α 3 151

Radiocarbon Dating

Radiocarbon Dating

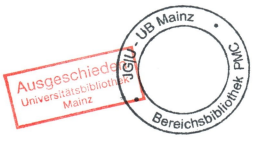

edited by

RAINER BERGER and HANS E. SUESS

Proceedings of the
Ninth International Conference
Los Angeles and La Jolla
1976

UNIVERSITY OF CALIFORNIA PRESS

BERKELEY   LOS ANGELES   LONDON

1979

University of California Press
Berkeley and Los Angeles

University of California Press, Ltd.
London, England

Copyright © 1979 by The Regents of the University of California

ISBN 0-520-03680-8

Library of Congress Catalog Card Number 78-54790

Printed in the United States of America

1 2 3 4 5 6 7 8 9

Contents

FOREWORD ix
PREFACE xi
Participants xiii

I. ARCHAEOLOGIC AND GEOLOGIC DATING

1. In Honor of Willard Frank Libby   *Kenneth Page Oakley*   3
2. Radiocarbon Dating and African Archaeology   *J. Desmond Clark*   7
3. Sea Level Variations and the Birth of the Egyptian Civilization   *Jacques Labeyrie*   32
4. An Appeal from the Consumer   *Lynn White, jr.*   37
5. A Radiocarbon Chronology for the Early Postglacial Stone Industries of England and Wales   *V. R. Switsur* and *R. M. Jacobi*   41
6. New Absolute Dates on Upper Pleistocene Fossil Hominids from America   *R. Protsch*   69
7. Were the Allerød and Two Creeks Substages Contemporaneous?   *Hans E. Suess*   76
8. A Second Millennium B.C. Radiocarbon Chronology for the Maya Lowlands   *Roy Switsur, Rainer Berger,* and *Norman Hammond*   83
9. The Effective Use of Radiocarbon Dates in the Seriation of Archaeological Sites   *Dwight W. Read*   89
10. Climate Controlled Uranium Distribution in Atlantic Ocean Core V29-179   *Emil K. Kalil*   95
11. The Neolithic Chronology of Switzerland by Underwater Archaeological Methods   *Gary Stickel* and *Rainer Berger*   102

II. RADIOCARBON METHODOLOGY

1. Correlation of ^{14}C activity of NBS Oxalic Acid with Arizona 1850 Wood and ANU Sucrose Standards   *H. A. Polach*   115
2. An Improved Procedure for Wet Oxidation of the ^{14}C NBS Oxalic Acid Standard   *S. Valastro, jr., L. S. Land,* and *A. G. Varela*   125

3. The Importance of the Pretreatment of Wood and Charcoal Samples
   Ingrid U. Olsson 135
4. Recent Progress in Low Level Counting and other Isotope Detection Methods *H. Oeschger, B. Lehmann, H. H. Loosli, M. Moell, A. Neftel, U. Schotterer,* and *R. Zumbrunn* 147
5. Measurement of Small Radiocarbon Samples: Power of Alternative Methods for Tracing Atmospheric Hydrocarbons *L. A. Currie, J. E. Noakes,* and *D. N. Breiter* 158
6. Further Improvement of Counter Background and Shielding
   S. Gulliksen and *R. Nydal* 176
7. The Effect of Electronegative Impurities on CO_2 Proportional Counting: An On-line Purity Test Counter *C. A. Brenninkmeijer* and *W. G. Mook* 185
8. Examination of Counting Efficiency during Measurements of Natural ^{14}C *W. Moscicki* 197
9. ^{14}C Dating to 60,000 Years B.P. with Proportional Counters
   M. Stuiver, S. W. Robinson, and *I. C. Yang* 202
10. A Thermal Diffusion Plant for Radiocarbon Isotope Enrichment from Natural Samples *Helmut Erlenkeuser* 216
11. Possibility of Measurement of ^{14}C by Mass Spectrometer Techniques
    H. W. Wilson 238
12. Specific Problems with Liquid Scintillation Counting of Small Benzene Volumes and Background Count Rate Estimation *Herbert Haas* 246
13. An Assessment of Laboratory Errors in Liquid Scintillation Methods of ^{14}C Dating *R. L. Otlet* 256
14. Radiocarbon Dating at the U. S. Geological Survey, Menlo Park, California *Stephen W. Robinson* 268

III. SOIL DATING

1. Soil Fraction Dating *H. W. Scharpenseel* 277
2. Radiocarbon Dating of Organic Components of Sediments and Peats
   John C. Sheppard, Syed Y. Ali, and *Peter J. Mehringer, jr.* 284

IV. ARTIFICIAL RADIOCARBON IN NATURE

1. Artificial Radiocarbon in the Stratosphere *Rainer Berger* 309
2. A Survey of Radiocarbon Variation in Nature since the Test Ban Treaty
   R. Nydal, K. Lövseth, and *S. Gulliksen* 313
3. The Uptake of Bomb ^{14}C in Humans *M. J. Stenhouse* and *M. S. Baxter* 324
4. Further Application of Bomb ^{14}C as a Biological Tracer
   M. J. Stenhouse 342

CONTENTS

    5. Radiocarbon Transmutation Mechanism for Spontaneous Somatic Cellular Mutations   *M. A. Tamers*   355
    6. The Origin of Drift-Gas Deposits as Determined by Radiocarbon Dating of Methane   *Dennis D. Coleman*   365
    7. Fossil Fuel Exhaust-Gas Admixture with the Atmosphere   *J. C. Freundlich*   388

V. SHELLS AND WATER PLANTS

    1. The use of Marine Shells in Dating Land/Sea Level Changes   *Joakim Donner* and *Högne Jungner*   397
    2. The Suitability of Marine Shells for Radiocarbon Dating of Australian Prehistory   *R. Gillespie* and *H. A. Polach*   404
    3. Fraction Studies on Marine Shell and Bone Samples for Radiocarbon Analyses   *R. E. Taylor* and *Peter S. Slota*   422
    4. Radiocarbon Activity in Submerged Plants from various South Swedish Lakes   *Sören Håkansson*   433

VI. OCEANOGRAPHY

    1. ^{14}C Activity of Arctic Marine Mammals   *Henrik Tauber*   447
    2. Environmental Effects on Radiocarbon in Coastal Marine Sediments   *Helmut Erlenkeuser*   453
    3. ^{14}C Routine Dating of Marine Sediments   *Mebus A. Geyh*   470

VII. TREE RINGS AND OTHER KNOWN AGE SAMPLES

    1. ^{14}C in Modern American Trees   *William F. Cain*   495
    2. ^{14}C Content of Nineteenth Century Tree Rings   *M. J. Stenhouse*   511
    3. Subsurface Radar Probing for Detection of Buried Bristlecone Pine Wood   *Henry N. Michael* and *Roger S. Vickers*   520
    4. Radial Translocation of Carbon in Bristlecone Pine   *A. Long, L. D. Arnold, P. E. Damon, C. W. Ferguson, J. C. Lerman,* and *A. T. Wilson*   532
    5. The ^{14}C Level during the Fourth and Second Half of the Fifth Millennium B.C. and the ^{14}C Calibration Curve   *Hans E. Suess*   538
    6. Composite Computer Plots of ^{14}C Dates for Tree-Ring-Dated Bristlecone Pines and Sequoias   *Elizabeth K. Ralph* and *Jeffrey Klein*   545
    7. Holocene Tree Ring Series from Southern Central Europe for Archaeologic Dating, Radiocarbon Calibration, and Stable Isotope Analysis   *B. Becker*   554

8. The Contribution of the Swiss Lake-Dwellings to the Calibration of Radiocarbon Dates *J. Beer, V. Giertz, M. Möll, H. Oeschger, T. Riessen,* and *C. Strahm* 566
9. Dendrochronology of Neolithic Settlements in Western Switzerland: New Possibility for Prehistoric Calibration *G. Lambert* and *C. Orcel* 585
10. Archaeological Evidence for Short-term Natural ^{14}C Variations *R. Burleigh* and *A. Hewson* 591
11. Radiocarbon Variations determined on Egyptian Samples from Dra Abu El-Naga *Ingrid U. Olsson* and *M.F.A.F. El-Daoushy* 601
12. The Radiocarbon Contents of Various Reservoirs *Ingrid U. Olsson* 613
13. Evaluation of Total Variability in Radiocarbon Counting *Stephen Meeks* 619

VIII. MODELING EXPERIMENTS

1. Prognosis for the Expected CO_2 Increase due to Fossil Fuel Combustion *H. Oeschger* and *U. Siegenthaler* 633
2. In Situ ^{14}C Production in Wood *C. J. Radnell, M. J. Aitken,* and *R. L. Otlet* 643

IX. RADIOCARBON AND CLIMATE

1. Isotopic Tree Thermometers: Anticorrelation with Radiocarbon *Leona Marshall Libby* and *Louis J. Pandolfi* 661
2. Causal Mechanisms in Climatic and Weather Changes as Revealed from Paleomagnetic Investigations of Samples Dated by the ^{14}C Method and from Geomagnetic Variations *Vaclav Bucha* 670
3. Sensitivity of Radiocarbon Fluctuations and Inventory to Geomagnetic and Reservoir Parameters *Robert S. Sternberg* and *Paul E. Damon* 691

X. RADIOCARBON AND OTHER DATING METHODS

1. Dating of Lime Mortar by ^{14}C *Robert L. Folk* and *S. Valastro, jr.* 721
2. Comparison of ^{14}C Ages and Ionium Ages of Corals *Kunihiko Kunihiko Kigoshi, Akira Tezuka,* and *Hiroshi Mitsuda* 733
3. The Dating of Fossil Bones using Amino Acid Racemization *Jeffry L. Bada, Patricia M. Masters, E. Hoopes,* and *D. Daling* 740
4. Amino Acid Racemization Dating of Fossil Shell from Southern California *Patricia M. Masters* and *Jeffrey L. Bada* 757

IX. APPENDIX

A Calibration Table for Conventional Radiocarbon Dates *Hans E. Suess* 777

INDEX 785

Foreword

For twenty-five years we have been coming together, working on problems, and doing our best to improve and perfect our method and to report new results. Two of the originals have passed on, Hessel de Vries and Richard Foster Flint, and we would say a word in memory of their contributions. Hessel did the first 70,000 year date, analyzed the residual background as due to neutrons generated in the shield by cosmic rays, and established the Groningen Laboratory. Dick led us hand in hand through the difficult field of glacial geology and helped us select the samples; in particular, he led us to Harlan Bretz and his Two Creeks forest bed. He served with Fred Johnson, Don Collier, and Froelich Rainey on our advisory committee for sample selection and general acquisition. His great prestige as a Pleistocene geologist helped us to receive consideration from geologists worldwide.

The method continues to enjoy improvements in two laboratory aspects: the counting and the chemisty of samples preparation. We have moved from the original black carbon method, with its low counting efficiency and susceptibility to radon contamination, to the gas counters (CO_2, C_2H_2, and CH_4) with the ultra-thin walls introduced by Oeschger greatly improving count rates relative to background. The next improvement may be the replacement of disintegration electron detection by direct counting of ^{14}C atoms. If successful, this method may reduce the required sample some one-hundred fold, opening up the field of valuable artistic and religious objects. Research to this end now is underway at Lawrence Berkeley Laboratory by Richard Muller and Luis Alvarez.

Improvements in sample chemistry include the hydrolysis of bone collagen protein and the subsequent isolation of the amino acids for burning and counting. As a result, Berger has been able to avoid tar contamination in dating the difficult La Brea Tar Pit bones. Cain and Suess have demonstrated ring-to-ring migration of sap and have developed a chemical method of purification for this material which otherwise mixes the outer twenty rings, at least in oak. This effect was discovered with the bomb ^{14}C of the 1962-63 tests. It is a small effect and would have been difficult to observe in any other way. It appears that the method of using the whole wood sample is slightly improved by separation of cellulose and lignin in cases where less than a limited number of rings per sample (twenty for oak) are to be used.

Because field collection techniques have improved over the years, the dates for a given site now are more likely to be consistent. This has occurred through the mutual education of the radiocarbon daters, the archaeologists, and the geologists.

Probably the most substantial advance in the method has been the determination of variations in the concentration of ^{14}C in the biospheric and marine and atmospheric carbon over the last 8000 years by ^{14}C dating and tree ring counting of a bristlecone pine forest in the Sierra Nevada. It has been proven conclusively that bomb ^{14}C injections in the atmosphere spread in a few years over the whole earth and biosphere. In addition no variation of biospheric ^{14}C concentration with latitude is evident despite the four- or five-fold variation in rate of cosmic ray neutron production. From these facts it is concluded that worldwide mixing probably has always been rapid and that the curve for the Schulman Grove in the White Mountains probably has worldwide application. It is our hope that the curve can be extended back to some 9400 years. Wes Fergesson has collected and conventionally dated the tree ring samples. Suess, Damon, and Ralph have done the ^{14}C dating.

What are our expectations? It has been reported at this Ninth Conference that the annual production of ^{14}C dates is about 15,000 on a worldwide basis. Most of these remain unpublished but provide practical guidance in field work. This output probably will increase. Berger and I are now engaged in an effort to date earthquakes in this way. We have some hope of success.

Much remains to be done in determining the rates of reactions of humus in soils and in marine organic sediments.

We expect ^{14}C dates to be important in determining the rate of climatic change in given regions during the onset and recession of ice ages. They will continue to play an important role in establishing climatic history during the last 40,000 years. These findings, combined with other evidence, may give us a worldwide record on which to base broad predictions of future climate, entailing major consequences for mankind and challenging us all to a mutual effort for survival.

The growing numbers of laboratories, of cooperating countries, and of daters is heart warming. I hope that one day every major university will have a laboratory and that archaeologists and earth scientists worldwide will have the opportunity to learn to date with radiocarbon.

W. F. Libby

University of California
Los Angeles

Preface

It is now nearly thirty years since Willard F. Libby announced the discovery of the first cosmic-ray-produced nuclear species in nature, that of the carbon isotope with mass 14. Certainly no other radioactive isotope has proven as fruitful for valuable research in so many fields of science as this carbon isotope. Willard Libby's ingenious idea to use it for measurements of the ages of carbon-containing substances has revolutionized our knowledge in several important fields. The early results of Libby's radiocarbon dating method obtained in collaboration with Drs. E. C. Anderson and James R. Arnold demonstrated the need for complete revision of the chronologies of the last stages of the great ice age and of the earliest civilizations of modern man. In his classic book, *Radiocarbon Dating*, Libby summarized his work, publishing the early dates which had necessitated the fundamental changes. The First International Radiocarbon Conference which, in 1954, was organized by Dr. Frederick Johnson and held in Andover, Massachusetts, has been followed by conferences devoted essentially to radiocarbon dating at two to three-year intervals. With each of these conferences, the scope of the results of radiocarbon measurements has been broadened. Natural radiocarbon is now being used in biochemistry, earthquake studies, geomorphology, meteorology, oceanography, and in other fields of geochemistry and geophysics as a source of valuable information.

In the course of time, great improvements have been made in counting instrumentation and in the accurate determination of radiocarbon. This has led to the discovery of hitherto unsuspected effects. For example, measurements of radiocarbon in samples of known age, primarily of tree ring dated wood, have shown that conditions on our planet do not remain entirely constant but undergo continuous change. The ^{14}C content of the carbon dioxide in our atmosphere has changed over the past ten-thousand years by approximately 10%. For this reason, the ^{14}C dates calculated in a conventional manner must be corrected by as much as a thousand years over what has previously been assumed. These corrections do not significantly change the results of measurements of Pleistocene age materials because one to three thousand years is not a large time interval in relation to ice age events. The corrections, however, have revolutionized Neolithic measurements. In particular, the relationship of the chronology of cultural developments in the near East and in Europe have had to be revised.

Variations in the ^{14}C content of atmospheric carbon dioxide may cause individual radiocarbon measurements to yield relatively imprecise dates in certain time ranges. Such short-range variations, the so-called "wiggles," may be of advantage in certain cases where sequences of samples of known age differences can be obtained, such as in series of samples from trees with some one-hundred annual rings.

The causes of the ^{14}C variations are still uncertain and present a fascinating geophysical problem. The more rapid variations occurring on a time scale of one-hundred or more years appear to correlate both with solar activity and with climatic events. Our global climate presently appears to undergo changes which underscore the importance of learning more about the causes effecting such changes in the conditions of our planet.

The scope of international radiocarbon conferences has increased in proportion to the number of scientific fields to which natural radiocarbon measurements contribute. Many thousands of radiocarbon determinations attest to this. It has become increasingly difficult to bring together all the experts interested in natural radiocarbon measurements and to discuss all the subjects of their fields of interest in a reasonably brief period of time. We hope that the subjects covered at this conference represent an appropriate compromise between extreme specializations and broad general interests.

Inasmuch as radiocarbon dating has increased our potential understanding of prehistory about tenfold for the time beyond the historical period, there are attempts underway to extend that range even further into the 70,000 year regime. In this regard, isotopic enrichment techniques may be helpful. Further advances in the techniques of radiocarbon measurement are now becoming feasible by sophisticated mass-spectrometric methods. Radiocarbon dating in the years to come promises many more existing results regarding the prehistory of man and his environment and consequently promises a far better understanding of man's condition today and tomorrow.

At the Ninth International Radiocarbon Conference the following resolutions were passed:

1. No change is recommended in the use of conventional ^{14}C years. A conventional ^{14}C year implies the use of the Libby half-life of 5568 years.

2. The reference standard remains 95% of the NBS oxalic acid activity, corrected for isotopic fractionation to a $\delta^{13}C$ value of -19.0 percent with regard to PDB. The year A.D. 1950 continues to be the reference year for conventional ^{14}C ages in years B.P. The use of lower case bp was rejected.

3. It is recommended that 1950 be no longer subtracted from conventional ^{14}C ages in order to arrive at a so-called A.D./B.C. age. A.D.*/B.C.* nomenclature is to be used after application of one of the available age correction curves or tables. The asterisk indicates a tree ring-calibrated age (e.g., 1250 B.C.*), whereas the text should specify the curve or table used.

4. It is recommended that a format for reporting radiocarbon dates compatible with computer-based retrieval systems be established.

The success of the conference was in large part due to the untiring and competent effort of Suzanne DeAtley. Moreover our gratitude extends to James Kubeck and Shirley Warren, of the University of California Press, to Gretchen Van Meter, who contributed much time in editing this book and to Roberta Smith who prepared the index.

Rainer Berger and Hans E. Suess

Participants

Jeffrey Bada	Scripps Institution of Oceanography, 3266 Sverdrup Hall, University of California, San Diego, La Jolla, California 92093, U.S.A.
Bernd Becker	Botanisches Institut, Universität Stuttgart-Hohenheim, 7000 Stuttgart-Hohenheim, Germany
Francesco Bella	Instituto di Fisica, Radiocarbon Dating Laboratory, Università di Roma, Città Universitaria, 001000 - Roma, Italy
Giorgio Belluomini	Centro di Studio per la Geochimica Applicata alla Stratigrafia Recente, Instituto di Geochimica, Università di Roma, Città Universitaria, 001000 - Roma, Italy
William Benson	Head, Earth Sciences Section, Division of Environmental Sciences, National Science Foundation, Washington, D.C. 20550, U.S.A.
Rainer Berger	Institute of Geophysics, Radiocarbon Laboratory, University of California, Los Angeles, Los Angeles, California 90024, U.S.A.
Junius Bird	American Museum of Natural History, Central Park West at 79th Street, New York, New York 10024, U.S.A.
Betty Lee Brandau	The University of Georgia, 4 GGS Building, Geochronology Laboratory, Athens, Georgia 30601, U.S.A.
Václav Bucha	Director, Geophysical Laboratory, Czechoslovakia Academy of Sciences, (Ceskoslovenska Akademie, Ved, Geofyzikalni Ustav), Bocni II - Cp, 1401 Praha, 4, Sporilov, Czechoslovakia
James Buckley	Teledyne Isotopes, 50 Van Buren Avenue, Westwood, New Jersey 07675, U.S.A.
Richard Burleigh	Research Laboratory, The British Museum, London WC1B 3DG, United Kingdom
William F. Cain, S. J.	Loyola Marymount University, Loyola Boulevard at West 80th Street, Los Angeles, California 90045, U.S.A.

J. Desmond Clark	Department of Anthropology, University of California, Berkeley, Berkeley, California 94720, U.S.A.
Dennis D. Coleman	Section of Analytical Chemistry, Illinois State Geological Survey, Natural Resources Bldg., Urbana, Illinois 61801, U.S.A.
Cesarina Cortesi	Instituto de Geochimica, Radiocarbon Dating Laboratory, Università di Roma, Città Universitaria, 001000 - Roma, Italy
Lloyd A. Currie	Head, Analytical Chemistry Division, National Bureau of Standards, U.S. Department of Commerce, Washington, D.C. 20234, U.S.A.
Garniss Curtis	Department of Geology and Geophysics, University of California, Berkeley, Berkeley, California 94270, U.S.A.
Emma Lou Davis	Research Associate, Los Angeles County Museum, 1236 Concord Street, San Diego, California 92101, U.S.A.
Suzanne P. de Atley	Institute of Geophysics, Radiocarbon Laboratory, University of California, Los Angeles, Los Angeles, California 90024, U.S.A.
Cesare Emiliani	University of Miami, Miami, Florida 33124, U.S.A.
Jonathon E. Ericson	Conservation Chemist, Los Angeles County Museum of Art, Los Angeles, California 90036, U.S.A.
Helmut Erlenkeuser	Institut für Reine und Angewandte Kernphysik der Universität Kiel, C-14 Labor, Olshausenstrasse 40/60, Geb. N20a, 2300 Kiel, Germany
C. W. Ferguson	University of Arizona, Laboratory of Tree-Ring Research, Tucson, Arizona 85721, U.S.A.
Gordon Fergusson	Director of Research, Scientific Research Instruments Corp., 6807 Whitestone Road, Baltimore, Maryland 21207, U.S.A.
John Fletcher	Research Laboratory for Archaeology and the History of Art, Oxford University, 6 Keble Road, Oxford OX1 3QJ, United Kingdom
Jürgen C. Freundlich	Universität Köln, Institut fur Ur- und Frügeschichte, C-14 Laboratorium, Weyertal 125, Köln - Lindenthal, Germany
Mebus A. Geyh	Director and Professor, Niedersächsisches Landesamt für Bodenforschung, D-3000 Hannover - Bucholz, Postf. 510153, Germany
Richard Gillespie	Department of Physical Chemistry, Sydney University, NSW, 2006, Australia
Marija Gimbutas	Department of Slavic Languages, University of California, Los Angeles, Los Angeles, California 90024, U.S.A.

PARTICIPANTS

Gayle Gittins	Radiocarbon Laboratory, Institute of Geophysics, University of California, Los Angeles, Los Angeles, California 90024, U.S.A.
Herbert Haas	Director, Radiocarbon Laboratory, Institute for the Study of Earth and Man, Southern Methodist University, Dallas, Texas 75275, U.S.A.
Sören Håkansson	Laboratory for C-14 Dating, University of Lund, Tunavägen 29, S-223 63 Lund, Sweden
Douglas Harkness	Scottish Universities Research and Reactor Center, East Kilbride, Glasgow G 75, OQU, United Kingdom
C. Vance Haynes	Department of Anthropology, University of Arizona, Tucson, Arizona 85721, U.S.A.
Patricia Helfman	Scripps Institution of Oceanography, University of California, San Diego, La Jolla, California 92093, U.S.A.
Jos Heylen	Institut Royal du Patrimoine Artistique, 1 Parc du Cinquantenaire, Brussels 4, Belgium
Carol Hutto	Radiocarbon Laboratory, Mt. Soledad, University of California, San Diego, La Jolla, California 92093, U.S.A.
E. M. Jope	Archaeology Department, The Queen's University of Belfast, Belfast BT 7 INN, United Kingdom
Högne Jungner	Radiocarbon Dating Laboratory, University of Helsinki, Snellmanikatu 5, SF - 00170 Helsinki, Finland
R. Kaldenberg	Department of Anthropology, California State University, San Diego, California 92101, U.S.A.
Emil K. Kalil	Uranium-West Labs, 15515 Sunset Blvd. B07, Pacific Palisades, CA 90272, U.S.A.
Aaron Kaufman	Geoisotope Laboratory, Department of Isotope Research, Weizmann Institute of Science, Rehovot, Israel
David Keeling	Scripps Institution of Oceanography, Ocean Research Division, 2314 Ritter Hall, San Diego, California 92093, U.S.A.
Kunihiko Kigoshi	Gakushiun University, Mejiro, Toshima-Ku, Tokyo, Japan
Serge Korff	Physics Department, New York University, 4 Washington Place, New York, New York 10003, U.S.A.
Jacques Labeyrie	Centre des Faibles Radioactivités, Laboratoire Mixte CNRS/CEA, 91190-Gif-sur-Yvette, France
G. Lambert	Laboratoire de Dendrochronologie, ler Mars 24, 2000 Neuchatel, Switzerland
Young Nam Lee	Department of Bacteriology, University of California, Los Angeles, Los Angeles, California 90024, U.S.A.

Juan Carlos Lerman	Isotope Geochemistry Laboratory, Geosciences Department, University of Arizona, Tucson, Arizona 85721, U.S.A.
Leona Marshall Libby	School of Engineering, University of California, Los Angeles, Los Angeles, California 90024, U.S.A.
Willard F. Libby	Director, Institute of Geophysics and Planetary Physics, University of California, Los Angeles, Los Angeles, California 90024, U.S.A.
Timothy Linick	Radiocarbon Laboratory, Mount Soledad, University of California, San Diego, La Jolla, California 92093, U.S.A. U.S.A.
Austin Long	Laboratory of Isotope Geochemistry, Geosciences Department, University of Arizona, Tucson, Arizona 85721, U.S.A.
Steven Meeks	Department of Statistics, Southern Methodist University, Dallas, Texas 75275, U.S.A.
Henry Michael	Department of Physics, DRL/E1, Philadelphia, Pennsylvania 19174, U.S.A.
A. Alan Moghissi	Office of Interdisciplinary Programs, Georgia Institute of Technology, Atlanta, Georgia 30332, U.S.A.
W. G. Mook	Natuurkundig Laboratorium der Rijks-Universiteit, Westersingel 34, Groningen, The Netherlands
W. Moscicki	Physical Institut C-14 Laboratory, Politechnika Slaska, Karolinki 11, 44 - 100 Gliwice, Poland
K. O. Münnich	Institut für Umweltphysik der Universität Heidelberg, D - 69, Heidelberg, Im Neuenheimer Feld 366, Germany
John Noakes	The University of Georgia, 4 GGS Building, Geochronology Laboratory, Athens, Georgia 30601, U.S.A.
Y. Nozaki	Department of Geology and Geophysics, Yale University, New Haven, Connecticut 06520, U.S.A.
R. Nydal	Radiological Dating Laboratory, 7034 Trondheim NTH, Norway
Kenneth Page Oakley	Subdepartment of Anthropology, British Museum (Natural History), London SW7 5BD, United Kingdom
Hans Oeschger	Physikalisches Institut, Universität Bern, Sidlerstrasse 5, 3012 Bern, Switzerland
Ingrid U. Olsson	University of Uppsala, Institute of Physics, Box 530 S - 75121 Uppsala, Sweden
Hain Oona	Physics Department, University of Arizona, Tucson, Arizona 85721, U.S.A.
C. Orcell	Laboratoire de Dendrochronologie, ler Mars 24, 2000 Neuchatel, Switzerland

PARTICIPANTS

R. L. Otlet	Carbon-14/Tritium Measurements Laboratory, Bldg. 10.46 AERE, Harwell, Didcot, Birkshire OX11 ORA, United Kingdom
Louis Pandolfi	Department of Chemistry, University of California, Los Angeles, Los Angeles, California 90024, U.S.A.
F. J. Pearson, Jr.	U.S. Geological Survey, Isotope Hydrology Laboratory, National Center, MS 432, Reston, Virginia 22092, U.S.A.
T. H. Peng	Lamont-Doherty Geological Observatory, Palisades, New York 10964, U.S.A.
Donald J. Piepgras	P.O. Box 249176, Department of Geology, University of Miami, Coral Gables, Florida 33124, U.S.A.
Henry A. Polach	ANU Radiocarbon Dating Laboratory, c/o Research School of Earth Sciences, The Australian National University, Canberra, ACT, Australia
Reiner Protsch	Anthropologisches Institut, J. W. Goethe Universität, Frankfurt/Main, Siesmayerstrasse 70, Germany
Christopher Radnell	Bldg. 10.46 AERE, Harwell, Didcot, Birkshire OX22 ORA, United Kingdom
Omar Rahmouni	Institute Ecudes Nucleaires, Bd. Frantz Fanon, Bp 1017, Algiers, Algeria
Elizabeth Ralph	Department of Physics, DRL/E1, University of Pennsylvania, Philadelphia, Pennsylvania 19174, U.S.A.
Dwight W. Read	Department of Anthropology, University of California, Los Angeles, Los Angeles, California 90024, U.S.A.
Stephen W. Robinson	U.S. Geological Survey, 345 Middlefield Road, Menlo Park, California 94025, U.S.A.
Meyer Rubin	U.S. Geological Survey, National Center, 971, Reston, Virginia 22092, U.S.A.
A. Rutherford	Saskatchewan Research Council, 30 Campus Drive, University of Saskatchewan, Saskatoon, Saskatchewan, Canada
H. W. Scharpenseel	Ordinariat für Bodenkunde, University of Hamburg, 2057 Reinbek, Schloss, Germany
John C. Sheppard	Department of Chemical Engineering, Dana 118, Washington State University, Pullman Washington 99163, U.S.A.
Irene Stehli	Radioisotope Laboratory, Dicar Corporation, 2040 Adelbert Road, Cleveland, Ohio 44106, U.S.A.
M. J. Stenhouse	Department of Chemistry, University of Glasgow, Glasgow G12 8QQ, Scotland
Robert Sternberg	Laboratory of Isotope Geochemistry, Department of

	Geosciences, University of Arizona, Tucson, Arizona 85721, U.S.A.
E. Gary Stickel	Department of Anthropology, California State University, Long Beach, Long Beach, California 90801, U.S.A.
J. J. Stipp	P.O. Box 249176, Department of Geology, University of Miami, Coral Gables, Florida 33124, U.S.A.
Christian Strahm	Institut für Ur- and Frühgeschichte, University of Freiburg, D-78 Freiburg i. Br., Adelhauserstrasse 33, Germany
Minze Stuiver	Quaternary Research Center, University of Washington Ak-60, Seattle, Washington 98195, U.S.A.
Hans E. Suess	Department of Chemistry, University of California, San Diego, La Jolla, California 92093, U.S.A.
V. R. Switsur	Department of Quaternary Research, Radiocarbon Research Laboratory, University of Cambridge, 5 Salisbury Villas, Station Road, Cambridge CB1 2JF, United Kingdom
Murray A. Tamers	Nova University, 3301 College Avenue, Fort Lauderdale, Florida 33014, U.S.A.
Henrik Tauber	Carbon-14 Dating Laboratory, National Museum, Ny Vestergade 10, DK-1471 Copenhagen, Denmark
R. E. Taylor	Radiocarbon Laboratory, Department of Anthropology, University of California, Riverside, Riverside, California 90650, U.S.A.
S. Valastro, Jr.	Radiocarbon Laboratory, Balcones Research Center, Rt. 4, Box 189, University of Texas at Austin, Austin, Texas 78757, U.S.A.
Gary Vecelius	Territorial Archaeologist for the Virgin Islands, Department of Conservation and Cultural Affairs, Charlotte Amalie, St. Thomas, Virgin Islands 00801, U.S.A.
John C. Vogel	Council for Scientific and Industrial Research, National Physical Research Laboratory, P.O. Box 395, Pretoria 0001, South Africa
Lynn White, jr.	Department of History, University of California, Los Angeles, Los Angeles, California 90024, U.S.A.
Evan T. Williams	Department of Chemistry, Brooklyn College of City University of New York, Brooklyn, New York 11210, U.S.A.
Henry W. Wilson	Scottish Universities Research and Reactor Center, East Kilbride, Glasgow G75, 0QU, United Kingdom
I. C. Yang	Geological Sciences, Ak-20, University of Washington, Seattle, Washington 98195, U.S.A.

PART I
Archaeologic and Geologic Dating

1

In Honor of Willard Frank Libby

Kenneth Page Oakley

When Willard Frank Libby was elected Fellow of the British Academy in 1969, his introduction of ^{14}C dating twenty years before was acclaimed as one of the great contributions of science to the humanities. The impace of this technique in my own domain, paleoanthropology, has been tremendous. The evolution and cultural emergence of our species *Homo sapiens* have had to be completely rewritten as a result of ^{14}C dating.

While recalling the immensity of Libby's direct contributions to dating the human past, we should also pay tribute to the ways in which he has stimulated individual research. I should like to record my deep appreciation of the practical encouragement he has given me in developing relative dating techniques applicable to skeletal materials.

In July 1950 on a brief visit to Chicago, at a party given by Al and Thelma Dahlberg, I was introduced to Libby. He invited me to the renowned Institute of Nuclear Physics for a demonstration of the new ^{14}C dating apparatus. Later in the year he visited me at the British Museum of Natural History and asked me about my application of Carnot's fluorine method for the relative dating of fossil bones. He asked me to consider ways in which radiocarbon dating might yield valuable results in the study of early man. Our resulting discussions and correspondence paid dividends.

In March 1953 at the invitation of the late Professor van Riet Lowe and with a grant from the Wenner-Gren Foundation, I made an advisory study-tour of Paleolithic sites in South Africa (Oakley 1954). Two months later, systematic excavations were made at the Cave of Hearths, Makapansgat, in the Northern Transvaal, by Revil Mason working under the guidance of van Riet Lowe. Undoubted hearths were encountered at ten levels and samples of these were accepted by Libby for ^{14}C dating. Preliminary tests showed that the carbon in some samples was contaminated, probably by roots, but there was a reasonably good sample from Middle Stone Age IV (Pietersburg culture) which he dated as 15, 100 ± 730 B.P. (C-925). This was the first radiocarbon date obtained on South African stone age material (Libby 1955).

In 1962 Libby asked if I would apply the fluorine method to one of two human bones dug out of Pleistocene deposits near Santa Barbara, California. Some of his colleagues doubted that they were contemporaneous with the containing deposits. If the material were fossil it would be a find of great importance. At that time I was developing, with a further grant from the Wenner-Gren Foundation, a series of techniques combining fluorine analysis with radiometric assay and nitrogen analysis for the relative dating of bones. Like fluorine, uranium increases in buried bone with the passage of time. Conversely, with the leaching out of soluble products of protein decay, the nitrogen content of bone decreases. I welcomed the opportunity to apply these combined techniques to the bones of dubious antiquity from California.

The two human femora had been discovered by P. C. Orr at a depth of 11m (37 ft) in Pleistocene deposits at Arlington Springs in Arlington Canyon near Santa Barbara (Orr 1960). During investigations by a team of geologists and archaeologists in 1960, sample of carbonaceous material found at nearly the same level was dated by W. S. Broecker in the Lamont Geological Laboratory as 10,000 ± 200 B. P. (L-650) (Orr 1962). If the femora were not part of an intrusive burial, they represented one of the very few Pleistocene men discovered in the Americas.

At Libby's suggestion a sample of one of the femora was sent to the British Museum of Natural History with the request that I organize relative dating tests. In the Subdepartment of Anthropology, Mrs. E. Gardiner made a radiometric assay counting beta-emissions as a means of assessing uranium content (described as eU_3O_8). In the government laboratory in London R. G. Cooper determined the fluorine and phosphate contents, and E. J. Johnson the nitrogen content. Although no other bones from the site were available for comparison, the fluorine content, the fluorine to phosphate ratio, and the equivalent-urania were high enough and the nitrogen content was low enough to indicate moderately high antiquity. Series of bone samples from Upper Pleistocene and early Holocene sites in North and Central America had been assayed for fluorine, uranium, and nitrogen (Oakley and Rixon 1958, Oakley and Howells 1961). When the composition of the Arlington Springs bone was considered in the light of these results (Oakley 1963) it appeared to be fossil rather than sub-fossil and, therefore, contemporaneous with the carbon sample dated by W. S. Broecker.

American Human Bone Analyses Compared in 1963

Arlington Springs	F = 1.2%	$100F/P_2O_5$ = 3.6	eU_3O_8 = 28 ppm,	N = 0.23%
Midland, Texas	F = 0.8%	$100F/P_2O_5$ = 7.7	eU_3O_8 = 13 ppm,	N = 0.03%
Natchez, Mississippi	F = 0.88%	$100F/P_2O_5$ = 3.4		N = 1.4%
Tepexpan, Mexico	F = 2.0%	$100F/P_2O_5$ = 6.8	eU_3O_8 = 1 ppm,	N = 0.06%
Compare modern bone	F = <0.1%	$100F/P_2O_5$ = 0.1	eU_3O_8 = <1 ppm,	N = >4.0%

Since 1963 the Pleistocene or near-Pleistocene age of several human bones from California has been established by ^{14}C dating under Libby's direction in the Institute of Geophysics, Los Angeles. Their F/U/N values have been recorded, so the analyses of these fossil bones may be added to the Arlington Springs comparative table.

Comparison of Fluorine, Nitrogen, Uranium Contents, and Radiocarbon Ages

	F (%)	N (%)	U (ppm)	
Diablo Canyon*	0.19	0.74		9,320 ± 140 (UCLA–1686A)
Laguna Beach**	0.30	0.26	60	17,150 ± 1470 (UCLA–1233A)
Los Angeles***	0.04–0.12	0.32		>23,600 (UCLA–1430)

*Berger and Libby 1976.
**Berger and Libby 1969.
***Berger et al. 1971.

Those of us in London concerned with applying F/U/N relative dating techniques to fossil bones were greatly encouraged to find that Libby had routinely measured the fluorine, uranium, and nitrogen contents of fossil bones submitted for ^{14}C dating in the UCLA laboratory. Rainer Berger, his successor as head of the UCLA radiocarbon laboratory, is continuing this practice. In the *Catalogue of Fossil Hominids* we have made both relative and absolute dating required information. We offer to provide contributors to the catalogue with facilities for having their fossil hominid material analyzed for fluorine, uranium, and nitrogen.

These notes illustrate that any research in association with Libby involves very rewarding interchange.

REFERENCES

Berger, R., et al.
 1971 *In* Stross, F. H. The Application of the Physical Sciences to Archaeology. Berkeley and Los Angeles, Univ. of Calif. Press, 12:43.

Berger, R. and W. F. Libby
 1969 UCLA Radiocarbon Dates IX. Radiocarbon, 11:194.
 1976 UCLA Radiocarbon Dates. Radiocarbon. (In press.)

Libby, W. F.
 1955 Radiocarbon Dating. 2d ed. Chicago, Univ. of Chicago Press.

Oakley, K. P.
 1954 Study Tour of Early Hominid Sites in Southern Africa. S. Afr. Archaeol, Bull. 9:75.
 1963 Relative Dating of Arlington Springs Man. Science. 141:1172.
 1967 Catalogue of Fossil Hominids. rev. ed. Africa. I. Oakley, Campbell, and Molleson, eds. Brit. Mus. (Nat. Hist.)
 1971 Ibid. Europe. II.
 1975 Ibid. The Americas, Asia, and Australasia. III.

Oakley, K. P., and W. W. Howells
 1961 Age of the skeleton from the Lagow sand pit, Texas. Am. Antiq. 26:543.

Oakley, K. P., and R. E. Rixon
    1958   The Radioactivity of materials from the Scharbauer Site, near Midland, Texas. Am. Antiq. 24:185.

Orr, P. C.
    1960   Late Pleistocene Marine Terraces on Santa Rosa Island, California. Bull. Geol. Soc. Am. 71:1113.
    1962   Arlington Springs Man. Science. 135:219.

2

Radiocarbon Dating and African Archaeology

J. Desmond Clark

It is well nigh impossible to measure the debt we owe to Dr. Willard Libby in regard to the ordering and understanding that the radiocarbon method of dating has given to African prehistory. All those who work on problems concerned with human society and behavior in prehistoric times and the paleoecological conditions under which man lived will agree that without the radiometric time scale that his research introduced, we would still be foundering in a sea of imprecisions sometimes bred of inspired guesswork but more often of imaginative speculation. In acknowledging our immense debt to the inventor of radiocarbon dating, I express, on behalf of all of us who work in Africa or on African problems, our very, very sincere and deep gratitude for providing us with the first reliable basis for the chronological framework within which African society in prehistoric and early historic times is now studied.

To one who has spent nearly forty years working on various problems in African prehistory, the changes in approach and understanding that the radiocarbon chronology has brought about appear all the more impressive, especially those of the last decade. Before 1950 there were already a number of excellent stratigraphic studies showing the relationships of succeeding archaeological occurrences, as well as taxonomic analyses of assemblages of prehistoric artifacts that together made possible the establishment of a succession of technological and typological stages or modes. However the generally accepted means of establishing the relative ages and relationships of these archaeological occurrences and any correlation between regional successions was by means of the "pluvial/interpluvial" hypothesis, an outcome of the Milankovitch theory of world glaciations. In 1958 and 1959 Flint's (1959) and Cooke's (1958) critical reexaminations of the evidence on which the pluvial/interpluvial framework was established showed this hypothesis to be without substantial foundation, especially in the type area of East Africa.

It seemed as if the props had been knocked from under both short-distance and long-distance correlations. Although the study of the African Quaternary thus suffered a seemingly severe setback, a beneficial effect was to turn prehistorians towards the natural sciences in attempts to make prehistory a more exact discipline.

From this impasse we had already been rescued — although at the time we did not realize it — by Dr. Libby's ^{14}C dating method. Results for sub-Saharan Africa were slow in making their appearance. Between 1951 and 1959 less than twenty dates were forthcoming. They were, however, spread throughout a large part of later Pleistocene and Holocene time so that it was possible to use them to establish a framework which, although it has undergone several significant readjustments since 1959, was our first glimpse of the true measure of the time depth in the continent where the oldest evidence for man's cultural abilities is now established. At the Leopoldville (Kinshasa) Pan-African Congress in 1959, a paper was read in which the collated results gave the chronology and correlations seen in figure 1 (Clark 1959). Surprisingly, this was the only paper at the conference concerned with radiocarbon dating. This may have been because Africa was then considered to be part of the prehistoric Third World and rather out of the mainstream of human biological and cultural development: a misconception that the range of radiometric dating methods and the unique archaeological discoveries of the past decade have done much to set right. With the increased number of laboratories processing samples and the establishment of laboratories in the continent itself — particularly those in Pretoria and Salisbury — radiometric chronology now plays a key part in the strategy of prehistoric research there.

The shortcomings of this 1959 chronology are obvious to us today with the large number of dates we now have available. But these first dates, most of them obtained by Dr. Libby in his Chicago laboratory, represent the cornerstones on which the present chronology stands. Before these early radiocarbon dates became available, the meteorological correlations of C. E. P. Brooks (1931, 1949) suggested that the Later Stone Age began ca. 850 B.C. and the beginnings of the Middle Stone Age were placed in the "Makalian Wet Phase" about 10,000 years ago (Söhnge, Visser and van Riet Lowe 1937). When Dr. Libby dated the first two samples from sub-Saharan Africa, from the site of Mufo in northeastern Angola, we obtained our first indication of the antiquity of the Later Stone Age (more than 6000 years as against 2500 years ago) (Libby 1951). The end of the Middle Stone Age looked as if it were about 14,000 years ago; its beginnings (if we include the First Intermediate industries) were put at more than 41,000 years and the late Acheulian was dated at a little over 57,000 years ago.

We now know that these dates must be extended back quite appreciably, but they gave us the first indication of the time depth in which we were working even while Sir Arthur Keith's estimate of ca. 500,000 years for the beginning of the Pleistocene still proved acceptable to many. In 1960 the Yale laboratory produced the first of a series of dates for the Later Stone Age microlithic Nachikufan Industry in Zambia, which indicated that its beginning stages were more than 10,000 years old (Stuiver and Deevey 1961). This illustrates well the early scepticism with which African prehistorians treated results that did not fall within accepted concepts of the age of an occurrence. In this case the excavator was myself. Because it seemed so highly unlikely to me that any Later Stone Age industry, especially a microlithic one, could be that old, I felt something had to have contaminated the samples! Later series of results from diverse parts of the continent have confirmed the reliability of the Yale dates, showing that the beginnings of Later Stone Age technology lie firmly within the later Pleistocene.

One of the most significant dates for African historians was that obtained by Libby for the age of a wooden beam (1361 ± 120 B.P.) from the base of a wall in the elliptical building at the Zimbabwe ruins in Rhodesia (Libby 1952, Summers 1955). Probably no other date from Africa has given rise to so much scientific (and nonscientific) discussion bound up with various political overtones. As the foundation stone of the chronology of the southern African Iron Age it has shown that instead of beginning ca. A.D. 1500, as many chose to think, the first farming populations were already established in southern Africa by the early years of the present era. The impetus given to Iron Age studies by the Zimbabwe date was phenomenal. Interest and research in this period throughout the continent took a tremendous leap forward and its importance for the African peoples can be readily appreciated.

The results that have become available in the intervening years appear to confirm the belief, based on Libby's dating of the Esh Shaheinab settlement on the Nile in the central Sudan to ca. 3200 B.C. (Libby 1952), that herding did not become significant in the prehistoric economies south of the Sahara until about that time. This date also initiated a series of studies devoted to documenting and trying to understand the causes underlying the development of agriculture in sub-Saharan Africa.

The radiocarbon dates provided a framework for correlation not only within Africa but also between that continent and Eurasia, particularly Western Europe (Movius 1959). Figure 1 shows the correlations suggested by the results in 1959. These were generally acceptable until 1972, when the first of a series of dates were published by Dr. Vogel's Pretoria laboratory (Vogel and Beaumont 1972). These have drastically changed our understanding of the age and length of the Middle Stone Age, particularly in southern Africa. The dates of ca. 40,000 years for the beginning and 10,000 years for the end of the Middle Stone Age, as the first radiocarbon dates suggested, indicated an approximate equivalence with the Upper Paleolithic industries of Western Europe and an earlier position for the technologically similar Middle Paleolithic stage vis-à-vis the Middle Stone Age. We now know that this seeming correlation is not correct, and that there is instead almost a one-to-one relationship between the Middle Paleolithic and the Middle Stone Age.

Today's refined laboratory methods, reliability assessments, and greatly increased number and runs of dates, sometimes with checks from other dating techniques, have made it possible to exclude many anomalous results and have provided us with the chronology set out in figure 2, showing the current state of knowledge in the continent north and south of the Sahara and its correlation with adjacent parts of Eurasia.

It is immediately apparent that even the latest Acheulian stage lies well beyond the lower limits of the radiocarbon method. The beginning, indeed the greater part of the

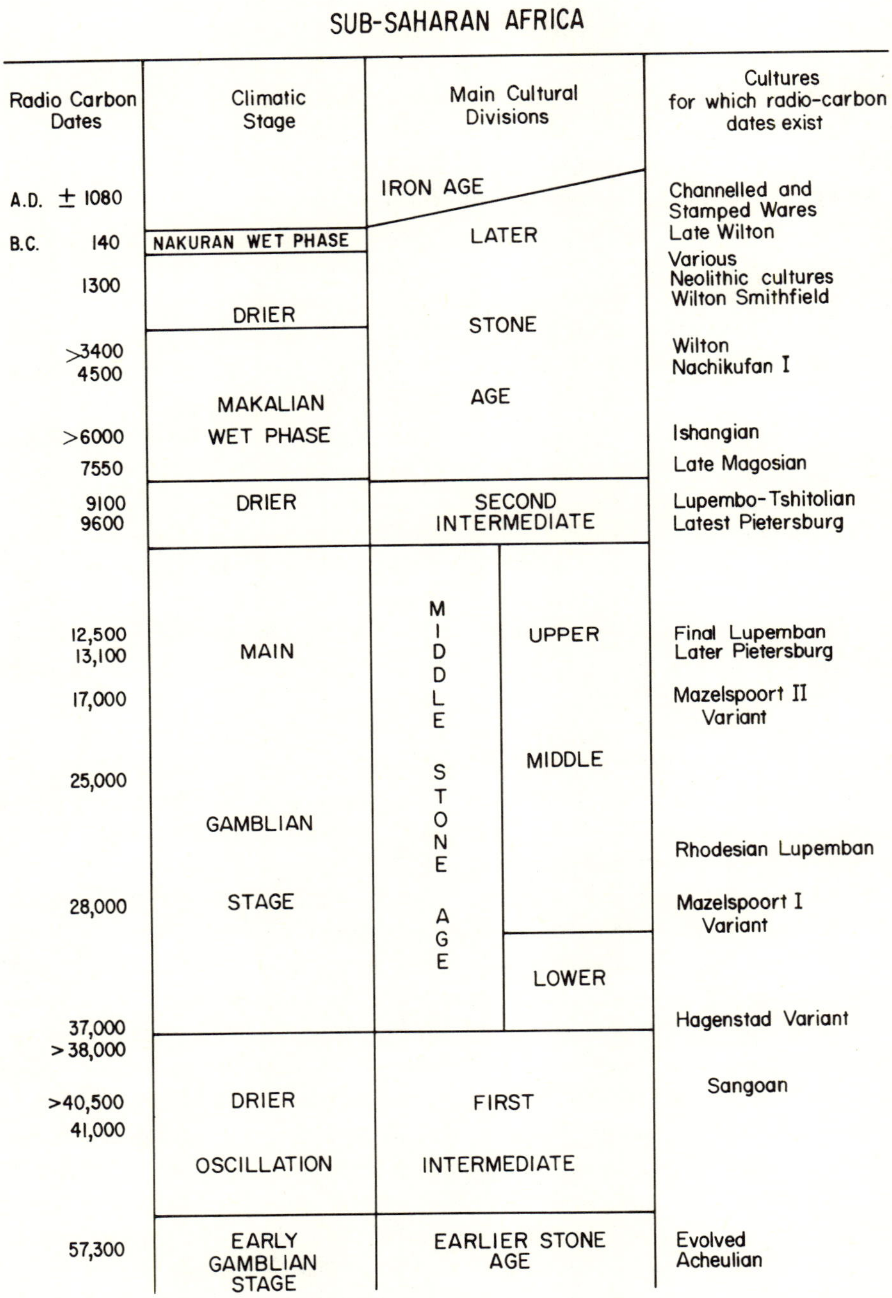

SUB-SAHARAN AFRICA

Radio Carbon Dates	Climatic Stage	Main Cultural Divisions		Cultures for which radio-carbon dates exist
A.D. ± 1080		IRON AGE		Channelled and Stamped Wares
B.C. 140	NAKURAN WET PHASE	LATER		Late Wilton
1300	DRIER	STONE		Various Neolithic cultures Wilton Smithfield
>3400 4500	MAKALIAN WET PHASE	AGE		Wilton Nachikufan I
>6000 7550				Ishangian Late Magosian
9100 9600	DRIER	SECOND INTERMEDIATE		Lupembo-Tshitolian Latest Pietersburg
12,500 13,100	MAIN	M I D D L E S T O N E A G E	UPPER	Final Lupemban Later Pietersburg
17,000				Mazelspoort II Variant
25,000	GAMBLIAN		MIDDLE	
				Rhodesian Lupemban
28,000	STAGE			Mazelspoort I Variant
			LOWER	
37,000 >38,000				Hagenstad Variant
>40,500 41,000	DRIER OSCILLATION	FIRST INTERMEDIATE		Sangoan
57,300	EARLY GAMBLIAN STAGE	EARLIER STONE AGE		Evolved Acheulian

FIG. 1. Radiocarbon chronology for sub-Saharan Africa compared with that for western Europe on the basis of dates obtained up to August 1959. (After Clark. 1959).

WESTERN EUROPE

Radio Carbon Dates	Climatic Stage	Main Cultural Division	Cultures	
A.D. 0- B.C.	POST GLACIAL	NEOLITHIC	Various	
		MESOLITHIC		
8000	LATE WURM		Magdalenian	
10,000-8850	ALLERØD OSCILLATION			
10,500	LATE WURM			
	BØLLING OSCILLATION			
11,500				
ca. 15,000	MAIN WURM : LATE PHASE	UPPER	Solutrian	
21,000				
25,000			PROTO-MAGDALENIAN	
	PAUDORF OSCILLATION		Aurignacian	Perigordian III-V
26,950	MAIN WURM : EARLY PHASE			
28,720		PALAEOLITHIC		
	GOTTWEIG INTERSTADIAL		Persisting Mousterian	Perigordian I
39,950				
	EARLY WURM STAGE	MIDDLE PALAEOLITHIC	Mousterian	

(MAIN WURM STAGE spans from ca. 15,000 through 28,720)

PL. 1. *The Hill Ruin or, Acropolis, at Zimbabwe near fort Victoria, Zimbabwe-Rhodesia.* The ruins of the massive stone structures at Zimbabwe are the most spectacular and most discussed buildings in the whole of sub-Saharan Africa. Built by the ancestors of the Shona peoples between about A.D. 1250 and 1450, these ruins are a record of the powerful, political, and religious institutions sometimes found in precolonial Africa based on a broad agricultural base, gold mining, and extensive internal and external networks. In 1951 Willard Libby's dating of a wooden beam from the Eliptical Building here to about A.D. 700 sparked the great interest and rapid development of Iron Age studies that is prevalent throughout the continent today.

Middle Stone Age as well as the Middle Paleolithic in North Africa, lies beyond this limit also. Only the latest part of this industrial mode is younger than 40,000 years.

Other dating techniques show that even the latest Acheulian, which has many characteristics in common with Middle Stone Age technology, is unlikely to be younger than ca. 200,000 years. Extrapolation from runs of radiocarbon dates and isotopic calibration between deep-sea cores and food shells stratified in the occupation deposits (e.g., at Haua Fteah cave and Klaasie's River Mouth caves at opposite ends of the continent) demonstrate that the early Middle Stone Age occurrences in sub-Saharan Africa, like the Pre-Aurignacian of Cyrenaica showing no remaining vestiges of the Acheulian, belong within the Last (Eem) Interglacial period and are 125,000 to 75,000 years old (McBurney 1975, Klein 1974).

Most of the early ^{14}C dates for the Middle Paleolithic or Middle Stone Age are not finite results and indicate that the archaeological occurrences they are dating lie beyond the limit of the method. This is the case with the early Middle Stone Age in the South African southern coastal caves, with the lower part of the Upper Pleistocene sequence in many of the cave sites in the interior, and with the desert oriented Aterian technocomplex in the Sahara and Maghreb. The latter is now established as being in part contemporary with the North African Mousterian (i.e., 40,000-50,000 years old) and has the distinction, through the invention of the tang, of providing the earliest indisputable proof for the hafting of stone working parts to form tools composed of two or more different materials (Clark 1975: 187-90). Evolved Middle Stone Age occurrences with smaller artifact dimensions and an increased element of non-Levallois blade technology appear to have lasted in some parts of the continent (e.g., in southern Central Africa) until shortly after 20,000 years ago, when they were replaced by technologies associated with the earliest of the industries that fall into the later Stone Age. The time for this event, suggested by the first ^{14}C results in 1959 (ca. 10,000 B.P.), is now more than doubled.

The dates in Figure 2 were selected to show only the earliest times at which the Middle Paleolithic/Middle Stone Age, Later Stone Age/Upper Paleolithic, and Neolithic food producing cultures made their appearance in each of four main regions of the Old World. Correlation shows some very interesting agreements and differences. The typical Middle Paleolithic of Europe and the Middle East begins $>$ 50,000 (more probably $>$ 60,000) years ago and the Middle Paleolithic in North Africa appears at about the same time.

In the African tropics it is now possible to see that the heavy-duty component of the early post-Acheulian tool kits, often described as Sangoan, belongs largely in the Last Interglacial period, the later stages only falling within the 40,000-50,000 year range on the evidence of a run of twelve dates from the long, almost continuous cultural sequence at Kalambo Falls (Clark 1969: appendix J).

In southern Africa the beginning of the Middle Stone Age is clearly within the Last Interglacial period, based on evidence of its relationship to the 6-8m high sea level in coastal caves and the use of marine food resources which this transgression made possible

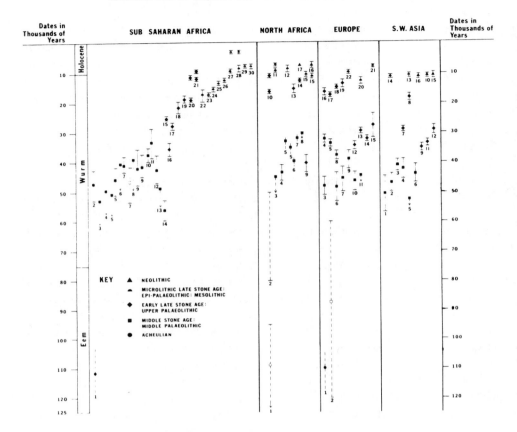

Site Lists, dates and Industrial Associations

A. *Sub-Saharan Africa*

Acheulian

1. Rooidam, Cape. 115,000 ± 10,000 B.P.

'Middle Stone Age'

2. Hout Bay, Cape. 47,100, +2,800, −2,100 B.P.
3. Bushman's Rock Shelter, Transvaal. > 53,000 B.P. (Pietersburg).
4. Florisbad, O.F.S. > 48,900 B.P. (Hagenstad).
5. Montagu Cave, Cape. 45,900 ± 2,100; > 50,800 B.P. (Howieson's Poort).
6. Mufo, Angola. > 40,000 B.P. (2 samples) (Lupemban).
7. Kalambo Falls, Zambia. 46,100 + 3,500, −2,400 (Lupemban).
8. Klassies River Mouth, Cape. > 38,000 (?Mossel Bay).
9. Red Cliff, Rhodesia. 41,800 ± 3,000; 40,780 ± 1,800 (Bambata).
10. Zombepata, Rhodesia. 37,290 ± 1,140 (Bambata).
11. Witkrans, Cape. 33,150 ± 2,500 (?Pietersburg).
12. Pomongwe, Rhodesia. 42,200 ± 2,300 (Bambata).
13. Border Cave, Ingwavuma, Natal. > 48,250 (Pietersburg).
14. Ndutu Beds, Olduvai Gorge, Tanzania. 56,000 ± 3,500.

FIG. 2. Radiometric dates with provenience for comparable Later Pleistocene techno-complexes in Africa and Eurasia. Assemblages selected to show the time range of each complex. (After Clark. 1975, fig. 3).

FIG. 2. (continued)

Early 'Later Stone Age'
15. Heuningsneskrans Shelter, Natal. 24,640 ± 300.
16. Border Cave, Ingwavuma, Natal. 35,700 ± 1,100.
17. Rose Cottage Cave, Ladybrand, O.F.S. 29,430 ± 520 ('Pre-Wilton').
18. Leopard's Hill Cave, Zambia. 21,550 ± 950 ('Proto-Later Stone Age').
19. Sahonghong, Lesotho. 20,900 ± 270 ('Howieson's Poort').
20. Nelson Bay Cave, Cape. 18,660 ± 110 (Robberg Ind.).
21. Nelson Bay Cave, Cape. 11,950 ± 150 (Albany Ind.).

Late Stone Age (microlithic)
22. Naisiusiu Beds, Olduvai Gorge, Tanzania. 17,550 ± 1,000 ('Kenya Capsian').
23. Leopard's Hill Cave, Zambia. 16,715 ± 95 (Nachikufu I).
24. Zombepata Cave, Rhodesia. 15,120 ± 170 ('Wilton').
25. Kamasongolwa, Zambia. 13,300 ± 250 ('Wilton').
26. Lake Nakuru: transgressive phase to ~ 60m Kenya. 12,140 ± 206 (Blade Ind.).
27. Nelson Bay Cave, Cape. 9,080 ± 180; 2,660 ± 150 (Wilton).
28. Wilton Rock Shelter, Cape. 8,260 ± 720 (Wilton).
29. Melkhoutboom, Cape. 7,300 ± 80. (Wilton).
30. Matjes River Rock Shelter, Cape. 7,750 ± 300 (Wilton).

All the above are radiocarbon dates except (14) which is an amino-acid racemisation age (Bada & Protsch 1973).

B. North Africa and the Sahara

Acheulian and 'Pre-Mousterian'
1. Harounian beach, Casablanca, Morocco ~145,000–~125,000 (Moroccan Acheulian, Stage 8).
2. Haua Fteah, Cyrenaica. ~80,000–~50,000 (extrapolation dates) (Pre-Aurignacian').

Middle Palaeolithic (Mousterian/Aterian).
3. Bir Sahara, Egypt > 44,680 (Aterian).
4. Haua Fteah, Cyrenaica, Libya. 43,400 ± 1,300 (Levalloiso-Mousterian).
5. Taforalt, Morocco. ±34,550
   ±32,350   (Mousterian/Aterian)
6. Bou Hadid, Algeria. ±39,900   (Aterian).
7. Berard, Algeria. ±31,800   (Aterian).
8. Dar-es-Soltan, Morocco. > 30,000 (Aterian).

Upper Palaeolithic: Epi-Palaeolithic
9. Hagfet ed Dabba, Cyrenaica, Libya. 40,500 ± 1,600 (Dabban).
10. Haua Fteah, Cyrenaica, Libya. 16,070 ± 100; 10,600 ± 300 (Eastern Oranian).
11. Haua Fteah, Cyrenaica, Libya. 8,400 ± 150; 7000 ± 110 (Libyco-Capsian).
12. El Mekta, Tunisia. 8,400 ± 400 (Capsian).
13. Gebel Silsila, Egypt. 15,200 ± 700; 16,000 ± 800 (Sebekian).
14. Wadi Halfa, Sudan. 14,000 ± 280 (Ballanan).
15. Taforalt, Morocco, 12,070 ± 400; 10,800 ± 400 (Ibero-Maurusian).
16. Khanguet el Mouhaad, Tunisia. 7,200 ± 120 (Upper Capsian).
17. Columnata, Algeria. 7,750 ± 300 (Capsian).

All radiocarbon dates except no. (1) which is an Th^{230}/U^{234} age.

C. Europe

Acheulian
1. Grotte du Lazaret, France. 110,000 ± 10,000.

Middle Palaeolithic
2. Weimer-Ehringsdorf, Germany. 60,000 – 120,000 (Mousterian).
3. Gorham's Cave, Gibraltar. 47,700 ± 1,500 (Mousterian).
4. Calombo Cave, Italy. 32,000 ± 680 (Mousterian).
5. Velika Pécina, Yugoslavia. 33,850 ± 520 (Mousterian).
6. Grotte aux Ours, France. 48,300 ± 230 (Mousterian).
7. Regourdoux, France. 45,500 ± 1,800 (Mousterian).
8. Les Cottes, France. 37,600 ± 700 (Mousterian/Lower Perigordian).
9. Combe Grenal, France. 39,000 ± 1,500 (Mousterian).
10. Broion Cave, Italy. 46,400 ± 1,500 (Mousterian).
11. Moldova, Ukraine, U.S.S.R. > 44,000 (Mousterian).

FIG. 2. (continued)

Upper Palaeolithic
12. Abri Pataud, France. 33,300 ± 760; 34,250 ± 675 (Basal Aurignacian).
13. Abri Pataud, France. 29,300 ± 450; 32,800 ± 450 (Aurignacian).
14. Willendorf, Austria. 32,060 ± 250 (Aurignacian).
15. Abri Facteur, Dordogne, France. 27,890 ± 2,000 (Aurignacian).
16. Puits de l'Homme, Lascaux Cave, France. 16,100 ± 500 (?Magdelenian).
17. Arka, Hungary. 17,050 ± 350 (Eastern Gravettian).
18. Angles sur l'Anglin, France. 14,160 ± 80 (Magdelenian).
19. Cueva Reclau, Spain. 13,200 ± 600 (Solutrean).

Mesolithic
20. Grotte di Ortucchio, Italy. 12,619 ± 410.
21. Shippea Hill. 7,610 ± 150 (Tardenoisian).
22. Starr Carr. 9,557 ± 210 (Maglemosian).

Neolithic
23. Nea Nikamedia, Greece. 6,100 ± 420.

All are radiocarbon dates except (1) dated by Th^{230}/U^{234} and (2) dated by Th^{230}/U. and Pa^{231}.

All are radiocarbon dates except (1) dated by Th^{230}/U^{234} and (2) dated by Th^{230}/U. and Pa^{231}.

D. South-West Asia

Middle Palaeolithic
1. Shanidar Cave, Iraq. 50,600 ± 3,000 (Mousterian).
2. Shanidar Cave, Iraq. 46,900 ± 1,500 (Mousterian).
3. Tabun Cave, Mt. Carmel, Israel. 40,900 ± 1,000 (Mousterian).
4. Geulah Cave, Mt. Carmel, Israel. 42,000 ± 1,700 (Mousterian).
5. Ras el Kelb, Lebanon. > 52,000. (Mousterian).
6. Ksar Akil, Lebanon. 43,750 ± 1,500 (Mousterian).

Upper Palaeolithic
7. Ksar Akil, Lebanon. 28,840 ± 380 (Aurignacian).
8. Ein Ager, Negev. 17,510 ± 290 (Final Upper Palaeolithic).
9. Rasaket. ± 34,600 (Aurignacian).
11. Shanidar, Iraq. 33,300 ± 1,000 (Baradostian).
12. Shanidar, Iraq. 28,700 ± 700 (Zarzian).

Mesolithic/early Neolithic
10. Zawi Chenu, Shanidar, Iraq. 10,800 ± 300 (Pre-Pottery Neolithic).
13. Shanidar, Iraq. 10,600 ± 300 (Mesolithic).
14. Jericho, Israel. 11,116 ± 107 (Natufian).
15. Jericho, Israel. 10,300 ± 500 (Pre-Pottery Neolithic).
16. Ganj-i-Dareh, Iran. 10,400 ± 150 (early Neolithic).

(Klein 1975a, 1975b). However, while these industries show the usual Levallois and disc core technology, they are characterized by a significant *blade* component which suggests a parallel with the Pre-Aurignacian industry from the lower part of the Haua Fteah Cave sequence in Cyrenaica and also with the "Amudian" blade occurrences in the Middle East. Whereas this early blade tradition is replaced by Mousterian and other Levallois (Levallois-Mousterian) or non-Levallois (Jabrudian) industries in the Mediterranean basin, in the African subcontinent the blade element becomes even more pronounced by 40,000-50,000 years ago (Klein 1972, 1974). The Levallois technology is still present, but the most significant tools are many utilized and retouched blades, some trimmed into truncated and backed blades and lunates. Figure 3 shows some of the backed blades and other artifacts from the Nelson Bay Cave on the south coast, excavated by Richard

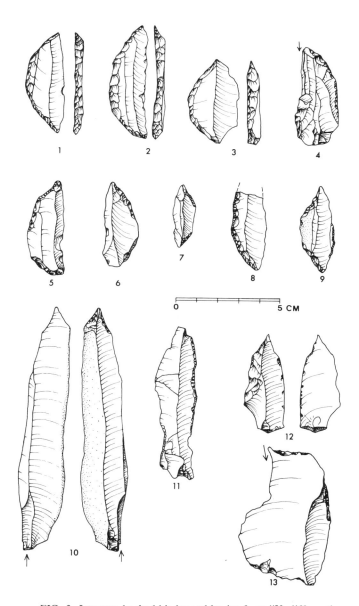

FIG. 3. Lunates, backed blades and burins from "Undifferentiated Blade Industry" occurrences from Nelson Bay Cave (nos. 1-3) and Montagu Cave (nos. 4-6) from South Africa dating to the late interglacial to early Würm, for comparison with "Libyan pre-Aurignacian" blade tools from the Haua Fteah Cave, Cyrenaica, dating to the last interglacial.

(Nos. 1-3 after Klein. 1972, fig. 7; nos. 4-9 below surface 7 ("Howieson's Poort") after Keller. 1973, figs. 11, 15, and 16; nos. 10-13 after McBurney. 1967, figs. iv 1 and iv 4).

Klein (1972:199), and from Montagu Cave (below surface 7) in the Cape Folded Mountain Zone (Keller 1973: plates 9, 15, 16), compared with Pre-Aurignacian blade artifacts from Haua Fteah (McBurney 1967:79, 83).

Since only preliminary reports of these blade industries are available, it is preferable to describe them as "undifferentiated blade industries." Though inferred similarity to the seemingly later occurrence at Howieson's Poort (ca. 18,000 B.P.) (J. Deacon 1966) has led to their ascription to that complex, they are separated from it if this date is correct by more than 30,000 years. Blade technology by direct percussion is thus seen to have its origins in the Last Interglacial period, not only in the eastern Mediterranean basin but also in the interior and the southern end of the African continent. The Upper Paleolithic *punched* blade traditions that appear in Europe and the Near East about 35,000 years ago can no longer be regarded as the revolutionary invention of modern man. They now appear as novel but well advanced stages in an ancient tradition, the appearance and development of which would seem to correlate with the use of the different animal and plant resources that accompanied climatic fluctuation.

The end of the Middle Stone Age is more difficult to determine. The radiometric dating for it is not always available and archaeological sequences are incomplete. In the Cape biotic zone it disappeared well before 30,000 years ago, and the hiatus in the cave sequences shows that they were unoccupied for about 25,000 years. At the same time, in the Sahara, the severe desiccation corresponding to the onset of the main glacial period brought human occupation of the greater part of the desert to an end about 30,000 years ago (Ferring 1975, Wendorf and Schild, in press). In the Valley of the Nile, however, and in much of tropical Africa this hiatus does not appear to have occurred or was much less significant. Here regionally adapted industries dating between ca. 25,000 and 15,000 B.P. combined a developed blade technology with a refinement of the characteristic elements of the Middle Stone Age mode. Thus, in the Nubian Nile, the Halfan appears to be a development out of the Khormusan (Marks 1975: 441-2, Wendorf and Schild, in press); in Rhodesia the Tshangulan appears to be an evolved form of the Bambatan (Cooke 1971, Sampson 1974: 236-42), and in South Africa, the Howieson's Poort occurrences had their origins in the undifferentiated blade industry occurrences in the early stages of the last glacial period. In North Africa, the culture pattern follows very much that of Europe and the Levant. An Upper Paleolithic is present in Cyrenaica by 38,000 B.C. (McBurney 1975:419). In Upper Egypt, Upper Paleolithic blade industries are present from ca. 20,000 B.P., and by ca. 14,000 B.P. there is a marked diminution in the dimensions of all artifacts which become of microlithic proportions (Marks 1975:442-3).

The earliest Later Stone Age industries of southern Africa, which appear ca. 25,000-18,000 years ago, represent a complete technological break with the Middle Stone Age, but they have little or nothing in common with the Upper Paleolithic industries of Europe or North Africa (Klein 1974:272-4, Sampson 1974: 258-91). Rather they appear to represent novel adaptations in the use of macrolithic tool forms following the adoption of more efficient strategies in hunting and the processing of meat and plant foods. They are essentially of autochthonous origin. However, in tropical Africa by 18,000-16,000 years ago, the earliest fully-developed microlithic industries make their appearance in Zambia, eastern Zaire, and the East African countries, and are more widespread in the tropics by 15,000 B.P. (Clark 1975:191). These small lunates and backed blade forms are usually considered to be the cutting parts of composite tools, in particular of arrows. It is possible that the bow and arrow may have been first an invention of the tropics or subtropics in response to the difficulties of hunting animals living in the thicker vegetation that became dominant there at the close of the last glacial period.

One of the most exciting aspects of archaeological research made possible by radiometric chronology is the study of the relationships between a number of environmental and cultural factors, such as climate and paleoecology and the biological changes that replaced the pre-neanderthal and neanderthal lineages by modern man; population distribution at different periods in the past; prehistoric economies, behavior patterns, and the raison d'être behind apparently contemporaneous but distinctive regional tool kits; record of the spread of domestication, and so to the sequence of events that led up to the movements of negroid populations bringing village farming to the subcontinent.

It is now evident that the appearance of modern man in the continent was an event that took place during Middle Stone Age times. In North Africa the radiometric chronology documents the sequence of pre-neanderthal stock (as represented in the Rabat, Temara, and Salé fossils), through the neanderthalers of Jebel Irhoud, Haua Fteah (Tobias 1968, Jaeger 1975), and perhaps also Dar-es-Soltan (Debénath 1975), and there are now reports that the late Aterian is associated with cranial fragments of more modern form (Roche and Texier 1976). Similarly, in sub-Saharan Africa, the modern lineage is represented by several dated fossils during the Middle Stone Age, some showing characteristics derived from a rhodesioid ancestor. Also it may well be that several of the innovations in the cultural equipment are associated with this evolutionary development. The Middle Stone Age populations were big game hunters who modified their hunting strategies to meet changes in the game population in response to the interaction of grassland and closed vegetation communities. However, well documented and dated evidence by McBurney (1967:54-9, 99), Voigt (1973), Klein (1974), Tankard and Schweitzer (1974), Bada and Deems (1975), and others, now shows that those groups living along the sea coasts also began to make use of marine resources as early as the Last Interglacial period; this is as yet the oldest unequivocal evidence for use of marine resources in the world. Shellfish were collected by people living along the North African shore of the Mediterranean and again by others at the south end of the continent. Here also they began to exploit seals and flightless birds (though not yet fish and flying birds) as long as the sea remained close to the beach of the 6-8m level on which they camped. After about 35,000 years ago, however, when the sea had receded with the onset of the main glacial period and had exposed large areas of the continental shelf, these coastal sites were deserted by the human populations who presumably moved after the game more than 50 km onto the continental shelf to sites that are now submerged. The base camps in the coastal cliffs were reoccupied only during the Later Stone Age, after the sea level was returning close to its former height. The occupational hiatus observed in the cave sites in northwest Africa between the Aterian and the epi-Paleolithic (Ferring 1975) may perhaps be explained in a similar way by the movement of a large part of the game resources, and the hunters with them, out onto the warmer and more favorable habitats on the continental shelf. Similarly the desiccation that brought the Aterian occupation of the Sahara to an end may have been responsible for stimulating southward movements out of the desert into the savannas. When systematic studies begin in Chad and the Central African Republic, we may expect dated assemblages that will show whether or not it was by this route that the use of the tang might have been diffused to the later Middle Stone Age populations of the Zaire basin.

Fluctuations in population densities and movements are well shown for the Later Stone Age in southern Africa in a study by Jeanette Deacon (1974) based on the patterning of a total of 223 radiocarbon dates. In her histogram, shown in figure 4, it is possible to identify the most favored regions and to note when they began to be consistently occupied (the number of dates are shown vertically and the ages are along the bottom; the different regions — mountains, inland plateau, Namibia arid zone, the south central African tropical savanna and the Cape biotic zone — are indicated by different hatchings). The two most intensely occupied regions are the winter rainfall zone of the Cape and the central African savanna. The desert zone can be seen to have become increasingly popular through time. Occupation of the escarpment mountains seems to have started later and to have fluctuated with climatic trends, with peaks around 10,000 B.P. and again after 3000 B.P. In particular, though, it is interesting that the interior high plateau does not appear to have been occupied to any important degree before 4000 B.P., and there is a significant gap between 9500 and 4600 B.P. Her second histogram (figure 5) shows cultural distributions and, though it is not as complete as one might have hoped, especially for tropical Africa, it does show how the less formalized assemblages of the late Pleistocene and early Holocene (Robberg, Albany, Pomongwe Industries) are gradually replaced by Wilton microlithic horizons, with the transition taking place between 9000 and 7000 B.P. There is a significant gap in the dates that relate to the other widely distributed complex in South Africa — the Smithfield — between 9500 and 4600 B.P. If the Wilton and the Smithfield were mutually exclusive complexes, which was the conventional view, it is inconceivable that the latter should have ceased to be made for nearly 5000 years before being resumed. This gap provides further proof that these traditions — the Wilton and Smithfield — are part of a continually developing system of tool sets. The lack of sites on the interior plateau during this period is probably explained by changes in environmental conditions and the unfavorable nature of the high veld at that time, while the tropics and the Cape both appear to have continued to provide the preferred resources and other favored living conditions. Similarly, the high incidence of microlithic backed segments in some Later Stone Age cultural traditions has been correlated with the hunting of large game animals, the remains of which preponderate in the food waste (H. Deacon 1972).

Without radiocarbon, prehistorians would have made little or no progress in elucidating the origins and spread of the domestication process in Africa. Shortcomings today, and there are many, are mostly the fault of the prehistorians themselves. The dates show

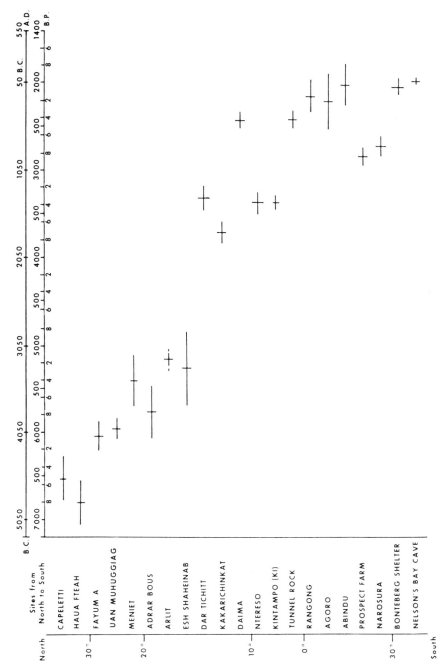

FIG. 4. Chronology of the spread of domestic stock in Africa, arranged geographically with the northernmost site at the top and the southernmost at the bottom. Radiocarbon dates from 21 sites associated with cattle and/or sheep/goat remains; shown to one standard deviation and based on the Libby half-life of 5568 years. Mean values have been used for sites with more than one dated sample. Number of dated samples recorded in brackets, Fayum A (4), Adrar Bous (2), Dar Tichitt (3), Karkarichinkat (6), Ntereso (3), Kintampo (3), Prospect Farm (2), Tunnel Rock (2), Narosura (3).

(after Clark 1972, In press.)

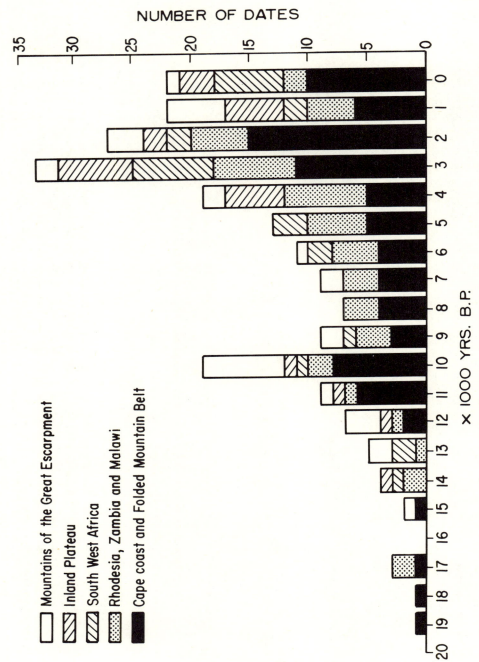

FIG. 5. Histogram of radiocarbon dates for the later Stone Age in southern Africa as distributed between the main geographical and biotic zones. (After J. Deacon. 1974, fig. 2.)

that between ca. 15,000 and 10,000 years ago there was a period of preadaptation to agriculture on the Upper Nile (Gautier 1968: 80-99, Greenwood 1968: 100-109, Wendorf and Schild, in press) when, from the food remains in the sites, it seems clear that some of the base camps were occupied all the year round (Churcher and Smith 1972). By preferential selection, and by seasonal use of terrestrial and freshwater animal protein and of wild cereal grains by hunting, fishing, fowling, and collecting, the populations there had been able to develop an enriched economy and to establish permanent settlements as an essential part of an ordered and limited system of transhumance. Failure of prehistorians to locate and investigate sites dating from the early Holocene, and dating before the fully agricultural settlements of the fifth millennium, is the reason why the origins of Egyptian agriculture still remain as obscure as they were before the Second World War. Not so, however, in the Sahara where the record of hunting/fishing communities in the sixth and seventh millennia reflects similar evidence for preadaptation as in the valley of the Nile (Clark, 1976). Domestic animals such as cattle, sheep, and goats are present in a sixth millennium context in the eastern Sahara (Wendorf et al. 1976) and were adopted more generally during the fourth and early fifth millennia on the evidence of remains from excavated camping places (Shaw 1972). Pictorial documentation in the art of the caves and rock shelters also reveals a unique record of these nomadic hunters/herders and grain collectors, showing their life-style to have been similar in many ways to that of the pastoral Hottentots at the south end of the continent. Dated settlements and records of fluctuating water levels in the Saharan lakes and stream systems (Butzer: N.D.) show that, after 3000 B.C., climatic deterioration and bad land management on the part of the pastoral peoples hastened the onset of desiccation. After 2000 B.C., the drought and widespread famine in the delicately balanced Saharan/Sahelian border zones initiated widespread movements of human and animal populations into the more favorable, better watered savannas to the south and east. The increased population densities caused by this imbalance and the consequent competition for resources among the hunting/gathering communities of the savanna and forest ecotone, can be seen as the catalysts that brought about the domestication of a wide range of food plants indigenous to these zones and the establishment of Neolithic farming settlements there during the second and late third millennia B.C. (Smith 1974, Flight, 1976).

Increased well dated archaeological evidence is needed before the dispersal details of the new economy can be known, but some indication is provided by radio carbon dating of the earliest appearance of domestic animal remains. As set out in figure 6, it is clear that the oldest sites with domestic animals are in the north and the youngest sites are in the south. The record begins in North Africa (top 3 sites), and those in the Sahara and on the Nile south of Nubia (next 9 sites) follow. There is a gap of about 1000 years (4800-3800 B.P.) before domestic animals are recorded in the Sahel and forest ecotone zones of West Africa (next 5 sites) and in Ethiopia (R. Gillespie, personel communication). In East Africa, all the sites with domestic animals are younger than 2000 B.C., the high altitude grasslands of the Rift being occupied before the more humid savanna in the Victoria basin (first 6 sites south of the equator). The last 2 sites, and there are 3 or 4

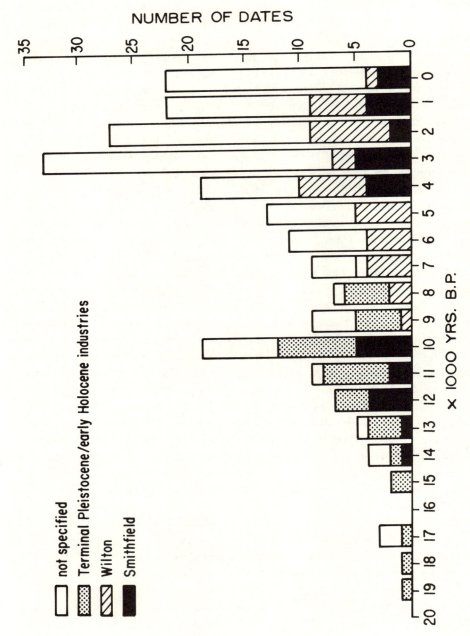

FIG. 6. Histogram of radiocarbon dates for the later Stone Age in southern Africa divided according to cultural associations. (After J. Deacon. 1974, fig. 6.)

more that could now be added, are where domestic sheep were associated with later Stone Age pastoral groups in the western Cape. This shows that sheep had been diffused to ancestral Hottentot peoples from some unknown source at least 2000 years ago, a few hundred years before the earliest Iron Age immigrants established themselves south of the Zambezi during the third or fourth centuries A.D.

There are now more than 400 published radiocarbon dates relating to Iron Age sites in Bantu-speaking Africa. Our understanding of the interrelationships between the various industries and groups, the cultural entities comprising the early and later Iron Age industrial complexes, rests almost entirely on the radiocarbon chronology. Using the radiocarbon dates, Phillipson (1975) has recently produced the first detailed study of the distribution in time and space of these industrial complexes, documenting the spread into the subcontinent of village farming and metallurgy as well as of the negroid physical stock, whose populations are assumed by most investigators to have been ancestral Bantu-speakers. Using a method developed by Ottaway (1973) for expressing the scatter of all grouped radiocarbon dates for a particular archaeological entity as the "interquartile range," Phillipson attempted to estimate the life period, the floruit time, of each of the main early and later Iron Age pottery wares in Bantu Africa. His study also made allowance for the shortcomings of the method, and he tested its reliability against one particularly well documented sequence.

The radiocarbon dates for early Iron Age times spread from 1020 B.C. to A.D. 1730, and the interquartile range is A.D. 610-A.D. 900. However, expansion of the data to include two-thirds of the known dates (the intersextile range) was found to give a closer approximation of the floruit period. Figure 7 shows the total and intersextile ranges for the early and later Iron Age entities in the main regions of Bantu Africa where investigations have been carried out. The results must be considered as general trends only, but they clearly show the older age of the early Iron Age wares and groups. These can be divided into two streams. An eastern stream begins earlier (270 B.C. - A.D. 390) in East Africa and spreads very rapidly during the fourth century down the eastern side of the continent to the northern parts of South Africa. The western stream (represented here by the groups in western and southern Zambia) does not begin until the fifth century (intersextile range A.D. 470, A.D. 920). The end of the early Iron Age complex comes as rapidly as its beginning. The later Iron Age complex begins generally in East, Central, and South Africa in the eleventh century. The absence, except in western Zambia, of any appreciable overlap between the early and later complexes, fully substantiates the sharp break that is exhibited by the archaeological evidence. Since many of the later Iron Age traditions continue to the present day, the latest dates are of little relevance and reflect only the frequency with which late sites have been dated.

The map, figure 8, illustrates well the results Phillipson obtained from this analysis. It shows that the East African material in the region of the Great Lakes (Urewe Ware), which was partly ancestral to the eastern stream, was the oldest (300-0 B.C.). The stream spread to the East African coast between A.D. 100 and 200 and between A.D. 300 and 400, very rapidly as far south as the Transvaal. Although the relationships between the

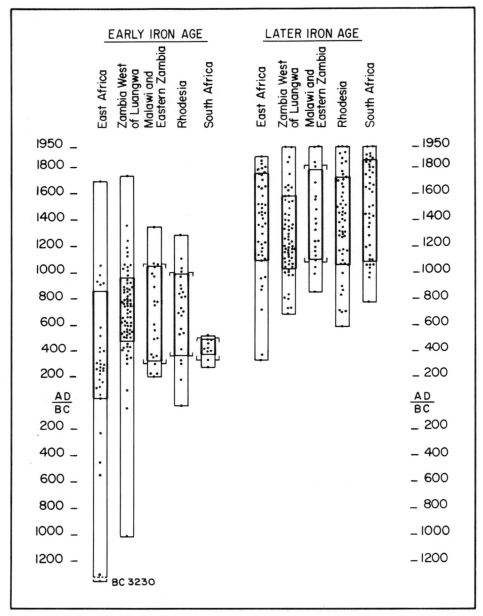

FIG. 7. Total (fine outline) and intersextile (bold outline) ranges of radiocarbon dates for early Iron Age and later Iron Age occurrences in various regions of Bantu Africa. For examples of less than 30 dates, the maximum limits for half the eventual dates are also given. (After Phillipson. 1975, fig. 1.)

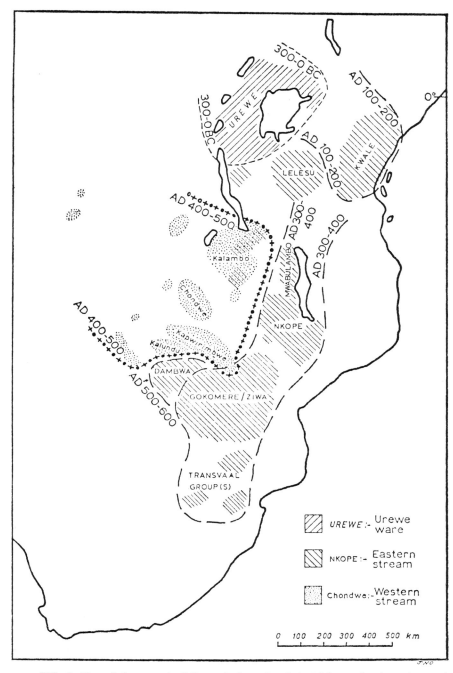

FIG. 8. Map of the spread of the early Iron Age industrial complex in eastern and southern Africa. The dated isochronal lines indicate the approximate date (on the radiocarbon calendar) of the commencement of the early Iron Age floruit in each area. (After Phillipson. 1975, fig. 2.)

eastern and the western streams are still uncertain, the isochronal map shows that the western is a later spread, though its origins are still obscure. This is of the greatest importance for historical linguistics and shows that the nuclear area in southern Zaire, from which Guthrie (1962) suggests that the Bantu-speakers spread, cannot have been connected with the main movements of early Iron Age culture as a whole, as others (Oliver 1966) have maintained, but only with the latest of these movements. The chronology and archaeological evidence are, therefore, now seen to be in agreement with the hypothesis of Greenberg (1972) that the homeland of Bantu was somewhere in the grasslands of Cameroun and to the east of Chad.

These, then, are some of the exciting developments in African prehistory made possible by radiocarbon chronology. There are still many uncertainties that derive from the method itself, and still more that derive from the gaps in archaeological surveys, from the lack of rigorous controls in the selection and collection of samples, and from the archaeologists' incomplete understanding and application of the results. Nevertheless, the method has been the most significant factor in revolutionizing knowledge of the age, of the relationships and behavioral patterning of prehistoric adaptations and cultural manifestations since the late Pleistocene. The new concepts, new goals, and new insights that characterize African archaeology today unquestionably owe their existence and their success to Dr. Willard Libby's inspired invention.

ACKNOWLEDGMENTS

I wish to thank the following authors and publishers for permission to reproduce the following figures that accompany this paper. For figure 1, the Musée Royal de l'Afrique Centrale: Actes du IV Congreès Panafricain de Préhistoire et de l'Etude du Quaternaire, 1962; for figure 2, the editors of *Man* 10 (2) 1975; for figure 3 and 5, Jenette Deacon and the editor of the *South African Archaeological Bulletin* (39) 1974, figures 2 and 6; for figure 6, Moutons. Origins of African Plant Domestication (1976); for figures 7 and 8, David Phillipson and the editors of *Journal of African History* 1975, (16) 334, 337.

REFERENCES

Bada, J. L., and L. Deems
    1975   Accuracy of dates beyond the ^{14}C dating limit using the aspartic acid racemisation reaction. Nature. 255:218-219.

Brooks, C. E. P.
    1931   The Correlation of Pluvial Periods in Africa with Climatic Changes in Europe. *In* Leakey, L. S. B. The Stone Age Cultures of Kenya Colony. Cambridge, Univ. Press., appendix B, 267-70.
    1949   Climate Through the Ages. London, Oxford Univ. Press.

Butzer, K. W.
    N.D.   Climatic changes in the arid zones of Africa during early to mid-Holocene times. Proc. Int. Symp. on World Climate from 8000-0 B.C. London, Oxford Univ. Press, 72-82.

Churcher, C. S. and P. Smith
  1972   Kom Ombo: Preliminary Report on the Fauna of Late Paleolithic Sites in Upper Egypt. Science 77:259-261.

Clark, J. D.
  1959   Carbon 14 Chronology in Africa South of the Sahara. Actes du IVe Congrès Panafricain de Préhistoire et de l'Etude du Quaternaire. Mortelmans and Nenquin, eds. 2:303-311.
  1969   Kalambo Falls Prehistoric Site. I. Cambridge, Univ. Press.
  1975   Africa in Prehistory: Peripheral or Paramount? Man (New Series). 10: 175-198.
  1976   Prehistoric Populations and Pressures Forming Plant Domestication in Africa. *In* Origins of Plant Domestication in Africa. Harlan, de Wet and Stemler, eds. The Hague, 67-105.

Cooke, C. K.
  1971   Excavation in Zombepata Cave, Sipolilo District, Mashonaland, Rhodesia. South African Archaeol. Bull. 26:104-127.

Cooke, H. B. S.
  1958   Observations Relating to Quaternary Environments in East and South Africa. Geol. Soc. South Africa. Annexure to Vol. 40.

Deacon, H. J.
  1972   A Review of the Post-Pleistocene in South Africa. South African Archaeol. Soc. Goodwin Series. 1:26-45.

Deacon, J.
  1974   Patterning in the Radiocarbon Dates for the Wilton/Smithfield Complex in Southern Africa. South Africa Archaeol. Bull. 29:3-18.
  1966   An annotated list of radiocarbon dates for sub-Saharan Africa. Ann. Cape Provincial Museums. 5:5-83.

Debénath, A.
  1975   Découverte de restes humains probablement atériens à Dar Es Soltan (Maroc). C. R. Acad. Sc. Paris. 281:875-876.

Ferring, C. R.
  1975   The Aterian in North African Prehistory. *In* Problems in Prehistory: North Africa and the Levant. Wendorf and Marks, eds. Dallas. 1:113-126.

Flight, C.
  1976   The Kintampo culture and its place in the economic prehistory of West Africa. *In* Origins of Plant Domestication in Africa. Harlan, de Wet, and Stemler, eds. The Hague 211-221.

Flint, R. F.
  1959   Pleistocene Climates in Eastern and Southern Africa. Bull. Geol. Soc. Amer. 70:343-74.

Gautier, A.
  1968   Mammalian Remains of the Northern Sudan and Southern Egypt. *In* The Prehistory of Nubia Wendorf, ed. Dallas. 1:80-99.

Greenberg, J. H.
  1972   Linguistic evidence regarding Bantu origins. J. African Hist. 13:189-216.

Greenwood, P. H.

    1968   Fish Remains. *In* The Prehistory of Nubia. Wendorf, ed. Dallas. 1:100-109.

Guthrie, M.
    1962   Some developments in the prehistory of the Bantu languages. J. African Hist. 3:273-282.

Jaeger, J. J.
    1975   The Mammalian Faunas and Hominid Fossile of the Middle Pleistocene of the Maghreb. *In* After the Australopithecines: Stratigraphy, Ecology, and Cultural Change in the Middle Pleistocene. Butzer and Isaac, eds. The Hague, 399-418.

Keller, C. M.
    1973   Montagu Cave in Prehistory: A Descriptive Analysis. Anthropological Records 28. Berkeley and Los Angeles, Univ. of Calif. Press.

Klein, R. G.
    1972   Preliminary report on the July through September 1970 excavation at Nelson Bay Cave, Plettenberg Bay (Cape Province, South Africa). *In* Paleoecology of Africa. van Zinderen Bakker, ed. 1969-1971. 6:177-210.
    1974   Environment and Subsistence of Prehistoric Man in the Southern Cape Province, South Africa. World Archaeol. 5:249-284.
    1975   Middle Stone Age Man-Animal Relationships in Southern Africa: Evidence from Die Kelders and Klasies River Mouth. Science. 190:265-267.
    1975b Ecology of Stone Age Man at the Southern Tip of Africa. Archaeol. 28:238-247.

Libby, W. F.
    1951   Radiocarbon Dates II. Science. 114:291-296.
    1952   Chicago Radiocarbon Dates III. Science. 116:673-681.

Marks, A. E.
    1975   The Current Status of Upper Paleolithic Studies from the Maghreb to the Northern Levant. *In* Problems in Prehistory: North Africa and the Levant. Wendorf and Marks, eds. Dallas. 1:439-458.

McBurney, C. B. M.
    1967   The Haua Fteah (Cyrenaica) and the Stone Age of the South-East Mediterranean. Cambridge, Univ. Press., 79-83.
    1975   Current Status of the Lower and Middle Paleolithic of the Entire Region from the Levant through North Africa. *In* Problems in Prehistory: North Africa and the Levant. Wendorf and Marks, eds. Dallas, 411-426.

Movius, H. L.
    1959   Radiocarbon Dates and Upper Paleolithic Archaeology in Central and Western Europe. Current Anthro. 1:355-391.

Oliver, R.
    1966   The Problem of the Bantu expansion. J. African Hist. 7:361-376.

Ottaway, B.
    1973   Estimating the Duration of Cultures. Antiquity. 47:231-232.

Phillipson, D. W.
    1975   The Chronology of the Iron Age in Bantu Africa. J. African Hist. 16:321-342.

Roche, J. and J. P. Texier
    1976   Découverte de restes humains dans un niveau atérien supérieur da la grotte des contrebandiers à Temara (Maroc). C. R. Acad. Sc. Paris. 282:45-47.

Sampson, C. G.
    1974   The Stone Age Archaeology of Southern Africa. New York.

Shaw, T.
    1972   Early Agriculture in Africa. Hist. Soc. of Nigeria. 6:143-191.

Smith, A. B.
    1974   Preliminary Report of Excavations at Karkarichinkat, Mali, 1972. West African J. Archaeol. 4:33-56.

Söhnge, P. G., D. J. L. Visser and C. van Riet Lowe
    1937   The Geology and Archaeology of the Vaal River Basin. Mem. Geol. Survey of the Union of South Africa. 35:1-184.

Stuiver, M. and E. S. Deevey
    1961   Yale Natural Radiocarbon Measurements 6. Radiocarbon. 3:134-135.

Summers, R. F. H.
    1955   The Dating of the Zimbabwe Ruins. Antiquity. 114:107-111.

Tankard, A. J. and F. R. Schweitzer
    1974   The Geology of Die Kelders Cave and Environs: a Paleoenvironmental Study. South African Science. 70:365-369.

Tobias, P. V.
    1968   Middle and Early Upper Pleistocene Members of the Genus *Homo* in Africa. *In* Evolution und Hominization. Kurth, ed. 2d ed. Stuttgart, 176-194.

Vogel, J. C. and P. B. Beaumont
    1972   Revised Radiocarbon Chronology for the Stone Age in South Africa. Nature. 237:50-51.

Voigt, E.
    1973   Stone Age Molluskan Utilization at Klasies River Mouth Caves. South African J. Science. 69:306-309.

Wendorf, F. and R. Schild
    (In press)   The Middle Paleolithic of the Lower Nile Valley and the Adjacent Desert. *In* Stone Age Technology and Culture. Tixier and Ghosh, eds.

Wendorf, F., R. Schild, R. Said, V. Haynes, A. Gautier, and M. Kobusciewicz
    1976   The Prehistory of the Egyptian Sahara, Science 193:103-114.

3

Sea Level Variations and the Birth of the Egyptian Civilization

Jacques Labeyrie

Recent measurements of past sea levels indicate that the profile of the lower Nile valley has varied considerably from 18,000 B.P. to the present, preventing the development of a flood-based civilization in lower Egypt during prehistoric periods.

According to all ^{14}C dates published before June 1975 in *Nature, Science,* and *Radiocarbon*, there is no human remain or artifact appearing in the lower Nile valley, including the Delta and the first 800 kilometers from the mouth, prior to 5000 B.P. (i.e., 3700 B.C. see Michael and Ralph 1974) (fig. 1). On the other hand, many artifacts dated between 5000 B.P. and 20,000 B.P. have been found in upper Egypt, Nubia, Fayum, in the desert surrounding the Nile valley in Jordan, and in Israël (fig. 1). (Note that the years B.P. are given in ^{14}C scale, which are converted to calendar years B.C. according to Michael and Ralph 1974.)

This absence of any remains of civilization in lower and middle Egypt before 5000 B.P. appears to be brought about by major variations in the general sea level during the Pleistocene and Holocene.

Measurements on the shores of Pacific Islands (Labeyrie et al. 1969, Bloom et al. 1974), on the western African Atlantic (Delibrias 1974) coast and on the Languedoc (Mediterranean) coast (Labeyrie et al., 1976) show that during the last 30,000 years — and probably during the last 100,000 years — the mean sea level has remained lower than its present position, with the exception of a short period culminating at +3 meters ca. 4500 B.P. (Delibrias 1971) (figs. 2 and 3).

During this long period of negative levels, the Nile cut its bed much deeper than at present, so its mouth remained constantly at sea level. The lowest sea level was attained in 18,000 B.P. at about 120 meters below the present level. It appears in figure 4 that the mean slope of the lower Nile profile was much steeper (26 cm/Km, profile c), than at present (9 cm/km as a mean value along the lowest 1000 km, profile a). Profile c starts from Sohaq (26°30'N, 700 km from the sea), because a shallow, rocky threshold seems to exist under the present bed in this region. Moreover, during the upper Würm and the early

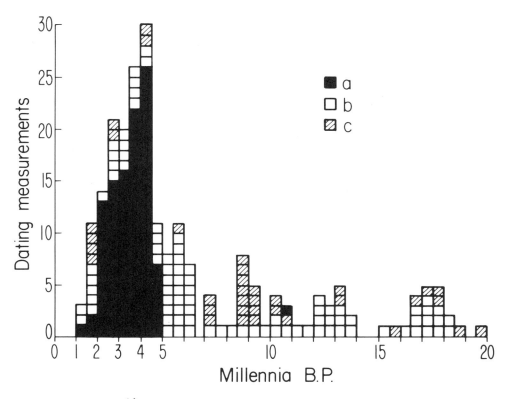

FIG. 1. Published ^{14}C dating measurements in *Nature, Science,* and *Radiocarbon* to June 1975. Delta and lower Nile valley (a), Upper Nile valley, Fayum, desert oasis (b), Israël, Jordan (c). The (a) type measurement at 10,500 B.P. are from 60 m depth in a boring at Tala-Betous, Nile Delta.

Holocene, a major flood period occurred (Fairbridge 1962, Wendorf et al. 1970), accelerating erosion of the valley.

The existence of a deep Nile bed during early Holocene is substantiated by the discovery of charred wood dated 10,580 B.P. (Nakhla and Mohammed 1974) in a drillhole at Tala-Betous, in the Delta, 100 km upstream from the present occidental mouth of the Nile. The charcoal samples were taken at a depth of 50 meters below present sea level, in accord with the past sea level curve in figure 2.

After its minimum at -120 meters the sea level rose continuously during 13,000 years at a mean speed of nearly 1 cm per year, invading the ancient valley and forming an estuary. This estuary probably remained shallow and boggy due to sedimentary deposition at each flooding of the river. At the time of maximum sea level (+3 m ca. 4500 B.P. or 3000 B.C.), which coincided with the early Old Kingdom, the southern end of the estuary was at about 28°30′N south of Beni-Suef and 450 km from the present Nile mouth (fig. 4, profile b). After attaining this maximum, the sea level decreased slowly, the estuary silted up quickly, and the valley began to fill with sediment. Figure 4 suggests

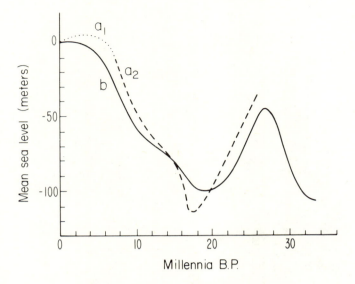

FIG. 2. Past mean sea level variations.
—a_2 from measurements on tropical West African coast.
—a_1 from Brazil coast (see fig. 3 for details).
—b from Languedoc (southern coast of France).

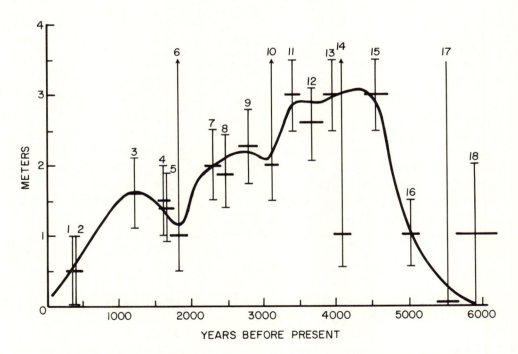

FIG. 3. General mean sea level during the past 7000 years measured on Brazil Atlantic coast.

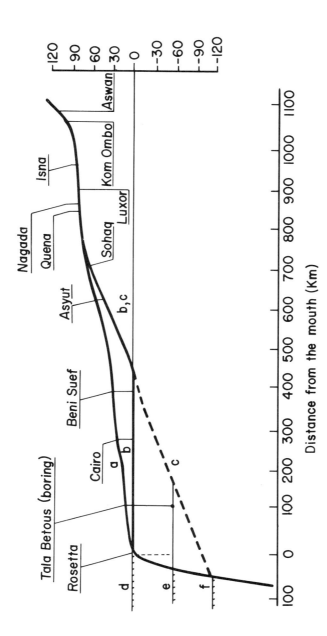

FIG. 4. Nile profiles: present (a), from 6500 B.P. to 4500 B.P. (b), ca 18,000 B.P. (c). Mean sea level: from 6500 B.P. to present (d), at 10,500 B.P. (e), at 18,000 B.P. (f). Profiels (a) from Carte OACI 1/1,000,000. ONC H-5. Defense Mapping Agency, Aerospace Center, Missouri 63118. Profiles (b) and (c) deduced from past sea levels (this paper).

that after the beginning of the sea's upward movement, layers of considerable thickness were deposited in the lower valley, ranging from 30 to 40 meters at Beni-Suef, to 50 meters at Cairo, to about 100 meters near the mouth. Since the time of the Old Kingdom, sedimentary deposits have attained about 15 m at Beni-Suef and 25 m at Cairo, but they have vanished gradually toward the south.

It is suggested that an important agricultural civilization making use of flood fertilization was unable to develop along the estuary which filled the lower valley during prehistoric times, because the level of the estuary was independent of the floods. Only after the beginnings of the Old Kingdom did the slope of the lower valley decrease, flattening its transverse profile and gradually increasing the area that was flooded each year. The upper valley south of Sohaq has, on the other hand, been periodically flooded independently of sea level fluctuations over a much longer period of time than the lower valley. It would seem probable that it is not in the Delta or in the lower valley that the birth of Egyptian civilization was prepared, but in the upper valley. It is, of course, possible that a pastoral or an "ordinary" agricultural civilization, not based on flood fertilization, developed during prehistoric times on the banks of the lower valley, with its remains now deeply buried under the sediments which deposited in the estuary.

REFERENCES

Bloom, A. L., W. S. Broecker, J. M. A. Chappell, R. K. Matthews, and K. J. Mesolella
    1974   Quaternary Research. 4:185-205.

Delibrias, G.
    1974   Colloque International du C.N.R.S. 219:127-133.

Delibrias, G. and J. Laborel
    1971   Quaternaria. 14:45-49.

Fairbridge, R. W.
    1962   Nature. 196:108.

Labeyrie, J., C. Lalou, and G. Delibrias
    1969   Cahiers du Pacifique. 13:59-68.

Labeyrie, J. C. Lalou, A. Monaco, and J. Thommeret
    1976   C. R. Acad. Sci. 282, D:349.

Michael, H. N., and E. K. Ralph
    1974   Radiocarbon. 16:198-218.

Nakhla, S. M., and F. Mohammed
    1974   Radiocarbon. 16:2-5.

Wendorf, F., R. Said, and R. Schild
    1970   Science. 169:1161-1171.

4

An Appeal from the Consumer

Lynn White, jr.

My brief paper will offer not information but exhortation. I am not a scientist, but rather an historian, one of the ultimate consumers of the dates arrived at by the physical methods that are the concern of most participants in this conference. What I want to say is that we humanists need from you more dates, more exact dates, and dates from all parts of the globe, not just as a convenience to our historical researches: we need them for a pressing moral reason.

The world of the late twentieth century is in many ways becoming unhinged, getting out of kilter. Nearly five-hundred years ago Columbus, Vasco da Gama, and Magellan — because of a concatenation of developments in medieval Europe that we are only beginning to understand — opened the great sea lanes and unified global communication for the first time. These last four-hundred-fifty years have been the era of oceanic empires, ruled by what today is called the First World; but in our own time we have seen these empires collapse, not least that of the United States. (The contemporary overland extension of the Tzarist-Soviet empire, today's Second World, is a related but rather different phenomenon.) The present human disarray is symbolized by the condition of the United Nations. In 1945 the UN was organized in San Francisco by statesmen whose outlook had been shaped by the age of Western imperialism which at that moment was disintegrating. It was a political mechanism for coping with the past, not the future. By 1976 the emergence of the post-colonial Third World had reduced the UN to a tragic incoherence which is merely the symptom of humanity's peril.

The danger to us all and to the continuity of civilization that is inherent in the present centrifugal forces is compounded by the spread of atomic weapons and the potential for making them. Our unhinging is the product of centuries of exploitation and oppression, but we may not have centuries in which to get hinged again. There is an element of time in our situation, although what it may be we can only guess.

Bitter memories have left a vast residuum of hatred in the Third World toward the First, including the United States. A tradition of dominance — euphemistically called "the white man's burden" — has made Westerners almost unconsciously contemptuous of non-Westerners. It is a psychic, if not a geographic, fact that in the twentieth century the

Mediterranean is far wider than it was, say, in the twelfth century when European intellectuals were feverishly translating Arabic scientific and philosophical works into Latin. Today — to speak simply of my own people — Islam is totally omitted from the education of most Americans, who get their image of the Muslim world largely from Shriners' conventions. Today, as a nation, we act like dolts in our relations with the Near East and we pay a catastrophic price for our ignorance. And, to most Americans, all Chinese *still* look alike simply because we know nothing in depth about China.

Having said this about my own tribe, I must add that most other peoples seem to have no clearer vision of groups unlike themselves. Our greatest peril on this shrinking, overpopulated, overarmed planet is mutual incomprehension. There may not be sufficient time to build and diffuse a global kind of education designed to reduce conflict. Nevertheless, it must be attempted because without such a common source of insight all clever devices for securing worldwide political and economic stability will be eroded and destroyed by the legacy of unreason.

In the late 1930's, Joseph Needham, a biochemist of genius, decided that mankind's greatest problem was less understanding nature than understanding the part of nature that is himself, that is, mankind. He abandoned biochemistry and since then has devoted himself to a masterly and monumental series of volumes on *Science and Civilization in China,* probably the most spacious and significant historical project now in process anywhere. On the surface, Needham's intent is to write the history of Chinese technology and science from the basic sources, and to show their relation to other aspects not only of East Asian but also of global culture. Below the surface, Needham is attempting something more. China — today one-quarter of mankind — has been isolated from the rest of humanity by ocean, deserts, and mountains. This isolation has made it defensive in attitude, yet its vast, cultural creativity has made it normally arrogant as well, contemptuous of anything alien. Very consciously, in his vast work, Needham is trying to reduce the barriers of ignorance and prejudice that for millennia have tended to separate the Middle Kingdom from most of the world. In all seriousness, I have proposed him for the Nobel Peace Prize. It has been awarded to some who have labored for peace in less basic ways.

I am an historian of medieval Europe. Thanks to what I call the "Marco Polo syndrome," considerable work has been done — not least by Needham — on cultural connections between China and Europe in the later Middle Ages. But what about India? Almost nothing has been learned, even though the evidence is ample that not a few Europeans reached India and the Indies before Vasco da Gama and that many Indic cultural elements reached medieval Europe and took root.

Here one runs into dating problems. Like the Mayas, the Chinese enjoyed dating things; and of course all the Semitic religions — Judaism, Christianity and Islam — with their dogmas of creation and a last judgment, have produced cultures in which it is axiomatic that the context of human destiny is a unique, linear time that had a beginning and will have an end. As the reading of Lord Kelvin's papers has shown, the second law of thermodynamics has been largely shaped by the Christian conviction of unidirectional and limited time.

The Indic religions, however — Hinduism, Buddhism, and Jainism — consider time cyclical and therefore not very significant: the cosmos palpitates through endless *kalpas* each of inconceivable duration, each with phases of birth, growth, decay, and death. One can be fairly confident that when modern Indian physics becomes autonomous, India will opt for a cosmology in which the red shift will at long last reverse itself, the universe will collapse into a single, black hole, and a new *kalpa* will begin. As our Western history shows, this sort of basic cultural preference is almost irresistible. (Incidentally, since Will Libby is present, let me say to him that I strongly suspect that if he had been a Hindu he would have become a superb physicist, but he would not have bothered to discover ^{14}C: time measurements and dating would not have seemed worth the trouble. He won his Nobel Prize for being a crypto-Christian.)

Indian nonchalance about time makes things inconvenient for historians working on global history. India did not produce, before the Muslim invasions, the sort of chronicles and annals to which European, Islamic, and Chinese scholars turn for dated information. Indian writings almost always have an abstract, mythic tone, a quality of timelessness. This attitude, common among the pre-Christian Greeks, expresses a view of experience that must be respected even if it is not shared. In the fourth century after Christ, Sallustius, one of the last great pagan intellectuals, said of myth that it "did not happen at any one time, but always is so: the mind sees the whole process at once: words tell part first, part second." Myth has its legitimate — even its essential — purposes; historians, however, deal not with eternal simultaneities but rather with time-bound sequences. India is hard on historians.

The imperviousness of India to historical happenings is indicated by the fact that no slightest reference has been found in India to Alexander the Great and his invasion of its northwestern region in 326 B.C. In the third century B.C., Aśoka, the greatest monarch of ancient India, set up some inscriptions that are roughly datable because they name Hellenistic monarchs with whom he had diplomatic realtions. Chinese Buddhist pilgrims went to India and returned to publish dated books telling what they had seen. The chronology of India and the further Indies is, to an amazing degree, derived from indications coming from cultures that took time seriously and had contact with India.

Let me offer a curious example of our resulting ignorance that is so serious an obstacle to writing global history. The first Christian millennium was a period of massive Indian expansion into Burma, Thailand, Malaya, Cambodia, and Indonesia. As an historical phenomenon it is comparable to the transplanting of Iberian culture to South and Middle America during the sixteenth century. Just as flamenco is danced in Lima and Mexico City, so today the *Mahābhārata* is performed in Bangkok and Djakarta. This Indic expansion is usually pictured as a peaceful penetration, the work of missionaries, merchants, and other kindly types. Considering the gory history of mainland India, I am skeptical; but the truth is that we know few details. Yet sometime between the first and the sixth century, a great migration of Indonesians populated Madagascar. There are linguistic reasons for believing that a large part of these emigrants took ship from the south coast of Borneo. Some of the earliest Indic inscriptions (they are in Sanskrit) thus far found in Indonesia come from Borneo and are of the fourth or fifth centuries. The Malagasy migration across six thousand kilometers of the Indian Ocean is one of the most spectacular and enigmatic movements of peoples in all history. Masses of families do not emigrate in such a way if all is going well in their homeland. Until, presumably by the new physical methods, we can date the Malagasy exodus much more exactly, and then, by means of carefully dated archaeological finds in Borneo, discover what was going on when it took to the sea, we shall be unable to make even educated guesses about the reasons for this incredible folkwandering.

In summary, the exigencies of the later twentieth century demand the construction of a global history to replace the disastrously myopic tribal histories now studied in all the world's schools. A new image of mankind must be provided, or there may be either no mankind or only residual, scattered barbarians.

Partly because some of the most interesting areas of human creativity either had no writing (like much of sub-Saharan Africa), or else had writing that we cannot yet read (like Maya Yucatan and Etruscan Tuscany), but more particularly because one of the largest, most seminal and literate cultures — India and the Indies — has not been much interested in time sequences, historians are badly handicapped in efforts to work out a global history that has coherence.

It follows that the perfecting and amplifying of new methods of dating, together with the provision of specific dates on a global basis, is a profound moral obligation. It is not simply a matter of scientific interest; it is far more than a game to be played in the laboratory. It is a scientific discipline to be pursued with passion, because in some measure the comity of peoples is at stake in a dangerous age.

5

A Radiocarbon Chronology for the Early Postglacial Stone Industries of England and Wales

V. R. Switsur and R. M. Jacobi

Clark, in his paper "A Microlithic Industry from the Cambridgeshire Fenland and other Industries of Sauveterrian Affinities from Britain" (Clark 1955), characterized what we now recognize to be but one of several variants of our later Mesolithic. Since 1955, over 150 radiocarbon determinations have become available for the Mesolithic of northern Europe, including Denmark, the Netherlands, Britain, and to a lesser extent, France. In 1973 it was suggested, considering the morphology of the stone industries and the relatively few (16) directly useful radiocarbon dates then available, that it was possible to divide our Mesolithic into an early and later stage. In elaborating on this concept through a discussion of the chronological evidence for the Mesolithic in Britain, we present herein the results of a series of radiocarbon measurements. A complementary paper on other aspects of the period is in press (Jacobi 1976).

The Mesolithic Age in Britain (Clark 1932) is still the largest and most useful archaeological source book. It has been followed, at the rate of about one major review article per decade, by publications seeking to interpret the particles of an ever-expanding corpus of information and data, covering just half of the postglacial period (Clark 1972; Radley and Mellars 1964; Clark 1955, 1954, and 1935).

It is helpful to divide Mesolithic artifactual material into two parts, (1) the microlithic component, comprising trimmed flint bladelets representing, primarily, the tipping and barbing of killing or wounding equipment and (2) the nonmicrolithic component, including scrapers, burins, and awls, associated with processing and tool renewal activities. Our discussion will be restricted to the microlithic component.

The microlithic shapes are divided into four main classes, as shown in figure 1, including (1) broad microliths, with a width greater than 7 mm and including simple, obliquely-blunted points, convex-backed pieces, isosceles triangles, and trapezes, (2) narrow microliths (classes 5-9) including 5 mm wide, scalene triangles, rods blunted along one or two sides, micro-rhomboids, lunates, boat-shaped, pear-shaped, and micro-tranchet (classes 5c–5d) microliths (3) Hollow based or Horsham points, and (4) a group of

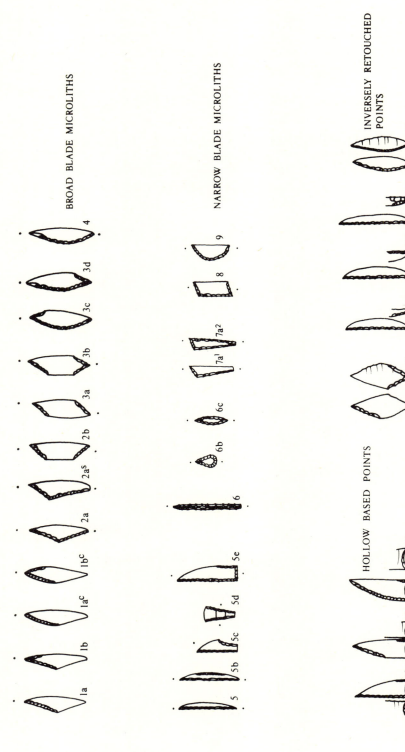

FIG. 1. Microliths from the Mesolithic.

inversely retouched points ranging from leaf-shaped to semicircular in outline, with the base brought to a pointed or rounded butt. The terms "geometric" and "nongeometric" are here replaced by the terms "narrow" and "broad," following the descriptive use by Francis Buckley (Petch 1924). In all thirty-one classes or subclasses of microliths were recognized.

The microlithic components of 115 sites in England and Wales were subjected to cluster analysis (see Table 1). The sites chosen possessed well recovered and unselected samples of either excavated or surface-collected material. In the case of earlier investigated sites, only collections with reliable documentation were incorporated into the present work. The sites were chosen to give the widest geographic coverage. Of the 115 sites, only the figures for Flixton site I (Moore 1950) and Misterton Carr (Buckland and Dolby 1973) were based on published data rather than on fresh counts.

It is appropriate to describe the process of cluster analysis, a procedure which has allowed us to link, on the basis of the microlith technology alone, a further eighty-three assemblages, which for various reasons are undatable or undated by other methods, to those dated by radiocarbon or pollen analysis, thereby assigning them to early or later Mesolithic (table 2). Considering cluster diagram 1 (figure 2), the groups formed as we work down the diagram are as follows: below the first division discriminating between early and later sites, a group of assemblages with a microlith component limited to narrow rod-like points, (classes 5, 5b, 6) is separated (cluster E) from the other, later sites. Dated to the early and middle fourth millennium B.C. at Rocher Moss south and at Dunford Bridge B (Q-799; 3430 B.C. ±80) on the Pennines, a total of some thirty identical sites can be recognized from this upland mass and the crests of the Cleveland Hills. In size these sites range from concentrations of worked flints and fire-crackled stone several meters across, probably indicative of a single tent, to substantial spreads 10 to 20 meters across, clearly the debris of multiple occupations. While isolated, rod-like microliths are frequently recorded as stray finds from low ground in northern England, no actual sites can be recognized below 1000 feet (300 meters).

Division into four groups continues in our diagram with the separation of the substantial cluster D at 0.218. Of the thirty sites comprising this cluster, all except two (Oakhanger VII and Broxbourne site 105) are northern English in distribution. Unlike the sites of cluster E, the sites recognized as making up this cluster are not confined solely to the Pennines and the Clevelands, but derive also from the lowlands around their flanks. This includes the coastal plain from Cheshire to Cumberland, and the eastern area from North Lincolnshire and the Yorkshire Wolds along the Durham coast. If one accepts the proposition that use of sites above 1,000 feet would have been largely restricted to the summer months, then it is of particular interest that their microlithic tool kit has linked so conclusively with that of the complementary sites at the lowest altitudes when populations of red deer, and hence their exploiting human groups, would have gathered during the winter months. Characteristic of this microlithic component, frequently referred to as Sauveterrian, is the absence of Horsham points (class 10) and leaf-shaped points (class 11), and the absence or extreme rarity of early microlithic shapes (classes 1-4). Instead, the assemblages are dominated by minute, scalene triangles (class 7) with a markedly lower proportion of classes 5, 5b, 6-6c, and 12a-12c.

TABLE 1

Cluster Analysis I
Later Sites

Cluster diagram number	Site name	County	Radiocarbon age B.C.
	CLUSTER B		
50	Farnham Pit 2	Surrey	
51	Selmeston Pit 1	Sussex	
52	Warnham	Sussex	
53	Stonewall Rock Shelter B	Kent	6110 ± 140 to 3820 ± 100
54	Broom Hill	Hants.	6590 ± 150 to 4585 ± 125
55	Wawcott Site 23	Berkshire	3910 ± 113
56	Tolpitts Lane	Herts.	6308 ± 120 to 4380 ± 80
57	Wicken Bonhunt	Essex	
58	Hullbridge	Essex	Before Zone VII B
59	Walton on the Naze	Essex	
60	Walton on the Naze	Essex	
62	Abinger Pit	Surrey	
63	Shippea Hill	Cambs.	5660 ± 150
64	West Keal Site I	Lincs.	
65	Sudbrook Sandpit	Lincs.	
66	Newton Cliff	Lincs.	
67	Honey Hill	Northants.	
68	Corley Rocks	Warwicks.	
69	Over Whiteacre	Warwicks.	
111	Hawkecombe Head	Somerset	
112	Wraxhall	Somerset	
113	Three Hole Cave	Devon	
114	Craig Y Llyn	Glamorgan	
115	Prestatyn	Flints.	Zone VII A
	CLUSTER C		
61	Wawcott	Berkshire	3310 ± 130
109	White Hill North	W.R. Yorks.	
110	Red Rotcher	W.R. Yorks.	
	CLUSTER D		
70	Broxbourne	Herts	Later than 5880 ± 520
71	Oakhanger Site 8	Hants,	
72	Risby Site V	Lincs.	
73	Risby VII	Lincs.	
74	Risby VIII	Lincs.	
75	Caistor	Lincs.	
76	Thorpe Common Shelter	W.R. Yorks.	4666 ± 220 to 3730 ± 150
77	South Rauceby	Lincs.	
78	Bagmore Mine No 2	Lincs.	

TABLE 1 (*continued*)

Cluster diagram number	Site name	County	Radiocarbon age B.C.
79	Holme on Spalding Moor	E.R. Yorks	
80	Mawdesley	Lancs.	VI/VII transition
81	Frodsham	Cheshire	
82	Crimdon Dene	Durham	
83	Filpoke Beacon	Durham	6810 ± 140
84	Eskmeals	Cumberland	
85	Drigg	Cumberland	
86	March Hill Top Site	W.R. Yorks.	
87	March Hill	W.R. Yorks.	
88	March Hill Top East	W.R. Yorks.	
89	March Hill Site 2	W.R. Yorks.	4070 ± 220
			3900 ± 80
90	Cupwith Hill N.W. Site	W.R. Yorks.	
91	Crooked Edge Clough	Lancs.	
92	Dunford A	W.R. Yorks.	
93	Broomhead Moor V	W.R. Yorks.	6620 ± 110
94	Mauley Cross	N.R. Yorks.	Zone VII A
95	Ecton Moor	N.R. Yorks.	
96	White Gill	N.R. Yorks.	
97	Coal Hill Whorlton	N.R. Yorks.	
98	Parci Gill	N.R. Yorks.	
99	East Bilsdale Main Site	N.R. Yorks.	
	CLUSTER E		
100	Cockayne	N.R. Yorks.	
101	Cow Ridge	N.R. Yorks.	
102	Pointed Stone	N.R. Yorks.	
103	East Bilsdale	N.R. Yorks.	
104	Bransdale Ridge	N.R. Yorks.	
105	Warcock Hill Site 4	W.R. Yorks.	
106	Cat Hill South Site I	W.R. Yorks.	
107	Rocher Moss South Site I	W.R. Yorks.	3880 ± 100
108	Dean Clough Site C	W.R. Yorks.	

The earliest sites in this cluster, indeed the earliest later Mesolithic sites in Britain, appear between 6800 B.C. at Filpoke Beacon, County Durham, and 6600 B.C. on the Pennines, Warcock Hill site III, and Broomhead site V. Dates within the fourth millennium from March Hill II (see figure 12) and Lominot site IV (Q-1189; 3660 B.C. ± 200) suggest that this tradition may have lasted from two and a half to three millennia. Although appearing to be similarly situated and partially contemporaneous with the rod-dominated assemblages of cluster E, these sites are rendered distinct on both the Pennines and Clevelands by typology and by the use of a completely different set of raw materials (Switsur and Jacobi 1975).

Two more clusters may be identified among our later Mesolithic sites. First, a group of three assemblages, cluster C, dominated by micro-rhomboids (class 8) of which only

TABLE 2

Cluster Analysis II
Early Sites

Cluster diagram number	Site name	County	Radiocarbon age B.C.
	CLUSTER A		
1	Broxbourne 104	Herts.	7660 ± 200
2	Star Carr	N.R. Yorks.	7610 ± 210
			7540 ± 350
3	Flixton Carr	N.R. Yorks.	Zone IV/V
4	Warcock Hill South	W.R. Yorks.	7265 ± 430
	CLUSTER B		
34	Rhuddlan Site M	Flints.	6580 ± 73
35	Aberystwyth	Cardigans.	
36	Nab Head	Pembs.	
37	Daylight Rock	Caldey Is.	
	CLUSTER D		
41	Broxbourne 106 B	Herts.	
44	Old Faygate	Sussex	
46	Bishop's Wood	Sussex	
47	Fox Hill	Sussex	
	CLUSTER E		
43	Roffey Halt	Sussex	
48	Old Bleeding Wood	Sussex	
49	Colgate	Sussex	
	CLUSTER C		
5	Broxbourne 102	Herts.	Prior to Zone VI c
6	Thatcham Site 1/5	Berkshire	7890 ± 160 to 7530 ± 160
7	Thatcham Site 2	Berkshire	Prior to 6626 ± 100
8	Thatcham Site 3	Berkshire	8415 ± 170 to 7715 ± 170
9	Greenham Dairy Farm	Berkshire	6830 ± 110
10	Marsh Benham	Berkshire	7740 ± 240 7350 ± 150
11	Oakhanger Site VII	Hants.	7275 ± 200 6935 ± 165
12	Downton (Patinated)	Wilts.	
13	Uxbridge	Bucks.	
14	Lackford Heath	Suffolk	
15	High Beech	Essex	
16	Kelling Heath	Norfolk	
17	Shedfield	Hants	
18	Hassocks: South Bank	Sussex	
19	Iping Common	Sussex	Zone VI
20	Wellocks Hill Basing	Hants.	

TABLE 2 *(continued)*

Cluster diagram number	Site name	County	Radiocarbon age B.C.
21	Dozmare Pool	Cornwall	
22	Ascott under Wychwood	Oxon.	
23	Brigham	E.R. Yorks.	
24	Willoughton A	Lincs.	
25	Misterton Carr	Notts.	
26	Deepcar	W.R. Yorks.	
27	Stanedge	W.R. Yorks.	
28	Warcock Hill N Site	W.R. Yorks.	
29	Pike Low Site I	W.R. Yorks.	
30	Hambleton Hill Site I	W.R. Yorks.	
31	Lominot Site 3	W.R. Yorks.	7615 ± 470
32	Aberffraw	Anglesey	6690 ± 150
33	Rhuddlan E	Flintshire	6790 ± 86
38	Buckland Corner, Reigate	Surrey	
39	Blackdown	Sussex	
40	Hassocks Sand Pit	Sussex	
42	Hastings Kitchen Midden	Sussex	
45	New Faygate	Sussex	

one (Wawcott site I in the Kennet Valley) has been dated, perhaps significantly, close to the local Mesolithic/Neolithic transition. The remaining pair (Red Rotcher and White Hill North) are both on the southern Pennines. It is clear from the large number of isolated finds of micro-rhomboids around White Hill that further sites must exist. Complementary to this Pennine material are groups of identical microliths in the same white flint from sites on the lower ground of the Lincolnshire Edge, most notably Manton Warren. It would be tempting to link these finds as further indicators of a summer/winter pattern of transhumance.

Only general comments will be made on the composition of the final cluster B made up of the residual twenty-four later sites incorporated in this analysis (numbers 50-69 and 111-115). Geographically, these sites are distributed over southern England, the Midlands, and Wales, the overall scatter being complementary to that of the sites forming clusters D and E. Chronologically these sites cover the time perior 6500 B.C. (Broomhill, Hants) to the end of the Mesolithic. In addition to possessing numerous later microlithic shapes (classes 5-8, occasionally 9), the sites retain a noticeable, often high, proportion of early microlithic shapes. It is this characteristic which helps to render them distinct from the sites of cluster D. In addition, in the Midlands and East Anglia sites contain many leaf-shaped points (class 11) and in southeastern England, sites contain Horsham points (class 10). In the southwestern peninsula the latter point type occurs sporadically, but it is absent from the assemblages considered in this analysis. These comments must be regarded as simplistic and highly generalized. The recovery of more "closed finds" and the application to these of radiocarbon dating shows that patterns of local evolution in

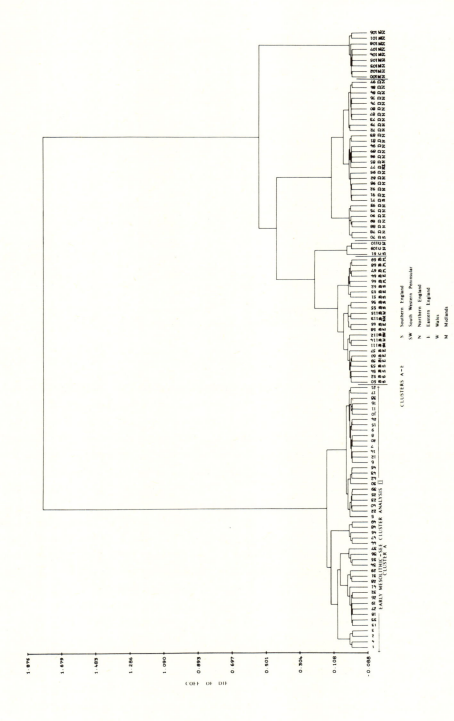

FIG. 2. Cluster diagram linking undated with dated sites.

the later Mesolithic and the occurrence of the more specialized point types are potentially more complex than has been hitherto supposed.

The next division is within the early Mesolithic group (at 0.127). Although it is archaeologically unsatisfactory, it is not unreasonable from the aspect of the tool frequencies. Subsequent divisions within the early group follow so rapidly that their differentiation cannot be treated with confidence. The number of variables (microlith shapes) is not so great among this early group of sites as among the later Mesolithic. Consequently, the distance coefficients are likely to be more distorted because of an excess of zero values among the sporadically represented classes 2b, 3a–d, and 4. The consequences of this are not discussed in this paper.

A separate examination was made of the early Mesolithic group of sites, producing a more refined division of the material (figure 3). The impressive feature of this diagram is that thirty-four of the sites have joined to form the single cluster C, spreading from Dozmare Pool on Bodmin Moor to the southern Pennines (Warcock Hill North, Pike Lowe) and from Kelling Heath on the northern Norfolk Coast west to Trwyn Ddu near Aberffraw on Anglesey. These sites, which could be described as Maglemosian, cover the same area that would in later times be divided between clusters B and E, with finds on the Clevelands and small sites in County Durham and Northumberland perhaps carrying their distribution northeast.

Separate from this major cluster are four sites, including Star Carr and Flixton Carr in Yorkshire, characterized by the rarity of convex-backed points (classes 3c, 3d, and 4) which comprise cluster A. A second group of four sites form cluster B, two of them on the present Welsh coast, the third on the Isle of Caldey, and the last at Rhuddlan (Flintshire). All are rendered distinct by rod-like points steeply trimmed down one or both margins, here associated with an otherwise early assemblage of microliths. The single date from Rhuddlan (site m) suggests that these sites may belong to the latter part of the seventh millennium B.C. On three of them there is significant evidence of coastal exploitation. The final clusters D and E incorporate all of Clark's classic Horsham sites except Warnham Lodge. These are characterized by a microlith tool kit dominated by obliquely-blunted points and isosceles triangles with a strong representation of Horsham points and asymmetric inversely-retouched points (classes 2a–c). The rare, later microlith shapes on these sites are suspected to be intrusive. As a combination of shapes, this assemblage appears confined within the Weald. There is the possibility, based on typological rather than on independent evidence, that it may fill a stage intermediate between such early sites as Iping site I (Sussex) and later sites such as Farnham, Warnham Lodge, or Stonewall, all in cluster B of figure 2, as described above.

Recapitulating the main observations from the cluster analysis, the dendrogram at distance coefficient 0.540 divides the assemblages into two prominent clusters. At this high level, each is comprised of numerous, distinct, smaller clusters from lower down on the diagram (a maximum of ten clusters at 0.073). The differences are striking. The first is made up of sites defined by the overwhelming preponderance of obliquely-blunted and simple microlith shapes and occasional, probably intrusive, narrow microliths. The second

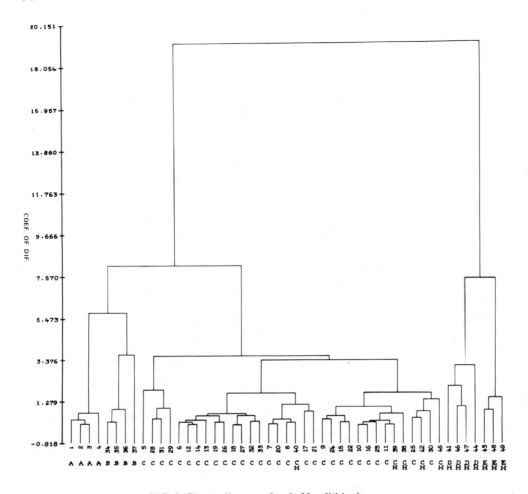

FIG. 3. Cluster diagram of early Mesolithic sites.

(and later) is defined by a preponderance of narrow forms, particularly rods and scalene triangles, with a sharp reduction in the importance of the simpler microlith shapes. Beyond this, variations in proportional representation of individual microlith types have yielded geographic groupings, particularly in the later Mesolithic, which are socially and economically significant.

Cluster analysis is an "agglomerative" technique gathering, in our case, more than one hundred individual sites or preexisting groups in order to unify new groups. It is more convenient to discuss the dendrogram produced as if it were the result of a "divisive" technique, that is, as if the original material were divided into two groups, then three, then four. This is particularly convenient, as here, where Ward's method has been used, for it results in neat clusters without stray sites, avoiding the necessity for discussing the early stages of the analysis when the groups formed may not be significant. Considering the groups as they increase numerically from the top downward, as the number of groups increases, so the difference between newly recognized groups decreases. Consequently, there comes a point when the differences between the relatively small groups is likely to become dependent on sampling error. Because of such factors as variation in sample size and change in the number of variables represented on each site, the definition of a minimum distance coefficient for recognition of valid groups is difficult to establish. In order to aid in the resolution of this, the data were examined at each division from the archaeological point of view and the process was arrested at an unsatisfactory division. This technique does not imply that subsequent divisions may not be valid.

Samples were taken from a representative number of sites and submitted to the radiocarbon dating process. Great care was exercised in the selection of samples to ascertain that the carbonaceous material was really associated with the site and the appropriate microlith assemblage. Usually the sample consisted of charcoal or burned wood, but in some cases bone was considered suitable. The chemical and physical pretreatment of the samples was meticulously performed prior to bomb combustion and gas counting of the radioactivity. In the instances where bone was used, it was first demineralized and the extracted collagen was used in preparation of the counting gas. The resultant measurements, in the form of conventional radiocarbon dates, together with some dates that were previously available, are depicted in figure 4. The majority of dates were obtained specifically in this project.

Pollen evidence from organic deposits (figure 5) confirmed that, while all except three (Broxbourne 102, Thatcham sites I, II) out of ten sites with a simple, microlithic assemblage were sealed by a Preboreal or early Boreal peat (Godwin's zones IV-VIa), no site with narrow, late microlith forms could be demonstrated to be earlier than zone VIb. Of those sites not yet dated by the radiocarbon method, all became covered by peat only during zones VIIa or VIIb. Where there is a temporal overlap, it is between the actual radiocarbon dated *occupations* of later sites and the pollen dated or radiocarbon dated organic deposits *overlying* early sites. Thus the flint industry of Broxbourne 102 is earlier by an unknown factor than the overlying late Boreal peat, whose relative position within zone VIc is also unknown. At Thatcham site I, the artifacts are overlain by late Boreal, zone VIc peat (Wymer 1958), but artifactual material representing debris from the site 4 to 5 meters away are found scattered into the swamp marl of site V, having entered the marl during the much earlier zones IV to VIa. The pair of dates from site II at Thatcham represents merely a *terminus ante quem* for the industry, since the material used for dating was a dark, humified substance derived from the peat which sealed and preserved the industry below (Barker and Mackey 1960).

From this list are omitted seven sites lacking these organic deposits, the artifacts from which are dated by their stratigraphic relationship to pollen grains within a mineral soil. In such a profile, only the overlying raw humus can be counted as "organic." Controversy has surrounded the interpretation of the diagrams, and while it appears that the relative depths of the pollen spectra within the soil represent changes in local ecology through time (Dimbleby 1962), the exact relationship of these spectra to other larger objects, such as flint artifacts, within the profile is unknown. Apart from Oakhanger VII,

Pl. 1. Profile in the bank of a gravel pit at Broxbourne (site 106A) Hertfordshire, showing thin bands of organic mud and redeposited algal marl overlying a compressed peat. Early mesolithic microliths and cores were recovered from this peat together with bone splinters and unburnt hazel nuts. These last were dated Q-1145 = 9,360 +/− 150 years B.P.

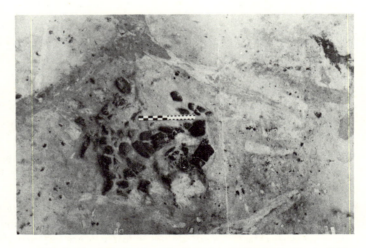

PL. 2. Carefully built wall of sandstone blocks in front of the Rock Shelter of Hermitage Rocks, near High Hurstwood, E. Sussex. Associated with the hearth was a rich later mesolithic stone industry with a wide range of microlithic shapes. Charcoal from the site was dated as Q-1312 = 6,800 +/− 110 B.P.

the ages suggested for these sites (Boreal for the early type assemblage from Iping Common [Keef et al. 1965], and Atlantic for the later type industry at Mauley Cross, White Gill site VIII (Simmons and Cundill 1970), and Addington Kent [Dimbleby 1963]) would appear to confirm the chronologies outlined.

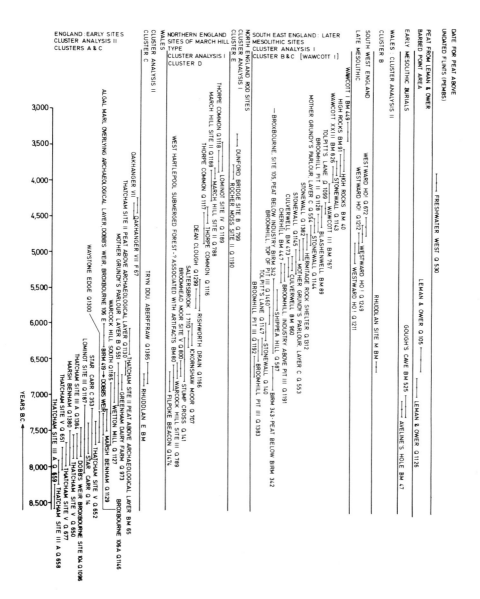

FIG. 4. English and Welsh Mesolithic radiocarbon dates.

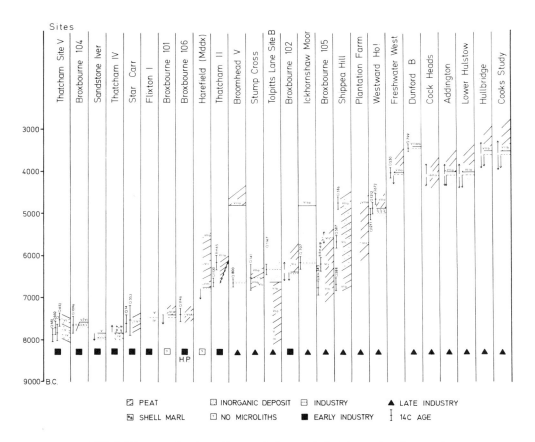

FIG. 5. Diagram of various sites correlating age, industries, and environment.

Oakhanger VII, level 2 (Rankine and Dimbleby 1960), which occurs in cluster analysis II as site 11 (figure 2), provides a single example of an industry, apparently later than 6800 B.C., where narrow microlith shapes are completely absent. This second analysis suggests that, of the four sites with which it is linked most closely, the two typologically closest microlith components would be Marsh Benham, which was dated by radiocarbon as 7350 B.C. ± 150 (Q-1129) and 7740 B.C. ± 210 (Q-1380), and Kelling Heath, Norfolk. The latter site has recently been completely destroyed and it is not possible to collect samples for radiocarbon or thermoluminescent dating. The original radiocarbon dates for the Oakhanger site were published in 1960 as 4430 b.c. ± 115 (F-68) and 4350 B.C. ± 110 (F-67), and were based on pine wood and hazel nut shells, respectively, using no pretreatment. The samples were collected at a depth of eight inches below the present surface of a much-disturbed *Calluna* heath, at the base of the A^1 horizon of a humus-iron podsol, rich in humic material. In order to check these problematic dates, so apparently out of sequence with the ages of this type of industry, further samples collected during the original excavation were obtained from Mrs. W. M. Rankine, who had carefully preserved them. The newly-dated samples were of pine charcoal and of hazel nut shells from the same horizon as the original specimens. They were thoroughly pretreated prior to radioactivity measurement. The results are given in table 1. Their clustering around 7000 B.C. is particularly striking (the pooled average is 7085 B.C. ± 66), and is more reasonable than the earlier datings mentioned above. For comparison, we present dates in the table for a later Mesolithic site which has artificial pits at Broomhill, near Romsey, also in Hampshire only twenty-four miles to the southwest. The bulk of over 2000 microliths from this latter site are narrow rods (class 5) or scalene triangles (class 7). The probability is strong for a typological succession identical with that which we document from elsewhere in Britain.

TABLE 3

Radiocarbon Dates for Oakhanger and Broomhill

Oakhanger VII		Broomhill	
Q-1489	7275 B.C. ± 170	Q-1128	4585 B.C. ± 125
Q-1490	7045 B.C. ± 160	Q-1191	5270 B.C. ± 120
Q-1491	7150 B.C. ± 160	Q-1460	5880 B.C. ± 120
Q-1492	7025 B.C. ± 160	Q-1383	6365 B.C. ± 150
Q-1493	7090 B.C. ± 160	Q-1192	6590 B.C. ± 150
Q-1494	6935 B.C. ± 160		

Support for the conclusion that industries with narrow-blade and geometric elements fall later in time than industries with only borad-blade microlith forms is given by the six sites or areas where it has been possible stratigraphically to relate industries of geometric and nongeometric types. In no case has a broad-blade, nongeometric industry been observed to overlie a narrow-blade, geometric industry.

At Broxbourne, in the valley of the river Lea, it has been possible to relate five occupation sites to various stages in the infilling of the late glacial or postglacial Nazeing channel. Two of these sites were investigated over forty years ago (Warren at al. 1934). At site 101, the artifacts lay below a peat of zone VIa age. At site 102, the industry contained obliquely and completely blunted microliths, together with a single, rod-like microlith. The industry lay toward the base of a leached sand, which was sealed by a peat, which began to accumulate during zone VIc. On site 104, artifacts from an occupation site could be traced passing out into deposits of the adjacent Nazeing channel. The peat, containing flecks of charcoal, was radiocarbon dated as 7660 B.C. ± 200 (Q-1096), and pollen dated to the base of zone V. The highest point of the ridge was sealed by peat during zone V-VIa. The tools bear visual similarity to those from Star Carr, with which they clustered in analysis II (cluster A), with wide, obliquely-blunted microliths, lacking retouch on the leading edge, and large scalene and isosceles triangles (figure 6). In their dimensions and degree of obliquity, these points differ markedly from the smaller, curved, backed microliths of site 102.

In sharp contrast is a later Mesolithic group (figure 7) with minute, narrow, scalene

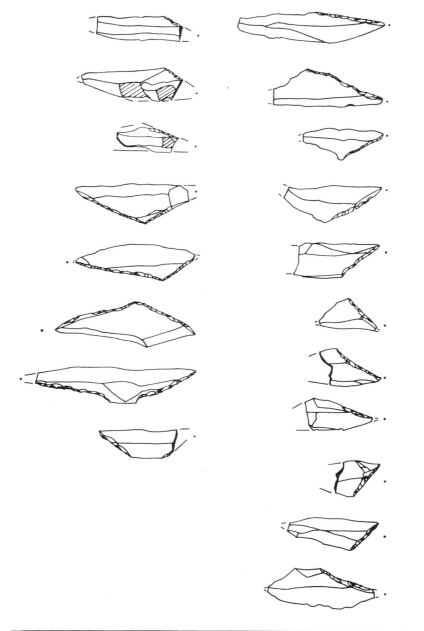

FIG. 6. Microliths from site 104 near Broxbourne, Valley of the River Lea.

FIG. 7. Stone tools from site 105.

triangles and rods from a small occupation site, 105, within the upper part of the peat infilling the same Nazeing channel mentioned in connection with site 104 and less than 100 meters to the east. This occupation has been pollen dated to late in zone VIc and lies above a wood peat that has been radiocarbon dated as 5880 B.C. ± 520 (Birm-342). The relative and absolute stratigraphy of this pair of sites is confirmed by a second radiocarbon date from halfway through the channel section Birm-343 = 6750 ± 170 B.C.

In the Colne Valley, Mesolithic finds are recorded from flood plain sites spread along seven miles of the river. Two of the sites, Sandstone Iver and Denham, have yielded only early microlith shapes. The group from Denham, near Uxbridge (Lacaille 1963), clusters most closely with Thatcham site I, whose suggested age lies within pollen zones IV to VIa. Samples of the overlying deposit collected by Haward, the original investigator, have now become too oxidized to yield pollen for analysis. However, comparison of the pollen diagram from Sandstone Ivor with that from Thatcham site V suggests that peat began to form over the artifacts as early as zone V, and certainly not as late as the late Boreal, suggested by Mitchell (figure 8); at Thatcham, zone V had ended before Q-652 = 7530 B.C. ± 160.

At Tolpitts Lane, Rickmansworth, seven miles northeast of Sandstone Ivor, a clay with Mesolithic artifacts passes into the backswamp deposits behind the site over a humified "channel-peat" of zone IV to VIb age. The similar ecological situation at the two sites suggests the relative synchroneity of the pollen zonations. The site itself is independently dated at two points: Q-1147 = 6310 B.C. ± 120 dates a sample of charcoal from the filling of a shallow scoop which contained much waste, cores, a backed flake, several thick, rod-like pieces, and small, scalene triangles, (figure 9), while the second date, Q-1099 = 4380 B.C. ± 80 dates scattered pieces of charcoal and nut shells associated with later microlith shapes and a wider range of nonmicrolithic equipment.

Along the third of the major tributaries of the Thames, the Kennet, the division of the sites into early and later groups is immediately apparent both from their microlithic content and their corresponding radiocarbon dates. A date of Q-1130 = 6630 B.C. ± 100 for the peat overlying Thatcham site II suggests that the underlying early Mesolithic industry must date, at the latest, from the earlier part of the seventh millennium B.C., a proposition which receives some support from a date of Q-793 = 6830 B.C. ± 110 for the collagen extract of roe and red deer bones associated with early Mesolithic material from a cutoff meander at Greenham Dairy Farm, Newbury (Sheridan et al. 1967). This date is the latest acceptable one for such an industry in England. In contrast, the earliest date for a late Mesolithic industry in the valley is that of BM-767 = 4170 B.C. ± 134 for Wawcott site III (Froom 1963).

This last site is stratified into a gradually accumulating, sandy flood-loam. It was excavated in a series of thirteen three-inch spits. Of these the lowest, spits J to M, contained only a range of simple microlith shapes (classes 1 to 4) identical with that of Marsh Benham (dated with two charcoal samples as Q-1380 = 7740 B.C. ± 240 and Q-1129 = 7350 B.C. ± 150), Thatcham, Greenham Dairy Farm, and Victoria Park, Newbury. All the higher spits, A to I, contained later shapes, their proportions relative to the simpler forms increasing toward the top of the section (Figure 10). The radiocarbon date BM-767 = 4170 B.C. ± 134 is for the contents of pit 2, at the level of spits E-F, which is roughly the center of the layers with the later microlith shapes, and it is in close agreement with the age BM-826 = 4129 B.C. ± 113 for charcoal from Wawcott XXIII, where 89 of the 105 microliths are of categories unrepresented on early sites. Once again, the evidence of stratigraphy and radiocarbon dating is in close agreement within the valley, and it is also in agreement with that from the valleys of the rivers Lea and Colne.

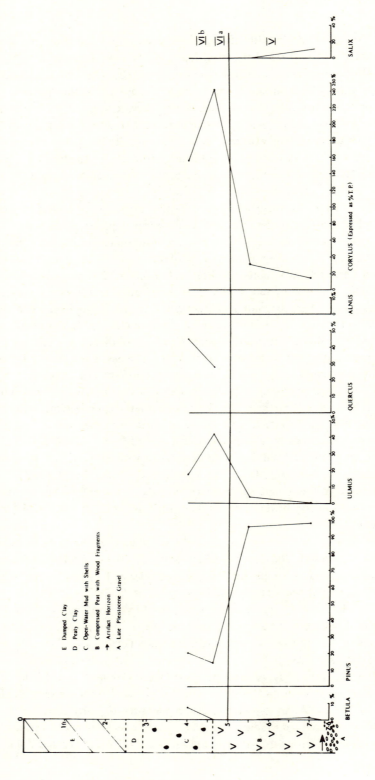

FIG. 8. Pollen diagram from Sandstone Ivor.

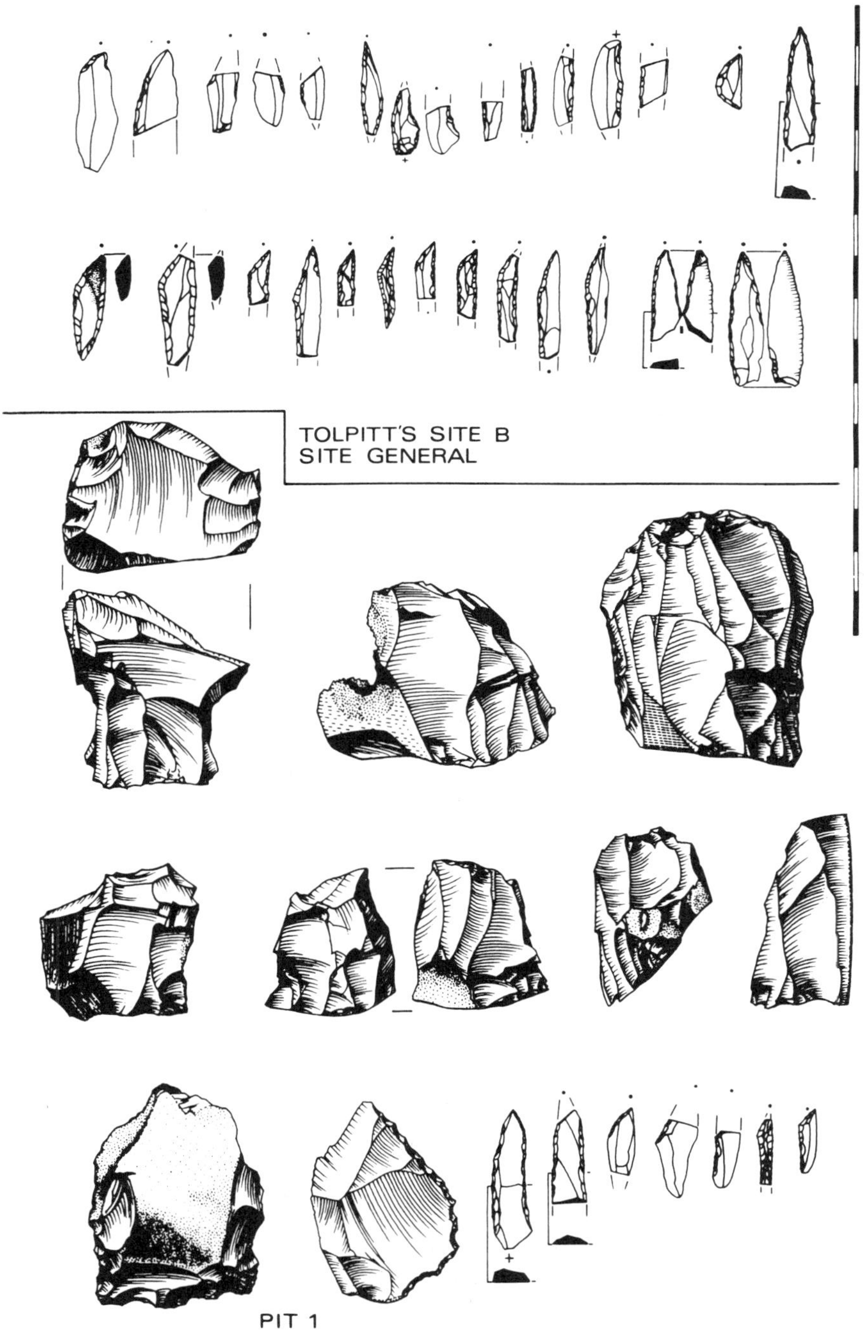

FIG. 9. Mesolithic artifacts from Tolpitts Lane, Richmansworth.

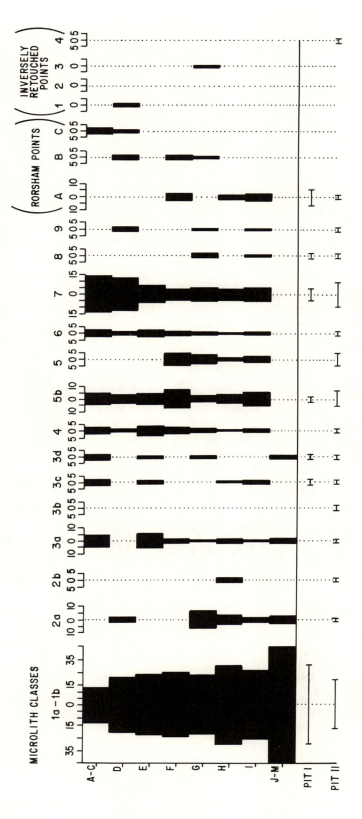

FIG. 10. Wawcott site 3. Vertical distribution of Microlith classes.

Further to the southwest of England, the site of Downton, Salisbury, has produced some 38,000 worked flints (Higgs 1959), of which 25,000 belong to a heavily patinated, early Mesolithic industry, clustered most closely with Thatcham I and Greenham Dairy Farm. Of the 132 microliths, 126 are simple, obliquely-blunted points with only isolated specimens of classes 2a, 3, 3c, and 4. Of the remaining 13,000 pieces, all unpatinated, the bulk is clearly Neolithic but does include a small group of microliths, with narrow rods, scalene triangles, and a small, hollow-based point, all of later Mesolithic type and identical with those from a northern Wiltshire site of Cherhill. Charcoal associated here with a later Mesolithic occupation was dated as BM-447 = 5280 B.C. ± 140, and this was incorporated into the base of a tufa deposit (Evans 1972) which itself overlay a burial soil. Within this soil occurred a group of simple, obliquely-blunted points of class 1. The sequence suggested by both the Downton and Cherhill sites, about thirty-two miles apart, is similar, if not identical, though based on the relative condition of the artifacts themselves in the first case and on stratigraphy in the second.

While in East Anglia proper there is no comparable stratigraphic sequence, and Plantation Farm (Clark et al. 1933) and Peacock's Farm (Clark et al. 1935) near Shippea Hill still remain the only independently dated late sites, the contents of the three sites, Kelling Heath, Hillwood Clay Pit (Essex), and Lackford Heath (Suffolk), are clearly early Mesolithic. At the last of these sites, Todd recorded a single rod and scalene triangle identical to examples from Peacock's Farm, which were within blown sand lying above a slightly sunken, tent-like structure with its early Mesolithic industry.

Reexamination of Sturge's collection from the Lakenheath site (figure 11) confirmed earlier observations (Sturge 1912) of two distinct groups of material: (1) A patinated group with a strikingly high proportion of simple microlith shapes, and (2) an unpatinated group dominated by narrow rods and scalene triangles. In an area, as at Downton, where the soil has been changing from a calcareous, brown earth to an acid podsol through the postglacial, it seems reasonable to interpret this observation as of some chronological significance. The difference, however, is between two microlith assemblages of later Mesolithic type, a simple cluster analysis omitted here suggesting that the patinated industry is closest in type to ones from Mother Grundy's Parlour, layers B/C and C. The unpatinated Lakenheath material, on the other hand, clustered most closely with that from the later Wawcott site XXIII (4129 B.C. ± 113). Further excavation of well stratified assemblages will be required to confirm any suggestion of the chronology that may be implicit in these data.

In northern England, the only direct, stratigraphic evidence of the relationship between the early and later Mesolithic industries is the record by Buckley that on his Warcock Hill North site, a group of seven, small, scalene triangles occurred ". . . within a square foot, rather above the White (i.e., Broad-Blade) industry. . . ." This observation is a complete reversal of that claimed by Woodhead (1929). On the southern Pennines, the chronological record is entirely dependent on radiocarbon dates since only in the case of Stump Cross (Walker 1956) has a Mesolithic flint scatter been traced into organic deposits. The charcoal for the measurements derives from excavations carried out by the late J. Radley (Radley et al. 1974) and by Francis Buckley.

LAKENHEATH SITE

	MICROLITH CLASSES																HORSHAM POINTS			INVERSE POINTS				SAMPLE SIZE
	①	2a	2b	3a	3b	3c	3d	4	5	5b	5c	5d	5e	6	7	8	9	A	B	C				
BRITISH MUSEUM W.A. Sturge Coll. "From one small patch of sandy soil on Lakenheath Warren" (Sturge to Gatty 5th Jan. 1901) Corresp.	UNPATINATED																							
%	7.65	1.1		.53	.26	.26	2.6	.79	10.3	8.2			1.06	1.32	57.2	3.4	2.64	.26			.26	.26	1.06	.26	.53	379
	PATINATED																							
%	4.84		3.3	1.1	1.1	2.2	6.5	5.4	4.4	5.4			2.2		15.1		2.2		2.04	2.2			1.1	1.1	93

	%
UNPATINATED	▬▬▬▬▬ — ▬ — ▬▬▬▬▬▬▬▬▬▬▬▬ — ▬ — ▬ ▬
PATINATED	▬▬▬ ▬▬▬▬▬▬▬▬ ▬▬ ▬ ▬ ▬
	SIMPLE MICROLITHS (CLASSES 1-4) / LATER MICROLITH FORMS (CLASSES 5-9) / HORSHAM POINTS / INVERSE POINTS

FIG. 11. Microlith distribution from Lakenheath site.

Buckley collected surface samples from over 250 find spots around Marsden in Yorkshire, and between 1921 and 1925 he excavated approximately twenty sites. The finished tools recovered from these excavations are now widely dispersed, often without their original proveniences, in museum collections throughout the British Isles and abroad. However, all are meticulously illustrated in a series of nine "drawing books," together with sketch plans and sections. Thus, while it has proven possible to trace the complete assemblages from only seven of the sites excavated (Badger Slacks 2, Dean Clough C, Lominot 4, Rocher Moss I, Warcock Hill III, Warcock Hill South, and White Hill North), it has been a relatively straightforward process to reconstruct the lost assemblages from these profuse illustrations. From charcoal collected during these excavations and stored in the Tolson Memorial Museum at Huddersfield, it has proven possible to date (Switsur and Jacobi 1975) two of his early broad blade sites (Lominot 3, and Warcock Hill South), and three of his later narrow blade sites (Lominot 4, March Hill 2, and Warcock Hill 3). These, together with six dates on more recently collected material, are plotted on figure 12 against dates for pollen zone boundary dates for Red Moss (Hibbert et al. 1971) and Scaleby Moss (Godwin et al. 1957).

Quite simply, the diagram shows the few early Mesolithic (broad blade) dates spread from close to, if not slightly before, the zone IV/V boundary to within zone V, while the earliest later Mesolithic (narrow blade) sites (Warcock Hill III and Broomhead V) belong to the early, but not to the earliest, part of zone VI. Use of the other narrow blade sites continues into zone VII.

The evidence taken from the North Riding of Yorkshire and County Durham gives the same picture, early Mesolithic sites at Star Carr and Flixton Site I (Moore 1950) being stratified into the muds of zone IV and IV/V ages respectively, occupation of Flixton ceasing during zone V, probably some centuries before the zone V/VI boundary, which is dated as Q-920 = 6840 B.C. ± 170 at Red Moss, and Q-161 = 161 = 7060 B.C. ± 194 at Scaleby Moss. The microlithic equipment of both these sites consists entirely of broad, obliquely blunted points, isosceles triangles, and trapezes.

In total contrast to these two assemblages is that from Filpoke Beacon (Coupland 1948) to the northwest of Hartlepool, where broken and charred hazel nut shells, clearly collected for food, have been dated at 6810 B.C. ± 140 (Q-1474) (see figure 13). The report is quite specific on the associations of these nuts, the artifacts coming from "beneath a compacted sandy mixture ... in or just above a 'black band' of broken hazel nut shells and blackened sand grains beneath which is a ... 'white layer' ... of bone ash and numerous bone fragments." Except for three pieces, two of them fragments and possibly interpretable as obliquely-blunted points, all the microliths are either narrow rods trimmed down one side or both sides, or scalene triangles of the narrow blade type. Clearly falling into zone VIa, this group represents the earliest later Mesolithic industry identified from England, and it is earlier than any from mainland Northern Europe. The dated sample consisted entirely of nut shells, so there can be no possibility of the age deriving from old or even semi-fossilized wood (Coles 1975). We feel that the date should certainly be accepted. The age is, however, statistically similar to determinations from North Wales at sites of Aberffraw (Q-1385; 6690 B.C. ± 150) and Rhuddlan Site E

PL. 3. Excavating an Early Mesolithic site at Pointed Stone II at 1,200 ft. on the Cleveland Hills, North Yorkshire. The artifacts are sealed by a narrow band of peat. The chances of charcoal for dating surviving in such a situation are minimal unless in an artifactual hollow.

PL. 4. *Stonewall Shelter B*. A later mesolithic rock shelter at Stonewall Park, Kent. Weathering and collapse of the soft sandstone has resulted in a deep accumulation of deposits in front of the shelter, which incorporate a series of hearths. The mesolithic material spans an age range between 8,000 and 5,750 years B.P.

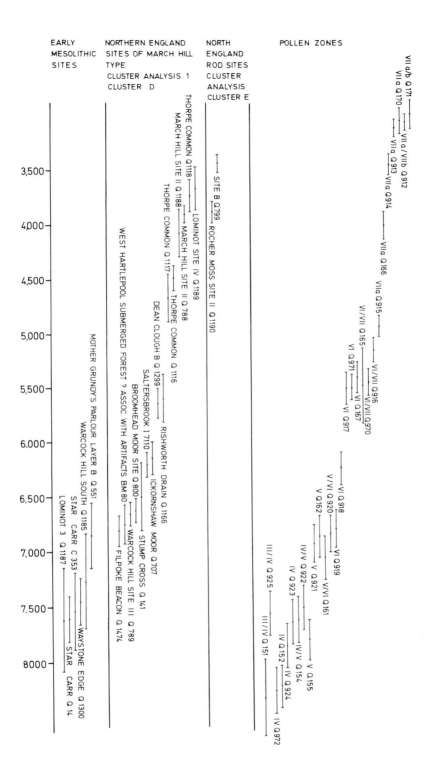

FIG. 12. Radiocarbon dates for sites and pollen zones from northern England.

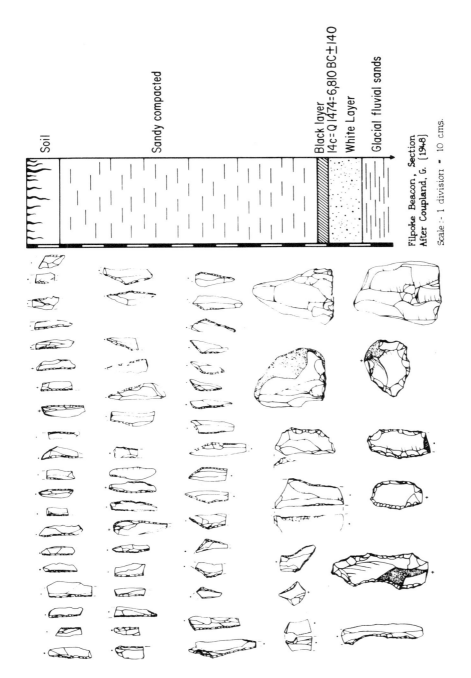

FIG. 13. Artifacts from site at Filpoke Beacon.

(BM-691; 6789 B.C. ± 86) which are of the early Mesolithic type. There is, of course, no reason why the technological changes involved in the early to later Mesolithic transition should be synchronous over the whole of England and Wales.

Just as on the Pennines, radiocarbon dating has confirmed the continued use of microlithic equipment with small, scalene triangles and rods into zone VIIa later than 5400 to 5200 B.C.), so the Atlantic zonation (Bartlett) of peat adhering to one of a group of scalene triangles from Cock Heads, Glaisdale (Radley 1970) confirms their continued use on the Clevelands.

On the Pennines, on the low ground of east Yorkshire, Lincolnshire, and Durham, and also on the crests of the Clevelands, it has been possible to identify early Mesolithic flint scatters identical in composition to Star Carr, Flixton, and Bingham, and also, on the other sites, with small, geometric microliths of the later Mesolithic type. Such an early assemblage at Money Howe Site I, at over 1200 feet (365 meters), has been dated by associated charcoal as Q-1560; 7480 B.C. ± 390. The weight of charcoal recovered from the site was very small, as it was from the Pennine sites of Warcock Hill and Lominot, and in each instance the sample had to be diluted with inert carbon prior to counting the radioactivity. The resulting dates inevitably possessed such uncertainties that they have been useful only in indicating the millennium of site usage. Nevertheless, recognition of the distinction between these site types denies the interesting supposition (Coles and Higgs 1969) that Sauveterrian (later) tool kits were a summer variant of the (early) Maglemosian. The difference can now be recognized as one of chronology, and what differences there are between upland and lowland tool kits are ones of proportion and not of types.

Evidence taken at both general and specific levels suggests the division of our Mesolithic into two parts: (1) an early Mesolithic, 8500 to 6700 B.C. with pollen zones IV to VIa, and (2) later Mesolithic, 6800 to 3500 B.C. with pollen zones VIa to VIIa. The two parts are differentiated by their microlithic content. Where isolated later microliths are recorded from otherwise early assemblages, as at Warcock Hill North, Lominot 3, and Pike Lowe I, there is reason to suspect that these later microliths are intrusive, particularly in the case of the first two sites, where on the same areas of hillside there is surface evidence for numerous, small, later Mesolithic sites.

Subsequent papers intend to present this information in greater detail, considering each site separately (Jacobi and Switsur, in preparation). Also, a penetrating study is intended, not simply of the technology of the two phases, but more particularly of their European connections (Jacobi, in press), for it was this first half of the postglacial which saw the greatest geographic and ecological changes. It will be important to seek, especially among the later Mesolithic material, for evidence of the social groupings that might be expected in an area of such extreme physiographic and economic variability as Britain. So far in the forty years of Mesolithic research, no serious attempt has been made to distinguish or search for any such patterning among the evidence, although these patterns are emerging and have been predictable even from the scant archaeological and the more abundant geographic literature.

REFERENCES

Barker, H., and C. J. Mackey
    1960   Brit. Mus. Natural Radiocarbon Meas. II. Radiocarbon. 12: 26-30.

Buckland, P. C., and M. J. Dolby
    1973   Mesolithic and Later Material from Misterton Carr, Notts: an Interim Report. Trans. of Thoroton Soc. LXXVII: 5-33.

Clark, J. G. D.
    1932   The Mesolithic Age in Britain. Cambridge, i-xxiii and 223.
    1954   Star Carr. Cambridge, i-xxiii and 200.
    1955   A Microlithic Industry from the Cambridgeshire Fenland and Other Industries of Sauveterrian Affinities from Britain. P.P.S. XXI: 3-20.
    1972   Starr Carr; a case study in bio-archaeology. Addison Wesley, 10: 1-42.
    1935   Report on Recent Excavations at Peacock's Farm Shippea Hill, Cambridgeshire. Ant. J. XV: 284-319.

Clark, J. G. D., H. Godwin, M. E. Godwin, and W. A. MacFadyen
    1933   Report on an early Bronze Age site in the South Eastern Fens. Ant. J. XIII: 266-296.

Coles, J.
    1975   Timber and Radiocarbon Dates. Antiquity. XLIX: 123-125.

Coles, J. and E. S. Higgs
    1969   The Archaeology of Early Man. London, Faber and Faber, 454.

Coupland, G.
    1948   A Mesolithic Industry at "The Beacon" S. E. Durham. Private printing, John Bellows Ltd. Gloucester, 7.

Dimbleby, G. W.
    1962   The Development of British Heathlands and their Soils. Oxford Forestry Memoirs No. 23, Clarendon Press, 120.
    1963   Pollen Analysis of a Mesolithic Site at Addington, Kent. Grana Palynologica. Uppsala. 4: 140-148.

Evans, J. G.
    1972   Land Snails in Archaeology. Seminar Press, i-xii and 436.

Froom, F. R.
    1963   An Investigation into the Mesolithic around Hungerford part II, the Wawcott District. Trans. Newbury Dist. Field Club. XI (2): 70-73.

Godwin, H., D. Walker, and E. H. Willis
    1957   Radiocarbon Dating and Postglacial Vegetational History: Scaleby Moss. Proc. Royal Soc. London, Series B. 147: 352-366.

Hibbert, F. A., V. R. Switsur, and R. G. West
    1971   Radiocarbon Dating of Flandrian Pollen Zones at Red Moss, Lancashire. Proc. Royal Soc. London, Series B. 177: 161-76.

Higgs, E. S.
    1959   Excavations at a Mesolithic Site at Downton, near Salisbury, Wiltshire. P.P.S. XXV: 209-232.

Jacobi, R. M.
    1976   Britain Inside and Outside Europe. P.P.S. 42 (In press.)

Keef, P. A. M., J. J. Wymer, and G. W. Dimbleby
    1965   A Mesolithic Site on Iping Common, Sussex, England. P.P.S. XXXI: 85-92.

Lacaille, A. D.
    1963   Mesolithic Industries beside Colne waters in Iver and Denham, Buckinghamshire. Records of Bucks. 17: 143-181.

Moore, J. W.
    1950   Mesolithic Sites in the Neighbourhood of Flixton, North East Yorkshire. P.P.S. XVI: 101-108.

Petch, J. A.
    1924   Early Man in the District of Huddersfield. Tolson Mem. Mus. Publ. 3: 1-95.

Radley, J.
    1970   The Mesolithic Period in North East Yorkshire. Yorks. Arch. J. XLII: 314-327.

Radley, J., and P. A. Mellars
    1964   A Mesolithic structure at Deepcar, Yorkshire, England, and the affinities of its associated Flint Industries. P.P.S. XXX: 1-24.

Radley, J., V. R. Switsur, and J. H. Tallis
    1974   The Excavation of three 'Narrow Blade' Mesolithic sites in the Southern Pennines, England. P.P.S. XL: 1-19.

Rankine, W. F., W. M. Rankine, and G. W. Dimbleby
    1960   Further Excavations at a Mesolithic Site at Oakhanger, Selborne, Hants. P.P.S. XXVI: 246-262.

Sheridan, R., D. Sheridan, and P. Hassell
    1967   Rescue excavation of a Mesolithic Site at Greenham Dairy Farm, Newbury, 1963. Trans. Newbury Dist. Field Club. XI: 66-73.

Simmons, I. G., and P. R. Cundill
    1970   Vegetation History during the Mesolithic in North East Yorkshire. Yorks Arch. J. XLII: 324-327.

Sturge, W. A.
    1912   Implements of the Later Paleolithic 'Cave' periods in East Anglia. P.P.S. E.A. I: 210-232.

Switsur, V.R., and R. M. Jacobi
    1975   Radiocarbon Dates for the Pennine Mesolithic. Nature. 256: 32-34.

Walker, D.
    1956   A Site at Stump Cross, near Grassington, Yorkshire, and the age of the Pennine Microlithic Industry. P.P.S. 22: 23-28.

Warren, S. H., J. G. D. Clark, H. Godwin, M. E. Godwin, and W. A. MacFadyen
    1934   An Early Mesolithic Site at Broxbourne Sealed under Boreal Peat. J.R.A.I. LXIV: 101-128.

Woodhead, T. W.
    1929   History of the Vegetation of the Southern Pennines. J. Ecology XVII: 1-34.

6

New Absolute Dates on Upper Pleistocene Fossil Hominids from America

Reiner Protsch

The first appearance of "anatomically modern man" in the Americas has been a long-debated question in paleoanthropology and archaeology. The anthropological facts are presented below, but problems related to materials found in association with fossil remains of anatomically modern man must be considered. These are restricted mainly to fauna, geology, relative and absolute dating, and foremost-archaeological remains.

The faunal extinction theory proposed by P. S. Martin (1973) proposed that the decline and extinction of some megafauna in America ca. 12,000 B.P. was due to a high population density attained by early hominid immigrants who practiced overkill. This theory does not negate the possibility that man may have migrated into America at an earlier time period, ca. 30,000 B.P., retaining a low population density till 12,000 B.P. Few sound radiocarbon dates were available in 1973 dating hominid remains to earlier than 13,000 B.P. Holmes and Hrdlicka, with earlier investigators less factual evidence, claimed that man could not have existed on the American continent during any glacial stage. Even though Hrdlicka allowed more than 10,000 years for postglacial time after the Wisconsin, it is doubtful that he thought of more than 20,000 to 30,000 B.P. (Hrdlička, 1907, 1912).

The existence and description of the Paleo-Indian "lithic complex" and its possible diffusion has been well explained by Müller-Beck (1966). He divided the American Paleolithic into two distinct technical traditions in which stone projectile points were used or in which they were not (Müller-Beck 1966 : 1191). The tools of the second tradition consist largely of crude stone implements dating to the earliest chronological phase. According to Müller-Beck, they do not date relatively and absolutely further back than 28,000 B.P. Claims of a greater antiquity lack the support of absolute dates. Some specialists claim that tools belonging to the second tradition date up to and even beyond 100,000 B.P. These claims are highly suspect since most artifacts are surface finds without other association and cannot be absolutely dated.

There are three basic problems which should be considered before dates obtained recently on very early hominid remains is accepted as a chronological marker for man's appearance on the American continent: (1) only one subspecies of *Homo sapiens* occupied, at any given time, the American continent; (2) the probable migrational route into the two American subcontinents is from Asia; (3) the time of first appearance of subspecies of anatomically modern man in America and his successors, the Indianoids, and the question of whether or not they are related to Mongoloids, remain uncertain.

It has been proven that no human genus and species but *Homo sapiens*, and only one subspecies, *Homo sapiens americanus*, has ever occupied the Americas. Fossil subspecies of *Homo sapiens*, such as *Homo sapiens neanderthalensis* or any other fossil hominids, have never been found on this continent. We can assume that the only member of the family Hominidae in the Americas is a subspecies of anatomically modern man, specifically, the geographical subspecies *Homo sapiens americanus* (Blumenbach 1775, 1798). It should be pointed out that the term *Homo sapiens americanus* merely designates one of the many different varieties of *Homo sapiens* in its specific geographical setting. The term is not meant to designate racial characteristics. At a time period prior to 30,000 B.P., the subspecies of *Homo sapiens* present on all continents are morphologically the same. Other such members are the European *Homo sapiens sapiens*, the Asian *Homo sapiens asiaticus*, the Australian *Homo sapiens australasicus*, and the African *Homo sapiens afer* (Blumenback 1798, Linnaeus 1758, St. Vincent de 1835). Early representatives of this basic anatomically modern man probably appeared on each continent between 35,000 to 30,000 B.P., after migration from Africa. Extensive dating by radiocarbon and supportive relative dating techniques (amino-acid dating and microanalysis) prove that a subspecies called *Homo sapiens capensis* formed the basis in Africa for all of the above-mentioned geographic subspecies, originating possibly ca. 90,000 to 80,000 B.P. (Protsch 1975).

In considering the possible migrational route taken by early immigrants into the Americas, it seems clear that the most suitable land route was across the Bering-Strait. Controversy surrounds the time period, however, during which immigration into the Americas could have occurred. Considering the different hypotheses of crossing this land bridge, a number of estimates exist ranging anywhere from 100,000 years to 10,000 years ago. These are relative geologic estimates and have as yet no supportive absolute dating. Most recent estimates (Butzer 1973) consider a land connection between the continents of Asia and America during the Altonian. There are several possibilities, ranging from 50,000 to 40,000 years ago and up to 28,000 to 10,000 years ago. The most elevated part of the now-submerged platform between Siberia and Alaska is about 35 to 40 meters below the present sea level (Hopkins 1959). Animals and man could have crossed during any of these time periods, but it is likely that man crossed only between 28,000 to 10,000 years ago, since anatomically modern man started to migrate out of Africa only about 35,000 to 40,000 years ago. The time of this first invasion of the American continent, based on climatic and archaeological evidence, is suggested by Müller-Beck as occurring between 26,000 to 28,000 years ago. A crossing of the Strait during the latter periods into central and northern Alaska was possible since they were free of ice. Further migration into the North American subcontinent was also possible along the area of the Mackenzie River.

The latter route was closed during the period between 14,000 to 20,000 years B.P. These hypotheses are valid in connection with a theory of crossing if early anatomically modern man needed a land bridge for migration, and if no crossing occurred other than by foot. Once migration had advanced beyond the ice and snow-covered areas of North America, a rapid infiltration into the two American subcontinents could have occurred. It is possible that within 5000 years after entering the American Continent, man could have arrived at the southernmost tip of South America.

The origin of Indians or Indianoids of the American continent is claimed by numerous specialists to be closely linked to Mongoloids. This claim is based on a superficial similarity of morphological characters, some of which are shared by man in Asia and the Americas. Ninety-five percent of early North American hominid remains, which are relatively estimated to date from 10,000 B.P. to about 2000 B.P., do indeed show many morphological characteristics not very different from those characteristics seen in Mongoloids (Protsch 1977). Those remains, however, which date relatively as well as absolutely older than 10,000 B.P., show morphological features which separate them clearly from those of Mongoloids. A more extensive morphological study of Laguna Beach Woman and Los Angeles Woman (not man) (Plhak 1975, 1977) indicates that these specimens are more closely aligned to what might be called a "basic anatomically modern man" or an "archaic-caucasoid" (Birdsell 1951). The latter term might seem to designate a racial category to these early American remains, but it merely intends to point out their morphological similarity to very early specimens of anatomically modern man in Europe about 30,000 B.P. It is based on the assumption of many specialists that fossil specimens of earliest anatomically modern man are the same as those found in Europe, *Homo sapiens sapiens*, and that they are the earliest representatives of any subspecies of *Homo sapiens* anywhere in the world.

Within progressively more recent time periods, from about 10,000 to 2000 B.P., typical mongoloid features become more prominent in hominid remains found in the New World. There is also some indication of a geographic gradient of these features; the further north in America one goes the more apparently pronounced these features become.

This could be linked to the following hypothesis. Migration of a basic anatomically modern man occurred out of Africa about 40,000 to 35,000 B.P., and the arrival of members of this migration into the different continents took place only a few thousand years later (Protsch 1975). After a few thousand years, each basic anatomically modern man developed its own racial characteristics on each continent after immigration. However, the typical racial features of each of the present-day races developed because of different environmental pressures at different time periods. The Bushman Rock Shelter mandible (Protsch and De Villiers 1975), for example, seems to indicate that Negroids had already developed about 29,500 B.P. in Africa. In Europe, Asia, and other continents, typical racial features developed at different times, somewhat later than those of Negroids in Africa. The above reasoning is supported by findings which indicate that typical mongoloid features are not found in anatomically modern man in Asia prior to 18,000 B.P.

If we assume that man migrated into America across the Bering Strait about 30,000 to 35,000 B.P., or as some claim (Bada et al. 1977) even earlier, we should not be able to recognize any racial features in those earliest remains. If man migrated into America before 18,000 B.P. and also displayed mongoloid features at that time, it could only mean that these features developed in America with subsequent migration back into Asia. Such an occurrence, considering all present evidence, is highly unlikely. Since we know that mongoloid features developed in Asia, we should look more closely at those remains dating earlier than 10,000 B.P. in the Americas. The hominid remains of Laguna Beach Woman and Los Angeles Man, as well as those of Otovalo, Punin, and others in South America prove, that remains of early hominids in America dating about 17,000 to 24,000 B.P. do not display mongoloid features and that typical mongoloid characteristics can not be seen in American hominids until about 8000 to 10,000 B.P.

With the absolute dating of the megafauna from Ayacucho, Highland Peru (MacNeish, Berger, and Protsch 1970), which were associated with crude unifacial tools and dated beyond 14,000 B.P. (UCLA-1464; 14,150 ± 180) and 19,000 B.P. (UCLA-1653A; 19,000 ± 2500), the presence of man in South America before 10,000 years B.P. seems to be well documented. Dates from North America, like those of Laguna Beach (UCLA-1233A; 17, 150 ± 1470) and Los Angeles (UCLA-1430; 23,600) (Berger et al. 1971), have indicated that anatomically modern man arrived on the American continent at least by 24,000 B.P. Recently, a number of hominid fragments from four different individuals were obtained through the courtesy of Prof. Byrne de Caballero of the University of Bolivia in Cochabamba. Each of these hominids were dated absolutely (A^1), and showed an age well before 10,000 B.P. A mandibular fragment from the site of Punin in Ecuador was absolutely dated to 10,600 B.P. and associated fauna were dated to 10,800 B.P. (Fra-A-5; Fra-A-6). A find from Altiplano near Cochabamba was absolutely dated to 13,050 (Fra-102) and 13,200 B.P. (Fra-A-7) (Protsch 1976). The latter two dates were on mandibular and maxillary fragments of an adult. The last specimen to be dated by amino-acids was a find from Sacaba which was found 15 km south of Cochabamba in Bolivia with an age of 18,000 years B.P. (Fra-A-10). This find is presently processed for radiocarbon dating.

The possibly oldest hominid, known as "Fred" from the site of Otovalo in Ecuador, was dated to 2,800 (Birm-360; and 22,800 ± 300) and even to 28,000 B.P. (Q-?; 18,000 ± 750) by radiocarbon. Information given to the author originally stated that these dates were processed on the collagen portion of the hominid. The above dates are, however, from the mineral portion of the bone, and these high ages could be due to contamination of older radiocarbon deposited secondarily after burial. These old dates are especially suspect since a date on the residual collagen comes only to 2670 B.P. (Shotton and Williams 1973). With a fairly good temperature reconstruction, it was possible to arrive at an amino-acid age of 27,000 B.P. (Fra-A-4). It should be pointed out that the radiocarbon dates run by the Birmingham laboratory were quite different and that they were processed on the mineral fraction and not on the collagen or organic fraction of the hominid.

With the above absolute dates we can postulate that anatomically modern man was already present in Central and South America by at least 20,000 B.P. We can also postulate that immigration of early Upper Pleistocene hominids into the American continent took place not much earlier than about 29,000 B.P. (Plhak 1975). The arguments of Müller-Beck (1966) on archaeological grounds are totally supportive of the anthropological and archaeometric evidence presented in this paper. Müller-Beck mentions that, after the appearance of man in America at about 28,000 to 26,000 years ago, these first invaders were isolated by the ice advances of the Wisconsin maximum and thus were separated from the continuing technological evolution of the Old World.

It has recently been claimed, through direct dating (A^1) of hominid remains in an area of southern California (Bada et al. 1974, 1977), that man in the Americas might date back to at least 50,000 B.P. or even to 70,000 or 100,000 B.P. These remains show, however, positively mongoloid features. Assuming that these dates are valid, it would mean, that the earliest Americans already displayed mongoloid features about 50,000 to 70,000 B.P., that they lost these mongoloid features about 25,000 to 10,000 B.P., and that they subsequently redeveloped these racial features about 8000 B.P.

If this were true, it would mean that early members of *Homo sapiens* showed the first racial features of any subspecies in America, and that basic anatomically modern man developed in America, not Africa or Europe. These dates, going back to earlier than 30,000 B.P., are peculiar because earliest anatomically modern man dates in Europe to no earlier than 32,000 to 34,000 B.P., in Australia to 34,000 B.P., and in Asia to around the same time. An exception is Africa with basic anatomically modern man dating to before 60,000 B.P.

To summarize, it is presently agreed that no remains of any other fossil hominid, such as Neanderthal or *Homo erectus*, have ever been found in America prior to anatomically modern man. Finds of fossil hominids other than modern man would be necessary to proclaim a theory of development of early Americans (i.e., anatomically modern man) from such fossil predecessors. If this picture is correct, dates earlier than 30,000 B.P. in the Americas must be viewed with scepticism.

All hominid finds belonging to anatomically modern man cannot date earlier than 30,000 to 28,000 B.P. and are merely manifestations of a population which migrated into America at around that time, developing racial features unique to the American Indian ca. 10,000 to 8,000 B.P. Subsequent migrations at more recent times brought mongoloid characteristics from Asia into the Americas. These features can clearly be seen in some Indian tribes of North America and in the Eskimos. These anthropological facts are supported by the findings of Müller-Beck (1966) who states that later contact between Asia and America existed after the inland ice barrier had melted and aurignacoid groups, such as the Mal'ta, invaded and expanded over North America for the first time.

REFERENCES

Bada, J. L., R. A. Schroeder, and G. F. Carter
    1974   New evidence for the Antiquity of Man in North America: Deduced from Aspartic Acid Racemization. Science. 184. : 791-793.

Bada, J. L.
    1977   Talk given at the SWAA-Conference April 7, 1977, at San Diego. Evidence

for a 50,000 Year Antiquity for Man in the Americas Derived from the Amino acid Racemization Analyses of Human Skeletons.

Berger, R., R. Protsch, C. Rozaire, and J. R. Sackett
  1971 New radiocarbon dates based on bone collagen of California Paleo-Indians. Contr. U. Calif. Berkely. Arch. Res. Fac. 12. : 43-47.

Birdsell, J. B.
  1951 The problem of the early peopling of the Americas as viewed from Asia. Pap. Phys. Anthr. Amer. Ind. The Viking Fund. New York.

Blumenbach, J. F.
  1775 De generis humani varietate nativa. Dissertation. Göttingen.
  1798 Über die natürlichen Verschiedenheiten im Menschengeschlechte. J.G. Gruber, Breitkopf und Härtel, Leipzig.

Butzer, K.
  1973 Environment and Archaeology: An Ecological Approach to Prehistory. Chicago, New York, Aldine-Atherton.

Hopkins, D. M.
  1959 Cenozoic History of the Bering Land Bridge. Science. 129 : 1519-1529.

Hrdlička, A.
  1907 Skeletal Remains Suggesting or Attributed to Early Man in North America. Bur. Amer. Ethnol. Bull. 33. Wash., D.C.
  1912 The Skeletal Remains of Early Man in South America. Hrdlička et al. Early Man in South America. Bur. Amer. Ethnol. Bull. 52. Washington, D.C., chap. 7.

Linnaeus, C.
  1758 Systemae Naturae. 10th ed. Stockholm.

Martin, P. S.
  1973 The Discovery of America. Science 179 : 969-974.

MacNeish, R.S., R. Berger, and R. Protsch
  1970 Megafauna and Man from Ayacucho, Highland Peru. Science. 168 : 975-977.

Müller-Beck, H.
  1966 Paleohunters in America: Origins and Diffusion. Science. 152. (3726) : 1191-1210.

Plhak, M.
  1975 Chronologie und Morphologie der frühesten fossilen Hominiden Amerikas. Diplomarbeit. J. W. Goethe Universität.
  1977 Die Morphologie der fossilen Hominiden von Laguna Beach und Los Angeles. Zeitschrift. Morphologie und Anthropologie. (In press.)

Protsch, R.
  1975 The Absolute Dating of Upper Pleistocene sub-Saharan Fossil Hominids and their Place in Human Evolution. J. Hum. Ev. 4 : 297-322.

Protsch, R., and H. De Villiers
  1975 Bushman Rock Shelter: Chronology and Morphology of a Child's Mandible. J. Hum. Ev. 3 : 387-396.

Protsch, R.
  1976 Proc. of 9th. Radiocarbon Conf. Univ. of Calif. Los Angeles and San Diego.

New Absolute Dates on Upper Pleistocene Fossil Hominids from Europe and South America. Proc. of 9th Int. Radiocarbon Conf. Univ. of Calif. Los Angeles and San Diego.

    1978   Catalog of Fossil Hominids of North America. Stuttgart, New York, Gustav Fisher Verlag.

Shotton, F. W. and R. E. G. Williams

    1973   Birmingham University Radiocarbon Dates VII. Radiocarbon. 15. (9): 467.

St. Vincent de Bory

    1825   Dict. Class. Hist. Nat. 8 : 269.

7

Were the Allerød and Two Creeks Substages Contemporaneous?

Hans E. Suess

Those who remember the days when the first radiocarbon dates were published (Arnold and Libby 1951) will perhaps remember that the most exciting and controversial result was the dating of wood from the Two Creeks forest bed, Wisconsin. The radiocarbon date obtained for this wood was 11,400 ± 350 B.P. Of course few people believed this. It took several years before people realized the tremendous impact this result must have upon archaeology, anthropology, and glacial geology. The date meant that ice was covering a large fraction of the North American continent only 11,000 years ago; the time elapsed between glacial time and the first Egyptian dynasty was just about as long as the time elapsed between the Egyptian dynasty and today. Human civilization must have developed much more rapidly after the time of the retreat of the ice than had been thought previously.

There was, of course, a large demand to correlate the date for the Two Creeks forest with European climatic records, especially with the glacial records in Scandinavia and with deGeer's varve chronology. Dr. Tauber (1964) in Copenhagen and Dr. Nydal (1970) in Trondheim dated such climatic fluctuations, finding them approximately synchronous with those recognized in North America. The beginning of the last major cold period in Europe, however, was dated as having occurred about 10,900 conventional radiocarbon years ago, considerably later than when the Two Creeks forest bed supposedly was overrun by advancing glaciers. Because samples from this time range contained only one-quarter of the ^{14}C present in modern wood, errors in background calibration could cause systematic differences of as much as several hundred years in the results of different laboratories. If the samples from America and Europe were dated in the same laboratory with the same sets of instruments, calibration errors in the dates for samples from the two continents would be identical and would cancel.

One of the most important problems of radiocarbon research is the radiocarbon level of atmospheric carbon dioxide during glacial time. Samples of precisely known age from late glacial time are not available. There is still a gap of more than 2000 years between the tree ring sequences from postglacial time and the ones predating the last glacial period, the Younger Dryas of Europe. It is hoped that this gap may be bridged within the next few years; in the meantime, the conversion of conventional radiocarbon dates from Pleistocene times into true calendar dates must be considered uncertain by perhaps as much as 2000 years.

The Euorpean samples of similar age as those from the classical Two Creeks forest came from a clay pit in the Dättnau valley near Winterthur in Switzerland (Kaiser 1972). This clay pit is presently being excavated for brick manufacturing. It contains large stumps of pine trees, which are being investigated by a group of scientists of the E.T.H., Zürich. The samples were obtained from Klaus Kaiser, who had established tree ring sequences of up to 400 years from these pine trees. Samples from such sequences afford opportunities to obtain well-defined, accurate dates for the times of initial growth, and also opportunities to recognize rapid fluctuations of the radiocarbon level of the atmospheric carbon dioxide during the extended growth time of the trees. Such rapid fluctuations have been considered as possibly causing distortion of the radiocarbon time scale at the very end of the last glaciation, perhaps explaining the unusually rapid sequence of climatic changes indicated by the conventional radiocarbon dates. Our results obtained so far do not show any indication of a rapid change in the radiocarbon level.

The results of the samples from Dättnau, Switzerland are listed in table 1 and are shown graphically in figure 1. This figure shows the conventional radiocarbon ages as a function of the tree ring number, which starts with the number one for the ring that formed in the year when the tree strated to grow. If the ^{14}C level of the atmospheric carbon dioxide did not change during the time when the tree grew, then the conventional dates obtained lie on lines with 45° slopes, provided experimental errors are negligible. In order to obtain information on the constancy of the ^{14}C level and also on the accuracy of the results, seventeen samples were measured from tree K212, which had almost three hundred rings. The results listed in table 1 show the deviations from a 45° line to be close to those expected from counting statistics and a least squares fit line of an angle somewhat smaller than 45°, indicating that it is slightly more probable than the ^{14}C level had been rising during the time when the tree grew than that it had remained constant or had been decreasing. Figure 1 shows that the oldest tree from this location started to grow about 12,200 B.P. and that the youngest died around 10,800 B.P.

Scandinavian climatic variations dated by the Copenhagen and Trondheim Radiocarbon Laboratories (Nydal et al. 1970, Tauber 1964) appear to have been contemporaneous, as shown in figure 2 and as discussed below.

A large number of samples from the Two Creeks forest bed, Wisconsin, were measured by various radiocarbon laboratories and discussed by Black and Rubin (1968). The dates obtained show a relatively large spread that may be due in part to systematic errors from uncertainties in the background counting rates encountered by some laboratories. In order to avoid such errors, we have measured four samples from Wisconsin. The results are listed in table 2 together with the results of previous measurements of the same samples by the Illinois State Geological Survey Radiocarbon Laboratory.

A comparison of the stratigraphic situations in Wisconsin, Switzerland, and in Scandi-

TABLE 1

Dättnau Samples
Pinus silvest

LJ No.	Tree	Rings	Libby age
3368	K-353	30-40	10 940 ± 67
3581	K-353	110-120	11 134 ± 67
3565	K-353	170-190	10 863 ± 67
3369	K-353	210-230	10 839 ± 67
3580	K-353	230-250	10 916 ± 67
3129	K-212	0-35	11 326 ± 67
3095	K-212	35-50	11 340 ± 67
3166	K-212	60-70	11 257 ± 67
3093	K-212	80-90	11 236 ± 67
3164	K-212	110-120	11 057 ± 67
3206	K-212	120-130	11 165 ± 67
3101	K-212	145-150	11 082 ± 67
3212	K-212	160-170	11 131 ± 67
3100	K-212	180-190	11 124 ± 67
3211	K-212	200-210	11 248 ± 67
3127	K-212	210-220	11 063 ± 67
3208	K-212	230-240	11 018 ± 67
3165	K-212	250-260	10 852 ± 67
3209	K-212	270-280	11 036 ± 67
3090	K-212	280-290	10 883 ± 67
3571	K-301	111-116	10 915 ± 91
3582	K-300	100-110	10 948 ± 66
3370	K-300	110-120	11 051 ± 67
3371	K-304	70- 80	11 468 ± 67
3130	K-203	160-185	11 553 ± 75
3094	K-209	57- 58	11 743 ± 67
3569	K-119	93- 97	12 104 ± 74
3570	K-260	90-100	12 106 ± 105
3372	K-110	30- 40	12 206 ± 73

navia is shown in figure 2. Stratigraphic horizons are shown here not as a function of depth, but of the conventional radiocarbon age B.P. The classical Two Creeks forest bed horizon is overlain by lake sediments and then by glacial till, which has been interpreted as indicating that an advancing ice front had caused the death of the trees. However, a comparison with the situation at Dättnau shows no conspicuous change in climate 11,800 years ago and shows a general warming up in Scandinavia at the end of the older Dryas, which appears synchronous with the time of the Two Creeks forest.

The radiocarbon dates from the three locations are sufficiently accurate so that they could be interpreted in the following way: The assumption, previously adopted in general, was that a date for the two Creeks forest of some 11,800 years indicated a time of decreasing temperatures and of an advance of the ice front. This would mean that the

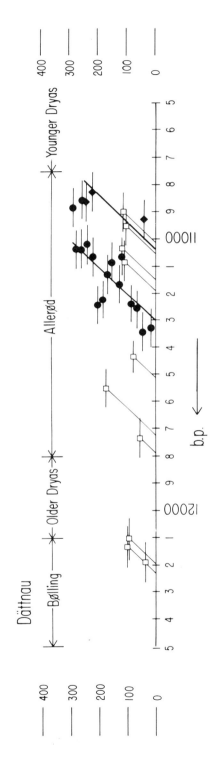

FIG. 1. Conventional radiocarbon dates of wood samples from Dättnau, Switzerland (Kaiser 1970) from different trees as function of the number of the annual tree-rings, from the center of the respective tree to the ring from which the sample was taken. The 45° lines extrapolate to the year when the respective trees started to grow. Filled-in points refer to trees from which a series of samples were measured. (See table 1).

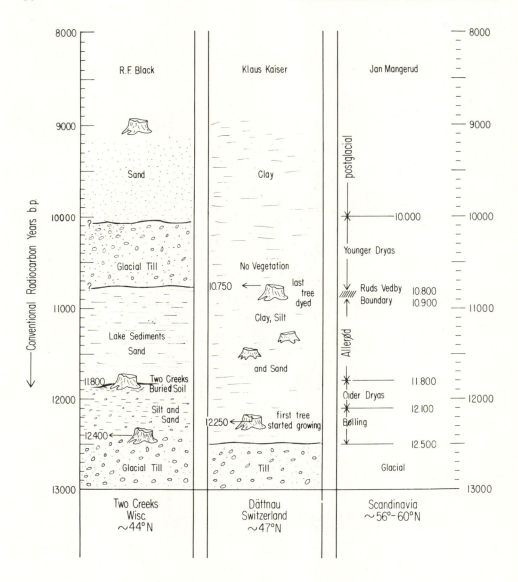

FIG. 2. Comparison of the stratigraphic situations at Wisconsin, Switzerland, and Scandinavia.

beginning of a cold period occurred in North America approximately three hundred years later than in Europe. The climatic oscillations of the two continents would have to be considered not synchronous, but out of phase by approximately three hundred years. It appears more probable now that the Two Creeks forest bed was drowned by a rising lake level at the beginning of a warm period synchronous with the end of the older Dryas in Europe. Accurate dating of the arrival of the ice by measurements on wood from the moraine and from other locations in North America appears most desirable.

TABLE 2

Wisconsin Samples

LJ no.	Conventional age B.P.	Illinois No.	Conventional age B.P.
3578	12 377 ± 73	ISGS 75	12 500 ± 120
3568	11 809 ± 71	–	–
3577	11 867 ± 71	ISGS 264A	11 790 ± 80
3579	9 004 ± 60	ISGS 73	9 270 ± 120

The samples from Switzerland agree perfectly with the climatic sequence as recognized in Scandinavia (Mangerud 1970). The only difference is that in Switzerland no indications for a cold period around 12,000 B.P. synchronous with the older Dryas of Scandinavia can be recognized. This is in agreement with observations by Oeschger (private communications), who found no indication from paleotemperature measurements in Swiss peat bogs.

In summary conventional radiocarbon dates for wood from the Dättnau forest bed (Wintertnur, Switzerland) were found to range from 10,750 to 12,250 B.P. These dates were compared with those for wood from Wisconsin using the same sets of instruments for the measurements. The results indicate that the classical Two Creeks forest bed did not form at a time when the climate in Europe turned colder.

ACKNOWLEDGMENTS

The technical operations of the La Jolla Radiocarbon Laboratory are carried out by Ms. Carol Hutto and are supervised by Dr. Timothy Linick. Thanks are due to Klaus Kaiser and Dr. Friedrich Schweingruber of the E. T. H. Zürich for supplying the samples from Switzerland and to Dr. William Farrand and Dr. Dennis Coleman for those from Wisconsin. Thanks are also due to Dr. Meyer Rubin for reading the manuscript and for valuable discussions. The operation of the La Jolla Radiocarbon Laboratory is made possible by grant DES 74-22864 from the National Science Foundation.

REFERENCES

Arnold, J. R., and W. F. Libby
    1951   Radiocarbon Dates. Science. 113 : 111-120.

Black, R. F. and M. Rubin
    1968   Radiocarbon Dates of Wisconsin. Wisconsin Acad. of Sci., Arts, and Ltrs. 56 : 99-115.

Black, R. F.
    1970   Glacial Geology of Two Creeks Forest Bed: Valderan Type Locality and Northern Kettle Moraine State Forest. Info. Circ. 13. Univ. Wisconsin Ext.

Black, P. R., N. F. Bleuley, F. D. Hole, N. P. Lasca, and L. J. Mahler, Jr.
    1970   Pleistocene Geology of Southern Wisconsin. Inf. Circ. 15, Univ. Wisconsin Ext.

Kaiser, K. R.
    1972   Ein eiszeitlicher Wald im Dättnau, Mittlg. d. Naturwiss. Ges. Winterthur. 34 : 25-41.

Mangerud, J.
    1970   Late Weichselian and Ice-Front Oscillations in the Bergen District, Western Norway. Norsk. geogr. Tidsskr. 24.

Nydal, R., K. Lovseth, and O. Syrstad
    1970   Trondheim Natural Radiocarbon Measurements V. Radiocarbon. 12 : 205-237.

Tauber, H.
    1964   Copenhagen Radiocarbon Dates VI. Radiocarbon. 6 : 215.

8

A Second Millennium B.C. Radiocarbon Chronology for the Maya Lowlands

Roy Switsur
Rainer Berger
Norman Hammond

The "Maya Lowlands" is a convenient, inclusive term for the Yucatan Peninsula of Mexico, the rain forest zone of Petén in Guatemala, and Belize, with part of the Mexican states of Tabasco and Chiapas adjoining to the west, and with the western margin of the Republic of Honduras on the east. The western boundary of the Maya Lowlands is around the Isthmus of Tehuantepec in Mexico, the eastern boundary is along the Ulua Valley in Honduras, and to the south the Maya Highland zone occupies the volcanic uplands of Guatemala, Chiapas, and western El Salvador (Hammond 1975, fig. 1). Much of this area is still occupied by Amerindians of Mayan speech, descendants of those who, in the first millennium A.D., built one of the world's greatest early civilizations; this Classic Maya civilization was centered on the forests of Petén, Belize, and the southern part of the Yucatán Peninsula, and it is known archaeologically from the ruins of such vast ceremonial centers as Tikal* (Coe 1967). Its period of florescence is formally A.D. 250-900, during which time the bulk of the stone monuments with dated inscriptions in the Maya Long Count calendar were carved and erected, but the recent discovery of a monument with a date of A.D. 126 at Abaj Takalik on the Pacific coast of Guatemala (Graham 1976) demonstrates the apparent Highland origins of both sculpture and writing, and it renders the designation "Classic" of little more than reference value. Similar violence has been done to the original concept of the Classic period as that when civilization first burgeoned in Mesoamerica by the discovery of the Olmec, whose complex and arguable "civilized" culture on the Gulf Coast of Mexico dates from the late second and early first millennia B.C.

*Place names in this paper can be located on the *Archaeological Map of Middle America* published by the National Geographic Society, Washington, D.C., except for Cuello, for which coordinates are given.

The period from 800-1500 B.C. was formally dubbed the "Early Formative" or "Early Preclassic." It was considered the period during which village farming communities established themselves in Mesoamerica, succeeding the Archaic period of hunters and gatherers (Willey and Philips 1958). At the time this concept was formulated, Willey and Phillips noted that the transition from Archaic to Formative took place at different times in various parts of the Americas, and that the proposed chronology was not immutable. Subsequently, the work of MacNeish (1964) on the origins of agriculture in the Tehuacan Valley of Highland Mexico led to acceptance of 2000 B.C. as a more realistic beginning for the Mesoamerican Formative. This acceptance came at a time when calendar and radiocarbon years were still considered equivalent for this period, but with the bristlecone pine calibration, this should probably be expressed as 2000 b.c./2500 B.C., using the British convention. We also feel that the Early Formative should be redated at its upper end, with the transition to Middle Formative (as defined) archaeologically and culturally occurring at ca. 1000 b.c./1200 B.C. (Hammond et al. 1976).

During the Early Formative much of Mesoamerica was occupied. Until the chronology reported here was obtained, there was no certain evidence that this occupation extended into the Maya Lowlands, although occupation of the Highland zone was known. At the time of a major conference on the origins of Classic Maya civilization in late 1974 (Adams, ed., in press), the earliest accepted archaeological evidence for settlement in the Lowlands was the Xe phase at the sites of Seibal and Altar de Sacrificios on the Rio de la Pasión in Guatemala. The former site had yielded a radiocarbon date of 660 ±75 b.c. (UCLA-1432b) for a cache that included a jade blood-letter of Olmec type, but it was disputed as to whether the first Lowland settlers had come up the Usumacinta basin from the Gulf Coast, down the Pasión and Chixoy from the Highland zone, or possibly both. There was, however, no doubt of the direction of migration and cultural influence into the Lowlands, nor of its occurrence at some time in the first half of the first millennium b.c.

This picture has been modified by a series of radiocarbon dates obtained from the site of Cuello, in northern Belize (approximately 18°05' N, 88°35' W), in the northeastern quadrant of the rain forest zone close to the ecological frontier due south of the karstic and scrub-covered Yucatan Peninsula. Excavations were carried out in 1975 and 1976 under the direction of N. Hammond and with the on-site supervision of D. C. Pring and S. Bishop, respectively. This was during the third and fourth field seasons of the Corozal Project, which began in 1973 as a joint venture of the British Museum and the University of Cambridge to investigate the prehistory and human ecology of northern Belize and the origins of Classic Maya civilization.

By the end of the project's second season in 1974, the regional chronology had been extended back to about 600 b.c. on the basis of ceramic typology, but a number of pottery types found in the earliest excavated levels at several sites did not seem congruent with those ceramic complexes. Some were found in disturbed layers at Cuello, but in the 1975 excavations they were found to be stratified below the earliest known ceramics. Four radiocarbon dates (Hammond et al. 1976) indicated a chronology spanning the second millennium b.c.. Excavations in 1976 enlarged the area of excavation to obtain further cultural and economic information and more samples for radiocarbon analysis. Four further dates have been obtained from the Cambridge laboratory and more are being processed, both there and at the Los Angeles laboratory; but occupation of the site during the second millennium b.c. (to the mid-third millennium B.C. in calendar years) is already substantiated.

The initial dates consisted of 1020 ± 1606 b.c. (Q1476) for the end of this previously unknown phase of Maya culture (now designated the Swasey Phase), and 1050 ± 160 b.c. (UCLA-1985a) for a stratigraphically close layer. A burial from another trench, dated 1230 ±195 b.c. (UCLA-1935g), is statistically inseparable from these. A sample from a midden, redeposited as fill for a low platform just above the old land surface in the first trench, yielded a date of 2050 ±155 b.c. (UCLA-1985e), which was not only the earliest acceptable date obtained for a Maya Lowland site by nearly one and a half millennia (dates of similar magnitude but associated with much later ceramics had been obtained and discarded for the Barton Ramie sequence; Willey et al. 1965:29), but was also one of the earliest for a sedentary, pottery-using community anywhere in Mesoamerica. Although it was stratigraphically consistent with the two dates of 1020 and 1050 b.c., the intervening, undated deposit and the uncertainty of a single date demanded confirmation through further selection and dating of samples.

The samples consisted of small pieces of dark-colored, charred wood in a black soil matrix in admixture with rootlets and whitish-lime plaster fragments; the samples were received in the laboratory as excavated. The latter contaminants were removed to obviate production of a false date: rootlets were removed manually under the microscope and plaster was removed by several applications of hot, dilute hydrochloric acid. An odor of rotten eggs during the latter procedure indicated the presence of sulphides, which were destroyed thereby. The yellowish, supernatant solution had appreciable iron content. The samples were also treated with very dilute, caustic, alkali solution to remove possible adsorbed humic acid contamination from decomposing vegetable matter leached by rainwater down through the upper, younger strata.

The radioactivity of the purified sample was measured as carbon dioxide in gas proportional counters shielded from cosmic and environmental radiation. Replicate measurements were performed to ensure statistical reliability of the results. The dates are presented in table 1 as conventional radiocarbon dates using the British Convention designation (b.c.) and are based on the Libby half-life (5568 years) for the ^{14}C isotope. The dates were further calibrated using the schemes of MASCA (Ralph et al. 1973) and of Clark (1975). These have been compared in table 1, which presents all the available results of measurements for this site, including a new date from the lowest stratigraphic level at Barton Ramie associated with the Jenney Creek Ceramic Complex (Gifford 1976). The Barton Ramie excavation, part of the Corozal Project, sought to obtain radiocarbon samples in stratigraphic association with pottery of the earliest phases at the site.

The four new dates, ranging from 1630 to 1950 b.c. with uncertainties of sixty-five to eight-five years, confirmed that this part of the Maya Lowlands was occupied during the first half of the second millennium b.c. The Barton Ramie date accords well with the ceramic evidence in that some characteristics of Jenney Creek pottery are found in the

TABLE 1

Radiocarbon dates from Cuello, Operation 17B and 17C (1975) and 17F (1976), and from Barton Ramie, Operation 20B (1976)

Operation and level	Ceramic complex	Age b.p.	Age b.c.	± yr	Calibrated according to:		Laboratory reference
					Ralph B.C.	Clark B.C.	
17B 6	Cocos	2125	175	65	390-70	370-80	Q-1558
17B 12	Lopez	2680	730	190	1100-750	1145-730	Q-1559
17B 14	Swasey → Lopez	2970	1020	160	1500-1020	1485-1030	Q-1476
17B 17	Swasey	3000	1050	160	1540-1030	1515-1065	UCLA-1985a
17B 28	Swasey	4000	2050	155	2910-2340	2870-2350	UCLA-1985e
17C 37	Swasey	3180	1230	195	1750-1270	1755-1285	UCLA-1985g
17F 176	Swasey	3670	1720	65	2180-2110	2280-1990	Q-1574
17F 189	Swasey	3580	1630	70	2140-2020	2150-1860	Q-1573
17F 239	Swasey	3760	1810	85	2460-2150	2410-2100	Q-1572
17F 230	Swasey	3900	1950	65	2580-2210	2610-2290	Q-1571
20B 9	Jenney Ck	3200	1250	110	1690-1400	1680-1410	Q-1575

In Op. 17F numerical and stratigraphic order of layers is not necessarily the same; 17F 239 ≃ 17B 28 stratigraphically.

latter part of the Swasey complex at Cuello (D. C. Pring, personal communication 1976).

Possible implications of this confirmed second millennium b.c. chronology are (1) that the Swasey Phase is long-lived, with a very slow, detectable rate of change in ceremic development; (2) that the Maya Lowlands relates to other subsequently civilized regions of Mesoamerica in having sedentary and probably agricultural communities by this date; (3) that the Maya culture was well established some centuries before the rise of the Olmec to the west, and the initial settlement of Yucatán was not part of an Olmec expansion; and (4) that Maya culture is as likely to have been the donor as the recipient in its relationship with the Olmec, and that the Olmec ceramic tradition in particular, may have derived from the Mayan contact.

A fifth implication has far-reaching geographic significance. In statistical terms of two standard deviations, the origin of the Early Formative Maya ceramic tradition is chronologically indistinguishable from the other early Mesoamerican ceramic tradition, that of the Purron-Pox pottery in Central Mexico (MacNeish et al. 1970, Brush 1965), but typologically dissimilar to it. The Purron-Pox tradition has a very restricted range of forms and decorative modes, and a clear local ancestry in stone bowl prototypes. The Barra-Ocos tradition (Lowe 1976) is more developed and begins by 1600 B.C., but is not similar to Swasey. The Swasey complex, from Cuello, unstratified as a separate complex from several other Lowland sites in and near northern Belize, has a wide range of vessel forms and of slip colors, including red, black, orange, buff, brown, cream, and gray. Decorative techniques including pre- and post-slip incision, bichrome painting and firing, negative painting, and pattern-burnishing. The tradition is developed and sophisticated, unlike a tentative ceramic tradition which must possess an antecedent phase of development. This phase may lie within the Maya Lowlands, in other parts of the Cuello site, or in sites elsewhere not yet excavated. Elsewhere, Central Mexico seems unlikely, while Barra at a slightly later date and the coeval pottery of Monagrillo in Panama (Willey and McGimsey 1954) are not sufficiently similar or exuberant to postulate a close relationship. The only areas of the Americas where pottery exists at an earlier date (back to 3000 B.C.) and possesses a similar range of variety are northwestern South America, the northern Andes of Ecuador and Columbia, and the circumambient coast from the Gulf of Guayaquil to the Gulf of Venezuela (Reichel-Dolmatoff 1965, Hill 1975).

As yet no close connections have been detected (although not enough work has been done in either area to make this serious negative evidence), but it is of some interest that maize, domesticated in Mesoamerica and not native to South America, has been found in coastal Ecuador in levels dated to about 3000 b.c. (Lathrap, personal communication). The possibility cannot be rejected that long-distance contact between Mesoamerica and northern South America may have occurred, by sea or overland, considerably earlier than has previously been supposed: "raising the ante" in the Maya Lowland chronology may have unexpectedly wide repercussions.

ACKNOWLEDGMENTS

The 1975 season was financed by the British Academy and the Crowther-Beynon Fund of the Cambridge University Museum of Archaeology and Ethnology. The 1976 season was financed by these and by the University of Texas at San Antonio, through the efforts of R. E. W. Adams and the generosity of private donors, by Belize Sugar Industries Ltd. and the Belize firm of G. A. Roe Insurance Services Ltd. and their London associates, Cooper Gay Ltd., by the City of Birmingham Museum and the University of Bradford, by the Society of Antiquaries of London, the University of London, the Royal Society, and *The London Times*. For permission to excavate we are indebted to the government of Belize, and especially to the minister responsible, the Hon. Santiago Perdomo, and to the archaeological commissioner, Mr. Joseph O. Palacio.

REFERENCES

Adams, R. E. W., ed.
    (In   The Origins of Mayo Civilization. Albuquerque.
  press.)

    1965   Pox pottery: earliest identified Mexican ceramic. Science. 149:194-195.
Clark, R. M.
    1975   A calibration curve for radiocarbon dates. Antiquity. 49:251-266.
Coe, W. R.
    1967   Tikal: Guide to the Ancient Maya Ruins. Philadelphia.
Gifford, J. C.
    1976   Prehistoric Pottery Analysis and the Ceramics of Barton Ramie in the Belize Valley. Memoirs of Peabody Mus. Harvard Univ.

Graham, J. A.
  1976 Mayas, Olmecs, and Izapans at Abaj Takalik. Proc. 42d Int. Cong. Americanists. Paris.
Hammond, N.
  1975 Lubaantun: A Classic Maya realm. Monographs of Peabody Mus. Harvard Univ.
Hammond, N., D. Pring, R. Berger, V. R. Switsur, and A. P. Ward
  1976 Radiocarbon Chronology for early Maya occupation at Cuello, Belize. Nature. 260:579-581.
Hill, B. D.
  1975 A new chronology of the Valdivia Ceramic Complex from the Coastal Zone of Guayas Province, Ecuador. Ñawpa Pacha. 10-12:1-32.
Lowe, G. W.
  1976 Papers of the New World Archaeological Foundation. Provo. 38.
MacNeish, R. S.
  1964 Ancient Mesoamerican Civilization. Science. 143:531-537.
MacNeish, R. S., F. A. Peterson, and K. V. Flannery
  1970 Prehistory of the Tehuacan Valley. 3. Ceramics. Austin, Univ. of Texas Press.
Ralph, E. K., H. N. Michael, and M. C. Han
  1973 MASCA Newsletter. 9:1.
Reichel-Dolmatoff, G.
  1965 Excavaciones arqueologicas en Puerto Hormiga (Dept. de Bolivar). Antropologia, Bogotá.
Willey, G. R., W. R. Bullard, jr., J. B. Glass, and J. C. Gifford
  1965 Prehistoric Maya Settlement in the Belize Valley. Papers of Peabody Mus. Harvard Univ.
Willey, G. R., and C. R. McGimsey
  1954 The Monagrillo Culture of Panama. Papers of Peabody Mus. Harvard Univ.
Willey, G. R., and P. Phillips
  1958 Method and Theory in American Archaeology. Chicago

9

The Effective Use of Radiocarbon Dates in the Seriation of Archaeological Sites

Dwight W. Read

The questions archaeologists wish to answer require fine control over the temporal relations among sites. Closer examination is needed, not only of analytical procedures for grouping and ordering sites (e.g., cluster analysis and seriation procedures) and for determining the archaeological meaning of these groupings, but of the role that dating must perform in establishing temporal relationships between such groups. The goal is to obtain the maximum amount of information from a limited number of radiocarbon dates.

The conceptual shift from a space-time to a processual framework in archaeology has increased the demand for accurate dating of a greater number of sites. The problem is particularly acute in regional studies where the focus is on topics such as trade networks or settlement systems and their change through time. The space-time framework has used a static categorization of sites into gross regional phases and has generally presumed that there is cultural homogeneity within a phase. Consequently, time control has largely been in terms of the beginning and ending of a phase. Since this framework tends to emphasize the regional phase as a unit, time control over individual sites is not so critical.

The processual framework, however, demands a much finer control over time. The fact that it rejects the uniformity of the regional phase and attempts to make sense of the variation among sites within a phase means that the cruder time estimates for phases are inadequate. Studies of settlement systems, trade networks, and so on, require that it be possible to indicate which of the sites in a region were occupied contemporaneously. In addition these studies will eventually need to provide occupational histories of individual sites to achieve their goal of explicating the development of a region. The difficulty stems from the fact that archaeological data are the sum total of such occupational histories, both at the level of the region and the site. Consequently, the archaeologist must be able to take these data in their accumulated form and separate them chronologically. At the regional level this ultimately requires assigning dates of founding and abandonment of individual sites. At the site level it implies reconstructing changes that occurred within the site.

With multi-featured sites, the latter task has been achieved primarily by assigning a construction date to each of the features (assuming there are datable materials) and assigning dates to artifactual material by association with a feature. Whether the artifactual material found in association with the feature is contemporaneous with it depends on the occupational history of the site. The use of abandoned rooms in pueblos for trash dumps is a well known example of lack of contemporaneity. Control over developmental histories is also partially achieved through excavation techniques, using the idea that superimposition is an ordering with respect to time. But mixing of levels does occur and so, ultimately, there is a need for time control at the smallest unit, the artifact.

Techniques are now being developed that permit assigning absolute dates to a wider variety of materials with increased accuracy, as some of the papers in this conference indicate. But even if the techniques are developed, cost must be considered, which is a limitation largely outside the control of the archaeologist. Regional studies involve hundreds of sites. In the Chevelon Archaeological Research Project (Plog et al. 1976) we have collected artifactual materials from some 300 sites, even though we are studying just one drainage in northeastern Arizona, and that is only a small fraction of the total number of sites in the drainage. The most generous of funding agencies would not underwrite obtaining absolute dates for all of these sites, even if datable materials were present in every site. Consequently, the archaeologist must continue to use indirect means of dating most sites.

Primarily these indirect means have involved certain assumptions about the role of material culture in the social system of a group of people. Thus, if one assumes cultural homogeneity throughout the social group, one can absolutely date the appearance of a particular artifact, pottery type, or design style in one site and use that date and the appearance of that artifact, pottery style, or design style in another site as a date for the other site. Or so the argument goes. Hence the lists of various pottery types and wares and their associated dates of appearance and disappearance. Given without reference to specific sites, these pottery sequences have presumed that cultures, including material culture, change in some monolithic fashion. It is a fiction that has not generated too many anomalies, so long as the regional phase has been the finest level of analysis. Clearly, though, it is an assumption that is in direct contradiction to the processual approach.

Now, the solution to this dilemma is (1) to be able to use properties of sites to order them in a relative time sequence, and (2) to use absolute dates to convert that ordinal time scale to an interval one. Two problems need to be resolved. First, how is that initial ordinal time scale devised? Second, what is the most efficacious way to convert the ordinal scale to an interval one? A solution to the first problem has traditionally been through using seriation techniques based on frequency of artifact types. The idea of the "battleship curve" as a paradigm for change in popularity of pottery types is sufficiently widely known not to need elaboration. The basic assumption is similar to that for pottery sequences — change is assumed to occur in a uniform fashion throughout a region. Here again regional development is assumed to be negligible, contradicting the processual approach.

Yet there is a valid assumption of homogeneity underlying the region-phase framework. To say that a set of individuals forms a cultural grouping is to imply that there are shared ideas, concepts, and meanings attached to objects and actions and that these are part of what defines that culture. The artifact, as Rouse (1939) has argued, is a consequence of those shared concepts and thus, somewhat imperfectly, is a physical representation of them. To the extent that there is a set of shared concepts, a change in those concepts will occur throughout the group, and presumably with minimal time delay. What the region-phase characterization has done is to assume that, in effect, there is no time delay for the diffusion of new or changed concepts within a region. The notion of regional development, however, presumes the contrary. It is these time delays which are precisely part of the explication of regional level variations.

A paradigm that is reasonable at the level of the site, the battleship-shaped curve for change in popularity of a type, becomes less accurate as it is extended beyond the site. While there are instances of accurate seriations, these are more likely to be accidental than to be a justification of the procedure as it is presently used.

When seriation is of sites that are from different phases, the actual trajectory of the popularity of a pottery type through both time and space can be collapsed into a single time trajectory which ignores space yet still remains reasonably accurate at the level of interphase analysis, even though distortion occurs at the level of intraphase analysis. To exemplify, if there are time lags in the spread of the popularity of a type, then, depending on the occupational length of sites in comparison to these time lags, the popularity as measured by the proportion of a type present in a site can give a seriation which distorts the actual time sequence (fig. 1). But if the time scale for change in popularity of a type and period of occupation of a site (i.e., interphase comparison) is substantially greater than the time scale for diffusion (i.e., intraphase comparison) then the distorting factor of diffusion will be small and sites from different phases will seriate correctly.

While seriation procedures use the criterion of similarity to order sites, the notion of diffusion implies that sites with similar characteristics need not be contemporaneous. Minimally, the model for devising a time sequence must be expanded to include a spatial dimension representing time lags in diffusion.

With these ideas in mind we can outline a first attempt at developing a more accurate model for seriation of sites. The model will also highlight the dating facility needed by the archaeologist. We will first indicate some specific problems with traditional seriation procedures using the Brainerd-Robinson technique, or some modification thereof.

The placement of all sites into a one-dimensional, ordinal sequence by such techniques is deceptive. Sites are occupied contemporaneously and the ordering procedure must be sensitive to this fact. Furthermore, relative ordering of sites within a phase may depend on small proportional differences between types. If these differences are small in comparison to sample and measurement error, the ordering is partially a consequence of such errors. Finally, the seriation procedure per se is not sensitive to the fact that the basic presumption underlying the relationship of material objects to culture in settlement studies is through hierarchical grouping of sets of individuals, such as is done with central place theory. Thus traditional seriation is an ordinal listing of sites, whereas the presumption of settlement studies is that sites should form a hierarchical classification corresponding to the properties of the settlement system as a system.

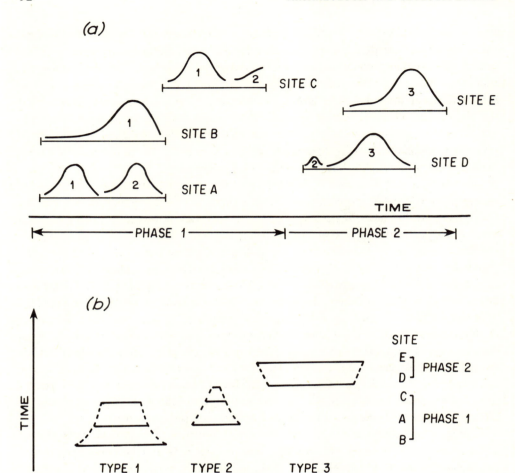

FIG. 1. (A) Hypothetical history of five sites. The curves indicate relative frequency of artifact types through time within a site. The number under the curve identifies the artifact type. The horizontal lines indicate the time and duration of occupation for each site. The sites are from different spatial locations so that there are time lags in the diffusion of the frequency of artifact types. (B) Seriation of the five sites using the "battleship" curve model. The length of the bar indicates the (cumulative) relative frequency of each artifact type. Sites within Phase 1 (intraphase comparison) are incorrectly seriated, but sites are correctly ordered with respect to time between phases (interphase comparison).

The group of procedures that goes under the name cluster analysis would seem to be a more appropriate analytical tool for seriation. This can be best understood by considering a model which views the totality of sites in a region as the superimposition of a series of time slices. Each time slice is composed of a series of sites interconnected by social and economic ties, with artifactual material distributed in a manner reflecting that system. The imagery of time slices is justified empirically, at least in the case of the Chevelon data, by the fact that the changes in the proportion of pottery types in sites is discontinuous (see fig. 3, Read 1973). Within a time slice, though sites may be occupied and abandoned, or though the geographical boundaries of the system may be expanding or contracting, there will be greater similarity among sites than between time slices. Thus cluster analysis should have the effect of sorting sites by time slices.

Groupings in cluster analysis based on the notion of similarity are not constrained by a priori presumptions, such as an ordering of sites in one dimension. Sites are grouped into primary clusters which can be hierarchically related as indicated by the data. Sampling and measurement error are not critical, since they mainly affect the tightness of a cluster. Though cluster analysis does not rank-order sites along a time scale, the clusters may be treated as units and may be seriated. Each cluster of sites plays the role of a phase in the earlier space-time framework, but it is not constrained in its properties by arbitrary external presumptions. Instead the partition of a collection of sites into clusters is determined analytically from the properties exhibited by that collection of sites, rather than paradigmatically as with the space-time, region-phase framework.

Evidence does seem to support the expectation that clusters represent time intervals. The site clusters based on proportion of black-on-white wares present in a site from the Chevelon data seriate to form a time sequence which is in accord with preliminary dating of the sites (Plog 1976).

Cluster analysis of sites would seem to represent the cultural dimension and, thereby, the time dimension possibly distorted by the effects of diffusion. Diffusion can be represented indirectly by spatial location.

In this fashion time can be measured along the two main dimensions through which it is expressed archaeologically. In effect the aim is to generate a stratified sample of the collection of sites that minimizes variation in each stratum. Consequently, if dating is to be done efficaciously, absolute dating of sites must represent those points in the classification scheme where time control is most critical.

At issue is whether or not the seriation of site clusters comprises a time sequence and, if it does, whether each cluster is disjoint in time. These questions can be answered by sampling randomly from within clusters and by dating the sites in the random samples to determine the distribution of sites in clusters with respect to time.

Another major issue is whether diffusion is a critical factor in ordering sites by similarity to obtain a time sequence. It is answerable by obtaining dates for sites classified both by cluster analysis and by location. In other words, is there regional change in time associated with sites within a cluster, or is the cluster homogeneous in terms of dates represented?

The main problem that arises in application of this procedure is one of simply being able to date sites. Traditionally, sites have been dated according to availability of a limited variety of materials. But the procedure being proposed here requires that the sites to be dated must be determined by the analysis. Hence, the range of datable materials needs to be expanded so that the archaeologist is able to date sites as needed to construct finer time frameworks for interpreting archaeological data.

In conclusion, the questions archaeologists are trying to answer demand more accurate control over time. So long as economics prohibits absolute dating of all sites, the archaeologist must make explicit the models being used for analysis of data and the implications of those models for establishing a chronology of sites. It has been argued that a combination of cluster analysis (classification by cultural remains) and regional analysis (classification by location) are a first step in this process. The model then indicates which sites need to be dated. Efficient dating of sites ultimately depends upon the archaeologist's ability to devise explicit models for the concepts involved in dating procedures and to expand the variety of material that can be dated.

REFERENCES

Plog, F.
    1976   Ceramic Analysis. *In* Chevelon Archaeological Research Project 1971-72. UCLA Archaeological Survey Monograph. II: 44-59.

Plog, F., J. Hill, and D. Read, Eds.
    1976   Chevelon Archaeological Research Project 1971-72. UCLA Archaeological Survey Monograph. II.

Read, D.
    1973   Some Comments on Typologies in Archaeology and an Outline of a Methodology. Am. Antiquity. *39*: 216-242.

Rouse, I.
    1939   Prehistory in Haiti: A Study in Method. Yale Univ. Pub. in Anthro. (21) New Haven.

10

Climate Controlled Uranium Distribution in Atlantic Ocean Core V29-179

Emil K. Kalil

ABSTRACT

Uranium content has been measured in eighty-nine samples from core V29-179 from the North Atlantic. Uranium analyses have been performed by nondestructive neutron activation and by the rapid and precise method of counting delayed neutrons. The sediments display an inverse relationship between uranium and calcium carbonate. When the uranium results are expressed on a carbonate-free basis, the depth distribution is not uniform but has six maxima at various depths which appear to correlate stratigraphically with glacial episodes. Presumably these values are the result of increased continental erosion. The six events occur at approximately 9000, 30,000, 60,000, 85,000, 130,000 and 150,000 B.P. This detailed uranium profile may prove to be a useful tool for stratigraphic correlations in Pleistocene chronology.

INTRODUCTION

Previous researchers have considered that, in general, uranium content of pelagic sediments has been constant with depth (Ku 1965, Mo et al. 1973). Where concentration changes with depth have been observed, some type of diagenetic reaction or diffusion process has been invoked (Bonatti et al. 1971, Veeh 1967). Sediments near midocean rises are suspect of having been affected by volcanic activity at the rise crests (Fisher and Boström 1969, Ku et al. 1972).

During periods of global glaciation, marine sediments display features which are a direct consequence of the change in erosion patterns. The most apparent features are an increase in sedimentation rate (Broecker et al. 1958), a decrease in calcium carbonate content (Ericson and Wollin 1964), and a change in faunal assemblages. Another is the increase in uranium content during glacial maxima as suggested by Boström and Fisher (1972).

Core V29-179 from the Lamont-Doherty Geological Observatory collection, and part of the CLIMAP Program, has been analyzed at one-cm increments taken approximately every 10 cm. V29-179 is a foraminiferal marl ooze, taken from the central North Atlantic at 44°00.6'N latitude and 24°32.4'W longitude in 3331 meters of water. The uranium profile will be compared to paleontological and radiometric age dates provided by the Lamont-Doherty Geological Observatory scientists.

Methods

Uranium analyses were performed by nondestructive neutron activation. Five to ten grams of dry sediment were irradiated at the UCLA Nuclear Reactor (Argonaut type) for one minute and then were counted for delayed neutrons after a 25-second delay (Amiel 1962, Gale 1967). Counting time was one minute. The automated system analyzed 24 samples per hour (Kalil 1976). Its precision was limited by counting statistics and for most samples was better than 3%. Samples with low uranium contents (<0.5 ppm) were run two or three times to increase the precision. All uranium concentrations were relative to NBL-76A standard.

Results

Figure 1 shows the depth distribution of uranium content expressed as ppm total weight sediment. This profile is complementary to the calcium carbonate profile.

Uranium content for most samples is inversely proportional to calcium carbonate, as shown in figure 2. This relation represents a two-component mixing system, where terrigenous material of approximately 2.0 ppm U is mixed with calcium carbonate of approximately 0.01 ppm U, as determined by a linear regression fit. This is consistent with data for similar models (Bertine et al. 1970, Ku et al. 1972), but there are numerous points in figure 2 that lie above the mixing line, which indicates that one of the mixing end markers must have a variable uranium content. When uranium is expressed on a calcium carbonate-free basis, as in figure 3, anomalously high uranium contents in the terrigenous material appear as several maxima above a relatively uniform background of about 1.7 to 2 ppm.

Discussion

Uranium and calcium carbonate contents show an inverse relationship. When calcium carbonate content is high during interglacial periods, uranium content is low and distinct minima occur. When uranium content is high during glacial periods, the $CaCO_3$ content is low, in particular the fine fraction coccoliths.

In figure 1, from the present to 10,000 B.P. and from 73,000 to 127,000 B.P., uranium content is low. Relatively high uranium values occur from 10,500 to 73,000 B.P. and from 127,000 to about 161,000 B.P. On a calcium carbonate-free basis (fig. 3), uranium maxima occur at about 10,555, at 26,000-37,000, at 62,000, at 86,000, at 127,000-140,000 and at 156,000 B.P.

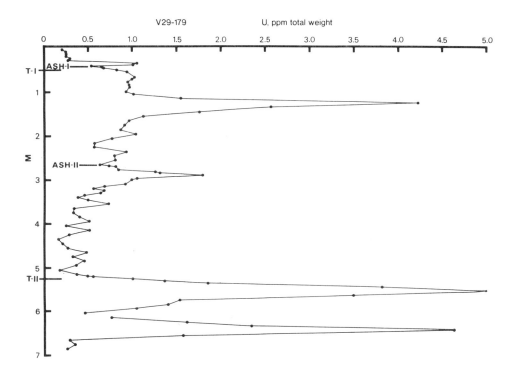

FIG. 1. Uranium concentrations of core V29-179 as ppm of total weight.

The first uranium peak (youngest) occurs at about 10,500 B.P., at the end of the Wisconsin, and falls in Emiliani's (1966) stage 2. The next peak is from 26,000-37,000 B.P. and could also be included in stage 2 when compared to the oxygen isotopic data of core 280 from Emiliani using Broecker and Von Donk's (1970) time scale. The peak at about 86,000 B.P. may relate to one of the cold periods associated within stage 5. There is some ambiguity to the events of stage 5 (Suggate 1974), and this may represent a cold period between Barbados I and II at 82,000 and 103,000 B.P., respectively. The peaks below 127,000 B.P. fall into stage 6 and the early Wisconsin.

The dates in V29-179 are based on paleontological evidence of Termination I at 13,000 B.P. and Termination II at 127,000 B.P. Dates younger than 40,000 B.P. are usually derived by ^{14}C methods. Ages older than 40,000 B.P. are determined radiometrically by $^{231}Pa/^{230}Th$ ratios.

The age of Termination II is substantiated by radiometric dates from raised coral reefs on Barbados (Broecker et al. 1968). The coral ages are based on the growth of ^{230}Th from ^{234}U and ^{231}Pa from ^{235}U, and are independent of concentration. This age marks a change in climate from cold to warm. The change in uranium content at 520 cm depth coincides with this change from glacial to interglacial and agrees with the age for T-II.

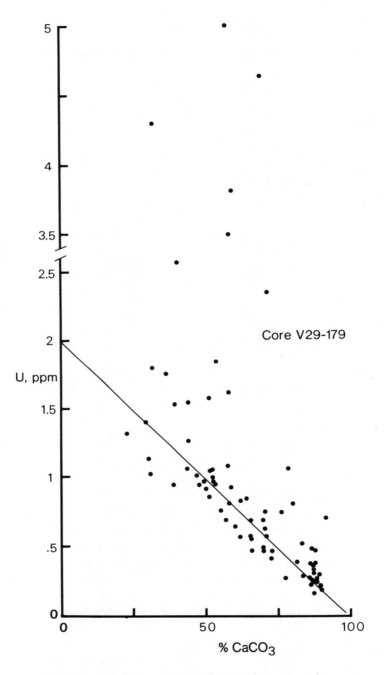

FIG. 2. Anticorrelation of uranium and calcium carbonate in core V29-179.

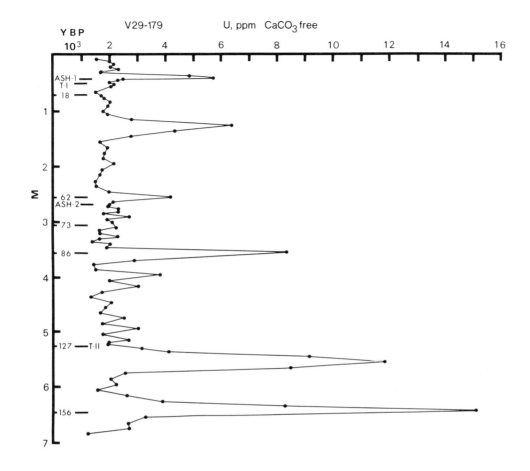

FIG. 3. Uranium concentration of core V29-179 on a calcium carbonate-free weight. Concentration is in ppm.

Once these dates are firmly established, the uranium content of V29-179 can serve at least two purposes. First, it denotes a uranium-glacial relationship which is independent of secondary diagenetic reaction. There is more uranium entering the ocean sediments during glaciations. Second, this uranium profile can be used to correlate with other cores. The uranium content of a core can be used stratigraphically to correlate different cores and to test the validity of the radiometric dates.

The episodic high uranium content in the sediment may be due to direct glacial input. Alternatively, it may be a result of a change in sea level, or a change in ocean circulation and sediment distribution, or a change in the rate of weathering on land, or it may be the result of volcanic activity from the mid-Atlantic Rise.

Another causal theory of increased uranium content is the formation of reducing conditions as postulated by Veeh (1967) and Bertine et al. (1970). Unfortunately, there is no evidence of reducing conditions in this core. Also, numerous cores from the southern California borderland show no increase of uranium due to reducing conditions (Kalil 1976). These cores also show a glacial occurrence of uranium.

Volcanic emanations appear doubtful as a source of uranium because the two ash layers contained in the core at 42 and 268 cm both have low uranium values.

CONCLUSIONS

Uranium content of pelagic sediments, as determined by counting delayed neutrons, is shown to vary directly with climatic fluctuations. These variations may be used stratigraphically to correlate different cores. Peaks of maximum uranium content coincide with glacial maxima at 10,500, at 30,000, at 62,000 and at 134,000-161,000 B.P.

ACKNOWLEDGMENTS

Support for the uranium analyses and for E. Kalil was provided by ERDA contract number ERDA E(04-3)-34, P.A. 134. I owe a special thanks to my thesis advisors, Dr. Willard F. Libby, Dr. Leona Marshall Libby, Dr. John T. Wasson, and Dr. Isaac R. Kaplan, for making this project possible. I would like to thank Dr. Andrew McIntyre for use of sediment samples from core V29-179 and Mr. J. Habib for his help in collecting the samples.

REFERENCES

Amiel, S.
    1962   Analytical applications of delayed neutron emission in fissionable elements. Anal. Chem. 34:1683-1692.

Bertine, K. K., L. H. Chan, and K. K. Turekian
    1970   Uranium determinations in deep-sea sediments and natural waters using fission tracks. Geochim. et Cosmochim. 34:641-648.

Bonatti, E., D. E. Fisher, O. Joensun, and H. S. Rydell
    1971   Post-depositional mobility of some transition elements, phosphorus, uranium and thorium in deep-sea sediments. Geochim. et Cosmochim. 35:189-202.

Boström, K. and D. E. Fisher
    1972   Lateral fluctuations in pelagic sedimentation during the Pleistecene glaciations, Boreas, 1:275-288.

Broecker, W. S., K. K. Turekian and B. C. Heezen
    1958   The relation of deep-sea sedimentation rates to variations in climate. Am. J. Sci. 256:503-517.

Broecker, W. S., D. L. Thurber, J. Goddard, T. L. Ku, R. K. Matthews and K. J. Mesolella
    1968   Milankovitch hypothesis supported by precise dating of coral reefs and deep-sea sediments. Science. 159:297-300.

Broecker, W. S. and J. von Donk
    1970   Insolation changes, ice volumes and the ^{18}O record in deep-sea cores. Rev. Geophys. Space Phys. 8:169-198.

Emiliani, C.
    1966   Paleotemperatures analysis of Caribbean cores P6304-8 and P6304-9 and a generalized temperature curve for the last 425,000 years. J. Geol. 74:109-126.

Ericson, D. B., and G. Wollin
    1964   The Pleistocene epoch in deep-sea sediments. Sci. 146:723.

Fisher, D. E., and K. Boström
    1969   Uranium-rich sediments on the East Pacific Rise. Nature 224:64-65.

Gale, N. H.
    1967   Development of delayed neutron technique as rapid and precise method for determination of uranium and thorium at trace levels in rocks and minerals, with application to isotope geochemistry. Symp. Radioactive Dating and Method of Low-Level Counting. IAEA. Vienna, 431-452.

Kalil, E. K.
    1976   The Distribution and Geochemistry of Uranium in Recent and Pleistocene Marine Sediments. Dissertaion. Univ. of Calif. Los Angeles.

Ku, T. L.
    1965   An evaluation of the $^{234}U/^{238}U$ method as a tool for dating pelagic sediments. J. Geophys. Res. 70:3457.

Ku, T. L., J. L. Bischoff, and A. Boersma
    1972   Age studies of Mid-Atlantic Ridge sediments near 42°N and 20°N. Deep-Sea Res. 19:233-247.

Mo, T., A. D. Suttle, and W. M. Sackett
    1973   Uranium concentrations in marine sediments. Geochim. et Cosmochim. 37:35-51.

Suggate, R. P.
    1974   When did the last interglacial end? Quat. Res. 4:246-252.

Veeh, H. H.
    1967   Deposition of Uranium from the Ocean. Earth Planetary Science Letters. 3:145-150.

11

The Neolithic Chronology of Switzerland by Underwater Archaeological Methods

E. Gary Stickel and Rainer Berger

INTRODUCTION

Exploratory dives in the Swiss Lakes have confirmed the presence of underwater sites related to the Swiss Lake Dwellers and have pointed to the feasibility of underwater excavations. It has been felt that the numerous, relatively undisturbed underwater sites in these lakes would provide excellent material for a major study of Swiss prehistory.

During the last twenty years, considerable development has been made possible in underwater archaeology with the availability of suitable equipment, such as flexible diving suits, air supplies, and ancillary instrumentation.* Many of the recovery techniques used in this study were gradually developed by U. Ruoff and his associates of the Zürich Stadtbüro für Archäeologie (Ruoff 1973). Dry diving suits supplied with air from a compressor in a boat above were used, since the water temperature of the Swiss Alpine lakes has proven too low for sustained underwater work of a few hours duration in wet suits.

The most widely held chronological scheme for Switzerland, originally devised by Vogt (1967) provides no calendar dates for the beginning and duration of each Swiss Neolithic culture. Sauter and Gallay (1969) suggest that the Swiss Neolithic began between 3000 and 2500 B.C. with two traditions, the Western and Danubian, coming into contact. The chronology by Ruoff (1973) extends the Neolithic period beyond 3300 B.C. and places the beginning of the Bronze Age at 1900-1800 B.C. with its termination at about 900-800 B.C. Although Ruoff is the first Swiss scholar to speculate about Neolithic terminal dates, he does not specify any definite Neolithic beginning dates. In this paper, new radiocarbon dates and other objective dating evidence suggest an updated chronology of the Swiss Neolithic.

*Exhibits on the development of underwater equipment can be seen at the J. Cousteau-designed museums on the Queen Mary, Long Beach, California, or in the Maritime Museum, Monaco, Principality of Monaco.

FIG. 1. Underwater archaeologist at work in Lake Zurich wearing dry suit. Water clouded by sediment is moved away from diver by slow water jets emerging from tube below. Seen in the background above the archaeologist is a dark occupation band which was above water level in antiquity.

New Research

In recent years the remarkable preservation of organic material in Swiss underwater archaeological sites, originally discovered by Keller in 1854, has prompted a continuing cooperative effort between the Zürich Stadtbüro für Archäologie and the Isotope Laboratory of the University of California at Los Angeles (Stickel 1974).

Even though some seventy ^{14}C dates were available for the Swiss Neolithic prior to this study*, much of the chronology, especially in the immediate study area of the environs of Zürich, was based on relative-stratigraphic and typological dating. For all the subsequent radiocarbon dates, the tree ring calibration of Suess was used to convert basic radiocarbon years into calendar years B.C. (Suess 1970).

Our initial chronological analysis focused on the Feldmeilen-Vorderfeld site located in Lake Zürich. This site was chosen for study because of its excellent stratigraphy and preservation of organic material which had been recently exposed by underwater archaeological excavations.

The settlement at Feldmeilen-Vorderfeld, on the right bank of Zürichsee, has been dated archaeologically by numerous remains of the characteristic ceramic forms of the Swiss Neolithic Pfyn and Horgen cultures. Also, there are some indications of the presence of the late Neolithic Schnurkeramik culture, although no extensive cultural stratum has been identified.

Both underwater and land excavations during 1970-71 exposed an excellent stratigraphical sequence of eleven cultural layers. The upper five belong to the older Horgen culture and the remaining six lower layers are identified as remains of the younger Pfyn culture. According to archaeological estimates, these remains should postdate the earliest Neolithic at Egolzwil, and should predate later groups such as the Schnurkeramik or Corded Ware culture. Thus, objective dating of these remains was of particular value. Nine samples from Feldmeilen were analyzed indicating an occupation for about 1400 years from 2950 to 4350 B.C. (table 1).

The oldest date of the Pfyn culture compares well with the beginning of the Cortaillod culture at the Egolzwil IV site. The date is based on a piece of hand-hewn wood which rested horizontally on the bottom of the lowest stratum. Great care was taken in the chemical treatment of the timber to exclude contamination by lake chalk and organic decay products.

In general the dates from the Pfyn levels are contemporaneous with those of the Enge site in Zürich identified as Michelsberg-Pfyn and perhaps earliest Horgen. On the other hand the Horgen culture from Feldmeilen overlaps with a series of dates from the Cortaillod settlement at Seeberg Burgäschisee-Süd in the Swiss Midlands, Canton Bern, ranging in age from 3000-3900 B.C. (Ferguson et al. 1966). The most recent date from Feldmeilen corresponds to the late Neolithic occupation at Auvernier, Lac de Neuchâtel.

On balance the Feldmeilen sequence dates from the earliest Neolithic of Switzerland to the beginning of the late Neolithic.

In 1854 Ferdinand Keller found a site which is now called Meilen-Schellen along with others which he later referred to as belonging to the famous Swiss Lake Dwellers. The site is located on the right bank of Zürichsee, 2.1 km south of the Feldmeilen site. The Swiss National Museum conducted excavations there in 1908-09 and found numerous Neolithic

*See Drack (1964); Feller (1970); Ferguson et al. (1966); Gfeller et al. (1961); Oeschger and Riesen (1966, 1967); Oeschger et al. (1959, 1970); Schwabedissen and Freundlich (1966); Suess and Strahm (1970).

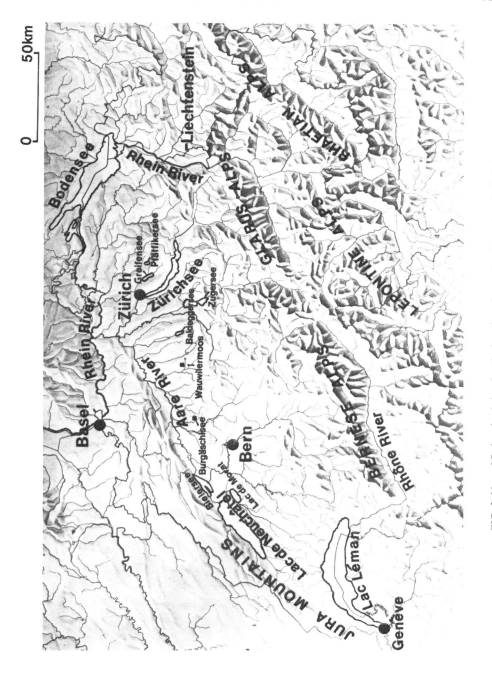

FIG. 2. Map of Switzerland showing the location of several major archaeological sites.

TABLE 1
Feldmeilen ^{14}C Measurements

Laboratory number	Stratum	Culture	Conventional radiocarbon years		Calibrated dates (B.C.)*	Material dated
			Underwater	Land		
1691 D	1x	Horgen	4250 ± 65		2950	Wood
1691 B	1y	Horgen	4500 ± 65		3400	Wood
1691 C	1	Horgen		4450 ± 65	3200 – 3400	Bark
1691 E	2	Horgen		4450 ± 65	3200 – 3400	Wood
1766 B	3	Horgen		4320 ± 65	3000 – 3350	Bone
1691 G	4	Pfyn		5200 ± 70	3950 – 4100	Wood
1691 H	6	Pfyn		4940 ± 70	3700 – 3900	Charcoal
1766 A	9	Pfyn		5100 ± 70	3900 – 3950	Bone
1691 F	9	Pfyn		5415 ± 60	4350	Charcoal

*Dates calibrated according to Suess 1970.

artifacts. It has been suggested that the various levels contain the remains of the older Cortaillod, Pfyn, Horgen, and Schnurkeramik cultures.

Recent underwater excavations exposed a clear sequence of eight strata which extends some three meters below the present lake bottom. In fact, this site provides, next to Feldmeilen, one of the clearest and most extensive stratigraphic sequences suitable for radiocarbon dating.

Ruoff's most recent archaeological assessment suggests that the site was not occupied by Cortaillod or Schnurkeramik populations as was previously thought. The topmost stratum belongs to the Bronze Age. Stratum two is characteristically Horgen. Strata three and four are listed as Horgen/Pfyn because of uncertainty regarding typologically determinative artifacts. Strata five through eight have been clearly identified as Pfyn, based on both pottery and axe forms.

All samples dated are the remains of beams and posts from former houses. The time span covering levels one through six represents almost two millennia from 2150 to 4250 B.C., exceeding that at Feldmeilen (table 2).

Comparison of the Meilen sequence with those of Feldmeilen and other sites provides a consistent chronological picture. The early dates in the Zürich area belong to the Pfyn culture beginning in the late fifth millennium B.C.

The Meilen dates of the Horgen culture match the Feldmeilen sequence. In fact they are among the first ^{14}C dates obtained for that culture. This culture must be placed later in time than the Pfyn on both stratigraphic and radiocarbon evidence.

Level one at Meilen may represent the Bronze Age with a calibrated radiocarbon age of 2150 B.C. This date is more recent than the age of the uppermost level for the

TABLE 2

Meilen ^{14}C Measurements

Laboratory number	Underwater stratum	Culture	Conventional radiocarbon years	Calibrated dates (B.C.)*	Material dated
1835 A	1	Bronze age?	3720 ± 65	2150	Charcoal
1835 B	2	Horgen	4040 ± 70	2500 – 2800	Charcoal
1835 C	3	Horgen	4390 ± 65	3000 – 3350	Charcoal
1835 D	4	Horgen	4480 ± 70	3200 – 3400	Charcoal
1835 E	6	Pfyn	5280 ± 60	4000 – 4250	Wood

*Dates calibrated according to Suess 1970.

Neolithic from the Auvernier type site. Therefore, it falls into a reasonable chronological place and is the first to mark the Swiss epoch of the major European cultural evolution now called the Bronze Age.

In addition to Feldmeilen and Meilen, sixteen of the twenty-six known Zürichsee sites were dated. Moreover, four of the eight known Neolithic Greifensee sites were also analyzed (Itten 1970). Each of these settlements is known either from past archaeological work, beginning with Keller in 1854, or from the latest surveys and test excavations being conducted by the Zürich Büro für Archäologie. All stratigraphic identifications are based on Itten (1970) with revisions by U. Ruoff (personal communication).

In order to determine the span of the Swiss Neolithic and its relationship to the major cultural periods which postdate it, samples from different Bronze Age contexts were analyzed. These samples are among the first to provide dates for the Bronze Age of Switzerland. Moreover, they allow the Swiss Bronze Age to be compared with that of the remainder of Europe.

The first of these sites analyzed was Zürich Alpenquai, which had an exclusive occupation by Bronze Age peoples. The topmost level, presently identified archaeologically as Late Bronze Age or Early Iron Age, dates to the end of the second millennium B.C.

Nearby Grosser Hafner site has a topmost level date of 1400 B.C. for the late Bronze Age, which may be the earliest epoch.

Whereas Horgen and Auvernier form the two major components of the Swiss late Neolithic, two Zürich Schnurkeramik dates of the first half of the third millennium B.C. should not be forgotten. The associated artifacts probably reflect cultural relations, such as trade, rather than actual occupations by Schnurkeramik peoples. It appears highly unlikely that three distinct cultures would have coexisted in such limited physical space. If one divides the Swiss Neolithic into an early and a late stage, then the latter is primarily composed of the Horgen and Auvernier sets of dates as well as a minor Schnurkeramik component.

The older is comprised of the Pfyn and Cortaillod cultures. In Vogt's Swiss chronology of 1967, the Neolithic is seen as a sequence beginning with cultures from Germany: (1) Linearbandkeramik (in cantons Basel and Schaffhausen), followed by (2) the Stichbandkeramik (in Schaffhausen and possibly Basel), followed by (3) the Rössen culture (in St. Gallen, Schaffhausen, Basel, Thurgau, and Zürich, with an extension into cantons Luzern, Zug, and Aargau). Vogt suggests that the first indigenous Swiss culture is the Egolzwil, coeval with the Rössen occupation in Switzerland. In addition he proposes that the Egolzwil culture developed in the cantons of Luzern, Zug, and Aargau, but at that time few radiocarbon dates were available. In fact the Egolzwil III site has three ^{14}C dates, but none of the actual Egolzwil levels has yet been dated. All other dates belong to an older Cortaillod cultural component. The only ^{14}C sample of possible Egolzwil context comes from the Zürich Kleiner Hafner site and dates to the middle of the fourth millennium (UCLA-1764A).

If there were direct progenitors of the Swiss Neolithic in Germany, the late Neolithic Michelsberg-Schüssenried cultural complex would be the likely candidate. The ^{14}C dating of this complex at Lautereck on the Upper Danube in Baden-Württemberg, Rieschachen, and Ehrenstein indicates a temporal span of this culture from 4490 to 3650 B.C., which is most coincident with the earlier stage of the Swiss sequence (Tringham 1971).

The later stage correlates best with the dates of the Eneolithic (Late Channeled Ware) culture at the site of Podolie, located in the Waag Valley in western Slovakia. The temporal span ranges there from 3400-2980 B.C. (Tringham 1971). Since the terminal Swiss date for the later stage is about 2500 B.C., it would appear that the Neolithic lasted considerably longer in Switzerland than it did in Central Europe.

Summary

The Swiss Neolithic appears about the middle of the fifth millennium B.C., according to presently available evidence. It presumably follows the Mesolithic, even though this transition has not yet been adequately dated. The beginnings are coeval with the Bandkeramik of Central Europe and the early Vinča culture of the Balkans (Tringham 1971) (table 3).

The fourth millennium B.C. (4000-3000 B.C.) sees the Neolithic incipiently developing in Switzerland contemporaneously with the late pre-dynastic cultures of Egypt, the later Neolithic of Greece (e.g., the Dhimini culture) (Theocharis 1973), the end of the important Vinča and Gumelnitsa cultures of southeastern Europe (each of which had a developed art of copper metallurgy) (Gimbutas 1974), the Trichterbecherkulture "A" and "B" of Denmank (Tringham 1971), and the Michelsberg-Schüssenried cultural complex of Germany (Tringham 1971). Also, evidence indicates that the early Swiss Neolithic was coeval with the most impressive Breton passage graves and the long barrows and megalithic tombs of Britain (Renfrew 1971).

The last ice age left its environmental effects longest in mountainous Switzerland and Scandinavia. Consequently, the Swiss region became one of the last settled in Europe. Moreover this resulted in a relatively late development of the Neolithic, which was then followed by a steady pace of cultural change.

TABLE 3

Comparative Neolithic chronology of Europe relative to Switzerland based in large part on treering calibrated radiocarbon dates and incorporating revisions by Renfrew (1971, 1973)

B.C.	SWITZERLAND	CENTRAL EUROPE	FRANCE	IBERIA/ITALY	NORTH EUROPE/BRITAIN	SOUTHEAST EUROPE
2000		EARLY AUNJETITZ VELUWE	BEAKER	EL ARGAR BEAKER + CASTELLUCCIO	HORIZON I CIST GRAVES + WESSEX I (STONEHENGE)	FÜZESABONY NAGYRE
2500	HORGEN SCHNURKERAMIK AUVERNIER	EARLY BEAKER	SEINE/OISE/ MARNE	LOS MILLARES + RINALDONE REMEDELLO	MIDDLE NEOLITHIC PASSAGE GRAVE + STONEHENGE I	VUČEDOL CLASSICAL BADEN
3000	PFYN CORTAILLOD	CORDED WARE	CHASSEY		TRICHTERBECHER C + NEOLITHIC	CERNAVODAEZERO EARLY BADEN EZEROVO
3500	EGOLZWIL	MICHELSBERG	LATE PASSAGE GRAVE	ALMERIAN + LAGOZZA	TRICHTERBECHER A + NEOLITHIC	TRIPOLYE C
4000	CORTAILLOD PFYN	RÖSSEN END OF BANDKERAMIK	EARLY CHASSEY	EARLY ALMERIAN + CHIOZZA	EARLY NEOLITHIC ERTEBÖLLE	GUMELNITSA VINČA
4500						

Considerable work lies ahead in surveying other Swiss lakes whose water levels have risen in the course of the last millennia, submerging habitation sites originally built on their shores. In particular the transitions from the Mesolithic to the Neolithic and from the latter to the Bronze Age need more exploration for a complete understanding of Swiss prehistory and the nature of its changes.

ACKNOWLEDGMENTS

We thank the National Science Foundation and the Wenner-Gren Foundation for Anthropological Research for their support, and we thank U. Ruoff and associates, J. and C. Bill, G. F. Wili, and J. Winiger for their invaluable advice and practical assistance. The late Emil Vogt was very helpful in consultations. This is publication No. 1730 of the Institute of Geophysics and Planetary Physics at the University of California, Los Angeles.

REFERENCES

Drack, W.
    1964   Enge, Zürcher Denkmalpflege. 2:125-132.
Felber, H.
    1970   Vienna Radiuminstitute Radiocarbon Dates I. Radiocarbon. 12:298-318.
Ferguson, C. W., B. Huber, and H. E. Suess
    1966   Determination of the Age of Swiss Lake Dwellings as an Example of Dendrochronologically Calibrated Radiocarbon Dating. Naturforschung. 21: 1173-1177.
Gfeller, C., H. Oeschger, and C. Schwarz
    1961   Bern Radiocarbon Dates II. Radiocarbon. 3:15-25.
Gimbutas, M.
    1974   Gods and Goddesses of Old Europe. Berkeley and Los Angeles, Univ. of Calif. Press.
Itten, M.
    1970   Die Horgener Kultur. Monographien zur Ur-und Frühgeschichte der Schweiz. 17. Basel.
Keller, F.
    1878   The Lake Dwellers of Switzerland and Other Parts of Europe. 2d ed. 1. 2. London, Longmans, Green.
Oeschger, H., U. Schwarz, and C. Gfeller
    1959   Bern Radiocarbon Dates I. Radiocarbon. 1:133-143.
Oeschger, H., and T. Riesen
    1966   Bern Radiocarbon Dates V. Radiocarbon. 8:22-26.
    1967   Bern Radiocarbon Dates VI. Radiocarbon. 9:28-34.
Oeschger, H., T. Riesen, and J. C. Lerman
    1970   Bern Radiocarbon Dates VII. Radiocarbon. 12:358-384.
Renfrew, C.
    1971   Carbon-14 and the Prehistory of Europe. Scient. Amer. 225(4):63-72.
    1973   Before Civilization. New York, Knopf.

Ruoff, V.
    1973   Palafittes and Underwater Archaeology. *In* Underwater Archaeology: A nascent discipline. Paris, UNESCO.

Sauter, M. R., and A. Gallay
    1969   Les Premieres Cultures D'Origine Mediterranienne. Archaologie der schweiz. 2:47-66.

Schwabedissen, H., and B. Freundlich
    1966   Köln Radiocarbon Measurements I. Radiocarbon. 8:239-247.

Stickel, E. G.
    974   A Temporal and Spatial Analysis of Underwater Neolithic Settlements in the Alpine Foveland of Switzerland. Dissertation. Univ. of California, Los Angeles.

Suess, H.
    1970   Bristlecone-Pine Calibration of the Radiocarbon Variations and Absolute Chronology. Olsson, ed. Stockholm, Almquist and Wiksell, 303-312.

Suess, H., and C. Strahm
    1970   The Neolithic of Auvernier, Switzerland. Antiquity. 44:91-99.

Theocharis, D.
    1973   Neolithic Greece. Athens, National Bank of Greece.

Tringham, R.
    1971   Hunters, Fishers, and Farmers of Eastern Europe: 6000-3000 B.C. London, Hutchinson Univ. Lib. Press.

Vogt, E.
    1967   Ein Schema des Schweizerischen Neolithikums. Germania. 45:1-20.

PART II
Radiocarbon Methodology

1

Correlation of ^{14}C activity of NBS Oxalic Acid with Arizona 1850 Wood and ANU Sucrose Standards

Henry A. Polach

INTRODUCTION

The twelfth Nobel Symposium on Radiocarbon Variations and Absolute Chronology recommended that radiocarbon dating results be related to NBS oxalic acid directly or, if indirectly, by using a substance of predetermined activity in relation to NBS oxalic acid (Olsson 1970:17). Undoubtedly the need for a *secondary laboratory standard* arose not only from the difficulty in obtaining CO_2 gas of reproducible isotopic composition on repeated oxidation (wet $KMnO_4$ oxidation, dry combustion, or electrolysis) of the oxalic acid radiocarbon dating standard (Craig 1961, Grey et al. 1969, Polach and Krueger 1972, Polach 1972), but also from the realization that the original supply of oxalic acid is quickly being exhausted.

In a study reported to the Eighth International Conference on Radiocarbon Dating (New Zealand, October 1972), Polach et al. (1972) proposed a three-way correlation of NBS oxalic, Arizona 1850 wood, and ANU sucrose. Bannister and Damon (1972) have prepared an Arizona 1850 wood primary radiocarbon dating standard (1846-1855 tree ring growth), and Polach has prepared an ANU sucrose which Polach and Krueger (1972) have evaluated for use as a secondary radiocarbon dating standard.

Selecting participants for this cross-correlation study was difficult because of the limited availability (about 880 g) of primary wood standard. The following rationale was used in selection. First, the number of participants was limited by the consideration that 40 g of wood was an adequate sample. Second, in the selection of participating laboratories priority was given to those laboratories engaged in dendrochronological cross-checking of the radiocarbon time scale. Coverage of a wide range of dating techniques was sought (gas proportional: CO_2, C_2H_2 CH_4, and liquid scintillation: C_6H_6), in order to achieve wide international participation, keeping in mind productivity and precision of determinations.

The following accepted invitations to participate: Agrawal (India), Berger (Los Angeles), Broecker (Lamont), Burleigh (London), Damon (Tucson), Geyh (Hannover), Gulliksen (Trondheim), Haas (Dallas), Kigoshi (Tokyo), Krueger (Geochron), Mook (Gröningen), Münnich (Heidelberg), Oeschger (Bern), Olsson (Uppsula), Ostlund (Miami), Polach (Canberra), Rafter (Lower Hutt), Ralph (Philadelphia), Stuiver (Seattle), Suess (La Jolla), Vogel (Pretoria), and Zavelsky (Moskva).

Since an overall precision of $\pm.5\,^o/_{oo}$ or better was sought, it was essential that not only multiple ^{14}C determinations of standards be carried out, but that supporting $^{13}C/^{12}C$ ratio measurements be made. Dr. Krueger agreed to carry out, free of charge, all $\delta^{13}C$ work related to this project, thereby assisting those laboratories without direct access to mass-spectrometric facilities. To further cross-check $\delta^{13}C$ determinations, Krueger, Damon, Polach, and Rafter proposed to carry out independent determinations on the same portions of CO_2.

ARIZONA 1850 WOOD Primary Radiocarbon Dating Standard

A sample of about 1.2 kg of wood consisting of precisely ten tree rings, corresponding to the growth season of A.D. 1846 to 1855 was prepared by the Tree Ring Research Laboratory, University of Arizona, from a Douglas fir (*Pseudotsuga menzieii* [Mirb.] Franco) known as the Hitchcock Tree. It grew twenty miles northeast of Tucson at an elevation of 6050 feet (Lat 32° 3'N, Long 110°41'W), and had sufficient ring width variability to allow positive cross-dating with other established tree ring chronologies of southern Arizona, permitting absolute identification of the A.D. 1846-1855 interval.

Chemical pretreatment to remove extractives while preserving as much as possible of the wood was carried out by the Laboratory of Isotope Geochemistry, University of Arizona. The match-stick-size raw wood was dried overnight in an oven at *ca.* 100°C, boiled in 1N HC1 (½ hour), rinsed in cold distilled water three to four times, then boiled in 2°/oo NaOH (½ hour), washed once thoroughly in cold 1N HC1 to neutralize alkaline treatment, rinsed with distilled water four to five times, and dried at *ca.* 100°C overnight (the total weight loss on treatment was 23.3°/oo). Following chemical treatment and drying, the wood sample was mechanically ground to sawdust, correctly proportioned, and packaged into 40 g lots for redistribution to selected laboratories (Bannister and Damon 1972). In order to obtain directly comparable results, the wood was ignited by participating laboratories, as supplied, withfurther pretreatment. The $\delta^{13}C$ value of the pretreated wood was established at $\delta^{13}C_{PDB} = -23.0 \pm 1\,^o/_{oo}$.

ANU Sucrose, Secondary Radiocarbon Dating Standard

A 1000 kg batch of high purity sucrose, suitable for this experiment and future dating needs, was donated by the Colonial Sugar Refining Company Ltd. (Sydney, Australia) and was manufactured at their New Farm plant, Brisbane (Queensland, Australia).

The analytical purity sucrose was recrystallized from a ten ton batch of raw sugar. The local provenience and growth period of the sugar cane was known to be the September 1969-June 1971 season. Standard purification techniques in the first instance were used. These involved treatment of the raw sugar with *milk of lime* ($Ca(OH)_2$; resulting in a final lime sugar solids ration of 0.5% CaO, pH 8.2). This treatment was followed by color bleaching in an acid pickle (sulfonic acid, pH 3), crystallization, centrifugation, redissolution, and filtering over *bone char* (jaw bones fragmented to 20 mesh, kiln roasted to produce activated charcoal with a large surface area). The filtered and clarified sugar syrup was then heated under partial vacuum with continuous stirring to evaporate excess water so that pure sugar would crystallize from solution.

Our batch was recrystallized under more stringent and precise conditions than is usual. To ensure uniform crystal size and high purity, only the mid, prime ton of the ten ton batch was collected for us and subjected to extra washing during centrifugation to ensure maximum purity and absence of fines. While cooking grade sugar is generally *ca.* 99% sucrose, (with .2% reducing sugars such as glucose and fructose, and .35% ash material such as $CaCO_3$, K^+, Na^+ and Ca^{++} and .3 to .4% of other organic matter), our specially produced *analytical reagent grade* ANU sucrose tested out as follows:

Sucrose	>99.95%
Specific rotation $[\alpha]_d^{20}$	+66.5
Insoluble matter	0.001%
Sulphated ash	0.004%
Chloride	< 0.0005%
Nitrogen	< 0.002%
Sulphate (SO_4'')	< 0.002%
Iron (Fe)	< 0.0001%
Heavy metal (Pb)	< 0.0001%

Reducing sugars (glucose & fructose)	0.001%
Loss of drying (100°C for 5 hours)	0.03%
Copper (Cu)	0.2 ppm
Potassium (K)	5.7 ppm
Sodium (Na)	1.4 ppm
Calcium (Ca)	1.5 ppm
Magnesium (Mg)	0.2 ppm

Each of these is superior, often very significantly, to laid down ANALAR (1967) specifications. The ANU sucrose was then packaged in heavy duty polyethylene bags which were closed using small-diameter, thick-wall "O" rings. These bags were inserted into specially manufactured, anti-corrosion-treated 20 kg steel drums with hermetically sealable lids. The drums were manufactured by their donor, Rheem Australia Ltd., who also arranged for free transport to Canberra. The drums are now stored at the Research School of Earth Sciences, and the distribution of sucrose will be undertaken by the ANU

Radiocarbon Dating Research Laboratory, on request, free of charge to all laboratories. Some 40 kg have been distributed so far.

Combustion of sucrose is not as straightforward as that of wood. We suggest that users try a few experimental runs, as experience in combustion provides the best procedural guide. During combustion, sucrose will swell, bubble, char, and run if combustion temperatures are excessive. Blockages of the combustion tube have been reported. We spread our sucrose sample into a thin layer within a silica boat inserted into a horizontal silica combustion tube of about 6 cm diameter. Ignition is accomplished by applying a small external flame at the far end of the boat from the pure oxygen inlet. On ignition the external heat source is removed and the low temperature ignition of *volatiles* (a clear blue flame, yield *ca.* 25% of CO_2) proceeds spontaneously against the gentle stream of oxygen (a backburn). The combustion of the black, tar-like residue follows spontaneously in the direction of the O_2, with a luminous flame (yield *ca.* 65% of CO_2) without application of external heat. As the flame passes over uncombusted *pyrolysis products*, thereby heating them, considerable sooting and tarring of the combustion tube will result, accompanied by some swelling and running. If the sucrose layer is too deep in relation to the diameter of the combustion tube, this swelling can result in a blockage of the tube. The combustion of residual *soot and coal* (yield *ca.* 10% of CO_2) then follows. A hot flame is applied externally, at the O_2 inlet end of the sample, and combustion proceeds normally. No residue, ash, or soot will remain; the tube will be clear.

We have tested the possibility of fractionation of sucrose on incomplete recovery and on portions of CO_2 gas derived from the *volatiles, tars,* and *coals* of partially burned sugar and found about ±1°/oo difference in $\delta^{13}C$ values (Polach and Krueger 1972). However, some of the deliberately unignited *volatiles* were precipitated in $AgNO_3$ and, upon recovery and ignition, their $\delta^{13}C$ value was some 2.5°/oo different from the other portions of the sample. We recommended that those laboratories using $AgNO_3$ bubblers between the combustion tube and the CuO furnace, bypass the $AgNO_3$ bubblers for their sucrose standard preparations; the ANU sucrose does not contain halides which would affect the purity of CO_2 for gas proportional counting. With normal care during combustion no significant fractionation is likely.

The same CO_2 measured by different mass spectrometers will give $\delta^{13}C$ results which can apparently differ by as much as ± 1.5°/oo. This is not due to fractionation during sucrose → CO_2 production and recovery, but is due to mass-spectrometer calibration using secondary standards in place of the original supply of PD belemite, which has become quickly exhausted. In our study, samples were prepared by Damon, Polach, and Rafter and measured by Krueger, Rafter (Institute of Nuclear Sciences, Lower Hutt, New Zealand), and Wilson (University of Waikato, Hamilton, New Zealand). The agreement between the two New Zealand laboratories was within ± 0.2°/oo with respect to PDB. The Krueger values were systematically displaced. Kureger has, since our measurements, recalibrated his instrument. Nevertheless, the possibility of $\delta^{13}C$ measurements deviating by ± 1.5°/oo (or possibly more) must be taken into consideration in the evaluation of the error of the cross-calibration of oxalic/sucrose/Arizona wood ^{14}C activities. The internal reproducibility for repeat combustions of ANU sucrose samples is likely to remain within ± 1°/oo, rendering repeat mass-spectrometric measurements less essential. This is particularly useful for laboratories without regular access to a mass-spectrometer. The $\delta^{13}C$ value of completely combusted sucrose should be within $\delta^{13}C = -11 \pm 1.5^o/oo$ with respect to PDB.

To assess completeness of combustion (e.g., in steel combustion furnaces or calorimetric bombs where combustion rates cannot be readily controlled) it is appropriate to use $C_{12}H_{22}O_{11}$ (molecular weight = 342.30 g) as the formula for the ANU sucrose. Stochiometric conversion to CO_2 will then yield 0.785ℓCO_2 at S.T.P. (or *ca.* 0.84ℓCO_2 at room temp.) per gram of ANU sucrose. In our case the yields ranged from 99.0 to 100.5%, due undoubtedly to error in manometer reading and gas temperature evaluation.

Cross-Calibration

Results of the international cross-calibration were collated by Polach and were obtained from all but two of the participating laboratories. Although the original request was for primary data (i.e., total β [^{14}C counts] for Ox + B, S + B, and W + B where Ox = oxalic, S = sucrose, W = Arizona wood, B = background, and T = counting time), only a few laboratories complied. Most preferred to express their results as ratios with respect to 95% oxalic. Complications arose, because questions relating to $\delta^{13}C$, age correction, and ^{14}C half-life (where age corrections were applied) were not resolved by all workers in the same way. Because it was acknowledged that minor deflections in computational pathways, dependent on personal interpretation, could give results not applicable to interlaboratory comparisons, discussions were held during the ninth International Radiocarbon Dating Conference (Los Angeles and La Jolla, July 1976) between participants of this cross-correlation study. Drs. Berger, Suess, Vogel, Stuiver, and Polach, in subcommittee, recommended establishing guidelines for expressing the cross-checking results in a significant, simple, and unique way, with one number defining each sample count rate with respect to 95% oxalic. It was further recommended that the study be enlarged to accommodate the Heidelberg Carbonate Solution Standard (Otto Münnich 1950's), already counted by some of the participants, and to keep results of each participating laboratory confidential.

Permutations of measurements of ^{14}C activity of ANU sucrose (S), Heidelberg Carbonate (H), 1850 Arizona wood (W), and 95% of ^{14}C activity of NBS oxalic (.950x), allow six ratios to be expressed:

$R_1 = S/H;$ $\quad R_2 = S/W_{1950};$ $\quad R_3 = S/.950x$
$R_4 = H/W_{1950};$ $\quad R_5 = H/.950x;$ $\quad R_6 = W_{1950}/.950x$

Of these $R_3(S/.950X)$, $R_5(H/.950x)$ $R_6(W_{1950}/.950x)$, and R_2 (S/W_{1950}) are of primary importance and will be reported in the final analysis of data. R_1 and R_4 can be calculated from:

$$R_1 = \frac{R_2}{R_4} \quad \text{and} \quad R_4 = \frac{R_2}{R_1}$$

The above, and other relationships (e.g., $R_6 = R_5/R_4$ and $R_6 = R_3/R_6$) will hold if the ^{14}C activity measurements are carried out on the same equipment (same detector) and will serve to indicate if any of the four standards (i.e., S, W, H, or Ox) are prone to erroneous measurements.

Expression of Data and Error

The error is expressed as ± 1 standard deviation (68.3%P) and it is based either on Poisson distribution of cont rate values around their mean, or on a combination of *actual* and *Poisson* distribution, depending on each laboratory's practice. If we let ± a be the error of A and ± b the error of B, then their combinations are given by

$$(A \pm a) + \text{or} - (B \pm b) = (A + \text{or} - B) \pm (a^2 + b^2)^{1/2}$$

$$(A \pm a)(B \pm b) = AB \pm AB \left[\frac{a^2}{A^2} + \frac{b^2}{B^2}\right]^{1/2}$$

$$\frac{(A \pm a)}{(B \pm b)} = \frac{A}{B} \pm \frac{A}{B}\left[\frac{a^2}{A^2} + \frac{b^2}{B^2}\right]^{1/2}$$

NBS OXALIC

95% of the net β count rate ± 1 σ normalized for isotopic fractionation to the base of $\delta^{13}C_{PDB} = -19.0^o/oo$.

ANU SUCROSE

Net β count rate ± 1 σ normalized for isotopic fractionation to the base of $\delta^{13}C_{PDB} = -25.0^o/oo$.

HEIDELBERG CARBONATE SOLUTION

Net β count rate ± 1 σ normalized for isotopic fractionation to the base of $\delta^{13}C_{PDB} = -25^o/oo$.

ARIZONA WOOD (1850)

Net β count rate ± 1 σ age corrected for ^{14}C decay from 1850 to the base year of 1950 ($T_{1/2}$ = 5730y : λ = 1.2097 × 10^{-4}), then normalized for isotopic fractionation to the base of $\delta^{13}C_{PDB} = -25^o/oo$.

Decay Correction of 1850 Arizona Wood

$$W_{1950} = W^1 e^{\lambda t}$$

Where W^1 = calculated* net β countrate of 1850 Arizona wood
λ = 1.2097 × 10^{-4} for 5730 y half-life
t = 100, i.e., base year 1950 minus year of growth 1850.

*Calculated net β incorporates such local corrections as temperature, pressure, dilution, and purity. Each asterisk in section refers to this Key.

TABLE 1

Summary of $\delta^{13}C$ and age correction of various standards used in the international cross-calibration program

Standard	Measured δ^{13}_{PDB} range	Defined** δ^{13}_{PDB}	Age correction
0.950 Oxalic	−14 to −22°/oo	−19.0°/oo	No
ANU Sucrose	−0 to −13°/oo	−25.0°/oo	No
Arizona Wood	−22 to −25°/oo	−25.0°/oo	Yes, 100 years
Heidelberg	−2 to −6°/oo	−25.0°/oo	No

**This is consistent with the proposal by Stuiver and Polach (1977) that $\delta^{13}C$ of all samples (with the exception of oxalic) be related to the base of −25‰ with respect to PDB. It is also consistent with the concept of the Conventional Radiocarbon Age BP where sample $^{13}C/^{12}C$ composition should be normalized to the base of $\delta^{13}C_{PDB} = -25.0°/oo$. (Olsson 1970:17).

$\delta^{13}C$ Normalization Formulae

Oxalic

$$0.950\, x = 0.95\, A_{ox}\left[1 - \frac{2(19 + \delta^{13}C_{ox})}{1000}\right]$$

where: A_{ox} = calculated* net count rate of NBS oxalic
$\delta^{13}C_{ox}$ = Measured (or estimated) $^{13}C/^{12}C$ ratio of oxalic acid CO_2 with respect to PDB in parts per thousand.

Sucrose

$$S = A_s\left[1 - \frac{2(25 + \delta^{13}C_s)}{1000}\right]$$

where: A_s = calculated* net β count rate of ANU sucrose
A_s = measured (or estimated) $^{13}C/^{12}C$ ratio of ANU sucrose CO_2 with respect to PDB in parts per thousand.

Wood

$$W_{1950} = A_w e^{\lambda t}\left[1 - \frac{2(25 + \delta^{13}C_w)}{1000}\right]$$

where: A_w = calculated* net β count rate of Arizona 1850 wood
λ = decay constant for 5730 y half-life, i.e., $\lambda = 1.2097 \times 10^{-4}$
$t = 1950 - 1850 = 100$

Example of Calculations

OXALIC

$$A_{ox} = 32.3858 \pm .0750 \text{ cpm}$$
$$\delta^{13}C_{PDB} = -18.3 \pm 0.1\text{°/oo}$$
$$T = 6000 \text{ Minutes}$$
$$.950x = 32.3858 \left[1 - \frac{2(19 + (-18.3))}{1000}\right] 0.95$$
$$= 32.3858 \times .99860 \times 0.95$$
$$= 30.7234 \pm .076 \text{ cpm}$$

SUCROSE

$$A_s = 47.3860 \pm 0.1000$$
$$\delta^{13}C_{PDB} = -11.5 \pm 0.2\text{°/oo}$$
$$T = 5000 \text{ Minutes}$$
$$S = 47.3860 \left[1 - \frac{2(25 + (-11.5))}{1000}\right]$$
$$= 47.3860 \times .9730$$
$$= 46.1066 \pm 0.110 \text{ cpm}$$

In absence of $\delta^{13}C$ measurements, if we had assumed the sucrose to have a value of $\delta^{13}C_{PDB} = -12.0 \pm 2\text{°/oo}$ then the calculated value for S would have been: $S = 46.1540 \pm 0.2$ cpm (i.e., essentially the same value but with an enlarged error.) In actual practice, the observed activity ratios (^{13}C uncorrected) of S/0.95 Ox is 1.0267 larger than the values reported here because of the ^{13}C correction for sucrose (from assumed value of $\delta^{13}C$ of -12°/oo to -25°/oo).

Wood

$$A_w = 30.4620 \pm .1112 \text{ cpm}$$
$$\delta^{13}C_{PDB} = -23.0 \pm 0.2\text{°/oo}$$
$$T = 3000 \text{ Minutes}$$
$$W_{1950} = 30.4620 e^{1 \cdot 01217 \times 10x^{-4} \; 100} \left[1 - \frac{2(25 + (-23))}{1000}\right]$$
$$= 30.4620 \times 1.01217 \times .996$$
$$= 30.7094 \pm .1115 \text{ cpm}$$

The ratios in these expressions are

$$R_3 (S/.950x) = 1.5007 \pm .0052$$
$$R_6 (W_{1950}/.950x) = .9995 \pm .0044$$
$$\text{and } R_2 (S/W^2{}_{1950}) = 1.5014 \pm .0065$$
$$\text{and the relationship } R_2 = \frac{R_3}{R_6} \text{ is upheld.}$$

Results of Cross-Calibration

When primary data were supplied, results were calculated using the above described procedures. When ratios were given, and it was indicated explicitly how these were derived, it was possible to assess the values received.

Fifteen results were compared; their distribution around their apparent mean value is shown in table 2.

TABLE 2

Distribution around apparent mean of ratios obtained for various dating standards

Distance from mean in terms of σ	S/0.950x	W_{1950}/0.950x	S/W_{1950}
+3 σ			
+2 σ	1	1	
±1 σ	13	13	9*
−2 σ	1	1	
−3 σ			

*The sucrose/wood ratio was given by all, but only nine of the given results could be resolved without ambiguity.

CONCLUSIONS

The calculated examples, taken directly from some of our ANU data, appear to fit within the ± 1 σ category so can probably be used as indicators of the order of magnitude of the actual relationship between the standards tested. Only confirmation of my calculations by the participants of this cross-checking program will allow us to define their precise relationship; yet it is not premature to indicate that there is excellent agreement so far. As only two laboratories have made the Heidelberg values available to me at this early stage, the Heidelberg/.950x ratio (and others) are not yet listed. From the available data, it appears to be H/.950x = 9.9 ± 0.1, with δ^{13}C normalization procedures the same as for sucrose, that is, $\delta^{13}C_{PDB}$ to the base of −25°/oo.

ACKNOWLEDGMENTS

I wish to thank all the participating radiocarbon daters for the tremendous effort made in supply precisely dated data, often requiring several months of counting time. It is appropriate to acknowledge once more the generous gift by Colonial Sugar Refining Company (Australia) and Rheem (Australia) Pty. Ltd., which made this study possible. My gratitude also goes to my colleagues at A.N.U., particularly to Professors Jack Golson and Donald Walker who encouraged me to proceed with this study.

REFERENCES

Bannister, B., and P. E. Damon
    1972   A dendrochronologically-derived primary Standard for Radiocarbon Dating. Proc. 8th Int. Conf. on Radiocarbon Dating. New Zealand. 676-685.

Craig, Harmon
    1961   Mass-spectrometer analyses of Radiocarbon Dating. Radiocarbon. 3 : 1-3.

Grey, D. C., P. E. Damon, and B. C. Haynes
    1969   Carbon-isotope fractionation during wet oxidation of oxalic acid. Radiocarbon. 11 : 1-2.

Olsson, Ingrid U.
    1970   Editorial preface. *In* Radiocarbon variations and absolute chronology. 12th Nobel Symp. Uppsala. 17.

Polach, Henry A.
    1972   Cross-checking of NBS oxalic acid and secondary laboratory Radiocarbon Dating Standards. Proc of 8th Int. Conf. on Radiocarbon Dating. New Zealand. 688-717.

Polach, Henry A., and Harold W. Krueger
    1972   Isotopic fractionation of NBS oxalic acid and ANU-Sucrose Radiocarbon Dating Standard. Proc. 8th Int. Conf. on Radiocarbon Dating. New Zealand. 718-724.

Polach, Henry A., Harold W., Bryant Bannister, Paul E. Damon, and Athol T. Rafter.
    1972   Correlation of C-14 activity of NBS Oxalic with Arizona-1850 Wood and ANU-Sucrose Radiocarbon Dating Standards: a preliminary report of investigations and results. Proc. 8th Int. Conf. on Radiocarbon Dating. New Zealand. 686-687.

Stuiver, Minze, and Henry A. Polach
    Reporting of C-14 data: A discussion. Radiocarbon 19 : 355-363.

2

An Improved Procedure for Wet Oxidation of the ^{14}C NBS Oxalic Acid Standard

S. Valastro, jr., L. S. Land, A. G. Varela

ABSTRACT

An improved procedure for wet combustion of ^{14}C NBS oxalic acid standard has been devised which gives consistent high carbon yields and an average δ^{13}C value of $-19.1‰$. The principal cause of fractionation in earlier attempts to prepare CO_2 by the wet oxidation method has been the inexact nature of the end point. The new procedure employs a chocolate-brown end point by adding 5 ml more of the sulfuric acid-potassium permanganate solution after the initial, reddish-brown end point is reached. The sulfuric acid-potassium permanganate solution is added to the NBS oxalic acid in a steady drop-wise flow, heat is applied to the generating apparatus, and a cycling technique is utilized to collect the CO_2.

Fifteen samples of the NBS oxalic acid have been processed. The carbon yields range from 97.8% to 100% with an average of 98.8% and an average δ^{13}C of $-19.12‰$. The results obtained by this procedure are much more consistent than previous results obtained by direct combustion.

Wet Oxidation of Oxalic Acid

Wet oxidation of the ^{14}C NBS oxalic acid standard to carbon dioxide is an accepted technique used by most radiocarbon laboratories. The procedure is simple and straightforward. After some years of use, however, investigators have discovered that fractionation does occur during CO_2 evolution because of difficulty in determining the proper end point of the reaction.

Fractionation during this chemical reaction has been documented by several investigators. Lindsay et al. (1949) reported as much as 3.5% change in the ^{13}C concentration in the CO_2 during the decomposition of oxalic acid, and Bernestein (1957), working with formic acid decomposition by sulfuric acid, stated that variations as much as 53%

occurred. Grey et al. (1969) in preparing CO_2 from the ^{14}C oxalic acid standard reported a $\delta^{13}C$ value of 25.5 ⁰/oo due to incomplete reaction.

Fractionation during oxidation of the NBS oxalic acid standard is of concern because it may offset counting rates, resulting in systematic errors in age determinations.

Craig (1961) determined the $^{13}C/^{12}C$ ratio of the standard gas samples from several laboratories, prepared by both wet oxidation and direct combustion in oxygen. $\delta^{13}C$ values by direct combustion with oxygen ranged from $-17.15$⁰/oo to $-22.72$⁰/oo with an average of $-19.3$⁰/oo (10 samples), while the wet oxidation values ranged from $-18.37$⁰/oo to $-31.37$⁰/oo. Rejection of the four most depleted values gave an average of $-19.6$⁰/oo (6 samples). According to Craig, depleted values were to be expected if the oxidation had not gone to completion.

Since the primary ^{14}C standard for the University of Texas radiocarbon laboratory is NBS oxalic acid, the personnel of the laboratory were compelled to improve the wet oxidation procedure so that no significant fractionation would occur. Thus, a new technique, which gives high yields and consistent $\delta^{13}C$ values that are close to the average value obtained by Craig (1961), is described.

Procedure

For the generation of approximately 6.77 liters of CO_2 from 3.31 gm of carbon, 17.35 gm of the ^{14}C NBS oxalic acid standard $(CO_2H)_2 \cdot 2H_2O$, is weighted on an analytical balance and transferred to a 2000 ml round-bottomed flask adapted for vacuum conditions. Degassed, distilled H_2O is added to the sample to immerse the contents. A separatory funnel with a 24/40 joint adapted for vacuum is attached to the flask (fig. 1). Vacuum is applied by a Welch Duo-Seal mechanical pump. All pumps used in the system are trapped, and the traps are refrigerated by a dry ice-acetone bath. The vacuum normally achieved is approximately 0.01 Torr. The separatory funnel is filled with sulfuric acid-potassium permanganate solution containing 16.25 gm of $KMnO_4$, 40 ml of concentrated H_2SO_4, and 250 ml of degassed, distilled H_2O. This solution is prepared by stirring and gently heating the contents until all the $KMnO_4$ dissolves.

Vacuum is also applied to the scrubbing and collecting apparatus, which consists of three gas wash bottles in series containing potassium-permanganate ($KMnO_4$), silver nitrate ($AgNO_3$) and potassium dichromate-sulfuric acid ($K_2Cr_2O_7$) + (H_2SO_4) solutions, respectively, a double water trap (E) cooled by dry ice-acetone mixture, and two CO_2 collecting traps (F and G) (fig. 2). One of the CO_2 collecting traps is a ten-foot glass coil (G) placed purposely at the end of the collection train to ensure total CO_2 removal. The function of the gas wash bottles is to scrub the CO_2 free of impurities. Gas scrubbing is applied to all samples, whether background, unknown, or NBS standard.

When both the generating apparatus and the collecting apparatus are evacuated to 0.01 Torr pressure, the sulfuric acid-potassium permanganate solution is added to the NBS oxalic acid sample in a steady, dropwise flow, while the contents of the flask are continually agitated by a magnetic stirrer. When enough CO_2 is generated to register approximately 38 cm Hg pressure on a monometer, the stopcock (H), which separates the

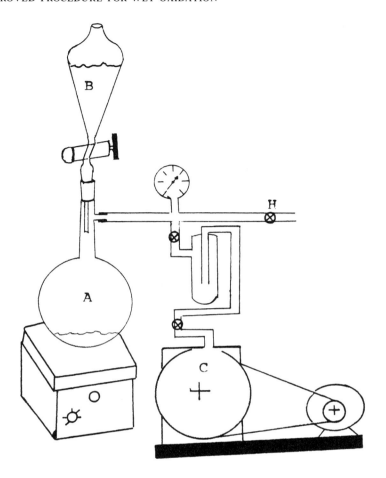

FIG. 1. CO_2 generating apparatus. 2000 ml round-bottomed flask and separatory funnel adapted for vacuum.

generating apparatus from the collecting apparatus is opened and the CO_2 is collected and frozen at liquid nitrogen temperature in the straight (F) and coiled (G) traps while the system is closed to the vaccum pump (I). The steady, dropwise flow of the sulfuric acid-potassium permanganate solution and the continual stirring of the contents are maintained throughout the entire procedure.

The apparent "end point" of the reaction between the sulfuric acid-potassium permanganate solution and the NBS oxalic acid sample is reached when the solution in the flask changes from clear to reddish-brown. This particular end point apparently does not satisfactorily ensure total evolution of CO_2 from the sample, according to Grey et al. (1969). Thus, to the contents of the flask, 5 ml more of the sulfuric acid-potassium permanganate solution is added until the solution changes to a chocolate-brown tint. At this point, heat from a stirrer-hot plate is applied and the solution is gently warmed. This serves to drive out residual CO_2 that may be trapped in the solution.

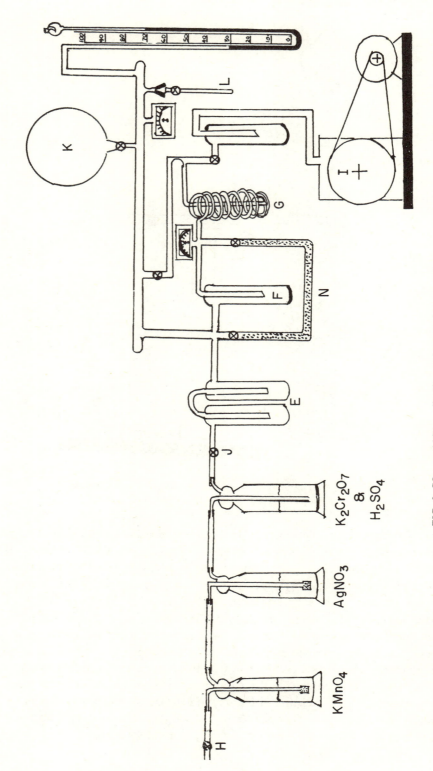

FIG. 2. CO_2 scrubbing, collecting, and measuring apparatus.

After the chocolate-brown end point is reached, a cycling procedure is started. The stopcock valve (H) separating the scrubbing and collecting apparatus is closed to the pump (I). The collecting apparatus is then evacuated for five minutes through the spiral trap (G) by pump (I). After five minutes the system is closed to the pump (I), and the stopcock (H) separating the collecting apparatus from the generating apparatus is reopened for five minutes, allowing the remaining CO_2 to expand from the round-bottomed flask (A) to the traps (F and G) where it is frozen out at liquid nitrogen temperature. This cycling is continued for thirty minutes to ensure that all the CO_2 generated from the NBS sample is trapped. At the end of the cycling procedure, when the pressure in the entire system is typically close to 0.02 Torr, the stopcock (J) is closed, isolating the collecting apparatus from the scrubbing apparatus, and the collecting system is evacuated through the spiral trap (G) to a pressure of 0.01 Torr. Finally, the CO_2 collected is expanded into a measuring flask (K) and into a collecting vial (L) for the measurement of the $\delta^{13}C$ via a silica gel desiccant tube (N).

Results

Fifteen consecutive NBS oxalic acid samples were processed by the method outlined under "PROCEDURE". These were analyzed by a dual collector mass-spectrometer for $\delta^{13}C$. The results of these samples are presented in table 1.

TABLE 1.

Yield in Grams of Carbon, Normalized percentage of Carbon, and $\delta^{13}C$ in $^o/oo$ Relative to PDB obtained by Wet Oxidation

Sample No.	Yield in gm of carbon	Normalized* % carbon yield	$\delta^{13}C$ in $^o/oo$ relative to PDB
NBS-1	3.34	100.00*	−19.11
NBS-3	3.32	99.43	−19.51
NBS-4	3.28	98.18	−19.05
NBS-5	3.27	97.94	−18.65
NBS-6	3.29	98.44	−19.01
NBS-7	3.29	98.50	−19.15
NBS-8	3.28	98.00	−18.71
NBS-9	3.34	99.91	−19.34
NBS-10	3.31	99.01	−19.31
NBS-11	3.33	99.61	−19.13
NBS-12	3.30	98.71	−19.30
NBS-13	3.28	98.24	−18.93
NBS-14	3.33	99.61	−19.35
NBS-15	3.27	97.82	−19.14
NBS-16	3.28	98.15	−19.08
	Av. 3.30	Av. 98.77 ± 0.76	Av. −19.12 ± 0.23

*Normalized % carbon yield = highest yield in gm of carbon expressed as 100% and the other yields expressed relative to it.

The average of -19.12%o for the fifteen NBS acid samples processed by wet oxidation is close to the average value of -19.3%o obtained by Craig (1961) by direct combustion.

To test the yields obtained by our wet oxidation procedure versus those obtained by direct combustion, we prepare six samples by direct combustion utilizing a micro-combustion system with a moderate vacuum, and a micro-furnace containing CuO maintained between 800-900°C. During combustion the generating gases were recycled with excess oxygen via a Topler pump to ensure total conversion to CO_2. The products were collected in a liquid nitrogen-cooled trap and transferred through dry ice traps prior to analysis.

The average of 18.69%o obtained from the six samples combusted in a stream of oxygen is close to the average value of 19.12%o obtained by the new wet oxidation procedure (table 2).

Plotting the percentage of carbon yields obtained for each sample of the NBS oxalic acid prepared by wet oxidation versus the $\delta^{13}C$ in parts per mil indicates that a small but significant trend still exists (fig. 3) between the two extreme values of -18.65%o and -19.51%o. The variation between samples is not as large as the difference of 5.57%o obtained by Craig (1961) in the analysis of samples by direct combustion.

A plot of the CO_2 yield versus the $\delta^{13}C$ ratio of the six samples processed by direct combustion, suggests that the depleted $\delta^{13}C$ values correlate with the larger CO_2 yields (fig. 4). This observation suggests that during the evacuation of the system prior to combustion, depleted oxalic acid preferentially sublimates. Thus combustion of large samples of oxalic acid may cause preferential sublimation and potential fractionation.

The new wet oxidation technique differs from the accepted wet oxidation technique (Kim 1970, Kolthoff and Sandell 1952) in the following ways: (1) The sulfuric acid-potassium permanganate solution is added to the NBS oxalic acid in a steady, dropwise flow while the contents of the flask are continually stirred and the system is closed to the

TABLE 2

CO_2 Yield and $\delta^{13}C$ in °/oo Relative to PDB
Obtained by Combustion in Oxygen

Sample no.	CO_2 yield in mm Hg mg oxalic acid	$\delta^{13}C$ in °/oo relative to PDB
5234	16.00	−18.65
5231	17.00	−18.87
5727	15.50	−18.45
5728	15.90	−18.71
5729	16.10	−18.84
5731	15.30	−18.59
		Av. −18.69 ± 0.16

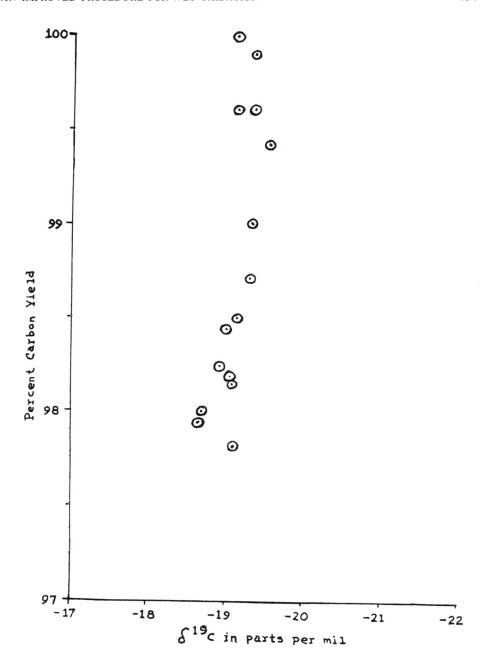

FIG. 3. Percent Carbon versus $\delta^{13}C$ in parts per mil.

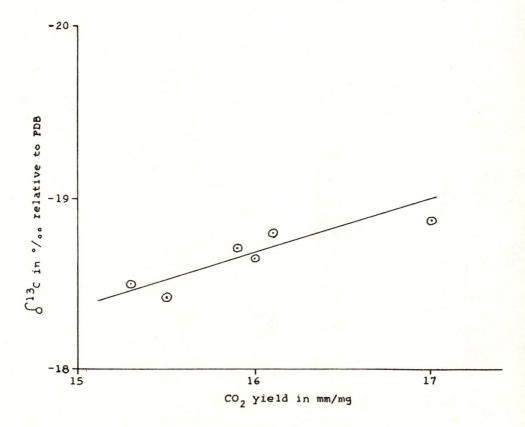

FIG. 4. CO_2 yield in mm/mg vs. $\delta^{13}C$.

vacuum pump; (2) the chocolate-brown end point is employed as a more accurate endpoint by adding 5 ml more of the sulfuric acid-potassium permanganate after the initial, reddish-brown end point is reached; (3) gentle heat is applied to warm the sample in order to drive off all residual CO_2 that may be trapped in solution or trapped in the flask or generating apparatus; and (4) a cycling technique is used by pumping the collecting apparatus while the generating apparatus is sealed for five minutes, and then by closing the system to the vacuum pump and reopening the generating apparatus.

This action allows the remaining CO_2 to expand from the flask into the liquid nitrogen traps where it is frozen out. This process is carried out for thirty minutes on an alternating basis.

Although this investigation directly concerns fractionation in the preparation of CO_2 by the wet oxidation procedure of ^{14}C NBS oxalic acid, it is also of interest to investigate whether further fractionation occurs when the CO_2 obtained by wet oxidation is converted to the counting liquid, benzene, via lithium-carbide and acetylene trimerization (Tamers 1965).

Six NBS oxalic acid samples were prepared and converted to benzene and the product was analyzed for $\delta^{13}C$ (table 3). The average $\delta^{13}C$ of $-19.42\,^{0}/_{00}$ did not differ

TABLE 3
Percentage Yield and $\delta^{13}C$ in °/oo Relative to PDB
of Benzene Obtained by Wet Oxidation

Sample no.	Overall % yield from CO_2	$\delta^{13}C$ of benzene in °/oo relative to PDB
NBS-1	98.3	−19.59
NBS-2	96.4	−19.48
NBS-3	96.0	−19.40
NBS-4	99.8	−19.44
NBS-5	99.7	−19.35
NBS-7	99.5	−19.23
		Av. −19.42 ± 0.12

significantly from our average value of CO_2 prepared from oxalic acid by wet oxidation. Correction for the activity of the counting liquid utilizing the formula proposed by Broecker and Olson (1961) accounts for a correction of approximately five years. These analyses substantiate that carbon isotopic fractionation during the acetylene and benzene syntheses are negligble (Noakes et al. 1967, Hubbs and Bein 1967).

CONCLUSIONS

If one follows the procedure outlined above, no significant fractionation in the isotopic composition of the CO_2 prepared from oxalic acid is observed.

The difference of 0.86°/oo obtained from the lowest and highest values is not as large as the difference of 5.57°/oo obtained by Craig (1969) in the analysis of the samples processed by combustion in oxygen. Small variations between the samples prepared by wet oxidation may be due in part to incomplete oxidation because of the inexact nature of the end point, or because of inexact following of procedural techniques.

The new oxidation technique is easily carried out and it eliminates the chances of sublimation while oxalic acid is under vacuum to expel air from the system.

The results obtained by the new procedure are much more consistent than the results previously obtained by direct combustion.

Since $\delta^{13}C$ fractionation by the wet method is insignificant, ^{14}C counting offset must also be insignificant.

REFERENCES

Bernestein, R. B.
    1957 Simple laboratory method producing enriched carbon-13. Science. 126:119-120.
Broecker, Wallace S., and Edwin A. Olson
    1961 Lamond Radiocarbon Measurements VIII. Radiocarbon. 3:176-204.

Craig, Harmon
    1961   Mass-spectrometer analyses of radiocarbon standards. Radiocarbon. 3:1-3.
Grey, D. C., P. E. Damon, C. V. Haynes, and Austin Long
    1969   Carbon-isotope fractionation during wet oxidation of oxalic acid. Radiocarbon. 11 (1):1-2.
Hubbs, C. L. and G. S. Beins,
    1967   La Jolla Natural Radiocarbon Measurements V. Radiocarbon. 9:261.
Kim, Stephen M.
    1970   Wet oxidation of oxalic acid used in radiocarbon dating and ^{14}C fractionation during the oxidation. Doehan Hwahak Hwoejee (J. Korean Chem. Soc.). 14. (1):49-50.
Kolthoff, M. I., and E. B. Sandell
    1952   Textbook of Quantitative Inorganic Analysis. New York, Macmillan, 564.
Lindsay, J. G., P. E. McElcheran, and H. G. Thode
    1949   The isotope effect in the decomposition of oxalic acid. J. Chem. Phys. 17:489.
Noakes, J. E., S. M. Kim, and L. Akers
    1967   Recent improvements in benzene chemistry for radiocarbon dating. Geochim. et Cosmochim. 31:1094-1098.
Tamers, M. A.
    1965   Routine carbon-14 dating using liquid scintillation techniques. Proc. 6th Int. Conf. on Radiocarbon and Tritium Dating. Clearing House for Federal, Scientific and Technical Information, U. S. Dept. of Commerce. 53-67.

3

The Importance of the Pretreatment of Wood and Charcoal Samples

Ingrid U. Olsson

INTRODUCTION

Insufficient pretreatment of samples for ^{14}C dating may lead to results that are seriously in error. A large error is usually obvious, permitting rejection of the resultant date, but small errors may be overlooked, and the resultant dates may wrongly be considered reliable. It is essential to understand the kinds of errors that can be made, whether they derive from treatment of the sample or from measurement of activity. For accurate measurements, such as calibration of the time scale with analysis of short-term variations, or of differences between the northern and the southern hemispheres, this can be critical. Statisticians assert that accuracy is usually less than that claimed by the laboratories.

The Standard Pretreatment

At the Uppsala laboratory, wood, charcoal, and similar samples always (unless some other procedure has been specified) have been boiled or treated for several hours with hot HC1 to remove any carbonate and to extract organic material. They have been washed repeatedly with distilled water to remove any dissolved material, treated with NaOH at +80°C to dissolve the humic acid, washed repeatedly with distilled water to remove the dissolved material, and then acidified to remove any CO_2 absorbed during the NaOH treatment. The humic acid has been precipitated with HC1. The strength of the solvents has been 1%. For silt, clay, and similar samples with little organic-carbon content, the fraction soluble in NaOH is probably most reliable, as has been shown in a sample contaminated with old carbon in the form of graphite (Olsson 1973). When dating very small grains of carbon (e.g., soot samples) from a soil rich in roots, another method has had to be used. The cellulose has been extracted from the sample by a method similar to that described by Haynes (1966). This method has been tried in Uppsala and will be detailed later.

The diagram in figure 1 illustrates the dating error when a recent contaminant remains in the sample. If the contaminant is 500 years younger than the sample and 10% re-

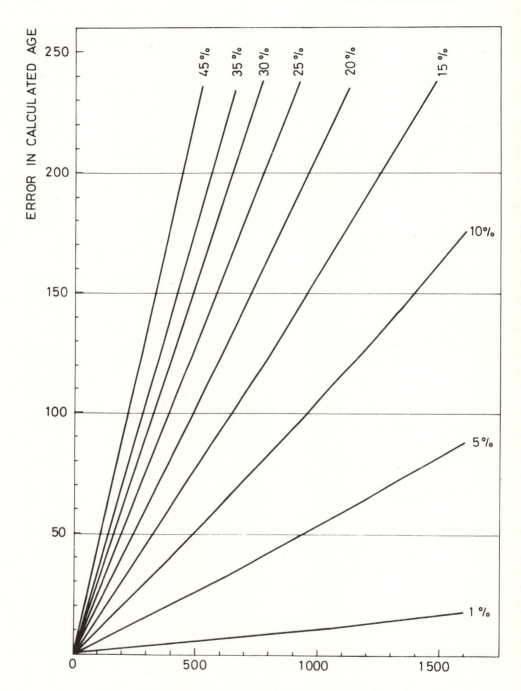

FIG. 1. The error in a date when a sample is contaminated to certain degrees by material somewhat younger than the actual sample.

mains in the sample, the error will total 50 years. If the contaminant is 1500 years younger than the sample and constitutes 10% of the sample, the error will be about 165 years. This is serious error, larger than the normal statistical error.

Another possible error should be mentioned which may be seen in certain mud samples or may derive from the oxygen used for combustion. Radon may occur in the gas as a daughter product from uranium present in the sample or in the commercial oxygen. In Uppsala the radon content has sometimes been so high that the sample has had to stand for more than a month for its level to become undetectable. The radon is routinely removed in Uppsala (Olsson 1958), but deliberate testing without its removal gave the results shown in figure 2. Forty days had to elapse after combustion before the sample could be dated. If the radon in this case has derived from the oxygen, 60 days would have been required before the sample could have been dated in a reliable way.

In the present paper a few cases will be discussed in which the different fractions obtained at the pretreatment yielded different ages, requiring further investigation into the sufficiency of the standard pretreatment. It has been found advisable to avoid extraction with organic solvents because of possible difficulty in removing all the solvents. It should be mentioned that in many other cases the different fractions obtained at the NaOH treatment have been found to be of the same age. Sometimes the standard treatment has been extended to include two or more extractions (fig. 3) with NaOH.

Pretreatment of Recent Wood

The dates for sapwood of *Pinus silvestris* from the period 1945-50 given by Olsson et al. (1972) indicate the difficulty inherent in removing from a wood sample all extractives with an age lower than that of the cellulose. These results and some on older rings (Olsson et al. 1969) from another tree indicate a transport of carbohydrates from the cambium layer through the rays. These conclusions are based on the fact that considerable excess of ^{14}C was found in the extractives of tree rings from 1935 to 1941 as well as in those from 1950 to 1955. The excess was larger than 100 $^o/oo$ and must be ascribed to transport of material affected by the excess of ^{14}C in the atmosphere caused by the testing of nuclear weapons (Olsson et al. 1972). Berger (1970, 1973) did not observe a similar increase. Jansen (1970), on the other hand, saw an age difference between the cellulose and the resin in a tree.

It is assumed that one of the $\delta^{13}C$ values published by Olsson et al. (1972) was in error, since it deviates so much from the normal value. When corrected to about -27 $^o/oo$, with the ^{14}C date recalculated using $\delta^{13}C = -27.4$ $^o/oo$, the excess of ^{14}C in the NaOH extract, U-2172, and in the extract obtained after a complicated chemical procedure (Olsson et al. 1969: 539-540), U-2170, were in agreement (see table 1). The wood obtained after the treatment with NaOH (U-2171) showed slightly higher activity than the wood after extraction with ethyl ether. To check this, another portion of the wood was dated (U-2585) with resultant agreement between the activities of these two portions of the pretreated wood. A third portion was treated one night with 1% NaOH at +80°C, washed, and treated in the same way for yet another night. The resultant soluble fraction

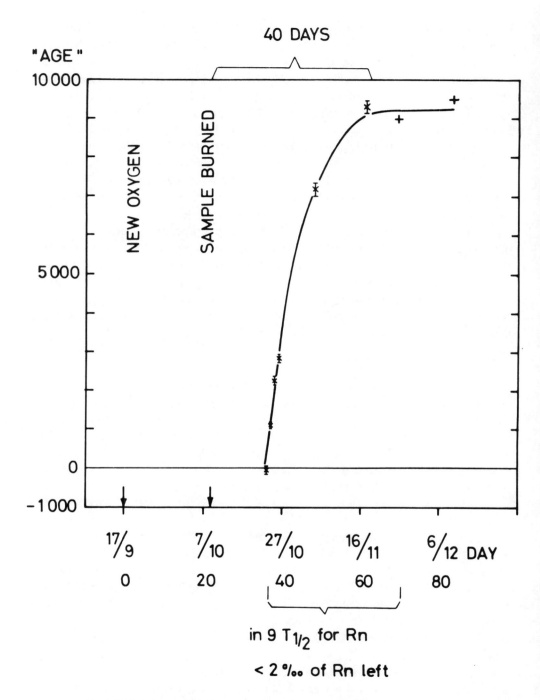

FIG. 2. The influence of radon on a sample. The radon may derive from the oxygen used or from uranium in the sample.

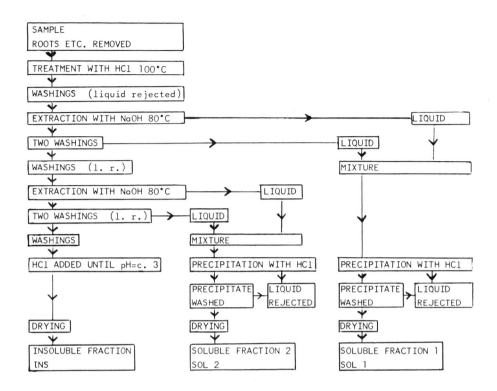

FIG. 3. The standard treatment of a wood or charcoal sample when two extractions with NaOH are made.

(U-2586) had an activity indicating that some contaminants had been removed, as was expected since the wood had probably been contaminated. The statistical uncertainty was great since the soluble fraction was small — only 4% of the starting material (after the first pretreatment). The activity of the insoluble fraction agreed very well with that of the holocellulose and the wood after extraction with ethyl ether. Since the soluble fraction (U-2586) probably contained some contaminant, a further extraction was performed on the wood remaining after a portion was used for U-2587. This time 3% of the sample was dissolved and the two fractions were dated. The insoluble fraction yielded slightly higher activity than expected, like U-2171 and U-2585. Results of the dating of all the different fractions are given in figure 4.

It has been shown by Jansen (1973) that it is extremely difficult to remove all organic solvents. The small differences of 2% to 3% between the activities of some fractions of wood samples as given by Olsson et al. (1969, 1972), may be due to residual amounts of solvents.

TABLE 1

^{14}C Dating Performed on Extracted Wood, Holocellulose, the Part of the Extractives Soluble in Ethyl-ether, the Part of the Wood Soluble in NaOH, the Part Insoluble in NaOH, and Two Fractions Mainly Consisting of Lignin and Cellulose

Wood formed during the years	Fraction	$\delta^{13}C$ °/oo	Δ °/oo	Dating no.
Pinus silvestris 1945-50	Extractives	−27.4	146 ± 10	U-2170
	Wood after ethyl-ether extraction	−23.1	−42 ± 7	U-2169
	Holocellulose	−22.8	−36 ± 8	U-2168
	Extractives (1 night in NaOH)	−27.4[a]	146 ± 23	U-2172
	Wood extracted with NaOH (1 night at +50°C)	−25.1	−14 ± 7	U-2171
	Wood extracted with NaOH (same as U-2171 but another portion)	−24.6	−21 ± 5	U-2585
	Extractives[b] (a further two extraction with NaOH at +80°C, each overnight)	−27.4[a]	19 ± 23	U-2586
	Wood extracted with NaOH[c] (total of 3 times, each overnight)	−23.6	−41 ± 6	U-2587
	Extractives (a further two extraction with NaOH at +80°C, each overnight)	−24.9	−63 ± 76	U-2599
	Wood extracted with NaOH[c] (total of 5 times, each overnight)	−25.0	−17 ± 7	U-2600

[a] Assumed value.
[b] Ratio of extractives to wood 6:94.
[c] Ratio of extractives to wood 3:97.

If 10% of the sample consists of extractives with an activity which is 2.5% higher than the cellulose, the resulting error will total 20 years. This is of very little importance in normal datings, but it must be considered in precise determinations. If the extractives have an activity which is 20% higher than the cellulose, 10% is needed to increase the activity of the unextracted sample by 2%.

Though the results from testing this wood sample do not give a definite answer as to whether one night's treatment with 1% NaOH at +80°C is sufficient to remove the extractives, the results indicate that some of the more recent extractives remained after the first night's treatment with NaOH.

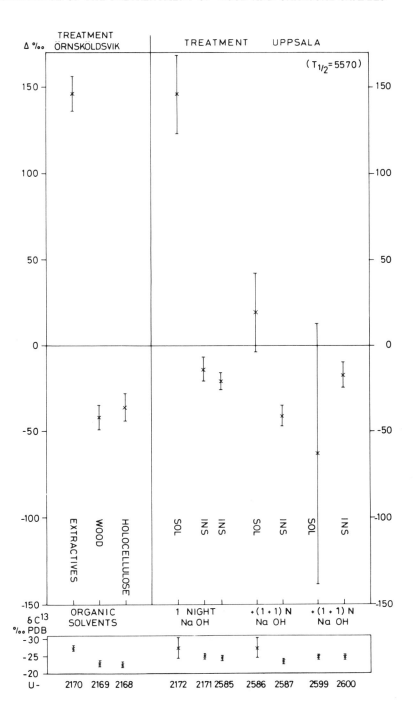

FIG. 4. The result of repeated treatments of a wood sample with 1% NaOH at +80°C. "SOL" indicates the fraction soluble in the NaOH and "INS" the insoluble fraction.

Pretreatment of Old Wood in Poor Condition

The sample was extracted twice with 1% NaOH at +80°C, each time overnight because of bad condition. During the first extraction, a fraction corresponding to one-quarter of carbon content was dissolved, but during the second, a fraction corresponding to only 3% of the carbon content was dissolved. All three fractions were dated (fig. 5), and the first extract was found to have an age of 1600 years less than the insoluble fraction. The second extract was found, within the limits of error, to be of the same age as the insoluble fraction.

Pretreatment of Charcoal

In dating some charcoal samples from Bua Västergård, Sweden, it was discovered that the fraction soluble in NaOH (figs. 6 and 7) in some cases yielded much too low an age. The samples were collected below sand and clay layers and seemed well protected against humic acid penetrating from above, but later discussions with geologists revealed that humic acid was probably penetrating through slightly sloping boundaries between the strata. Very different amounts of humic acid were extracted by the treatment at 80°C with NaOH. As much as 50% of the carbon could be extracted by the NaOH. In some cases a second treatment was performed, since all of the soluble fraction could not have been extracted at a single treatment, but there was no indication that a second treatment was necessary in this case. It is worth estimating what the dating error might be were no pretreatment carried out. Since the age difference may be about 1000 years and the amount of soluble carbon may be about 30% in a typical case in this investigation, it can be seen from figure 1 that the errror would be 315 years.

ACKNOWLEDGMENTS

Sincere thanks are due to Mr. A. Assarsson and Miss G. Åkerlund for performing the extractions, to Dr. R. Ryhage for the $^{13}C/^{12}C$ determinations, and to those of my collaborators in the ^{14}C laboratory who assisted in this work. I am indebted to Professor K. Siegbahn, the head of the institute, for making available the facilities of the Institute of Physics in Uppsala. The financial support received from the Swedish Natural Science Research Council is gratefully acknowledged.

REFERENCES

Berger, Rainer
    1970   Ancient Egyptian radiocarbon chronology. Phil. Trans. Roy. Soc. Lond. A 269:23-36.
    1973   Tree ring calibration of radiocarbon dates. Proc. 8th Int. Conf. on Radiocarbon Dating. New Zealand. A97-A103.

Jansen, H. S.
    1970   Secular variations of radiocarbon in New Zealand and Australian trees. In Radiocarbon Variations and Absolute Chronology. Olsson, (ed.) 12th Nobel

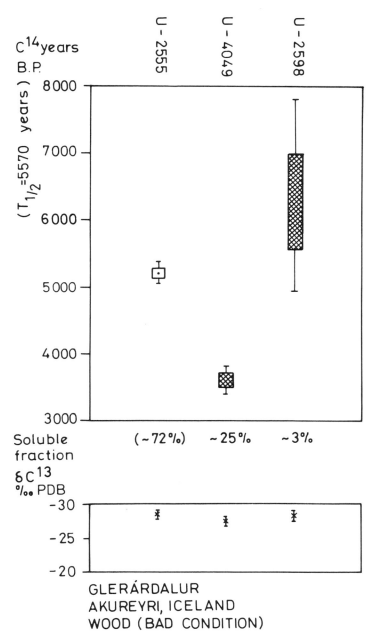

FIG. 5. Three fractions of a wood sample after the standard treatment with two extractions with NaOH. The first of these yielded a sample containing 25% of the carbon of the sample (U-4049) and the second of these yielded a sample containing only 3% of the carbon. The first soluble fraction was significantly lower than the insoluble fraction, whereas the second soluble fraction seemed to be free from younger contaminants.

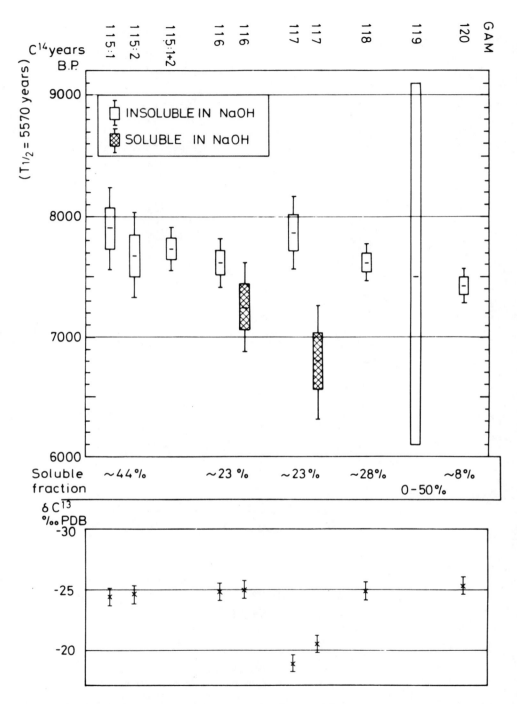

FIG. 6. The first set of dates from Bua Västergård, indicating an appreciably lower age of the soluble fraction than for the insoluble fraction of some samples.

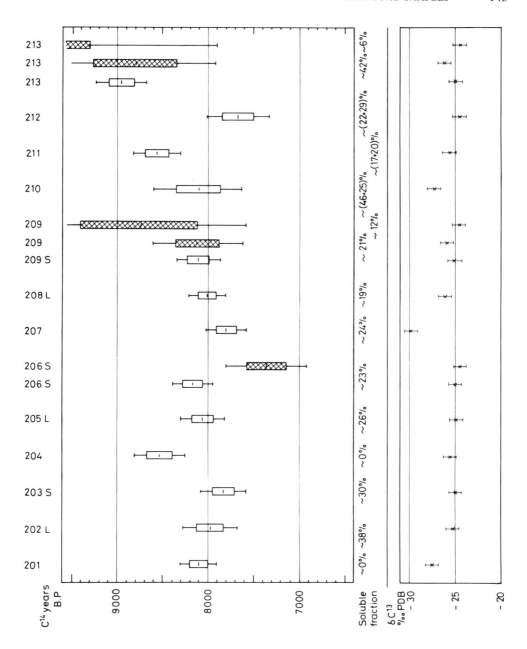

FIG. 7. The second set of dates from Bua Västergård, indicating that most probably the second fraction obtained after the NaOH treatment is free from younger contaminants.

Symposium. Uppsala. Stockholm, Almqvist and Wiksell. New York, John Wiley and Sons, 261-274.

Jansen, H. S.
    1973   Transfer of carbon from solvents to samples. Proc 8th Int. Conf. on Radiocarbon Dating. New Zealand. B63-B68.

Olsson, Ingrid
    1958   A ^{14}C dating station using the CO_2 proportional counting method. Arkiv för Fysik. 13: 37-60.
    1973   A critical analysis of ^{14}C datings of deposits containing little carbon. Proc. 8th Int. Conf. on Radiocarbon Dating. New Zealand. G11-G28.

Olsson, I. U., Shawky El-Gammal, and Yeter Göksu
    1969   Uppsala natural radiocarbon measurements IX. Radiocarbon. 11: 515-544.

Olsson, I. U., Martin Klasson, and Abdalla Abd-El-Mageed
    1972   Uppsala natural radiocarbon measurements XI. Radiocarbon. 14:247-271.

4

Recent Progress in Low Level Counting and other Isotope Detection Methods

H. Oeschger, B. Lehmann, H. H. Loosli, M. Moell, A. Neftel, U. Schotterer, and R. Zumbrunn

ABSTRACT

To increase the spectrum of radioisotopes available for environmental studies and to expand the spectrum of ^{14}C applications, we have studied possible improvements of the radioactivity detection methods.

When operated underground ($\simeq 70$ m water equivalent below surface) the background of low level proportional counters is a factor of 2 to 4 lower than that obtained in the previously-used laboratory ($\simeq 7$ m water equivalent below surface). We believe that the residual background is mainly due to radioactive contamination of the counter's construction material. It shows little counting-gas pressure dependence.

For ^3H, the rise-time discrimination technique promises considerable background reduction and the results are comparable to those obtained with counters with internal anticoincidence.

We have attempted to measure isotopic ratios by means of IR laser spectroscopy. At present molecules with abundances of 10^{-6} to 10^{-7} can be detected.

INTRODUCTION

The scientific impact of ^{14}C studies on natural samples goes far beyond radiocarbon dating of the history of man, animals, plants, and the landscape during and after the last glaciation. ^{14}C measurements on atmospheric, oceanic, and biospheric samples help us to understand the dynamics of the CO_2 exchange in the atmosphere – hydrosphere – biosphere system and, therefore, are the basis for predictions of atmospheric response to the future fossil fuel CO_2 input. The ^{14}C variations observed in tree rings probably reflect fluctuations of solar activity in the past and enable us to check for solar influence on climate.

Perhaps the most fascinating aspect of ^{14}C work is the fact that through ^{14}C analyses, scientists of earth and of historical sciences are confronted with the simultaneity of historic events; thus interdisciplinary research is stimulated and the history of man is understood in the framework of environmental history (climate, flora, fauna).

The importance of ^{14}C studies demands improvement of ^{14}C measuring techniques with respect to (1) accuracy of measurement of up to $\simeq 10,000$ year-old samples. Increased precision in this range would augment information about solar effect on the atmospheric $^{14}C/C$ ratio; (2) dating of relatively old ^{14}C dating range, and (3) determination of the $^{14}C/C$ ratio in relatively small samples.

New Developments in the Measurement of Isotopic Ratios

The classical measuring technique of natural ^{14}C by determining the decay rate of a sample of given size can be improved by increasing the ratios of sample count rate (S) to background count rate (B). In our laboratory, we have investigated possibilities of further reduction of the background count rate of proportional gas counters. Considerable background reduction has been obtained by operating our counters in an underground laboratory (70 m of water equivalent below the surface). The ratio S/B could be further increased by working at high counting-gas pressure. We have also studied the effect of rise-time discrimination on background count rates of proportional gas counters.

Radioactive isotopes with half-lives similar to that of ^{14}C or shorter are usually analyzed by measuring the radioactivity, that is, the number of decays per unit time of a given sample. Until now all measurements of natural ^{14}C samples have been done this way. However, during a measuring period of three days only one-millionth of the ^{14}C atoms present in the sample decay. This means that the radioactivity measurement utilizes only a very small portion of the available information. For this reason, other techniques which permit measurement of isotopic ratios should be investigated. Promising methods are mass-spectrometry and laser infrared spectroscopy, by which a much greater fraction of the physically available information is used. A high power, infrared laser beam is able to excite several times per second each molecule containing a ^{14}C atom. Despite their great advantages, to our knowledge these methods are not yet ready for natural ^{14}C studies, largely due to small $^{14}C/C$ ratios ($\simeq 10^{-12}$ for a "modern" sample). We will report on our attempts to measure isotopic ratios by infrared laser spectroscopy.

Background Reduction in an Underground Laboratory

In summer 1975 we moved part of our counting equipment from the cellar of our institute ($\simeq 7$ m water equivalent below surface) to a new underground laboratory ($\simeq 70$ m of water equivalent below surface). The 40 cm thick walls of this laboratory are constructed from specially selected concrete (Pulfer 1974). Based on γ-measurements we chose serpentine (gravel and sand) from Torino, Italy, and Danish cement. γ-measurements indicated a reduction factor of >30 for sand and gravel and of $\simeq 4$ for the cement when compared with the concrete components commonly used in our region.

In the following we restrict ourselves to a summary of the most important results. For more details see Oeschger and Loosli (1975, in press). With an old lead shield, 10 cm thick, we observed the following reduction factors for the different background-generating radiation components:

1. The muon flux is lower by a factor of 11, the muon stop rate (according to literature) is lower by a factor of 40 than in the old laboratory.
2. Measured with a 60 cc Geli-detector in the interval between 239 and 1460 keV, the γ-count rate shows a reduction of about 6.
3. Neutron measurements with a BF_3-counter and a He-counter (Neftel 1976) indicate that the n-flux is lower by at least a factor of 8 than in the old laboratory.
4. The measured integral background values for different counters in the old and in the underground laboratory are given in table 1. Unfortunately the Pb shields used in the two laboratories are not identical. Based on our experiences with the different lead qualities this should not greatly influence the results. The ^{14}C and ^3H counters have an internal anticoincidence system; the others are operated in an external guard counter system with a 2 cm thick, pure, old lead shield between guard and actual counter. As can be seen from table 1, moving into the underground laboratory results in background reductions by factors of 2 to 4.

The strong reduction factors for the muon-, γ- and n-fluxes, and the availability of background values for counters varying considerably in size has enabled us to estimate the relative contribution of the different background components. We consider the following four background components (Oeschger and Loosli 1975, Oeschger 1963):

1. α- and β-particles from the inner counter wall; flux $S_{\alpha\beta}$ (cpm cm^{-2})

TABLE 1

Measured Integral Background Values for Different Counters in Two Laboratories

Counter	Anticoincidence	Gas	Pressure	Background (cpm) Old laboratory	Underground
10 cc	external	CH_4	4 at	0.12	0.04
50 cc	external	Ar + 5% CH_4	36 at	–	0.08
100 cc	external	Ar + 5% CH_4	18 at	0.27	0.12
1000 cc	external	Ar + 2% CH_4	4.5 at	1.70	0.9
2.8 lt with $\simeq$1.5 lt internal volume	internal, foil	CH_4	1.5 at	1.25	0.35
	internal, wire grid	CH_4	1.3 at	0.48	0.11
	internal, wire grid ^3H window	CH_4	1.3 at	0.24	0.06

2. muons not detected by the anticoincidence system; rate c (cpm)
3. recoil nuclei from collisions with neutrons; source strength S_{np} (cpmg^{-1})
4. γ-induced secondary electrons; source strength $S_{\gamma e}$ (cpm g^{-1})

The background count rate BG(cpm) can be approximated by the equation:

$$BG = S_{\alpha\beta} \cdot 0 + c + S_{np} \cdot \left(V \cdot \rho + \frac{O\overline{R}_\rho}{4}\right) + S_{\gamma e} \cdot \left(V \cdot \rho + \frac{O\overline{R}_e}{4}\right)$$

with V = volume (cm^3),
O = surface (cm^2) of counter
ρ = density of counting gas (g cm^{-3})
$\overline{R}_\rho$ and $\overline{R}_e$ = average range of recoil nuclei and secondary electrons, respectively, (g cm^{-2}).

Table 2 shows the parameters adopted for the background discussion which permitted most of the counters with thick walls to obtain agreement within 10% to 20% between measured and calculated background values. For $S_{\alpha\beta}$ we had to adopt values between 8 and 13 · 10^{-4} cpm/cm^2. Considering the fact that the counters compared varied by a factor of 100 in volume and the filling pressure by a factor of up to ten the observed agreement is very satisfactory.

TABLE 2

Parameters Adopted for Background Discussion

Parameter	Old laboratory	Underground laboratory
$S_{\gamma e}$ (cpm/g)	0.036	0.006
R_e (mg/cm^2)	70	80
S_{np} (cpm/g)	0.08	0.01
R_p (mg/cm^2)	2	2

The results of this analysis can be summarized as follows:
1. In the old laboratory the highest background contributions probably were from radioactive contamination of counter material (30% to 45%), from γ-induced background in counter walls, and from high pressures in the gas (together 40% to 55%). Not neglibible were the contributions by residual mouns ($\simeq$10%) and of n-induced background for the CH$_4$-component of the counting gas.
2. In the new underground laboratory the background due to radioactive contamination of the counter material dominates (70% to 85%), whereas the component induced by γ-rays is only 15% to 20%, and the contributions by n-induced effects and residual muons are negligible.

Though this background discussion needs confirmation by additional measurements, it appears that in many highly-developed, low-level gas counting systems the background induced by muons and neutrons dominates, and that considerable background reduction can be obtained by operating the counting systems underground. The high proportion of background contribution from the counter wall suggests that even purer material for counter construction should be sought. Since the counters show only a very low background increase with raised pressure, the operation of proportional counters at very high pressures would seem promising.

To summarize the most significant background results:

1. For ^{39}Ar measurements, by operating a 50 cc proportional gas counter filled with Ar and CH_4 (5%) at 36 atmospheres pressure, background (full spectrum) of 0.025 cpm/g of counting gas is obtained. The background of 0.08 cpm compares with a modern net effect of 0.14 cpm, i.e., in the underground laboratory the background for ^{39}Ar measurements is now smaller by a factor of 1.7 than the modern net effect (specific activity of modern Ar: 0.112 ± 0.012 dpm/lt Ar). If both background and sample are counted for 10,000 minutes each, the statistical error for a 540-year-old sample becomes about 50 years. In addition, the necessary amount of sample can be reduced from 5 lt of Ar (first measurements) to 2 lt of Ar.
2. In the undergound laboratory our routine ^{14}C counter, for which the modern net effect is $\simeq 15$ cpm, shows a background of 0.35 cpm, i.e., the ratio S/B is 43. We are reconsidering our standard ^{14}C measuring techniques to take full advantage of the new laboratory. We plan to operate the counters at higher pressures.
3. For our 10 cm^3 counter for "small" ^{14}C samples, filled up to a pressure of 5 atm of CH_4 ($\simeq 60$ cm^3 NTP), the values for S and B are 0.31 cpm and 0.04 cpm, respectively. This enables us to expand the dating range of polar ice to about 20,000 years.
4. Our routine ^{3}H counter, with a sensitivity of 74 T.U./cpm filled with CH_4 at 1000 Torr, shows in the ^{3}H-window a background of (0.06 ± 0.015) cpm, corresponding to 5 ± 1 T.U.

The Rise-time Discrimination Method

The method of rise-time-discrimination (RTD) has been successfully applied to LLC proportional systems in the last few years (Wahlen et al. 1972, Frommer 1973, Oeschger and Wahlen 1975, Davis et al. 1972). Thus in gaseous detectors for α particles, X rays, and low to moderate energy β particles, considerable background reduction can be achieved without substantial loss in counting efficiency and with consequent improvement in sensitivity and precision. The method makes use of the fact that in a proportional counter the rise time of a pulse observed at the central anode depends on the extension of the initial ionization track produced by a charged particle interacting with the counter gas. Low energy X rays or low energy electrons following electron capture or β decay cause very short ionization tracks, and the consequent signal at the anode is of short rise-time. On the other hand, "background events" such as charged cosmic ray particles or energetic Compton electrons produce extended ionization tracks, resulting in signals with considerably longer rise-times although the same energy may have been deposited. If, in identifying a certain isotope, one measures the pulse rise-time along with the pulse height, one can to distinguish effectively between source events and background events.

Experimental realization of this principle is preferably done in a two-parameter mode: pulse height versus rise-time. The signals from the proportional counter are amplified by a fast charge-sensitive preamplifier and then routed in parallel into a "fast branch" and a "slow branch". In the fast branch the pulse is fed into a timing filter amplifier (TFA) where it is differentiated with a short time constant of 10-100 ns (depending on counter geometry, pressure, and filling gas), then stretched and digitized by an analog-to-digital-converter (ADC). The digital information (rise-time) is displayed on the x-axis of a two-parameter plot. In the slow branch the signal is amplified and differentiated with a time-constant of 200-300 ns and subsequently digitized by a second ADC (as in usual pulse height analysis). This information (pulse height) is plotted on the y-axis. One then finds the signals originating, for example, from an X ray source, in an area determined by the corresponding energy and rise-time. Background events of the same energy are displaced toward longer rise-time, that is, they give smaller signals from the fast differentiating branch.

The electronic circuit is given schematically in figure 1. Figure 2a shows a spectrum obtained for a ^{3}H labeled methane standard sample in a proportional counter (without anticoincidence), indicating rates versus pulse height (slow branch) and rise-time (fast branch). The pulses of ^{3}H appear close to the diagonal at relatively small pulse heights. The energy deposited by muons is higher on the average than by β-particles from ^{3}H. Since, in addition, the rise-time of the muon pulses are longer, they are shifted towards the y-axis. Figure 2b shows count rates versus rise-time obtained from the first diagram by a central projection. The ^{3}H-pulses and the background pulses (mainly muons) are clearly separated. If the counter is operated in an anticoincidence system, the background peak is strongly reduced. The spectrum given in figure 1 is also from a ^{3}H standard gas sample, but with anticoincidence enabled.

Frommer 5 applied the RTD method to a tritium counter with internal anticoincidence, with the inner counter and the anticoincidence separated by a wire grid. By removing the internal anticoincidence and operating the counter in an external anticoincidence shield, the RTD method enabled similar background values to be obtained as in the conventional mode, though the sensitive volume doubled. As an example of the additional background obtainable by the use of RTD, we refer to our 0.5 lt copper proportional counter filled with CH_4 at 5.5 atmospheres, operated in an external anticoincidence counter, surrounded by 10 cm Pb. Use of RTD gives an additional background reduction in the ^{3}H energy window by a factor of 7.

Unfortunately, using counting-gas pressures below a few atmospheres, the RTD method does not give a significant improvement for ^{14}C measurements because of the extended tracks of the β-particles from ^{14}C decay. We are investigating the possible application of the RTD method for ^{14}C by working at higher pressures and with admixtures of Ar.

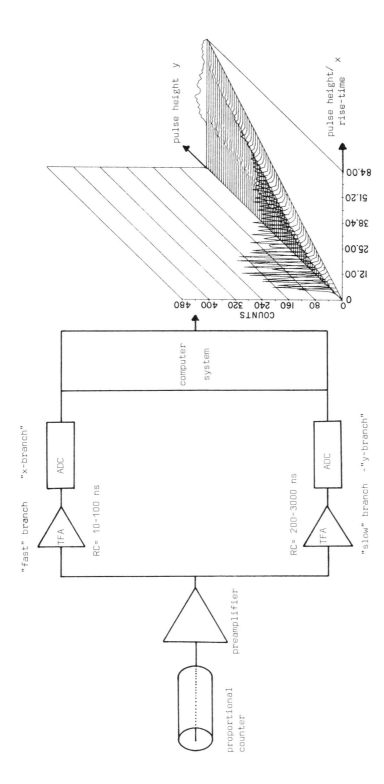

FIG. 1. Electronic circuit schematically for rise-time discrimination.

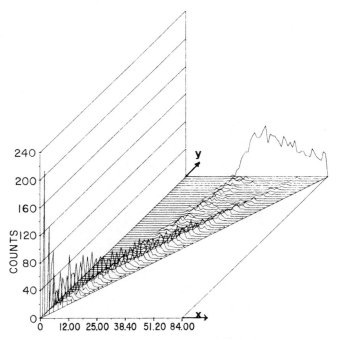

FIG. 2a. Counts ³H-standard versus pulse height and pulse height/rise-time (without anticoincidence system)

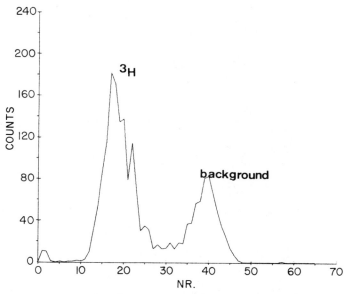

FIG. 2b. Counts ³H-standard versus rise-time (without anticoincidence system)

Determination of Isotopic Ratios and Abundances by Means of IR Laser Spectroscopy

In our laboratory we have been attempting to determine isotopic ratios, like $^{13}C/^{12}C$ or $^{18}O/^{16}O$, and abundances of ^{13}C, ^{12}C, ^{18}O, ^{17}O, and ^{16}O in small samples (≈ 0.1 cm^3 NTP) of CO_2 with a tunable diode laser (Lehmann et al. 1976, Zumbruan 1976).

The experimental arrangements is as follows:
The light emitted from a continuous, tunable PbS diode laser (spectral range for quasi-continous current tuning: 2310 cm^{-1} to 2330 cm^{-1}) is focused and sent through an absorption cell containing the gas sample. The transmission signal as a function of wave length, changed by current tuning, is determined with a PbSe detector and processed by a lock-in amplifier. Line frequencies are measured with a Czerny-Turner spectrometer.

This technique makes use of the vibrational isotope-effect which produces a frequency shift when one or more atoms are substituted by atoms with different mass numbers. The extremely small laser band width (less than 10^{-5} cm^{-1}) and the quasi-continuous tunability facilitate resolution of doppler-broadened absorption lines. A typical measured transmission spectrum from part of the very strong 4.3 μm absorption band of CO_2 is shown in figure 3. At present, molecules with abundances of 10^{-6} to 10^{-7} and $^{18}O/^{16}O$ ratios can be detected with an accuracy of 1°/oo to 2°/oo. With a long path, absorption cell (white cell) sensitivities of 10^{-8} to 10^{-9} should be within reach. Further improvements may be possible by use of high power lasers, which facilitate direct detection of absorption in the cell by means of an opto-acoustical detection technique.

Though the sensitivities presently reached are still considerably below those needed for the detection of long-lived environmental radioisotopes such as ^{14}C and ^{36}Cl, it is possible that in a few years techniques such as mass-spectroscopy may become competitive with classical techniques for measurement of isotopic ratios such as low level counting.

CONCLUSIONS

Considerable background reduction in low level gas proportional counting has been obtained by operating the counters in an underground laboratory, where the γ- and muon-fluxes are considerably reduced. Backgrounds as low as 0.025 cpm per g of counting gas have been measured. Most of the remaining background is probably due to α- and β-particles from the counter wall. A successful search for radioactively purer construction material may permit considerably greater background reduction. Since in the underground laboratory the backgrounds of proportional counters are minimally dependent on counting-gas pressure, the operation of relatively small counters at high filling pressures promises excellent ratios of sample count rate to background count rate.

Rise-time discrimination permits the distinguishing of short track from long track events. Its use for background reduction shows results similar to the internal anticoincidence systems. Since counting efficiency is higher for counters with external anticoincidence, rise-time discrimination is superior to internal anticoincidence counting, though the electronic effort is considerably.

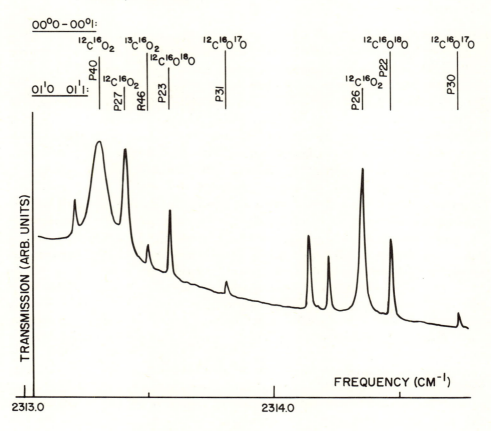

FIG. 3. Vibration-rotation spectrum from CO_2.

For long-lived radioisotopes, determinations by other analytical techniques, such as mass-spectrometry, n-activation, and optical spectroscopy, are already or may become competitive with low level counting.

ACKNOWLEDGMENT

This work has been supported by the Swiss National Science Foundation.

REFERENCES

Davis Jr., R., J. C. Evans, U. Radeka, and L. C. Rogers
    1972   Report on the Brookhaven Solar Neutrino Experiment. BNL Report 16937.
Frommer, H.
    1973   Lizentiatsarbeit. Univ. of Bern.

Lehmann, B., M. Wahlen, R. Zumbrunn, and H. Oeschger
    1976   Isotope Analysis by Infrared Laser Absorption Spectroscopy. Applied Optics. (In press.)

Neftel, A.
    1976   Der Nulleffekt im Tieflabor. Diplomarbeit. Univ. of Bern.

Oeschger, H., and H. H. Loosli
    1975   New Developments in Sampling and Low Level Counting of Natural Radioactivity. Proc. Int. Conf. on Low Radiocarbon Measurements and Applications. Czechoslovakia. (In press.)

Oeschger, H., and M. Wahlen
    1975   Low Level Counting Techniques. Ann. Rev. of Nuclear Sci. 25:448-449.

Pulfer, P.
    1974   Lizentiatsarbeit. Univ. of Bern.

Wahlen, M., et al.
    1972   Proc. 3d Lunar Sci. Conf. Geochim. et Cosmochim. 2(2):1773.

Zumbrunn, R.
    1976   IR Absorptionspektroskopie mit einem kontinuierlich abstimmbaren PbS-Diodenlaser für Isotopenanalysen an CO_2-Molekülen. Diplomarbeit. Univ. of Bern.

5

Measurement of Small Radiocarbon Samples: Power of Alternative Methods for Tracing Atmospheric Hydrocarbons

L. A. Currie, J. E. Noakes, and D. N. Breiter

ABSTRACT

Measurement problems with very small radiocarbon samples arise whenever such samples have a high specific "cost," that is, are limited in supply or are difficult to collect. One such problem, due to collection difficulties, involves the measurement of carbonaceous species in the atmosphere to determine the relative contributions of man and nature. The characteristics of miniature gas proportional and liquid scintillation counters especially developed for this work have been evaluated and have been judged adequate for the measurement of 10 mg of contemporary carbon. A graphical means of assessing the relative performance of an entire set of small counters has been introduced. It has been used to deduce the equimerit curve for the most sensitive gas and liquid scintillation counters. Preliminary results on the carbonaceous fraction of an urban dust sample have been compared with the results of earlier research.

INTRODUCTION

The present investigation has a two-fold objective: (1) to evaluate the capabilities of state-of-the-art gas proportional and liquid scintillation counters for measuring small natural radiocarbon samples; and (2) to make a quantitative assessment of the applicability of such small counters to outstanding questions concerning the sources and residence times of carbonaceous species in the atmosphere. The earliest measurements of ^{14}C in an atmospheric hydrocarbon were carried out by Libby. (Libby's results for the radiocarbon content of CH_4 samples, collected in 1949 and 1950, appear in Ehhalt's review [1973] of the methane cycle.)

As shown by the work of Oeschger et al. (1965) conventional counters used for radiocarbon dating, requiring gram quantities of carbon, are inappropriate for the measurement of very small samples. Because of the low concentration of CO_2 in ice (~15 ml CO_2/ton of ice) and the technical problems associated with extracting several tons of ice, Oeschger was limited to samples having less than 100 mg of carbon. Dilution with dead CO_2 and use of a large counter with its attendant larger background rate would have needlessly reduced the sensitivity.

A number of additional situations where only small amounts of carbon are available can be cited: limited or precious materials (archaeological, geological, or biological specimens such as tree rings); substances having relatively small concentrations of carbon (alloys, ceramics); and mixtures in which sampling difficulties preclude the collection of very much carbon (glaciers, precipitation, the atmosphere). Our efforts to measure ^{14}C in atmospheric hydrocarbons led to our interest in developing and evaluating small counters having good performance.

The applicability of ^{14}C to the identification of sources of carbonaceous contaminants rests upon the dichotomous origin of organic species in the atmosphere: they may be either "dead" or "alive." To quite a good approximation, the ^{14}C content of atmospheric molecules derives from living matter or from the use (or storage) of fossil fuel.* This was first noted by Suess (1955) during his investigation of the dilution of natural CO_2 by that resulting from the combustion of fossil fuel. Measurement of $^{14}C/^{12}C$ ratios in atmospheric species thus can provide direct information on the fraction of a particular substance which arises from natural sources and that which derives from man's use of fossil fuels. This matter is of considerable importance because of the role of hydrocarbons in producing urban and rural photochemical oxidants. There is some question as to whether man's activities or natural sources are primarily responsible for atmospheric hydrocarbons (Maugh 1975).

POLLUTANT LEVELS AND RADIOCARBON MEASUREMENTS

Some Outstanding Problems

Pollutant evaluation via detailed examination of sources, sinks, reaction kinetics, atmospheric transport, and observed concentrations and their variability does not always lead to definitive conclusions because of uncertainties in both the data and the geophysical models. Three atmospheric components of interest will serve as examples:

1. Non-methane hydrocarbon (HC) production rates (Mton/yr):

anthropogenic	~60
natural	~440

2. CCl_4 concentrations (tropospheric, parts per billion):

anthropogenic	~0.070
total observed	~0.120

*The ^{14}C currently present in living matter itself has a dual origin: cosmic ray production from atmospheric nitrogen and nuclear testing. The resulting specific activity is relatively uniform, however, although it is slowly decreasing with time (Jansen 1972, Nydal et al. 1976, Zimen 1972).

3. CO residence time (years):

   deduced from ^{14}CO   ~0.1
   deduced from stable CO   ~1. to 3.

For these components of interest, the following findings illustrate the uncertainties:

1. The HC production rates, reported by Covert et al. (1975), indicate that natural contributions far outweigh anthropogenic emissions on a global basis. This fact, together with observations that significant rural air pollution episodes repeatedly occur and that forests yield photochemically reactive HC's (terpenes; R. Rasmussen 1972) has raised the issue of whether the offensive HC's are, in fact, primarily due to man's activities.

2. Carbon tetrachloride, which has a tropospheric concentration comparable to that of the freons, may have considerable significance with respect to stratospheric ozone (Rowland and Molina 1975). Yet as the above data show (Altshuller 1976), the discrepancy between the reported global (tropospheric) concentration and that computed from man's emissions, leaves room for a significant natural source.

3. The very large difference between estimated residence times (τ) as deduced (a) from the ratio ^{14}CO/^{12}CO (Weinstock and Chang 1976), and (b) from known sources and sinks, plus observed concentrations (Robinson and Robbins 1968, Junge et al. 1971), is of some importance, for τ is a key parameter in deducing the steady-state concentrations of CO at increased rates of energy production from fossil or synthetic (organic) fuels. Radiocarbon measurements of CO should be repeated in remote locations in order to reassess the τ-discrepancy. Similarly, measurements of ^{14}C/^{12}C in HC and in CCl_4, if feasible, could provide unambiguous evidence as to the relative source strengths.

Prior Radiocarbon Measurements

Within the past decade and a half, a number of investigators have measured ^{14}C/^{12}C ratios in atmospheric components (other than CO_2), and in bodies of water. ^{14}C/^{12}C in atmospheric particles was first measured by Clayton et al. (1955) and more recently by Lodge et al. (1960). The results of these measurements, to be discussed at the end of this paper, demonstrated the utility of the technique as well as the limitation due to the large sample size normally required. MacKay et al. (1963) reported the specific activity of atmospheric CO; a result later employed by Weinstock (1969, 1974) to estimate the residence time (~0.1 year) of CO in the troposphere. As noted above, a considerably longer time was suggested by Junge et al. (1971) in their detailed analysis of the global CO budget. Radiocarbon in atmospheric CH_4 was first measured by Libby. Libby's results, together with subsequent measurements by Bainbridge et al. are discussed in Ehhalt's review article (1973).

The ^{14}C content of dissolved organic carbon (DOC) in surface or groundwater has also been successfully employed to assay the fraction of petrochemical pollutants (Spiker and Rubin 1975). Here again, the ability to measure natural radiocarbon in small amounts of carbon would be quite beneficial because of the low concentrations encountered.

Atmospheric Concentration

In order to assess the feasibility of resolving contemporary (natural) and fossil (anthropogenic) contributions, we must first consider the stable and radiocarbon concentrations to be sampled and measured. The contemporary ("living") concentration of ^{14}C is roughly 8.6 pCi/g-C. It is approximately 40% higher than that of the modern radiocarbon standard, defined as 0.95 × NBS oxalic acid specific activity (Standard Reference Material No. 4990-B). (For more precise information, see Rafter and O'Brien 1972, and Nydal et al. 1976)

Observed concentration ranges for carbon in certain atmospheric gases and particles and in rainwater have been collected in table 1, together with maximum air quality standards. In order to permit rapid estimation of sampling requirements, concentrations have been expressed approximately in terms of mg of carbon per m^3 of air (or per liter of rainwater). It is immediately evident that for most species, collection of gram quantities of carbon is difficult if not impossible.

Turning to small radiocarbon sample techniques, and anticipating the results of the next section, let us assume that 10 mg of carbon will permit us to adequately distinguish fossil from contemporary carbon. Collection feasibility may be reevaluated from table 1. It is clear that dissolved organic carbon (in rainwater) presents no problem, nor is the collection of 10 mg of particulate carbon in a polluted environment very difficult. With respect to the gaseous species in table 1, collection of adequate amounts of HC (non-methane hydrocarbons) is clearly more difficult than CO, and any laboratory-scale proce-

TABLE 1

Concentrations of Selected Atmospheric Constituents[a]

	Approximate range (mg – C/m^3)	Air quality standard (mg – C/m^3)
Hydrocarbons[b] (non-methane)	<0.001 – 2.	0.14
Particles[c]	<0.001 – 0.1	0.052
CO[d]	0.06 – 20.	17.
CCl_4[e]	$2 \times 10^{-5} - 8 \times 10^{-4}$	–
DOC[f]	2 – 12 (mg/l)	–

[a] All values are approximate and expressed in mg-carbon/m^3 of air except for DOC (dissolved organic carbon) whose units are mg-carbon/liter of rainwater.
[b] K. Rasmussen et al. (1975), Lonneman et al. (1974).
[c] Limits estimated from data reported by Shaw (1975), Clayton et al. (1955), and Lodge et al. (1960). Mass of carbon was taken to be 20% of the particle mass (see table 4).
[d] K. Rasmussen et al. (1975); Colucci and Begeman (1969).
[e] Galbally (1976). The concentration of CH_3Cl in rural air (Grimsrud and Rasmussen 1975) is equivalent to 2.5×10^{-4} mg – C/m^3.
[f] Swinnerton et al. (1971).

dure for collecting CCl_4 is quite impossible. If we wished to collect HC at a concentration of 0.07 mg/m³, an oil-free pump with a throughput of 1 liter/sec would require about forty hours for the collection of 10 mg of carbon. The hydrocarbon experiment thus becomes feasible by reducing the required mass of the counting sample by a factor of 100 (1g→10 mg).

MEASUREMENT OF VERY SMALL RADIOCARBON SAMPLES

Counter Characteristics

Before discussing existing state-of-the-art gas and liquid scintillation counters, let us consider measurement feasibility in terms of the performance of an "ideal" counter — that is, one which has zero background and 100% detection efficiency. In order to reduce the Poisson counting error (ϕ) of the ideal counter to 10%, 100 counts must be collected. As the specific activity of contemporary carbon (including the residual increment from nuclear testing) is about 8.6 pCi/g, the required counting time (t) would be,

$$t = 100/[8.6 \times 10^{-3} (2.22)] = 5263 \text{ min/mg-C}.$$

Obviously a 1 mg sample is too small, because half of a week is rather long to wait, especially for an ideal counter. A more reasonable lower limit for sample size is 10 mg, for which the ideal counter would require about nine hours. If a contemporary sample could be measured with a standard deviation of 10%, samples having increased fossil carbon content would exhibit fewer counts and, therefore, decreased standard deviation for the same counting time. The standard deviation of a sample which is 80% fossil (anthropogenic) for example, would be reduced to $\sqrt{20}$ which equals 4.5 counts, that is, 4.5% of contemporary carbon.

Because zero background and 100% counting efficiency are not achievable, let us ask what limiting values should be sought in designing a real counter. Having decided upon 10 mg as a reasonable, achievable mass and taking 90% as a reasonable, achievable counting efficiency, we may next consider the point (R_B^o) beyond which background reductions have little consequence. For paired measurements (equal counting times for sample and background),

$$\phi_{PS}^2 t = (R_S + 2R_B)/R_S^2 \tag{1}$$

where ϕ_{PS} equals the relative (Poisson) standard deviation of R_S; t, the counting time; and R_S and R_B, the net sample and background counting rates, respectively. Clearly, the relative importance of background diminishes once $R_B < R_S/2 \equiv R_B^o$. A factor of two reduction from R_B^o, for example, would result in only a 25% reduction in t or a 13% reduction in ϕ_{PS} (The maximum reduction in ϕ_{PS}, obtained with the ideal counter, would be only 33%). 10 mg of contemporary carbon would have a counting rate of 0.17_2 cpm; therefore, $R_B^o = 0.086$ cpm.

The analysis can be carried a step further in designing a gas counter. The approximate size and pressure dependence of background, for a relatively clean and well-shielded counter, may be deduced from Oeschger's semiempirical theory (Oeschger 1963; Oeschger and Wahlen 1975). For small masses of counting gas, the background is largely due to residual contamination and γ-induced electrons in the wall of the counter. Based upon observed values for these terms (Oeschger and Loosli 1975), together with a length to diameter ratio of 6:1, we estimate $R_B^o \approx 0.013\ V^{2/3}$, where V is the counter volume (ml). The volume corresponding to $R_B^o = 0.086$ cpm is thus 17 ml.

Lacking an ideal counter, two real counters (one for liquid scintillation counting, and one for gas proportional counting) have been specially constructed for this work. Both counters have been constructed of high purity quartz in order to minimize radioactive contamination, and both have been made as small as feasible to minimize background arising from the interaction of external radiations with the counter walls. Characteristics of the counters are as follows:

	Volume (ml)	Background rate (cpm)	σ [10 mg-C]
Liquid Counter Ⓑ	1.0	0.42	0.41
Gas Counter E	15.0	0.17	1.0

The last column refers to the reduced activity ($\rho = R_S/R_B$) for 10 mg of contemporary carbon in each counter. It corresponds to 8.6 pCi/g and 90% detection efficiency. (Labels Ⓑ and E have been assigned to these two counters for subsequent identification and graphic use.) The counters differ significantly in carbon capacity. Counter Ⓑ can contain up to 800 mg of carbon (as C_6H_6), while E holds only 7.5 mg-C (as CH_4) at 1 atm. Counting times for 10 mg of carbon with $\phi_{PS} = 0.10$ derive from equation (1): 57 hours for Ⓑ, and 29 hours for E – approximately six and three times the period calculated for the ideal counter.

The small volume (1 ml), low level liquid scintillation cell Ⓑ is shown in figure 1. The cell is constructed of quartz with cylindrical outside dimensions of 7 mm length and 22 mm diameter with 1 mm thick walls. The scintillation counting mode of the cell is to position it between two 180° photomultiplier tubes which count only coincidence events. Counter background is reduced by an electronic guard and massive shielding. The electronic guard is constructed of a $NaI(T_l)$ annulus which surrounds the cell and counts in the anticoincidence mode. Massive shielding covers the cell and guard, and is constructed of 5000 lbs of lead bricks (~15 cm thick walls) with an inner graded shield of cadmium and copper sheeting. The electronics are of NIM BIN construction with measurement of fast-pulse time intervals replacing the conventional pulse height analysis mode of counting. A more detailed review of the counter is given by Noakes et al. (1973).

The small volume, low level gas proportional counter E is shown in figure 2. It consists of a quartz body, 0.05 mm tungsten wire, and a high-purity copper cathode. The shielding consists of a section of a naval gun from the early twentieth century (First World War) having a wall thickness of about 20 cm of iron. A muon anticoincidence counter surrounds the sample counter. Neither neutron shielding nor internal γ shielding

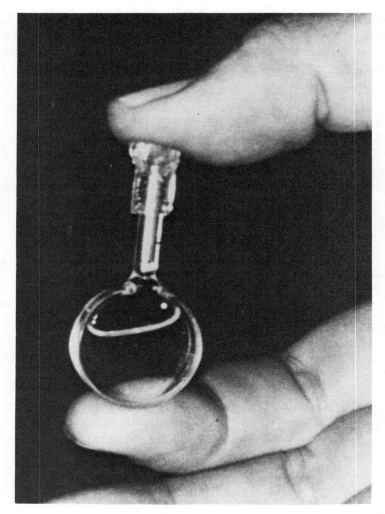

FIG. 1. 1 ml Quartz liquid scintillation cell (B).

(within the anticoincidence counter) were in place at the time of the background measurements. This may account, in part, for a background rate of about twice that estimated from Oeschger's semiempirical theory. Further description of the counting system may be found in Currie and Lindstrom (1975).

BACKGROUND REPRODUCIBILITY

The ultimate performance of any counter, as well as the comparative performance of alternative counters, is governed by the stability of the background. The quantitative

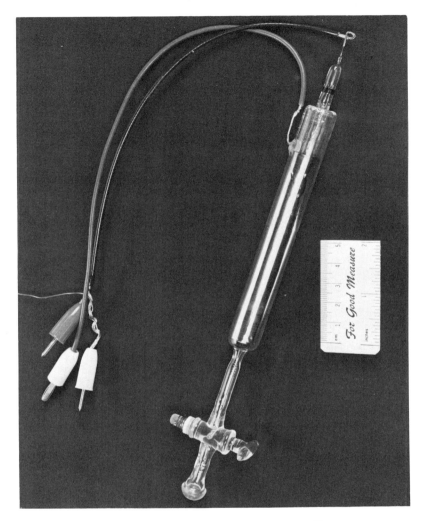

FIG. 2. 15 ml quartz gas proportional counter (E).

effect of (non-Poisson) background fluctuations is an added relative standard deviation (RSD) term which appears on the right side of relation (2). A necessary condition for Poisson errors to predominate is

$$\phi_{PS} > \phi_B (\sqrt{2}/\rho) \qquad (2)$$

where ϕ_B represents the intrinsic variability of the background (RSD) and ρ equals the reduced activity (i.e., R_S/R_B). Thus for 10 mg of carbon in Ⓑ ($\rho = 0.41$), the extra term equals $3.5\phi_B$, whereas for E ($\rho = 1.$) it equals $1.4\phi_B$. Taking ϕ_{PS} to be 10% (eq. 1), we see that random errors arise principally from counting statistics only if ϕ_B is kept within 2.9% for the liquid scintillation counter Ⓑ and 7.1% for the gas proportional counter E.* The *real* advantage of decreased background is thus a less stringent requirement on its stability.

Non-Poisson random errors are not easy to determine. Simple detection of such variations requires forty-seven (background) replicates if ϕ_B equals the Poisson value (Currie 1972), where the detection limit is defined as that level for which the errors of the first and second kinds (false positives and false negatives) each equals 5%. In the above reference, it is shown also that the *minimum* detection limit for ϕ_B occurs when the total available (background) counting time is subdivided into six to fifty intervals. This limit is shown to be about six times the *overall* Poisson standard error (ϕ_{PT}) based on the total accumulated counts.

Background reproducibility for the miniature liquid scintillation counter was assessed by means of three sets of ten 100 minute observations. Thus the total counting time was 3000 minutes, yielding 1260 counts; so ϕ_{PT} is $1/\sqrt{1260}$ which equals 2.8%. For twenty-seven degrees of freedom (three sets of ten observations), one can show (Currie 1972) that the detection limit for background fluctuations is 6.5 ϕ_{PT}, which equals 18.2% for a single 100 minute count, or $18.2\%/\sqrt{30}$ which equals 3.3% for the mean of the full set of thirty observations.**

Bounds for the background variations may be set by comparing the observed ratio s^2/σ^2 with χ^2/ν (s^2 equals the statistical estimate of the total background variance based upon ν degrees of freedom, and σ^2_{PB} is the variance derived from Poisson statistics). For the three sets of ten observations each, where $\nu = 27$, the pooled $s^2/\sigma^2_{PB} = 1.072$. This is not statistically significant, but it provides an upper limit (2.5%)*** for the relative standard deviation (RSD) due to background fluctuations for the mean of thirty observations (the detection limit (3.3%) is conveniently viewed as a *maximum* upper limit). Thus the background fluctuations of B lie within the bound (2.9%) previously set for Poisson predominance.

A second example, encompassing a broad range of background rates, is given in table 2 or a 100 ml gas proportional counter C used at NBS for low-level measurements of ^{37}Ar. Variance has been analyzed as above for four different background rates: the net

---

*Relation (2) may be transformed to yield the minimum reliable value for ρ, given ϕ_B. Thus for quantitative measurements ($\phi_{PS} = 10\%$), $\rho > 0.14\phi_B$; if $\phi_B = 5\%$, then $\rho > 0.71$.

**The total counting time (T) required to establish a detection limit for ϕ_B equal to its permissible limit, may be computed by equating 6.5 $\phi_{PT}/\sqrt{n}$ to ϕ_B as derived from relation (2), where $\phi_{PT} = 1/\sqrt{R_B T}$ and n = number of replications *normally* included in a background (and sample) measurement. The result is T = $8400 \cdot R_B/(nR_S^2)$. For 10 mg of contemporary carbon, $R_S = 0.17$ cpm; for counter Ⓑ ($R_B = 0.42$ cpm) and n = 30, T = 4070 min. For counter E ($R_B = 0.17$ cpm), the result is T = 1650 min.

***The upper limit derives from Eq. (9) in Currie (1972), that is,

$$2.5\% = \phi_{PT} \left(\frac{s^2/\sigma^2_{PB}}{F_L} - 1 \right)^{1/2}$$

where $\phi_{PT} = 2.8\%$, $s^2/\sigma^2_{PB} = 1.072$ and F_L, the fifth percentile of the χ^2/ν distribution, equals 0.60 for twenty-seven degrees of freedom.

TABLE 2

Background Reproducibility, Counter C
(100 ml, gas proportional)[a]

	^{37}Ar-Window	Integral background	Meson rate (coincidences)	Guard rate
$\bar{R}$ (cpm)	0.040	1.40	29.5	927
ϕ_{PT} (%)[b]	6.4	1.08	0.24	0.042
s^2/σ_{PB}^2	0.79	4.6*	3.0*	35.1*
$\phi_{\bar{B}}$ (%)[c]	(0, 8.0)	(1.2, 4.0)	(0.16, 0.69)	(0.17, 0.44)

[a]Eight observations with a total counting period of 4.2 days.
[b]Detection limit for background variability (RSD) is 6.6 $\phi_{PT}/\sqrt{10}$.
[c]Background variability bounds (90% C.I.) for mean of 10 observations. For a single observation, these would be increased by $\sqrt{10}$.
*Significant at ≤0.5% significance level.

background in an energy — pulse-shape "window" selected for ^{37}Ar events, the integral (total anticoincidence) background, the coincidence background due to muons, and the gross rate of the guard counter. The latter three, for different reasons, each show the presence of a non-Poisson error component. Significant deviations of coincidence counts from Poisson statistics were due to the (expected) dependence of the muon intensity on the barometric pressure. A correlation plot (not shown) exhibited a slope of −2%/cm-Hg. The guard counter errors were strongly influenced by a software oversight (improper rounding). The source of the integral background variations has not yet been identified.

COMPARATIVE PERFORMANCE OF DIFFERENT COUNTERS

Planning the measurements and assessing the small sample potential of alternative counters can be rapidly accomplished by means of a reduced activity plot similar to that published by Currie (1973) for radiocarbon dating. Figure 3 depicts the performance of two liquid scintillation counters (circled letters) and four gas proportional counters. Characteristics of the six including the three already discussed (Ⓑ, C and E) are given in table 3.

Figure 3 comprises four primary features: the two coordinates, the contours, and the −45° trace. The ordinate represents reduced activity or signal/background; the abscissa, background counts; the contours, Poisson imprecision (RSD) and detection limit; and the −45° trace, the locus of all counters having 90% efficiency and containing the same mass of contemporary carbon (here, 20 mg). The lettered points, which represent counters which differ simply in volume and background, except for Ⓐ, lie along this line. Counter Ⓐ lies slightly below the line because its efficiency is only 70%.

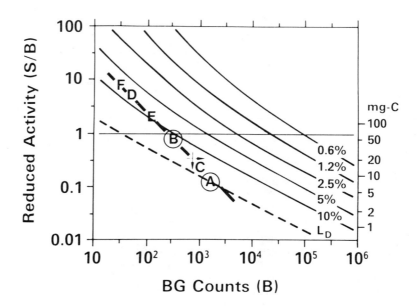

FIG. 3. Reduced activity diagram (20 mg-C). Reduced activity ($\rho = R_S/R_B$) vs. Background Counts (B) and Poisson counting error (ϕ_{PS} contours). (Dashed contour is the detection limit.) The $-45°$ trace is the locus of all counters for 90% detection efficiency and 20 mg of contemporary carbon. Normalization arrows indicate the time in minutes (abscissa) and the mass of carbon in mg (right ordinate). Lettered points correspond to the counters listed in table 3.

Of the additional three counters characterized in table 3, only D exists. Counter Ⓐ is taken to represent a typical commercial liquid scintillation counter, well shielded and finely tuned for the counting of radiocarbon as benzene. Counter F, similarly, is taken for asymptotic purposes. It represents the smallest gas proportional counter likely to be applied to the problem under discussion; its background was deduced from Oeschger's semiempirical theory. Smaller counters would yield inconsequential gains from background reduction, and gas pressure and dead volume difficulties might appear. Counter D, on the other hand, is real. Its characteristics for P-10 gas and radioargon were reported by Oeschger and Loosli (1975). For the purposes of this comparison, we assumed 90% efficiency and the same background (0.06 cpm) for radiocarbon. The outstanding background rate was due in part to an underground counting facility.*

The trace in figure 3 may be used to deduce counting times and Poisson precision. Its vertical normalization is set at 20 mg-C, and it has been translated horizontally in the figure so that the 15 ml quartz gas counter E intersects the $\phi_{PS} = 10\%$ contour. The time index (vertical arrow) shows that about 600 minutes are required. (The exact time

---

*A counter having still better characteristics for radiocarbon has been reported by Oeschger et al. (1976). 12 ml in volume, it has a background of 0.04 cpm (underground), and a net rate of 0.31 cpm for "modern" carbon at 5 atm. (CH_4).

TABLE 3

Counter Characteristics[a]

Counter	Vol. (ml)	mg – C[b]	Background rate (cpm)	(R_S/R_B) 10 mg – C[c]
Ⓐ	5.0	4000	2.2	0.061
Ⓑ	1.0	800	0.42	0.41
C	100	50	1.40	0.12
D	20	10	0.06	2.9
E	15	7.5	0.17	1.0
F	5	2.5	(0.04)	(4.3)

[a] Ⓐ, Ⓑ, are liquid scintillation counters, all others are gas proportional. ^{14}C counting efficiency is taken to be 70% for Ⓐ, 90% for all others.
[b] Mass of carbon (as benzene for Ⓐ and Ⓑ; as CH_4 for the remainder) required to fill the counter at 1 atm.
[c] Reduced activity ρ (signal/background) for 10 mg of contemporary carbon (8.6 pCi/g – C).

value is 579 minutes.) 20 mg of carbon in E would, of course, require an internal pressure of 2.7 atm, if it were mono-carbonic as CO_2 or CH_4. Counting precisions and background counts for the remaining counters, for 20 mg-C and 579 minutes, are indicated by the positions of the corresponding points. Precision thus ranges from about 8% (D and F) to beyond the detection limit (>30%, Ⓐ).

The time required to quantitate 10 mg of contemporary carbon with the 1 ml quartz liquid scintillation counter Ⓑ is deduced by a vertical translation of the trace to 10 mg followed by horizontal translation until point Ⓑ intersects the 10% contour, as shown in figure 4. The time index shows that 3400 minutes are now required, and locations of the remaining points are indicative of Poisson errors ranging from 5% to just above the detection limit.

The asymptotic character of the trace for the smaller counter background rates is evident from the figure. The approximate parallelism between the 5% contour and the 45° trace to the left of point D confirms that precision is relatively insensitive to further background reduction. In fact, the precision for a zero-background counter shows only about a 20% improvement over that of counter F (4.1% vs. 5.0%).

Because of background variability, the apparent precision for the larger background counters in the figure may be misleading. Inequality (2) indicates the point at which such variability becomes important. Taking $\phi_{PS} = 10\%$ and $\phi_B = 3\%$, for example, we see that background fluctuations must be reckoned with once ρ decreases below 0.42. We therefore anticipate significant effects for Ⓑ, C, and Ⓐ. The total random error may be

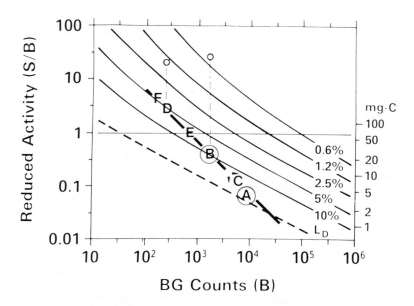

FIG. 4. Reduced activity diagram (10 mg-C). Same as figure 3, but normalized to 10 mg-C. Open circles correspond to maximum carbon capacity of D (80 mg) and Ⓑ (800 mg).

easily estimated by combining the Poisson component, as given by figure 4, with the background variability component, as given by the right side of relation (2). The results follow:

	ZBC[a]	F	D	E	Ⓑ	C	Ⓐ
Poisson error (RSD)	4.1%	5.0%	5.4%	7.1%	10%	17%	27%
Total error (RSD)	4.1%	5.2%	5.6%	8.2%	14%	38%	75%

[a] Zero-background counter; efficiency = 90%

Selection of the best counter for a specific task does not necessarily follow from the application of a simple figure-of-merit (FOM) expression, such as E^2/R_B or EV (E = counting efficiency, V = sample volume). Counters Ⓑ and D, which represent the state-of-the-art for small sample liquid scintillation and gas proportional counting, even have *different* asymptotic FOM expressions: for D, $R_S > R_B$, and for Ⓑ, $R_S < R_B$. It is clear from figure 4, however, that as long as these two counters lie on the same 45° trace, the Poisson error for D will be smaller than that of Ⓑ. The effect of background variability will make the difference even more pronounced, as shown above.

Reversal of merit will come about, however, because of the different counter capacities. The open circles above D and Ⓑ in figure 4 give the coordinates of the two counters at full capacity, 80 mg-C (8 atm. CH_4) and 800 mg-C (1 ml C_6H_6), respectively (changes of background rate with sample size have been ignored). The result is clear. If 800 mg of carbon are available the precision with the liquid scintillation counter will be considerably better than that of the gas counter, rather than the reverse.

Obviously there must be some mass of available sample between 80 mg-C and 800 mg-C for which the performance of the two counters is identical. This crossing point, which can be deduced from figure 4 or computed from equation (1), is 107 mg-C. For this mass of available carbon, if the errors are governed entirely by Poisson counting statistics, the points will be located on the same contour and the counters must therefore have equal merit. Background variations distort this simple picture. Counter Ⓑ, having the larger background rate, is more susceptible to such variations. The effect is to increase gradually the equal-merit mass as counting time increases. The quantitative effect of background variability, taking $\phi_B = 5\%$ for illustrative purposes, is presented in figure 5.

In conclusion, the equi-merit curve of figure 5 provides a definitive answer with regard to determining the better counter from the standpoint of counting precision. Though its capacity (m_D) is but 80 mg of carbon, the gas counter will always yield better precision if the total available sample is 107 mg-C or less. The region above the curve is the domain of the liquid scintillation counter. Taking the time as given in figure 4 (3400 min.), we see that Ⓑ is the counter of choice if more than 160 mg of carbon is available.

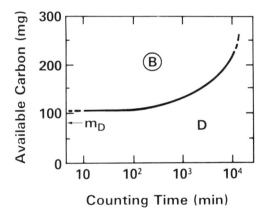

FIG. 5. Equimerit curve, counters Ⓑ and D. Curve indicates the mass of available contemporary carbon for which the liquid scintillation (Ⓑ) and gas proportional (D) counters will exhibit equal precision. Ⓑ yields better precision for the region above the curve, and the converse. Variation with counting time is due to assumed background variability of 5%. m_D represents the capacity of counter D (80 mg-carbon).

Contemporary Carbon in Urban Dust

A preliminary assay was made of the natural (contemporary carbon) contribution to a recent sample of atmospheric particles collected over a period of two years in the St. Louis area. The sample is not necessarily representative of fresh particulates nor entirely free from contamination, such as collection bag fibers, though considerable effort was expended to remove foreign material from the sample prior to analysis. At this time only an upper limit is available from our analysis, but the result is worthwhile for comparison with previous measurements by Clayton et al. (1955) and Lodge et al. (1960). Results for the three sets of analyses are given in table 4. It is interesting to note that liquid scintillation counting was employed by the first group of workers, and gas proportional counting by the second group.

Although there is little doubt concerning the potential of radiocarbon measurements to help identify carbon sources in urban particles, previous work and the data in table 4 point up certain difficulties. First, as noted by both sets of previous researchers, carbon does not make up a major fraction of the air particulate load. As a result the collection of an adequate sample for conventional radiocarbon measurements requires large capacity pumps ($\gtrsim 1000$ ft^3/min.) operating for a period of about a week per sample. If the samples are further fractionated chemically, even with such large collections, chemically interesting fractions result which have too little carbon for normal radiocarbon assay.

Following the first work in this field by Clayton et al. (1955) demonstrating the applicability of the technique, Lodge et al. (1960) took the important step of examining separated chemical fractions. By this approach they sought to eliminate the large amount of scatter in gross ^{14}C assays — presumably due to innocuous "resuspended organic dusts such as ground paper, leaves etc." The percentage of contemporary radiocarbon in the several organic fractions, moreover, ought to give clearer insight as to the primary source (natural or fossil) of compounds having deleterious health effects. As suggested by the

TABLE 4

Natural Carbon in Urban Particles

Collection area	Total mass (g)	Percentage carbon in total sample	Contemporary carbon as percentage of total carbon
Clayton, Arnold and Patty (1955)			
Detroit (II)	~18.	~16.	13.
Los Angeles (I)	31.	19.	26.
Lodge, Bien and Suess (1960)			
St. Louis (filter 7)	9.8	60. / 17.	27 (total) / 20 (ether extract)
Los Angeles	38.	48. / 10.	40 (total) / 14 (ether extract)
This Work (1976)			
St. Louis	10.	12.	≤20.

two ether extract results quoted in table 4, Lodge et al. succeeded in eliminating much of the original variability. They demonstrated also that nearly all of the paraffin material (≳95%) and a large percentage of the aromatic hydrocarbons (~90%) were fossil in origin.

With the possibility of measuring radiocarbon in samples as small as 10 to 20 mg-C, one should be able to extend the above studies to quantify man's contribution to several of the more important classes of organic compounds. With respect to atmospheric particulates, an early and critical application will be the evaluation of chromatographic fractions derived from insoluble organic constituents. Radiocarbon measurements of these species should be especially fruitful, for it has been shown by high-vacuum pyrolysis gas chromatography (Kunen et. al. 1976) that they tend to reflect particle origins.

ACKNOWLEDGMENTS

Thanks are due to J. K. Taylor of the National Bureau of Standards for providing the sample of urban dust and to E. K. Ralph and B. Lawn of the University Museum, University of Pennsylvania, for conversion of a portion for gas counting. Other colleagues providing assistance at NBS included E. Wolle, J. Ritter, W. Schmidt, W. Dorko, and J. Andrews. J. Spaulding of the Geochronology Laboratory, University of Georgia, assisted in determining the characteristics of the 1 ml liquid scintillation counter. Partial support by the Office of Air and Water Measurement of the National Bureau of Standards is gratefully acknowledged.

REFERENCES

Altshuller, A. P.
    1976   Average Tropospheric Concentration of Carbon Tetrachloride Based on Industrial Production, Usage, and Emissions. Env. Sci. and Tech. 10:596.

Clayton, George D., James R. Arnold, and Frank A. Patty
    1955   Determination of Sources of Particulate Atmospheric Carbon. Science. 122:751.

Colucci, Joseph M., and Charles R. Begeman
    1969   Carbon Monoxide in Detroit, New York, and Los Angeles Air. Env. Sci. and Tech. 3:41.

Covert, D. S., R. J. Charlson, R. Rasmussen, and H. Harrison
    1975   Atmospheric Chemistry and Air Quality. Rev. of Geophysics and Space Physics. 13:765.

Currie, L. A.
    1972   The Limit of Precision in Nuclear and Analytical Chemistry. Nucl. Instr. Meth. 100:387.

Currie, L. A.
    1973   Inherent Statistical Limitations in Age Resolution. Proc. 8th Int. Conf. on Radiocarbon Dating. Royal Society of New Zealand.

Currie, L. A., and R. M. Lindstrom
    1975   The NBS Measurement System for Natural ^{37}Ar. Proc. Noble Gases Symp. EPA and Univ. of Nevada. Las Vegas.

Ehhalt, Dieter H.
    1973   Methane in the Atmosphere. Carbon and the Biosphere. Woodwell and Pecan, eds. Conf. 720510, AEC.

Galbally, I. E.
    1976   Man-made Carbon Tetrachloride in the Atmosphere. Science 193:573.

Grimsrud, E. P., and R. A. Rasmussen
    1975   The Analysis of Fluorocarbons in the Troposphere by Gas Chromatography-Mass Spectrometry. Washington State University.

Jansen, H. S.
    1972   Extending the Use of Bomb Carbon in the Life Sciences. Proc. of the 8th Int. Conf. on Radiocarbon Dating. New Zealand.

Junge, C., W. Seiler, and P. Warneck
    1971   The Atmospheric ^{12}CO and ^{14}CO Budget. J. of Geophysical Res. 76:2866.

Kunen, S. M., M. F. Burke, E. L. Bandurskii, and B. Nagy
    1976   Preliminary Investigations of the Pyrolysis Products of Insoluble Polymer-like Components of Atmospheric Particulates. Atmospheric Environment. 10:913.

Lodge, James P. jr., George S. Bien, and Hans E. Suess
    1960   The Carbon-14 Contents of Urban Airborne Particulate Matter. In. J. Air Poll. 2:309.

Lonneman, W. A., S. L. Kopczynski, P. E. Darley, and F. D. Sutterfield
    1974   Hydrocarbon Composition of Urban Air Pollution. Env. Sci. & Tech. 8:229.

MacKay, Colin, Mary Pandow, and Richard Wolfgang
    1963   On the Chemistry of Natural Radiocarbon. J. Geophysical Research. 68:3929.

Maugh, T. H.
    1975   Air Pollution: Where Do Hydrocarbons Come From? Science. 189:277.

Noakes, John E., Michael P. Neary, and James D. Spaulding
    1973   Tritium Measurements With a New Liquid Scintillation Counter. Nucl. Inst. & Meth. 109:177.

Nydal, R., K. Loveseth, and S. Gulliksen
    1976   A Survey of ^{14}C Variations in Nature Since the Test Ban Treaty. Proc. 9th Int. Radiocarbon Conf., Univ. of Calif. Los Angeles and San Diego.

Oeschger, H., B. Alder, H. Loosli, C. C. Langway, jr., and A. Renaud
    1965   Radiocarbon Dating of Ice. Proc. 6th Int. Conf. Radiocarbon and Tritium Dating. Pullman. Washington. 53.

Oeschger, H. and M. Wahlen
    1975   Low Level Counting Techniques. Ann. Rev. of Nucl. Sci. 25:423.

Oeschger, H., B. Lehmann, H. H. Loosli, M. Moell, A. Neftel, U. Schotterer, and R. Zumbrunn
    1976   Recent Progress in Low Level Counting. Proc. 9th Int. Radiocarbon Conf. Univ. of Calif. Los Angeles and San Diego.

Oeschger, H., and H. H. Loosli
    1975   "New Developments in Sampling and Low Level Counting of Natural Radioactivity." Proc. Int. Conf. on Low Radioactivity Measurement and Applications. Czechoslovakia.

Oeschger, H.
 1963 Low Level Counting Methods. Proc. Int. Conf. on Radioactive Dating. Athens. IAEA.:13.

Rafter, T. A., and B. J. O'Brien
 1972 ^{14}C Measurements in the Atmosphere and in the South Pacific Ocean: a Recalculation of the Exchange Rates between the Atmosphere and the Ocean. Proc. 8th Int. Conf. on Radiocarbon Dating. New Zealand.

Rasmussen, Karen H., Mansoor Taheri, and Robert L. Kabel
 1975 Global Emissions and Natural Processes for Removal of Gaseous Pollutants. Water, Air and Soil Poll. 4:33.

Rasmussen, Reinhold A.
 1972 What Do the Hydrocarbons from Trees Contribute to Air Pollution? J. Air Poll. Control Assoc. 22:537.

Robinson, E., and R. C. Robbins
 1968 Sources, Abundance, and Fate of Gaseous Atmospheric Pollutants. Final Report. PR-6755. Stanford Research Institute.

Rowland, F. S., and Mario J. Molina
 1975 Chlorofluoromethanes in the Environment. Rev. of Geophysics and Space Physics. 13:1.

Shaw, G. E.
 1975 The Vertical Distribution of Tropospheric Aerosols at Barrow, Alaska. Tellus. 27:39.

Spiker, E. C., and M. Rubin
 1975 Petroleum Pollutants in Surface and Groundwater as Indicated by the Carbon-14 Activity of Dissolved Organic Carbon. Science. 187:61.

Suess, H. E.
 1955 Radiocarbon Concentration in Modern Wood. Science. 122:415.

Swinnerton, J. W., R. LaMontagne, and V. Linnenbom
 1971 Carbon Monoxide in Rainwater. Science. 172:943.

Weinstock, B.
 1969 Carbon Monoxide: Residence Time in the Atmosphere. Science. 166:224.

Weinstock, Bernard, and Tai Yup Chang
 1974 The Global Balance of Carbon Monoxide. Tellus. 26:108.
 1976 The Steady-State Concentration of Carbon Monoxide in the Troposphere. In Environmental Biogeochemistry. Nriagu, ed. Ann Arbor, Ann Arbor Science Publ., 1: Chap. 4.

Zimen, E. E.
 1972 The Future CO_2 Burden of the Atmosphere and Carbon-14 in the Ethanol from Wines. Proc. 8th Int. Conf. on Radiocarbon Dating. New Zealand.

6

Further Improvement of Counter Background and Shielding

S. Gulliksen and R. Nydal

ABSTRACT

During recent years efforts have been made at our ground level laboratory to obtain lower and more stable backgrounds for our counters by improving the shielding against cosmic radiation. Investigations have been made concerning the significance of lead shielding inside the anticoincidence guard counter. The effect of adding extra counting shells to the guard system has also been examined. The background level due to radioactive contaminations in counter material and lead has been measured for one particular counter (1.2 liters) by moving it 380 m below ground. Our best ground level arrangement, including three guard shells and 6.5 cm old lead, gives a background exceeding this level by 0.15 c/min. The considerable improvements obtained by using multiple guard shells indicate that a closer approximation of ideal underground conditions should be possible through careful design of the shielding system.

INTRODUCTION

The history of radiocarbon dating (Libby 1955) is also a history of the struggle to date greater age. Each laboratory has its own experience in this field, and literature describes the search for pure material for ^{14}C counter and shield, the problems with stability in the electronic apparatus, and the efforts to eliminate that part of counter background due to cosmic radiation. Different from radioactive contamination in counter and shield, cosmic radiation gives irregular fluctuations in the counter background, depending on solar activity, atmospheric pressure, temperature, and humidity. The maximum age which can be dated with the ^{14}C method is, due to the half-life of 5730 ± 40 years and the low concentration in nature, limited to about ten half-lives without enrichment of the samples. Even with enrichment the limitation is around 75,000 years.

DeVries (1957), who made the first comprehensive study of the background in ^{14}C counters, suggested that the only safe method of avoiding the neutron component in cosmic radiation was to place the counters below five to ten meters of earth. DeVries and his collaborators reached an impressively low background as early as the 1950's, and they were able to date samples up to 70,000 years with isotope enrichment (Haring et al., 1958). During recent years, very low counter backgrounds have been obtained in underground laboratories established in Bern (Oeschger 1975) and Seattle (Stuiver et al. 1976).

At our laboratory we are currently investigating means of reducing counter background and are studying the various components of cosmic rays. This paper contributes further to the study, indicating a way to approach an "underground" level of counter background even in a ground level laboratory.

Analysis of Previous Shielding

The shielding of our counters, shown in figure 1, has basically been unchanged since 1962 (Nydal 1962). For counter 2 it consists of 3.8 cm old lead between the CO_2 counter and the C_3H_8 ring counter, all inside a 22 cm thick iron chamber. Above the counting room, a four-story building provides additional shielding of 1 m concrete. The iron guard counter consists of a closed counting shell filled with propane to 1.3 atm pressure. With this shielding, the background of a 1.5 liter counter (counter 2) filled to 2 atm pressure was decreased to 0.83 c/min. Although this result was considered satisfactory, background measurements (Gulliksen 1972), showed a barometric effect of 2.2% – 2.3%/cm Hg of total background (Nydal et al. 1975). This effect, important when dating very old samples, inspired further investigation into our shielding efficiency. Unlike several other laboratories, we had no special shield against neutrons, and the walls and roof of our iron chamber were not so thick as those proposed by DeVries (1957).

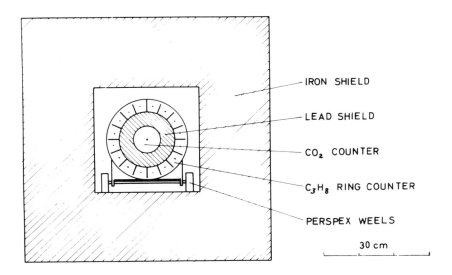

FIG. 1. The counter arrangement in the iron shield.

THE IRON CHAMBER

We first studied the roof thickness of our iron chamber. We also examined the fraction of background due to star production and showers occurring in the iron shield. For the latter purpose we applied a flat guard counter on top of the shield in order to detect charged particles before they might decay, initiating showers in the iron shield. The background of the counter was observed both with and without this extra flat guard counter in anticoincidence. The resultant measurements (Nydal et al. 1975) indicated the great complexity of the shielding problem. For a ^{14}C counter filled with CO_2 and with 5.3 cm thickness of old lead close to the counter, an iron shield of about 20 cm thickness seemed to be an optimum value. When the counter was filled with C_3H_8, the background decreased with greater thickness of the roof. After replacing the roof material of the chamber with lead, a nearly constant background was obtained in CO_2 for roof thickness exceeding 12 cm. The flat guard counter on top of the shield reduced the background by 20% − 25%. The shielding effect of this guard counter would most certainly have been improved had it been larger and had guard counters covered the sides of the chamber.

The shielding experiment showed (especially for C_3H_8) that showers occurred in the iron shield, and we considered it important to put most of the shielding material within guard counters for further improvement.

SHIELDING WITH OLD LEAD

The γ-radiation within the iron shield was studied with a Harshaw NaI crystal (2" x 2½") and a multichannel analyzer (Intertechnique, 200 channels). Shown in figure 2 are measurements inside the iron chamber with and without 5.3 cm additional lead shielding. Most of the γ-radiation behind 5.3 cm lead is due to contamination in the NaI crystal. The area between the two curves describes the γ-spectrum inside the iron chamber. The main part of the γ-radiation is found below 0.5 MeV; this shows that even a thin layer of lead inside the guard counter gives great reduction of the counter background. γ-radiation of 0.5 MeV can be reduced by 95% with a layer of 2 cm lead.

In the γ-spectrum certain peaks can be recognized as being due to contamination in the scintillation crystal, such as Potassium-40 with an energy of 1.46 MeV. Two small peaks between 0.5 and 0.7 MeV are also probably related to contamination in the crystal.

The actual reduction in background of counter 2 with various lead shields is seen in figure 3. Measurements on this curve at lead thickness 0 − 3.8 cm were done in 1962 (Nydal 1962) when the background of this counter was higher. These data have now been adjusted to the present background with repeated measurements at 3.8 cm lead. The curve has been extended to 5.3 cm lead. A verification of the whole curve has not yet been possible. For a lead thickness between 0.5 cm and 2.5 cm, the curve shows the effect of showers and star production in the lead. The assumed shower effect is illustrated by the dotted curve III on the figure. Even though the main part of the γ-radiation is absorbed with 3 cm lead, the shape of the curve indicates that further reduction of the background could be obtained with old lead thicker than 5.3 cm. The background obtained for counter 2 (1.5 l) with 5.3 cm lead is 0.59 c/min.

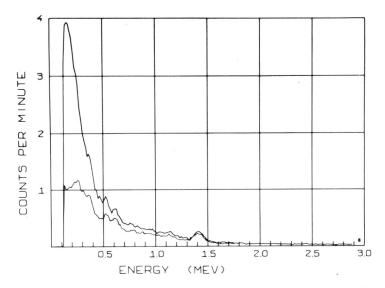

FIG. 2. Gamma-radiation inside iron shield, lower curve with 5.3 cm lead shielding.

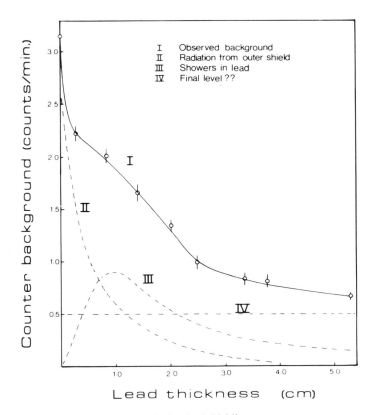

I   Observed background
II Radiation from outer shield
III Showers in lead
IV Final level??

FIG. 3. Old lead shielding.

Further Shielding Experiments

Our Experience showed that counter background was reduced by placing as much as possible of the shielding material inside guard counters. We also wanted to test the efficiency of our guard counters against charged particles and γ-rays. We found it necessary to build a larger guard counter and a new iron chamber. The dimensions of the new guard counter were so large that even the largest of our other guard counters could be placed inside it.

GUARD COUNTER CONSTRUCTION

The new guard counter shown in figure 4 consists of 2 coaxial iron cylinders with a clearance between them of 6.0 cm and a total length of 130 cm. The inner diameter of the inner cylinder is 29 cm. Aside from its big size, this construction has a major advantage over the three previous iron guard counters (Nydal 1962, 1975) because of its two separate counting shells. Separated by a thin iron foil (1.2 mm thick), these counting shells have 33 and 39 counting sections respectively, with effective lengths of 105 cm and 110 cm. The inner shell is not visible in figure 4. There are no shielding foils between the individual anode wires, which are 0.01 cm thick. The wires in each layer are continuous and are supported by two aluminum rings, each insulated from the cathode wall by three glass insulators. There is only one terminal (not shown) for outside connection to each wire system. Like our three previous iron guard counters, this one is sealed by tin soldering. The glass to metal seals are made of araldite, polymerized at 190°C for three hours. The counter is filled with 1.3 atm propane.

THE EFFICIENCY OF OLD LEAD AND GUARD COUNTERS

Figure 5 shows an experimental setup for testing the efficiency of guard counters as well as the influence of lead thickness on counter background. Our small guard counter

FIG. 4. Guard counter 5.

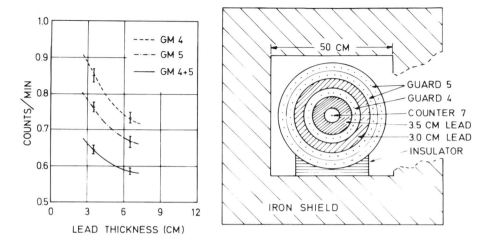

FIG. 5. Counter background for various shieldings.

4 and two layers of old lead were sandwiched inside the big guard counter 5. For this experiment, counter 7 was used (Nydal et al. 1975) with an effective volume of 1.2 liters, filled with 2 atm CO_2. Unfortunately the new iron chamber resulted in about 0.1 c/min higher background (with counter 7 inside counter 4 and 3.5 cm lead) than was obtained in the chamber previously used. The reduction in background which occurred with additional lead (figure 5) accords with the result in figure 3, but the curves drawn are very tentative. Most surprising was the marked decrease in background with additional guard counters. The efficiency of the guard counters had not previously been tested in our laboratory. From various literature, we has assumed nearly 100% efficiency against charged particles, and about 1% against γ-radiation. As shown in figure 5, we obtained a significant 10% to 15% drop in counter background by adding a new guard shell to the shielding. This reduction was surprisingly high. We wonder if it was due to detection of γ-radiation or muons.

In the present experiment, the background of counter 7 was reduced from 0.85 c/min to 0.59 c/min (30%) by adding 3 cm old lead and two extra guard shells.

Upper Level Discriminator

The effect of an upper level discriminator on counter background was investigated during the first operative years of this laboratory (Nydal and Sigmond 1957). It was found to be negligible compared to the total background. The possibility of excluding radiation from the minimum background level for a counter, such as that caused by radioactive impurities in the counter materials, must be examined.

Recent measures show that approximately 10% of the background is counted in the energy range exceeding the maximum energy of ^{14}C. With our operating conditions for the counter, an upper/lower threshold ratio of 70:1 was found to be an adequate value. A more comprehensive study concerning the influence of counter gas gain and threshold ratio must be made before optimum values can be determined.

Methods of pulse shape discrimination should also be considered, but the improvement to be obtained thereby may not compensate for the complexity of the necessary electronic circuits.

Counter Background Below Ground

In order to determine the background of the counter in the absence of any cosmic radiation, in the winter of 1976 we established our counting apparatus in a pit 380 m below ground level. Unfortunately the γ-intensity from minerals in the walls was very high and the counter had to be carefully shielded.

Figure 6 shows the shielding arrangement. The walls of the iron chamber were approximately 15 cm thick and the counter was surrounded with 0.5 to 9.5 cm old lead. A flat guard counter was used on top of the shield in order to ensure the presence of minimal cosmic radiation. The curve shown in figure 6 indicates that the counter background at maximum lead thickness approached a level corresponding to the background due to contamination in the counter materials and the old lead. This level can be found by careful analysis of the γ-spectrum inside the iron chamber, its absorption in lead, and the counter detection efficiency for different γ-energies. A rough determination has been made by assuming that the curve in figure 6 is composed of four components:

$$B = B_s \cdot e^{-(\mu/\rho)_s \cdot d} + B_h \cdot e^{-(\mu/\rho)_h \cdot d} + B_{muon} + B_{min}$$

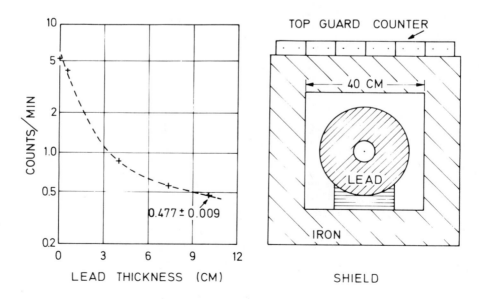

FIG. 6. Counter background 380 m below ground.

B_s = the background component due to soft γ-radiation that is absorbed with increasing lead thickness; $(\mu/\rho)_s$ = the corresponding mass absorption coefficient; B_h = the hard γ-component; B_{muon} = the contribution from muons traversing the counter; and B_{min} = the minimum background due to contamination; d = the surface density of lead shielding. For $(\mu/\rho)_h$ is used 0.04 cm²/g, the minimum absorption coefficient found for lead (Bleuler and Goldsmith 1952). Best fit to the data is then obtained for

$$B_s = 4.90 \text{ c/min}$$
$$B_h = 2.01 \text{ c/min}$$
$$B_{muon} + B_{min} = 0.45 \text{ c/min}$$
$$\text{and } (\mu/\rho)_s = 0.16 \text{ cm}^2/\text{g}$$

With the top guard counter working, we found coincidences indicating a muon component (B_{muon}) of approximately 0.015 c/min. This is approximately 0.1‰ of the muons counted in the chamber at ground level, and it agrees with the meson flux that exists below 950 m water equivalents (Heisenberg 1953).

By subtracting the muon contribution, we find that B_{min}, the minimum background for counter 7, is 0.44 ± 0.02 c/min.

CONCLUSIONS

Our shielding experiments demonstrated the feasibility of closely approximating underground background conditions in a ground level laboratory. The use of efficient guard counters and the inside application of a thick layer of old lead were found to be of critical importance. With one of the experiment systems, shown in figure 5, counter 7 reached a background of 0.59 ± 0.01 c/min, about 0.15 c/min above the ultimate level, as estimated from measurements at 380 m below ground. As shown in figure 5, further reduction can be obtained using more lead and better guards. As demonstrated by the design of guard counter 5 (fig. 4), sensitivity can be increased by adding one or more counting shells inside the same counter. The sensitivity of the guard counters can also be increased by higher pressure and greater thickness of the counting shells.

Future investigations will include consideration of an alternative system with a top guard counter (Nydal et al. 1975). We also will determine whether background can be reduced by adding a boron/paraffin shield inside the iron chamber.

It is important that a final shielding system should not be more complicated and expensive than necessary.

ACKNOWLEDGMENTS

This work was supported by the Norwegian research Council for Science and the Humanities (NAVF). The authors are indebted to Chief Ingenieur Ole Nordsteien and his staff in the Astrup pit at Lökken Verk for their valuable assistance in the underground measurements.

REFERENCES

Bueuler, E., and G. J. Goldsmith
    1952   Experimental Nucleonics. New York, Rinehart.

DeVries, Hl.
    1957   Further analysis of the neutron component of the background of counters used for ^{14}C age measurements. Nuclear Physics. 3:65-68.

Gulliksen, Steinar
    1972   Low cost electronics and twin counter assembly. Proc. 8th Int. Conf. on Radiocarbon Dating. New Zealand. 1:B69.

Haring, A., A. E. deVries, and Hl. deVries
    1958   Radiocarbon dating up to 70,000 years by isotope enrichment. Science. 128 (3322):472-473.

Heisenberg, W.
    1953   Kosmische Strahlung. 2d ed. Berlin, Göttingen, Heidelberg, Springer-Verlag.

Libby, W. F.
    1955   Radiocarbon Dating, 2d ed. Univ. of Chicago Press, Chicago.

Nydal, R., and R. S. Sigmond
    1957   Radiocarbon dating in Trondheim. Appl. Sci. Res. 6 (B):393-400.

Nydal, Reidar
    1962   Proportional counting technique for radiocarbon measurements. Rev. Sci. Instr. 33 (12):1313-1320.

Nydal, R., S. Gulliksen, and K. Lövseth
    1975   Proportional counters and shielding for low level gas counting. Proc. Conf. on Low Radioactivity Measurements and Applications. Czechoslovakia.

Oeschger, H.
    1975   Limits for low level counting. Proc. Int. Conf. on Low Radioactivity Measurements and Applications. Czechoslovakia.

Stuiver, M., S. W. Robinson, and I. C. Yang
    1976 ^{14}C dating up to 60,000 years with low background proportional counters. Proc. 9th Int. Radiocarbon Conf. Univ. of Calif. Los Angeles and San Diego.

7

The Effect of Electronegative Impurities on CO_2 Proportional Counting: An On-line Purity Test Counter

C. A. M. Brenninkmeijer and W. G. Mook

ABSTRACT

The effect of electronegative impurities on the CO_2 counter performance was studied quantitatively by determining the attachment coefficients for O_2, NO_2, NO, H_2O and SO_2. From the reasoning presented, some ideas have been deduced about the optimization of CO_2 counters.

An on-line purity test counter was constructed and put into operation. The optimal design followed from the theory and the measured values of the attachment coefficients.

INTRODUCTION

For the use of CO_2 as a counting gas in proportional counters, a high degree of gas purity is essential (de Vries and Barendsen 1953). Impure counting gas can result from extremely contaminated CO_2, or from an operational failure in the purification system.

In routine CO_2 counter operation, the gas purity is determined in the counter itself but, changing samples in the dating counters is a relatively laborious procedure. In order to avoid the introduction of impure gas into the dating counters, a simple and on-line purity check directly after the production of the gas is preferable. Moreover, a direct determination of the gas purity indicates in a quick and simple way the efficiency of the purification system.

The influence of impurities on counter performance depends on the specific counter used; therefore, the use of previous measurements resulting from particular counter set-ups is limited (Sdroc and Sliepcevic 1963, Freundlich and Rutloh 1972). Useful results are obtained when expressing the influence in terms of attachment coefficients, as was reported for O_2 only (Zastawny 1974).

The Effect of Impurities on $^{14}CO_2$ Dating

CO_2 proportional counters are used to compare the activities of CO_2 samples under identical conditions. The sample material is combusted and the CO_2 is thoroughly purified to remove traces of electronegative gases. Electronegative gases can bind the slow primary electrons released by the ionizing radiation. The so-formed negative ions, having roughly a thousand times smaller drift velocities than the electrons, do not participate in the buildup of the pulse, which results in smaller pulses and apparently lower gas amplification. As indicated by the slope of the beta plateau, a fraction of the pulses is just above the fixed discriminator level, so that the decrease in pulse size lowers the counting rate (CR). When working with fixed voltage, discriminator level, and gas pressure, the degree of purity thus influences the measured ^{14}C age.

A mathematical correction for the impurity of the counting and sample gas is possible without changing the counting conditions. Decrease of pulse height results in a shift of the beta plateau toward higher voltages. This shift is related to a similar shift of the *meson* characteristic curve (Barendsen 1955). The latter is measurable by the decrease in counting rate (CR) at the steep part of the curve. The effect of attachment can be corrected for by converting the decrease in meson counting rate to that in β counting rate, through a comparison of the respective slopes; the measured β counting rate A_m is related to the corrected value A by the following equation (cf. figure 1):

$$A = A_m \left[1 + (z_o - z) \left(\frac{\partial V}{\partial CR}\right)_{ms} \left(\frac{\partial CR}{\partial V}\right)_{\beta p} / A_o \right] \text{cpm} \qquad (1)$$

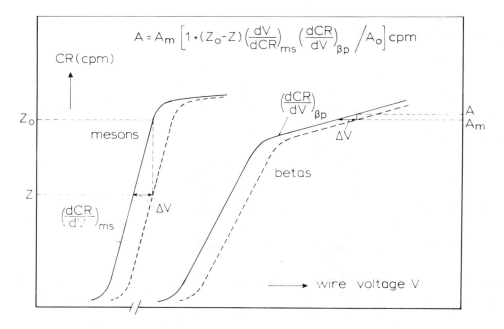

FIG. 1. The beta counting rate is corrected for impurities in the CO_2 by comparison of their respective influences on the meson and beta characteristic curves.

where A_o is the beta counting rate and $(\partial CR/\partial V)_{\beta p}$ the corresponding beta plateau slope as measured in the counter for standard recent $^{14}CO_2$; ms refers to meson steep slope. The voltage shift and the quantitative effect mainly depend on the applied pressure, the tube radius, and the nature of the impurity.

The Electron Attachment

An ionizing event will release a low density cloud of electrons. The electrons gain energy by drifting against the field direction and lose energy by colliding with the gas molecules. The low energy rotational and vibrational levels of CO_2 molecules keep the electron energy low, as long as the electrons are drifting under moderate field strength/pressure ratios as prevailing in proportional counters, except in a region several millimeters wide near the wire. Thus in CO_2 proportional counters the electrons are nearly in thermal equilibrium with the gas, and the drift velocity is simply given by $v_d = \mu \cdot E/p$. When E is the electric field strength in volt/cm and p is the pressure in torr, the mobility coefficient $\mu = 5.4 \times 10^5$ cm² torr/volt.sec at 20°C. The path traveled by the electrons to the wire is minimally affected by diffusion and can be considered a straight line. If electro-negative molecules are present, electron attachment will occur, decreasing the number of free electrons. The rate of decrease of n_e electrons is given in terms of reaction rates:

$$dn_e/dt = -n_e n_i (K_1 + n_i K_2 + n_c K_3) \tag{2}$$

The respective densities of the electrons (e), the impurity (i) and the CO_2 molecules (c) are indicated by n (cm⁻³). The reaction constant K_1 refers to the process in which a free electron is bound to an impurity molecule after a collision. The reaction constants K_2 and K_3 refer to three-particle collisions in which a second collision with a molecule $I(n_i K_2)$ or CO_2 $(n_c K_3)$ stabilizes the negative ion, preventing the ejection of the electron within a short time. When starting with n_r electrons at a distance r from the wire, the number arriving at the wire will be:

$$n = n_r \exp[-(K_1 + n_i K_2 + n_c K_3) n_i \Delta t]$$

where Δt is the time during which the electrons are drifting from a distance r to the anode wire. This time can be found from $\Delta t = -\int_r^a dr/v_d$ where a is the wire radius and the drift velocity $v_d = \mu E/p$. Using the field distribution $E = V/r \ln b/a$ where V is the wire voltage and b the counter radius, we have (for $r \gg a$):

$$\Delta t = \frac{p r^2 \ln b/a}{2\mu V}$$

The number of electrons arriving at the wire is therefore:

$$n = n_r \exp\left[-(K_1 + n_i K_2 + n_c K_3) \frac{cfp^2 r^2 \ln b/a}{2\mu V}\right] \tag{3}$$

where f is the fractional impurity concentration n_i/n_c and $c = n_c/p \simeq 3.3 \times 10^{16}$/torr cm³. Because the life time of the metastable negative ion I⁻ is very short compared to the time needed for stabilizing collisions, most ions I^- lose their additional electron. The primary electrons are permanently "lost" only by transfer of excitation energy to other gas molecules during collisions before the region of gas multiplication close to the wire is reached. We expect that the first term K_1 can be neglected. Moreover, n_i is orders of magnitude smaller than n_c, while K_2 and K_3 will not differ much. As a result the second term with K_2 is negligibly small. By inserting the proper values of c and μ, this expression becomes:

$$n \simeq n_r \exp\left[-\frac{c^2 f p^3 K_3 r^2 \ln b/a}{2\mu V}\right] \qquad (4)$$

where $c^2/2\mu \simeq 10^{27}$, if f is in ppm, p in torr and r in cm.

Method and Apparatus

Equation (4) shows that in order to achieve a measurable attachment, the ionization must occur at a well defined spot. In the case of extended ionization tracks, the pulse size distribution will depend on attachment. Furthermore the number of electrons generated during primary ionization should not be too low. The attachment is a statistical process and can lead to broad pulse size distributions subject to large errors when the initial number is too low. Moreover, a small number of electrons arriving at the wire would require the use of a high gas amplification which might interfere with the attachment process. Alpha particles can produce enough ionization within a few millimeters. On the other hand, the electron density along the ionization track is sufficiently low not to cause direct or indirect interaction of the electrons. A suitable alpha emitter is Americium-241 (half-life 458 y) emitting alphas with an energy of 5.5 MeV by 98%. The used source consists of Americium sintered on silver and coated with gold alloy. In addition the source is covered with mylar foil to exclude contamination. The range of the alpha particles within the counter, after passing through a mylar counter window, is thus reduced to less than 1 mm at STP. The counter itself is made of commercial copper tubing with an inner diameter of 60 mm and a wire diameter of 50 microns is used.

Information on the effect of attachment on the number of electrons collected at the wire is obtained as follows: For pure CO_2 the counting rate (CR) of the pulses above a certain discriminator level is measured as a function of the wire voltage (viz. the characteristic curve of the counter). For a given CR, voltage V_o, and gas multiplication M_o, n_o electrons arrive at the wire. After adding the impurity the measurement of the characteristic curve is repeated, where the same CR is now reached at a higher voltage V and gas multiplication M, corresponding to the collection of n electrons. The loss of primary electrons is compensated for by an increase of the gas multiplication, so that $M_o n_o = Mn$. The ratio M/M_o is accessible to measurement by determining the CR-voltage relation for several discriminator settings. An analogous method was recently applied by Chatterjee (1976).

The CO_2 gas as prepared for dating purposes shows no attachment at pressures up to 3.5 atm when purified by repeated circulation over hot copper (temperature above

450°C). This has been investigated by replacing the ^{241}Am source by a ^{55}Fe source which emits 5.9 keV gammas. These generate short-range (~1 mm in CO_2 at STP) photoelectrons at all possible distances from the wire. Any attachment would affect the shape of the narrow spectrum and give it a tail in the low energies, but no deformation has been detected, which confirms the purity of the gas and the fact that CO_2 itself does not attach electrons (Smith and Conway (1962) undoubtedly used impure CO_2). Assuming three-particle attachment only, in the case of primary electrons generated at the counter-wall we have:

$$\ln \frac{n_o}{n} = \ln \frac{M}{M_o} = \frac{c^2 f K_3 p^3 b^2 \ln b/a}{2\mu V} \tag{5}$$

Measurements of the Attachment

The effect of 1.6 ppm (by volume) of O_2 in a counter filled with CO_2 at 2380 torr on the alpha counting rate is illustrated in figure 2. The CR-characteristic curve is shifted to higher voltages by 240 volt. The plateau CR does not change over its whole length. This indicates that no pulses are generated by negative ions. The meson CR is affected in a different way (figure 3). The low energy pulses are affected more than the high energy pulses: a relatively large part of the low energy pulses comes from short ionization tracks near the counter wall. Electrons released here suffer most attachment.

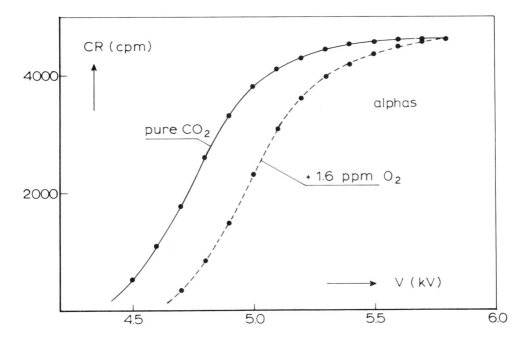

FIG. 2. The shift of the alpha characteristic curve by adding 1.6 ppm of O_2 to the pure CO_2.

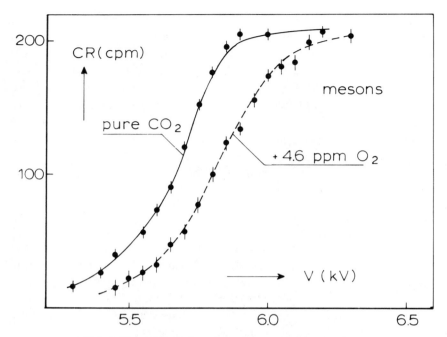

FIG. 3. The shift and distortion of the meson characteristic curve by adding 4.6 ppm of O_2 to the pure CO_2.

From equation (5) and from the shift for alpha particle ionization, one obtains a calculated shift for mesons of at most 460 volts. This value is corrected for the higher voltage as well as for the slightly different voltage dependence of the gas multiplication. Because only a very small fraction of the meson tracks is at the maximum distance from the wire and the attachment rapidly drops approaching the wire, the measurable maximal shift is lower. The influence of impurities on the characteristic meson curve depends on the electronics. For extended ionizations, the time distribution of an electron swarm striking the wire at an angle is affected by attachment. Pulse heights are affected when they are combined with differentiating and integrating time constants that are not matched with drift time differences up to 25 μsec (3 atm CO_2). This may be one of the reasons why the voltage shift of the meson plateau is not simply equal to the shift of the beta plateau. It is also clear, considering end effects, that the voltage shift for the plateau itself is larger than that of the steep part of the CR-V characteristic curve for mesons and beta ionizations.

The three-particle attachment coefficient K_3 has been measured applying equation (5). While the distance to be traveled by the electrons is constant (b = 30.5 mm), the pressure varies between 750 and 2400 torr. Changing the amount of O_2 added affords the opportunity to measure the influence of the voltage V. The measured value of $\ln(M/M_o)$ is plotted as a function of $p^3 f/V$ (fig. 4). When $p^3 f/V = 13$ torr^3 Volt^{-1}, the number of surviving electrons is 16% only. The linear relationship agrees well with equation (5) and confirms our expectations about K_1 being zero. The resulting value of $K_3 =$

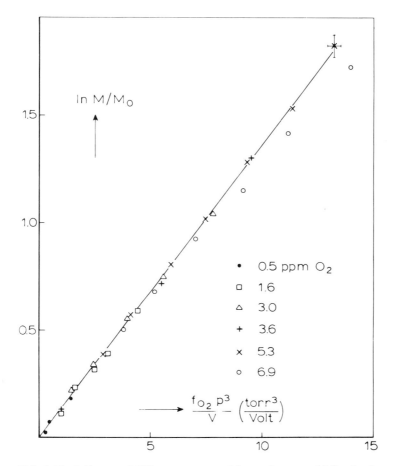

FIG. 4. The influence of different amounts of O_2 on the gas multiplication in a proportional CO_2 counter (cf. Eq. 5).

$(2.0 \pm 0.2)10^{-30}$ cm^{6}sec^{-1}. Our experiments and others with proportional counters (Zastawny 1974, Smith and Conway 1962) indicate that attachment is influenced by more processes and possibly by nonuniformity of the field (Pack and Phelps 1966a; 1966b). Experiments with electron drift tubes yield $K_3 = (3.0 \pm 0.3)10^{-30}$ cm^6 sec^{-1}. Therefore it is important to know if attachment is influenced by the counting rate, the amount of ionization, and the gas amplification. CR values up to 9000 cpm combined with alpha energies up to 700 keV and gas amplifications between 10 and 1000 give the same results. Care must be taken that the counter operates in the proportional region. Additional measurements with 5.9 keV gammas have resulted in $K_3 = (2.0 \pm 0.4) 10^{-30}$ cm^{6}sec^{-1}. Literature values of K_3 for O_2 are (in 10^{-30} cm^{6}sec^{-1}): 1.7 (Smith and Conway, 1962), 3.1 ± 0.3 (Pack and Phelps 1966), 3.18 ± 0.3 (Zastawny 1974).

Other gases, such as NO, NO_2, SO_2 and H_2O vapor, are likely to be present in CO_2 counting gas. Therefore, small amounts of these pure gases have been added to pure CO_2.

Each of them shows three-particle attachment. Our results are shown in figure 5 and in the following table:

Impurity gas	K_3 in $10^{-30} cm^6 sec^{-1}$
O_2	2.0 ± 0.2
NO	2.0 ± 0.2
NO_2	0.8 ± 0.1
SO_2	0.8 ± 0.2
H_2O	$(0.6 \pm 0.1) \times 10^{-2}$

The general conclusion is that other impurities are not more effective in binding electrons than O_2. The electronegativity of a gas does not uniquely determine the electron binding effect that leads to lower pulse sizes. This is because the *effective* electron attachment is a multiple process as can be shown schematically:

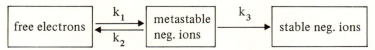

It can be shown that the three-particle collision-attachment coefficient K_3 in fact equals $k_1 k_3 / k_2$.

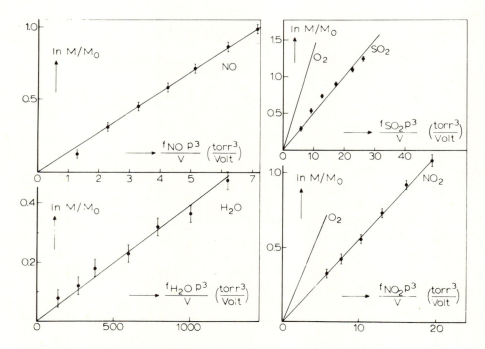

FIG. 5. The influence of different amounts of NO, SO_2, H_2O and NO_2 on the gas multiplication in a proportional CO_2 counter.

Optimization of an Existing CO_2 Counter

Knowledge of the influence of electro-negative impurities can be used to improve the figure-of-merit of a counter if enough counting-gas is available. As a certain loss of electrons due to attachment can be tolerated,

$$\ln \frac{n_o}{n} = \frac{Cp^3 b^2 \ln b/a}{V} \qquad (6)$$

has a given value (cf. eq. 5).

The constant C is determined by the purity of the used counter gas and the vacuum properties of the counter. An increase in the amount of gas in a given counter will, despite a slight increase in background, improve the figure-of-merit. The problem of obtaining a sufficient degree of gas purity does not limit this method in the case of CO_2 counting-gas, because a decrease in wire diameter can compensate for the higher pressure, according to the relation (Rossi and Staub 1949):

$$M = M\left(pa, \frac{V}{\ln b/a}\right) \qquad (7)$$

The maximum operating voltage, however, depends on the insulating properties of the counter and has in practice an upper limit of 10kV. With this voltage limit not yet reached, the pressure can be increased, without losing more electrons by attachment, through increasing V according to equation (6). The obtainable operating pressure p_{max} thus is related to the existing operating pressure p_o and to the obtainable voltage V_{max} by:

$$p_{max} = p_o \left(\frac{V_{max}}{V_o}\right)^{1/3} \qquad (8)$$

The wire diameter must be increased in order to keep M unchanged. An explicit relation for the gas amplification M has been derived by Zastawny (1966). The pressure range of this relation, however, is limited. It will also not necessarily give a satisfactory relation for the pulse height behavior, because it is based on pure multiplication yield, but, the measured pressure-voltage relation at constant M can be used together with equation (7) to calculate the necessary increase in wire diameter.

The Purity Test Counter

Ionization by alpha particles near the counter wall provides a highly sensitive means of detecting attachment, because the ionization happens in a limited space at a maximal distance from the wire where the electric field is weak and, consequently, the electron drift velocity is low. This principle is applied in the construction of a purity test counter combining the following features (fig. 6):

1. An enlarged counter diameter; $\ln M/M_o$ is proportional to the square of the diameter (b). Because both the voltage and the relation between $\ln M/M_o$ and the

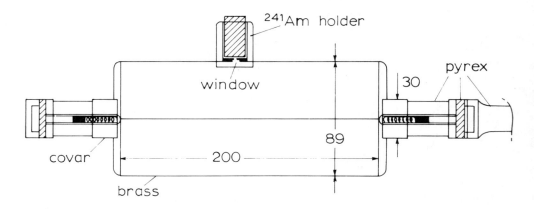

FIG. 6. Design of the purity test counter for CO_2.

voltage shift $(V - V_o)$ is minimally affected by an increased value of b, the voltage shift increases proportionally to the square of the diameter.

2. A decreased wire diameter; although the resulting voltage shift is lowered because the dependency of gas amplification on the voltage is increased, the effect of a lower voltage on attachment dominates.
3. A high counting rate; short measuring times will yield sufficiently accurate results.

For routine operations it is practical to operate the purity test counter at a moderate pressure so that it can be installed directly on-line to the CO_2 purification system. This simplifies the pressure reading and voltage supply. The counter diameter is fixed by the amount of gas to be tested and by the length-diameter ratio, which is taken, in this instance as 200 mm/90 mm; the wire diameter is 25 μ. The decrease in sensitivity because of the rather low pressure of 90 cm mercury (compared to 210 cm Hg for the 6 cm diameter dating counter) is compensated for by the better geometry. The ^{241}Am source is covered by 0.75μ nickel foil, and the counter window consists of 3.75 μ Havar metal foil, giving both source and window virtual permanence. A counting rate of 1600 cpm simplifies the required electronics. The purity check counter shows roughly the same voltage shift as the dating counters for a certain degree of gas purity: the relative voltage shift is, therefore, much higher. The sensitivity expressed in relative changes in counting rate is about 2.5 times higher (figure 7). At a working voltage of 2000 volts the impurity limit allowed for ^{14}C counters is a decrease in CR at the meson slope of 20%, equivalent to a drop in CR of about 50% (800 cpm) in the purity counter.

ACKNOWLEDGMENTS

We gratefully acknowledge many fruitful discussions and the assistance of D. J. Groeneveld, P. M. Grootes, and P. P. Tans during the project.

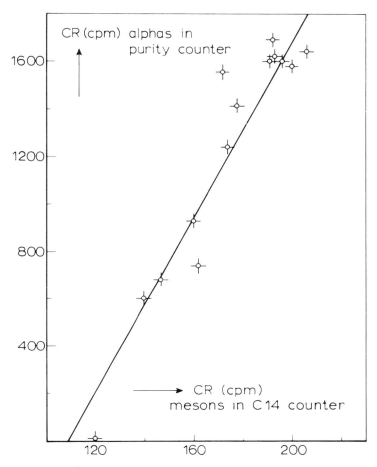

FIG. 7. Comparison of the effects of different impurities on the counting rates of the purity test counter and a routine proportional CO_2 counter for ^{14}C.

REFERENCES

Barendsen, G. W.
    1955 Ouderdomsbepaling met radioactieve koolstof. Thesis. Univ. of Groningen.

Chatterjee, M. L.
    1976 A method to obtain the unchanged energy calibration in the pulse height analyses for a gas-filled DE-proportional counter at different filling pressure. Nucl. Instr. Meth. 133:275-277.

De Vries, Hl., and G. W. Barendsen
    1953 Radiocarbon dating by a proportional counter filled with carbon dioxide. Physica. 19:987-1003.

Freundlich, J. C., and M. Rutloh
    1972 Radiocarbon dating by carbon dioxide method, influence and removal of

known impurities. Proc. 8th Int. Conf. on Radiocarbon Dating. New Zealand. 134-144.

Pack, J. L., and A. V. Phelps
    1966a Electron attachment and detachment. I. Pure O_2 at low energy. J. Chem. Phys. 44:1870-1883.
    1966b Electron attachment and detachment. II. Mixtures of O_2 and CO_2 and of O_2 and H_2O. J. Chem. Phys. 45:4316-4329.

Rossi, B. B., and H. H. Staub
    1949 Ionization chambers and counters. New York, McGraw Hill.

Sdroc, D., and A Sliepcevic
    1963 Carbon dioxide proportional counters: effect of gaseous impurities and gas purification methods. Int. J. Appl. Rad. Isot. 14:481-488.

Smith, C. F., and D. C. Conway
    1962 Distortion of proportional counter spectra by counter Poisons. Rev. Sci. Instr. 33:726-729.

Zastawny, A.
    1966 Gas amplification in a proportional counter with carbon dioxide. J. Sci. Instr. 43:179-181.
    1974 Measurement of the attachment of thermal electrons to oxygen molecules in the proportional counter with mixture $CO_2 + O_2$. Acta Phys. Pol. A46:39-46.

8

Examination of Counting Efficiency during Measurements of Natural ^{14}C

W. Moscicki

INTRODUCTION

The precision achieved in modern ^{14}C laboratories make it possible to evaluate the ^{14}C activities of recent samples with an accuracy as high as 0.2% – 0.3%. Some results raise questions, however, as to their reproducibility. One frequently finds that when a laboratory publishes several results obtained on the same sample (e.g., on background [B] or standard activity [S_o]), the standard error σ_p of a single measurement calculated with the assumption of a simple Poisson distribution of ^{14}C decays is significantly smaller than the standard error of a single result calculated on the basis of the mean square error σ_s (Lowdon et al. 1970). Occasionally one finds drastic differences amounting to $5\sigma_p$ and more between the results of interlaboratory checks (Reeburgh and Young 1976). Even results bearing major significance for further development of ^{14}C dating may be suspect. Out of 579 results, Damon et al. (1972) found 37 which may be suspect. This number of suspect results is about three times more than statistical expectation when the Cauvenet criterion is applied.

The methods applied in laboratories using proportional counters to examine the efficiency of counting from sample to sample are usually indirect. These methods may be divided into two groups: one in which the external source of soft X ray is used (Srdoc et al. 1971); and one in which the counting characteristics are studied. In cases when the former is applied, one determined by standardized electronic parameters whether or not the pulses are counted. This is, however, an uncertain test. With the presence of electronegative ions producing impurities (ENII) the decrease of gas amplification depends not only on the concentration of the ENII but on the location of the initial ionization. The gas amplification of ionization occurring in the vicinity of the anode will be the same as in the case when no ENII are present. Consequently, the pulse amplitudes from this ionization will be unchanged. To obtain reliable information, one would have to repeat a careful examination of the pulse-amplitudes of the X ray ionization spectrum from sample to sample.

An example of those methods relying on the examination of counting characteristics is one in which the pulse-amplitudes spectrum from muons is divided by a suitable discrimination threshold (D). The ratio (k) of the number of pulses above D to those below D is calculated. If, for fixed K, the value of HV is placed in arbitrarily chosen limits, the gas is believed to be pure. In all such methods, a uniform influence on the counting characteristics for every possible kind of ENII is assumed a priori. In fact, however, little is known about influences on the counting characteristics of any ENII, except for a few, such as O_2 and SO_2. Several purely physical factors suggest the possibility of ascertaining compositions of ENII by which particular properties of counting characteristics will remain fixed even if the counting efficiency is diminished. For detailed discussion see Moscicki et al. (1976).

In the Gliwice ^{14}C laboratory experiments have been carried out on the immediate measurement of the efficiency of muons counts (η_m) and on the evaluation of efficiencies of background (η_B) and ^{14}C (η_c) countings from η_m.

General Theory of η_m Measurements

The scheme of the arrangement of AC shield counters around the proportional counter (P) in the Gliwice laboratory is shown in figure 1. The AC shield consists of two separate units: the multi-electrode-ring (GM) counter (R), and the tray (T) of seven GM counters. Separately counts are made of the sum (L) of coincidences (P,T) + (P,R) and the number (N) of coincidences (T,R).

The C = L/N random number supplies immediate information on the counting efficiency of muons crossing P: The N random number is fully independent of the performance of P. It is only a matter of statistical fluctuation around the instantaneous mean value being influenced by atmospheric pressure and other factors causing changes of the muon flux.

The L random number, however, depends not only on the muon flux but also on the

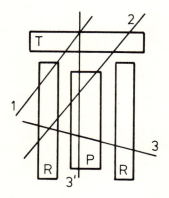

FIG. 1.

EXAMINATION OF COUNTING EFFICIENCY

efficiency of P in muon counting. Because the ratio of the mesons crossing the AC shields in a given time interval to the number of those crossing P depends only on the geometric distribution of T,R, and P, then, if the efficiency of P is constant, the number C will also be constant within the limits of statistical fluctuation. When N and L become great enough, the probable fluctuations of C become minimal and make it possible to reveal even small differences in P performance from sample to sample.

Probability Distribution of C

The possible tracks by which muons can activate coincident pulses (P,T), (P,R), and (T,R) are shown in figure 1. If x_1, x_2, and x_3 are the numbers of pulses caused by muons moving along the 1, 2, and 3 tracks, then $N = x_1 + x_2$ and $L = x_2 + x_3$ and, consequently,

$$C = \frac{x_2 + x_3}{x_1 + x_2} \tag{1}$$

Since the expected values of $\bar{x}_i (i = 1, 2, 3)$ are sufficiently high, x_i may be regarded as normally distributed. In this case the distribution obtained by Geary (1930) may be used (see also Moscicki 1958). By sufficiently high mean value $\bar{N}$, this distribution may be approximated by a normal distribution with the dispersion of

$$\sigma_c = \left[\frac{C_o (1 - 2\gamma C_o^{1/2} + C_o)}{\bar{N}} \right]^{1/2} \tag{2}$$

where

$$\gamma = \frac{\bar{x}_2}{(\bar{x}_1 + \bar{x}_2)(\bar{x}_2 + \bar{x}_3)} \tag{3}$$

and

$$C_o = \bar{L}/\bar{N} \tag{4}$$

In table 1a few numbers of σ_c for some r, C_o, and $\bar{N}$ are given.

TABLE 1

r	N	C_o	1.0	C_o	1.5	C_o	2.0
0.0	10^4		0.0142		0.0195		0.0247
	10^5		0.0045		0.0061		0.0078
0.2	10^4		0.0127		0.0175		0.0222
	10^5		0.0040		0.0055		0.0070
0.6	10^4		0.0090		0.0125		0.0163
	10^5		0.0028		0.0039		0.0051

Experimental Results

The preliminary examination of gas purity in the Gliwice laboratory employed the previously defined k criterion. As one of the counters, gas was accepted as pure for k = 0.7 ± 0.1 and the corresponding HV values 6200 ± 40. Within the limits of k and HV thirty-one measurements were made each ca. 3000 minutes of different samples. The mean value of C obtained was C_o = 1.4118, and the mean square dispersion of C was σ^E_{3000} = 0.0068. Applying σ^E, according to the Chauvenet criterion, four results were suspect as obtained with decreased efficiency of muon-counting. The results for three suspect measurements are listed in table 2 together with the results obtained after repurification of gas.

TABLE 2

Lab no.	HV	k	C	η_m	η_B	η_c	T_m	T_c
Gd 336a	6210	0.652	1.374	97.2	98.1	97.8	3120 +150	2930 +200
336b	6220	0.616	1.407	99.8	99.9	99.9	2980 +135	2970 +140
Gd 337a	6200	0.668	1.375	97.3	98.2	97.9	5390 +180	5050 +240
337b	6230	0.780	1.405	99.5	99.7	99.6	5180 +160	5150 +170
Gd 339a	6190	0.657	1.370	97.0	98.0	97.6	4440 +160	4160 +220
339b	6240	0.707	1.411	100.0	100.0	100.0	4120 +150	4120 +150

The successive columns of table 2 represent the laboratory number (Lab No.), HV at which the measurement was performed, k, C, efficiency of muon counting ($\eta_m = C/C_o$), calculated efficiency of background (η_B), and ^{14}C counting (η_c), age (T_m) calculated immediately from the results and (T_c) after correction for efficiency. Symbols (a) and (b) denote, respectively, results obtained before and after the second gas purification. For 1.36<C<1.425 calculations of η_B and η_c may be made by use of the experimentally obtained correlations: $B(C) = B_o + \alpha(C-C_o)$ and $S(C) = S_o + \beta(C-C_o)$.

In the arrangement ($\bar{x}_1 \simeq 85, \bar{x}_2 \simeq 40, \bar{x}_3 \simeq 135$) the value of (r) was estimated to be r = 0.25 + 0.05. Using this value together with C_o = 1.4118, N = 38·10⁴ (3000 minute measurement) the value of dispersion of C in single measurement was obtained σ^T_{3000} = 0.0026. In the 1 σ^T limit the uncertainty of η_m evaluation amounted to 0.18%. The series consisting of about twenty-two to sixty-six results of 100 minute measurements (each series on the same gas) furnished a number of results for σ^E_{100}. Individual values of σ^E_{100} fell within the limits from 1.07 σ^T_{100} to 0.92 σ^T_{100} with an average of σ^T_{100}. This result confirmed the above evaluation of statistical error of η_m in 3000 minute measurements. The mean square dispersion σ^E_{3000} used by applying the Chauvenet criterion reflects carelesness in reproduction of reference values of k and HV from sample to sample. σ^E_{3000} may be regarded as an effect of the superimposition of two random events: C and HV variations (σ_v) from sample to sample. From $(\sigma^E)^2 = (\sigma^T)^2 + (\sigma_v)^2$ one can evaluate the dispersion of HV from the reference value. A series of thirty-one measurements produced a result of σ_v = 28V, in very good agreement with the mean square value of $\sigma_{v,s}$ = 22V.

DISCUSSION

The most striking feature of Table 2 consists in the values of k and HV falling both before and after second purification within the limits of the reference values for k and HV. The small differences of HV for samples before and after the second purification cannot explain the changes in efficiency of muon-counting. From the slope of characteristics, one obtains the maximum changes of η_m after second purification amounting successively to 0.16%, 0.40%, and 0.64% in contrast to 2.6%, 2.2% and 2.9% found from corresponding C values.

The precision which can be achieved by means of C values in evaluating the efficiency of the ^{14}C decay-counting is limited by statistical fluctuation of the random value of C. The dispersion of C depends on the value of r. Under favorable conditions, great precision in η_m evaluations can be achieved because for r tending to 1 (only muons having tracks 2 [fig. 1] participate in L and N counts) the shape of distribution resembles δ-Dirac. It is, however, not the η_m, but η_B and η_c that influences the results of ^{14}C measurements. Except for managing r as high as possible, other requirements must be fulfilled to make the C method work successfully. To evaluate η_B and η_c from η_m, the correlations B = B(C) and S = S(C) are needed. These can be obtained experimentally with sufficient precision only when (1) the distribution of tracks of muons crossing P is sufficiently variable and (2) the shape of the spectrum of the primary ionization produced by muons is apparently similar to that produced by ^{14}C decays. These requirements apparently have been met by the arrangement of counters in the Gliwice laboratory.

REFERENCES

Damon, P. E., A. Long, and E. I. Wallick
    1972   Dendrochronology Calibration of the Carbon-14 Time Scale. Proc. 8th Int. Conf. on Radiocarbon Dating. New Zealand. 57.

Geary, R. C.
    1930   J. Stat. Soc. 93:442.

Lowdon, J. A., R. Wilmeth, and W. Blake, Jr.
    1970   Geological Survey of Canada Radiocarbon Dates X. Radiocarbon. 12(2): 473.

Moscicki, W.
    1958   Acta. Phys. Polon. XVII(5):32.

Moscicki, W., M. Pazdur, and A. Walanus
    1976

Reeburgh, W. S., and M. S. Young
    1976   University of Alaska Radiocarbon Dates I. Radiocarbon. 18(1):1.

Srdoc, D., B. Breyer, and A. R. Sliepcevic
    1971   Rudger Boskovic Institute Radiocarbon Measurements I. Radiocarbon. 13(1):135.

9

^{14}C Dating to 60,000 Years B.P. with Proportional Counters

M. Stuiver, S. W. Robinson, and I. C. Yang

ABSTRACT

A special counter design gives overall ^{14}C counting efficiencies of 89% to 98%, depending on size. Background activities appear approximately proportional with counter volume at 1.5, 5, and 25 mwe of shielding above the counters (1 mwe = one meter water equivalent = 100 g/cm^2). Special selection of counter materials results in a figure-of-merit of more than 5000 for 4 liter counters with 25 mwe of shielding.

INTRODUCTION

A sizable fraction of samples submitted for ^{14}C dating contains less than a few grams of carbon. For such small samples a higher counting efficiency gives increased precision as well as an increased age range. Even with abundant sample material, improved counting efficiency is favorable, as it results either in shorter counting periods or in smaller samples for the same precision. The special counter construction discussed in the following report increases the already comparatively high counting efficiency of proportional counters.

Counter backgrounds depend strongly on the type of materials used for construction. However, secondary radiation produced by cosmic rays appears to be the principal contributor to background activity for most ^{14}C laboratories. The influence of increased shielding on counter backgrounds is discussed in the concluding portion of the paper.

Counter Design

Considerable variation in design is possible for proportional counters. Leakage currents, across or through the insulating materials used to support the wire, can be troublesome. The "classical" proportional counter therefore incorporates a guard tube at ground potential to prevent the arrival of disturbing pulses at the wire. Such an arrangement contributes to the distortion of the electric field near the ends of the counter. This so-called "end effect" causes incomplete collection of emitted electrons. The field distortion can be compensated for by incorporating field tubes in the counter design (Curran 1958).

Four different kinds of counters were used for our ^{14}C dating projects. They are the following:

1. Groningen counters. These are quartz counters with a copper envelope. This counter type has been discussed previously (deVries et al. 1959).
2. A counter entirely constructed of quartz, as in figure 1. This counter is a variant of the Groningen counter and employs similar guard tubes.
3. UW (University of Washington) counters with minimal dead volume. These can be made entirely of copper, or they may contain a quartz liner, as shown in figure 2(A). Neither guard nor field tubes are used in this design. Quartz tubing, about 10 cm long and with wall thickness of 2-1/2 to 3 mm, is used as an insulator. The quartz tube can be shortened to insulate a six or seven kilovolt voltage differential. The extra length, however, reduces the possibility of surface leakage currents and eliminates the need for guard tubes.

    The end effect is kept minimal through the shape of the end plate. High voltage is applied to the end plate, with the inner curvature near the counter wire partly compensating for the field disturbance caused by the end tubes.

    The counter end plate in figure 2(A) has both a gas inlet and a high voltage contact. The end plate at the opposite side of the counter is similar in design without the gas inlet.

    The dead volume around the quartz tube is 10 cm^3 for each end plate. Where the end tubes extend into the counter cylinder dead volume is also encountered. This volume is small because the extension into the counter tube can be as short as 2 mm.

    The outer parts of the Groningen counters are electrically grounded because the voltage is applied to the tin oxide-coated inner quartz walls only. For our UW counters we traditionally apply a negative voltage on the cylinder wall and keep the wire near ground potential. A 0.25 mm thick teflon sheet provides insulation between sample and guard counter.
4. Counter end plate B, figure 2, was used to simplify construction. The fairly long quartz end tube increases dead volume. The end tubes perhaps could be made shorter, at the risk of increasing spurious pulse activity. The basic appeal of this counter is its extreme simplicity of construction and design. The electrical field configuration at the counter ends is far from ideal, however, and results in shorter plateaus.

Counter Materials and Construction

Quartz and copper are low in radioactive impurities and are favored materials for low level counter construction. Our quartz components are made from a high commercial grade natural quartz (Amersil-Engelhard Company). The variability in wall thickness and

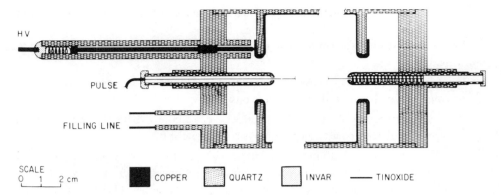

FIG. 1. All quartz counter of 2.4 liters volume. The tin oxide coated quartz end tubes are grounded and act as guard tubes.

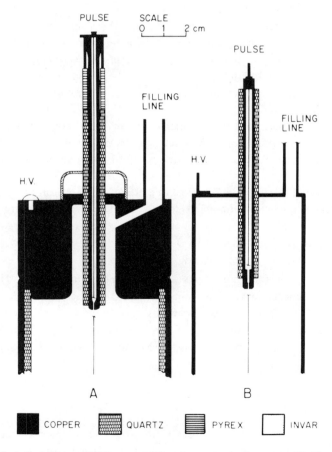

FIG. 2. End plates of UW counter (A) and experimental counter (B). The UW counter can be either all copper, or with a quartz liner.

tube diameter induced us to try synthetic quartz because of its smaller dimensional variability. However, the synthetic quartz (General Electric Company), increased background activity twenty-fold. This material, furthermore, had a tendency to leak at higher pressures. The synthetic quartz end tubes were leak proof when subjected to a few atmospheres of overpressure, but small holes "formed" at pressures of 4 to 5 atmospheres.

Wall thicknesses of at least 2.5 to 3 mm are required for the quartz endtubes. Insulating problems were encountered with 1 mm wall thickness (T. H. Peng, personal communication).

Insulators may exhibit a "shock" effect when a sudden voltage is applied. Spurious pulses occur immediately after application of the voltage. The number of spuriour pulses diminishes exponentially with time, and may suggest radioactive contamination. This effect is particularly pronounced in our guard counters. The counters have a simple end plate construction, similar to plate B in figure 2, but are made with commercial high voltage leads (Carborundum Company). A sudden voltage application induces initially a spurious counting rate of up to 100 counts per minute. Fortunately this behavior is not critical for guard counter operation because voltages are changed very infrequently.

In general, the quartz end tube insulators are not subject to this effect, which was measurable in only one of our counters. Here the application of the counter working voltage adds about 50 spuriour pulses in the first hours of counting. This is a small effect as it increases the counting rate for a standard 1000 minute run by 0.05 counts per minute only.

The copper parts are all made from oxygen-free high-conductivity (OFHC) copper. Although several batches of copper have been purchased over a ten year interval, we have never encountered contamination problems.

Counter wires are either 1 mil (0.025 mm) tungsten (Sylvania Company), or 1 mil stainless steel (Cohn, N.Y.).

Invar is used because it has the same thermal expansion coefficient as quartz. Epoxy makes a vacuum-tight joint between the invar insert and quartz end tube.

Commercial copper-pyrex seals are joined to the quartz end tubes by way of a graded seal. In the older counters, metal parts were joined either with silver or tin solder. The parts of our newer counters are all joined with epoxy. Clear Hysol epoxy is used, with an occasional dash of conducting epoxy when electrical contact has to be made. Overpressures of at least six atmospheres can be applied to such an "all epoxy" counter without causing leaks or ruptures.

One of the first steps in UW counter assembly is to epoxy the copper end tube inside the quartz insulator (copper to copper seal). This unit is then epoxied to the end plate (invar to quartz seal). An essential part of this operation is keeping the 0.25 mm hole in the end tube centered in the end plate.

The copper end tubes keep the wire in the center of the counter. Wire tension is provided in the Groningen counter by a small spring inside the quartz insulator that acts as a guard tube. In UW counters the wire is subjected to a weight of about 15 grams during counter assembly and is epoxied in the end tube with the aid of small copper plugs. This procedure appears to supply sufficient tension on the wire. No springs are used in this type of counter.

The UW end plate in figure 2(A) has been used by us for 2.3 liter counters. Different counter sizes, in our case from 0.5 to 4.5 liters, can be used with the same design. Inside diameters for these counters vary between 44 mm and 77 mm. The radius of the curvature of the end plate near the 0.025 mm hole in the end tube is 7 mm for all counters.

Counter Performance

UW counter plateaus are at least 400 volts long for CO_2 counting at 3 atmospheres. Longer plateaus are found for lower filling pressures. For most counters oxalic acid counting rates change about 0.8% per 100 V change in plateau counting voltage. In our 4.5 liter counter, the slope is as low as 0.4% per 100 V at 3 atmospheres of CO_2 pressure. The counting characteristics of a 2.3 liter UW copper counter are given in figure 3.

The counting stability of UW counters is excellent. Figure 4 gives background and oxalic acid activities for a 2.3 liter counter, as recorded over a twelve-month interval, with 25 mwe of shielding. For oxalic acid the standard deviation around the mean (σ_x) is 2.7 per mil for a total of twenty determinations. The standard deviation in the number of counts of a single determination (σ) is 2.5 per mil. The scatter of the values around the mean, and the variability in a single determination, evidently are nearly equal. Thus only statistical variability in the number of counts plays a role in the oxalic acid variations.

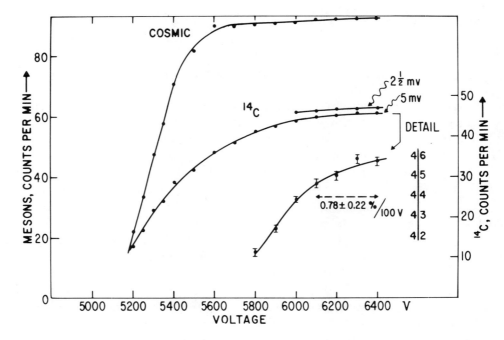

FIG. 3. Counting plateaus of a 2.3 liter copper UW counter. The β plateaus are for 2.5 and 5 mV thresholds.

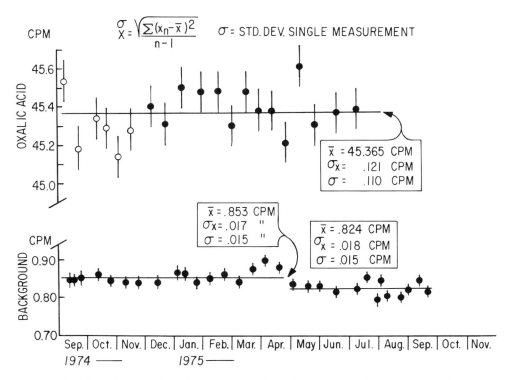

FIG. 4. Long term oxalic acid and background counting rates for a 2.3 liter UW counter.

Similarly, background activities have a normal distribution, except for an 0.03 count per minute shift that is possibly associated with a change in guard-counter efficiency (figure 4).

Table 1 lists the dimensions of some of our counters. In addition, the following properties are listed:
1. Total volume. The gas filled volume of the counter. This volume has been calculated from the counter dimensions.
2. Effective length. The length of the counter wire between the end tubes.
3. Counted volume. The volume of gas actually counted with 100% efficiency.
4. Dead volume. The difference between total and counted volume.
5. Counting efficiency. The ratio of counted and total volume, in percentage.

The counted volume can be calculated using the A.D. 1950 specific activity of oxalic acid, as determined by Karlén et al. (1964). This activity, corrected for decay to A.D. 1975, is 14.23 ± 0.07 cpm/gC. From the measured oxalic acid counting rate the amount of CO_2 actually counted is determined, thus yielding counted volume after correction for pressure and temperature. The CO_2 pressure is 2 to 3 atmospheres for all counters for which the above calculations are made.

Counting efficiencies of the UW counters increase with size from 89% to 98%. The Groningen counter, because of its design, has a lower counting efficiency of 83%. Dead

TABLE 1

COUNTER TYPE	CYLINDER LENGTH cm	DIAMETER cm	TOTAL VOLUME cm³	EFFECTIVE LENGTH cm	COUNTED VOLUME cm³	DEAD VOLUME cm³	COUNTING EFFICIENCY %
UW QUARTZ	35.3	4.40	557	34.8	497	60	89.1
UW QUARTZ	60.0	5.40	1394	59.0	1317	77	94.5
UW COPPER	98.4	7.07	3880	97.4	3752	127	96.7
UW COPPER	97.0	7.72	4560	96.4	4481	79	98.3
GRONINGEN QUARTZ	70.0	5.60	1724	62.0	1432	296	83.1

volume increases for the UW counters with size, except for the 4.5 liter counter. The dead volume calculations, however, are sensitive to possible variability in counter tube diameter. A change in average diameter from 7.72 cm (as measured at the counter ends) to 7.76 cm increases the calculated dead volume of the 4.5 liter counter to 126 cm³.

Background Activities and Shielding

Counter backgrounds were measured at three localities:
1. The Yale Geochronometric Laboratory. The shield consisted of 20 cm of iron on all sides. Underneath the "roof" of the shield was placed a 5 cm thick neutron absorber (85% paraffin, 15% boric acid by weight). Total cosmic ray absorber thickness above the counters was about 1.5 mwe for this arrangement (one meter of water equivalent = 1 mwe = 100 g/cm²).
2. The Yale Radiocarbon Laboratory. The shield was similar to the above one, except for an additional 30 cm of iron placed on top. This laboratory was housed in a three-story concrete building. The cosmic ray absorber thickness, including the building and the geometry of the sub-basement counting room, was estimated at ca. 5 mwe.
3. The University of Washington Quaternary Isotope Laboratory. The shield consisted of 30 cm of selected low level lead (St. Joseph, Missouri) on all sides. The counting room was 11 meters underground, and total absorber thickness for cosmic radiation was estimated at 25 mwe. The inner walls of the shield are currently lined with 10 cm of the paraffin-boric acid mixture. Backgrounds, however, were also measured (a) without the neutron absorber, and (b) with the absorber underneath the roof only.

Some additional changes were made when moving from the geochronometric to the radiocarbon laboratory. The guard counters were replaced, and their mode of operation was changed from Geiger to proportional counting. The iron shield between guard and sample counters also was replaced by a lead shield. These changes must have caused some improvement in counter backgrounds, but we feel this to be a second order effect in comparison to the changes caused by the differences in shielding. In the further discussion we consider shielding thickness to be the basic parameter influencing backgrounds.

A quartz counter of 1.5 liter volume of the Groningen type was used at all three localities. A second quartz counter of 0.5 liters was used in both Yale laboratories. The background for this counter at the University of Washington was estimated from the backgrounds of other counters, as will be discussed later.

Backgrounds of the 1.5 and 0.5 liter counters are given in figure 5 as a function of shielding in meters water equivalent and in cosmic ray flux. A neutron absorber is used underneath the roof of the shield. The biggest reduction in background occurs when shielding above the counters increases from 1.5 to 5.0 mwe. This is caused by the strong reduction of secondary radiation associated with the cosmis ray flux. The intensity of star formation, directly related to secondary radiation, is given in figure 6. In this figure, shielding thicknesses include the atmosphere. When shielding is increased from 0 to 5 mwe (10-15 mwe when the atmosphere is included), star intensity is reduced by a factor of 20. A subsequent factor of 20 reduction in star intensity is obtained only if shielding is increased to values in excess of 100 mwe. For the first 95% attenuation of the secondary radiation intensity, a 2% reduction is obtained on the average for each 0.1 mwe increase in shielding. To reduce the last 3% of the secondary radiation to 1%, however, takes 40 mwe.

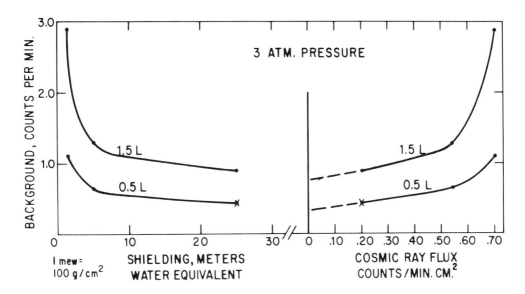

FIG. 5. Background activities of a 1.5 and 0.5 liter quartz counter at 3 localities. The shielding includes a neutron absorber above the counters, but not along the side walls. Observed backgrounds are given as dots; the background for the 0.5 liter counter at 25 mwe was obtained indirectly.

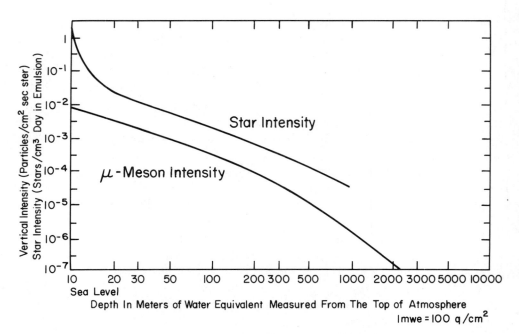

FIG. 6. Star formation and meson intensity measured underground. From Condon and Odishaw, Handbook of Physics (1958). Total shielding in mwe includes the atmosphere.

The secondary neutron and γ radiation may cause large changes in background when atmospheric pressure changes. Figure 7 gives the barometric effect of the 1.5 liter Groningen counter in the 1.5 mwe shield. Extreme changes in barometric pressure (ca. 3 cm Hg) result in a large change in background. The correction to be applied for the barometric effect limits the maximum age range to values less than those calculated from purely statistical considerations of the number of counts accumulated.

A barometric effect is not noticeable for the 1.5 liter counter in the 5 mwe shield. Beyond the 10 mwe range an increase in shielding reduces backgrounds by limited amounts only (figure 5). For most ^{14}C projects, 10 mwe of shielding would appear to be sufficient. It should be noted that the backgrounds in figure 5 are not the lowest possible because the backgrounds given are for shields with neutron absorption at the top only. Increased guard counter efficiencies also will yield lower backgrounds than those in figure 5 for 1.5 mwe of shielding (R. Nydal, this volume).

The background activities for counters of various sizes are given in figure 8. For each specific cosmic ray absorber thickness the backgrounds vary approximately linearly with volume. Quartz and copper counters have nearly identical backgrounds with 25 mwe of shielding. The counters listed for 1.5 mwe of shielding include two counters at the Yale Geochronometric Laboratory, one counter (IS) tested in the shield of Isotopes Inc., and one counter (ML) tested at the Miami Tritium Laboratory. All these shields have approximately 1.5 mwe of cosmic ray shielding.

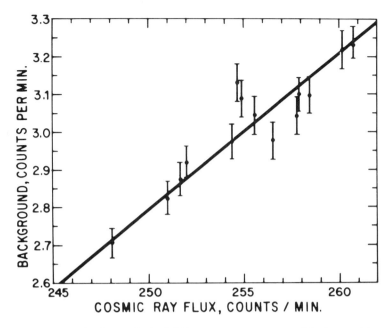

FIG. 7. The change in background of a 1.5 liter quartz counter caused by barometric pressure changes. Total cosmic ray shielding is 1.5 mwe.

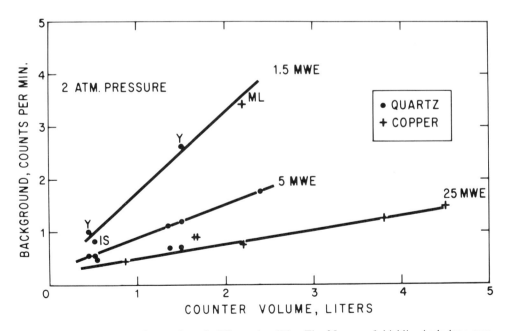

FIG. 8. Counter backgrounds at 3 different localities. The 25 mwe of shielding includes a complete 10 cm paraffin-boric acid neutron absorber inside the shield. The 5 and 1.5 mwe of shielding contain a 5 cm paraffin-boric acid absorber above the counters only.

Background contributions can be separated into two major components:

1. Amount effect A is the contribution to the background caused by γ and neutron interaction with the counter gas.
2. Wall effect W is the contribution of the counter walls to the background. This can be through α and β radiation from radioactive impurities, through ray-induced secondary electrons, and through recoil nuclei from collisions with neutrons.

The amount effect A, in cpm per gram of CO_2, can be calculated from the change in background for different filling pressures. The extrapolated zero pressure background activity, divided by the surface area S of the counter, gives wall effect W in cpm per cm². Both amount effect and wall effect are given in table 2 for 1.5, 5 and 25 mwe of shielding. For the 1.5 and 5 mwe of shielding, neutrons are absorbed by a 5 cm thick paraffin-boric acid mixture placed underneath the roof of the shield. For 25 mwe of shielding two sets of numbers are given; one set without any special neutron absorber; the other set with a 10 cm paraffin-boric acid mixture placed along the inside walls of the shield. Backgrounds are appreciably lowered by the neutron absorber; for instance, the background of our 3.8 liter counter at 3 atm CO_2 pressure reduces from 2.1 cpm to 1.3 cpm in the 25 mwe shield.

The concept of wall and amount effect is useful in calculating background activities. This can be seen from the background calculations of five counters in table 3, for a 25 mwe shield with complete neutron absorption. The backgrounds are calculated by substituting the W and A values given in the last row of table 2 in the equation $B = S \times W + G \times A$, where S is surface area of the counter, W is wall effect, G is the weight of CO_2

TABLE 2

SHIELDING	COSMIC RAY FLUX cpm/cm²	WALL EFFECT W cpm/cm²	GAS EFFECT A cpm/gCO₂
1.5 mwe NEUTRON ABSORBER AT TOP ONLY	.70	18×10^{-4}	.107
5 mwe NEUTRON ABSORBER AT TOP ONLY	.54	8.7×10^{-4}	.039
25 mwe WITHOUT NEUTRON ABSORBER	.20	4.2×10^{-4}	.054
25 mwe 10 CM NEUTRON ABSORBER ON ALL SIDES	.20	4.2×10^{-4}	.018

gas in the counter, and A is the amount effect. Although the counters vary in size by a factor of five, the differences between calculated and observed backgrounds is less than 20%.

The backgrounds seem to change linearly with volume (figure 8), but the linearity is only approximate. For constant filling pressure, the amount contribution is indeed proportional with volume V. However, the wall contribution $S \times W = 2V/r \times W$ (r is the radius of the counter) is not entirely proportional with volume because the radius r changes with counter size (from 2.4 to 3.9 cm). As a result the wall contribution increases less with size than the amount contribution.

The backgrounds given in table 3 are the lowest obtained thus far by us with standard anticoincidence techniques. Attempts to reduce the background further by risetime discrimination methods will be made, and this technique may possibly reduce the backgrounds by an additional few tenths of a count per minute. The lowest attainable backgrounds for a 1.5 liter counter, with 70 to 100 mwe of shielding, probably will be around 0.3 at 3 atmospheres of filling pressure. The remaining background is mainly caused by radioactive impurities in counter materials (α, β, γ radiation) and shield (γ radiation).

The interaction of γ radiation with the CO_2 gas, assuming full neutron absorption, results in the amount effect $A = 0.018$ cpm/g for 25 mwe of shielding. The source strength of the γ radiation interacting with the counter walls will be identical to the strength of the radiation inside the counter. To obtain a wall effect of 4.2×10^{-4} cpm/cm² the amount of wall material involved, if 100% of the emitted radiation is

TABLE 3

MATERIAL	EFFECTIVE COUNTER VOLUME Liter	WALL SURFACE AREA S cm²	WALL CONTRIBUTION S × W cpm	GAS CONTRIBUTION (3 ATM.) cpm	CALCULATED BACKGROUND (3 ATM.) cpm	OBSERVED BACKGROUND (3 ATM.) cpm
COPPER	.85	720	.30	.08	.38	.48
QUARTZ	1.5	1090	.46	.17	.63	.66
COPPER	2.2	1385	.58	.22	.80	.83
COPPER	3.8	2160	.91	.38	1.29	1.29*
COPPER	4.5	2340	.98	.45	1.43	1.55*

*Actual OBSERVED VALUE 0.1 cpm higher due to mesons not detected by the ANTI-COINCIDENCE SYSTEM.

counted, would be $4.2 \times 10^{-4}/0.018$ g/cm^2 = 23.3 mg/cm^2. This calculation assumes that γ radiation is the predominant cause of the wall effect. The effective range of the γ radiation contributing to the background is 4×23.3, or about 90 mg/cm^2, when the geometry of the emitted electrons from the counter walls is taken in to consideration.

The effective range of recoil nuclei is much less because the reduction in wall effect in the 25 mwe shield is about zero after addition of the neutron absorber, indicating a range close to 0 mg/cm^2.

Figure 9 gives the maximum ages of some of our counters for 25 mwe of shielding and full neutron absorption. A 5-fold increase in ^{14}C concentration in our thermal

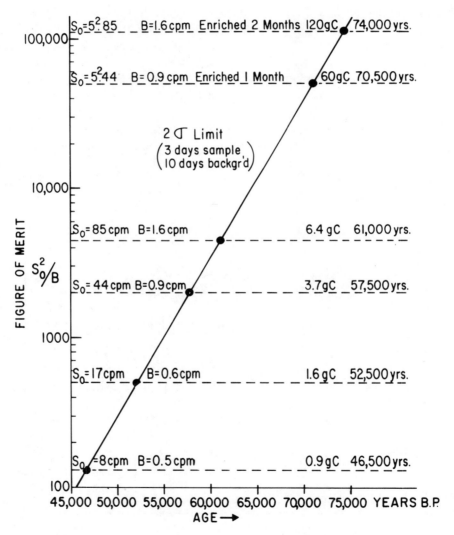

FIG. 9. Maximum age range of a set of counters with 25 mwe of shielding. S_0 is zero age activity (95% of oxalic acid activity) and B is background. ^{14}C isotope enrichment is five times for the upper lines.

isotope enrichment system increases the maximum age to 74,000 years. Much larger sample amounts are needed because depletion by more than 30% of the source gas used for enrichment is very inefficient. The enrichment process is slow, and only a limited number of samples can be processed each year. Finite conventional ^{14}C ages up to 75,000 years have been obtained so far. These dates clearly demonstrate that wood samples exist with ^{14}C contamination levels of less than 1 part in 10,000 (Stuiver et al. 1978).

ACKNOWLEDGMENT

This work has been supported by the National Science Foundation through grant DES-72-01712.

REFERENCES

Condon, E. U., and H. Odishaw
    1958   Handbook of Physics. New York, McGraw-Hill, 9-233.

Curran, S. C.
    1958   The proportional counter as detector and spectrometer. Ency. of Physics (Springer Verlag). 45(II):174-221.

de Vries, Hl., M. Stuiver, and I. V. Olsson
    1959   A proportional counter for low level counting with high efficiency. Nuclear Instruments and Methods. 5:111-114.

Karlén, I., I. V. Olsson, P. Kallberg, and S. Killicci
    1964   Absolute determination of the activity of two ^{14}C dating standards. Arkiv för Geofysik. 4(22):465-471.

Stuiver, M., C. J. Heusser, and I. G. Yang
    1978   North American glacial history extended to 75,000 years ago. Science 200: 16-21.

10

A Thermal Diffusion Plant for Radiocarbon Isotope Enrichment from Natural Samples

Helmut Erlenkeuser

ABSTRACT

The concept of a thermal diffusion plant, separation tube dimensions, and optimum operating conditions for enrichment of radiocarbon from natural samples are outlined.

Use of methane gas is recommended for operation of the enrichment plant. Concentric tube columns are preferable to hot wire columns because of shorter run times due to greater production rates. Isotope depletion in the storage volume at the negative end of the enrichment column determines minimum costs and run times obtainable and should be kept as low as possible. No sampling bulb should be used at the positive column end, but the enriched gas should be withdrawn from a sampling section of appropriate length separated from the enrichment column itself.

Optimum operating conditions with respect to minimum costs and minimum run times are calculated for a one-stage separation column of different dimensions on the basis of thermal diffusion column theory. Hot tube radii of a few centimeters and separation gap widths of 5 to 6 mm are found convenient. At hot wall temperatures of 400°C and with operating pressures falling between 1 and 2 atm, 1.8g of methane enriched in ^{14}C by a factor of 12 is obtained with a separation tube of 10m length within 8 to 9 days. The isotope yield is as high as 80%. Construction details of an all-metal concentric tube column are shown and measurements illustrating the validity of the mathematical model are presented.

INTRODUCTION

The radiocarbon dating method is usually restricted to samples younger than 50,000 years. Several attempts have been made to extend the range beyond this limit in order to date interstadial periods up to 70,000 to 80,000 years ago, and to improve the statistical accuracy for samples younger than 50,000 years.

By conventional techniques Geyh (1965) obtained a dating limit of 60,000 years. Sample size was about 20g of carbon. He used low background counters, counting gases with several atoms per molecule, and high filling pressures.

A more promising approach to extending the dating range appears to be radiocarbon isotope enrichment, particularly by thermal diffusion in gaseous medium. Concentrating ^{14}C by a factor of 2 extends the dating range by one half-life of the radiocarbon nuclide. To gain 20,000 years, radiocarbon must be enriched by a factor of 12 (fig. 1).

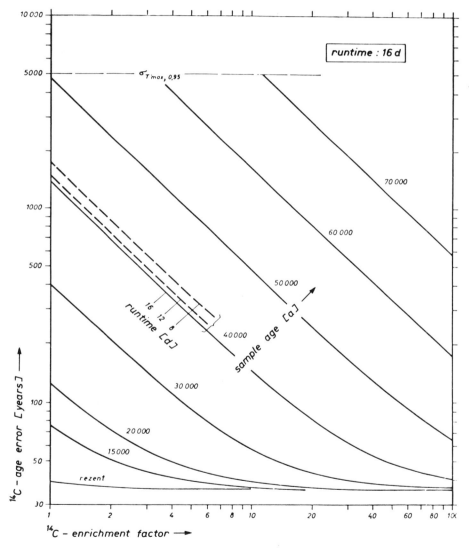

FIG. 1. ^{14}C-age error vs. enrichment factor (^{14}C-counter data assumed: background n = 8 cpm, $g_1 = \sigma_n/n = 2\ ^o/oo$ = residual uncertainty of background, recent activity without enrichment r = 21 cpm, $g_2 = [(\sigma_r/r)^2 + (\sigma_p/p)^2]^{1/2} = 5\ ^o/oo$, p = enrichment factor).

In an early attempt, Haring et al. (1958) used a cascade of hot-wire thermal diffusion columns working on carbon monoxide. They obtained 1.9g C, 12-fold enriched in ^{14}C, from an initial sample size of 160 g, in about one month. Recently Grootes et al. (1975) revived this equipment and produced 3.4g of carbon, 13-fold enriched from a total sample size of 130g C, within 44 days.

Pak (1970) and Felber and Pak (1973) constructed a thermal diffusion plant for methane consisting of a one-stage column of the concentric type. With a column of 4 m of length they produced 2.5g C, 23-fold enriched, from an initial sample of 90g C, within 35 days, and their measurements would indicate that they could get 2.5g C at an enrichment of 12-fold within 11 days.

Kretner (1973) also used a thermal diffusion column of the concentric type operating on methane. With a column length of 10.9 m and 122g C of sample weight he obtained 0.95g C, 20.6-fold enriched, or 3.7g C at an enrichment of 11.4-fold, within 45 days.

A concentric type thermal diffusion column for enrichment of ^{14}C-methane was also developed in Kiel (Erlenkeuser 1971 a,b). Our main purpose was to study in detail the behavior and performance of a thermal diffusion column. Although the plant was not operated at optimum conditions, the agreement achieved between experiment and theory to give a numerical approach to the optimum conditions for ^{14}C-enrichment from finite sample size was encouragingly close, suggesting that the effectiveness of the enrichment process could be considerably increased with respect to run time and initial sample size.

The Thermal Diffusion Process

The molecular theory of isotopic gases predicts a separation effect between any two of the different isotopic molecules in a temperature gradient. For binary mixtures, at constant pressure, the diffusion equation is

$$c_1 c_2 (v_1 - v_2) = -D_{12} \operatorname{grad} c_1 + a_{12} D_{12} \frac{\operatorname{grad} T}{T} \qquad (1)$$

with c_1, c_2 = concentrations (molar fractions) of isotope 1 and 2, respectively
v_1, v_2 = drift velocities of the isotopes
D_{12} = binary diffusion coefficient
a_{12} = thermal diffusion factor

There is an isotope net transport in the direction of the temperature gradient resulting in a small concentration gradient which is counteracted by ordinary concentration diffusion.

The thermal diffusion effect is utilized in the Clusius-Dickel separation tube, where the microscopic isotope separation effect in a horizontal temperature gradient is combined with a vertically circulating convection loop with the gas ascending at the heated wall and descending at the cold wall (fig. 2), thus establishing a counter-current process which concentrates the light and the heavy isotopes at opposite ends of the separation tube. This macroscopic separation effect along the z-axis of the tube is counteracted, finally becoming balanced by ordinary concentration diffusion and isotope back-

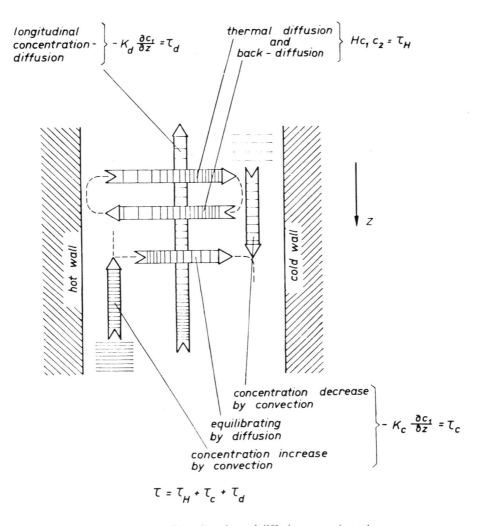

FIG. 2. Isotope fluxes in a thermal diffusion separation tube.

transport by the convection currents. In most isotopic mixtures, but not necessarily in all, the heavier isotope migrates to the cold wall and is consequently enriched at the lower end of the column. This situation is illustrated in figure 2. The different isotope fluxes add up to the so-called "transport equation" (Furry et al. 1939).

$$\tau_1 = H c_1 c_2 - (K_c + K_d) \frac{\partial c_1}{\partial z} \qquad (2)$$

where z is taken in the direction of increasing concentration of the isotope considered. H, K_c, and K_d are the "column transport coefficients" which are combinations of the gas properties a (thermal diffusion factor), D_{12} (binary diffusion coefficient), η (viscosity), λ

(thermal conductivity), and ρ (density) averaged over the horizontal temperature distribution within the space between the hot and the cold walls.

The mode of entry of the gas properties into the transport coefficients of the thermal diffusion column can be seen from the limiting case of small temperature difference between the hot and the cold walls.

$$H = \frac{a\rho^2 g(2w)^3 B}{96\,\eta}\left(\frac{\Delta T}{\overline{T}}\right)^2$$

$$K_c = \frac{\rho^3 g^2 (2w)^7 B}{2304\,\eta^2\,D}\left(\frac{\Delta T}{\overline{T}}\right)^2 \qquad (3)$$

$$K_d = 2w\,b\rho D$$

with $2w = R_K - R_H$ = separation gap width
$\qquad\qquad\quad g$ = gravity acceleration
$\qquad B = 2\pi R$ = horizontal length (circumference) of the separation gap
$\qquad R_K, R_H$ = cold and hot wall radii, respectively.

Note that $D \sim 1/\rho$. Introducing the equation of continuity,

$$-\mu\,\frac{\partial c}{\partial t} = \frac{\partial \tau}{\partial z} \qquad (4)$$

μ = mass of the gas filling per length unit of the separation tube.

and replacing $c_1 = c$, $c_2 = 1 - c$, the "separation tube equation" is obtained:

$$-\mu\,c_t = H\,c_z\,(1 - 2c) - (K_d + K_c)\,c_{zz} \qquad (5)$$

c_t, c_z, c_{zz} = 1st and 2nd order partial derivatives with respect to time t and longitudinal coordinate z.

The initial condition is

$$c(z,o) = c_0 \qquad (6a)$$

and the border conditions for a one-stage column are

$$-M_-\,c_t(o,t) = \tau(0,t) = H\,c(0,t)\,[1 - c(0,t)] - (K_c + K_d)\,c_z(o,t)$$
$$+M_+\,c_t(L,t) = \tau(L,t) = H\,c(L,t)\,[1 - c(L,t)] - (K_c + K_d)\,c_z(L,t) \qquad (6b)$$

$\qquad L$ = active column length
M_-, M_+ = mass of the gas in the reservoirs attached to the negative ($z = 0$) and positive ($z = L$) ends of the column, respectively.

The tube equation is a quasi-linear partial-differential equation of the parabolic type and must be solved numerically for the general case of border conditions as given by equations (6).

As seen from the tube equation, two characteristic scale factors can be defined:

1. the characteristic column length $\Lambda = \dfrac{K_c + K_d}{H}$

   Λ is the scale factor of the exponential concentration distribution in the final stationary state which is obtained in the limit of infinite run time.

2. the internal relaxation time $\tau_e = \dfrac{(K_c + K_d)}{H^2} \cdot \mu$

   The external relaxation time* as it appears from the time variation of the isotope concentration is based on τ_e, but is influenced to a greater extent by the border conditions, especially by the amount of gas in the end reservoirs.

It can readily be seen from equation (2) that the concentration in the separation tube is being built up from the ends of the tube. To illustrate, figures 3 and 4 give two examples, one for a column with both ends closed off and the other for a column with an infinite reservoir attached to the negative end (i.e., at $z = o$).

In the final steady-state when thermal diffusion transport is balanced by longitudinal backward-diffusion and convectional back-transport, the transport coefficients H, K_c, and K_d determine the ratio of the isotope molar fractions at the column ends rather than their concentrations. The latter are determined by mass conservation of the isotopic species under consideration. As a consequence, with finite end reservoirs a depletion at the negative column end occurs which reduces the enrichment obtainable with respect to the initial isotope concentration (compare fig.'s 3 and 4). Compensation for this depletion effect demands an effort in run time and column length, both of which increase rapidly as the negative end volume becomes exhausted. This feature makes optimization of enrichment plants with finite sample size very complex because operation conditions (pressure and temperatures), transport coefficients, column length, size of the reservoirs, desired enrichment factor, and size of the enriched sample are intimately interwoven.

The Concept of the Thermal Diffusion Plant

THE OPERATING GAS

For radiocarbon dating, enrichment in gaseous medium is highly preferable because appropriate vacuum techniques minimize the risk of sample contamination during the different steps of sample-to-gas conversion and the subsequent enrichment procedure. Two carbon-bearing gases are suitable for thermal diffusion enrichment: carbon monoxide and methane. The thermal diffusion factors are about the same. Carbon monoxide can be operated at higher temperatures than methane, which dissociates at noticeable rates at

---

*Mathematically, there is an infinite number of relaxation times, but it appears from experiments reported in literature, and it may be proved by theory, that for times much greater than τ_e one single time constant is sufficient to describe the time variation of isotope enrichment at the ends of the column.

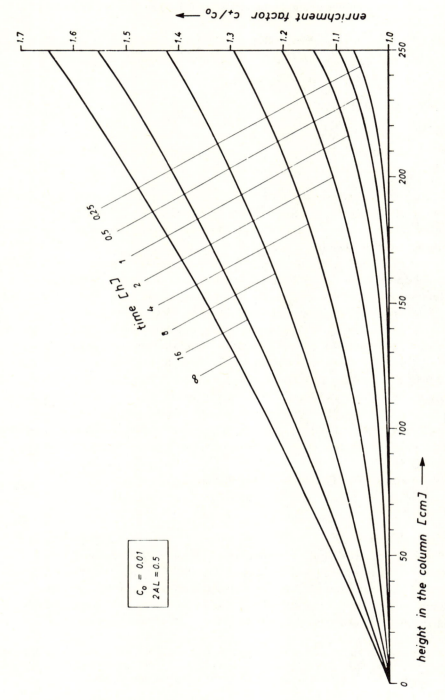

FIG. 3. Time variation of concentration distribution of $^{13}CH_4$ in methane in a one-stage thermal diffusion column closed at the positive end and open at the other ($^{13}CH_4 - {}^{12}CH_4$ binary mixture, $c \ll 1$, $H = 1_{10}-5$ g/sec, $K = K_c + K_d = 5_{10}-3$ g cm/sec, $\mu = 5_{10}-3$ g/cm, $L = 250$ cm).

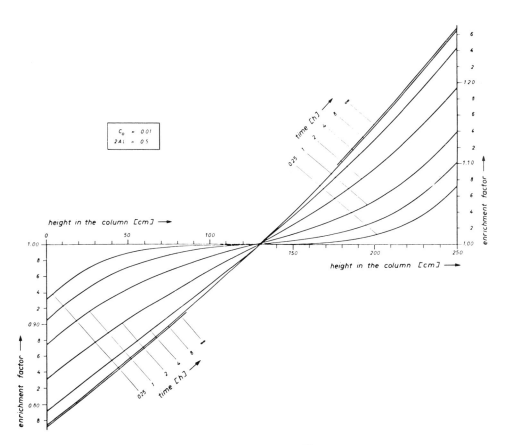

FIG. 4. Time variation of concentration distribution of $^{13}CH_4$ in methane in a one-stage thermal diffusion column closed at both ends (same parameters as in fig. 3).

temperatures above 400°C.* This allows steeper temperature gradients to be used, and the isotope net transport which depends on the term $(\Delta T/\overline{T})^2$ (see eq. 3) might be slightly higher for carbon monoxide operated columns. However, the chemical preparation techniques to convert the sample into a suitable counting gas seem much easier for methane than for carbon monoxide, which must be converted again after enrichment for mass spectrometric analysis and radiocarbon measurement. Methane, on the other hand, is easily prepared from carbon dioxide obtained from sample combustion. It is nonreactive, can be easily handled in conventional vacuum systems, and is readily analyzed for stable carbon and radiocarbon isotopes without further chemical pretreatment.

The isotope variation during the enrichment process may be observed from the carbon-13 isotope. However, the relations between the different isotopic molecules of different thermal diffusion factors are very sensitive to the operation conditions of the

*At 700°K and 1 atm CH_4 the partial pressure of H_2 from the thermal dissociation reaction $C + 2H_2 = CH_4$ is about 0.33 mmHg (Tipton 1960).

thermal diffusion plant. The thermal diffusion factors of $^{14}CH_4$ and $^{13}CH_4$ vary more than those for the isobaric molecules $^{14}C^{16}O$ and $^{12}C^{18}O$ available in carbon monoxide, and the requirements as to column performance and reproducibility of the enrichment process might be more severe for methane in order to allow calibration of radiocarbon enrichment on the basis of the stable isotope ratio.

THE OPTIMIZATION PROBLEM

The concept of the enrichment plant and the dimensions of the separation tube are governed by demands arising from the intended application in radiocarbon dating. For economic reasons the costs for producing a given amount of gas enriched by a given factor should be as low as possible. For routine operation the time for processing a single sample should be small as well.

The costs for providing the radiocarbon sample are too variable to be taken into account here. The costs for chemical treatment of the sample are minimized by operating the enrichment plant at high isotopic yield in order to have small samples for processing at the outset. The costs of enrichment derive from consumption of electric power and cooling water and will be proportional to the product of the total active length of the column (L), run time (t), and heating power per unit of column length (q). The costs for cooling water may be included in the heating costs.

The problem is to minimize the length-time-power-product and simultaneously to find a minimal run time with enrichment factor, final sample size, and isotopic yield as high as possible. Principally, the existence of optimal operating conditions can be seen from figure 6. With too short a column, the run time to achieve a given enrichment will be infinite. On the other hand, for long columns, costs increase with column length without further gain in run time because maximum isotope net flux through the column is independent from column length (see eq.'s 2 and 3).

Before numerical results of the optimization calculations are presented some general findings should be discussed. According to equations (2) and (3), the isotope net transport is proportional to the circumference of the annular separation gap. Concentric type columns thus appear to allow achievement of much lower run times than can be obtained with hot wire columns. The greater thermal capacity and the larger surface of the hot wall probably increase the stability of the gas convention loop in the separation gap. Convenient hot tube diameters are a few centimeters.

The thermal diffusion plant must be a closed one because of the finite sample size available. The major part of the sample gas will be stored in a reservoir at the negative column end. The enriched gas, then, may be sampled in a small volume at the positive end or may be withdrawn from an appropriate section of the active length separated and closed off from the other parts of the system after enrichment has become sufficiently large.

However, the numerical calculations show that in order to keep the effects arising from isotopic depletion of the negative end reservoirs as low as possible, the isotope to concentrate should be confined to the negative end reservoir in order to maintain the isotope concentration there as high as possible throughout the enrichment process. Using collection bulbs at the positive column end seems inappropriate because an isotope mass has to be transported to fill the positive part of the column up to an amount necessary for producing the enrichment factor desired and additionally to fill the sampling bulb. This greater transport has to be established under conditions which are less favorable because of the greater depletion of the storage volume at the negative column end. The effort of concentrating the isotope within the column is lost. When the enriched gas, on the other hand, is withdrawn from an appropriate section of the active column itself the column length must be enlarged, but because of the steep concentration gradients at the positive column part, the total amount of isotope to be transported is lower than with a passive bulb attached. Costs and run times become reduced, at least for high isotope yield, sample size, and enrichment factor as studied in this paper. Accordingly, all inactive dead volumes for gas sampling should be avoided at the positive column end.

The third essential constructional parameter of a thermal diffusion column is the distance between the cold and the hot walls. There is an optimum filling pressure for operating a thermal diffusion column which results from the fact that the isotope transport by thermal diffusion increases as p^2 (p = operation pressure; see eq. 3), while remixing effects by convection-flow increase as p^4 and are dominating at high pressures. The optimum pressure to give maximum enrichment in the final stationary state is related to the separation gap width according to

$$P_{opt} = \frac{1}{\sqrt[4]{\frac{K_c^*}{K_d}}} \tag{7}$$

$$\approx (2w)^{-1.5} \, \varphi(T_K, T_H) \left(\text{for } 2w \ll \frac{R_K + R_H}{2} \right)$$

T_K, T_H = temperature of cold and hot walls, respectively
$K_c = K_c^* p^4$; $2w$ = separation gap width.

φ is a slowly varying function of the temperatures only, ranging between $\varphi(13°C, 417°C)$ = 13.6 atm mm$^{1.5}$ and $\varphi(12°C, 295°C)$ = 11.8 atm mm$^{1.5}$ in case of methane. At $2w$ = 6 mm, p_{opt} is about 0.8 to 0.9 atm.

For the transient state of operation, however, the optimum pressure is found to be higher by a factor of about 1 to 2 (see fig.'s 7-10). Working pressures appreciably exceeding 1 atm may be inconvenient for laboratory work. Moreover, the column performance depends strongly on the relative inaccuracy of the separation gap width which probably can not be made better than 0.1/2w mm/mm. It appears that separation gap widths of about 5 to 6 mm are convenient, with operation pressures of about 1 to 2 atm.

The concept of the diffusion plant as presently worked out is shown in figure 5.

Results of Optimization Calculations

The optimization calculations reported here are performed on the basis of the tube equation (eq. 5) and the appropriate initial and border conditions (eq. 6), for a one-stage column with a negative end reservoir but without a sampling bulb at the positive column

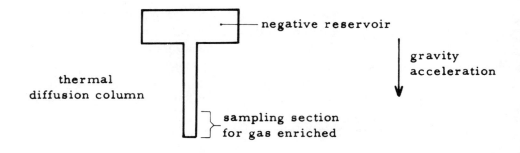

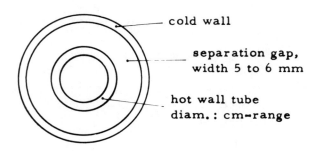

FIG. 5. Concept of a thermal diffusion plant for enrichment of ^{14}C from natural samples.

end. The calculations are based on numerical solution of the tube equation and on binary enrichment of $^{14}CH_4$ in $^{12}CH_4$. Run times and tube lengths calculated appear to be accurate within 1% or 2%.

Some parameters of the diffusion plant were preset and were not varied:

cold wall temperature	T_K = 14°C
size of enriched sample, withdrawn from a column section of appropriate length	ΔM = 1.8g $\hat{=}$ 2.6 l atm STP
enrichment factor, with respect to the initial ^{14}C concentration	m = 12

The following parameters were varied as shown

hot wall temperature	T_H =	350	400	450°C
hot wall radius	R_H =	15	20	30 mm
separation gap width	2w =	4	6	8 mm
operation pressure factor	f_p =	0.5	......	2.0
isotopic yield	ϵ =	40	60	80 %

The operation pressure factor f_p is defined as

$$f_p = \frac{p_B}{p_{opt}} \quad (p_B = \text{actual operation pressure})$$

The isotopic yield is calculated according to

$$\eta = \frac{\Delta M(m-1)}{M_{total} - \Delta M}$$

M_{total} = total amount of sample gas, m = enrichment factor

η compares the amount of isotope obtained by column action with the additional amount of sample required for the enrichment process.

The computer calculates, for a given set of R_K, R_H, T_K, T_H, η, and p, the optimal value of the column length (L) and run time (t) for which the length-time-power-product is minimal and the enrichment factor and sample size are achieved as preset. The above parameters are varied, then, and their influence on the Ltq-product is studied visually from the results obtained (fig.'s 7 to 9).

The costs as measured by the length-time-power-product (Ltq-product) increase with isotopic yield η approximately as $(100 - \eta)^{-0.81}$ (fig. 7). A reasonable compromise between costs and total sample size is at η of about 60% to 80%.

It is seen from figure 7 that the length-time-power-product decreases with increasing hot wall temperature. This feature results from the coefficient H (see eq. 3) of the thermal diffusion transport which increases with temperature (see fig. 11) while the remixing coefficient $K = K_c + K_d$ decreases at higher temperatures due to the decrease of K_c (fig. 12). Thus the column scale factors $\Lambda = (K_c + K_d)/H$ and $\tau_e = \mu \cdot (K_c + K_d)/H^2$ become smaller with increasing hot wall temperature. In the Ltq-product, however, the reduction of length and run time is almost compensated for by growing power demand. A small gain remains, suggesting selection of a hot wall temperature as high as thermal stability of methane will allow.

The calculations further indicate that the Ltq-product is strongly affected by the operation pressure factor f_p. Increasing pressures allow smaller sampling sections. Because of the steep concentration gradients at the positive part of the column, this effectively reduces the expenditure of energy and time during enrichment. The other parameters, however, become dominating at conditions when the negative end volume is strongly depleted in the isotope considered. Steeper concentration gradients have to be built up in the column in order to achieve the desired enrichment. These parameters then determine the location of the minimum Ltq-product on the f_p-scale and its numerical value. The strong influence of the depletion effect on enrichment costs becomes most pronounced when the isotope yield is varied (fig. 7).

Concerning the dimensions of the column cross-section, a larger hot wall radius allows shorter sampling length. The column scale factors Λ and τ_e remain unchanged and the column length L is nearly independent on hot wall radius R_H (fig. 10) when pressure factor f_p and separation gap width 2w are fixed, as long as the depletion effect is

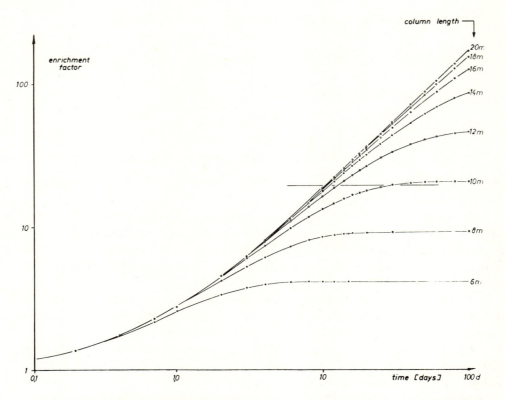

FIG. 6. Variation of enrichment factor with run time at the positive end of a one-stage thermal diffusion column closed at the positive end and open at the other (binary mixture, theoretical calculations).

negligible. The reduced collection length at a higher hot wall radius results in effective reduction of the run time. In the Ltq-product, however, the gain in run time is nearly compensated for by an increase in heating power consumption with the hot wall circumference. When R_H is increased, the depletion effect becomes important at lower pressures (fig. 8), and because power consumption is high the minimum value of the Ltq-term is comparatively large. Smaller hot wall radii and higher temperatures prove more opportune with respect to the costs although the run time may be larger with operating pressure at minimum Ltq-conditions.

When the separation gap width (2w) is enlarged, the column scale factors Λ and τ_e increase and softer concentration gradients along the column axis result. With the hot wall radius and pressure factors fixed, the column length must be increased (fig. 10). The run time, however, remains nearly unchanged because the effect of greater τ_e is canceled out by a reduced sampling length. Simultaneously the Ltq-product is nearly independent of the separation gap width 2w (fig. 8) because the effect of greater column length on heating power is overcompensated by smaller temperature gradients between the hot and cold wall. However, the depletion effect becomes important at lower pressure factors if

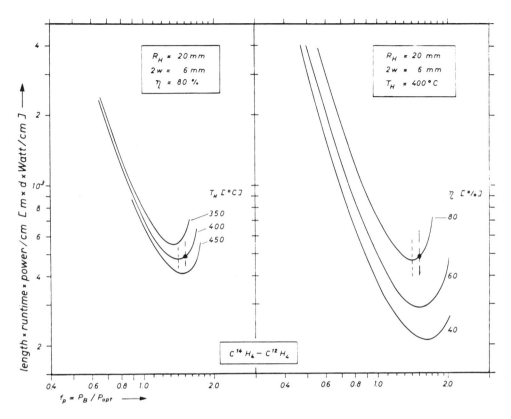

FIG. 7. Variation of the length-time-power-product with operation pressure factor f_p for different isotope yield η and hot wall temperature T_H (cold wall: 14°C, sample withdrawn: 1.8g CH_4 12-fold enriched).

2w is high, and the Ltq-values are relatively high at large separation gap widths in spite of low heating power.

Summing up the above discussion, we find costs to be low at small hot wall radii and small separation gap widths (i.e. high operating pressures), but when selecting the tube parameters it should be considered that operating pressures greatly exceeding one atmosphere are inconvenient for vacuum systems, demand increased accuracy in the constructional parts of the column, and result in higher run times at optimum length-time power-conditions.

The variations of the column transport coefficients with temperature are presented in figures 11-13. They can be approximately converted to other column dimensions on the basis of equation (3) if $2w = R_K - R_H \ll 0.5 (R_K + R_H)$. Note that $H \sim a \sim (M_1 - M_2)/(M_1 + M_2)$ (M : molecular weights).

The Kiel columns were designed before optimization was studied. A hot wall radius of 20 mm and a separation gap width of 6 mm were chosen (black circles in fig.'s 7-10). During operation, pressure is at 1.25 atm. With an isotope yield of 80% and a hot wall

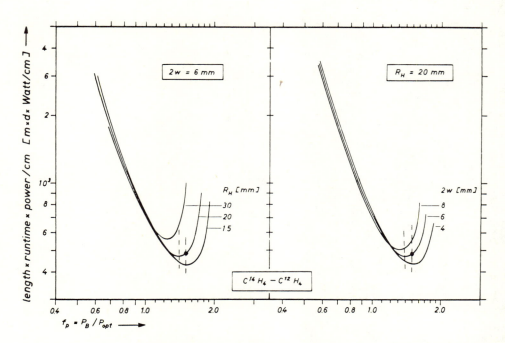

FIG. 8. Length-time-power-product vs. operation pressure factor f_p for different hot tube radii and separation gap widths 2w (cold tube: 14°C; hot tube: 400°C; total sample size: 26.6 g; withdrawn: 1.8g, × 12).

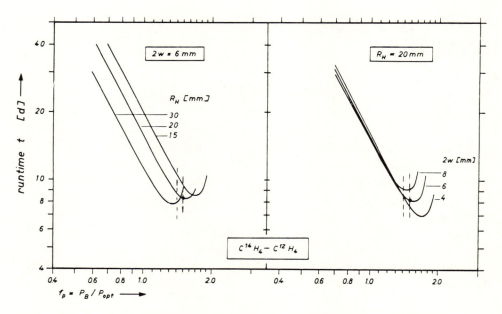

FIG. 9. Run time (at optimum ltq-product) vs. operation pressure factor f_p for different hot wall radii R_H and separation gap widths 2w (parameters as in fig. 7).

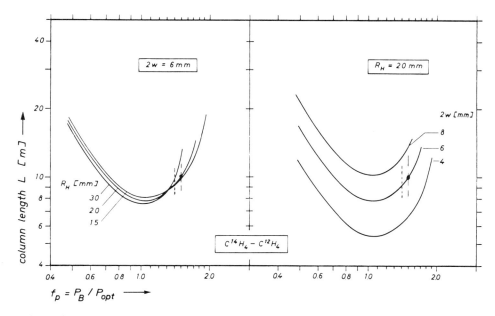

FIG. 10. Column length (at optimum ltq-product) vs. operation pressure factor f_p for different hot wall radii R_H and separation gap widths $2w$ (parameters as in fig. 7).

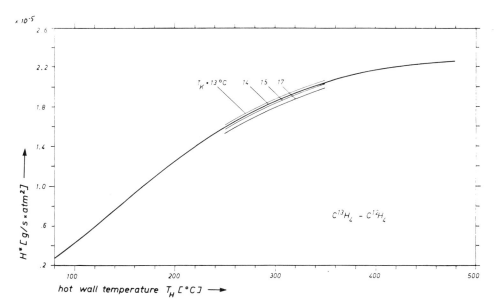

FIG. 11. Transport coefficient $H^* = H/p^2$ vs. hot wall temperature T_H for $^{13}CH_4 - ^{12}CH_4$ binary mixture (parameter: cold wall temperature T_K; separation gap width $2w = 6$ mm; hot wall radius $R_H = 20$ mm).

Approximate conversion to other column dimensions and methane isotope mixtures may be performed on the basis of $H^* \sim (2w)^3 \cdot R_H \cdot (M_1 - M_2)/(M_1 + M_2)$, $2w \ll R_H$, M: molecular weights.

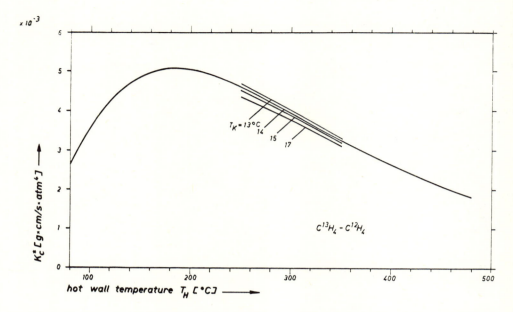

FIG. 12. Transport coefficient $K_c^* = K_c/p^4$ vs. hot wall temperature (parameters as in fig. 10). Approximate conversion to other column dimensions on the basis of $K_c^* \sim (2w)^7 \cdot R_H$, $(2w \ll R_H)$.

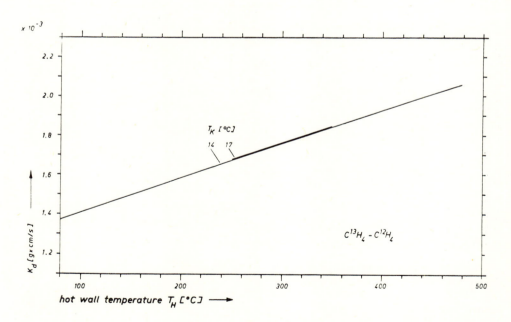

FIG. 13. Transport coefficient K_d vs. hot wall temperature T_H (parameters as in fig. 10). Approximate conversion to other column dimensions on the basis of $K_d \sim 2w \cdot R_H$, $(2w \ll R_H)$.

temperature of 400°C, the run time for producing 1.8g of methane, 12-fold enriched, is 8.5 days. The optimum column length is 9.24 m, the sampling section is 4 m, and the total sample gas amounts to 26.6 g.

With $\eta = 60\%$ the operation pressure is 1.34 atm, run time is 5 days, column length is 8.1 m, sampling length is 3.73 m and total gas amount is 34.8 g.

The Mechanical Design of the Thermal Diffusion Column

The theory of the thermal diffusion column, as outlined in the excellent papers of Furry, Jones, and Onsager in 1939 to 1946, is based on a few fundamental principles. This theory should succeed in accurately describing the enrichment process. Many of the discrepancies between theory and measurement reported in the literature appear to be caused by improper constructional concepts, inaccurate data of gas properties, which are often purely known even today, and insufficient theoretical approximations and numerical treatments. Concerning the mechanical design, it must be realized that the enrichment process in a thermal diffusion column appears to be extremely sensitive to temperature inhomogeneities on the walls of the separation gap and to the accuracy of the mechanical structure.

Maximum mechanical precision can be attained only with all-metal construction. In the Kiel columns (fig. 14) hot wall and cold wall tubes have been made from high precision tubes (wall thickness: 3 mm) of oxygen-free copper in order to minimize temperature inhomogeneities on the walls. Stainless steel tubes appear to give markedly lower column performance, probably because thermal conductivity is insufficient. The cold tube is suspended in the cooling water jacket tube and is kept under tension of about 900 kp by means of six stretching bolts at one end of the column. The cold tube is aligned by optical methods (laser beam for reference, Henseler 1973) by means of three adjusting screws placed at mid-height on the tube. Absolute deviation from a perfect cylinder is estimated to be less than 0.2 mm. The hot tube is centered in the cold tube by radially-arranged conical spacing bolts, three at every 400 mm of column length. The spacers are fixed in the cold wall. The hot wall is heated by a bundle of eight heating rods (diam. 8 mm, 200 watt max. each) which are guided in ceramic multiple hole tubes precisely fitted to the inner hot tube diameter (fig. 15). Henseler (1973) simplified the heating system by using a commercial heating rod (diam. 25 mm, 2000 watt max.) and placing it in a copper-made temperature equalization tube centered between heater and hot tube. The tube arrangement is filled with nitrogen. An additional heating element is placed at the upper end of the hot tube to balance heat losses through the hot tube suspension rod. Temperatures are measured by very thin thermocouples placed directly within the hot wall at different levels on the column.

The active length of a single separation tube is about 2.5 m. Four tubes are connected in series by thermal convection loops. The tubes are mounted on a bearing shelf, well protected against shock and vibrations from the floor.

The reproducibility of the enrichment factor is generally better than 1%, for steady-state operation and for transient operation, for ^{14}C isotopes and ^{13}C enrichment, at the

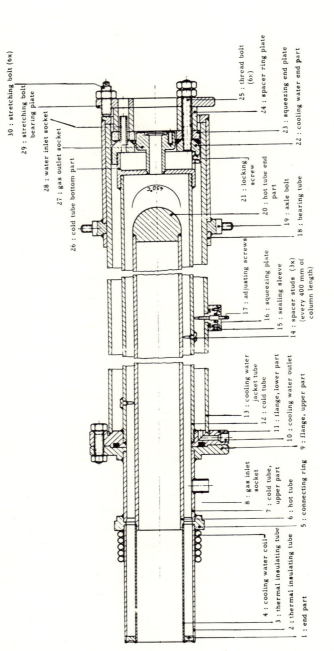

FIG. 14. Longitudinal cross section of Kiel thermal diffusion column.

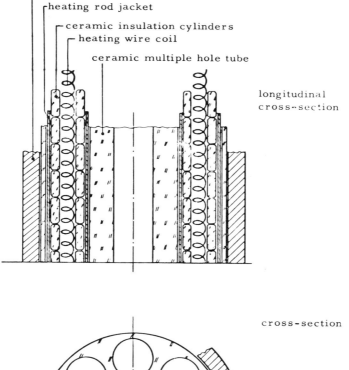
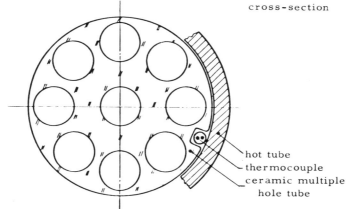

FIG. 15. Thermal diffusion column heating system.

ends of the column and for the average concentration in the sampling section. Maximum difference found between prediction and experiment was 1.5%. For illustration, the results of three runs are presented in figure 16.

Henseler (private communication) recently connected all tubes, their total length being about 10 m. The first results he obtained on stable carbon isotopes met the theoretical prediction as expected.

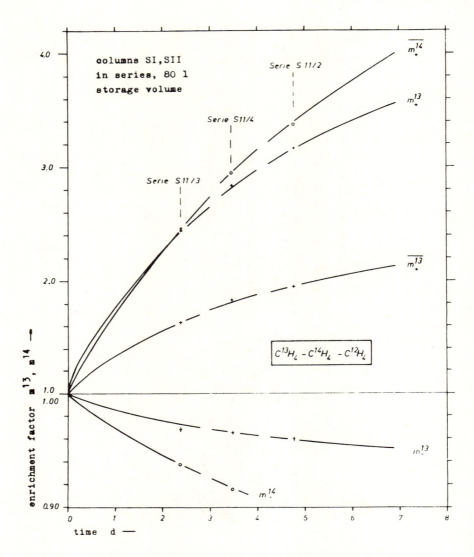

FIG. 16. Time variation of enrichment factors m of ^{13}C-methane and ^{14}C-methane as compared to theory (solid curves), at the positive column end (m_+), at the negative column end (m_-), and average over the total column length ($\bar{m}$). (column length: 4.9 m; reservoir at the negative column end: 80 l; column closed at the positive end; hot wall temperatures for the different runs are between 271°C and 298°C and operation pressures between 0.768 atm and 0.829 atm; cold wall temperature: 13°C).

REFERENCES

Erlenheuser, H.
    1971a Aufbau einer Thermodiffusionsanlage zur Anreicherung von Methan-Isotopen in Hinblick auf die Verwendung bei Altersbestimmungen nach der ^{14}C-methode, Dissertation. Univ. of Kiel.

1971b Predictable low enrichment of methane isotopes by Clusius-Dickel thermal-diffusion columns for use in radiocarbon dating technique. Z. Naturforsch. 26a:1365-1370.

Felber, H. and E. Pak
1973 Erweiterung der ^{14}C-Altersbestimmungsmethode durch quantitative Isotopenanreicherung im Trennrohr. Sitzungsberichte der Österreichischen Akademie der Wissenschaften, Mathem.-Naturw. Klasse. II (180) 8:10. Springer Verlag, Wien.

Furry, W. H., R. C. Jones, and L. Onsager
1939 On the theory of isotope separation by thermal diffusion. Phys. Rev. 55:1083.

Furry, W. H., and R. C. Jones
1946 Isotope separation by thermal diffusion: the cylindrical case. Phys. Rev. 69:459.

Geyh, M. A.
1965 Proportional counter equipment of sample dating with ages exceeding 60,000 B.P. without enrichment. Proc. 6 Int. Conf. Radiocarbon and Tritium Dating. Pullman, Washington (29).

Grootes, P. M., W. G. Mook, J. C. Vogel, A. E. deVries, A. Haring, and J. Kistemaker
1975 Enrichment of radiocarbon for dating samples up to 75,000 years. Z. Naturforsch. 30a:1-14.

Haring, A., A. E. deVries, and H. deVries
1958 Radiocarbon dating up to 70,000 years by isotope enrichment. Science 128:472.

Henseler, T.
1973 Aufbau einer Thermodiffusionsanlage zur Anreicherung von Methan-Isotopen und experimentelle Untersuchung des Trennverhaltens. Diplomarbeit. Univ. of Kiel.

Jones, R. C., and W. H. Furry
1946 The separation of isotopes by thermal diffusion. Rev. Mod. Phys. 18:151.

Kretner, R.
1973 Quantitative Anreicherung von $^{14}CH_4$ mit dem Clusius-Dickelschen Trennrohr zur Anwendung in der ^{14}C-Datierungs-methode. Doctoral thesis. Univ. of München.

Pak, E.
1970 Erweiterung der ^{14}C-Alterbestimmungsmethode durch Isotopenanreicherung im Trennrohr. Doctoral thesis. Univ. of Wien.

Tipton, C. R. jr., ed.
1960 Reactor Handbook. 2d ed. I: Materials. New York, Interscience Publ.

11

Possibility of Measurement of ^{14}C by Mass Spectrometer Techniques

H. W. Wilson

Carbon-14 is traditionally measured by radioactive counting methods. This is certainly the simplest technique although it is still not simple. It is inefficient in one sense because we must wait for a ^{14}C atom to decay before we know it exists. A considerable apparent increase in efficiency would result if one could measure the ^{14}C atoms directly. The difficulty is that the amount of ^{14}C in samples for carbon dating, only 1 in 10^{12} even when modern, drops by a factor of 126x if the age of the sample is 40,000 years and 422x if 50,000 years (based on $\tau_{1/2}$ = 5730 yr), i.e., ~ 1 in 10^{14} and 2 in 10^{15}, respectively. Even if direct measurement were possible one would hope to measure milligram samples in periods on the order of a few hours or less.

There are various ways to attempt measurement of this ratio (other than by radioactivity). One obvious technique which has been considered in recent years is mass spectrometry. I have thought about it on and off for quite a long time, particularly after being involved in high abundance sensitivity mass spectrometry at the AWRE, Aldermaston. Abundance sensitivity denotes the ability to measure a very small peak near a very large one. Of various isotopic ratio measurement techniques, mass spectrometry is the most promising.

A typical, single-stage, mass spectrometer of 30 cm magnet radius has an abundance sensitivity of 1 in 10^4 to 1 in 10^5 at mass 180, a convenient mass at which to measure, since Ta has two isotopes (^{180}Ta and ^{181}Ta) with an abundance ratio of 1 to 10^4. Measurement is limited because the large ^{181}Ta peak has a tail on each side. The tail on the lower mass side is higher than that on the high mass side. These tails are due to scattering of ions by the residual gas in the mass spectrometer, to scattering from the walls and slits of the mass spectrometer, to collisions in the source region, and to aberrations caused by the angular spread of the beam and the energy spread of the ions. Scattering of ions is usually the major cause and can be reduced by use of a better vacuum. Sacttering from the walls can perhaps be reduced by careful design, although I am not aware of attention paid to this. Aberrations are not usually significant in spectrometers of reasonable size and good design in the mass range required for carbon dating.

A good way of improving abundance sensitivity is to build a two stage or "tandem" mass spectrometer as shown in figure 1 (Inghram and Hess 1954, White and Collins 1954, Wilson et al. 1961). The small beam as it passes through the slit between the magnetic stages (fig. 2a) is accompanied by ions scattered from the large beam. Much of the scattered ion beam is resolved out in the second stage. In principle one might expect 1 in 10^4 abundance sensitivity to be increased to $(1 \text{ in } 10^4)^2$ or 1 in 10^8. In practice it depends on design and would not be quite so good as that. In general the abundance sensitivity is better on the high mass side. This is convenient because when measuring ^{14}C against ^{12}C, the small peak is on the high mass side.

The large MSY spectrometer (fig. 2b) at Aldermaston (radius 63.5 cm) gives abundance sensitivities in the uranium region at mass 239 due to ^{238}U of 1 part in 7.7×10^5 (single stage), 1 in 2.8×10^7 (double stage), and $\sim 10^9$ (triple stage). The triple stage is an electrostatic stage energy filter which improves abundance sensitivity because scattered ions lose energy. A cheaper solution uses a suppressor electrode just in front of the detector set at a voltage slightly lower than those of the ions. (Ridley et al. 1964, Freeman et al. 1967).

Although the abundance sensitivities quoted above appear well below those required for carbon dating, it must be remembered that the mass separation is very much greater and, in fact, rough calculations and extrapolation show that it should be possible to attain sufficient abundance sensitivity to measure the ratios in the mass regions we have considered. Indeed, the experiments of Schnitzer, et al. (1974) have confirmed that the required abundance sensitivity can be attained.

Let us now do some arithmetic. If, using a multiplier type collector, we could measure 1 ion per second of a beam containing the ^{14}C atoms and there were no background, 10,000 seconds or 2.8 hours would suffice to measure the ratio to 1%. If the sample were modern, this would involve a total beam current from the source of 10^{12} ions/second, or $\sim 10^{-7}$ amps, larger than in a normal mass spectrometer. A 40,000 year old sample with a $^{14}C/^{12}C$ ratio of 1 in 10^{14} requires a current of $10\mu A$, which definitely requires a different type of source. The duoplasmatron source used in magnetic isotope separators would be the most appropriate. This can give an additional benefit in that source efficiencies can be as high as 50%. We shall assume 10% source efficiency in the following discussion.

It is clear that, in principle at least, the times of measurement are much shorter than in the counting method. Also the size of sample required is smaller. For example, if we assume a 10% efficient source and no further losses in the spectrometer, the number of atoms of ^{14}C consumed in the source would be 10^5 and the number of ^{12}C atoms in a 40,000 year old sample would be $\sim 10^{19}$ (i.e., about 2×10^{-4} gm), much less than for counting.

The presence of background, whether continuous or consisting of a peak at the ^{14}C position, increases these figures considerably. A background of ten times the sample beam current would increase the counting time (or beam current required) at least thirty times even in the favorable circumstance of an absolutely steady and well determined background. This is optimistic, since the background is likely to be much more variable than when using counting techniques. We shall return later to the question of background.

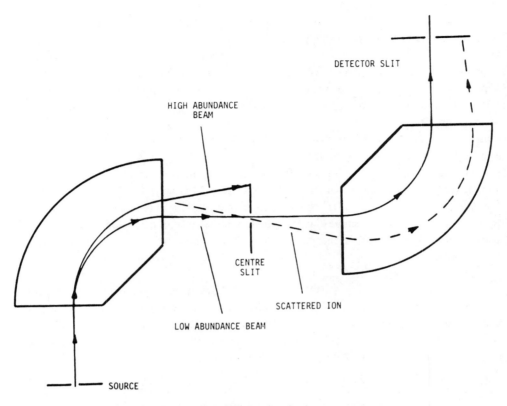

FIG. 1. Principle of the two stage (tandem) mass spectrometer.

It seems likely that the required abundance sensitivity and beam current could be achieved. The next question concerns the compound of carbon one should use. An obvious choice is CO_2 which provides, in a mass spectrometer, CO_2^+ as the major peak, but in the duoplasmatron source the major peaks are likely to be those corresponding to CO^+. In practice, because of the isotope composition of oxygen, one would measure $^{14}C\ ^{18}O^+$ (mass 32) against $^{14}C\ ^{17}O^+ + ^{13}C\ ^{18}O^+$ (mass 31). This means, however, that the main beam $^{12}C\ ^{16}O^+$ would be increased by the ratio of $^{16}O : ^{18}O$ (i.e., by a factor of 490 times). Thus the major ion beam corresponding to 1 ion per second of 40,000 year old carbon would be approximately 5mA and, with background present, would be considerably higher still, leading to unacceptable values. The problem could be overcome by the use of oxygen enriched in ^{18}O, if at some cost. The cost of ^{18}O of adequate enrichment (it does not have to be particularly high) is $230 – $430 per liter, depending on the enrichment (U.K. prices). This would lead to an additional cost of $25 – $50 for a 50 mgm carbon sample.

Another possibility would be to measure ions from CD_4. Deuterium is considerably cheaper than ^{18}O, and the cost of the deuterium for a 50 mgm C sample would be between 10 and 37 cents, depending on the quantity of deuterium ordered (at U.K. prices). In a mass spectrometer, the major peak would in fact be CD_3^+, but in a plasma source further dissociation of the molecule would be likely; in fact C^+ itself might be the main peak.

The isotopic problem could be overcome by converting the carbon to CF_4, since F is monoisotopic to begin with. It is not certain what the major peaks would be, but with an electron impact source the major peaks would probably be $^{12,\ 13,\ 14}CF_3$ (i.e., masses 69, 70, 71) and one would measure $^{14}CF_3$ relative to $^{13}CF_3$. At these higher masses the background tends to be less but, on the other hand, the abundance sensitivity for adjacent masses decreases as the mass goes up and it is not certain that it would be adequate in this region. It is very likely that the ions from a duoplasmatron source would have fewer F atoms, probably CF^+ or even C^+, and one would therefore operate at lower masses.

The most serious problem of the spectrometer approach is background. Figure 3 shows the residual background, in ions/second on a logarithmic scale of the well-pumped Aldermaston MSG UHV mass spectrometer. For example, background at $^{14}C\ ^{18}O^+$ is over 1000 ions/second, at $^{14}C^+ \sim 10^4$ ions/second, and at $^{14}C\ D^+$ almost 10^5 ions/second. These are all considerably higher than in the example quoted earlier. (It is worth pointing out that, if CF^+ turns out to be the major ion from CF_4, then the background at $^{14}CF^+$ is only 3 ions/second.) To make the situation worse, a spectrometer fitted with a duoplasmatron source seems bound to have a much higher pressure, since the source itself operates at 10^{-2} to 10^{-1} torr while the mass spectrometer source of the AWRE MSG is 10^{-9} torr (measured at a point below a gate valve directly below the source and above a nitrogen trap). Naturally one would take steps to reduce the pressure in the spectrometer as much as possible, but the count rates at the detector are bound to be much higher unless some ingenious method of reducing background can be devised.

One possibility is to resolve out the background peaks. If these are mainly due to hydrocarbons (as is often the case with mass spectrometers) the resolving power required is in the region of 1000, which is within the capabilities of the tandem mass spectrometers mentioned earlier. The contributions from the tails of the hydrocarbon peaks would be higher, of course, than at one mass unit removed, but calculation suggests these might still be sufficiently small. The large spectrometers would score here, but some other possible background peaks would be much harder to resolve. For example, if the major ion is C^+, then background peaks at mass 14 could include $^{12}CH_2^+$, $^{12}CD^+$, and ^{14}N. The corresponding resolving powers required are 1130, 1290, and 8.8×10^4. The last is impossibly high for the kind of spectrometer considered.

There is room for more ingenious solutions to the background problem, perhaps involving techniques from nuclear physics, but there is no doubt that the problem using conventional mass spectrometry is extremely difficult. Before leaving this topic, it is worth mentioning that Schnitzer et al. (1975) have sought to reduce background by use of a hollow cathode source (with reduced input power), purification of the sample gases from hydrocarbons, and background reduction by photo detachment of wanted ionic species.

The double and triple stage spectrometers discussed earlier are large and expensive and would require considerable development for this particular purpose. I believe they

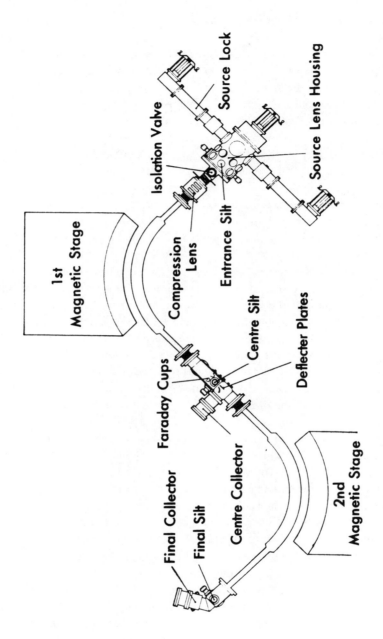

FIG. 2(a). Diagram of MSY two stage mass spectrometer. A third electrostatic stage was added later.

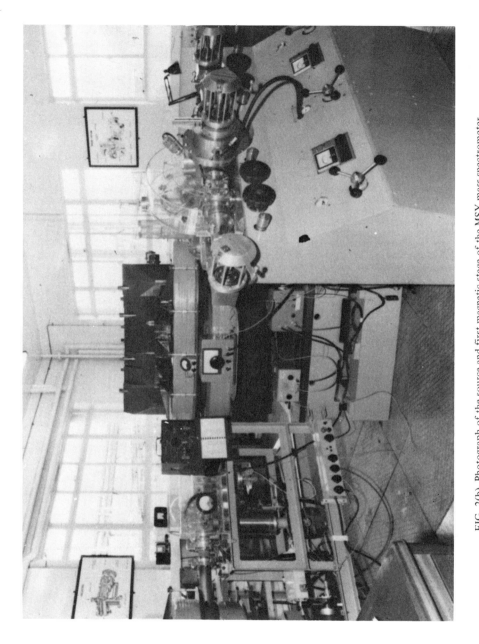

FIG. 2(b). Photograph of the source and first magnetic stage of the MSY mass spectrometer.

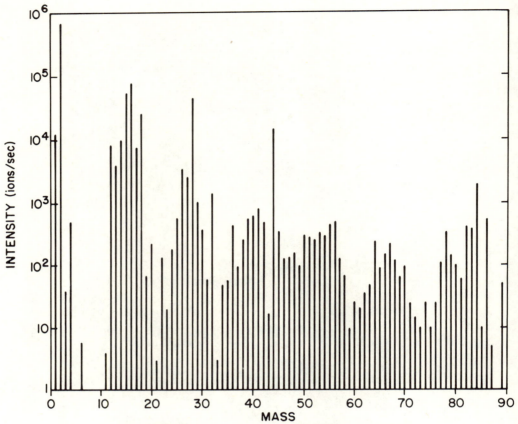
FIG. 3. Background spectrum from the AWRE UHV mass spectrometer.

could be smaller, although size does help a good deal — the dispersion of the peaks is greater and the tails of the peaks contribute correspondingly less to adjacent masses. Nevertheless it is hard to imagine building such an instrument for under $200,000 – $300,000.

A number of people have suggested the possibility of enrichment of the sample by thermal diffusion or by laser action. In theory the latter in a single step could enhance the ^{14}C count rate by a factor of 10^3 or even more. In practice, use of a frequency-doubled tunable-dye laser using rhodamine 6G dye-illuminating carbon in the form of formaldehyde has given enrichment factors of the order of 80, as shown by Moore et al. (Clark et al. 1975), and higher enrichments seem very possible. With an enrichment factor of 126X, a 40,000 year sample would have the activity of a modern sample. It might be asked why, if this is done, the sample is not used for counting. In the mass spectrometry method the amount required for analysis is much smaller and the ease of producing 10 mgm C which has been enriched in ^{14}C is much greater than that of producing 1 gm of C for counting.

Calculations suggest that, using a 1 mW laser, 1 mgm of formaldehyde could be enriched in ^{14}C by a factor of 100 in half-a-minute (Hedges, personal communication

1976). More powerful lasers are possible and a preliminary enrichment by this method could make a considerable difference to the practicability of the method. Also, even if the theoretical enrichment values could not be approached, a two-stage process might be practicable. Other laser techniques using different compounds have been tried, but that described above seems the most promising.

Another interesting possibility considered by M. Dunn (personal communication), and also mentioned during this conference, is optical measurement of ^{14}C concentration from the isotope shift. It is too early to assess its feasibility, although it appears very difficult.

To summarize, the mass spectrometer method gives promise of measuring old carbon samples in very much shorter times and with fewer samples than the counting method. The background problems will be very severe and the equipment will be much more expensive than that associated with the counting method. Preliminary laser enrichment may make the difference between success and failure of the technique.

ACKNOWLEDGMENTS

I had useful discussions with a number of people and would particularly mention Mr. M. Dunn of St. Andrews University and Dr. D. D. Harkness of the Radiocarbon Laboratory at the Scottish Universities Research and Reactor Center. Dr. N. R. Daly of AWRE, Aldermaston, kindly provided the background spectrum (figure 2).

REFERENCES

Clark, J. H., Y. Haas, P. L. Houston, and C. B. Moore
    1975   Chem. Phys. Letters. 35:82.

Freeman, N. J., N. R. Daly, and R. E. Powell
    1967   Rev. Sci. Inst. 38:945.

Inghram, M. G., and D. C. Hess
    1954   Nuclear Sci. Series. Inghram and Haydon, eds. Report 14:24.

Ridley, R. G., R. Munro, W. A. P. Young, R. Hayes, R. W. D. Hardy, and H. W. Wilson
    1964   Adv. in Mass Spectrometry. 3:553.

Schnitzer, R., W. H. Aberth, H. L. Brown, and M. Anbar
    1974   Proc. ASMS Ann. Conf. Mass Spectrometry and Allied Topics. Philadelphia. 64.

Schnitzer, R., W. H. Aberth, and M. Anbar
    1975   Proc. ASMS Ann. Conf. on Mass Spectrometry and Allied Topics. Houston. 479.

White, F. A. and T. L. Collins
    1954   App. Spectroscopy. 8:169.

Wilson, H. W., R. Munro, R. W. D. Hardy, and N. R. Daly
    1961   N. R. Nuc. Instr. and Meth. 12:269.

12

Specific Problems with Liquid Scintillation Counting of Small Benzene Volumes and Background Count Rate Estimation

Herbert Haas

Because the S.M.U. Radiocarbon Laboratory has, since its inception, dealt with a steady influx of small samples, we have needed to optimize our methods of benzene synthesis to a degree where high benzene yields can be achieved predictably for every conversion. Similarly, the counting procedures have had to be adapted to obtain narrow error fields with these small samples. Furthermore, it is in the nature of many sample series received here that they pertain to exploratory archaeologic work and therefore range over the whole age spectrum datable with conventional radiocarbon equipment. We must be prepared to process in sequence samples which will eventually be dated as modern or as very old (i.e., around 40,000 B.P.). This recurring possibility requires a system totally devoid of any memory effects.

Such problems are not new, of course, and many articles are available in the literature on possible solutions. Other investigators using the benzene dating method may be interested in our methods of detecting existing problems with statistical evaluation of accumulated data and notes, and in our solutions to these problems.

In attempting the catalytic trimerization of acetylene to benzene, we made a major effort to find a catalyst which would give, with a suitable pretreatment, a 95% or better yield with a fast reaction rate and a long useful life. We tried some of the procedures described in the literature (Polach 1972, Coleman 1972) but did not obtain satisfactory results. The most promising results have been obtained with the Noakes and Harshaw catalysts, which can be reused many times but seldom more than fifty times. At the end of the "useful life" of a catalyst charge, and this may be as early as after twenty uses, the average yield drops by about 2%.

In our early experiments we heated the catalyst to 300° *in vacuo* between conversions. 93% yields were frequently reached, but after twenty uses the reaction rate declined to the point where 75 g of catalyst pellets would barely convert 2 liters of acetylene in four hours. It was suspected that the organic by-product produced in the catalyst and described by Coleman (1972) might be cumulative for each use of the catalyst. To test this hypothesis, 60 g of catalyst which had been used nineteen times were combusted with oxygen. 6.1 liters of CO_2 were collected from this operation, which represents approximately 70% of the cumulative carbon loss that occurred in the catalyst. The CO_2 from this catalyst combustion was synthesized to benzene and then counted for activity.

From each of the nineteen converted samples, the ^{14}C activity and the amount of unrecovered sample was entered into the calculation of a theoretical activity which is 5.808 cpm per g carbon. The measured activity from the combusted catalyst was 5.739 ± .045 cpm per g carbon. This result strongly supports the assumption that the cumulatively retained by-product cannot be extracted from the catalyst by vacuum baking at 300°C. A statistical study (Haas et al. 1976), undertaken later on memory effects from the catalyst alone, has shown a correlation of .00704 from the previous sample activity to the net activity of a background sample passed through the same catalyst in the following conversion. The memory effect disappeared as soon as a different regeneration process using a hot oxygen-containing gas mixture was used. A similar study with modern standard activities shows the raise in activity measured once the memory effect from previously converted low activity samples was removed (fig. 1).

During the continuing search for a satisfactory reconditioning procedure, we brought the oxidation state of the vanadium catalyst to its highest state, the yellow V_2O_5, by baking in pure oxygen, and also to a reduced state, dark gray, with hot hydrogen. While the observed results were usable for the first few reactions, the reaction rate declined soon to an annoyingly low value and yields became highly variable and unpredictable.

In the course of further research we tried mixed gases for the bakeout process, that is, a nitrogen-oxygen mixture to lower the fugacity of oxygen, and an argon-oxygen mixture to prevent vanadium nitride from forming. Each of these steps resulted in a slight improvement in the catalyst performance, but the problem with occasional low yields remained. A discussion with chemists alerted us to the fact that hydrogen-vanadium bonds are an important part of the catalytic reaction in cycling acetylene to benzene. High temperature treatment of the catalyst with argon, nitrogen, and oxygen is likely to reduce the hydrogen content in the catalyst. In the initial stage of catalytic reactions with acetylene, the depleted hydrogen is replaced by breaking the C-H bonds from acetylene and leaving carbon as residue. The result is a low yield and carbon combustion during the subsequent bakeout-process with presence of oxygen.

A new and final step in the conditioning of the catalyst was added to the procedure: exposing the catalyst to an argon-hydrogen mixture at 200°C for approximately 20 minutes, then cooling the catalyst to room temperature in this gas mixture before vacuum pumping. 200°C was chosen because at this temperature the reduction of vanadium does not proceed. With the introduction of this treatment the yields varied between 94% and 98% and, most important, no more low yields were observed.

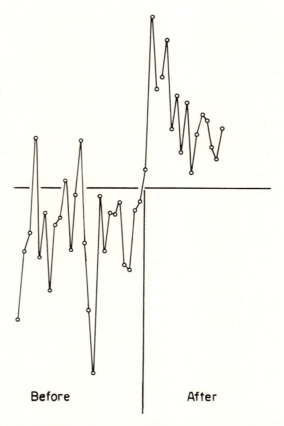

FIG. 1. Activity of modern standard benzenes synthesized before and after change in catalyst cleaning procedure

The complete regeneration procedure presently adopted at our laboratory is the following:

1. Baking temperature 440°C, minimum of 4 hours in about 50%/50% argon-oxygen mixture at about atmospheric pressure. Minimum of 6 pumping and refilling cycles.
2. Cooling of vessel to 200°C.
3. Filling with about 50%/50% argon-hydrogen left for ½ hour at 200°C; pumping and refilling with argon-hydrogen.
4. Cooling to room temperature.
5. Maximum of 1 minute evacuation to vacuum.
6. Immediately before admission of acetylene, short evacuation to remove residual pressure.

An additional advantage of the hydrogen treatment is that any free oxygen which may remain in the catalyst even after prolonged high-vacuum pumping is thoroughly removed and can be extracted together with the benzene. Oxygen dissolved in a scintillation counting solution can lead to quenching. In order to extract oxygen and traces of unconverted acetylene from the sample benzene, we have developed the benzene extraction system shown in figure 2. Acetylene is admitted to the catalyst vessel for conversion to benzene and left overnight. On the following day the benzene is collected in a freeze bulb of 100 ml size, into which a side arm has been fitted which leads either to the vacuum pump or to a removable extraction vial which is actually a threaded centrifuge tube screwed vacuum-tight to the system (fig. 2).

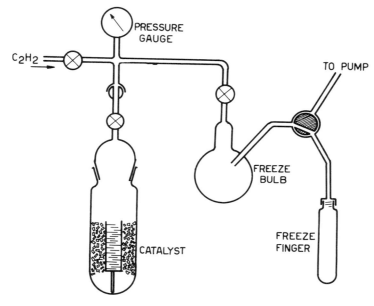

FIG. 2.

The freeze bulb is cooled with liquid nitrogen, the catalyst is heated to 120°C, and the extraction process is allowed to proceed for three hours very little is collected after the first hour. In the next step the frozen bulb is brought up to dry ice temperature while the vacuum pump is connected through the side arm to the bulb for ten to fifteen minutes. Residual pressure of acetylene is removed in this way, as well as possible traces of oxygen and hydrogen.

The final transfer brings the benzene from the bulb into the removable extraction vial. The vial is cooled with dry ice. The benzene transfers rapidly into it since all residual gas pressure has previously been pumped off.

After the transfer the extraction vial is removed. The vacuum is broken at the threaded connection, which is protected during this step from direct contact with the atmosphere by a stream of argon gas. The vial is immediately closed with a screw cap. The benzene in the vial is stored in a freezer for about a week prior to counting.

COUNTING PROCEDURES

We did not expect to synthesize more than 3.5 cc of benzene from most of the small samples we processed, and quite often much less was available for counting. The counter we had selected, an Intertechnique LS 20, offered low background characteristics without cooling of the photomultiplier assembly. This means that no "antifreeze solution" like toluene needed to be added to the counting benzene. The largest vial we expected to need in the routine dating work was of 3 cc usable volume. The characteristics we sought in a vial were low background, good reproducibility of counting efficiency and background levels, and availability of smaller vials of similar design.

We were concerned about reducing the counting background level, originating in most liquid scintillation counters from four different and independent sources:

1. Cosmic radiation striking the counter tubes, the vial, and the benzene inside.
2. Cross talk between counter tubes.
3. Nuclear decays within the vial material or in the benzene which are not due to sample ^{14}C.
4. Electronic noise.

Outside or cosmic radiation, the largest contributor to background, can be reduced only by installing more shielding material around the whole counter, or better, near its sensitive elements, the counter tubes and the vial containing the sample benzene.

Cross talk between counter tubes, the second largest contributor, is reduced by operating the tubes at the low voltage end of the plateau. Further reduction is achieved by choosing a suitable counting vial design. We tested round counting vials made of quartz and various other materials and a square quartz vial. The reference level for cross talk was established by counting on the empty sample chamber, that is, with no vial inserted. The square quartz vial showed the lowest cross talk level; apparently it provides for the weakest optical link between the two opposite counter tubes. In contrast, a round vial acts like a wide angle focusing lens placed between the photomultiplier tubes, facilitating the transmission of light impulses from one tube to the other and thus enhancing the cross talk effect. We filled round vials with various solutions and then measured their approximate optical angle. The background count rate for 3 cc fillings was also determined for each case. A positive correlation between the optical angle and the background count rate was suggested from the experimental data shown in the following table.
The lowest background was achieved with the square cell set in its lead frame. This background was later lowered to 2.8 by masking the margins of the counter tube faces.

The square vial is a commercially available spectrophotometric cell purchased from Precision Cells, Inc., Hicksville, New Jersey, with a tapered teflon stopper and a lightpath of 10 mm. For small samples we have used a smaller, rectangular vial with a lightpath of 5 mm.

There are further reasons for selecting the square vial design. Such vials are manufactured from flat quartz plates of equal thickness on all four sides. Optical properties and lightpaths through the vial are clearly defined and reproducible. This is not the case with

Type of Vial	Filling	Refr. Ind.	Optical Angle	
round clear	benzene	1.5011	127°	6.04 ± .05
round clear	toluene	1.4961	127°	6.16 ± .09
round clear	H_2O dist.	1.332	100°	5.76 ± .09
round clear	air	1.000	43°	4.95 ±
round frosted	H_2O dist.	1.332	–	8.69 ± .08
round clear	benzene + scint.	1.5 approx.	135°	4.07 to 4.25
square clear	benzene + scint.	1.5 approx.	57°	3.77

a round vial made from fused quartz. The uneven wall thicknesses cause changes in optical property as the vial is rotated. Variations up to 1% in sample count rates have been reported by Garfinkel et al. (1965).

Counters with automatic changers use round vials which are inserted into the counting chamber in random position. Small and unpredictable changes in count rates can occur as the sample is returned several times to the counting position.

In our LS 20 counter the square vial rests inside a tightly fitting lead frame which is in turn mounted into the manually operated sample drawer of the counter (fig. 3). This design has the advantage of forming an additional radiation shield directly surrounding the vial. Possible radiation coming from the hardware of the sample drawer, mostly brass and steel, can be greatly reduced. The inside surfaces of the lead frame are lined with aluminized mylar as a reflective surface. This material was found most suitable for increasing the counting efficiency.

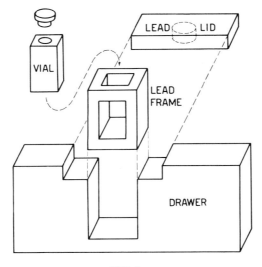

FIG. 3.

Filling the vials with exactly the same amount of benzene for each counting run is important. An experimental series of background counts with over- and underfilled vials has demonstrated that for every 10 mg sample benzene *over* the norm quantity, the background decreases by ¼% or .007 counts per minute (fig. 4). This is nearly half of the background standard deviation entered into the sample age calculations. The negative correlation is explained with the observed fact that the benzene-scintillator mixture in the vial is light-absorbing. This is shown in the preceding table where pure benzene has a higher background level than benzene containing .91% scintillator.

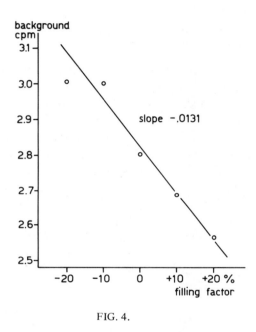

FIG. 4.

Another experimental series with modern standard benzene solutions demonstrates how net count rates are also affected by changes in the fill factor of the vials (fig. 5).

For every 10 mg added modern benzene, the count rate per gram carbon decreases by 1%-10% or approximately .009 counts per minute. Again this represents a significant part of the standard deviation of the "modern" calibration. The negative correlation in this case is caused by the filling level's close proximity to the upper margin of the lead frame window. Added sample may not be seen by the counter with full efficiency.

PREPARATION OF SAMPLE BENZENE FOR COUNTING

After the sample benzene has been extracted from the catalyst it is stored for about a week in the freezer. During this time it is thawed and refrozen several times to expel dissolved gases such as traces of unconverted acetylene or residuals from the catalyst regeneration process. Further preparation varies depending on the sample size.

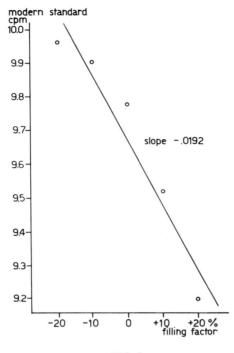

FIG. 5.

Benzene samples which are larger than 3 cc are transferred into a screw-cap glass vial with a teflon liner in the cap. The transfer is performed with an all-glass syringe with a metal needle. Transfer losses are low in this way and air contact is minimal. The amount of benzene in the screw-cap vial is determined by weight. The necessary quantity of scintillator powder, .91% per weight, is added to the benzene in the vial. The benzene is then stored in the vial until the counting procedure. With a calibrated syringe the sample is transferred to the square counting vial.

Benzene samples which are smaller than 3 cc can either be counted in undiluted form in the 1.5 cc vial or can be diluted to bring the volume to 3 cc. Very small samples of less than 1.5 cc must always be diluted. In order to keep transfer losses at a minimum, the sample is transferred directly with a syringe into the counting vial With practice, one can keep the benzene from touching the plunger, in which case the overall transfer loss can be kept at less than 50 mg benzene. The dilution benzene (Baker No. 9155 Spectrophotometric grade) and the scintillator powder are added. Shaking of the vial is not recommended because of possible capillary leaks between vial and stopper. Mixing can be achieved easily by rubbing the vial with nylon gloves. The induced electrical charges cause vigorous convection inside the vial.

For this reason the vial should not be touched with nylon gloves during insertion into the counter or be brought into friction with nonconducting sample holders. The induced electrical charges cause an increased count rate which declines within ½ to 3 hours, depending on the magnitude of the effect and on air humidity. In our laboratory the counting vials are touched at this stage of procedure only with long, metal tweezers.

Mixing and stirring the solution leaves the scintillator molecules in an excited state, which produces an artificially high count rate in the counter if counting is commenced immediately after sample preparation. This effect dies out within 24 hours. It is therefore necessary to store the sample for a full day in a dark, dry cabinet kept at the temperature prevailing inside the counter. This room-temperature storage of a sample prior to counting is also important when the sample has been stored in a refrigerated state for a prolonged period.

Cold storage over several weeks and subsequent warmup in the counter will sometimes cause a high initial count rate which we interpret as a thermoluminescence effect. We have observed that the intensity of the initial count rate increases with the length of cool storage. A sample stored for several months will display an elevated count rate for several days at room temperature. This effect is minimal and may have gone unnoticed in some radiocarbon laboratories. Sensitive statistical tests performed by the Department of Statistics of Southern Methodist University allowed detection of this and other effects. More details on the statistical models used in these investigations are given by Meeks (1976).

A final note on the use of spectrophotometric grade benzene for sample dilution is in order. We have run sensitive chemical tests with gas chromatography and nuclear magnetic resonance equipment, and have not been able to detect any organic impurities, such as acetone, which would have a quenching effect on the counting properties. Dissolved gases accumulate in this benzene, especially when the bottle has been opened many times. In order to purify the benzene, one can alternately freeze and thaw the benzene in a screw-cap vial until gas bubbles in the benzene ice no longer form and the thawed benzene freezes into a clear, bubble-free ice showing only tension cracks. Flushing of the vial with an argon gas stream after each thawing is necessary. The resulting loss of spectrophotometric benzene by evaporation is acceptable. This is not so with the small quantities of sample benzene in which a probable benzene contaminant is unconverted acetylene or traces of gases used in the catalyst reconditioning process. For this reason, the loss-free method of vacuum pumping on the frozen benzene was introduced.

ACKNOWLEDGMENTS

I wish to thank Vance Haynes for his many suggestions for improving our laboratory procedures. Also, I wish to thank Steve Meeks, Nancy Neubert, Cathy Campbell, and the many assistants in our laboratory for their valuable contributions.

REFERENCES

Coleman, D. D. et al.
    1972   Proc. 8th Int. Conf. Radiocarbon Dating. New Zealand. 158-166.
Garfinkel, S. B. et al.
    1965   Int. J. Appl. Radiation and Isotopes. 16:27-33.

Haas, H. et al.
    1976   Report on Preliminary Investigation into Adequacy of ^{14}C Counting Procedures. Report to Inst. for Study of Earth and Man. Southern Methodist Univ.

Meeks, S. L.
    1976   Evaluation of Total Variability in Radiocarbon Counting. Proc. 9th Int. Radiocarbon Conf. Univ. of Calif. Los Angeles and San Diego.

Polach, H. et al.
    1972   Proc. 8th Int. Conf. Radiocarbon Dating. New Zealand. 145-157.

13

An Assessment of Laboratory Errors in Liquid Scintillation Methods of ^{14}C Dating

R. L. Otlet

INTRODUCTION

The proper evaluation of the ± error term associated with a ^{14}C measurement is recognized as a vital part of the dating determination. Apart from archaeological or geologic implications, it is important that the significance of the error terms quoted should be consistent when results from different laboratories are compared, especially if they are to be combined mathematically, as in the calculation of a weighted mean.

Examination of recent issues of *Radiocarbon* raises doubt as to whether consistency of meaning in error terms can be relied upon in published dates. Sometimes it is stated that "counting statistics" of the basic parameters (counting rates of the sample, background, and modern calibration standard) are all that have been considered, but derivation of the overall error term quoted is seldom explicitly described.

The procedure by which to assess the error term is necessarily the business of those in the measuring laboratory. It appears to be the view in many laboratories employing gas counting with relatively simple chemical preparation processes that the errors associated with counting are all that need be considered. Whatever assessment procedure is used, it is asserted that the estimate should be consistent with the variability observed in practice for the results of replicate samples, that is repeat check measurements with samples undergoing the full laboratory process. It is the purpose of this paper to present an assessment procedure applicable to replicate sample reproducibility which is specifically oriented toward a practical liquid scintillation counting process.

Practical Considerations

In exception to the generalization made above, Callow (1965) set out a full error term derivation adopted by the National Physical Laboratory staff for its gas counting system. A statistical combination of all parameter variances was made. This is basically the procedure followed here. In addition the question is raised of whether the errors associated with the preparation of replicate scintillation counter samples are more significant than is frequently assumed, in comparison with the errors normally associated with counting statistics alone.

The principal difficulty in trying to establish possible sample-to-sample differences is that it can be done only from observations of counting experiments. In theory, real differences between replicate samples should be demonstrated from a comparison of the gross counts accumulated from each of a number of such samples counted sequentially, many times around, over a long period. With this procedure the uncertainty in the counting of each sample eventually becomes relatively small. Any variations in count-rate due to instrumental drifts should tend to cancel one another out since, in the long term, all samples are similarly affected. The logistics of such an experiment largely preclude its use as a practical proposition for a working service laboratory. This type of continuous monitoring is given incidentally to the groups of "background" and "modern" samples used in routine batch counting sequence in this laboratory (Harwell). An examination has been made of two years of counting data to determine what sample-to-sample differences may occur in these types.

The batch counting sequence used should be described in detail. It employs the kind of sample-train system described by Polach (1969) which allows quasi-simultaneous counting of a complete batch of sample vials and makes optimum usage of the sample exchanging automatics generally found in liquid scintillation counting instruments. A typical batch includes one set of background samples, one set of modern samples, and a number of samples to be dated (see fig. 1).

The instrument controls are preset to achieve either 100 minutes counting time, or 2000 total counts, if that occurs earlier. The vial is taken out of the counting chamber and the next one moved into its place. This continues until the last sample in the train has been counted; the automatics then move on to begin the first sample again for twenty cycles over a period of about the same number of days for a typical batch.

At the end of the full period, the counting statistics of each sample are related either to 40,000 total counts or to 2000 minutes total count time from each individual unknown sample. The statistics are related to factors of 4 to 5 times more than this for the "background" and "modern" values, as determined from the means of the representative samples of each group included in the batch. Over a period of one to two years of repeated counting periods, real sample-to-sample differences might be expected to emerge unobscured by counting statistics or instrumental drifts.

Figure 2 shows a histogram summarizing the results of eighteen separate background samples derived from long term data from two instruments. The experimental values fall short of the ideal requirements in that not all the samples could be counted simultaneously in the period. Only a small group is included at one time, but new samples have been introduced periodically, with others being removed and interchanged between the two instruments. This has given long, overlapping periods of about a year for any one

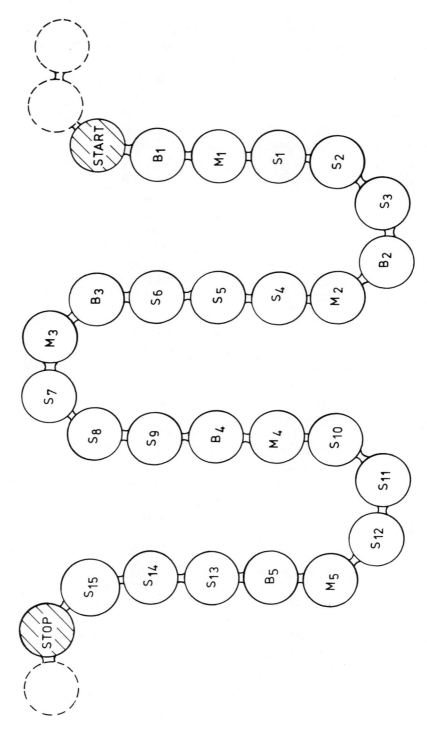

FIG. 1. Schematic make up of a typical sample train for batch counting. S1-15 represent samples to be dated, M1-5 the modern calibration standards and B1-5 the background samples.

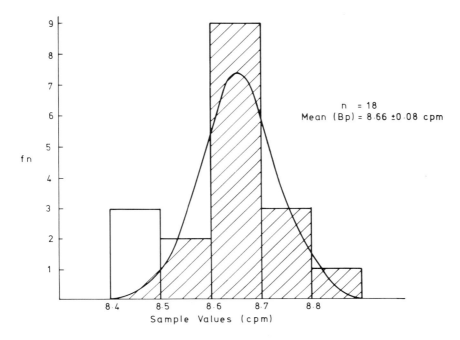

FIG. 2. Distribution of replicate background samples

sample relative to two or more others. In this way all eighteen have been normalized to a single, horizontal scale. The overall mean is 8.63 ± 0.12 cpm (1σ) for the separate samples. The best fit Gaussian to the fifteen closest set values about the mean is also shown. Three rejections were made to achieve this theoretical curve, which was computed from the revised mean 8.66 ± 0.08 cpm.

Figure 3 shows the comparable histogram for sixteen "modern" samples counted in the same period. This gives a mean of 12.49 ± 0.05 with only one rejection, which lies more than three times the standard deviation of the others from the mean. The best fit Gaussian is also drawn in.

These results demonstrate that sample-to-sample differences are not insignificant. On counting statistics alone, each sample was counted to an expected standard deviation of better than ± 0.02 cpm on the backgrounds and ± 0.02 dpm/g on the moderns.

It has been suggested (Polach 1974) that modern samples prepared from NBS oxalic acid should not be used as indicators of stability owing to fractionation problems in their preparation to carbon dioxide. In contrast it has been found that with the system of dry combustion used in this laboratory, the mean and standard deviation on 15 $\delta^{13}C$ values of NBS oxalic was −19.25 ± 0.19 °/oo. Of sixteen samples tested to date, only one (at −21.49 °/oo) lay beyond the 3σ criterion. A further eleven samples prepared from oxalic acid of a different source gave −24.88 ± 0.24 °/oo with no rejections.

Because the variation in the stable isotope ratio is so small, no correction is applied to the results shown.

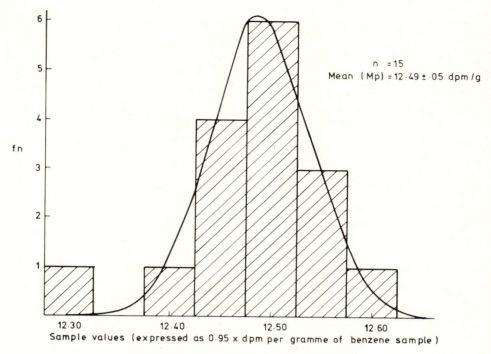

FIG. 3. Distribution of replicate NBS oxalic samples.

Error Term Derivations

Ignoring fractionation effects, the basic age relation is,

$$t = \tau \ln \frac{A_o}{A_t}$$

τ is the mean life for ^{14}C decay (8033 y)
A_o is the original, modern, ^{14}C specific activity of the sample
A_t is the residual ^{14}C specific activity t years later when it is measured.

To calculate the residual specific activity, it is necessary to know the sample gross counting rate (S), the actual background count rate (B) applicable to the given sample, its weight (w), and its counting efficiency (ϵ). The ratio A_o/A_t is then found by substituting the appropriate modern specific activity (m):

$$\frac{A_o}{A_t} = \frac{mw\epsilon}{S - B}$$

Values of (m) and (ϵ) need not be specified in absolute terms. If their derivations are strictly comparable, uncertainties in absolute calibration will cancel out in the ratio calculation. The occurrence of distributions of background and modern replicate samples raises other problems. To obtain a valid A_o/A_t ratio, it follows that the values to be substituted must be the appropriate members of the distributions which exactly correspond to the preparation history and the counting quality of the given unknown sample.

In practice one cannot exactly specify the sample count rate, background, and modern specific activity of a single sample. The best that can be done is to measure S by counting (to a 1σ value of $\pm \delta S$), assume that B will be the mean of the observed background sample population distribution, B_p (and hence carry the uncertainty of the 1σ of the population mean, $\pm \delta B_p$), and assume that m will be the mean of the observed modern sample population, M_p (and hence carry the uncertainty δM_p), thus

$$\frac{A_o}{A_t} \text{ becomes } \frac{(M_p \pm \delta M_p)(w \pm \delta w)(\epsilon \pm \delta \epsilon)}{(S \pm \delta S) - (B \pm \delta B_p)}$$

By partial differentiation it can be shown that the error term relationship in t (ignoring errors in (w) and (ϵ) just for the moment) is

$$\sigma_t = \tau \sqrt{\frac{\delta M_p^2}{M_p^2} + \frac{\delta S^2}{(S - B_p)^2} + \frac{\delta B_p^2}{(S - B_p)^2}}$$

This is the minimum possible value of σ_t since it assumes that the long term means for M_p and B_p correctly apply to the particular counting sequence in which the count on the unknown sample was carried out. Opinions differ as to whether long term or short term values apply. In the system employed in this laboratory the short term value determined from the means of the representative sets of background and modern samples included in every counting batch is always used. Variations in these means are occasionally large enough to preclude the general use of long term means. Some of these are seen in figures 4 and 5, in which means of the appropriate groups are plotted over a two-year period. The hatched sections represent the range of $\pm 1\sigma$ calculated on the basis of Poisson counting statistics. In some cases, the steps can be related to instrumental difficulties, for example, after breakdowns, but at other times there has been no apparent explanation.

If we substitute the short term, batch means B_c and M_c in place of either B_p or M_p, then the added uncertainty due to the shorter term counting statistics must be allowed for. Thus, for example, the background, $B_p \pm \delta B_p$ is replaced by

$$B_c \pm \sqrt{\delta B_p^2 + \delta B_c^2}$$

The error relation for σ_t becomes,

$$\sigma_t = \tau \sqrt{\frac{\delta M_p^2 + \delta M_c^2}{M_c^2} + \frac{\delta S^2}{(S - B_c)^2} + \frac{\delta B_p^2 + \delta B_c^2}{(S - B_c)^2}}$$

To a first order, the substitution ignores the interdependence of the value M_c upon the prevailing background B_c on which its calculation is made. Its omission leads to a slightly exaggerated estimate of the overall error.

If the uncertainties in all the relevant parameters (appendix I), such as sample weight

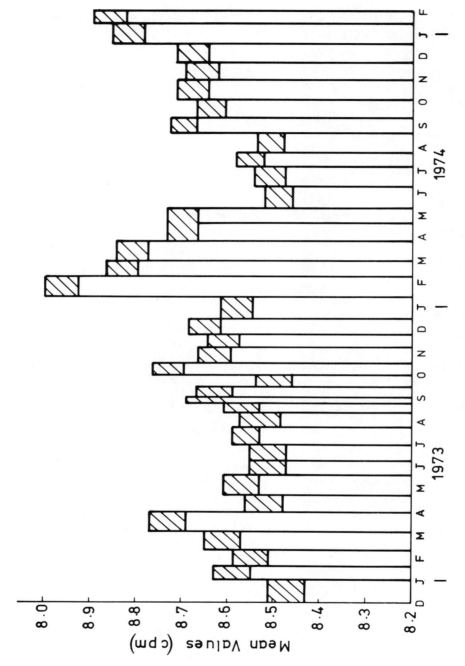

FIG. 4. Variations observed from the 'background' group of samples in routine batch counting sessions over a 2 year period.

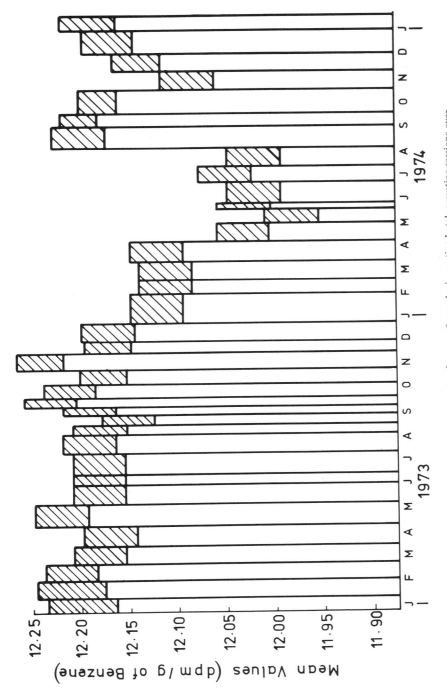

FIG. 5. Variations observed from the 'modern' group of samples in routine batch counting sessions over a 2 year period.

(δw), sample counting efficiency correction ($\delta\epsilon$), and sample stable-isotope correction ($\delta\Delta$), are included, the overall error term is

$$\sigma_t = \tau \sqrt{\frac{\delta M_p^2}{M_c^2} + \frac{\delta M_c^2}{M_c^2} + \frac{\delta S^2}{(S - B_c)^2} + \frac{\delta B_p^2}{(S - B_c)^2} + \frac{\delta B_c^2}{(S - B_c)^2} + \frac{\delta \epsilon^2}{\epsilon^2} + \frac{\delta w^2}{w^2} + \left(\frac{2\delta\Delta}{1000}\right)^2}$$

Using the data obtained from the replicate sample analyses and incorporating allowances for the estimated errors in the mass, counting efficiency, and stable isotope errors, a good approximation to the overall error can be made by substituting typical values in the above equation. The ± error term derived is the precision or reproducibility of the full laboratory measurement expressing the one standard deviation, the 68% probability that the result of a replicate sample determination will fall within the stated range. Two examples are given in table 1.

TABLE 1

Age Equivalents of Terms Contributing to the Overall Error
(Values given in conventional-radiocarbon years)

	Term	Case 1: t = 3000 y			Case 2: t = 10,000 y		
Counting	$\frac{\delta M_c}{M_c}$	$\frac{0.03}{12.5}$	19.3	52.2	$\frac{0.03}{12.5}$	19.3	73.3
	$\frac{\delta B_c}{S - B_c}$	$\frac{0.03}{30}$	8		$\frac{0.03}{12.3}$	19.6	
	$\frac{\delta S}{S - B_c}$	$\frac{0.19}{30}$	50.9		$\frac{0.104}{12.3}$	67.9	
Replicate samples	$\frac{\delta M_p}{M_c}$	$\frac{0.05}{12.5}$	32.1	45.4	$\frac{0.05}{12.5}$	32.1	84.7
	$\frac{\delta B_p}{S - B_c}$	$\frac{0.12}{30}$	32.1		$\frac{0.12}{12.3}$	78.4	
Others	$\frac{\delta\epsilon}{\epsilon}$	$\frac{0.05}{68}$	5.9	10.1	←	←	10.1
	$\frac{\delta w}{w}$	$\frac{0.01}{12}$	6.7		←	←	
	$\frac{2\delta\Delta}{1000}$	$\frac{0.6}{1000}$	4.8		←	←	
				72.2			113

DISCUSSION AND CONCLUSIONS

This type of assessment raises questions about what is to be gained simply by extending counting on an individual sample. In making a longer count it is noted that only the counting statistics terms are reduced. Thus in case 1 it can be shown that the improvement in σ_t by exactly doubling the counting time is small, changing ± 72 years to ± 64 years; case 2, σ_t becomes ± 103 years (previously ± 113 years), a questionable improvement. The argument is strengthened, however, for doubling up the entire measurement process with a repeat sample. If the two samples are counted in separate batches, normal association of the overall error term is justified because all error terms are independently assessed in each measurement. Thus in case 1,

$$\sigma_t = \pm \tfrac{1}{2} \sqrt{72^2 + 72^2} = \pm 51 \text{ years}$$

and case 2 reduces to ± 80 years. A more detailed comparative summary is given in table 2.

TABLE 2

Summary of Error Assessments (y)

Method	Age of sample	
	3000 y	10,000 y
1. Counting statistics only	±55	±73
2. Full error analysis as this paper	±72	±113
3. As 2, but given two counting sessions	±64	±103
4. Combination of errors (full error analysis as 2) on two replicate samples	±51	±80

In conclusion, it is emphasized that contributions to the overall error due to background and modern sample reproducibility are not insignificant in this assessment. Summation of only the counting statistics terms would lead to a considerable underestimate of the overall measurement error.

This conclusion is highly relevant to the way in which results are quoted by this laboratory. The main purpose of this paper, however, is to present a method of error assessment which can be applied more generally to liquid scintillation counting systems. It is hoped that this analysis may stimulate further thought about methods of error assessments and may contribute to the attainment of a more uniform definition of what is quoted as conventional radiocarbon measurement error.

APPENDIX 1

Assessment of Typical Individual Error Terms for a Liquid Scintillation Counter System.

COUNTING ERRORS

The text describes the batch counting method employed. Each sample train includes 5 moderns and 4 backgrounds. Samples with count rates < 20 cpm count for 2000 mins, or $\geqslant 20$ cpm for 40,000 total counts, per session. In the table below all samples are assumed to be 5 g benzene.

Parameter	Count rate (cpm)	Net count rate (cpm)	Total counts per session	Counting error from $\sqrt{N}$ (%)	Related parameter error
Background B	8.5	–	68K	0.383	0.033 cpm
Modern Spec. Act. M (12.5 dpm/g C^6H^6 at $\in$ = 68.4%)	51.5	43	200K	0.224	0.030 dpm/g
Sample S (t = 3000 y)	38.5	30	40K	0.500	0.193 cpm
Sample S (t = 10,000 y)	20.8	12.3	40K	0.500	0.104 cpm

Analysis of actual results shows that the standard deviation on the means of repeated counts (e.g., 100 min counts) is generally larger than $\sqrt{N}$. This may simply indicate that the distribution is not entirely Poisson. This effect has also been observed and examined in detail by Hall and Hewson (1975) using a similar liquid scintillation counter. For this reason

$$\sqrt{\frac{\Sigma(\overline{x} - x)^2}{n(n - 1)}}$$

which assumes a normal or Gaussian distribution is taken as the counting error in dating calculations.

OTHER PARAMETERS

Efficiency ($\epsilon + \delta\epsilon$). — The efficiency is estimated from the test reading provided by the Packard Tricarb automatic external standardization (AES) facility. The average counting efficiency is ~68 ± 2%. The variation is due to small but variable amounts of quenching in the synthesized samples. The residual error in the corrected efficiency term (using an AES-efficiency calibration relationship) is estimated to be ~ ± 0.05%.

Weight (w ± δw). — Although all weighings are done on an accurate beam type balance to better than ± 0.001 g, an allowance of ± 0.01 g relative to 12 g total contents, benzene + benzene scintillant, is made in this paper. Some losses are inevitable in the dispensing process and complete sealing is still difficult to achieve in every case despite the special techniques developed in this laboratory (Otlet and Slade 1974). Losses using standard vial caps, even with a final weighing correction allowance, may be as large as ± 0.1 g in 12 g total since the loss of weight is obscured by gain in weight of the plastic tops in the cooler, more humid, environment of the counting chamber.

Stable isotope correction $\delta^{13}C$ ($\Delta \pm \delta\Delta$). — The basic $\delta^{13}C/^{14}C$ relationship is described by Craig (1954). In the terminology of this paper the modification to (A_o/A_t) is

$$\frac{A_o}{A_t \left\{ 1 - \frac{2(\Delta + 25)}{1000} \right\}}$$

where Δ is the stable isotope measurement relative to PDB of the unknown sample, given its usual meaning in °/oo as described by Craig (1954).

In conclusion of this modification in the full age formula and partial differentiation with respect to Δ gives the error term

$$\frac{2\delta\Delta}{1000}$$

Measurements of Δ for all samples is made to ∼ ± 0.3 °/oo.

REFERENCES

Callow, W. J. et al.
    1965   National Physics Laboratory Radiocarbon Measurements III: Radiocarbon. 7:156-161.

Craig, H.
    1954   Carbon-13 in plants and the relationships between Carbon-13 and Carbon-14 variations in nature. J. Geology. 62 (2):115-149.

Hall, J. A., and A. D. Hewson
    1975   On-line Computer Processing of Radiocarbon Date Results: Its Use in Dating and Research. Proc. Symp. on Archaeometry and Archaeol. Prospection. Oxford. (not published)

Otlet, R. L., and B. Slade
    1974   Harwell Radiocarbon Measurements I. Radiocarbon. 16 (2):178.

Polach, H. A.
    1969   Optimizations of Liquid Scintillation Radiocarbon Age Determinations and Reporting of Ages. Atomic Energy in Australia. 12 (3):21-28.
    1974   Application of Liquid Scintillation Spectrometers to Radiocarbon Dating. Liq. Scint. Counting: Rec. Dev. Stanley and Scoggins, eds. Academic Press, London.

14

Radiocarbon Dating at the U.S. Geological Survey, Menlo Park, California

Stephen W. Robinson

INTRODUCTION

The directive authorizing establishment of a radiocarbon laboratory at the western headquarters of the U.S. Geological Survey in Menlo Park, California, has provided an opportunity for the design of a new laboratory free from constraints imposed by preexisting equipment and facilities. The new laboratory will utilize the gas proportional counting of carbon dioxide. This counting gas was selected for its simplicity of synthesis and its demonstrated potential for yielding low background levels with counters of simple design.

The basic design concepts were derived directly from the radiocarbon laboratory at the Quaternary Research Center, University of Washington, and are indirectly a consequence of the pioneering work of Professor Hessel De Vries and his students at Groningen, Netherlands (De Vries and Barendsen 1953). The design objectives for the new laboratory are:

1. Attainment of low background levels

2. Annual measurement capacity of 500 samples

3. Counters for samples as small as 0.25 gm carbon

4. Large counters for old samples and high-precision measurements.

Laboratory Description

The development of the Menlo Park Radiocarbon Laboratory began with the design and construction of a specialized building. It is situated about 70 m from the nearest existing structure to minimize electrical interference and the possibility of laboratory contamination with radioactive materials. Other papers in this volume stress the importance of the contribution of cosmic radiation to counter background levels (Stuiver et al. 1977, Oeschger et al. 1977). To attenuate the effect of cosmic rays, the room containing the counters and electronics is located 9.7 m below the ground surface (fig. 1). The vertical shielding against cosmic radiation is estimated at 1.8 kg/cm^2. The added shielding of earth fill is expected to reduce the cosmic ray flux and the associated counter background by a factor of 3 compared with a ground-surface laboratory. The underground counting room is equipped with apparatus for maintaining constant temperature and low humidity.

Sample combustion and preliminary purification follow the method of De Vries (1956), and final purification is accomplished by repeated passage of the sample gas over copper and silver at 420 C. A second vacuum line will be used for preparation of carbonate samples and wet combustions. This system will permit the processing of up to five samples per day. The counting electronics were constructed according to a design based on that developed at Trondheim, Norway (Gulliksen 1972), and they combine simplicity of construction and electrical maintenance with compact size and low cost. Each of the two desktop units contains counting circuitry for five counters in a single anticoincidence ring.

The counter shield is contained in a small chamber connected to the underground room. The shield of selected low-radioactivity lead bricks is 20 cm thick on the sides and bottom and 30 cm thick above the counters. Inside the lead is a borated paraffin neutron absorber, 5 to 10 cm thick. The shield contains two anticoincidence rings, 90 cm long, with an inside diameter of 15 cm. The anticoincidence rings are composed of discrete

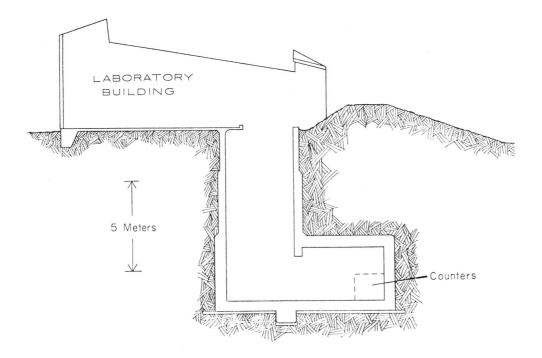

FIG. 1. Vertical cross section of Menlo Park U.S. Geological Survey Radiocarbon Laboratory

counter tubes, 5 cm in diameter, made of high-purity copper with quartz insulators. The guard ring is operated in the proportional mode with a filling of commercial counting gas (90% argon, 10% methane) at about one atmosphere pressure. Each guard ring contains four or five sample counters, each wrapped in a teflon sheet for electrical insulation.

Two designs of sample counters are in use (fig. 2). Each design utilizes only quartz and high-purity copper and is assembled with epoxy cement. The center wire is 25-μm-diameter stainless steel. Since high voltage is applied to the outer wall of the counter, the segment of the filling tube adjacent to the counter is of quartz to provide insulation. Figure 2a shows a simplified version of the Stuiver-type counter (Stuiver et al. 1977). The other counter design (fig. 2b) utilizes a quartz disc end wall epoxied into the copper tube. The electric field configurations in the two types of counter differ radically. In one the end wall is held at high voltage, while in the other the end wall is a dielectric. However, there is no significant difference in the performance of the two designs, either in background level, efficiency, or plateau slope. The Stuiver-type counter is more durable but requires more elaborate machining.

A wide array of counter volumes will permit convenient measurement of undersized samples and the high-precision analysis of geophysical samples, in addition to routine dating and the achievement of a dating limit of about 55,000 B.P. for sufficiently large samples. The proposed array of counters is detailed in table 1. The effective volume is based on the exposed length of the center wire, and for this volume a counting efficiency of 92%-95% is anticipated. Background levels in the underground counting chamber are estimated on the basis of experience in the Seattle laboratory using 0.03 cpm/g carbon and 5.0×10^{-4} cpm/cm^2 of counter wall area. Plans call for the installation of two each of the 2.47- and 3.00-liter counters.

Completion of the facility is scheduled for January 1977.

TABLE 1

Counter Array for Menlo Park Laboratory

Effective volume (liters)	Inside diameter (cm)	Effective length (cm)	Net modern count rate, 3 atm (cpm)	Estimated background (cpm)	Amount of carbon to fill system (g)
0.12	2.29	29.1	2.2	0.13	0.23
0.25	3.56	25.1	4.7	0.18	0.44
0.40	3.56	40.2	7.5	0.29	0.66
0.65	3.81	57.0	12.2	0.44	1.04
1.04	4.75	58.7	19.6	0.57	1.61
1.22	4.75	68.8	23.0	0.67	1.90
2.47	6.67	70.7	46.6	1.00	3.76
3.00	7.14	74.9	56.6	1.14	4.55

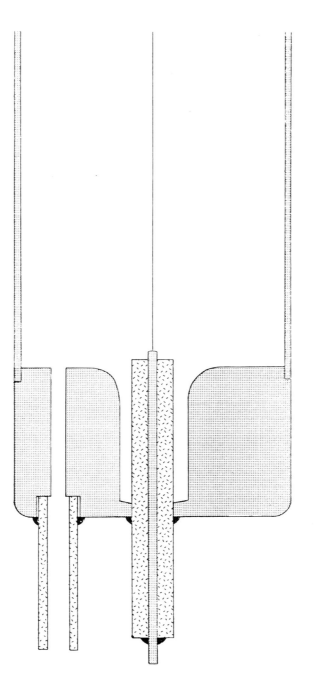

FIG. 2. End geometry of two counter designs used in Menlo Park laboratory. (a) "Stuiver-type counter" with contoured copper end-plate.

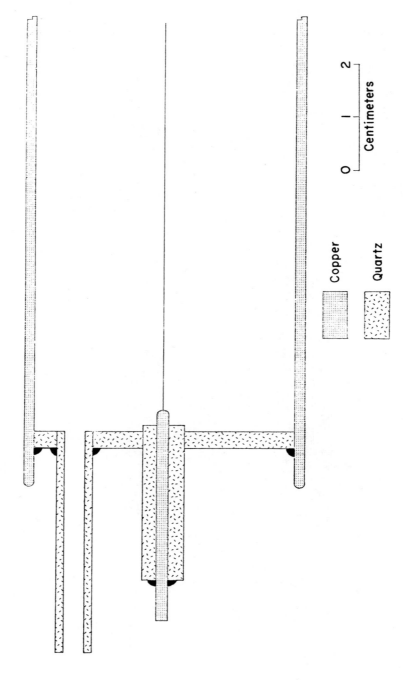

FIG. 2. End geometry of two counter designs used in Menlo Park laboratory. (b) Counter with quartz end-plate.

REFERENCES

De Vries, Hl.
    1956   Purification of CO_2 for use in proportional counter for C-14 age measurements. Appl. Sci. Research. 5B:387-400.

De Vries, Hl., and G. W. Barendsen
    1953   Radiocarbon dating by a proportional filled with carbon dioxide. Physica. 19:987–1003.

Gulliksen, S.
    1972   Low cost electronics and a twin counter assembly. Proc. 8th Int. Conf. on Radiocarbon Dating. New Zealand. 178-188.

Oeschger, H., B. Lehmann, H. H. Loosli, M. Moell, A. Neftel, U. Schotterer, and R. Zumbrunn
    1977   Recent progress in low level counting (this volume).

Stuiver, M., S. W. Robinson, and I. C. Yang
    1977 ^{14}C dating up to 60,000 years with high efficiency proportional counters (this volume).

PART III
Soil Dating

1

Soil Fraction Dating

H. W. Scharpenseel

The application of ^{14}C dating techniques to fossil soils is adequate and its feasibility depends entirely on the precision of site description and interpretation. Considerable constraints are associated with recent soil dating. They have been analyzed concurrently in a review paper on soil radiocarbon dating (Scharpenseel and Schiffmann 1976).

In order to enhance the significance of the minimum age produced from recent soil's organic carbon dating, three alternatives exist to minimize the rejuvenating influence of root growth, animal transport, and chelate percolation:

1. exclusive use of C, derived from clay organic complexes, which should consist in soil formations not too young and mainly of older carbon species;

2. selective use of a C-fraction, be it on the basis of classical humus fractionation techniques, or of particle size separation by gel permeation, or be it as a product of gradient extraction from the soil matrix; and

3. insistence on careful sampling at the ultimate, deepest fringes of the solum that still contain 0.2-0.3% of organic carbon, enough to support a full-size benzene sample.

One can integrate and condense these boundary conditions into the following: take a texture sample of (1) coarse-silt to clay (comprising the clay organic complexes) from (2) the relatively oldest chemical fraction out of (3) the relatively oldest (deepest) zone of the soil profile, still containing 0.2% to 0.3% organic carbon.

Results of fraction dating (table 1) in carefully selected soil samples indicate distinct differences between the individual fractions, such as superiority in age of humic acid compared to fulvic acid, of gray humic acid to brown humic acid, of Sephadex fraction 2 (delayed fraction) versus Sephadex fraction 1 (passing fraction) (fig. 1). Organic, solvent-extracted fractions are less reliable due to the danger of incorporating young, solvent-derived C from ethanol, as in the case of the hymatomelanic acids or of old C, which, despite thorough vacuum drying, incorporates the considerably older dimethylformamid- and dimethylsulfoxid-extracts (table 1, column 10). A marked difference exists between 6 N HCl hydrolysate and hydrolysis residue, which Paul et al. (1964) observed in the

TABLE 1

Soil Fraction Dating

Fraction	Podzol Scherpenseel (Ah) 15-25 cm	Podzol, Scherpenseel near Geilenkirchen, Bh, 40-55 cm	Podzol Flaesheim (O/Ah) 10-20 cm	Chernozem, Aseler forest Ah, 40-50 cm	Chernozem Söllingen (Ap) 15-25 cm	Pseudogley-Chernozem Adlum(SwA) 40-50 cm	Fossil paparendsina below Trachyt tuff of Laach, fA, 200 cm	Fossil Chernozem Michelsberg (fA) 140 cm	Nethermoor Koislhof, overlaying Isargravel Hn 80 cm	Kalkarer Nethermoor Hn 80-90 cm
Total soil		2470 ± 70		2470 ± 70			10,600 ± 120		7,200 ± 110	
Petroleumbenzene extract									3,290 ± 320	
Benzone extract				3220 ± 80			4,130 ± 100		3,920 ± 120	
Methanol extract									6,380 ± 90	
Acetone extract									4,020 ± 360	
Acetonitril extract									2,240 ± 440	
Dimethylformamid extract									10,760 ± 130	
Dimethylsulfoxide extract									13,140 ± 200	
Hydrolysis residue				3160 ± 70			11,360 ± 150		9,730 ± 170	
6 H HCl hydrolysate				104.4 0.6% mod			2,510 ± 100		7,270 ± 140	
Fulvic acids (after organic extraction)	2930 ± 40		140.1% mod	370 ± 70	104.3% mod	1800 ± 60	1,140 ± 200	4310 ± 210	6,860 ± 250	4270 ± 80
Fulvic acids dialyzed				380 ± 70					7,060 ± 110	

Hymatomelanic acid	1580 ± 80								
Total humic acid		2940 ± 60	114.1% mod	2100 ± 70		6,970 ± 210		8,810 ± 120	
Humic acids Sephadex 50 fraction 1 (passing fraction)			2230 ± 70	1480 ± 60		6,110 ± 100		7,590 ± 120	
Humic acids Sephadex 50 fraction 2 (delayed fraction)			5410 ± 90	2940 ± 90		6,830 ± 130		7,820 ± 90	
Brown humic acids		2530 ± 60		920 ± 50	1560 ± 70	4890 ± 50	7600 ± 220		5380 ± 80
Grey humic acids		2980 ± 70		1140 ± 70					5970 ± 40
Humine		2850 ± 70	117.2% mod.	2460 ± 60	2275 ± 80	2910 ± 80	10,320 ± 140	6930 ± 80	7,110 ± 110	3490 ± 70
Humus coal		Traces		2920 ± 70		2810 ± 60	9,940 ± 140	6830 ± 100	7,230 ± 110	4460 ± 80
Charcoal							10,330 ± 120			

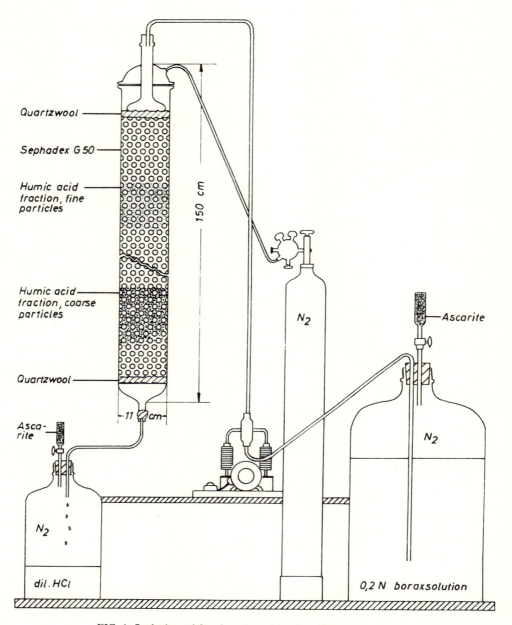

FIG. 1. Sephadex gel fractionation of humic acids under inert gas.

early days of natural ^{14}C measurements in soil organic matter. Repeated hydrolysis would seem to be the most promising approach to possible isolation of a biologically inert C, which may exist in recent soils (Gerasimov 1971, 1974) protected by clay organic interaction and by polycondensation.

Decay studies with uniformly ^{14}C-labeled organic matter in soils (Jenkinson 1971,

Sauerbeck and Johnen 1973) would seem to indicate a C residence half-life of about 6 years, a mean lifetime of 9 years, a residue of 1/1024 after $10 \times T^{1/2} = 60$ years, and $1/10^9$ after $30 \times T^{1/2} = 180$ years. Instead, in the deepest zones of A-horizons in our loessic soils, we find minimum ages (apparent mean residence time, AMRT) of 5000 to 6000 years. In a soil of about 8000 years true age (TA), holocene in origin, this means that in the sample, taken from the deepest fringes of organic substance subjected to ^{14}C-dating, old carbon species have dominated overwhelmingly.

This lends support to the claim of Gerasimov that there may exist a biologically inert C which has been chelated by the clay minerals at the early stages of clay formation and which has since been protected in this clay organic linkage against microbial interactions and interchange with younger carbon.

Acid hydrolysis as a means of extracting from soils large portions of the younger carbon species was selected from the host of fractionation methods (table 1) and was applied repeatedly to several soils. This rather tedious procedure is illustrated in figure 2. Beginning with sixteen repetitions, 1 kg of soil was hydrolyzed over a period of 48 hours in 3 liters of 6 N HCl. The hydrolysate was filtered and the residue was washed, dried, and weighed. The hydrolysis residues derived from two runs of each hydrolysis step were combined, combusted in a quartz tube under O_2 and converted into a benzene sample.

Table 2 shows the results obtained with two soils, a loess-chernozem from Hildesheim and a gley-chernozem from the lower, loess-covered terrace of the Isar river in Bavaria (both samples from the FRG). A tendency was revealed of increasing fraction ages from the unhydrolyzed soil sample toward the higher replications of 6 N HCl-hydrolysis. Coking pretreatment by combustion in a quartz tube under N_2 flow produced no age increment, either in the terrestrial soil sample from Asel (Argiudoll) or in the hydromorphic, gley soil from Landshut (Argiaquoll). The unhydrolyzed sample, the coked sample, and the first hydrolysis fraction from Asel soil indicated bomb carbon influence and dated $> 100\%$ modern. The disappearance of this effect under conditions of continued hydrolysis would suggest that most of the bomb carbon, containing younger organic C had been removed by the action of the 6 N HCl. The Landhut soil indicated a gradient of increasing age parallel to the repetitions of hydrolysis.

The age increase of the fractions would seem almost steady. This suggests that, by a continuous process of 6 N HCl hydrolysis with constant replacement of the acid and solutes, a method for pretreatment of recent soil samples might permit production of a sample carbon close to the quality hypothesized by Gerasimov (1971, 1974) in his discussions of a possible biologically inert C. The dates of these C-species would be near the true age of the soil if sampling were carried out near the depth boundary of organic carbon.

SUMMARY

Rejuvenation of C in recent soils makes radiocarbon dating a priori meaningless. Root growth, chelating, lye-produced carbon transport, and the actions of meso- and macrofauna are responsible for the rejuvenation. A reduction of rejuvenating interaction is

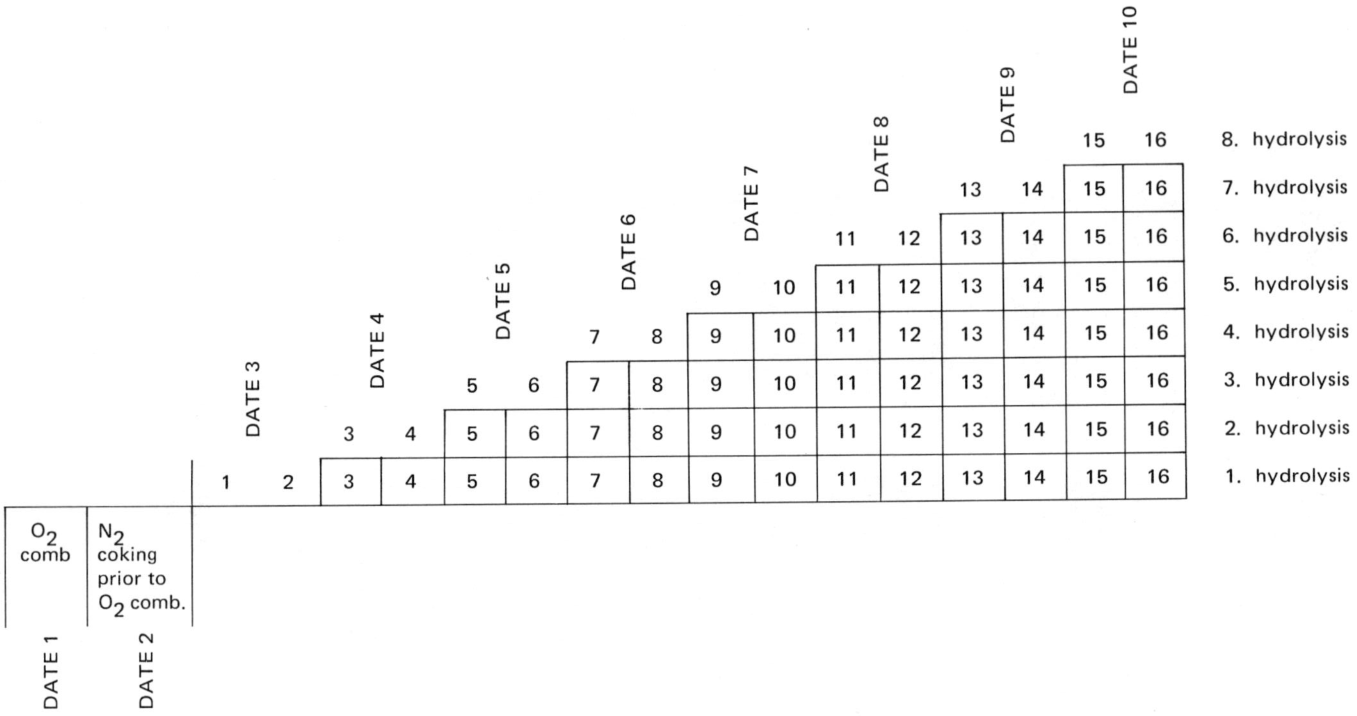

FIG. 2. Repeated hydrolysis of 1 kg soil in 3 liters of 6 N HCl.

TABLE 2
Fraction Dates from Hydrolysis Replications

Treatment	Soil of Asel, Argiudoll		Soil of Landshut, Argiaquoll	
O_2 – combustion	>100% modern		4640 ± 90	56% modern
N_2 – coking + O_2 – combustion	>100% modern		4570 ± 110	56.6% modern
1st hydrolysis step	>100% modern		5610 ± 80	49% modern
2nd hydrolysis step	2530 ± 80	73% modern	5820 ± 110	48.5% modern
3rd hydrolysis step	2410 ± 70	74% modern	5700 ± 100	49.2% modern
4th hydrolysis step	2340 ± 80	74.7% modern	5540 ± 110	50.1% modern
5th hydrolysis step	2770 ± 80	70.8% modern	6110 ± 90	46.8% modern
6th hydrolysis step	2560 ± 90	72.8% modern		
7th hydrolysis step	2960 ± 80	69.2% modern		
8th hydrolysis step	3260 ± 100	67.2% modern		

obtainable by sampling near the deepest boundary of organic C in the soil, and by producing the dating sample from the texture and organic matter fraction, representing the relatively oldest carbon. Gerasimov's claim for existence of a biologically inert C-fraction in recent soils was tested by fraction dating within the organic matter pool. Consistent age increase in 6 N HCl hydrolysis residues led to fraction dating following repeated acid hydrolysis. Two soil samples from a terrestrial Argiudell material as well as of a hydromorphous, gley Argiaquoll revealed almost steady increment of fraction age, parallel to hydrolysis replications. Continuous 6 N HCl hydrolysis over a longer time period, with concurrent replacement of acid with solutes, could enhance the quality of dating samples derived from the organic matter of recent soil.

REFERENCES

Gerasimov, I. P.
    1971   Nature and originality of paleosols. In Paleopedology: Origin, nature and dating of paleosols. Jerusalem, 15-27.
    1974   The age of recent soils. Geoderma. 12 (1/2):17-25.

Jenkinson, D. S.
    1971   Studies on the decomposition of ^{14}C labelled organic matter in soil. Soil Sci. 111 (1):64-70.

Paul, E. A., C. A. Campbell, D. A. Rennie, and K. J. McCallum
    1964   Investigations of the dynamics of soil humus utilizing carbon dating techniques. Proc. 8th Int. Cong. Soil Sci. Bukarest. 3:201-208.

Sauerbeck, D., and B. Johnen
    1973   Radiometrische Untersuchungen zur Humusbilanz. Landw. Forsch. Sonderheft 30 (II):137-145.

Scharpenseel, H. W., and H. Schiffmann
    1976   Radiocarbon dating of soils. Pflanzenernährung u. Bodenkunde.

2

Radiocarbon Dating of Organic Components of Sediments and Peats

John C. Sheppard, Syed Y. Ali, and Peter J. Mehringer, Jr.

ABSTRACT

Lipids extracted from sediments and peats are potential radiocarbon dating materials. Paired-t analysis of humic acid-peat residue, humic acid-sediment residue, peat-charcoal, humic acid-charcoal, bone collagen-charcoal, and shell-charcoal pairs did not reveal large systematic differences in radiocarbon ages. Soil radiocarbon dates can suggest contradicting conclusions about the usefulness of soil-organic materials.

INTRODUCTION

The search is constant for suitable radiocarbon dating materials that are minimally subject to age-altering contamination. Examples of such materials are bone collagen and apatite. The former has proven to be a very useful radiocarbon dating material (Olsson et al. 1974) while the latter is of questionable value (Haynes, pers. comm. to Olsson). Similarly, organic components like humic acid and lipids have been investigated both as contaminants and as radiocarbon dating materials for sediments, peats, and soils.

This research was directed toward the evaluation of humic materials such as humic acid and lipids as alternative radiocarbon dating materials for soils, sediments, and peats. To evaluate the utility of these materials, directly associated charcoal, wood, and peat and gyttja residues were dated. These pairs of dates (i.e., charcoal-humic acid) were subjected to the paired-t test (Youden 1951, Bennett and Franklin 1954) to establish whether the sets of dates might differ significantly. To obtain a larger data base, additional paired dates were obtained from the literature, mainly *Radiocarbon*. In the process of obtaining the paired data for sediments, soils, and peats, shell-charcoal and bone collagen-charcoal data also were collected and analyzed. These results are included for comparison.

EXPERIMENTAL METHODS

Extraction of humic acid from soils, sediments, and peats followed standard radiocarbon pretreatment procedures (Olsson 1952). To minimize solvent contamination problems, lipids were extracted before the humic acid extraction. Grant Taylor (1972) extracted lipids from peats using benzene as a solvent in a soxhlet extractor and demonstrated that lipids are potential radiocarbon dating materials. The soxhlet extraction method is slow and requires the use of heated solvents. As an alternative we used the very rapid, ultrasonic method (Murphy 1969) utilizing a common ultrasonic cleaner and solvent mixtures, such as methanol-chloroform, to extract lipids from soils. Complete extraction of the lipid fraction requires about one hour. The solvent is removed from the lipid using a roto-evaporator and a vacuum oven. Subsequently the lipid is converted to CO_2 and then to CH_4 for counting. A flow chart for the extraction procedure for lipids, humic acid, and humine is shown in figure 1.

Radiochemical purity is a major problem when establishing the utility of lipids as radiocarbon dating materials. When large samples are available, they can be split into two fractions and subjected to varying degrees of pretreatment. If the two samples have the same specific activity after pretreatment, radiochemical purity and the minimum pretreatment can be established. If significantly different specific activities are obtained after pretreatment, additional purification steps must be performed or a new purification method must be used.

Another approach, the one used for this investigation, is to compare the lipid or humic acid radiocarbon ages to radiocarbon ages of reference materials like charcoal, wood, or a well-characterized volcanic ash, such as the 6700-year-old Mt. Mazama ash (Kittleman 1973), which is found throughout the Pacific Northwest. This method is well suited for the small samples encountered in this research. Examples of the paired sample approach are given in table 1.

Analysis of Data

Use of paired sets of radiocarbon dates provides the opportunity to evaluate whether significant systematic differences exist between various materials used for radiocarbon dating. The paired-t test (Youden 1951, Bennett and Franklin 1954) is one way to determine systematic differences between radiocarbon dating materials. An important advantage of the paired-t test is that the limits for the average difference and the significance for the observed difference are based only on the variation in the differences. Most importantly, factors which have equal effects on paired radiocarbon dates do not influence the accuracy of the comparison. Thus constant factors such as isotope effects, type of contemporary standard (i.e., nineteenth century wood or NBS oxalic acid) used, and the counting gas used are canceled out leaving only the difference between radiocarbon dating materials.

To make the statistical analysis of the paired data more significant, standard deviations for the individual radiocarbon dates should be approximately equal. In practice this

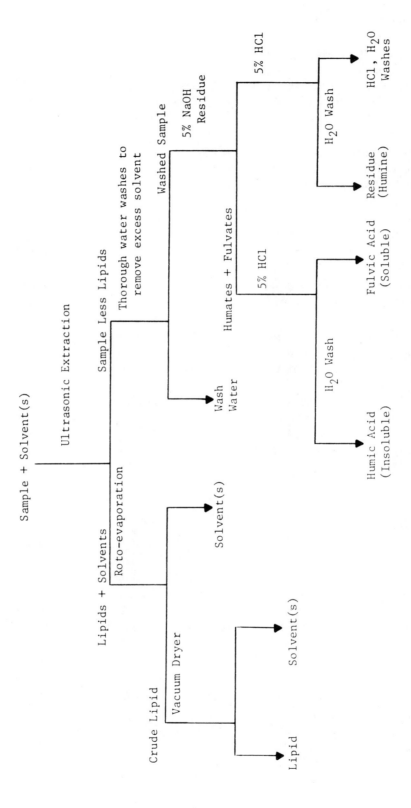

FIG. 1. A flow chart of the procedure used to extract lipids, humic acid, and humin from sediments and peats.

TABLE 1
Comparison of Paired-t Radiocarbon Dates

Material	Sample No.	Radiocarbon Age	Reference Material	Sample No.	Radiocarbon Age
Lipid	WSU-1552	6750 ± 220	Mt. Mazama[a] Ash		6710 ± 100
Lipid	WSU-1520L	5990 ± 120	Peat Residue	SI-1757	6030 ± 295
				WSU-1520R	5870 ± 120
Humic Acid	WSU-1484	4690 ± 125	Charcoal	WSU-1491	5120 ± 100
Humic Acid	WSU-1451H	3690 ± 90	St. Helens[b] Ash in Sediment	WSU-1519	3600 ± 100
				WSU-1451	4180 ± 100
Humic Acid	WSU-1552H	7100 ± 225	Mt. Mazama[a] Ash in Sediment		6710 ± 100
Humic Acid	WSU-1573H	4830 ± 180	Wood	WSU-1573W	5010 ± 180

[a]Average of large numbers of radiocarbon ages of Mt. Mazama tephra (Kittleman 1973). Our bracketing ages of gyttja just above and below the Mt. Mazama tephra places rather narrow limits on the date for the eruption of Mt. Mazama.

[b]The 3890 ± 300 average of two bracketing dates, WSU-1519 and WSU-1451, is in agreement with other dates for St. Helens Y eruptions (Crandell and Mullineaux 1973, 1975).

is rather difficult because the standard deviation of a radiocarbon age increases with increasing age, but a fair approximation can be obtained by using paired dates of less than 10^4 years. In spite of this limitation the paired-t test was applied to each class of paired samples with the hope of obtaining useful comparisons. Where sufficient numbers of paired data were available for a single radiocarbon laboratory, the paired-t test was performed also and compared to the combined results.

Data from other radiocarbon laboratories were obtained by exhaustive review of the current literature. Care was taken to include only those data involving minimal uncertainty as to the paired data, and an effort was made to verify that pairs of dates were for material with a common source or stratum.

Consider a set of paired residue and lipid, radiocarbon age data from two normal populations with unknown means μ_L, μ_R. In this case, we tested the null hypothesis

$$Ho: \mu_L = \mu_C.$$

The paired radiocarbon dates are:

$$\begin{array}{c|ccc} L_i & L_1 & L_2 - L_n \\ \hline R_i & R_1 & R_2 - R_n \end{array}$$

Where L_i = the radiocarbon age of the i-th lipid sample
and R_i = the corresponding radiocarbon age of the paired charcoal sample

Let
$$D_i = L_i - R_i$$

and
$$\overline{D} = \frac{1}{n} \sum_{i=1}^{n} D_i$$

The standard deviation for the i-th pair is

$$S_D = \sqrt{\frac{\Sigma D_i^2 - \frac{1}{n}(\Sigma D_i)^2}{n-1}}$$

and the standard deviation of average difference, $\overline{D}$ is

$$S_{\overline{D}} = \frac{S_D}{\sqrt{n}},$$

where n is the number of pairs.

The test statistic is

$$t = \frac{\overline{D}}{S_{\overline{D}}}$$

and has n − 1 degrees of freedom.

The paired-t test was applied to the set of lipid-residue pairs derived from peat and gyttja samples obtained from Lost Trail Pass, Montana; Twin Lakes, Colorado (Petersen and Mehringer, in press); and Wildcat Lake, Washington (Davis et al. 1976). The lipid data are shown in table 2, and the results of the t test are summarized in table 3.

At the 5% level, the critical value of t for 9 or 10 degrees of freedom is approximately 2.3; thus the paired-t test indicates that the lipid-residue pairs are statistically indistinguishable and suggests that $\overline{D}$ lies within the errors of the radiocarbon dating method and that lipids extracted from peats and gyttjas are suitable radiocarbon dating materials. Figure 2 shows, graphically, that these lipid radiocarbon dates are concordant with the residue dates.

To obtain supplementary information, humic acid, residues (humine), charcoal, and wood associated with the peat and gyttja samples were also radiocarbon dated. Since rather few data were obtained, supplementary data from the literature were used to provide a larger and more significant sample for the paired-t analysis. Table 4 contains a summary of the paired-t analysis for the combined results, as well as the bone-charcoal and small-charcoal data.

Discussion

Examination of the data in table 4 reveals that differences of radiocarbon ages of the paired materials are, on the average, comparable to an average standard deviation of about 100 years. If comparisons are made for those samples with radiocarbon ages less than 10^4 years, only the humic acid-charcoal and the shell-charcoal sample sets have t values

TABLE 2

Paired Lipid-residue Radiocarbon Dates

Lipid		Residue		
Lab No.	Radiocarbon Age	Lab No.	Radiocarbon Age	Material
WSU-1550L	3400 ± 170	WSU-1550R	3410 ± 90	Gyttja
WSU-1560L	5270 ± 200	WSU-1560R	4560 ± 130	Gyttja
WSU-1551L	5415 ± 180	WSU-1551R	5340 ± 100	Gyttja
WSU-1552L	6750 ± 220	WSU-1552R	6700 ± 100	Gyttja
WSU-1553L	7680 ± 240	WSU-1553R	6720 ± 100	Gyttja
WSU-1570L	8795 ± 160	WSU-1570R	8800 ± 160	Gyttja
WSU-1558L	8860 ± 260	WSU-1558R	9220 ± 230	Gyttja
WSU-1554L	10,670 ± 310	WSU-1554R	11,300 ± 230	Gyttja
WSU-1523L	500 ± 200	WSU-1448S	900 ± 100	Gyttja
WSU-1520L	5990 ± 120	WSU-1520S	5870 ± 120	peat
WSU-1526L	6650 ± 140	WSU-1526R	6760 ± 120	peat

TABLE 3

Summary of t Test of Lipid-residue Pairs

Data Pairs	d.f.	t	$\overline{D}$, years
All, WSU	10	0.26	40
$<10^4$ years	9	0.76	100

exceeding the 5% critical value. However, it should be noted that their $\overline{D}$ values are not 500 or 1000 year values, which might be considered excessively large. Large $\overline{D}$ values are noted for data taken from individual laboratories and may be indications of intralaboratory biases; but even here the $\overline{D}$'s are not much more than twice an average standard deviation of 100 years. But examination of the data in appendix 1 reveals several pairs of dates differing by as much as 10^3 years. Two explanations can be given for these large differences. The first is that the differences may be real and may represent dynamic chemical and physical processes taking place in the environment from which the pair of samples were taken. This is particularly evident for soils where $\overline{D}$ was 1100 years for humic acid-residue pairs. An example might be large moisture and concentration gradients capable of rapid transport of soil components like humic acid. Young, mobile soil components certainly have the potential to yield systematic errors (Martel and Paul 1974) and rootlet contamination must be considered. A less likely factor might be complications of

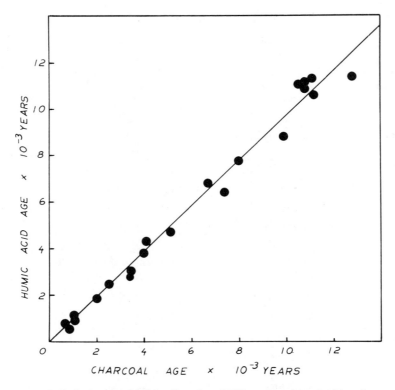

FIG. 2. A plot of liqid radiocarbon MRTs against the paired humin age, which suggests that liqid radiocarbon ages are close approximations to the humin radiocarbon ages.

the hardwater effect (Deevey et al. 1954) but such factors must always be kept in mind, especially when anomalous data are obtained. Second, the large $\overline{D}$'s may be "flyers" due to erroneous factors related to the counting and calculation of the radiocarbon age. Such discrepancies can be resolved only by reexamination of the original data or by more age determinations. If the flyer hypothesis is accepted for the data in appendix 1, the t and $\overline{D}$ values become much smaller and the agreement between all paired comparisons becomes much better. But this hypothesis can be resolved only by further experiment.

Comparison of the paired data for organic components derived from peats, sediments, and soils reveals that the more mobile components (i.e., humic acid and lipids) do not, on the average, have mean residence times (MRTs) that differ greatly from the discrete radiocarbon ages of the reference materials — charcoal, wood, and residues. The largest $\overline{D}$'s of about 225 years are found for humic acid-charcoal pairs. As shown by figures 2 and 3, these are not large, so humic acid and lipid MRTs are close approximations to the radiocarbon age of their peat and sediment environment. Thus, humic acid and lipids can be used to obtain a "second best" date of a stratum if more desirable material is unavailable. These second best dates must be used with caution. The humic

TABLE 4

Additional Paired-t Analysis Results

Pair		d.f.	t	Crit. Value (0.05)	D, Years
Humic Acid Peat Residue	all	79	−2.25	1.99	−130
	<10^4 years	60	−1.22	2.00	−40
	LU	28	−1.81	2.05	−60
	UB	21	−2.70	2.08	−280
Humic Acid Gyttja Residue	all	25	−0.38	2.06	−30
	<10^4 years	23	0.03	2.07	0
	Lu	9	−2.55	2.26	−150
Peat Charcoal	all	15	−0.71	2.13	−50
	<10^4 years	13	−1.27	2.16	−90
Humic Acid Charcoal	all	25	−1.98	2.06	−225
	<10^4 years	16	−2.67	2.12	−245
	NZ	5	−1.77	2.57	−270
Bone Collagen Charcoal	all	64	−1.71	1.99	−130
	<10^4 years	48	0.60	2.00	30
	Ly	5	−2.46	2.57	−100
Shell Charcoal	all	49	3.16	2.00	140
	<10^4 years	44	2.97	2.01	125
	Tx	18	2.32	2.10	150
Humic Acid Residue	all	18	3.08	2.10	1100

acid-residue data are much less encouraging. Examination of these data yield contradictory conclusions, suggesting in one instance that soils can yield usable radiocarbon dates (Ruhe et al. 1971) and, in another, that radiocarbon dates can yield only the mean residence time of carbon in soils (Gejh et al. 1971). The charcoal-humic acid pairs support the former conclusion (the concordance of pairs of soil-wood radiocarbon dates). The research of Scharpenseel (1972) and Paul (1974) has demonstrated that organic components of soils have many MRTs and supports the latter conclusion. The situation with regard to the radiocarbon dating of soils is confused, and it is not surprising that conflicting results are frequently obtained. Clearly, additional research is necessary to clarify or establish the utility of soils as sources of radiocarbon dating materials.

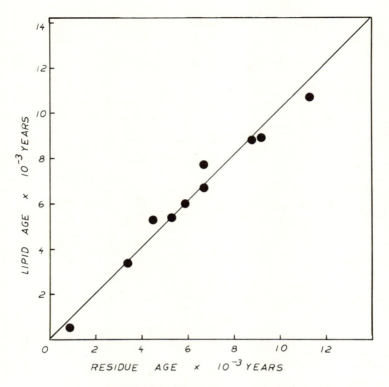

FIG. 3. A plot of humic acid MRTs against paired charcoal radiocarbon ages. This plot further demonstrates that, on the average, humic acid MRTs yield good approximations of the radiocarbon age of the soil environment adjacent to the charcoal.

The bone collagen-charcoal data suggest that, even with large differences in the collagen extraction and purification processes used by radiocarbon laboratories, collagen-derived radiocarbon ages closely approximate charcoal ages and that, on the average, sample contamination is minimal. Large differences between collagen-charcoal pairs do exist and may have explanations like those mentioned by Haynes (pers. comm. to Olsson). To resolve the occasional differences requires reexamination of the data or additional radiocarbon dating.

Shell-charcoal pairs have a small but statistically significant $\overline{D}$ which suggests that some $CO_3^=$ exchange has occurred and that the usual pretreatment practice may be inadequate. The $\overline{D}$ is not large and shell dates may be useful for the radiocarbon dating of archaeological sites.

Although the statistical analyses above have shown average differences to be small between radiocarbon ages of pairs of directly and closely associated materials, large differences are not uncommon and are caused by numerous factors. Interpretation of the radiocarbon ages of samples from specific localities requires further understanding of stratigraphic relationships, depositional environments, post-depositional history, and the species dated. Radiocarbon dates from desert aquatic habitats of eastern California and adjacent Nevada serve as examples. While peats and other organic sediments usually produce reliable age determinations, desert salt marsh sediments often are contaminated by the modern roots of a common sedge (*Scirpus olneyi*) and possibly by salt grass (*Distichlis spicata*). For example, whole core, residue, and humate fractions of buried organic sediments from Warm Sulphur Spring, Panamint Valley, California, have given modern ages (A-848, A-849) while charcoal and sedge seeds from the same sediments have been dated at 3400 ± 500 B.P. (A-952) (Haynes et al. 1971). Conversely, radiocarbon dating of sediments and peats of a core taken from nearby Little Lake, California, yielded two sets of parallel ages (Mehringer and Sheppard, in press) (see table 5) one for the peats, which was clearly reasonable, and another for lake sediments, which was displaced by about 5500 years. One explanation for the differences between the Little Lake peat and sediment data is that peats were formed from *Typha* sp. and *Scirpus acutus*, which metabolize undiluted atmospheric carbon dioxide and ^{14}C, while the lake sediments were largely composed of macrofossils of *Chara* sp., *Potamogeton pectinatus*, and *Ruppia maritima* which metabolize "old" bicarbonate ion formed by the reaction

$$Ca^{12}CO_3 + {}^{14}CO_2 + H_2O = Ca^{2+} + H^{12}CO_3^- + H^{14}CO_3^-.$$

Little Lake's sediments are expected to have older radiocarbon ages than the peats. Similar observations have been made by Deevey et al. (1954).

Because of the possible unreliability of peat dates, only charcoal and *Scirpus* seeds were dated in a study of the marsh, dune, and archaeological chronologies at Ash Mead-

TABLE 5

Little Lake Radiocarbon Dates

Lab No.	Depth (meters)	Material	Apparent Radiocarbon Age, B.P.
WSU-1460	--	Modern *Chara* sp.	3660 ± 120
1473	--	Modern *Potamogeton*	3020 ± 80
1461	1.17–1.25	*Chara* Ooze	7160 ± 100
1490	1.17–1.25	Carbonate	7060 ± 100
1462	3.39–3.49	*Chara* Ooze	6030 ± 150
1462H	3.39–3.49	Humic Acid	6030 ± 100
1465	4.91–5.02	*Chara* Ooze	6380 ± 150
1474	5.91–6.01	Peat	3020 ± 120
1466	7.86–8.06	Peat	3980 ± 120
1463	8.31–8.41	Chara Ooze	9480 ± 200
1464	11.21–11.28	Peat	5050 ± 140
1464H	11.21–11.28	Humic Acid	5000 ± 140

ows, Nevada (Mehringer and Warren 1976). In the process of this study, snails also were dated. Again the ages differ with the depositional environments indicated by the living and fossil assemblages. Shell carbonate of living *Tryonia* sp., collected from the outlet channel of a large spring fed by old water (Grove et al. 1969), gave an age of 20,000 ± 630 B.P. (A-1204). Assemblages representing a spring-fed marsh with perennially flowing water, 2 km from a large spring source, give ages in closer agreement with the bracketing dates on seeds and charcoal (table 6). These assemblages are dominated by *Fontelicella* sp. a perennial-water hydrobiid. The maximum and minimum errors given in table 6 were derived by assuming that the snails were the same age as the lower and upper bracketing dates. The "real" errors probably vary from 800 to 3500 years.

TABLE 6

A Comparison of *Scirpus* Seed and *Prosopis* Charcoal Dates with Snail Dates from Ash Meadows, Nevada

Snail Locality	Depth (meters)	Date (years, B.P.)	Lab No.	Material Dated	Error (years)	
					Minimum	Maximum
Trench 1	2.00–2.20	3970 ± 120	A-1178	Charcoal		
	2.68–2.83	8210 ± 560	1240	Snails	2890	4140
	3.55–3.65*	5320 ± 70	1177	Seeds		
Trench 5	2.57–2.67	3980 ± 130	A-1174	Seeds		
	2.77–2.87	5020 ± 150	1203	Snails	570	1040
	2.90–3.00	4450 ± 110	1175	Seeds		
Trench 8	1.60–1.70*	2940 ± 100	A-1064	Seeds		
	1.75–1.82	6610 ± 260	1239	Snails	2890	3670
	2.00–2.10*	3720 ± 200	1172	Seeds		
Locality II	1.44–1.48	3640 ± 100	A-1069	Seeds		
	1.70–1.80	6290 ± 300	1163	Snails	1550	2650
	1.98–2.04	4740 ± 110	3766	Seeds		

*Depths are adjusted to correspond to the stratigraphic position at the snail collection locality (Mehringer and Warren 1976).

SUMMARY

This research has demonstrated that lipids are potential radiocarbon dating materials for sediment, gyttja, and peat. Use of the paired-t test indicates, with the possible suggestion of humic acids derived from soils, that bone collagen, shells, and humic acid derived from peats and sediments are useful radiocarbon dating materials. Most of the differences between reference materials (wood and charcoal) are within the standard deviations of most radiocarbon age determinations. Radiocarbon dating of soils is still a questionable practice, except when mean residence times are being determined.

ACKNOWLEDGMENTS

Acknowledgment is made to the donors of the Petroleum Research Fund, administered by the American Chemical Society, for partial support of this research. Ms. Yvonne Welter pretreated, converted, and counted many of the samples. Mr. Bernard Au assisted in the pretreatment of samples. The desert salt marsh studies were supported by National Science Foundation grant GB-8646, and the mollusks were identified by Mr. D. W. Taylor.

APPENDIX

TABLE Ia

Humic Acid — Sediment/Gyttja Residue

Humic acid		Sediment/Gyttja residue	
Lab no.	Age, Years	Lab no.	Age, Years
Birm-485b	9550 ± 290	Birm-485a	8980 ± 450
486b	9420 ± 190	486a	9760 ± 230
487b	9470 ± 190	487a	8760 ± 310
488b	10,060 ± 380	488a	9630 ± 240
489b	10,080 ± 320	489a	10,160 ± 260
LU-302A	6450 ± 75	LU-302	6640 ± 70
306A	5770 ± 85	306	6000 ± 70
307A	5520 ± 90	307	5700 ± 70
309A	4640 ± 85	309	5030 ± 80
310A	4340 ± 70	310	4470 ± 80
311A	3930 ± 65	311	4270 ± 60
312A	3870 ± 65	312	3950 ± 60
313A	3610 ± 65	313	3820 ± 65
565A	4360 ± 60	565	4080 ± 65
641A	2250 ± 60	641	2280 ± 55
U-748	4360 ± 220	U-2248	4170 ± 80
749	3990 ± 90	2249	3950 ± 90
2205	2540 ± 80	707	2650 ± 150
2207	2620 ± 80	2206	2660 ± 80
UB-227C	10,795 ± 140	UB-227A	10,440 ± 110
299C	11,225 ± 160	299A	12,360 ± 165
WSU-1462H	6030 ± 100	WSU-1462R	6030 ± 150
1550H	3480 ± 100	1550R	3410 ± 90
1551H	5540 ± 125	1551R	5340 ± 100
1552H	7100 ± 225	1552H	6700 ± 100
1451H	3690 ± 90	1451R	4180 ± 90

TABLE Ib

Humic Acid – Peat Residue

Humic acid		Peat residue (humine)	
Lab no.	Age, Years	Lab no.	Age, Years
Birm-494b	8480 ± 170	Birm-494a	8030 ± 170
495b	1780 ± 110	495a	2090 ± 170
496b	1500 ± 120	496a	1600 ± 120
507b	1610 ± 110	507a	1360 ± 280
508b	920 ± 130	508a	860 ± 130
GrN-5390	5315 ± 65	GrN-5348	5265 ± 75
5391	4630 ± 90	5349	4370 ± 40
BSC-233	7820 ± 140	GSC-233	7680 ± 140
240	10,640 ± 150	240	10,550 ± 160
253	6510 ± 150	253	6230 ± 150
300	2210 ± 130	300	2330 ± 150
305	270 ± 130	305	220 ± 130
310	9510 ± 150	310	9620 ± 150
I-3536	17,250 ± 600	I-3535	20,500 ± 450
L-391B	4700 ± 150	L-391B	4650 ± 150
368	8350 ± 200	368	7350 ± 650
LU-116A	4890 ± 100	LU-116	4840 ± 100
135A	1860 ± 55	135	1850 ± 60
182A	8100 ± 100	182	8200 ± 100
183A	5740 ± 100	183	5860 ± 100
184A	4520 ± 100	184	4630 ± 100
185A	4050 ± 100	185	4030 ± 100
239A	8440 ± 90	239	8560 ± 90
240A	3880 ± 65	240	4130 ± 65
258A	2980 ± 60	258	3100 ± 60
259A	1720 ± 55	259	1590 ± 50
272A	360 ± 50	272	330 ± 50
273A	1090 ± 55	273	1200 ± 50
287A	7090 ± 80	287	7600 ± 155
288A	3300 ± 60	288	2930 ± 60
289A	3270 ± 60	289	3380 ± 70
291A	170 ± 50	291	220 ± 50
292A	5010 ± 70	292	4900 ± 95
293A	4880 ± 65	293	5110 ± 70
331A	5070 ± 65	331	5240 ± 90
332A	1890 ± 75	332	2070 ± 55
334A	3370 ± 85	334	3420 ± 60
433A	1850 ± 50	433	1860 ± 55
434A	760 ± 60	434	650 ± 75
435A	1200 ± 50	435	1110 ± 50
613A	3480 ± 60	613	3410 ± 50

TABLE Ib (*continued*)

Humic acid		Peat residue (humine)	
Lab no.	Age, years	Lab no.	Age, years
LU-642A	1190 ± 55	LU-642	1160 ± 50
782A	5590 ± 70	782	5540 ± 70
805A	8600 ± 90	805	9550 ± 95
805:2	10,010 ± 100		
806A	2400 ± 55	806	2850 ± 55
832A	8890 ± 90	832	8960 ± 140
N-955	1460 ± 140	N-953	1480 ± 110
Q-1029	4250 ± 60	Q-1028	4230 ± 60
U-750	9120 ± 180	U-2224	9460 ± 100
UB-153C	3245 ± 70	UB-153F	3890 ± 110
158C		158F	3930 ± 105
193C	2970 ± 85	193F	3135 ± 75
195C	2850 ± 60	195A	3220 ± 65
196C	2625 ± 90	196A	2745 ± 60
227C	10,795 ± 140	227A	10,440 ± 110
		227D	10,805 ± 125
		227F	10,945 ± 145
299C	11,225 ± 160	299A	12,360 ± 165
		299D	12,060 ± 125
		299F	12,470 ± 125
399C	10,430 ± 150	399A	10,835 ± 165
		399D	10,190 ± 145
		399F	9995 ± 145
400C	10,380 ± 80	400A	10,730 ± 145
		400D	11,140 ± 155
		400F	10,400 ± 175
433C	5690 ± 85	433A	6610 ± 115
		433F	5690 ± 85
WSU-1464H	5000 ± 140	WSU-1464R	5050 ± 140
1520H	5760 ± 130	1520R	5870 ± 120
		SI-1757	6030 ± 295
1526H	6600 ± 120	WSU-1526R	6760 ± 120
		SI-1555	5575 ± 185
1571H	3690 ± 110	WSU-1571R	3560 ± 110
1572H	3960 ± 110	1572R	4030 ± 120
1573H	4830 ± 180	1573R	4510 ± 110

TABLE Ic
Peat − Charcoal/Wood

Peat		Charcoal	
Lab no.	Age, years	Lab no.	Age, years
Gif-1991	9700 ± 200	Gif-1990	9400 ± 200
GSC-1161	7480 ± 150	GSC-1069	7470 ± 140
IGS-C14/55	5300 ± 100	IGS-C14/55	5205 ± 105
LE-810	6580 ± 70	LE-811	6960 ± 80
750	4670 ± 150	748	4520 ± 120
LU-22	8860 ± 100	LU-21	9060 ± 100
N-276-2	1530 ± 110	N-276-1	1340 ± 110
SU-180	1770 ± 100	SU-179	1610 ± 100
182	1420 ± 100	181	1370 ± 100
184	1980 ± 100	183	2090 ± 100
189	2810 ± 100	188	3220 ± 100
W-2020	15,840 ± 700	W-2039	13,280 ± 1200
1737	11,090 ± 300	1736	10,630 ± 300
WSU-1571R	3560 ± 110	WSU-1571W	3710 ± 110
1572R	4030 ± 120	1572W	4510 ± 190
1573R	4510 ± 110	1573W	5010 ± 180

TABLE Id

Humic Acid – Charcoal

Humic acid		Charcoal	
Lab no.	Age, years	Lab no.	Age, years
A-410B	950	A-410A	1,130
415B	560	415A	820
425B	1,170	425A	1,020
A-525B	11,080	A-525A	10,550
615B	3,050	615A	3,440
805B	11,300	805A	11,150
S102/625	4,330	S102/624	4,100
S74/662	6,810	S74/654	6,720
S74/659	800	S74/655	640
NZ-1545	940	NZ-1542	1,050
1547	1,870	1543	1,860
1604	1,840	1553	2,000
1399	7,760	1452	8,030
1401	6,430	1453	7,440
1403	2,520	1454	2,600
1606	11,400	1554	12,850
1608	19,400	1555	21,100
1610	26,800	1556	26,100
SMU-20	3,800	SMU-21	4,000
27	11,210	28	10,890
42	11,160	43	10,840
56	2,770	57	3,390
158	10,600	140	11,200
195	10,860	168	10,800
201	8,800	197	9,900
WSU-1484	4,690	WSU-1491	5,120

TABLE Ie
Bone Collagen – Charcoal/Wood

Bone collagen		Charcoal/wood	
Lab no.	Age, years	Lab no.	Age, years
BM-237	3720 ± 110	BM-236	3840 ± 65
357	4700 ± 135	356	4760 ± 130
358a*	4620 ± 140		
358a*	4530 ± 110		
395*	3900 ± 90	396	3950 ± 90
397	3850 ± 90		
400*	4000 ± 90	398	3930 ± 90
F-53	24,210 ± 410	F-54	23,750 ± 390
Gif-2295	250 ± 90	GrN-6111	375 ± 50
GrN-4661	20,000 ± 190	GrN-4660	22,300 ± 185
5446	24,430 ± 400	5425	25,500 ± 200
5606	5540 ± 65	5443	5300 ± 40
4948	4130 ± 40	2480	4190 ± 70
5566	3630 ± 40	5705	3635 ± 60
K-1078*	1040 ± 100	K-1099	1020 ± 100
1080*	1010 ± 100	1100	1040 ± 100
1487*	1120 ± 100	1505	930 ± 100
1488*	1150 ± 100	1506	880 ± 100
LU-330	2100 ± 55	LU-329	1960 ± 55
436	3390 ± 60	471	2870 ± 55
472	3720 ± 60		
473	4230 ± 80		
641A	2250 ± 60	653	2130 ± 55
		654	2060 ± 55
653	2030 ± 50		
		655	1930 ± 55
		656	2070 ± 55
Lv-342	460 ± 100	Lv-341	350 ± 120
Ly-301	3970 ± 130	Ly-302	4060 ± 220
303	6140 ± 140	304	6300 ± 140
305	7780 ± 250	306	7890 ± 170
307	7750 ± 340	308	7770 ± 410
309	22,900 ± 600	310	24,150 ± 550
314	17,150 ± 300	315	16,740 ± 300
318	11,500 ± 380	319	11,750 ± 300
321	12,350 ± 200	322	17,320 ± 600
633	5300 ± 130	412	6050 ± 120

TABLE Ie (*continued*)

Bone collagen		Charcoal/wood	
Lab no.	Age, years	Lab no.	Age, years
M-1970	12,750 ± 500	M-1971	13,300 ± 600
2189	1080 ± 120	2188	1040 ± 120
2191	1180 ± 120	2190	1080 ± 120
N-1212-1	1120 ± 165	N-1212-2	1090 ± 140
S-820	3890 ± 160	S-821	3725 ± 95
		823	3615 ± 95
T-447	3300 ± 100	T-925	2950 ± 90
881	2900 ± 90		
951B	800 ± 80	915A	1250 ± 120
1076	4920 ± 90	1005	5150 ± 130
		1001	5020 ± 100
1326	1020 ± 100	1325	1040 ± 110
1210	990 ± 70	1156	890 ± 150
UCLA-689	2500 ± 80	UCLA-690	2510 ± 80
		692	2590 ± 80
1292E	14,500 ± 190	Y-354B	14,640 ± 115
		LJ-55	14,400 ± 300
1708A	3845 ± 60	UCLA-1208	3980 ± 60
		1389A	4000 ± 60
1764B-2	9230 ± 160	GrN-5872	9580 ± 85
1746	10,100 ± 190	5871	10,030 ± 55
1746A	10,120 ± 200	L-336G	10,500 ± 400
1744	35,630 ± 2500	UCLA-1235	35,000 ± 2400
1745B	38,680 ± 2000	1745C	38,550 ± 3800
1754C	32,400 ± 2500	Pta-424	35,700 ± 1100
1754D	34,800 ± 2500	433	36,100 ± 900

*Antler or tusk

TABLE 1f

Shell — Charcoal/Wood

Shell		Charcoal/wood	
Lab no.	Age, years	Lab no.	Age, years
Gif-2108	4910 ± 120	Gif-2109	4990 ± 120
GrN-5061	9780 ± 60	GrN-5871	10,030 ± 55
5886	9450 ± 55	5872	9580 ± 85
GSC-38	12,360 ± 140	GSC-24	12,150 ± 250
741	3380 ± 170	743	2680 ± 140
906	2510 ± 180	894	2870 ± 140
ISGS-45	21,350 ± 320	ISGS-39	20,000 ± 500
L-391E	12,350 ± 250	L-391D	12,150 ± 250
LJ-84	1060 ± 150	LJ-85	960 ± 150
611	1800 ± 300	645	1660 ± 200
925	2600 ± 200	922	2200 ± 200
924	6140 ± 250	923	5480 ± 200
1035	1480 ± 80	UCLA-187	1500 ± 80
1034	1630 ± 80		
MC-210	7000 ± 150	MC-209	6800 ± 150
N-366-1	5090 ± 130	N-365	4920 ± 195
366-2	5090 ± 140		
318	3960 ± 120	317	3960 ± 175
320	3260 ± 120	319	3370 ± 135
SI-450	2460 ± 100	SI-449	1750 ± 90
T-1017	4910 ± 80	T-1014	5250 ± 80
TX-341	1990 ± 100	TX-343	1890 ± 100
342	1840 ± 90	344	1560 ± 100
345	2280 ± 90	356	2070 ± 110
388	2150 ± 60	397	2540 ± 110
389	2240 ± 90	396	1900 ± 90
390	2010 ± 90	399	1740 ± 100
392	2040 ± 90	400	1880 ± 90
393	1890 ± 90	401	1780 ± 100
394	1810 ± 90	402	1400 ± 110
455	1950 ± 70	456	2010 ± 90
457	2180 ± 90	449	1950 ± 80
460	2220 ± 80	450	2020 ± 80
521	820 ± 50	520	780 ± 40
946A	1100 ± 70	946B	720 ± 90
947A	1300 ± 70	947B	1120 ± 110

TABLE If (continued)

Shell		Charcoal/wood	
Lab no.	Age, years	Lab no.	Age, years
TX-949A	1740 ± 70	TX-949B	1120 ± 70
951B	1310 ± 60	951A	1060 ± 90
968A	1500 ± 70	968B	2170 ± 180
1259B	1870 ± 70	1259A	1650 ± 90
U-2194	2920 ± 80	U-753	2560 ± 90
VRI-127	3460 ± 90	VRI-36	3650 ± 130
W-26	2540 ± 200	W-27	2500 ± 200
154	580 ± 200	155	600 ± 200
WIS-325	3650 ± 65	WIS-321	3550 ± 65
WSU-349	1890 ± 170	WSU-348	1670 ± 160
354	1740 ± 170		
404	5675 ± 175	403	5600 ± 175

TABLE Ig

Humic Acid – Soil Residue

Humic acid		Soil residue	
Lab no.	Age, years	Lab no.	Age, years
A-715B	7990 ± 500	A-715A	7390 ± 400(R)
	7890 ± 420		
746B	10,100 ± 700	746A	11,000 ± 1000(R)
755B	8050 ± 400	755A	6450 ± 300(R)
969B	9270 ± 800	969A	8900 ± 400(R)
970B	10,150 ± 600	970A	7800 600(R)
ANU-144B	24,500 ± 600	144A	20,300 ± 2400(R)
145B	22,190 ± 600	145A	17,740 ± 1500(R)
146B	24,960 ± 580	146C	25,360 ± 580(R)
Bonn-362	5380 ± 80	Bonn-364	3490 ± 70(R)
363	5970 ± 40		
368	2530 ± 60	370	2850 ± 70(R)
369	2980 ± 70		
399	4890 ± 50	401	2980 ± 70(R)
671	1560 ± 70	672	2275 ± 60(R)
674	7600 ± 220	675	6930 ± 80(R)
I-1419B	19,000 ± 4500	I-1419A	16,500 ± 500
Campbell et al. Soil Sci. 104, 81 (1967)	1235 ± 60		1140 ± 50
	195 ± 50		485 ± 70

REFERENCES

Bennett, C. A. and N. L. Franklin
    1954   Statistical Analysis in Chemistry and the Chemical Industry. New York, John Wiley and Sons, 180-182.

Crandell, D. R. and D. R. Mullineaux
    1973   Pine creek volcanic assemblage at Mount St. Helens, Washington. U.S. Geol. Surv. Bull. 1383A.

Davis, O. K., D. A. Kolva, and P. J. Mehringer, jr.
    1976   Pollen analysis of Wildcat Lake, Whitman County, Washington: the last thousand years. Northwest Sci. 8:275.

Deevey, E. S. jr., M. S. Gross, and G. E. Hutchinson
    1954   The natural C^{14} contents of materials from hard-water lakes. Proc. Nat. Acad. Sci. 40:285.

Geyh, M. A., J. H. Benzler, and G. Roeschmann
    1971   Problems of dating plastocene and holocene soils by radiometric methods. *In* Paleopedology. Vaalan, ed. Jerusalem, Israel Univ. Press, 63.

Grant-Taylor, T. L.
    1978   The extraction and use of the plant lipids as a material for radiocarbon dating. Proc. 8th Int. Conf. on Radiocarbon Dating. New Zealand. 2 (E58).

Grove, G. D., M. Rubin, B. B. Hanshaw, and W. A. Beetem
    1969   Carbon-14 dates of ground water from a Paleozoic carbonate aquifer, south-central Nevada. U.S. Geol. Surv. Prof. Paper. 650-C:215-218.

Haynes, C. F., jr., D. C. Grey, and A. Long
    1971   Arizona radiocarbon dates VIII. Radiocarbon. 8:1.

Kittleman, L. R.
    1973   Minerology, correlation, and grain-size distribution of Mazama tephra and other postglacial pyroclastic layers, Pacific Northwest. Geol. Soc. of America Bull. 84:2957.

Martel, Y. A. and E. A. Paul
    1974   The use of radiocarbon dating of organic matter in the study of soil genesis. Soil Sci. Soc. Proc. Am. 38:501.

Mehringer, P. J., jr., and J. C. Sheppard
    Holocene History of Little Lake, Mohave Desert, California. *In* Ancient Californians, PaleoIndians of the Lakes Country. Davis, ed. Nat. Hist. Mus. of Los Angeles County. Sci. Series 29 (In press).

Mehringer, P. J., jr. and C. N. Warren
    1976   Marsh, dune and archaeological chronology, Ash Meadows, Amargosa Desert, Nevada. *In* Nevada Archaeol. Res. Paper No. 6. Elston, ed. Univ. of Nevada, 120-150.

Mullineaux, D. R., J. H. Hyde, and M. Rubin
    1975   Widespread late glacial and postglacial tephra deposits from Mt. St. Helens Volcano, Washington. J. Res. U.S. Geol. Surv. 3:329.

Murphy, M. T. J.
    1969   Analytical methods. *In* Organic Geochemistry. Longman. Berlin, Springer-Verlag, 75.

Olsson, I. U.
    1972   The pretreatment and the interpretation of the results of ^{14}C determinations. *In* Climatic Changes in the Arctic Areas During the Last Ten-thousand Years, Vasari et al., eds., Ser. A. Geol. No. 1, Oulu University, 9-34.

Olsson, I. U., et al
    1974   A comparison of different methods for pretreatment of bones I. GFF. 96:171.

Petersen, K. L. and P. J. Mehringer, jr.
    In Press   Postglacial timberline fluctuations, La Plata Mountains, southwestern Colorado. Arctic and Alpine Research.

Ruhe, R. V., G. H. Miller, and W. J. Vreeken
    1971   Paleosols, Loess sedimentation and soil stratigraphy. *In* Paleopedology. Yaalan, ed. Jerusalem, Israel Univ. Press, 41.

Scharpenseel, W. H.
    1972   Messung der naturlichen C-14 Konzentration in der organischen substanz von rezenten boden: Eine zwischenbelanz. A. Pfanz, Bodenkunde. 133:241.

Youden, W. J.
    1951   "Statistical Methods for Chemists," New York, John Wiley and Sons, 28.

PART IV
Artificial Radiocarbon in Nature

1

Artificial Radiocarbon in the Stratosphere

Rainer Berger

The UCLA radiocarbon laboratory has monitored tropospheric $^{14}CO_2$ levels since 1960 at its field station at China Lake, California (35°37'N Lat, 117°41W Long) (Berger and Libby 1965). This desert location, away from large urban areas emitting quantities of fossil CO_2, ensures the measurement of concentrations of ^{14}C representative of average tropospheric carbon dioxide. Samples are collected once monthly over a period of one week by exposing 2N carbonate-free sodium hydroxide to air. After acidification and purification the resulting $^{14}CO_2$ is measured in a gas proportional counter overnight.

The maximum tropospheric radiocarbon concentration was observed in 1963, a fact corroborated by other laboratories (Nydal et al., this volume). At that time radiocarbon levels were twice the natural production rate. In the laboratory the natural rate is equivalent to 95% of the count rate of NBS oxalic acid and is the same as the concentration of biospheric radiocarbon in A.D. 1890, prior to the large scale exhaust of fossil CO_2 from the world's industrial plants. Since 1963 there has been observed a steady decline of radiocarbon in the troposphere. In 1970 only one-half remained and in 1977 about one-third. Thus the mean residence time of CO_2 in the troposphere near the surface of the earth can be approximated by 7/0.693 or about ten years.

In the course of the years of tropospheric monitoring summer maxima were initially observed that later gave way to occasional annual spikes of at least 10% above the normal flow of the data (fig. 1). Inasmuch as the large nuclear detonations of the early 1960's had deposited large quantities of radiocarbon in the stratosphere, this locale was the obvious source region for radiocarbon mixing with the air in the lower troposphere. In April 1962, before massive Russian injections a year later, the ^{14}C level at 12.2 km altitude (36°N Lat) was already approaching three times the normal production rate of A.D. 1890 (Fergusson 1963a). Consequently it was very interesting to attempt direct measurements of stratospheric CO_2. The only obstacle was lack of a convenient way to collect air samples above 12-15 km altitude.

With the advent of the civilian NASA Earth Observation Program, two U-2 high-altitude aircraft became available in 1972 for the gathering of data. In order to collect CO_2

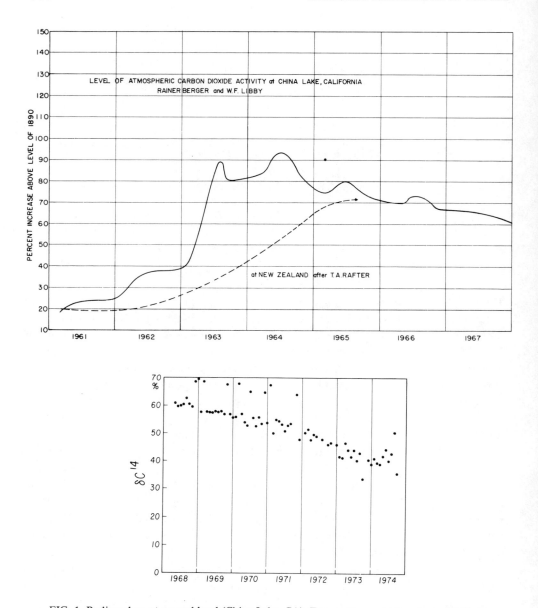

FIG. 1. Radiocarbon at ground level (China Lake, CA). Top, summer maxima during 1963-66. Only one spike was observed in Spring 1965. Bottom, spikes observed during 1968-1974.

at an altitude of some 20 km a modified version of the technique developed by Fergusson was used (1963b).

The collection unit is a high volume sampler fabricated from a 36 cm long aluminum tube containing a volume of about 5ℓ. At each end pilot-operated, electric, air-tight gate valves permit a flow of air through the collector. It is filled with a synthetic zeolite, Linde molecular sieve 4A, which adsorbs both CO_2 and moisture. The unit is mounted behind the pilot in the "canoe", or a backbone-like structure on top of the aircraft. Air enters a special scoop and passes through the gate valve over the sieve to the discharge port. Typical absorption times are on the order of a few hours.

After arrival of an exposed collector at the UCLA laboratory, the molecular sieve is heated under vacuum for several hours to release its CO_2 which is then purified, stored, and counted. After overnight baking the sieve can be reused for new collections without exhibiting a significant memory effect.

So far all collections at altitude were made over the western United States as listed in the table (fig. 2). When these data are plotted (fig. 3), it is immediately apparent how

UCLA lab no.	Date	Altitude	Location	$\delta^{14}C$ in %*
2020	13 June 75	19800 m	N. Sierras	+127.4
2022	25 Sept 75	19800 m	California	+90.7
2030	3 Dec 75	18-21 Km	San Francisco-Phoenix	+79.1
2031	30 March 76	19800 m	San Francisco-Oregon	+107.0
2033	2 June 76	18-21 Km	San Francisco-Salt Lake City	+85.6
2042	23 Sept 76	19800 m	San Francisco-Denver	+99.2
2044	29 March 77	19800 m	Pacific near Los Angeles	+199.6

*In percentage above 0.95 NBS oxalic acid uncorrected.

FIG. 2. Stratospheric $^{14}C_2$ collections

much greater numerically the radiocarbon concentrations are than those in the troposphere. Since all measurements were made at roughly similar altitudes, the data can be compared. In the lower stratosphere there seem to appear spring peaks not unlike the summer maxima clearly observed in tropospheric air of the mid-1960's. This implies that, during the summer, pumping action through the tropopausal break siphons off ^{14}C-rich air into the troposphere where it is quickly dissipated. At present the experimental data are too few to allow a firm and final determination of the stratospheric residence time of carbon dioxide. Since the dates are the only modern measurements in the world today, a somewhat quantitative assessment ought to be made. The residence time can be expressed as the mean-life of CO_2 in the stratosphere and hence is equal to the half-life of CO_2 in the air divided by ln 2. Using the available measurements, a residence time on the order of about twice that in the troposphere is found.*

Continued measurements should allow us to improve this estimate significantly. A firm residence value would permit assessment of the effects of massive stratospheric air travel and would aid in determining if certain worldwide limitations should need to be enforced to reduce the danger of a Greenhouse Effect with its far-ranging climatic impact on man and nature (see Cont. No. 1748, Inst. of Geophysics and Plan. Physics, UCLA).

*A residence time of 15-20 years would be a maximum estimate, based on the time period of April 1962 through the mean of 1975-1976. There were, however, further weapons tests after 1962 resulting in stratospheric additions of radiocarbon which complicates this assessment.

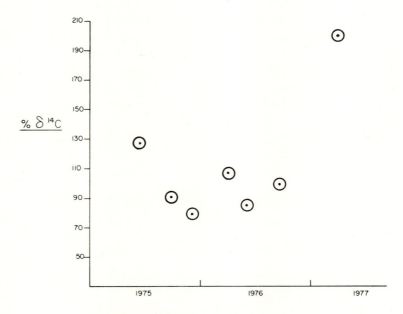

FIG. 3. Stratospheric radiocarbon concentrations.

ACKNOWLEDGMENTS

Many of these data would not have been available without the unfailing contributions of Dr. G. Plain, Ridgecrest, CA. My thanks go to the pilots and personnel of NASA Moffett Field. The collectors were designed with the aid of Lockheed-Burbank scientists and engineers and constructed at UCLA. W. Ferguson has been very helpful in arranging shipments. The entire project was supported by the National Aeronautics and Space Administration and the National Science Foundation.

REFERENCES

Berger, R., and W. F. Libby
    1965   UCLA date lists. Radiocarbon.
    to
    Present

Fergusson, G. J.
    1963a   Upper Tropospheric Carbon-14 levels during Spring 1962. J. Geophys. Res. 68:3933-3941.
    1963b   High Volume Sampler for Atmospheric Carbon Dioxide. Rev. Sci. Instr. 34:403-406.

Nydal, R., K. Lövseth, and S. Gulliksen
    1979   A survey of radiocarbon variation in nature since the Test Ban Treaty. Proc. 9th Int. Radiocarbon Conf. Univ. of Calif. Los Angeles and San Diego (this volume)

2

A Survey of Radiocarbon Variation in Nature since the Test Ban Treaty

R. Nydal, K. Lövseth, and S. Gullicksen

ABSTRACT

^{14}C variation in the atmosphere and ocean surface has been studied since the early 1960's. This paper presents ^{14}C data from the troposphere during the last twelve to fourteen years between Spitsbergen (78°N) and Madagascar (21°S). At present the ^{14}C content in the troposphere constitutes about 40% above normal level, and there are only small differences between various latitudes. Measurements of ^{14}C in the surface water of the ocean have been performed at some locations for a period of only six to seven years. ^{14}C concentration in the ocean surface sometimes shows an oscillating pattern which is not quite understood. In some cases these oscillations appear to be related to temperature changes which, for instance, could be caused by internal mixing of water. Even though there is great scattering in the ^{14}C data, they show that the ^{14}C concentration in the ocean surface has reached an upper level which has been almost constant for the last seven to eight years.

INTRODUCTION

In 1963 Great Britain, the Soviet Union, and the U.S.A. negotiated a general agreement on prohibition of nuclear tests in the atmosphere and ocean. This decision was welcomed because of concern about possible hazards to humanity from these tests. Underground tests were, however, not included in this agreement, but such tests produce little radioactivity in the atmosphere. France and China did not sign the treaty and then continued small scale atmospheric testing after 1962. The end of the major tests provided an opportunity to use bomb-produced isotopes as tracers in order to study exchange between various reservoirs in nature. Bomb-produced ^{14}C was found extremely valuable for studying the carbon cycle.

A program for studying the distribution of ^{14}C from nuclear tests was started at the Trondheim laboratory in 1962. The initial purposes were to obtain exact data on terrestrial ^{14}C activity for estimating the hazard to man and to examine transport processes in the atmosphere (Nydal 1963). Our interest later included the carbon cycle in nature. Calculation of CO_2 exchange rates between atmosphere and ocean became important in this respect (Nydal 1967, 1968; Nydal and Lövseth 1970; Gulliksen et al. 1972).

In the present paper, data from the atmosphere and the ocean surface from the early 1960's until the present time are presented in graphs. Some aspects concerning ^{14}C variation are discussed, but further calculations on the CO_2 exchange in nature have not been performed.

ATMOSPHERE

During the first half of 1962, a sampling program for atmospheric ^{14}C was initiated at ten locations between Spitsbergen and Madagascar (fig. 1).

In 1967, the ^{14}C concentration of the lower atmosphere was nearly uniform, and calculation of exchange between various parts of the atmosphere became less important. Further measurements were, therefore, reduced to the following four stations: Nordkapp (71°N); Mas Palomas (28°N); Fort-Lamy (10°N); and Madagascar (21°S). The sampling

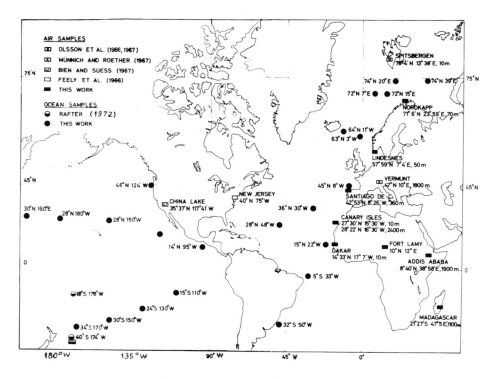

FIG. 1. Collection of ^{14}C samples.

on Spitsbergen continued until 1968. Most of the samples were not measured, but were stored as calcium carbonate. Measurements at this location have been performed since 1961 by Dr. Ingrid Olsson, University of Uppsala. Parallel measurements at the same location were found unnecessary (Olsson 1966). Presently the station at Mas Palomas is no longer in operation. Sampling has continued at Observatorio Astronomico, Izana, Tenerife since the early days of the program (Nydal 1968). In addition to the four stations, samples are still collected at Lindesnes (58°N) and Dakar (15°N). These samples are stored as calcium carbonate and are available.

Ground level samples are collected by direct adsorption of CO_2 in NaOH. One liter of 1/2N NaOH is placed in an open dish (450 cm²) and exposed to the atmosphere for one week. After exposure the solution contains dissolved Na_2CO_3 and H_2CO_3. Only distilled water or rain water is used in the solution, in order to avoid contamination by fossil carbon. The solution is transferred to a plastic bottle and is sent to the laboratory, where HCl is added under vacuum. A CO_2 volume of approximately 5 liters is thus recovered, sufficient for radioactive measurements in a 1.5 liter proportional counter (Nydal et al. 1975).

OCEAN

Sampling from the ocean surface was begun in 1965. The number of sampling locations increased during the following five years. Figure 1 shows all the locations for regular collection of ocean samples. It has been difficult to perform regular measurements for long periods at all these locations. Sampling in the Norwegian and Barents Seas lasted for only a short period during 1965 and 1966 (Nydal 1967).

There have been annual changes in shipping routes, so each sampling locality has not always been well defined. Some lines also have been cancelled. This applied to the lines to Seattle and to South America in 1970 and 1972, respectively. Collection of samples in the Atlantic Ocean ceased in 1972 after a measuring period of six to seven years. Sampling continued in the Pacific Ocean until spring 1976. After that time ships to the Far East passed through the Mediterranean and the Suez Canal. There are presently no sample collections in the Pacific Ocean.

The sampling procedure to date for the ocean samples has been the following: A steel drum (free of CO_2) is filled with 200 liters of sea water collected from a depth of 6-8 m. The temperature in the water at the collection depth is measured, and the drum is sealed for transport to the laboratory. The time lag between collection of samples and treatment varies according to changes in shipping routes. Storage time for samples from the Atlantic Ocean has generally been between one and two months, and from the Pacific Ocean from two to four months. Some exceptions, with storage time up to six months, have occurred. On reaching the laboratory, the drums are opened and 1 liter of concentrated sulphuric acid is added. The pH value in the sea water is then reduced to below 3.0. A circular metal tube with small holes is plunged to the bottom of the drum, and nitrogen (free of CO_2) is flushed through the drum and an ammonia trap. About two-thirds of the CO_2 amount is flushed out, which produces about 5 liters of CO_2. The flushing time is approximately two days, and about 500 liters of nitrogen is consumed.

Measurements and Calculation

The major part of the ^{14}C measurements of both the atmospheric and the ocean samples were carried out in a 1.5 liter CO_2 proportional counter, working at 2 atm pressure. The background of this counter has recently been lowered to 0.59 c/min (Nydal et al. 1975) but has generally ranged between 0.80 c/min and 0.90 c/min. The recent standard net count (oxalic acid) is 19.0 c/min.

The increase in ^{14}C concentration is calculated in percentage of excess above normal ^{14}C level, represented by 95% of the activity in the NBS oxalic acid standard (Broecker and Olson 1959). This activity represents the normal activity in twentieth century wood when corrected for age until 1950.

The $^{13}C/^{12}C$ ratios have been measured in the majority of the samples in order to make corrections for isotopic fractionation. Samples from the ocean and the atmosphere are corrected to $-25^{o}/oo$, which is regarded as the normal ^{13}C value in terrestrial vegetation. The corrected ^{14}C excess ($\Delta^{14}C$) per mil above normal level is calculated from the formula

$$\Delta^{14}C = \delta^{14}C - 2(\delta^{13}C + 25)\left(1 + \frac{\delta^{14}C}{1000}\right)$$

In this formula $\delta^{14}C$ represents the uncorrected ^{14}C excess per mil relative to the NBS standard, and $\delta^{13}C$ represents the deviation per mil in the $^{13}C/^{12}C$ ratio relative to the PDB standard (Craig 1961). The limit of error in each $\Delta^{14}C$ value is one standard deviation, which is generally between 10 and 20 per mil and includes only statistical errors.

Results show that atmospheric CO_2 samples, which are absorbed in NaOH, have a mean $\delta^{13}C$ value close to -25 per mil at both higher southern and northern latitudes. The $\delta^{13}C$ value of the samples increases toward the equator.

Results and Discussion

ATMOSPHERE

Figure 2 shows ^{14}C variation in the troposphere above Norway and Spitsbergen. After two years of measurements at three locations in Norway (58°N, 63°N, 71°N) and one on Spitsbergen (78°N), we realized that the troposphere was in such rapid circulation at higher latitudes that measurements were confined only at Nordkapp (71°N, 24°E). This station was chosen because of its northerly location and because of a regular mail connection to our laboratory throughout the year. During the period 1963 to 1967, the ^{14}C variation was quite regular, mainly because the U.S.A., the Soviet Union, and Great Britain conducted no atmospheric tests. China and France have not joined the treaty, but they have tested only a few, small fission bombs during this period. The latter two nations tested their fission bombs in 1967 and 1968 respectively. The irregularity of the curve after 1967, especially in 1970 and 1971, reflects further injections of ^{14}C into the atmosphere.

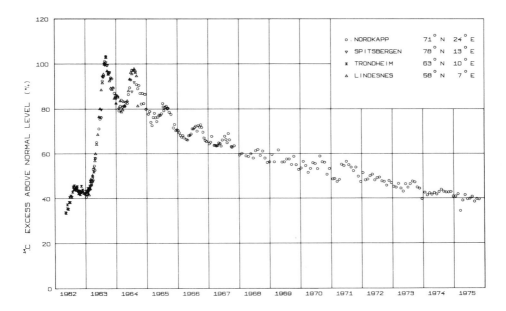

FIG. 2. Radiocarbon in the troposphere.

Figure 3 shows a graph of ^{14}C variation at Tenerife and Gran Canaria. Collections at both localities took place only during 1963 and 1964. ^{14}C variation was later monitored only on Gran Canaria, while the samples on Tenerife were collected and stored for later use. Different meteorological conditions at the two stations indicate differences in ^{14}C concentration during 1963 and 1964. The two stations are separated by the trade wind inversion layer, and this layer may be responsible for the deviations. The Canary Islands are situated on the eastern side of the Azores anticyclone, which is periodically renewed by colder air masses from northern latitudes. At the upper level of the cyclone there is also advection of warmer air masses from southern regions where the circulation is slower. The ^{14}C curve from Tenerife and Gran Canaria is for this reason not as smooth as the curve farther north. Any further injection of ^{14}C in the atmosphere during 1970-1971 is, therefore, not clearly discernible.

The ^{14}C variation seen in figure 4 clearly shows that seasonal variations in Africa at Fort-Lamy and Addis Ababa are normally much lower than variations farther north. The special situation at Dakar differed from this picture when air from the north produced a local increase in ^{14}C concentration, mainly in the period 1965-1967. There are some very low data in Dakar in winter 1966-1967, and in Addis Ababa in winter 1968-1969. The values are lower than corresponding data observed at Madagascar. The very low values are probably due to contamination from fossil carbon during the time of collection. Since the end of 1969 the measurements have been performed only at Fort-Lamy.

Figure 5 shows ^{14}C measurements from Madagascar performed from 1964 to 1975, with the exception of 1974. The ^{14}C level at Madagascar in 1975 is, as for higher

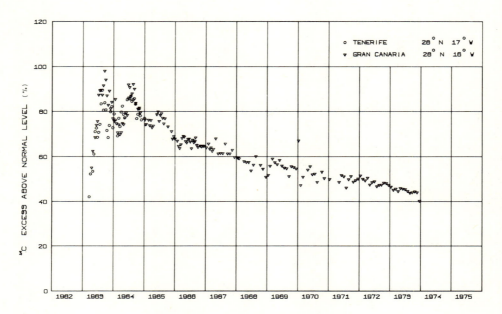

FIG. 3. Radiocarbon in the troposphere.

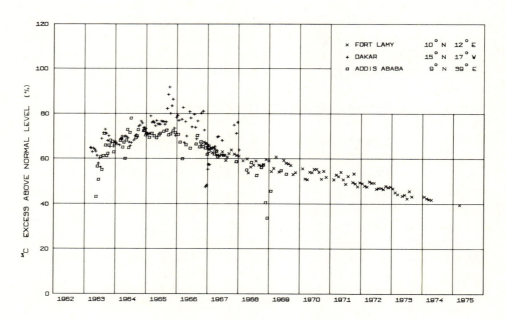

FIG. 4. Radiocarbon in the troposphere.

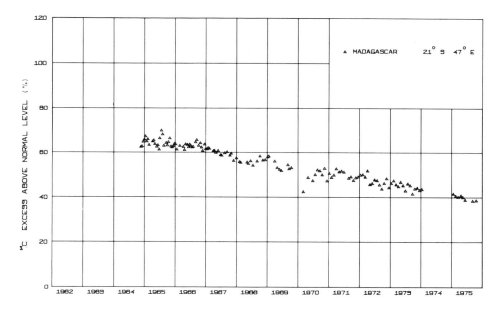

FIG. 5. Radiocarbon in the troposphere.

latitudes, close to 40% above normal. The curve from Madagascar also shows seasonal variations during the whole measuring period.

In figure 6, the ^{14}C data from four locations in the troposphere are collected in one graph. This shows that important differences in ^{14}C concentration between various parts of the troposphere nearly disappeared during four years (1963-1967). In the period 1967 to 1975, the whole troposphere can be regarded as a single and well mixed reservoir. This simplifies the calculations of the CO_2 exchange between the atmosphere and the ocean.

OCEAN

Figure 6 also shows ^{14}C observations in sea water from the Atlantic and Pacific Oceans. The graph was constructed in order to obtain an approximate mean value for the ocean, at least in the region 0°-30°N. Because we had no observations at our laboratory in the period from 1962 to the end of 1965, the data from this period are from papers by Münnich and Roether (1967), and Bien and Suess (1967).

The scattering in ^{14}C data from the ocean surface is great, but the mean value indicates that the maximum ^{14}C level had already been reached in 1967-1968. Since then it has been almost at a constant level or has been slightly decreasing (Gulliksen et al, 1972).

We have recently studied variation in ^{14}C concentration and surface temperature at different locations. There tends to be a positive correlation between ^{14}C concentration and temperature at 28°-31°N (fig. 7). A similar trend is also seen in the Pacific Ocean at

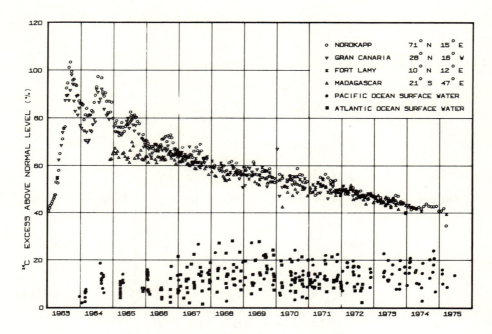

FIG. 6. Radiocarbon in the troposphere.

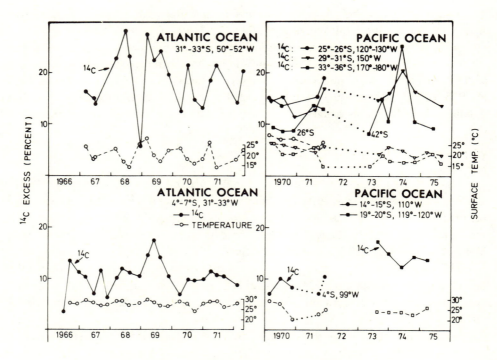

FIG. 7. Radiocarbon in the ocean surface.

25°-34°N. If this trend is real, the most convincing explanation is a greater exchange of water in winter time between ocean surface and deeper layers. When water is cooled, it sinks and is replaced by deeper water with lower ^{14}C concentration.

In figure 8 a great and regular oscillation in ^{14}C concentration north of the Canary Islands is noted. The regular temperature curve indicates a normal yearly change in surface temperature. There tends to be a positive correlation between ^{14}C concentration and temperature at this locality, and also in the Pacific Ocean at the same latitude. In the lower graphs in figure 8 the temperature variations are smaller, and it is difficult to find any convincing correlation.

In figure 9 the ^{14}C concentration in the ocean surface is given for some locations in the southern hemisphere. These graphs have certain features similar to those from the corresponding (approximately) northern latitudes (fig. 8), but it is difficult to see any systematic temperature correlation with the ^{14}C variations.

There have been objections to the ^{14}C variation found in our ocean samples. We have been informed during the Los Angeles conference (personal communications) that there have been adverse experiences with storage of sea water in polyethylene bottles. The bacterial action in sea water continues after collection, and a possible exchange of carbon with the polyethylene material of the bottle itself has been observed. The method of avoiding contamination has been to pretreat the drums with mercuric chloride.

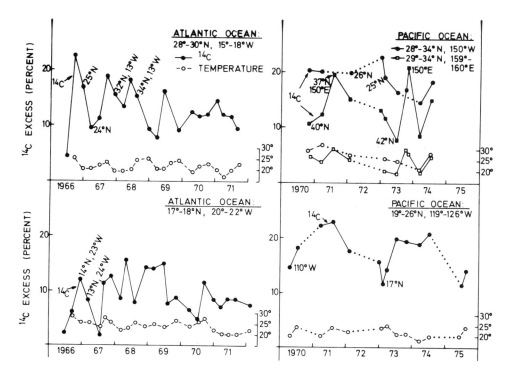

FIG. 8. Radiocarbon in the ocean surface.

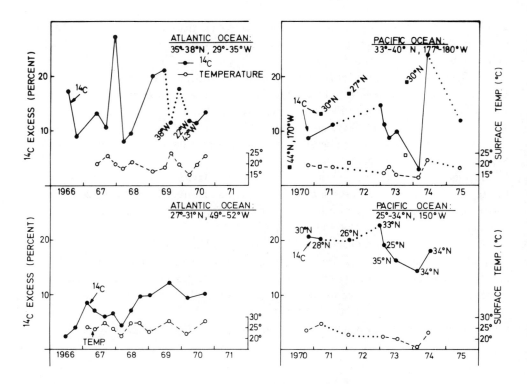

FIG. 9. Radiocarbon in the ocean surface.

In our work the containers have been new oil drums direct from the factory. In addition they have been cleaned and the air has been blown out with nitrogen. The sea water inside has been kept in complete darkness. One possibility of contamination could be oil in the sea water itself, a contingency which cannot be ignored. An exchange with oil carbon depresses the ^{14}C concentration. In the Atlantic Ocean between 5°N and 45°N we obtained mean ^{14}C values, regarded as too high (Oeschger et al. 1974). Even if some data may be wrong, we believe that errors are infrequent. It is reasonable to expect some fluctuation in ^{14}C in the ocean surface during the year.

Very recently the flushing time for the sea water samples has been reduced to one hour. It has thus become possible to process them on shipboard. We intend to collect parallel samples, storing them in the former way in order to check the reliability of the previous data. A further program for the ocean surface is now in progress.

ACKNOWLEDGMENTS

This work has, since 1962, been financially supported by the Norwegian Research Council for Science and the Humanities (NAVF). We are indebted to all persons who have been collecting samples, most of whom have been mentioned previously (Nydal 1968). Special thanks to Captain K. Hornburg, inspector in Wilh. Wilhelmsen's shipping company, Oslo, who has most kindly supported this work during the past seven years. Thanks also to the laboratory staff, especially to Fred Skogseth, for careful work in processing and counting the samples.

REFERENCES

Bien, G., and H. Suess
    1967   Transfer and exchange of ^{14}C between the atmosphere and the surface water of the Pacific ocean. Symp. on Radioactive Dating and Methods of Low-Level Counting. Vienna, Atomic Energy Agency. UN Doc. SM-87/55.

Broecker, Wallace S. and Edvin A. Olson
    1959   Lamont radiocarbon measurements VI. Am. J. Sci. Radiocarbon Supp. 1:111-132.

Craig, Harmon
    1961   Mass-spectrometer analysis of radiocarbon standards. Am. J. Sci. Radiocarbon. Supp. 3:1-3.

Gulliksen, S. R. and K. Lövseth
    1972   Further calculations on the C-14 exchange between the ocean and the atmosphere. Proc. 8th Int. Conf. on Radiocarbon Dating. New Zealand. 1:C63.

Münnich, K. O., and W. Roether
    1967   Transfer of bomb ^{14}C and tritium from the atmosphere to the ocean. Internal mixing of the ocean and the basis of tritium and ^{14}C profiles. Symp. on Radioactive Dating and Methods of Low-Level Counting. Vienna, Int. Atomic Energy Agency. UN Doc. SM-87/22.

Nydal, R.
    1963   Increase in radiocarbon from the most recent series of thermonuclear tests. Nat. 6:212-214.
    1967   On the transfer of radiocarbon in nature. Symp. on Radioactive Dating and Methods of Low-Level Counting. Vienna, Int. Atomic Energy Agency. UN Doc. SM 87/29.
    1968   Further investigation on the transfer of radiocarbon in nature. J. Geophys. Res. 73:3 617-3635.

Nydal, R., and K. Lövseth
    1970   Prospective decrease in atmospheric radiocarbon. J. Geophys. Res. 75:(12) 2271-2277.

Nydal, R., S. Gulliksen, and K. Lövseth
    1975   Proportional counters and shielding for low level gas counting. Proc. Conf. on Low Radioactivity Measurements and Application. Czechoslovakia.

Oeschger, H., U. Siegenthaler, and A. Gugelmann
    1974   A box diffusion model to study the carbon dioxide exchange in nature. Tellus. 17 (2):167-191.

Olsson, I. U., I. Karlen, and A. Stenberg
    1966   Radiocarbon variations in the atmosphere. Tellus. 18 (2-3): 293-297.

Rafter, T. A., and B. J. O'Brien
    1972 ^{14}C measurements in the atmosphere and in the south Pacific Ocean: a recalculation of the exchange rates between the atmosphere and the ocean. Proc. 8th Int. Conf. on Radiocarbon Dating. New Zealand. 1: C17.

3

The Uptake of Bomb ^{14}C in Humans

M. J. Stenhouse and M. S. Baxter

ABSTRACT

Human soft tissues were analyzed for ^{14}C content and compared with theoretical tissues ^{14}C curves predicted from environmental ^{14}C levels. In this way estimates of the mean residence times of body organ carbon were obtained. A value of 6 ± 4 years for the mean residence time of soft tissue carbon was determined. Dose rates from bomb ^{14}C were evaluated for those years with an excess contribution over the pre-bomb era, thus assessing the human radiation burden from ^{14}C released in nuclear weapon tests.

INTRODUCTION

Artifical ^{14}C, released in nucelear weapon tests and injected mainly into the stratosphere, has been used extensively as a global tracer for determining rates of exchange of carbon in nature — especially for CO_2 exchange between atmosphere and ocean (Bien and Suess 1967, Young and Fairhall 1968, Rafter and O'Brien 1970, Walton el at. 1970, Gulliksen and Nydal 1972). Within the terrestrial biosphere, which has a relatively small carbon reservoir, this technique has been applied to studies of the transfer of bomb ^{14}C to humans. $^{14}CO_2$, of both manmade and natural origins, is assimilated in plant material during photosynthesis and is incorporated in human tissue via transfer through the food chain. As expected with the advent of the nuclear era, early measurements of the ^{14}C activity of soft tissue confirmed that uptake of bomb ^{14}C has occurred (Broecker et al. 1959). Since these initial results, further investigations on a similar theme (Libby et al. 1964, Harkness and Walton 1969, Nydal, et al. 1971) have utilized the major additional pulse of ^{14}C produced by the powerful detonations of the early sixties. With the continuing existence of nonequilibrim ^{14}C levels in nature, the opportunity is still available for determination of residence times of carbon within the human body. On the other hand, the resultant increase in ^{14}C concentration in humans has been cause for concern over possible genetic and somatic damage due to bomb ^{14}C irradiation to future as well as present generations. An extimated 9×10^{28} ^{14}C atoms, released in nuclear explosions during the past three decades, doubled the natural ^{14}C concentration in the troposphere in late 1963. Although ^{14}C is a relatively weak β-emitter, its long half-life and its ease of incorporation in all biochemical molecules merit consideration of this radioisotope as a long-term radiological health hazard (Pauling 1958, Sakharov 1968, Leipunsky 1958, Totter et al. 1958).

This paper summarizes the major research effort at the University of Glasgow Radiocarbon Laboratory initiated for purposes of (1) estimating the residence time of carbon in different tissues of the human body from knowledge of their ^{14}C content, and (2) on the basis of these estimates, assessing the human radiation burden from bomb ^{14}C, both present and potential.

Experimental Method

PREPARATION AND PURIFICATION OF CO_2

Details of the collection and pretreatment of tissues are given elsewhere (Farmer et al. 1972, Stenhouse and Baxter 1976). It is sufficient to note here that tissues were obtained unfixed from postmortem examinations and where therefore free from contamination from noncontemporary carbon. In order to monitor environmental ^{14}C levels, several dietary samples (meat and fish) were analyzed for ^{14}C content, while atmospheric ^{14}C level was measured on CO_2 collected by absorption in 2M KOH at three locations in the United Kingdom (fig. 1).

CO_2 was produced from tissue and dietary samples by combustion in a stream of oxygen in a quartz tube and was absorbed in 4M KOH solution. Subsequently, for all samples, CO_2 was precipitated as $BaCO_3$ by addition of $BaCl_2$ solution; the resultant solid was filtered, washed with hot distilled H_2O, and dried in an oven at 110°C. Thereafter, CO_2, released by acid hydrolysis of $BaCO_3$ using 95%H_3PO_4, was purified in a high vacuum system by absorption on CaO at 700°C, followed by evacuation of the furnace, cooled to 400°C, to $< 2 \times 10^{-5}$ torr. Sample gas was regenerated once more by heating the furnace to >900°C; partial removal of radon was achieved by passage of CO_2 through a charcoal trap at 0°C. The sample CO_2 was stored for at least three weeks prior to counting.

CO_2 from NBS oxalic acid was produced via wet oxidation using $KMnO_4$ in 0.5M H_2SO_4, while activity measurement was performed on inactive CO_2 derived from several sources (anthracite, marble chips, and crude petroleum) in order to check each state of the processing system for contamination.

COUNTING OF CO_2

Sample ^{14}C activities were measured using a Johnston Laboratory, Inc. Beta-Logic Gas Proportional Counting System. The characteristics of the counters were as follows:

1. *guard ring*: Gas-flow (Q-gas) Geiger type, 12 independent anode wires; working voltage 1300-1400 volts, count rate ~600 c.p.m.

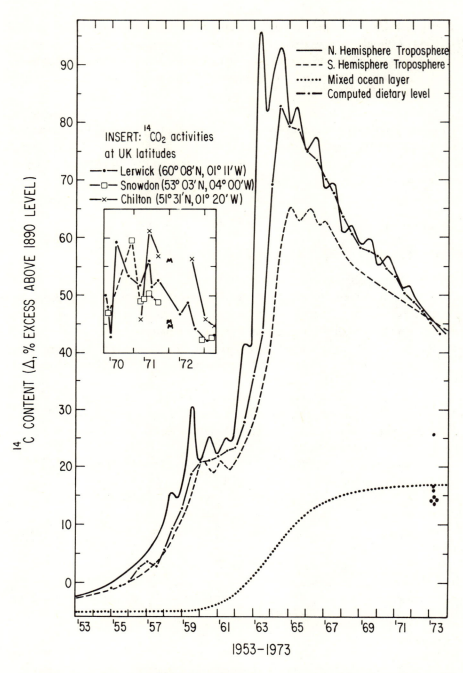

FIG. 1. Environmental ^{14}C levels: 1953-1973. M ≡ meat samples, obtained fresh in January 1972; • ≡ fish obtained fresh in March 1973. The anomalously high ^{14}C concentration in the fish samples belongs to plaice.

2. *sample detector*: 2.6 liters, stainless steel 0.001" diam. center wire; ^{14}C plateau: >500 volts with slope <1% per 100 volts change in working voltage; filling pressure 760 torr at 15°C; working voltage 3.45 ± 0.03 Kv.; background ~4.6 c.p.m., net ^{14}C activity of NBS oxalic acid ~14 c.p.m.

Two channels were available for analysis of ^{14}C pulses, while a third channel was set up for detection of radon. Distribution of meson pulses in the ^{14}C channels was used to establish the working voltage.

Net sample ^{14}C activities were compared with 0.95 × activity of the NBS ^{14}C standard, corrected for fractionation by normalization to $\delta^{13}C = -19.0°/oo$, and for radioactive decay since 1958. A small aliquot of sample gas was retained for $^{13}C/^{12}C$ ratio measurement; thus the final values, Δ in %, quoted in table 1 have been corrected for isotopic fractionation via $\delta^{13}C$, calculated relative to PDB Belemnite. Several intercalibration samples were analyzed throughout this study as a check on the overall counting performance (Stenhouse and Baxter 1976).

Discussion

The results contained in table 1 represent two years of measurement of tissue ^{14}C levels. Despite the numerous series involved, no significant trends were apparent in Δ values as a function of age of donor or the particular organ from which each tissue was sectioned. Similarly, no major difference in ^{14}C content between normal and cancerous tissue was evident, although in this case the sampling frequency was low. For the bone collagen sample, M/72, the low ^{14}C activity reflects a relatively slow turnover of carbon in this system, a conclusion in agreement with previous research findings (Libby et al. 1964, Berger et al. 1965, Berger and Libby 1967, Harkness and Walton 1972).

For a theoretical treatment of the uptake of bomb ^{14}C in the human body, a box model approach was selected. Whereas previous studies using box models have employed at least two (Nydal et al. 1971) compartments to depict the transfer of ^{14}C within the atmosphere-biosphere-ocean system, a single box model treatment was considered appropriate here. This decision was based on the homogeneous nature of the material analyzed and on the decrease, since 1970, of the marked disequilibrium in environmental to latitude. Using the model shown in figure 2, the ^{14}C content of any tissue, Δ_T^{ni}, at any time, was expressed in terms of dietary ^{14}C level and of the residence time, γ years, of carbon in that tissue, namely:

$$\Delta_T^{ni} = \Delta_T^{o} + \sum_{n=1}^{n} \left\{ \Delta_D^{ni} - \Delta_T^{(n-1)i} \right\} \left\{ 1 - e^{-kT^i} \right\} \quad (1)$$

$\gamma = 1/k_T$, the mean residence time of tissue carbon, is the time required for replacement of a quantity of carbon equivalent to the organ being considered. The above equation is representative of mature tissue in which an essentially steady state prevails, that is, carbon input = carbon output. For the special situation of a tissue/organ under growth, certain assumptions are necessary in order to derive a similar expression and are found in the

TABLE 1
^{14}C Content of Human Tissues: (Δ, %)

Donor/date of death	Liver	Kidney	Spleen	Heart	Lung	Miscellaneous
M/21:9/71	54.6 ± 0.9	51.2 ± 1.6	46.0 ± 0.8	42.4 ± 1.3	53.9 ± 1.6	Pancreas: 52.2 ± 1.1 Testes: 56.0 ± 1.3 Thyroid: 43.3 ± 1.6 Muscle: 55.9 ± 1.3
F/21:9/71	55.2 ± 1.0	53.1 ± 1.2	54.8 ± 1.6	54.5 ± 1.3	51.1 ± 1.1	Pancreas: 54.6 ± 1.1 Ovaries: 50.6 ± 2.0 Thyroid: 50.7 ± 1.1 Muscle: 51.7 ± 1.3
M/37:4/72	48.4 ± 0.9	–	–	47.3 ± 1.0	–	Testes: 49.8 ± 1.8
M/39:8/71	42.3 ± 1.2	–	38.6 ± 1.0	39.8 ± 1.2	–	Pancreas: 46.0 ± 1.2 Thyroid: 37.9 ± 1.5 Brain: 48.1 ± 0.9
M/42:9/71	47.7 ± 1.0	33.9 ± 0.9	43.5 ± 2.5	51.3 ± 1.5	37.8 ± 1.0	Thyroid: 43.3 ± 0.9 Brain: 51.9 ± 1.0
M/50:1/71	43.3 ± 0.8	44.6 ± 0.8	–	52.3 ± 1.9	–	Testes: 37.8 ± 1.0 Muscle: 50.2 ± 1.3
M/61:10/72	48.5 ± 1.0 P45.0 ± 1.1	45.3 ± 0.8 P43.7 ± 1.0	43.4 ± 0.9 P46.2 ± 1.0	50.4 ± 0.8 P44.9 ± 1.0	– –	–
M/64:11/72	47.5 ± 0.8 P48.7 ± 0.9 L47.2 ± 1.1	46.3 ± 0.9 P48.4 ± 1.0	43.2 ± 1.0 P42.8 ± 0.9	– P40.9 ± 0.9	44.0 ± 1.2 P46.5 ± 2.2	–
F/64:4/71	–	49.5 ± 1.0	45.6 ± 1.2	41.5 ± 1.5	–	Muscle: 48.2 ± 1.0 Brain: 49.1 ± 0.9 Ovaries: 47.5 ± 1.5 Adipose tissue: 46.9 ± 1.2
M/67:1/73	49.4 ± 0.8 P49.4 ± 1.0 L48.7 ± 1.2	46.6 ± 0.8 P48.5 ± 0.9 –	– – –	46.8 ± 0.8 P45.4 ± 0.8 L51.1 ± 0.8	– – –	–
M/72:3/71	53.2 ± 1.0	55.2 ± 0.9	51.5 ± 1.0	50.7 ± 1.0	51.2 ± 1.0	Pancreas: 54.7 ± 0.9 Testes: 49.8 ± 1.0 Bone (mf): 48.9 ± 0.8 Bone (col): 10.4 ± 2.7 Muscle: 51.9 ± 1.1 Brain: 50.4 ± 1.0
F/87:1/73	48.0 ± 0.9 P49.1 ± 1.0	46.2 ± 0.9 P47.9 ± 0.8	– P51.1 ± 0.9	–	–	–
M/5:9/71	52.8 ± 0.9	50.4 ± 1.5	44.9 ± 0.8	–	–	Pancreas: 53.5 ± 1.2 Muscle: 52.5 ± 1.0 Brain: 50.5 ± 0.9

M/66:3/73 Liver (normal): 50.0 ± 0.9; Liver (cancered): 55.2 ± 1.4
M/63:2/73 Spleen (normal): 51.0 ± 0.9; Liver (cancered): 53.0 ± 1.2
M/59:2/73 Kidney (cancered): 48.1 ± 0.8

Key: Age and sex of donor, month and year of death are given.
P = protein-rich; L = lipid-rich; mf = marrow fat, col = collagen; muscle analyzed forearm

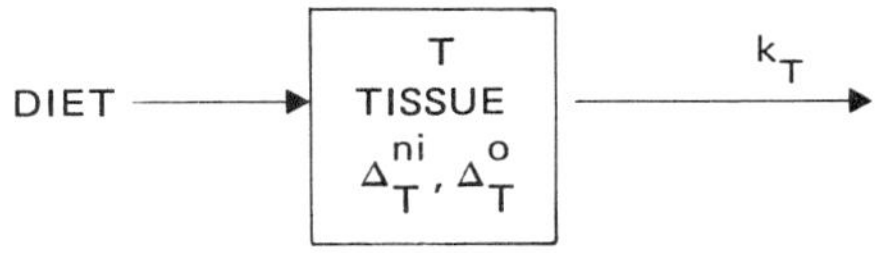

```
                    ┌──────────┐
                    │    T     │        k_T
  DIET  ─────────▶  │  TISSUE  │ ─────────────▶
                    │ Δ_T^{ni}, Δ_T^o │
                    └──────────┘
```

T, tissue carbon, with dietary input as the source of 'new' carbon, turns over with rate constant k_T, that fraction of T replaced in unit time (1 year).

Definition of symbols: Δ_T^o = ^{14}C content of tissue at 't_o', computed zero time

Δ_T^{ni} = ^{14}C content of tissue 'ni' years after 't_o'

'i' = a suitable time increment, viz. 0.5 year

Δ_D^{ni} = mean ^{14}C content of diet over the nth 'i' period after t_o.

For steady-state conditions, it can be shown that

$$\Delta_T^{ni} = \Delta_T^o + \sum_{n=1}^{n} \left\{1 - \exp(-k_T i)\right\} \left\{\Delta_D^{ni} \Delta_T^{(n-1)i}\right\}.$$

while for a tissue under growth, the corresponding equation is

$$\Delta_T^{ni} = \Delta_T^o + \sum_{n=1}^{n} \left\{1 - \frac{N_T^o + (n-1)(N_T^i - N_T^o)}{N_T^{ni}}, \exp(-k_T i)\right\} \left\{\Delta_D^{ni} - \Delta_T^{(n-1)i}\right\}$$

assuming conditions of linear growth.

N_T^{ni} = no. of carbon atoms in tissue, 'ni' years after t_o.

N_T^o = no. of carbon atoms at t_o, (i.e., at birth).

FIG. 2. Single compartment model of carbon turnover.

growth curve of most body tissues. The shape of this curve may be considered, to a first approximation, to be a combination of two consecutive periods of linear growth — from birth to 11 years, and from 11 to 20 years of age (Spiers 1958). Hence the equation corresponding to (1) is:

$$\Delta_T^{ni} = \Delta_T^o + \sum_{n=1}^{n} \left\{1 - \frac{N_T^o + (n-1)(N_T^i - N_T^o)}{N_T^{ni}} \cdot \exp(-k_T i)\right\} \left\{\Delta_D^{ni} - \Delta_T^{(n-1)i}\right\} \quad (2)$$

The number of carbon atoms at birth or at 11 years of age (N_T^o) is expressed as a fraction of the mature tissue size; for the former situation, however, N_T^o is estimated using the underlying assumption of linear growth extrapolated to birth. This procedure is adopted since the first three years of growth are better described by an exponential function; thus the true N_T^o is smaller than the extrapolated version.

Obviously a one-compartment approach for each tissue oversimplifies the true situation within the body. The existence of at least one metabolic pool, whereby interaction/exchange of carbon between different organs can occur, has long been recognized (Sprinson and Rittenberg 1949). So there is a certain interdependence of tissues with respect to ^{14}C content. On the other hand, no single source of carbon analyzed in this study, with the exception of bone collagen, consistently possesses a Δ value sufficiently different in magnitude to influence other tissue ^{14}C activities. This observation is particularly true of muscle protein, which comprises $\sim 50\%$ total body protein.

Both equations derived from the model shown in figure 2 are based on a transfer of carbon from diet to tissue. The ^{14}C content of the diet in terms of atmospheric and oceanic ^{14}C levels — environmental activities which have been monitored regularly over the past two decades (fig. 1) — is expressed by the following equation:

$$\Delta_D^{ni} = 0.24\Delta_{TN}^{ni} + 0.19\Delta_{TN}^{(n-1)i} + 0.16\Delta_{TN}^{(n-2)i} + 0.28\Delta_{TN}^{J} + 0.02\Delta_{TS}^{ni}$$
$$+ 0.035\Delta_{TS}^{(n-2)i} + 0.035\Delta_{TS}^{J} + 0.03\Delta_{M}^{ni} + 0.01\Delta_{M}^{(n-1)i} \quad (3)$$

Equation (3) was established via a survey of the dietary habits of the U.K. population (Stenhouse 1975) coupled with information presented in table 2 concerning lag times between foodstuffs and atmospheric/oceanic ^{14}C activities.

TABLE 2
Dietary Components: Environmental ^{14}C Relationships

Main food category (D)	Contribution to typical diet (%)	Environmental relationship ($\Delta_D^{ni} \equiv$)
Milk/dairy products	23.0	Δ_T^{ni}
Cereals/flour/sugar	31.5	Δ_T^{J}
Meat/poultry	34.0	$\Delta_T^{(n-1)i}, \Delta_T^{(n-2)i}$
Fruit/vegetables – fresh	3.4	Δ_T^{ni}
– other	3.7	$\Delta_T^{(n-1)i}$
Fish – fresh/frozen	3.3	Δ_M^{ni}
– canned	1.1	$\Delta_M^{(n-1)i}$

T = tropospheric CO_2; M = mixed surface-ocean layer: i = 6 months

The above relationships have been determined using previous relevant ^{14}C data (Tauber 1967, Scharpenseel et al. 1968, ibid. 1969, Baxter and Walton 1971, Harkness and Walton 1972, Nydal and Lovsëth 1965; also Chatters, personal communication) together with ^{14}C measurements of food items performed in this study.

Comparison of theoretically-determined curves of Δ^{ni} vs. time (fig. 3 and 4) with experimentally measured activities yields estimates of the residence time, γ years, of carbon in each tissue (table 3). The 5-year-old series has not been subjected to model treatment owing to the breakdown of linear growth assumption applied to the first few years after birth. The measured ^{14}C activities of these tissues are similar to dietary ^{14}C levels over the 18-month period prior to death, with the exception of spleen. The latter organ, a small carbon reservoir because of its relatively small size, may be more responsive to dietary ^{14}C content. The different dietary pattern of a child appears to favor a more direct correlation between atmospheric $^{14}CO_2$ and tissue ^{14}C activities.

For many tissues a choice of γ values/range of γ values is possible. This ambiguity is attributed to the proximity of some Δ curves at the time of sampling and does not imply that both choices exist. More than 25% of all tissues have been shown to possess a γ value $\geqslant 8$ years; thus these tissues support the longer of the two choices of residence time obtained for other samples. The 21-year-old series, considered in its entirety, contains values significantly higher than those of other series; since the $\gamma = 0\text{-}2$ year curves in figures 3 and 4 show little variation, this observation also favors a residence time > 2 years.

The uncertainty in ^{14}C activity measurement ($\pm 1\sigma$) and in dietary-environmental Δ relationships, estimated at ± 1 year, were included in the estimates shown in table 3. Thus the ^{14}C tracer method is sensitive only in terms of years rather than days/months. Within this limitation, and in the context of the model used in this study, a representative value for the mean residence time of soft tissue carbon is taken as 6 ± 4 years.

Such a conclusion must be considered in the knowledge that early tracer investigations, using ^{15}N for example, demonstrated a fast metabolism within tissues, with residence times of the order of days/months (Schoenheimer 1942). With respect to protein, Neuberger and Richards (1964) recognized three types of turnover, the extend of each varying with different cell/tissue, namely (1) replacement of whole cells by mitosis — cell division, (2) renewal of subcellular protein, and (3) turnover of the soluble constituents of the cell without renewal of microscopic structures such as nuclear protein. The bomb ^{14}C tracer method is not sensitive enough to distinguish between these three cases. Thus, in the context of bomb ^{14}C as a biological tracer, there are essentially two carbon entities: (a) a metabolically "active" fraction, with residence time $\gamma = 0\text{-}1$ years, and (b) a metabolically "inactive" fraction, with $\gamma > 10$ years. The terms "active" and "inactive" are purely relative qualifications. The former component, by far the major contributor, is identified readily by normal kinetic experiments. Furthermore, within this group, there is a degree of metabolic heterogeneity; for example, the biological half-life ($t_{1/2} = 0.693 \times \gamma$) of the long-lived component of brain is estimated to be 150 days, while bone contains a protein fraction with $t_{1/2} > 300$ days (Thompson and Ballou 1956).

Type (b) carbon is notable for its relatively small incorporation of bomb ^{14}C. Although identification of this source is not possible from the composite measurements performed in this study, the nuclear protein mentioned in (3) above or structural protein may be the source(s) responsible. Discrete proteins and lipids must be analyzed for bomb ^{14}C content in order to resolve this question.

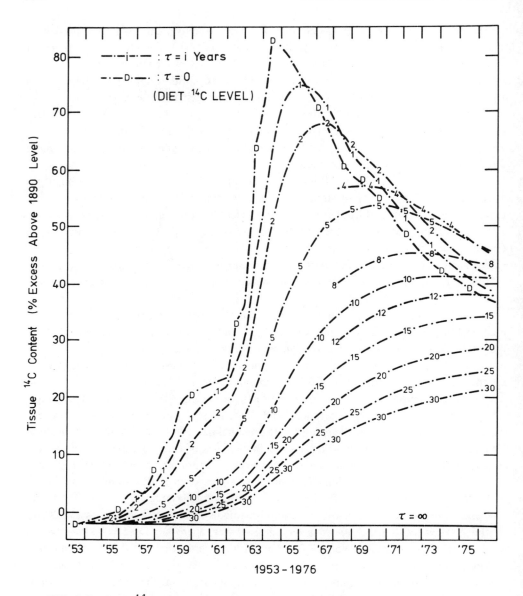

FIG. 3. Predicted ^{14}C content of mature tissue, 1953-1976. Each curve corresponds to a different mean residence time of tissue carbon. Steady-state conditions and first-order kinetics of carbon turnover are assumed to exist. $-i-\tau = i$ year; $-D-\tau = 0$.

Human Radiation Burden

With a knowledge of tissue ^{14}C levels it is possible to evaluate the radiation hazard due to this radioisotope. The annual absorbed dose, D_T, to any tissue from bomb + natural ^{14}C irradiation is determined by the equation:

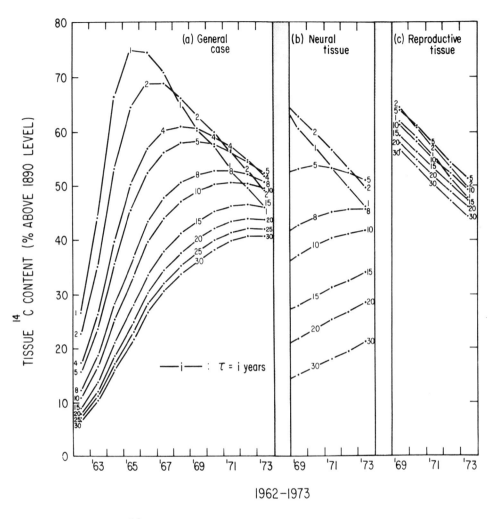

FIG. 4. Predicted ^{14}C content of tissues under growth, 1953-1973. Non-steady state conditions exist: (a) the general series of curves was based on 2 linear growth periods – from birth to 11 years, and from 11-20 years of age; (b) for brain tissue, curves were computed for linear growth from birth to 5 years; thereafter steady-state conditions were applied; (c) for reproductive tissue, steady-state conditions were assumed from birth to 15 years, followed by linear growth until 20 years. $-i-\tau = i$ years.

$$D_T = D_N \{1 + \Delta(\%)/100\} \qquad (4)$$

where D_N represents the annual absorbed dose attributable to natural ^{14}C (i.e., ^{14}C level of 1980 wood). The latter dose rate is evaluated at 0.69 mrad and 0.74 mrad administered to bone marrow/soft tissue and bone-lining cells, respectively. Thus the dose-level curves shown in figure 5 have been calculated using the $\gamma = 6$ years Δ_T plot for bone marrow/soft tissue and $\gamma = 30$ years ($\equiv$ mean residence time of bone collagen carbon) for

TABLE 3

Most Probable Residence Times of Tissue Carbon (τ)

Donor	Liver	Kidney	Spleen	Heart	Muscle	Lung	Testes/Ovaries	Others	
M/21	1-6	8-10	15	20-30	1-6	5-8	1-10	Pancreas:	7-9
								Thyroid:	15-25
F/21	1-6	6-8	1-7	1-7	8-10	8-10	10-15	Pancreas:	1-7
								Thyroid:	9-10
								Brain:	5-6
M/37	7	–	–	7	–	–	5-6	–	
M/39	9	–	11	10-11	–		Pancreas:	7-8
						–	–	Thyroid:	11-12
								Brain:	7
M/42	7	13-14	8-10	5-6	–	11-12	–	Brain:	5-6
								Thyroid:	9
M/50	8-9	8	–	5-6	6	–	10-12	–	
M/61	6-7	8	9	6					
	P 8	P 9	P 7-8	P 8-9	–	–	–	–	
M/64	7	7-8	9	–	–	8-9		–	
	P 6-7	P 6-7	P 9-10	P 10-11			P 7-8		
	L 6-7								
F/64	–	6-7	8	9-10	7	–	7	Brain:	6-7
M/67	6-7	7-8		7-8					
	P 6-7	P 6-7	–	P 8	–	–	–	–	
	L 6-7			L 5-6					
F/72	5	1-4	6	6	5-6	6	6	Brain:	6
								Pancreas:	4-5
								Bone (mf):	6
								Bone (col):	⩾30
F/87	7	7-8	–	–	–	–	–	–	
	P 6-7	P 7	P 5-6						

M/66 Liver (normal): τ = 5-6; Liver (cancerous): τ = 3-4.
M/63 Spleen (normal): τ = 5-6; Liver (cancerous): τ = 2-4.
M/59 Kidney (cancerous): τ = 6-7.

bone-lining cells. In the same way, future absorbed dose rates have been estimated using predictions of future environmental ^{14}C activities (Gulliksen and Nydal 1972) and hence predicted future tissue ^{14}C are included for comparison and are based on a theoretical assessment of the Suess effect applied to atmospheric ^{14}CO$_2$ activities (Baxter and Walton 1970). As shown in figure 6, there remains a substantial contribution from bomb ^{14}C to the total dose absorbed over the next 50 years, although environmental ^{14}C levels are likely to reach pre-bomb concentrations by the first half of next century in the absence of further nuclear weapon testing.

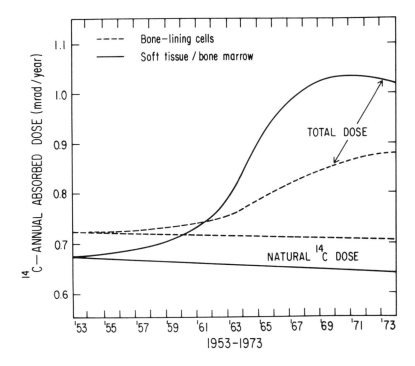

FIG. 5. Annual absorbed doses attributable to ^{14}C, 1953-1973. Total doses (natural + bomb ^{14}C) were calculated using tissue ^{14}C curves of Fig. 3; $\tau = 6$ and $\tau = 30$ years were assigned as mean residence times of soft tissue/bone marrow and bone-lining cell carbon respectively. Natural ^{14}C dose rates were determined via theoretical consideration of the Suess effect.

For evaluation of nonthreshold biological effects of radiation, the absorbed dose accumulated over the organ's critical years of function is of greater relevance than annual absorbed dose rates. Of particular interest is the 30-year dose to reproductive tissue in consideration of possible genetic damage, and a 60-year dose to bone-lining cells for assessment of late somatic effects (e.g., leukemia). Thus table 4 contains several examples of the accumulated dose to soft tissue and bone-lining cells from bomb ^{14}C irradiation; values represent the excess dose over that from the natural ^{14}C level depicted in figures 5 and 6.

The total absorbed dose attributable to bomb ^{14}C is obtained by summation of all annual absorbed doses in excess of that delivered by the corresponding pre-1890 ^{14}C level. Resultant totals calculated in this way are $\sim$ 10 mrad to bone-lining cells delivered over $\sim$ 100 years (until the year 2050) and $\sim$ 8.5 mrad delivered to soft tissue/bone marrow over $\sim$ 80 years. These values, though well below the corresponding figures calculated in the Unscear report (1972a), take account of the Suess effect and, therefore, are more realistic for assessment of the additional radiation burden from bomb ^{14}C. With the exception of this long-lived radioisotope, essentially all of the total dose, due to irradia-

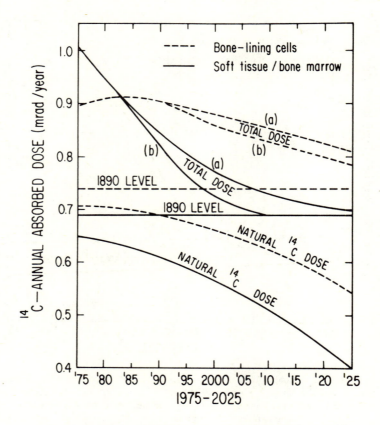

FIG. 6. Predicted annual absorbed doses attributable to ^{14}C, 1975-2025. Total doses to soft tissue/bone marrow and bone-lining cells due to ^{14}C were calculated from predicted tissue ^{14}C levels which are related to environmental ^{14}C activities. The latter data in line (a) were based on predicted future atmospheric and oceanic ^{14}C levels (Gulliksen and Nydal 1972), and in line (b) were based on an atmospheric decrease in ^{14}C content of 2%/year.

Natural ^{14}C dose rates were assessed as in Fig. 5.

Linear extrapolation of upper curves (a) indicates that the total absorbed doses to soft tissue/bone marrow and bone-lining cells will return to natural levels by the years 2035 and 2050 respectively.

tion by radioactive fallout produced in nucelar weapon tests carried out piror to 1970, will have been delivered to the population by the year 2000 as a result of radioactive decay and environmental process which will have effected removal of access of fission products to man, (e.g., erosion and dispersion). Radioactive decay, on the other hand, is responsible for the loss of < 1% of the total bomb ^{14}C; in addition, this radioisotope is retained in the dynamic carbon cycle where it is available for transfer to man. Thus the dilution effect of fossil-produced CO_2 is the major influence responsible for the attenuated radiation dose attributable to ^{14}C of nuclear weapons origin. By the year 2000,

TABLE 4

Accumulated Absorbed Doses Due to Bomb ^{14}C

Period under consideration	Accumulated dose (mrads)	
	Soft tissue/ bone marrow	Bone-lining cells
pre-1954	–	–
1955-1984	7.26	3.16
1964-1993	8.81*	4.95
1970-1999	8.31	5.69
1985-2014	6.65	6.57*
1965-2024	Bone growth during initial 20 years	24.80* (60-year dose)

*maximum doses

therefore, the predicted impact of ^{14}C released in nuclear discharges from the nuclear power industry will be the predominant source of ^{14}C irradiation in the environment (Kelly et al. 1975).

Prediction of the genetic damage to the world's population resulting from the 30-year accumulated doses included in table 4 is difficult, since no critical lesions at the cellular level have as yet been identified. Although chemical effects of ionizing radiation on mammalian cells are known (Cerutti 1974), quantitative assessment of this damage in terms of DNA-strand/chromosome breakage is complicated by the existence of repair mechanisms which are possible for some species within the cell (Painter 1970, Rasmussen et. al. 1970, Cleaver and Burki 1974).

In addition to damage from β-irradiation alone, consideration must be given to the transmutation effect accompanying ^{14}C decay, that is, to those processes associated with the daughter nucleus ^{14}N. Definitive experimental conclusions in this field of research are inhibited for low energy β-emitters owing to difficulty in differentiating between a possible mutagenic effect and damage caused by β-ionization alone. There is conflicting evidence with regard to a transmutation effect of ^{14}C, a topic which is reviewed in detail by Krisch and Zelle (1969). The most recent experimental findings, based on a general ^{14}C labeling technique, do not support a transmutation effect (Lee et al. 1972).

For the most relevant evaluation of the social implications of bomb ^{14}C, reference is made to the results of experimental observations on different types of mutation caused by radiation. This genetic damage is expressed in terms of congenital malformation, embryonic losses, and spontaneous abortions. Estimates of the incidence of genetic disorders in the population, calculated with respect to irradiation damage to spermatogonia and oöcytes — the germ cells most at risk (Unscear 1972b) — are combined with the maximum 30-year dose to the gonads. This procedure yields an estimate of $\sim$ 1100 persons who will possess severe, deleterious traits as a result of bomb ^{14}C; $\sim$ 40-70 first generation offspring will be affected.

Late somatic effects — in particular, an increase in the number of leukemia and cancer victims — are also worthy of consideration in terms of the excess human radiation burden from bomb ^{14}C. In this case, the total absorbed dose to bone-lining cells from artificial ^{14}C (60-year dose) is combined with an induction rate of 100 cases/10^6 persons/rad $\sim$, a crude estimate of the total risk of leukemia and different types of bone tumors (Spiers 1973). In this way, a total of $\sim$ 2000 cases is predicted as a result of bomb ^{14}C irradiation delivered to the world's population over three to four generations. It should be stressed that risk estimates derived here, both genetic and somatic, have been determined on the basis of a dose-effect relationship which is linear, even to the low doses determined in this study. Leukemia and tumor incidence rates were determined under conditions of high dose and acute irradiation; hence resultant risk estimates must be regarded with caution. For consideration of somatic damage due to ^{14}C, the transmutation effect may be the predominant factor involved with this low energy β-emitter; such a possibility is discussed elsewhere (Tamers 1974).

CONCLUSIONS

Residence times of tissue carbon, determined via the bomb ^{14}C tracer method, are expressed realistically in terms of years rather than days/months. Within this sensitivity, the following comments may be made:

1. There is no difference in the residence time of carbon in protein and lipid fractions relative to whole tissue.

2. There is no significant difference between γ values of cancerous and corresponding healthy tissues.

3. No apparent relationship between γ and age of donor is apparent.

4. Many tissues possess carbon with a residence time $>$8 years.

5. With respect to soft tissue, a representative value for the residence time is taken as 6 ± 4 years.

6. Results of this study imply the presence of a tissue component with carbon of γ value greater than ten years.

The human radiation burden due to bomb ^{14}C, in terms of severe genetic and somatic damage, does not exceed 5000 persons over $\sim$ 3 generations. Although minor effects are anticipated in excess of this figure, the total number of cases is likely to remain undetected within the time span involved and in the world population. This situation is directly attributable to the dilution of atmospheric ^{14}C concentrations by fossil-fuel produced CO_2.

POSTSCRIPT

With respect to identification of tissue components containing carbon of slow turnover rate, subsequent ^{14}C assay of the structural proteins, elastin and collagen, extracted from lung tissue, implied a residence time $\gamma = 20$ years with Δ values ranging from 25% to 30% in donors who died in 1976 (Stenhouse and Saugier, unpublished data).

ACKNOWLEDGMENTS

The authors are indebted to the many suppliers of samples analyzed throughout this project. Technical aid was supplied by A. Miller. This work was generously supported by a grant from the Medical Research Council. One of us (MJS) thanks the National Environmental Research Council for financial assistance.

REFERENCES

Baxter, M. S., and A. Walton
    1970   A theoretical approach to the Suess effect. Proc. Roy. Soc. London. Series A. 318:213-230.

Baxter, M. S., and A. Walton
    1971   Fluctuations of atmospheric carbon-14 concentrations during the past century. Proc. Roy. Soc., London. Series A. 321:105-127.

Berger, R., G. J. Ferguson, and W. F. Libby
    1965   UCLA Radiocarbon Dates IV. Radiocarbon. 7:336-371.

Berger, R., and W. F. Libby
    1967   UCLA Audiocarbon Dates VI, Radiocarbon. 9:477-504.

Bien, G., and H. E. Suess
    1967   Transfer and exchange of ^{14}C between the atmosphere and the surface water of the Pacific Ocean. Proc. Symp. Radioactive Dating and Methods of Low-Level Counting. Vienna. IARA, 105-1150.

Broecker, W. S., A. Schulbert, and E. A. Olson
    1959   Bomb carbon-14 in human beings. Science. 130:331-332.

Cerutti, P.
    1974   Effects of ionizing radiation on mammalian cells. Die Naturwissen. 61:51-59.

Cleaver, J. E., and H. J. Burki
    1974   Biological damage from intranuclear carbon-14 decays. DNA single-strand breaks and repair in mammalian cells. Int. J. Radiat. Biol. 26:399-403.

Farmer, J. G., M. J. Stenhouse, and M. S. Baxter
    1972   Glasgow Univ. Radiocarbon Measurements VI. Radiocarbon. 14:326-330.

Gulliksen, S., and R. Nydal
    1972   Further calculations on the C-14 exchange between the ocean and the atmosphere. Proc. 8th Int. Radiocarbon Conf. New Zealand. C58-C72.

Harkness, D. D., and A. Walton
    1969   Carbon-14 in the biosphere and humans. Nature. 223:1216-1218.

1972 Further investigation of the transfer of bomb ^{14}C to man. Nature. 240:302-303.

Kelly, G. N., J. A. Jones, and P. M. Bryant, et al.
1975 The predicted ratiation exposure of the population of the European community resulting from discharges of krypton-85, tritium, carbon-14, and iodine-129 from the nuclear power industry to the year 2000. Report to the Comm. of the European Comm. Study-Contract No. 079-74-8 PST GB.

Krisch, R. E., and M. R. Zelle
1969 Biological effects of radioactive decay: the role of the transmutation effect. Adv. Radn. Biol. 3:177-213.

Lee, W. R., G. A. Sega, and E. S. Bensin
1972 Transmutation of carbon-14 within DNA of *Drosophila Melanogaster* spermatozoa. Mutation Res. 16:195-201.

Leipunsky, O. I.
1958 Harmful effects of the radioactivity from explosion of pure hydrogen bombs and ordinary atomic bombs. Atomic Energy (U.S.S.R.) Eng. trans. 3:1413-1425.

Libby, W. F., R. Berger, and J. F. Mead, et al.
1964 Replacement rates for human tissue from atmospheric radiocarbon. Science 146:1170-1172.

Neuberger, A., and F. F. Richards
1964 Protein biosynthesis in mammalian tissues. Part II: Studies on turnover in the whol animal. *In Mamalian Protein Metabolism*. Munro and Allison, eds. New York and London, Academic Press, 1:Chap. 7.

Nydal, R., and K. Lövseth
1965 Distribution of radiocarbon from nuclear tests. Nature. 206:1029-1032.

Nydal, R., K. Lövseth, and O. Syrstad
1971 Bomb ^{14}C in the human population. Nature. 232:418-421.

Painter, R. B.
1970 Repair of DNA in mammalian cells. In Current Topics in Radiation Res. Amsterdam, North Holland Publishing Co., 7:45-70.

Pauling, L.
1958 Genetic and somatic effects from atmospheric radiocarbon. Science. 128:1183-1186.

Rafter, T. A., and B. J. O'Brien
1970 Exchange rates between the atmosphere and the ocean as shown from recent ^{14}C measurements in the Pacific Ocean. In Radiocarbon Variations and Absolute Chronology. Proc. 12th Nobel Symp. Olsson ed. Stockholm, Almquist and Wiksell, 355-377.

Rasmussen, R. E., B. L. Reisner and R. B. Painter
1970 Normal replication of repaired human DNA. Int. J. Appl. Radn. Biol. 17:285-290.

Sakharov, A. D.
1958 Radioactive carbon from nuclear explosions and non-threshold biological effects. Atomic Energy (U.S.S.R.) Eng. trans. 4:757-762.

Scharpenseel, H. W., F. Pietig, and M. A. Tamers

1968 University of Bonn Natural Radiocarbon Measurements I. Radiocarbon. 10:8-28.

1969 Univ. of Bonn Natural Radiocarbon Measurements II. Radiocarbon. 11:3-14.

Schoenheimer, R.
1942 The Dynamic State of Body Constituents. Cambridge, Harvard Univ. Press.

Spiers, F. W.
1958 Radioisotopes in the Human Body. New York and London, Academic Press, Chap. 2.
1973 Radiosensitivity and dosimetry of the tissues of bone: implications for setting maximum permissible levels of beta-emitters. Proc. 439-447. 3d Int. Cong. of Int. Radn. Prot. Asscn. Washington D.C. 439-447.

Sprinson, D. B., and D. Rittenberg
1949 The rate of interaction of the amino acids of the diet with the tissue proteins, J. Biol. Chem. 180:715-726.

Stenhouse, M. J.
1975 Artificial ^{14}C from Nuclear Weapon Tests and its Uptake by Man. Ph.d. thesis. Univ. of Glasgow.

Stenhouse, M. J., and M. S. Baxter
1976 Glasgow Univ. Radiocarbon Meas. VII. Radiocarbon. 18:161-171.

Tamers, M. A.
1974 Somatic mutation by endogenous radiocarbon. Int. J. Nucl. Med. Biol. 1:227-228.

Thompson, R. C., and J. E. Ballou
1956 Studies of metabolic turnover with tritium as tracer. J. Biol. Chem. 223:795-809.

Totter, J. R., M. R. Zelle, and H. Hossister
1958 Hazard to man of carbon-14. Science. 128:1490-1495.

Tauber, H.
1967 Copenhagen Radiocarbon Measurements VIII: geographical variations in atmospheric ^{14}C activity. Radiocarbon. 9:246-256.

Unscear
1972a Report of the United Nations Scientific Committee on the Effects of Atomic Radiation. 1:41-42, 95.

Unscear
1972b ibid., 2:429.

Walton, A., M. Ergin, and D. D. Harkness
1970 Carbon-14 concentrations in the atmosphere and carbon dioxide exchange rates. J. Geophys. Res. 75:3089-3098.

Young, J. A., and A. W. Fairhall
1968 Radiocarbon from nuclear weapon tests. J. Geophys. Res. 73:1185-1220.

4

Further Application of Bomb ^{14}C as a Biological Tracer

M. J. Stenhouse

ABSTRACT

The ^{14}C content of lipid extracted from arterial tissue was measured and compared with the ^{14}C activity of corresponding tissue sections of these aortas. Theoretical treatment of the results suggests that, whereas carbon of the artery layers is metabolically inert, carbon of intimal plaques is as active metabolically as that of ordinary tissue. Complementary measurements of the ^{14}C content of cholesterol extracted from gallstones have been shown to provide a unique method of determination of the minimum formation time of the stone.

INTRODUCTION

It has been demonstrated (see Stenhouse and Baxter, preceding paper) that bomb ^{14}C is, at present, best applied as a biological tracer to detection of carbon reservoirs with slow turnover rate/long residence time ($\tau \geqslant 10$ years). For shorter residence times, the sensitivity of this tracer method is restricted by the overlapping of Δ curves corresponding to different γ values; in this situation, a choice of γ values/range of values is possible for one tissue ^{14}C activity. Within this limitation, the development of atherosclerosis within the arterial system afforded a suitable topic for investigation, first suggested by Prof. W. A. Harland of the University of Glasgow. Preliminary ^{14}C measurements performed at the University of Glasgow Radiocarbon Laboratory (table 1a) provided a basis for further research on this theme.

Atherosclerosis, a common disease in humans, involves abnormal accumulation of cholesterol and other lipids on the arterial wall. Narrowing and progressive hardening of the arteries due to calcification are subsequent stages of the process, leading, in extreme cases, to severe lesions, total blockage of an artery, then death. Postmortem examinations reveal that initial stages of this disease are present from about 20 years of age (Enos et al. 1955). Almost all males in the United Kingdom have grossly visible coronary atheroscle-

TABLE 1a
^{14}C Content of Artery Samples*

Donor information	Sample	Δ (%) ± 1σ	Mean residence time (years) τ
M/68: died 1963	Intimal lipid	4.8 ± 0.6	10-30
M/67: died 1964	Intimal lipid	9.7 ± 0.9	5-25
M/69: died 1964	Intimal lipid	17.9 ± 0.8	2-10
F/86: died June 1973	Intimal lipid	47.1 ± 0.9	0-5
	Adventitia	31.6 ± 0.8	15

*The above data were obtained at the University of Glasgow Radiocarbon Laboratory, and are discussed elsewhere (Stenhouse and Baxter 1976 1977). These results are included for comparison with the ^{14}C activity measurements performed in this work.

rosis by the age of 60 (French 1970). Atherosclerotic lesions are known to accumulate lipids of the three major classes: cholesterol and its esters, phospholipids, and neutral fat/free fatty acids. Cholesterol and its esters predominate (Insull and Bartsch 1966). As the disease progresses, the proportion of total arterial lipid composed of free cholesterol and its esters increases and the phospholipid content decreases; the remaining component is essentially unchanged (Böttcher and Woodford 1962, Insull and Bartsch 1966). The exact mechanism by which lipid accumulates in the vessel wall is unresolved, but it is throught to occur either by (a) deposition of blood components, either soluble lipoprotein or formed cells, or (b) in situ synthesis at the arterial wall (Bowyer and Gresham 1970). Development of the disease is influenced by diverse factors, although the relative importance of each is difficult to assess (Harland 1971). In this respect similar causal factors exist between atherosclerosis and gallstone disease (cholelithiasis), particularly with reference to those dietary components which are known to enhance the prevalence of each disease (Portman et al. 1975). In addition to artery samples several gallstones were analyzed for ^{14}C content. Comparison of the measured ^{14}C activities with dietary ^{14}C levels permitted some assessment of the residence times of carbon in tissue and lipid fractions, and the formative time of gallstones.

Experimental Method

PRETREATMENT

Sections of aorta were collected fresh from postmortem examinations (Veterans Administration and University Hospitals, San Diego). Each artery was visibly marked with lipid deposits, ranging from fatty streaks and fibrous plaques (thickening of the artery wall) to lesions which extended beyond the inner wall (intima) to the middle (media) and, in some cases, to the outer (adventitia) layers. Lipid composition typically would vary with extent of damage but, in the arteries analyzed in this work, the bulk of lipid

(>60%) was estimated to be cholesterol/cholesterol esters. Each artery was opened longitudinally and washed with distilled H_2O to remove adhering blood. The three layers were identified easily, and media and adventitia were stripped from the intima using scissors; subsequent pretreatment of each sample is shown diagrammatically in figure 1. Evaporation of solvent was performed under vacuum and, toward the latter stage of the first evaporation, ~5 ml. benzene was added to reduce fromthing and to aid removal of water (Foote and Coles 1968). The backwash with warm saline solution was carried out to ensure removal of nonlipid constituents (Entenman 1961). Yields of lipid obtained in this way were in the range 200-500 mg. Treatment of the extracts discussed in table 1a was identical to that described here.

The gallstones provided for this study were rich in cholesterol and had been removed surgically from three patients. Three of the stones were large enough for separate analyses of outer and inner layers. Each stone was dissolved in warm dioxane

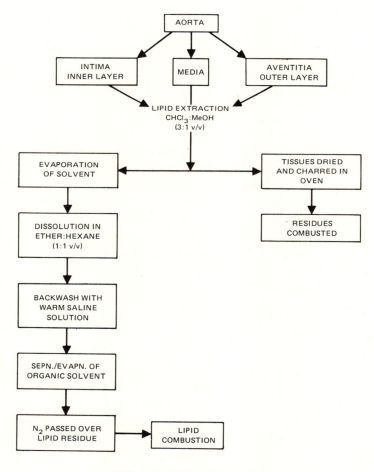

FIG. 1. Arterial pretreatment.

(25 ml.) and the residue was filtered; on cooling, the filtrate yielded crystals of cholesterol which, in turn, were filtered and purified by recrystallization from chloroform to give long, white crystals (m.p. 136°C).

PREPARATION AND COUNTING OF CO_2

Samples were combusted in a stream of oxygen in a silica tube. The subsequent purification of this gas, which is necessary for gas-proportional counting, is shown schematically in figure 2. After regeneration of CO_2 from the furnace at 900°C, partial removal of radon was achieved by passing the gas through a charcoal trap at 0°C. Sample CO_2 was stored for a minimum of four weeks prior to ^{14}C activity measurement.

Samples were counted in a 100 cm³ quartz-shell counter since the total quantity of CO_2 produced was small (<0.2 liter at STP). Background CO_2 was prepared from marble chips or anthracite and processed in the normal manner, while CO_2 from NBS oxalic acid was produced by wet oxidation using acidified $KMnO_4$. A secondary standard — wood

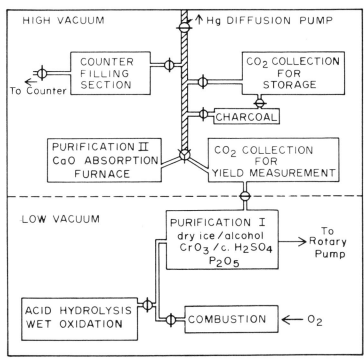

FIG. 2. CO_2: Preparation and pretreatment.

from the 1871-1880 section of Douglas fir (Linick 1977) — was employed as an independent check on counter performance throughout this work. Background count rate and the ^{14}C activity of NBS oxalic acid are included in figure 3; these measurements were accumulated in the course of the research project only. Although the background activity was high in relation to the NBS ^{14}C standard, the characteristics of this counter were considered adequate for routine measurement of bomb ^{14}C standard, the characteristics of this counter were considered adequate for routine measurement of bomb ^{14}C activities.

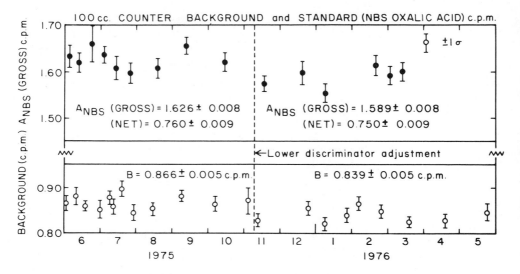

FIG. 3. 100 cm^3 Quartz-shell counter: background and NBS oxalic acid values.

Each sample CO_2 was counted for at least four days, usually for two separate counting periods of up to four days. Net sample ^{14}C activities were compared with 0.95 × activity of NBS oxalic acid using the Lamont VIII formulae (Broecker and Olson 1961); the latter count rate was corrected for isotopic fractionation via normalization of $\delta^{13}C = -19.0‰$, and for radioactive decay since 1958.

Results and Discussion

Tables 1b and 2 contain the results of ^{14}C activity measurement of the arterial samples and gallstone cholesterol, respectively. Δ values are quoted in percentages, in consideration of the larger differences between sample activities and the 1980 ^{14}C level. Errors of ^{14}C analysis are normally in the range 2%-3%, and table 1a results are included for comparison.

ARTERY RESULTS

The arterial system, in common with most tissues in the human body, is complex kinetically as well as biochemically; therefore it is not surprising that the ^{14}C contents of

TABLE 1b

^{14}C Content of Arterial Fractions

Donor information	Fraction analyzed	δ^{14}C ± 1σ (%)	δ^{13}C (‰)	Δ ± 1σ (%)
Male/78 Died 6/75	A	30.5 ± 2.2	−13.0	27.4 ± 2.2
	M	28.4 ± 2.3	−16.4	26.2 ± 2.3
	I	29.0 ± 2.8	(−18.0)	27.2 ± 2.8
	M(lipid)	42.4 ± 2.3	−16.5	40.0 ± 2.3
	I(lipid)	41.7 ± 2.3	−19.9	40.3 ± 2.3
Male/56 Died 8/75	M/A	21.1 ± 2.1	−18.0	19.4 ± 2.1
	I	26.9 ± 4.9	(−18.0)	25.1 ± 4.9
Male/65 Died 8/75	M/A	19.6 ± 2.1	−18.7	18.1 ± 2.1
	I	29.5 ± 2.0	(−18.0)	27.7 ± 2.0
	I_n	25.8 ± 2.0	−19.2	24.3 ± 2.0
	A	18.8 ± 1.5	(−18.0)	17.1 ± 1.5
	M	21.6 ± 2.2	−16.8	19.6 ± 2.2
Male/61 Died 10/75	M(lipid)	47.6 ± 3.5	−18.1	45.6 ± 3.5
	I	32.4 ± 2.2	−17.2	30.4 ± 2.2
	I(lipid)	37.3 ± 2.3	(−18.0)	35.4 ± 2.3
	IP extract	52.6 ± 3.1	(−18.0)	50.5 ± 3.1
Male/64 Died 3/76	IP extract	34.3 ± 4.2	(−18.0)	32.4 ± 4.2
Male/51 Died 4/76	IP extract	47.1 ± 3.5	(−18.0)	45.1 ± 3.5
Male/78 Died 4/76	IP extract	48.1 ± 3.2	(−18.0)	46.0 ± 3.2

A = adventitia; M = media; I = intima; I_n = relatively healthy areas of intima; M/A = media/adventitia IP = intimal plaques. Where no δ^{13}C measurement was performed, a mean value is included in parentheses.

various components of aortas analyzed in this project exhibit a wide range of Δ values — from 17.1% to 50.5% above the 1890 ^{14}C level. Although complicated mathematical expressions are required to describe interaction/exchange rates of carbon between different tissue fractions, relatively simple treatments are adopted here. Within the limitations of the bomb ^{14}C tracer method discussed previously, single-box models are considered adequate to distinguish between short (τ = 0-6 years) and long ($\tau \geqslant$ 10 years) residence times. $\tau, = 1/k_T$, where k_T is the rate constant for turnover of carbon (in units of fraction of total tissue per year), is defined as the average time required by a reservoir for replacement of an amount of carbon equivalent to the reservoir itself.

TABLE 2
^{14}C Content of Gallstone Cholesterol

Donor	Date of removal		$\delta^{14}C \pm 1\sigma$ (%)	$\delta^{13}C$ (º/oo)	$\Delta \pm 1\sigma$ (%)
(A) Born in 1931	8/75	* 1 2a 2b	46.5 ± 4.0 47.6 ± 2.9 57.2 ± 5.1	−18.0 (−18.0)† (−18.0)	44.4 ± 4.0 45.5 ± 2.9 55.0 ± 5.1
(B) Born in 1936	7/75	1 2a 2b	55.5 ± 4.0 51.7 ± 4.3 67.5 ± 4.2	−17.8 (−18.0) −17.3	53.2 ± 4.0 49.6 ± 4.3 65.2 ± 4.2
(C) Born in 1892	8/75	1 2 3a 3b	1.9 ± 3.7 2.5 ± 3.0 −5.7 ± 2.7 −3.1 ± 3.2	−18.2 −18.5 (−18.0) (−18.0)	0.5 ± 3.7 1.1 ± 3.0 −7.0 ± 2.7 −4.5 ± 3.2

*Numbers refer to individual gallstones

 a = surface layer of stone

 b = centre of gallstone

†Where no $\delta^{13}C$ measurement has been performed, a value of −18.0, in parentheses, is assumed.

Accordingly the steady-state model discussed in the preceding paper (figure 2), together with its equation for solution, is utilized here to describe the buildup of bomb ^{14}C in arterial tissue — intima, media, and adventitia. The equation which relates dietary ^{14}C content to environmental ^{14}C levels, together with the resultant family of curves of Δ as a function of γ, residence time, are identical to those derived previously (figure 3, preceding paper).

By contrast the lipid carbon reservoir is nonstationary, so a modified version of the original single-box model treatment is necessary, as depicted in figure 4. In this case a linear increase in the amount of carbon which accumulates on the artery wall is assumed in order to derive the corresponding model equation. Such an assumption is justified on the basis of numerous pathological findings which indicate a linear incidence in the number of atherosclerotic lesions with time from 20-25 years of age (Eggen and Solberg 1968). Incorporation of bomb ^{14}C in an arterial lipid deposit, with respect to time, is represented by the following expression:

$$\Delta_L^{ni} = \Delta_L^{(n-1)i} + \sum_{n=2}^{n} \left\{ 1 - \frac{(n-1)}{n} \cdot \exp(-k_L i) \right\} \left\{ \Delta_S^{ni} - \Delta_L^{(n-1)i} \right\}$$

$\gamma, = 1/k_L$, is the mean residence time in years of a carbon atom in the deposit. Subscript S denotes the source of this carbon, of which ~70% originates from lipid synthesized within the body (Kaplan et al. 1963); the remaining 30% is derived directly from the diet. Furthermore, liver is known to be a major site for synthesis of cholesterol and cholesterol

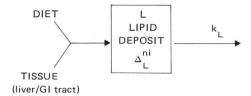

L, lipid deposit carbon, turns over with rate constant k_L, that fraction of L replaced in unit time (1 year).

Definition of symbols: Δ_L^{ni} = ^{14}C content of lipid 'ni' years after t_o

'i' = 0.5 year

Δ_S^{ni} = mean ^{14}C content of source carbon (30% dietary and 70% endogenous)

Under assumed conditions of linear growth of lipid carbon, it can be shown that

$$\Delta_L^{ni} = \Delta_L^{(n-1)i} + \sum_{n=2}^{n} \left\{ 1 - \frac{(n-1)}{n} \cdot \exp(-k_L i) \right\} \left\{ \Delta_S^{ni} - \Delta_L^{(n-1)i} \right\}$$

FIG. 4. Single compartment model of carbon turnover in lipid deposit.

ester, two major components of a lipid extract (Ho et al. 1970). Thus each Δ_S^{ni} term is subdivided as follows:

$$\Delta_S^{ni} = 0.30 \Delta_{D2}^{ni} + 0.70 \Delta_{Liver}^{ni}$$

In the computation of Δ_S^{ni} data from 1953-1976, Δ_{Liver}^{ni} values are obtained from the $\gamma=6$ years curve (figure 3, preceding paper) while dietary emphasis is focussed on those factors which appear to influence serum lipid levels and, in turn, the development of atherosclerosis, namely, intake of cholesterol and total fat (Connor et al. 1964), and of saturated fatty acids (Hardinge and Stare 1953, Bronte-Stewart et al. 1956). Coefficients of the original dietary equation are altered, therefore, to give the following relationship:

$$\Delta_{D2}^{ni} = 0.58 \Delta_A^{ni} + 0.09 \Delta_A^{(n-1)i} + 0.31 \Delta_A^{(n-2)i} + 0.02 \Delta_M^{ni}$$

With a knowledge of environmental ^{14}C levels (figure 1, preceding paper), a family of curves was computed to give the ^{14}C content of lipid deposit carbon, expressed as the excess ^{14}C activity over the 1890 level, as a function of γ, residence time. The fact that the ^{14}C activity of lipid source carbon does not exceed 55%, as shown in figure 5, is due to the major contribution of liver carbon, for which the $\gamma=6$ years Δ curve itself is never higher than 55%.

Comparison of theoretically predicted ^{14}C activities with measured ^{14}C activities at the time of death yields estimates of the mean residence time of carbon in each sample — arterial layers and lipid deposits (table 3). Included in each comparison is the error of

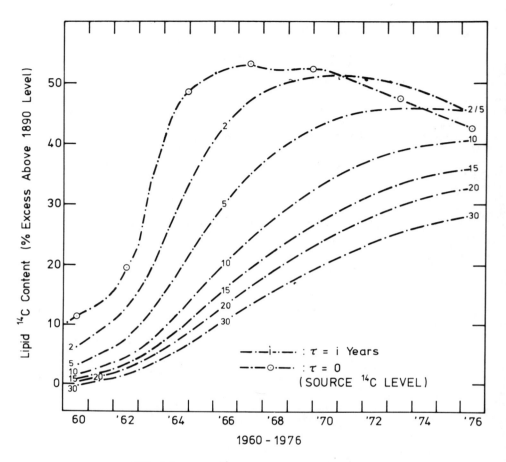

FIG. 5. Predicted ^{14}C content of lipid, 1960-1976.

activity measurement (±1σ) together with the estimated uncertainty of ±1 year in dietary-environmental ^{14}C level relationships. Given the additional limitations inherent in the model treatment, values of γ estimated here are suitable for comparative purposes only.

Included in table 3, in the last column, are estimates of the formation time of each total lipid extract in the absence of carbon turnover, that is, deposition is assumed to be irreversible. By this alternative approach, quantities determined here are obtained by direct comparison of source and ^{14}C levels.

It is evident from the results in table 3 that carbon in each arterial layer residue (adventitia, media, and intima, after lipid extraction) turns over very slowly with residence time, $\gamma = \sim 20$ years. This conclusion is expected since the bulk of this residue is composed of collagen which, from previous research findings on the same component, is known to be metabolically inert (Kao et al. 1961, Thompson and Ballou 1956). On the other hand it is difficult to make definite conclusions based on the γ values evaluated for the lipid extracts owing to the sparsity of data and to the greater spread of γ estimates. In

TABLE 3

Estimates of the Residence Time of Carbon in Artery Components

(a)

Donor		Mean residence time (years)	
M/78:	A	~20	
	M	~20	
	I	~20	
M/56:	M/A	>30	
	I	~25	
M/65:	M/A	>30	Tissue Model
	I	20-25	
	I$_n$	~25	
M/61:	A	>30	
	M	>30	
	I	15-20	

(b)

Sample		Mean residence time (years)	Deposition time (years)*	
M/78:	M(lipid)	0/8-12	10-15	
	I(lipid)	0/8-12	10-15	
M/61:	M(lipid)	0-10	~15	
	I(lipid)	10-20	15-25	Lipid Model
	IP(lipid)	0-5	≤10	
M/64:	IP(lipid)	15-20	~25	
M/51:	IP(lipid)	0-10	~15	
M/78:	IP(lipid)	0-10	~15	

*The values in the last column have been calculated assuming irreversible conditions during lipid accumulation, and on the basis of a linear increase in deposit carbon up to the time of death.

general, the mean residence times of lipids are considerably shorter than those of the corresponding tissue from which each was extracted. In addition the carbon in the intimal plaques appears to be as active metabolically as that in ordinary soft tissue, with one exception (male/64). This conclusion supports the hypothesis inherent in the model treatment, that the precursors of the lipid are deposited directly from the bloodstream, in agreement with the observations of Newman and Zilversmit (1959). Comparison with the previous study (table 1a) indicates agreement with the recent 1973 samples. Agreement with the earlier (1963-1964) data is not so good, although this group itself exhibits a range of values.

For the situation in which lipid accumulation is assumed to be irreversible, resultant figures in the last column of table 3b range from a deposition time of less than 10 years to one of ~25 years. With respect to the male/61 donor, with $\Delta=50\%$, a corresponding deposition time $\leqslant 10$ years seems unlikely on the basis of the age of donor and the postmortem findings, discussed previously, which demonstrated the presence of plaques in males of 20-25 years of age.

The balance of the results of this study favor a state of flux at the arterial wall, where lipids undergo dissolution and deposition. The fact that different interpretation is possible in some cases may be attributed to the inhomogeneous nature of the lipid deposit, that is, the presence of at least three classes of lipid, each with its individual turnover properties. Unfortunately in this study the quantity of total lipid extracted from each sample was insufficient for isolation and ^{14}C analysis of separate fractions. It should be considered essential, therefore, in any subsequent investigation of this carbon system, that the ^{14}C activity of each lipid component be measured independently so that ambiguity in interpretation of the ^{14}C results may be avoided.

GALLSTONE CHOLESTEROL

Interpretation of the ^{14}C data of gallstone cholesterol is more straightforward and is discussed separately in terms of each donor. Direct comparison of Δ values with the source, $\gamma=0$, curves (figure 3, preceding paper; figure 5, this paper) yields a minimum estimate of the mean age of each stone (time since the cholesterol was deposited). These curves may be considered boundary lines for the ^{14}C content of source carbon in which two extremes in ^{14}C activity are represented. The raw data suggest that the $\gamma=0$ curve of figure 3, preceding paper, is more representative of the true ^{14}C level of the source of cholesterol carbon. The latter line is favored since a Δ value of 65% (donor B) is meaningless in figure 5. It can be explained using the same lipid accumulation model only if the liver source of carbon has a shorter residence time ($\gamma \ll 6$ years). Such a condition may be justified satisfactorily since the carbon utilized by the liver in the biosynthesis of cholesterol is to be a part of a metabolically active pool.

The most striking results shown in table 2 are found for donor C. In this case, it is evident that all three gallstones were formed prior to the nuclear era, since virtually no bomb ^{14}C was detected in these stones. Since these obstructions had been removed surgically only in 1975, they had remained inert within the patient for at least 20 years. In the case of donor A, the ^{14}C content of stone 1 may be regarded as contemporary, within counting error. On the other hand, the difference in Δ values between the outer and inner layers is statistically significant, implying a formation time for the stone of at least 3 years. It should be apparent that this estimate is minimal since no consideration has been given to the possibility of partial dissolution and subsequent recrystallization of cholesterol, a known feature (Goldstein and Schoenfield 1975). Reference to either figure should make the previous statement self-evident.

For donor B, born in 1936, a similar but even more apparent situation exists; the results indicate that the gallstones have been present within the patient since 1973 at the latest, while stone 1 data imply a time of formation >4 years. It is encouraging that so few ^{14}C measurements have yielded considerable information. Analysis of gallstones by electron microscopy can reveal concentric deposition layers (Osuga et al. 1974) but give no indication of the time-scale involved in the formation of a stone. Assessment of formation times by cholecystography (X ray picture of the gallbladder) may also be subject to complications caused by stone movement or dissolution.

Application of bomb ^{14}C to such a study is a unique and definitive method of investigation of gallstone deposition rates, even long after the stones have been removed from the patient.

ACKNOWLEDGMENTS

The author is indebted to the many suppliers of artery material from the Veterans Administration and University Hospitals, San Diego. Henry Mok, M.D., supplied the gallstones analyzed in this study and offered advice in interpretation which was greatly appreciated. Patricia Case performed the extraction of cholesterol and participated in the analysis of the ^{14}C content of these samples. Ellen Druffel performed the ^{14}C activity measurements of the lipid extracts. Both these contributions are gratefully acknowledged. Financial support for this work was supplied by generous grants GA-25952 and DES75-05519 from the National Science Foundation.

REFERENCES

Böttcher, C. J. F., and F. P. Woodford
    1962   Chemical changes in the arterial wall associated with atherosclerosis. Fed. Proc. 21:15-19. Suppl. 11.

Bowyer, D. E., and G. A. Gresham
    1970   Arterial lipid accumulation. In Atherosclerosis. Proc. 2d Int. Symp. Chicago. Jones, ed. New York, Springer-Verlag.

Broecker, W. S., and E. A. Olson
    1961   Lamont Radiocarbon Measurements VIII. Radiocarbon 3:176-204.

Bronte-Stewart, B., A. Antonis, L. Eales et al.
    1956   Effects of feeding different fats on serum-cholesterol level. Lancet. 1:521-526.

Connor, W. E., R. E. Hodges, and E. Bleiler
    1964   The serum lipids in men receiving high cholesterol and cholesterol-free diets. J. Clin. Invest. 40:894-901.

Eggen, D. A., and L. A. Solberg
    1968   Variation of atherosclerosis with age. Lab. Invest. 18:571-579.

Enos, W. F., J. C. Byer, and R. H. Holmes
    1955   Pathogenesis of Coronary Disease in American soldiers killed in Korea. J. Amer. Med. Ass. 158:912-914.

Entenman, C.
    1961   The preparation of tissue lipid extracts. J. Amer. Oil Chem. Soc. 38:534-538.

Foote, J. L., and E. Coles
    1968   Cerbrosides of human aorta: isolation, identification of the hexose, and fatty acid distribution. J. Lipid. 9:482-486.

French, J.E.
    1970   General Pathology. 4th ed. Florey, ed. London, Lloyd-Luke (Medical Books), 459.

Goldstein, L. I., and L. J. Schoenfield
    1975   Gallstones: Pathogenesis and medical treatment. Adv. Intern. Med. 20:89-119.

Hardinge, M. G., and F. J. Stare
    1953   Nutritional studies of vegetables 2: dietary and serum levels of cholesterol. Amer. J. Clin. Nutr. 2:83-88.

Harland, W. A.
    1971   The problems of atherosclerosis. The practioner. 206:321-329.

Ho, K. J., C. B. Taylor, and K. Biss
    1970   Overall control of sterol synthesis in animals and man. In atherosclerosis. Proc. 2d Int. Symp. Chicago. 271-273.

Insull, W. jr., and G. E. Bartsch
    1966   Cholesterol, triglyceride, and phospholipid content of intima, media, and atherosclerotic fatty streak in human thoracic aorta. J. Clin. Invest. 45:513-523.

Kao, K. Y., D. M. Hilker, and T. H. McGavack
    1961   Connective tissue IV: synthesis and turnover in proteins in tissues of rats. Proc. Soc. Exptl. Biol. Med. 106:121-124.

Kaplan, J. A., G. E. Cox, and C. B. Taylor
    1963   Cholesterol metabolism in man. Arch. Pathol. 76:351-359.

Linick, T. L.
    1977   La Jolla Radiocarbon Measurements VII. Radiocarbon. 19(1):19-48.

Newman, H. A. I., and D. B. Zilversmit
    1959   Quantitative aspects of cholesterol flux in rabbit atheromatous lesions. J. Biol. Chem. 237:2078-2084.

Osuga, T., K. Mitamura, S. Miyagawa et al.
    1974   A scanning electron microscopic study of gallstone development in man. Lab. Invest. 31:696-704.

Portman, O. W., T. Osuga, and N. Tanaka
    1975   Biliary lipids and cholesterol gallstone formation. Adv. Lipid Res. 13:135-194.

Stenhouse, M. J., and M. S. Baxter
    1976   Glasgow Univ. Radiocarbon Measurements VIII. Radiocarbon 18:161-171.
    1977   Bomb ^{14}C as a biological tracer. Nature 267:828-832.

Thompson, R. C., and J. E. Ballou
    1956   Studies of metabolic turnover with tritium as tracer. J. Biol. Chem. 223:795-809.

5

Radiocarbon Transmutation Mechanism for Spontaneous Somatic Cellular Mutations

M. A. Tamers

ABSTRACT

Naturally occuring radiocarbon is present in the human body in sufficient quantities to provide a significant mechanism for somatic cell mutation. This is due to a chemical transmutation, carbon becoming nitrogen, that occurs during the radioactive disintegration. The transformation can take place in a strategic position of a DNA base in a living cell. The particular base need not be dislodged from its location on the chain and the resulting stable nitrogen remains in place of the decayed carbon atom in approximately half the reactions. The change in the DNA base, undetected by repair enzymes, can cause a copy error during the ensuing replication. If one of the initiating phases of neoplasia is a specific somatic mutation, endogenous radiocarbon transmutation provides a reasonable mechanism for a primary etiology of the disease.

This theory of chemical transmutation etiology of cancer can be tested experimentally. Food of low radiocarbon concentration, grown in the special atmosphere from which the carbon dioxide was obtained is fed to laboratory rodents. The few isotopes in the diet and animals are monitored by a radiocarbon dating system. In the second generation, the effect of the reduction of the radiocarbon burden on neoplasia incidence and longevity can be determined. Increased radiocarbon concentrations can be examined for any augmented incidence of malignant disease in experimental animals, but results can be ambiguous, since naturally occurring radiocarbon concentrations might be more than sufficient to initiate neoplasia. Higher activities of the isotope would, in such cases, be superfluous.

Of the principal radioisotopes in the human body (summarized in table 1), ^{40}K and ^{14}C present the greatest radioactive disintegration activities. However, the potassium isotope has high energy beta and gamma rays, and its capacity for ionizing radiation damage is approximately ten times that of radiocarbon. The other radioisotopes have

TABLE 1

Principal Radioisotopes in the Human Body*

Isotope	Product	Energy (Mev)	Half-life (years)	Body radioactivity (Curie)	Origin
^{14}C	^{14}N + β	0.156	5730	10^{-7}	Cosmic radiation produced
^{40}K	^{40}Ar + β + γ	1.32(1.46)	1.26×10^9	10^{-7}	Naturally occurring, long-lived
^{87}Rb	^{87}Sr + β + γ	0.275	6.2×10^{10}	10^{-9}	Naturally occurring, long-lived
^{90}Sr	^{90}Y + β	0.54	28.1	10^{-9}	Nuclear devices fallout
^{3}T	^{3}He + β	0.0186	12.3	10^{-11}	Cosmic radiation produced
^{226}Ra	^{222}Rn + α + γ	4.6	1620	10^{-12}	Naturally occurring from ^{238}U decay chain

*See Eisenbud 1973

orders-of-magnitude lower activities. It should be noted that ^{90}Sr, a recent arrival in our environment, and ^{226}Ra, a principal radioactive daughter of the uranium series, are accumulated in bone. These would be in a particularly propitious location for damage to hematopoiesis tissues and, therefore, are suspect of being one of the causes of some leukemias. They can not be major factors though, because of the minimal amounts of natural ^{226}Ra and because the disease occurred at approximately the same incidence before nuclear weapon tests began to produce significant amounts of ^{90}Sr. Tritium and ^{87}Rb, likewise, are present in low concentrations compared to the two major endogenous radioisotopes.

The actual radiations (α, β, γ) emitted during disintegration of the radionuclides are not efficient mutagenic agents. Although large doses of radioactivity will cause pathological changes, normal levels of ionizing radiation appear to have a negligible effect on longevity. The higher energy of the potassium-40 radiation is not as significant as at first supposed.

Naturally occurring radiocarbon is extensively diluted with stable carbon (Anderson et al. 1947). Its normal concentration is only 1.17×10^{-12} gm ^{14}C/gm carbon or, with respect to its radioactivity, 13.56 disintegrations/ minute/gm carbon for materials in equilibrium with the atmosphere (Karlen et al. 1964). Concentrations have been augmented during the past two decades because of thermonuclear weapons testing, but contamination is presently only 40% above normal level and is decreasing (Nydal et al. 1976).

Chemical Transmutation of Endogenous Radiocarbon

There is significant probability that radiocarbon decay could occur in a strategic location in a living cell (Tamers 1974). This would be the situation if the carbon isotope involved were part of a DNA base in a critical position on a gene. The radioactive transformation would not usually be lethal to the cell and the particular DNA base would not inevitably be dislodged from its position on the chain. The stable ^{14}N product could assume the place of the decayed atom in a significant precentage of the transformations. Wolfsberg (1956) demonstrated theoretically that there is a high probability of nonrupture of the bond in the ^{14}C $\rightarrow$ ^{14}N + β transformation. Wolfgang (1956) has shown experimentally that, in the disintegration of a ^{14}C in the ethane molecule, 47% of the reactions produce a methyl amine in which the molecule entity is conserved.

This phenomenon would not be important in the case of ^{40}K, since it changes into a noble gas element and does not occur in the body in a bound state; rather, it exists in the dissolved ionic form. ^{226}Ra and ^{3}T also produce noble gases in their disintegration processes.

The possible occurrence of radiocarbon transmutation in a specific position in one of the genes in a human somatic cell is significant. Taking the spleen as representative, the whole cell contains about 0.77 mg DNA-P/gm fresh tissue (Davidson 1947). A 70 kg person would have 99 gm DNA base carbon in his cells. Of this 1.39×10^{-10} gm or 7.0×10^{12} atoms would be radiocarbon, taking into account an estimated 20% excess due to thermonuclear weapon contamination. During a lifetime of 70 years, there would be 5.9×10^{10} DNA base ^{14}C decays.

Since the human cell has approximately 1.1×10^{10} DNA nucleotides per cell (Lehninger 1970), and since there is a weighted average of 4.8 carbon atoms/DNA base, it is seen that a carbon atom of any chosen DNA base in at least one of the cells would undergo the ^{14}C $\rightarrow$ ^{14}N transmutation in 70 years. If 47% of the decays allowed the ^{14}N to take the exact place of the original ^{14}C, there would be a 53% chance of this type of modifications occurring in any particular carbon atom of a specific DNA base in a gene. Actually it would be expected that any one of numerous positions would result in loss of function of the produced enzyme or enzymes that control cell growth. This would mean a much higher probability of a pathological effect due to radiocarbon transmutations. In tissues that have a constant turnover, the change in the DNA base could cause a copy error during the proceeding replication. This would result in a somatic mutation of the affected cell that would be perpetrated in successive mitoses.

This endogenous radiocarbon mechanism is distinct fro that of nuclear radiation-induced mutations. Pauling has stressed the significance of additional radiation burden from fallout radiocarbon on the human population (Pauling 1958). It now appears that nuclear radiation is a relatively inefficient mutagenic agent. A recent study shows that a doubling of the normal radiation dose would produce, at most, a 3% increase in the frequency of genetic defects (Grahn 1972). An analysis of epidemiological evidence for the effects of radiation on American radiologists led to the conclusion that doses forty times the normal background increased the mortality rates only 1.02 times the norm for the 39-70 years age range (Burch 1971). Naturally occurring background radiation has not been observed to be the principal cause of any neoplastic disease. This is seen from medical statistics in areas of high natural radiation, for example, in specific regions of Brazil (Freire-Aiai 1971). In the United States there are considerable variations in the levels of natural sources of external ionizing radiation. This comes from long-lived terrestrial radioisotopes, principally ^{40}K, ^{238}U, ^{232}Th, and the daughters of the latter two as well as cosmic radiation. For example, inhabitants of Denver, Colorado, receive two to three times higher doses than those of Dallas, Texas (Eisenbud 1973); yet there is no appreciable difference in the incidence of neoplasia between residents of these cities. All evidence points to the conclusion that naturally occurring ionizing radiation cannot be a principal cause of cancer.

Oncogenesis by Somatic Mutation

The somatic cell mutation theory of cancer and certain other diseases of old age is attractive for numerous reasons. Chromosomes can be damaged either spontaneously or by various outside agents, but the principal exogenous factor considered, ionizing radiation, has been shown to be an inefficient mutagen. The Proceedings of the Fifth International Symposium on Molecular Biology (1972) indicate that this is due to the highly sophisticated DNA repair enzymes present in the cell. Even the pyrimidine dimer, the principal mutation mechanism in UV radiation, is recognized immediately by the excision-repair enzymes. The damaged portion is removed from the DNA chain and replaced by a healthy piece. The ionizing radiation effect, which causes multiple bond breakage, is likewise rapidly repaired.

Another mechanism for mutation is that due to tautomerism (see Stent 1971 for general discussion). Mutations could arise in the situation where a DNA base at the moment of replication happens to be in its rare enol or imino tautomeric form. At such an instant, an incorrect nucleotide could be included in the growing replica polynucleotide. The mutation would be propagated in succeeding replications. It would escape detection by the repair enzymes which, presumably, detect errors by ionizations and distortions in the double helices (5th Symp. Mol. Biol. 1972). However, there is a doubt that the tautomeric forms exist as such. The resonance theory of organic chemistry would describe the structures of the bases as intermediate (but weighted toward the keto form) between the two tautomers.

The possible relevance of chemical transmutation to the problem of the etiology of cancer is evident. Naturally occurring endogenous radiocarbon could offer a mechanism for the particular irreversible alteration of a cell gene influencing growth regulation. This could be the first step in the production of cancer. The 53% probability of a specific, nondisruptive mutation is greater than the one-in-four chance of contracting clinical cancer for man. But this presents no contradiction, since the somatic mutation would be only an initiating step. Ensuing conditions would have to become operative before the cancer could manifest itself.

Naturally Occurring Radiocarbon as a Mutagen

The transmutation of a carbon atom into a nitrogen atom in virtually any position in the DNA bases would result in a stable molecule. This type of mutation could cause numerous replication errors. One example, indicated in figure 1, is as follows: if the number 4 carbon atom on a thymine base were transmuted to a nitrogen atom, the keto tautomeric form would convert to enol in this position. In this way, it would combine preferentially with guanine instead of with adenine.

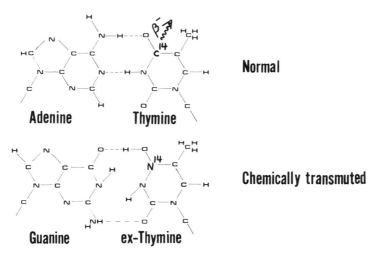

FIG. 1. Chemical transmutation causing a DNS replication error. In this case, a radiocarbon atom transformed into a nitrogen in the 4 position of a thymine ring causes a coupling of that base with quanine instead of with the normal adenine.

Another example would be the number 5 carbon atom in an adenine base transmuting to nitrogen, as shown in figure 2. In this case the aromatic character of the ring would be lost and the electron balancing could lead to a hydrogen addition and a rearrangement to an imino form. After this the ex-adenine would combine preferentially with cytosine instead of with thymine.

Many similar modifications could be devised from chemical transmutations on almost any other position in the four principal bases. It is unnecessary to speculate about the relative importance of each for somatic mutation probability, since the previous argument has shown that the frequency of human cancer could be explained even if only one of all the possibilities caused the specific DNA mutation that initiated the neoplasm.

Experimental Tests of Radiocarbon as an Ocogenic Agent

The theory of somatic mutation by endogenous radiocarbon has an important advantage over the tautomeric mechanism theory because experiments are feasible to test the

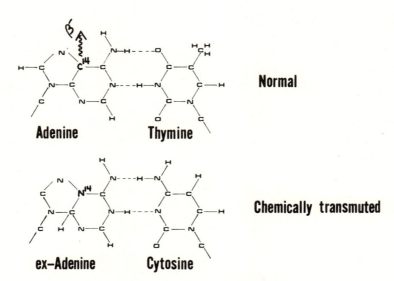

FIG. 2. Chemical transmutation as an error source. The radiocarbon atom in the 5 position on an adenine base transmutes to nitrogen. In this case, when the aromatic character of the ring is lost, the electron balancing could cause a hydrogen addition and rearrangement to an imino form. Subsequently, the ex-adenine would pair with cytosine instead of thymine.

role of this chemical transmutation as an oncogenic agent or a factor in other diseases of aging. In the tautomeric mutation mechanism, no experiments have yet been devised to test the theory.

It would be feasible to examine the radiocarbon transmutation effect by producing animals with either increased or deficient radiocarbon levels. Although the first alternative is considerably simpler, the generation of radiocarbon-free rodents would lead to a less ambiguous result. Previous calculations have demonstrated that the naturally occurring radiocarbon level is sufficient to be a primary cause of neoplasia, even in the most conservative case when a single, specific locus is the only site whose modification has led to the disease. It would be more reasonable to assume, however, that numerous types of DNA changes would produce defective enzymes that could lead to abnormal cell proliferation. The naturally occurring ^{14}C level would therefore be sufficient to cause the initial phase of neoplasia early in the life of most animals. An addition of artificial radiocarbon would not noticeably increase the cancer incidence.

On the other hand, if radiocarbon-free animals were developed and compared with controls, a decrease in the tumor appearance rate would have to occur if the chemical transmutation theory is correct. If the ^{14}C content were less than 1% of modern and no change occurred, the theory could be discarded.

It is recognized that an increase in chromosome aberrations is associated with aging. In addition to observing cancer incidence, it would be interesting to determine whether a reduction in the burden of endogenous radiocarbon has an effect on the longevity of animals, since aging could also be related to somatic mutation (Curtis 1966).

Production of Low Radiocarbon Level Rodents

A rodent diet with low levels of radiocarbon was investigated in the project using petroleum-derived yeast material (Elzinga and Laskin 1971). In trial runs it was possible to maintain mice on a mixture of 2/3 of this yeast and 1/3 normal soybean base feed mixture, with added vitamins and essential amino acids. The results, presented in figure 3, show the reduced ^{14}C burdens that can be achieved in living animals. After several months on this diet the radiocarbon content of these mice was lowered to almost half the normal value, but the theoretical minimum concentration, represented by the dotted line in the figure, was not attained here. It will be necessary to approach this with second generation animals.

In practice, protein-rich supplemental yeasts derived from petroleum or synthetic methanol are kept at dietary levels below 20% (Elzinga and Laskin 1971). The materials cannot be used more extensively because of the high nucleic acid content of the fast-growing organisms. One difficulty would be the metabolism of purine bases to uric acid. Most animals, but not man, convert this allantoin with uricase. The capacity to do this is

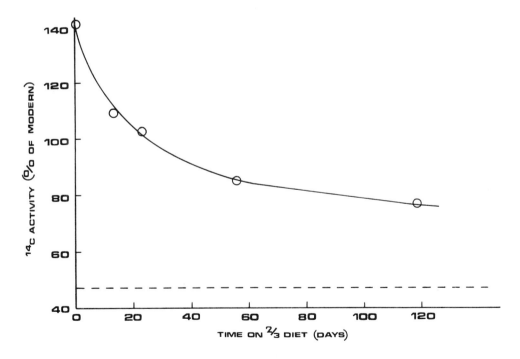

FIG. 3. Preliminary trials of BALB/c mice on two-thirds radiocarbon-free diet consisting of 67% synthetic methanol-derived yeast material and 33% Purina Chow feed (soybean base). The results are expressed as % of the levels of natural pre-nuclear-weapon-testing materials. This is 95% of the activity of the NBS oxalic acid. Errors are of the order of ±1% for one standard deviation criterion. Radiocarbon analyses employed the benzene method (Tamers 1975).

not unlimited and excessive concentrations of nucleic acids lead to urinary system problems.

As another approach, plants can be grown in enclosed systems with carbon dioxide generated from radiocarbon-free sources. Carbon dioxide available commercially contains only minimal amounts of the isotope, as it is generally produced from industries using petroleum feedstock. These natural foods, despite their being the most difficult to produce, would have to form the bulk of the animals' diet.

Synthetic food materials have been used as radiocarbon-free nutrients. Two nonbiogenic food supplements, acetate and propionate from petrochemical sources, have been investigated in our laboratory (Young 1976). Biochemical pathways for their metabolism in the mammalian digestive system are known and chronic toxicity is low. The compounds have been fed to mice in the form of starch esters, as shown below. The other alcohol positions on the glucose unit are also acetylatable.

$$\text{STARCH (GRAIN)} + \text{CH}_3\text{COOH (PETROCHEMICAL ACETIC ACID)} \rightarrow \text{STARCH ACETATE} + H_2O$$

This method permits the introduction of the radiocarbon-free compounds to concentrations of 5%–25%. It has been found that the uptake of the acetate, for example, is greater than 90%. The starch carriers would have to come from grains grown in a fossil carbon dioxide atmosphere.

CONCLUSIONS

The irreversible nature of malignant transformations indicates a change in one or more genes of the cell. The basic biochemical lesion should be in a DNA molecule. There is strong evidence that the initial transformation occurs in a single cell (Linder and Gartler 1965) which then propagates itself, producing malignancies with a bias towards uncontrolled growth. Since cancer is not hereditary, its etiology involves a somatic cell.

The age dependency of neoplasia can be explained by the hypothesis that there are two or more stages in the production of clinical cancer (Curtis 1966). The initial step is the formation of the first transformed cell and its limited propagation. The microscopic tumor remains dormant, prevented from uncontrolled growth by immunological systems of the body. There then follows a latent period which is interrupted by a promoting phase. At this time the cancer becomes malignant.

If one of the initiating phases of cancer is a specific somatic mutation, naturally occurring endogenous radiocarbon transmutation provides a reasonable mechanism for a primary etiology of the disease. Significant increases in atmospheric ^{14}C due to thermonuclear tests began in 1952. The spontaneous clinical cancer incidence has been increasing in the past few years at an accelerating rate. Since there is a twenty to thirty years latent period between the initiation and rapid growth phases of tumors, the radiocarbon theory would be in agreement with this finding.

REFERENCES

Anderson, E. C., W. F. Libby, S. Weinhouse, A. F. Reid, A. D. Kirshenbaum, and A. V. Grosse
    1947   Natural radiocarbon from cosmic radiation. Science. 105:576.

Burch, P. R. J.
    1971   Ionizing radiation and man: the biological hazards. Int. J. Env. Studies. 2:211.

Curtis, H. J.
    1966   A Composite Theory of Aging. The Gerontologist. 6:143.

Davidson, J. N.
    1947   The distribution of nucleic acid in tissues. Cold Spring Harbor Symp. Quart. Biol. 12:50.

Eisenbud, M.
    1973   Environmental Radioactivity. 2d ed. New York, Academic Press.

Elzinga, E. R., and A. I. Laskin
    1971   Proteins from Petroleum. *In* Enc. of Chem. Tech. 2d ed. Suppl. Vol. 836.

Freire-Aial, A.
    1971   Human genetics studies in areas of high natural radiation. II. First results of an investigation in Brazil. Cienc. 43:457.

Grahn, D.
    1972   Genetic effects of low level radiation. Bioscience. 22:535.

Karlen, I., U. Olsson, P. Kallberg, and S. Kilicci
    1964   Absolute determination of the activity of two C-14 dating standards. Arkiv Geofysik. 4:465.

Lehninger, A. L.
    1970   Biochemistry. New York, Worth.

Linder, D., and S. M. Gartler
    1965   Glucose-6-phosphate dihydrogenase mosaicism: utilization as a cell marker in the study of Kiomyomds. Science. 150:67.

Nydal, R., K. Loveseth, and S. Gulliksen
    1976   A survey of variations in nature since the test ban treaty. Proc. 9th Int. Radiocarbon Conf. Univ. of Calif. Los Angeles and San Diego.

Pauling, L.
    1958   Genetic and somatic effects of carbon-14. Science. 128:1183.

Stent, G. S.
    1971   Molecular Genetics. San Francisco, W. H. Freeman and Co.

Tamers, M. A.
    1974   Somatic mutation by endogenous radiocarbon. J. Nucl. Med. Biol. 1:227.
    1975   Chemical yield optimization of the benzene synthesis for radiocarbon dating. Int. J. Appl. Radiation Isotopes. 26:676.

Wolfgang, R. L., Anderson, R. C., and R. W. Dodson
    1956   B rupture and nonrupture in beta decay of carbon-14 studied by double isotope labeling. J. Chem. Phys. 24:16.

Wolfsberg, M. J.
    1956   Excitation and dissociation of molecules due to B-decay of a constituent atom. J. Chem. Phys. 24:24.

Young, M. W.
    1976   The use of synthetic organics as non-biogenic feed extenders. MS thesis. Nova Univ.

6

The Origin of Drift-Gas Deposits as Determined by Radiocarbon Dating of Methane

Dennis D. Coleman

ABSTRACT

Drift-gas is natural gas that occurs in glacial drift. Previous studies have suggested that its formation is due to bacterial decomposition of organic materials within the glacial drift. Stratigraphic positions of some of the deposits suggest, however, that the gas could have accumulated as a result of upward migration from coal or petroleum in the underlying bedrock. This study purports to ascertain the origin of drift-gas, determining the geologic conditions under which drift-gas deposits form.

Samples were collected from twenty-two drift-gas wells in Illinois, and radiocarbon ages were determined for the methane fractions. The results indicate that all of the samples were formed by bacterial decomposition of organic materials and that the most significant source of methane is organic material within the Robein Silt unit, which ranges in age from about 22,000 to 28,000 B.P.

A comparison of the radiocarbon ages of the samples to the stratigraphic positions of the deposits shows that in each case the gas is the same age or younger than the material in which it occurs. This observation, along with the results of radiocarbon analyses of eight samples of gas collected from water wells, indicates that methane commonly exists far below any possible source material of equivalent age and that some mechanism is necessary to facilitate downward migration of drift gas.

It is concluded that movement of the gas as a solute, with ground water, is the most likely mechanism of drift-gas migration, and that the rate and direction of migration are controlled by groundwater movement. Several mechanisms are proposed by which methane (even from the modern soil) could be carried downward in solution and then released from solution (as a result of a decrease in pressure) to form gas deposits well below the source material.

INTRODUCTION

Drift-gas is composed primarily of methane, but contains small amounts of nitrogen and carbon dioxide. Although drift-gas deposits are not generally of commercial significance, many produce sufficient gas for private use. Studies by Meents (1960) and Wasserburg et al. (1963) have indicated that some of the gas is formed by bacterial decomposition of organic materials buried within the glacial drift. The possibility also exists that some deposits could have formed by upward migration and entrapment of natural gas resulting from the formation of coal or petroleum. The author's interest in this subject began in spring 1974 when W. F. Meents suggested studying the origin of drift-gas. Because the most prominent peat deposits in Illinois are in the Wisconsinan drift and have ages within the range of determination by radiocarbon dating, it was anticipated that this technique could be used to ascertain the source of the gas.

Most of the drift-gas wells in Illinois are within the region that was covered by the Wisconsinan glaciers. Meents (1960) reported that approximately three-fourths of these wells produce from the Sangamon Soil "zone" where it is covered by Wisconsinan drift. The Sangamon Soil itself is generally interpreted as having formed earlier than 75,000 B.P., but in his definition of the Sangamon Soil zone, Meents included the silts, sands, and gravels directly above and below the Sangamonian surface. It is now known that throughout much of the area covered by the Wisconsinan glaciers, the Sangamon Soil is directly overlain by the Robein Silt, which generally dates from 22,000 to 28,000 B.P. It is this unit, rich in organic matter, that is a more likely source of bacterially produced drift-gas than is the Sangamon Soil itself. The sands and gravels directly above and below provide a reservoir, and the overlying till serves as a cap. Most of the gas wells in glacial drift from areas that were not covered by the Wisconsinan glaciers produce from sands in buried valleys.

In addition to the gas that exists as a free phase, large quantities of gas are contained in the groundwaters throughout the state. The gas is in solution within the aquifer, but when the water is pumped to the surface and the pressure is released, the solution becomes supersaturated and the gas escapes. Water from wells completed in bedrock and also from those completed in glacial drift commonly contains dissolved gas. The relationship between this gas and that existing in drift-gas deposits has not been previously studied.

Description of Samples

The gas wells in glacial drift that were sampled are described in table 1. Well locations are given in table 5. All of the wells for which detailed well logs are available are completed in sand or gravel, and the remainder probably are completed in similar materials. The ages of the reservoir materials reported in the table are estimates based on well logs, local geologic maps, and the contour map of the Sangamon surface by Horberg (1953). For those wells that appear to be completed within 10 feet of a contact, both the upper and the lower units are noted. As shown on the table, the wells sampled range in depth from 72 to 324 feet. The maximum gas pressure encountered was 64 psig. Several wells

TABLE 1
Description of Gas Wells in Glacial Drift

Sample number	County	Surface elevation (ft)	Well depth (ft)	Depth to base of casing (ft)	Gas pressure (psig)	Geologic age of reservoir material
1-74	Champaign	787	130	126	10	Wis.–Ill.
2-74	Macon	685	95	95?	10.5	Ill.
3-74	Macon	702	100	100	10.5	Ill.
4-74	McLean	860	137	132	15	Wis.–Ill.
5-74	Champaign	742	125	125	9.5	Wis.–Ill.
8-74	Champaign	735	100	93?	3.5	Wis.–Ill.
9-74	De Witt	771	120	120?	~0	Ill.
10-74	Peoria	835	324	322	64	Ill.–Kans.–Penn.
11-74	Marshall	735	144	140	28	Wis.–Ill.
12-74	Tazewell	675	120	120?	<0	Ill.
13-74	McLean	842	163	140	38	Ill.
18-75	Fayette	558	102	102	12	Kans.
19-75	Shelby	718	72	72?	0.6	Wis.–Ill.
32-75	Kane	920	200	194	43	Ill.
39-75	Champaign	735	155	105	8.5	Ill.
40-75	La Salle	648	137	137	8	Ill.
54-75	Bureau	780	150	150?	6	Wis.–Ill.
55-75	Bureau	692	200	200	~0	Kans.
56-75	Bureau	661	174	165	~0	Kans.
57-75	Bureau	695	85	85	1.5	Ill.
60-75	Piatt	735	95	95?	7.5	Ill.
61-75	Woodford	810	145	145	21	Ill.

had pressures very near atmospheric pressure; gas is produced from these low pressure wells by the use of vacuum pumps. In one of the wells (12-74), there is a shut-in vacuum of 3.4 inches of mercury in the reservoir.

In addition to the samples that were collected from drift-gas wells, eight samples of gas from freshwater wells were collected for radiocarbon dating. Descriptions of these wells are given in table 2. Four of the samples were from water wells completed in glacial drift. The remaining four samples were from wells completed in bedrock, but the geologic, chemical, and isotopic data reported by Coleman (1976) suggests that this gas originated in the glacial drift.

Chemical and isotopic compositions for all of the samples were reported by Coleman (1976). The samples considered here had methane contents ranging from 45% to 97%; the remainder of the gas was nitrogen and carbon dioxide. The $\delta^{13}C$ values for the methane from these samples ranged from -70 $^o/_{oo}$ to -91 $^o/_{oo}$ relative to the Chicago PDB standard.

TABLE 2

Description of Water Wells Sampled

Sample number	County	Surface elevation (ft)	Well depth (ft)	Depth to base of casing (ft)	Aquifer material	Geologic age of aquifer
2-75	Piatt	786	>162	162	drift	Ill.
3-75	De Witt	705	105	93	drift	Ill.
4-75	De Witt	725	172	152	drift	Kans.?
15-75	McLean	730	210	?	drift	Kans.
22-75	Iroquois	645	193	170	ls.	Sil.
46-75	Vermilion	600	82	45	ls. & sh.	Penn.
52-75	Livingston	615	55	50	ls.?	Penn.
72-75	Kane	878	220	210	dol.	Sil.

Experimental Techniques

SAMPLE-COLLECTION TECHNIQUES

The techniques used for collecting samples from gas wells have been described by Meents (1958). Most samples were collected at the well heads, but in some cases it was necessary to take samples from storage tanks or from gas outlets in private homes. Steel tanks were used of different sizes, ranging in capacity from 4.6 to 9.6 liters depending on the gas pressure in the well. Each tank was evacuated prior to use. The tank was connected to the well with high pressure tubing, and the tubing was purged by allowing gas to flow out through a venting port located immediately above a valve on the tank. The tank was then filled to the pressure in the well or the pressure system.

Samples from water wells were collected in 2-gallon polyethylene bottles by water displacement. All samples from water wells were from private or municipal pressure systems, care being taken to ensure that the water had not undergone any prior treatment. The bottle was filled with water and then inverted and immersed in a pail of water. A hose was connected to the water system, and the water was allowed to run for several minutes. After the flow rate was adjusted to about 3 gallons per minute, a copper U-tube was inserted into the inverted bottle. Because of the decrease in pressure as the water flowed out of the tube, gas came out of solution and collected in the bottle, gradually displacing the water. When sufficient gas was collected, the U-tube was removed and the top of the bottle was fitted with a rubber stopper forced tightly into position by a screw-top lid.

ANALYTICAL TECHNIQUES

The techniques used to analyze the gas have been described in detail by Coleman (1976). Only a brief description is given here. The sample gas was passed through a dry ice trap to remove water vapor and through a column filled with Ascarite to remove CO_2; the gas was then adsorbed on activated charcoal in a 150-cm³ stainless steel pressure cylinder cooled with liquid nitrogen. After the cylinder warmed to room temperature, it was attached to the vacuum system shown in figure 1. In this system the gas was passed through a NaOH solution to remove any remaining CO_2 and was then mixed with N_2. The gas mixture was allowed to enter the combustion tube and was burned at the tip of the inlet tube. The last fraction of sample gas was removed from the pressure cylinder by freezing it into a trap filled with silica gel. Residual gases remaining on the silica gel were removed by flushing with N_2. The CO_2 formed was synthesized to benzene following the procedures outlined by Kim et al. (1969), Coleman et al. (1972), and Coleman (1973).

^{14}C concentrations of synthesized benzene were determined in a modified Packard Tri-Carb Model 3375 liquid scintillation spectrometer operated at 1°C, using Butyl PBD dissolved in toluene as a scintillator. All samples were counted in a 10-ml low-potassium glass vial. Background count rates were determined before and after counting each sample, the total counting times ranging from 3 to 9 days. With this system the background was about 5.2 counts per minute and the efficiency was about 71%. The reference activity used was 95% of the activity of the NBS oxalic acid standard.

In accordance with the decision of the Fifth Radiocarbon Dating Conference, Cambridge University, 1962, all dates were calculated using the Libby half-life of 5568 years. The dates reported in the text have been corrected for the effects of isotopic fractionation. The error (1σ) associated with uncorrected ages accounts only for uncertainties in activity measurements of the sample, standard, and backgrounds. An additional error of at least ± 10 years has been added to the errors of the corrected ages to compensate for possible errors in the determinations of the isotopic compositions.

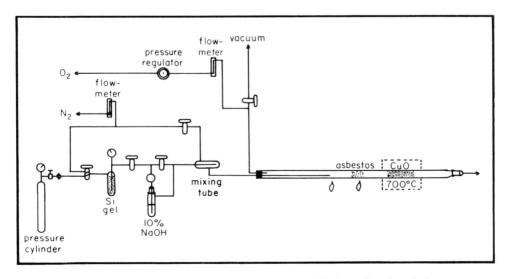

FIG. 1. System for combustion of methane samples for radiocarbon dating.

Experimental Results

If the methane in drift-gas is derived from buried organic material, the methane should contain the same proportion of ^{14}C (after correction for isotopic fractionation) as the source material, and the radiocarbon age of the methane should be the same as that of the source material. It should be emphasized, however, that the radiocarbon age of a methane sample is indicative *only* of the age of the source material and is independent of the time of gas formation. If the methane in a particular deposit originates from more than one source, the radiocarbon date obtained on that gas will be a "composite" age. Many of the radiocarbon dates that will be given in the following sections are only apparent ages. The errors that are given indicate only the precision of the ^{14}C determination and imply nothing about the range in age of the source material(s).

The results of the age determinations on samples from drift-gas wells are given in table 3, and those for samples from water wells are given in table 4. Numerous organic zones have been reported in the glacial drift of Illinois, but the Robein Silt is the most prominent. It was therefore anticipated that most of the methane samples from gas wells in glacial drift would have radiocarbon ages between 22,000 and 28,000 B.P., the range of ages commonly determined for the Robein Silt (Willman and Frye, 1970, table 1). Ten of the samples do have ages between 20,000 and 30,000 B.P., but nine of the samples are younger and three are older. The value E-E' given in tables 3 and 4 is an estimate of the difference in elevation between the base of the well casing and the probable position of the Robein Silt. A negative value of E-E' indicates that the well extends below the Robein Silt. Those wells having very small E-E' values may be completed within the Robein Silt, but it is more likely that they are completed in the underlying Illinoian sands and gravels. To indicate this uncertainty, the age of the reservoir material is followed by a question mark in the tables.

For samples that have radiocarbon ages in the range of 20,000 to 30,000 B.P. and that are from wells completed near the position of the Robein Silt, there is little question that most of the methane formed from organic material in the Robein Silt. The remainder of the samples, however, require consideration of the possibility that local geologic and hydrologic conditions may account for the apparently anomalous radiocarbon ages of methane from some of these wells.

Discussion of Results

PEORIA-MARCHALL COUNTY LINE AREA

Two of the samples (10-74 and 11-74) are from drift-gas wells located about 2 miles apart near the Peoria-Marshall County line. Sample 10-74 is from a well on the Providence Moraine of Wisconsinan age. The well extends 324 feet into the drift, almost to the bedrock. The base of the well is approximately 180 feet below the position of the Robein Silt, yet the radiocarbon age of the methane is 21,290 ± 160 B.P. Sample 11-74 is from a well east of the moraine (toward the Illinois River Valley). The base of this well is estimated to be about 10 feet below the Robein Silt. The radiocarbon age of this sample is 24,040 ± 230 B.P.

TABLE 3

Radiocarbon Ages of Samples from Drift-Gas Wells

Sample number	Well Depth (ft)	Depth to base of casing (ft)	Geologic age of reservoir material	E-E'* (ft)	Date number (ISGS-)	Radiocarbon age (years B.P.)
1-74	130	126	Wis.?	+5	287	16,770 ± 220
2-74	95	95?	Ill.	−50	291	14,630 ± 230
3-74	100	100	Ill.	−40	293	14,240 ± 130
4-74	137	132	Ill.?	−10	301	20,920 ± 280
5-74	125	125	Ill.	−20	302	23,890 ± 340
8-74	100	93?	Ill.?	−10	304	15,610 ± 270
9-74	120	120?	Ill.	−30	305	15,590 ± 190
10-74	324	322	Kans.	−180	306	21,290 ± 160
11-74	144	140	Ill.?	−10	308	24,040 ± 230
12-74	120	120?	Ill.	−40	309	10,230 ± 130
13-74	163	140	Ill.	−60	310	30,820 ± 460
18-75	102	102	Kans.	†	335	39,000 ± 1900
19-75	72	72?	Ill.?	−5	330	21,470 ± 360
32-75	200	194	Ill.	−60	339	30,720 ± 510
39-75	155	105	Ill.	−20	341	23,580 ± 780
40-75	137	137	Ill.	−30	340	23,230 ± 510
54-75	150	150?	Ill.?	−5	355	21,510 ± 610
55-75	200	200	Kans.	−110	342	16,200 ± 200
56-75	174	165	Kans.	−100	345	16,040 ± 200
57-75	85	85	Ill.	−30	349	11,890 ± 460
60-75	95	95?	Ill.?	−5	362	20,200 ± 1100
61-75	145	145	Ill.	−35	353	25,120 ± 450

*E-E' is the difference in elevation between the base of the well casing and the probable position of the Robein Silt. The values have been estimated with the help of J. P. Kempton of the Illinois State Geological Survey. For areas for which well logs and local geologic maps (many of which are unpublished) were available, the values were estimated from those logs and maps; for other areas, the estimates were made from the Sangamon plain contour map of Horberg (1953). Negative values indicate that the wells extend below the Robein Silt.
†This well is outside the margin of Wisconsinan glaciation.

Figure 2 is a topographic map of the area from which these samples were taken; the locations of the two wells are shown. A comparison of this map with the contour map of the bedrock surface by Horberg (1957) shows that these wells are over bedrock tributary valleys of the Ancient Mississippi Valley. The Sankoty Sand Member, which is believed to be Kansan in age, occurs in the deepest part of the Ancient Mississippi Valley and adjacent tributaries (Willman and Frye 1970). This sand is the reservoir material for several gas wells both in this area and farther to the north, in the Tiskilwa drift-gas area.

Figure 3 is a cross section from point A to point A' on figure 2, and includes the wells represented by samples 10-74 and 11-74. This figure and the other cross sections, to be shown later, were drawn to illustrate the relationship between the sand units, which are potential reservoirs of gas, and materials that are rich in organic matter (mainly

TABLE 4
Radiocarbon Ages of Samples from Water Wells

Sample number	Well depth (ft)	Depth to base of casing (ft)	Geologic age of reservoir material	E–E'* (ft)	Depth into bedrock (ft)	Date number (ISGS–)	Radiocarbon age (years B.P.)
2-75	162	162	Ill.	−25		311	25,540 ± 880
3-75	105	93	Ill.	−115		314	33,770 ± 1,900
4-75	172	152	Kans.?	−80		315	31,880 ± 1,200
15-75	210	200?	Kans.	−170		323	>32,600
22-75	193	167?	Sil.		26	360	30,600 ± 1,100
46-75	82	45	Penn.		50	361	16,810 ± 200
52-75	55	50?	Penn.		5?	365	29,620 ± 800
72-75	220	210	Sil.		10	364	23,560 ± 550

*See footnote (*) on Table 3.

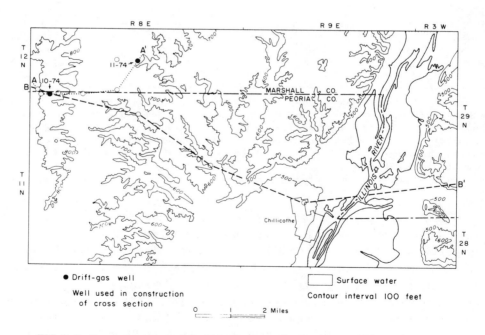

FIG. 2. Surface contour map of the Peoria-Marshall County-line area. (See figs. 3 and 4 for cross sections A to A' and B to B'.)

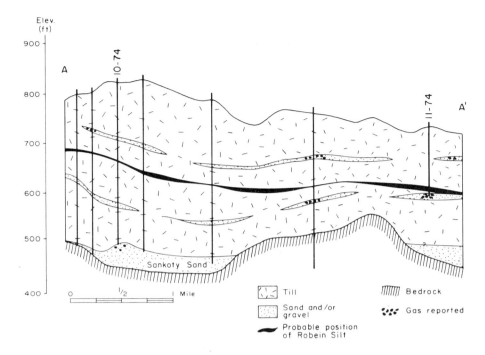

FIG. 3. Cross section from A to A' in the Peoria-Marshall County-line area (fig. 2). The cross section was constructed from well records and sample descriptions on file at the Illinois State Geological Survey, and from the Sangamon plain contour map by Horberg (1953) and the bedrock surface contour map by Horberg (1957).

Robein Silt); they are not intended to be precise representations of all available stratigraphic data.

The dark band in figure 3 represents the probable position of the Robein Silt. Despite the continuity implied in the cross section, this unit is not continuous throughout the area and in many wells is weakly developed or absent. The sand lenses may not be as continuous as they are shown to be in the figure, but they do appear in many of the wells at approximately equivalent positions, and many of them contain gas.

The well represented by sample 11-74 is completed in one of the sand lenses. In the vicinity of this well, the Robein Silt (where present) is very close to the position of the sand unit and, as is suggested in figure 3 may in some places directly overlie the sand. The Robein Silt therefore provides the most likely source for the gas; that it is the source is confirmed by the radiocarbon date on the gas of 24,040 ± 230 B.P.

Sample 10-74 is from a well in the Sankoty Sand, but the radiocarbon date on this sample, 21,290 ± 160 B.P., implies that most of this methane is also from the Robein Silt. Figure 4 is a cross section from point B to point B' in figure 2. This cross section is approximately parallel to a bedrock tributary valley of the Ancient Mississippi Valley.

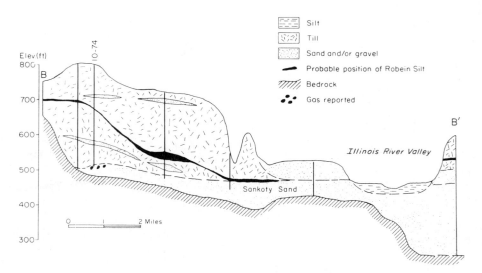

FIG. 4. Cross section from B to B' in the Peoria-Marshall County-line area (fig. 2); modified from Horberg (1953).

The figure shows that the Robein Silt (where present) drapes down into the valleys and in some areas probably directly overlies the Sankoty Sand. The Sankoty Sand is generally saturated with water in this area. The cross section suggests that methane could form in the Robein Silt and then migrate, as a gas, along the top of the water and collect in high areas. The water is reported to contain dissolved methane, however, so movement in or with the ground water also appears to be a likely mechanism of migration. Groundwater movement in this area is to the east, toward the Illinois River Valley (K. Cartwright, Illnnois State Geological Survey, personal communication, 1975). If migration in ground water is the mechanism by which the gas deposit represented by sample 10-74 has accumulated, it appears that either (1) the gas has migrated in a direction opposite to that of the flow of ground water, or (2) there are areas in which the Robein Silt directly overlies the Sankoty Sand at elevations higher than those shown in figure 4, or (3) the gas has migrated *down* through the tills into the Sankoty Sand.

TISKILWA AREA

The Tiskilwa drift-gas area, approximately 20 miles north of the Peoria-Marshall County line area, is also located along the Ancient Mississippi Valley. Meents (1958) reported that there were 51 drift-gas wells in the area, all but two of which produced from the top of the Sankoty Sand. Because coal is present about 55 feet below the base of the Sankoty Sand, Meents concluded that the gas could be coming either from the coal or from bacterial decomposition of organic material in the glacial drift. Three wells from the area were sampled for the present study. Two of the wells (55-75, 56-75) are in the Sankoty Sand, and one (57-75) is in a sand unit within the glacial tills that overlie the Sankoty Sand.

An origin from Wisconsinan or younger materials is verified by radiocarbon dates on the three samples. The radiocarbon ages of the two samples from the wells in the Sankoty Sand are 16,200 ± 200 B.P. (55-75) and 16,040 ± 200 B.P. (56-75), in close agreement with each other. Sample 57-75 has an even younger radiocarbon age, 11,890 ± 460 B.P.

A cross section has been constructed (fig. 5) through these three wells to show the relationship between the gas deposits and the subsurface geology. Although the sand units within the drift may not be continuous, as shown, they do appear at approximately equivalent positions in practically all of the wells for which records are available. It was possible to confirm the presence of the Robein Silt in only a few of the well records, and therefore its exact position is hypothetical. The Sankoty Sand is a valley fill and does not extend up onto the bedrock highs. Meents (1958) reported that the static water level here is about 40 feet below the top of the Sankoty Sand (that is, there is no head), and that gas occupies the space above the water. The gas pressure is about equal to atmospheric pressure. The Sankoty Sand crops out along the Illinois River, and therefore gas can flow out of the unit during times of low atmospheric pressure and air can flow in during times of high atmospheric pressure. Meents compiled a contour map showing the variation in "gas gravity" of samples from wells in the area. Air has a gas gravity of 1.00, whereas pure methane has a gas gravity of 0.55. The lower the gas gravity, the higher the methane content for an air-methane mixture. Figure 6 is a modification of Meents's map; it shows that the gas having the lowest gas gravity, that is, the highest methane content, is in the southwestern part of the area and that the gas gravity increases to the north and east.

Several possible sources of the methane in the Sankoty Sand need to be considered.

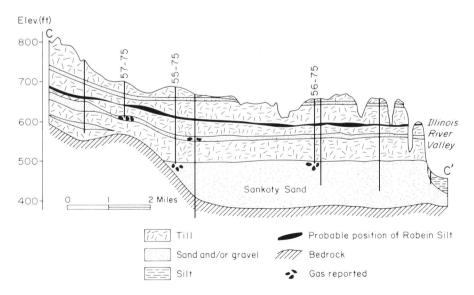

FIG. 5. Cross section from C to C' in the Tiskilwa drift-gas area (fig. 6); constructed from well records and references to the works of Horberg (1957) and McComas (1959).

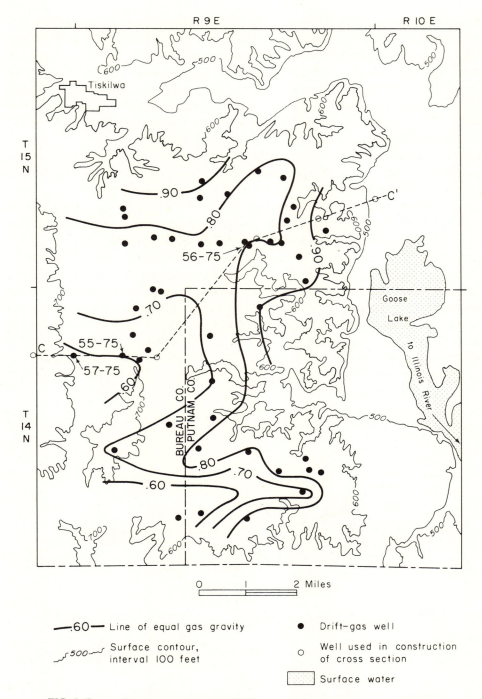

FIG. 6. Gas-gravity contour map of the Tiskilwa drift-gas area (after Meents 1958). (See fig. 5 for cross section C to C'.)

If the gas is a mixture of "old gas" formed in the glacial deposits and "young gas" formed in the recent sediments in the Illinois River Valley, one would expect that those wells nearest the valley might contain a higher proportion of young gas than those farther away. Yet, samples 55-75 and 56-75 have nearly identical radiocarbon ages (table 3).

Meents (1958) interpreted the gas-gravity contours (fig. 6) as evidence that atmospheric dilution has occurred where the Sankoty Sand crops out along the Illinois River Valley. These contours also suggest, however, that the source of methane may be in the regions where the gas gravities are lowest. Sample 57-75 is from the low gas-gravity region in the western part of the area. This well is one of the few gas wells in the area that is not in the Sankoty Sand and that contains gas under pressure. Figure 5 shows that this well is in a sand unit considerably above the Sankoty Sand and that it is within the glacial drift. A similar sand unit was reported in a water well just to the west, only a foot below the Robein Silt. The occurrence of 57-75 is similar to that of sample 11-74 in the Peoria-Marshall County line area. One would expect the radiocarbon age of the methane to be similar to the age of the Robein Silt, but the radiocarbon age of sample 57-75 is 11,890 ± 460 B.P., considerably younger than anticipated. Although part of this gas probably is from the Robein Silt, it apparently is diluted with methane from a much younger source. It is possible that recent bog deposits provide much of the methane. Although no such deposits are known in the immediate area, MacClintock and Willman (1959) reported extensive peat and muck deposits a few miles to the northeast. These deposits were interpreted as being postglacial and associated with stagnant ice features.

The similarity of the isotopic and ^{14}C compositions of sample 57-75 and the two samples from the Sankoty Sand (55-75, 56-75), along with the gas-gravity contour map (fig. 6), suggests that the well represented by sample 57-75 may be within a source region for the gas in the Sankoty Sand. The only other gas well in the area reported by Meents (1958) that is not in the Sankoty Sand is in a sand unit similar to that in which the well represented by sample 57-75 is completed, and it is located in the low gas-gravity region in the southwestern corner of the area. This is another possible source region for the gas in the Sankoty Sand. If the methane in this region is older — for example, the same age as the Robein Silt — a mixture of the gases from the western and southwestern source regions could result in a gas having a radiocarbon age of 16,000 B.P., similar to that obtained on samples from the two gas wells in the Sankoty Sand. There may, of course, be other sources of methane. Regardless of the exact source, the data suggest that the gas in the Tiskilwa drift-gas area has accumulated as a result of some mechanism of downward migration. Mechanisms of drift-gas migration will be discussed later.

CHAMPAIGN COUNTY AREA

The locations of three of the gas wells sampled in Champaign County (1-74, 5-74, 39-75) are shown on figure 7. A generalized cross section developed in conjunction with a groundwater study by the Illinois State Geological Survey was made available by J. P. Kempton and is shown in figure 8. As was true for the previous cross sections, the sand lenses shown may or may not be continuous; their positions relative to the probable position of the Robein Silt are somewhat hypothetical.

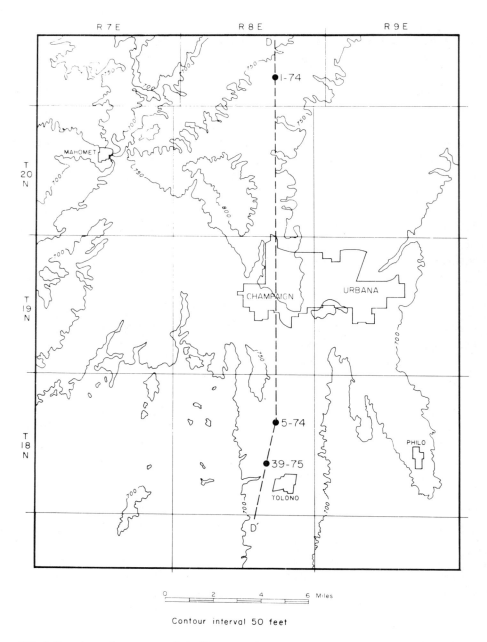

FIG. 7. Locations of some gas wells in Champaign County. (See fig. 8 for cross section D to D'.)

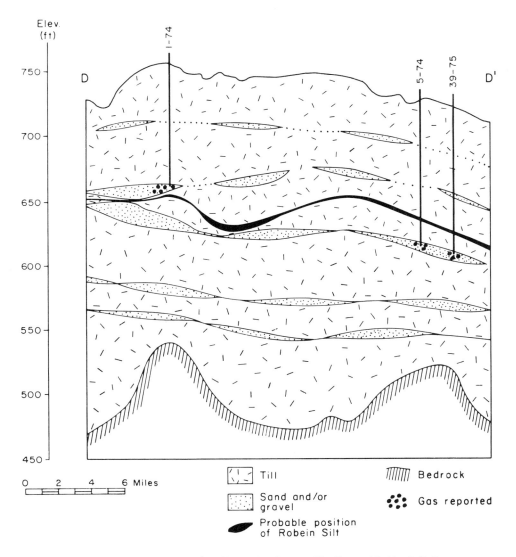

FIG. 8. Cross section from D to D' in Champaign County (fig. 7); provided by J. P. Kempton.

The two gas wells represented by samples 5-74 and 39-75 appear to be in a sand unit 10 to 20 feet below the Robein Silt. There is evidence that in some areas this sand unit directly underlies the Robein Silt. The radiocarbon ages on these samples are 23,890 ± 340 B.P. and 23,580 ± 780 B.P., confirming that most of the methane did form in the Robein Silt.

The well represented by sample 1-74 appears to be in a sand unit immediately above the Robein Silt, but since there is no log on record for this well, the exact position of its base relative to the Robein Silt is uncertain. The radiocarbon age of this sample, 16,770 ± 220 B.P., indicates that material younger than the Robein Silt has been a significant source of methane. This well is located near the crest of the Rantoul Moraine, but even the lowland areas beyond the margin of the moraine are at an elevation of about 700 feet, considerably above the base of this well. Simple lateral migration of younger gas into this area does not seem likely.

Approximately 8 miles to the south, there is a peat deposit that has a radiocarbon age of 13,980 ± 200 B.P. (ISGS-69). The peat has resulted from accumulation of organic material in a small postglacial lake on top of the Champaign Moraine (W. H. Johnson, *in* Coleman 1973). It is highly probable that similar deposits are present at other places in the area and may be actively producing methane. If deposits of this type are the source of the methane, it again seems that some mechanism of downward migration exists.

NATURALLY OCCURRING METHANE FROM OTHER AREAS

Of the remaining samples shown on table 3, some are younger than the Robein Silt and some are older. However, no sample is older than the reservoir material in which the gas occurs. This observation suggests that upward migration of gas has not been a major factor in the formation of drift-gas deposits.

Radiocarbon dates were obtained on four samples of gas from water wells completed in glacial drift. The results of these determinations are shown in table 4. Sample 2-75 is from Illinoian sand and gravel, approximately 25 feet below the position of the Robein Silt, but the radiocarbon age (25,540 ± 880 B.P.) suggests that the gas originated from organic material in the Robein Silt. Sample 3-75 is from Illinoian materials (>75,000 B.P.) approximately 115 feet below the Robein Silt, yet the radiocarbon age of the methane is 33,770 ± 1900 B.P. Sample 4-75 is from approximately 80 feet below the position of the Robein Silt in what are probably Kansan deposits, but it has a radiocarbon age of 31,800 ± 1200 B.P. The dates for samples 3-75 and 4-75 do not necessarily imply that the methane formed from organic materials approximately 33,000 radiocarbon years old. It is more likely that the gas is a mixture of methane from different sources. For example, a sample having a radiocarbon age of 33,000 B.P. could be obtained if 37% of the methane were derived from source material 25,000 radiocarbon years old (Robein Silt) and 63% from source material more than 50,000 radiocarbon years old. Sample 15-75, from approximately 170 feet below the Robein Silt, in Kansan drift, contained no detectable ^{14}C; thus formation of methane does occur from organic material within older, Pre-Robein glacial deposits. The absence of radiocarbon in sample 15-75 also shows that routine sampling and analytical procedures do not introduce significant amounts of modern carbon.

The radiocarbon ages on these samples indicate that methane does exist in solution in groundwater, well below the source material. This point is confirmed by analysis of four samples from water wells completed in bedrock that were considered, on the basis of their chemical and isotopic compositions, to have originated in the glacial drift (Coleman 1976).

Sample 22-75 is from a well that extends 26 feet into the top of a Silurian-limestone dome covered by glacial drift. Since there is very little organic material in the limestone to serve as a source of methane, the water migrating down through the glacial drift, may have become saturated with bacterially produced methane. The radiocarbon date for this sample, 30,600 ± 1100 B.P., supports this explanation.

The well from which sample 46-75 was taken extends 50 feet into Pennsylvanian limestone and shale. Organic matter in the shale provides a possible source of methane, but the radiocarbon date for this sample is 16,810 ± 200 B.P. Since this well is located only a few feet from a small lake, it is possible that methane produced by bacteria in the lake sediments is carried downward in solution in ground water. In the aquifer, this methane can mix with any gas that forms in the shale. As little as 8% methane from

TABLE 5

Locations of Wells Samples

Sample number	Name of well or owner	Location						Type*
		County	¼	¼	¼	Sec.	T.	R.	
1-74	T. Whetzel	Champaign	SE	SE	SE	27	21N	8E	G,D
2-74	M. Hizer	Macon	NE	SE	SE	27	17N	3E	G,D
3-74	C. E. Maxey	Macon	SW	SE	NW	26	17N	3E	G,D
4-74	T. Moffit	McLean	NW	NE	SW	34	23N	3E	G,D
5-74	W. M. Joyce	Champaign	NE	NE	NW	14	18N	8E	G,D
8-74	D. Clapper	Champaign	NE	NW	NW	25	20N	7E	G,D
9-74	W. Klemm	DeWitt	NE	NE	SE	8	20N	1E	G,D
10-74	Holt	Peoria	NE	NE	NW	6	11N	8E	G,D
11-74	G. Shimp	Marshall	NW	NW	NW	34	12N	8E	G,D
12-74	H. Yordy	Tazewell	SE	SW	SE	34	25N	3W	G,D
13-74	E. Bane	McLean	SW	SW	SE	36	24N	4E	G,D
2-75	V. Kammeyer	Piatt	SW	NW	NW	22	21N	6E	W,D
3-75	C. Wendell	DeWitt	SE	SW	SE	19	21N	4E	W,D
4-75	Farmer City No. 6	DeWitt	NW	NE	NE	28	21N	5E	W,D
15-75	R. Ringer	McLean	SW	NE	NE	31	26N	3E	W,D
18-75	M. Miller	Fayette	NW	NE	NE	21	7N	2E	G,D
19-75	M. Rohlf	Shelby	SE	SE	SE	2	11N	5E	G,D
22-75	N. Ill. Gas. Co.	Iroquois	SW	SE	SW	34	27N	13W	W.B
32-75	D. Van Haelst	Kane	NW	NW	NW	10	40N	7E	G,D
39-75	W. Maxwell	Champaign	SE	SE	SE	22	18N	8E	G,D
40-75	M. Walter	La Salle	NE	SE	SE	15	32N	3E	G,D
46-75	H. Schroeter	Vermilion	SE	NW	NE	1	19N	13W	W,B
52-75	Camp Aramoni	Livingston	SE	SE	NE	3	29N	4E	W,B
54-75	E. Schiall	Bureau	SE	SW	NE	23	18N	10E	G,D
55-75	E. Kane	Bureau	NW	NW	SW	9	14N	9E	G,D
56-75	R. Albrecht	Bureau	NE	NE	NW	34	15N	9E	G,D
57-75	R. Workman	Bureau	NE	NW	SE	7	14N	9E	G,D
60-75	L. Dobson	Piatt	NW	NW	SW	25	17N	4E	G,D
61-75	Marquordt	Woodford	SE	NE	NW	20	25N	1E	G,D
72-75	Big Timber B.S.	Kane	NW	SW	SW	31	42N	8E	W,B

*G - gas well; W - water well; D - completed in glacial drift; B - completed in bedrock.

modern organic material mixed with methane formed in the shale could result in an apparent radiocarbon age of 16,800 B.P. Since the aquifer is covered by glacial materials, a significant amount of the methane could also be derived from these materials. Another possible explanation for the results on this sample is that dissolved organic matter may be carried down with the groundwater. If methane formation is taking place in the shale, this dissolved organic matter could be reduced to methane and mixed with the methane from the shale. A more detailed discussion of the mechanism of gas migration in groundwater is given later.

Sample 52-75 is from a very shallow well (55 feet) which extends only about 5 feet into the Pennsylvanian bedrock, probably limestone. Here again the geologic setting is favorable for the incorporation of the methane from the overlying glacial drift. The radiocarbon date of 29,620 ± 800 B.P. confirms that the gas is from this source. Sample 72-75 is from a similar geologic setting in Silurian dolomite covered by drift. The radiocarbon date on this sample is 23,560 ± 550 B.P. Table 5 charts the location of all wells sampled.

Mechanisms of Drift-Gas Migration and Accumulation

It has been shown in the previous sections that drift-gas commonly exists some distance from the probable source material. When a gas phase forms in porous materials that also contain water, buoyancy causes the gas to move upward. Free flow, however, requires a continuous network of channels. Carman (1956) reported that this network of channels does not form (and thus free flow of gas does not occur) until approximately 15%–20% of the water is displaced by gas. Once this condition is met, gas can migrate freely, either vertically or laterally, to the surface or to a point where a structural, stratigraphic, or hydrodynamic trap prohibits further movement. Flow of a free gas phase may be the mechanism responsible for the formation of some of the driftgas deposits encountered, but many of the deposits occur well below the source material. These deposits cannot be accounted for by migration of a gas phase unless the rate of downward movement of water is greater than the rate of upward movement of gas caused by buoyancy (Hubbert 1953). This condition is unlikely. It appears that movement of methane as a solute in or with groundwater is the most probable mechanism for drift-gas migration.

The solubility of a gas in water at constant temperature can be approximated by Henry's law:

$$X_1 = \frac{P_1}{K_1} \qquad (1)$$

where K_1 is the Henry's law constant for the particular gas, P_1 is the partial pressure of the gas, and X_1 is the mole fraction of the gas in solution. If n_1 is the number of moles of gas and n_2 is the number of moles of water, then for dilute solutions

$$X_1 = \frac{n_1}{n_1 + n_2} \approx \frac{n_1}{n_2} \qquad (2)$$

The Henry's law constant for methane at 12°C (average groundwater temperature) is approximately 3.3×10^4 atm (Hougen et al. 1954). If C_1 is the methane concentration in the solution in moles per liter of water, and P_1 is expressed in atmospheres, then

$$C_1 = 1.68 \times 10^{-3} P_1 \qquad (3)$$

If methane is the only dissolved gas, its concentration can increase until its partial pressure (P_1) is equal to the hydrostatic pressure. At equilibrium, the concentration of methane cannot exceed this value, and any additional methane will go into a gas phase. If more than one gas is present, a gas phase will form when the sum of the partial pressures of the various gases is equal to the hydrostatic pressure.

Theoretical consideration of the relative importance of diffusion and convection as mechanisms of migration of methane in solution indicates that diffusion is probably negligible unless the rate of groundwater movement is extremely slow (Coleman 1976). Groundwater movement appears to be the primary factor governing the direction and rate of migration of dissolved methane, but for the release of gas from solution and the subsequent formation of a gas deposit, it is necessary that the water encounter a zone in which the hydrostatic pressure is less than the sum of the partial pressures of the dissolved gases. The following sections discuss some of the natural situations in which such a decrease in pressure could be encountered at depths greater than those of the source materials.

THE EFFECT OF A PERCHED WATER TABLE

The radiocarbon data on the samples from the Tiskilwa drift-gas area suggest that methane is transported downward in groundwater and released at a depth greater than that of the source material. The water table in the drift that overlies the Sankoty Sand is perched (fig. 5). Water movement in the Sankoty Sand is towards the Illinois River Valley (from left to right in fig. 5) (K. Cartwright, personal communication, 1975), and only during times of flooding does water from the river move into the aquifer. The water in the Sankoty Sand in this area must be coming from the overlying drift or from the highlands to the west. If the water moves slowly down through the glacial tills, silts, and sands, towards the Sankoty Sand, it will undoubtedly pass through one or more zones rich in organic matter in which methane production is taking place. Hydrostatic pressures within the glacial drift overlying the Sankoty Sand can be quite high. For example, in the well to the left of 57-75 on figure 5, the water table is 124 feet above the base of the well. This head corresponds to a pressure at the base of the well of about 4.7 atm (54 psig). Equation (3) predicts that, at this pressure, water can contain up to 7.9×10^{-3} moles of CH_4 per liter of water (assuming that CH_4 is the only gas present). The pressure within the Sankoty Sand, however, is only about 1 atm (0 psig). If methane-saturated water from the overlying drift flowed into the Sankoty Sand, a decrease in pressure could occur and the solution would then become oversaturated. At 1 atm pressure, the solubility of methane is about 1.7×10^{-3} moles per liter. Almost 80% of the gas held in solution could be released.

A similar, but less dramatic, situation may exist in the Champaign County area. Although a true perched water table does not appear to be present in that area, apparent restrictions in flow between the Wisconsinan and Illinoian drifts result in what may be considered a "partially perched" water table in the Wisconsinan units. Well tests have shown that hydrostatic pressures within the Wisconsinan drift (above the Robein Silt) are higher than those in the underlying Illinoian and Kansan aquifers. Differences in water table levels corresponding to differences in pressures of up to 2 atm are common (K. Cartwright, personal communication, 1975). These differences are enhanced, and may even be caused, by modern pumpage. When water moves from the Wisconsinan drift into the underlying Illinoian materials or other low pressure zones, there is a decrease in pressure that may be large enough to cause the release of dissolved gases.

THE EFFECT OF A HYDRAULIC GRADIENT

Although perched or partially perched water tables provide an environment in which a decrease in pressure could occur at depth, there are other possible mechanisms. Consider, for example, a confined aquifer such as that illustrated in figure 9. Groundwater flow within the area would be governed by hydraulic gradients, controlled in turn by the regional hydrology. The potentiometric surface, as used by Hubbert (1953), is the level to which the water would rise if it were not restricted. In the aquifer shown in the figure, flow would be from left to right since the potential decreases from left to right. At any point in the aquifer, the hydrostatic pressure is proportional to the vertical distance from that point to the potentiometric surface. If ϕ is the elevation of the potentiometric surface and E is the elevation of any chosen point, then the hydrostatic pressure (P) at that point, expressed in terms of feet of water, is given by

$$P = \phi - E. \qquad (4)$$

Consider two points in the aquifer. If point 2 is at a lower elevation than point 1 (that is $E_1 > E_2$), it can be seen that $P_1 > P_2$ if $(\phi_1 - \phi_2) > (E_1 - E_2)$; that is, the pressure at point 2 will be lower than that at point 1 if the difference in elevation between the two points is less than the difference in potential. If the water is saturated with gas at point 1, gas will be released from solution at point 2 and may remain trapped at that position.

OTHER MECHANISMS

In the preceding discussion it has been assumed that gas is formed at one point, then carried in solution to another point and released. It is also possible that gas accumulates in ground water from a number of different sources. Assume, for example, that water within the soil is saturated with methane at 1 atm pressure. If this water then migrates downward and encounters the Robein Silt at a depth of 100 feet, additional methane could be taken into solution at this point. If the water again becomes saturated with methane at this point, the radiocarbon age of this hypothetical gas mixture would be approximately 12,000 B.P. Although this amount of "modern contamination" is possibly greater than what would normally be encountered in nature, it is apparent that downward

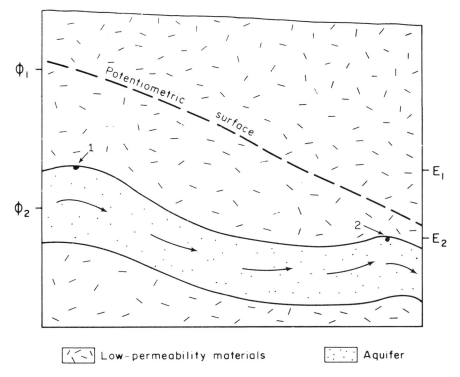

FIG. 9. Hypothetical cross section showing migration under the effects of a hydraulic gradient.

migration of methane derived from modern soil could be responsible for many of the anomalously young radiocarbon ages encountered.

"Anomalous" radiocarbon ages could also be explained by the dissolution and downward migration in groundwater of carbon dioxide or organic matter from modern soil zones and the subsequent degradation of this material in the subsurface by methane-producing bacteria. It is doubtful, however, that this type of process is a major source of methane. The bacteria that form methane are restricted to highly reducing environments; for the reduction of CO_2 to CH_4, free hydrogen is needed. Zones in which active organic decomposition is occurring are the only likely sites of such highly reducing environments. In these zones, the quantity of methane produced from in situ organic material is probably much greater than the quantity of potential source material transported in ground water. In particular, the quantity of CO_2 carried in with groundwater is likely to be negligible relative to the quantity produced by bacteria.

SUMMARY AND CONCLUSIONS

Radiocarbon dating of methane demonstrates that, in Illinois, organic material within the Robein Silt is the most significant source of drift-gas, even though samples both younger and older than the Robein Silt were encountered. In all cases, the gas had a radiocarbon age equal to or less than the age of the reservoir material in which it occurred. This evidence suggests that upward migration from the source and later entrapment are not major factors in the formation of drift-gas deposits.

Deposits of drift-gas commonly exist well below the most probable source materials. In some instances the lateral migration of a free gas phase can account for these deposits. In many cases, however, some mechanism of downward migration is required. The fact that methane commonly occurs dissolved in groundwater suggests that movement in or with groundwater is a likely mechanism for the migration of drift-gas.

Theoretical considerations suggest that the transport of dissolved methane by diffusion is negligible compared to the transport resulting from movement of groundwater. Water that moves downward can become saturated with methane (as well as with other gases) as it passes through zones rich in organic matter where bacterial decomposition occurs. If this water encounters a zone in which the hydrostatic pressure is less than the quilibrium vapor pressure of the dissolved gas, a free gas phase will form. This mechanism provides the most likely explanation for the accumulation of gas in many of the deposits of drift-gas in Illinois.

It is possible that methane generated in the modern soil or in other near-surface organic deposits could be carried down in solution and mixed with gas formed in deeper deposits. This mechanism may be responsible for many of the anomalously young radiocarbon ages encountered. Carbon dioxide and dissolved organic material that originated in the modern soil could also be converted to methane at depth and thus affect the radiocarbon ages determined on drift-gas, but this effect is probably of secondary significance in most cases.

ACKNOWLEDGMENTS

The successful completion of this research would not have been possible without the help of a number of people. W. F. Meents either collected or helped to collect all of the gas samples that were analyzed in this study. He also supplied the pertinent geologic information for most of the wells sampled. C. L. Liu helped with much of the routine laboratory work. J. P. Kempton assisted in the interpretation of well records and in the construction of some of the geologic cross sections. I would also like to express my appreciation to K. Cartwright and T. F. Anderson for their helpful discussions and suggestions. This research is part of a dissertation completed at the University of Illinois (Coleman 1976).

REFERENCES

Carman, P. C.
    1956   Flow of Gases Through Porous Media. New York, Academic Press.
Coleman, D. D.
    1973   Illinois State Geological Survey radiocarbon dates IV. Radiocarbon. 15:75-80.

1976 Isotopic characterization of Illinois natural gas. Ph.D. dissertation. Univ. of Illinois.

Coleman, D. D., C. L. Liu, D. R. Dickerson, and R. R. Frost
1972 Improvement in the trimerization of acetylene to benzene for radiocarbon dating with a commercially available vanadium oxide catalyst. Proc. Int. Radiocarbon Dating Conf. 1:B50-B62.

Horberg, L.
1953 Pleistocene deposits below the Wisconsin drift in northeastern Illinois. Ill. State Geol. Survey. Report of Investigations 165:61.
1957 Bedrock surface of Illinois. Ill. State Geol. Survey Map.

Hougen, O. A., K. M. Watson, and R. A. Ragatz
1954 Chemical Process Principles. Part I. Material and Energy Balances. New York, John Wiley and Sons.

Hubbert, M. K.
1953 Entrapment of petroleum under hydrodynamic conditions. Bull. Amer. Ass. of Petroleum Geologists. 37(8):1954-2026.

Kim, S. M., R. R. Ruch, and J. P. Kempton
1969 Radiocarbon dating at the Illinois State Geological Survey. Ill. State Geol. Survey. Env. Geol. Note 28:19.

MacClintock, P., and H. B. Willman
1959 Geology of Buda Quandrangle, Illinois. Ill. State Geol. Survey, Circular 275:29.

McComas, M. R.
1959 Pleistocene geology and hydrogeology of the Middle Illinois Valley. Ph.D. dissertation. Univ. of Illinois.

Meents, W. F.
1958 Tiskilwa drift-gas area. Bureau and Putnam Counties, Illinois. Ill. State Geol. Survey. Circular 253:15.
1960 Glacial-drift gas in Illinois. Ill. State Geol. Survey. Circular 292:58.

Wasserburg, G. J., E. Mazor, and R. E. Zartman
1963 Isotopic and chemical composition of some terrestrial natural gases. In Geiss and Goldberg, eds., Earth Sciences and Meteoritics. Amsterdam, North Holland Press, 219-240.

Willman, H. B., and J. C. Frye
1970 Pleistocene stratigraphy of Illinois. Ill. State Geol. Survey. Bull. 94:204.

7

Fossil Fuel Exhaust-Gas Admixture with the Atmosphere

J. C. Freundlich

ABSTRACT

Environmental pollution of the atmosphere by automobile exhaust-gases has been studied using natural radiocarbon in carbon dioxide as a tracer. Hence carbon dioxide from automobile exhaust can be clearly distinguished from atmospheric carbon dioxide. Near a heavily-traveled highway, a decrease of exhaust CO_2 by at least a factor of 100 has been observed.

INTRODUCTION

The admixture of fossil fuel exhaust-gases with the atmosphere using natural radiocarbon as a tracer was initially studied by Suess (1955), Münnich and Vogel (1958), and DeVries (1958); more recent work has been summarized by Vogel (1972), by Houtermans et al. (1967), by Libby and Libby (1972), and by Zimen (1972). Due to the very rapid mixing rate of the atmosphere, local variations in concentration of fossil (i.e., ^{14}C-free) CO_2 are scarcely detectable unless the local source of fossil CO_2 is sufficiently strong. Chatters et al. (1969), investigating fumaroles, reported appreciable local deficiencies. Our study is intended as a further investigation of this question for three reasons: (1) a methodical interest in a quantitative study of apparent "age" deviations caused by sources of ^{14}C-free CO_2, such as volcanism (fumaroles), or exhaust-gases of fossil fuel combustion; (2) the need to investigate the spread of exhaust CO_2 and its attenuation into the atmosphere by tracers (although in this case the "marked" substance is free from ^{14}C, whereas the atmosphere contains it), and (3) a need to determine the amount of actual environmental pollution by exhaust-gas on and near to a heavily-traveled highway.

Extent of the Effect

On a well-traveled road (one car every ten seconds), CO_2 is generated at a rate of 10^{-5} mol/(cm. sec). That is about fifty times the natural CO_2 content (1.7×10^{-8}

mol/cm^3) of the atmosphere contained in a cross section over the road 10^6 cm^2. If these exhaust-gases should not undergo quick dilution into the atmosphere, the local CO$_2$ content after one day would exceed the CO$_2$ MAK (1971) value (= 0.5%) by a factor of 3. Assuming only 5% CO in the exhaust-gas, the local CO content after one day would be more than ten times the CO MAK (1971) value (= 0.01%). If a local variation of radiocarbon content of that degree were possible, it would be readily detected by radiocarbon analysis. A ratio of

$$\frac{\text{exhaust CO}_2}{\text{natural CO}_2} = \frac{50}{1}$$

would result in an apparent local ^{14}C content of about 2% as compared to the natural atmospheric value of 100% (there is always a wake of forced turbulence along the outline of the moving car).

Diffusion Considerations

Considering stationary diffusion in a system of cylindrical symmetry (using the road line as the axis of the cylinder) into the space above the road, we have

$$c_{100} - c_1 \doteq \frac{1}{D} \cdot 500 \cdot c_N$$

where

c_N = natural CO$_2$ concentration of the atmosphere = 1.7×10^{-8} mol/cm^3

c_1 = exhaust CO$_2$ concentration at 1 meter off the road

c_{100} = exhaust CO$_2$ concentration at 100 meters off the road

D = diffusion (resp. turbulent exchange) coefficient

500 is the approximate product of (a) primary CO$_2$ exhaust production of ca. 10^4 motor vehicles per day (= ca. 10^{-5} mol/(cm. sec)), (b) $(1/\pi) \cdot \ln (100 \text{ meter}/1 \text{ meter}) \doteq 1.5$, and (c) $1/c_N \doteq 6 \times 10^7$ cm^3/mol. Expressing the result in percentage of c_N (the undisturbed natural CO$_2$ content), we would expect the following range of exhaust-gas spread (table 1) if we assume a variety of diffusion coefficients from 1 cm^2/sec (as for molecular gas diffusion) up to 3.10^{10} cm^2/sec (tubulent exchange coefficient in the higher atmosphere (cf. Münnich 1963)). As can be inferred from table 1, radiocarbon analyses are promising only when the diffusion constant is 10^6 cm^2/sec or lower.

Experimental Investigation

Samples for measurement were taken from a heavily-traveled highway, the Autobahn Koeln-Oberhausen-Berlin, road-kilometer no. 532.8, between the overpass of Bundesstrasse 8 and the overpass at the Opladen exit, just south of a third intermediate overpass at the local Muchlenweg Road. All samples taken were plant material grown during the 1973 growth period. The sample locality was a road trench cut about 4 meters below the

TABLE 1

Diffusion Considerations

Assumed diffusion coefficient	D	10	10^3	10^5	10^7	10^9	cm^2/sec
Exhaust – CO_2 concentration at 1 m off the road	c_1	2	50	98	99.98	(99.9998)	% modern ^{14}C
Exhaust – CO_2 concentration at 100 m off the road	c_{100}	90	95	99	100	100	% modern ^{14}C

normal land surface. Both slopes bordering the road trench were densely covered with bushes and trees (fig. 1).

In a ditch bordering the road we found a natural CO_2 content of as high as 86%, hence the observed ratio

$$\frac{\text{exhaust } CO_2}{\text{natural } CO_2} = \frac{100\% - 86\%}{100\%} = \frac{1}{7}$$

From this we inferred a dilution of the exhaust CO_2 by a factor of 300 directly off the road edge. Only a few meters farther we found almost 90%, and some 100 meters farther on, 98% of the natural, undisturbed CO_2 value (table 2).

According to Haxel (1958), an investigation in 1953 by Münnich showed about 97% natural CO_2 in plants grown between the lanes of a highway, whereas values as low as

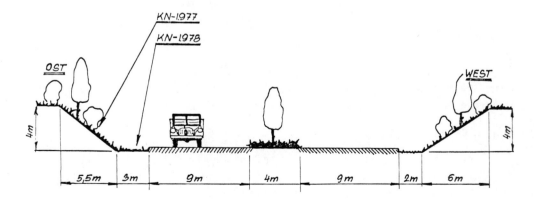

FIG. 1. Sampling Site

TABLE 2

Exhaust Analysis Samples

Laboratory number	KN-1978	KN-1977	KN-2014
Location	Opladen, ditch	Opladen, slope	Opladen, garden
Sampling date	1 Nov 73	1 Nov 73	13 Jan 74
Road edge distance	0.5 meter	4 meter	120 meter
Material: leaves from	grass	oak	oak
$\delta^{13}C$ (°/oo, PDB)	-26.9	-26.7	-27.6
% ^{14}C (relative to 95% NBS exalic acid)	122.9 ± 0.4	128.0 ± 0.5	140.0 ± 0.4
% ^{14}C (relative to 1973 reference	86.0	89.5	98.0
Local ^{14}C defect	14.0	10.5	2.0

80%-85% are found in places with very little ventilation (Haxel, personal communication, 1974).

Reference samples for the uncontaminated atmosphere were taken from three different sites near hilltops far removed from any highway in calm, rural areas not densely populated (table 3).

CONCLUSION

Figure 2 has been drawn from the data of tables 2 and 3. It shows a pattern well in accord with steady-state diffusion of cylindrical symmetry. A formal diffusion coefficient on the order of

$$D \doteq 10^4 \text{ cm}^2/\text{sec}$$

can be calculated from these data. Comparing this value with the higher atmospheric turbulent exchange coefficient of 3×10^{10} cm²/sec, it has to be presumed that (1) close to the earth's surface, all turbulent exchange is fairly restricted as compared to the higher atmosphere, (2) this restriction is especially effective in a groove or trench, and (3) considering smaller distances of exchange (some tens of meters here versus tens of kilometers in the higher atmosphere), the turbulent exchange may be appreciably lower. Moreover, it is known that in the earth's general atmosphere, horizontal exchange occurs more readily than vertical exchange.

Even in a highway site with a fairly high fossil CO_2 source (about one metric ton of CO_2 yearly per meter length), one would expect to find no remarkable ^{14}C deviation outside the immediate vicinity of the gas source. Fossil CO_2 originating from volcanic sources (fumaroles) is accompanied by SO_2 gas, which is poisonous to plants. This prevents the finding of plant samples near fumaroles with a fossil-CO_2-induced contamination.

TABLE 3
Reference Samples (1973)

Laboratory number	KN-1975	KN-1976	KN-2013	Mean
Location	Urft	Münstereifel	Dürscheid	–
Sampling date	28 Oct 73	11 Nov 73	20 Jan 74	–
Distance	67 km	58 km	15 km	–
Material: leaves from	beech	oak	oak	–
$\delta^{13}C$ (°/oo, PDB)	– 27.6	– 26.4	– 25.1	–
% ^{14}C (relative to 95% NBS oxalic acid)	143.2 ± 0.6	144.1 ± 0.4	141.8 ± 0.4	143.0
% ^{14}C (relative to 1973 mean)	100.1	100.7	99.1	100.0

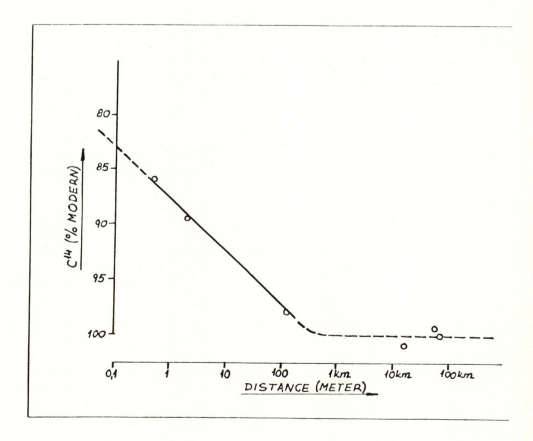

FIG. 2. Diffision Pattern

In localities where the turbulent exchange conditions are good, the local fossil CO_2 contamination is very low because all fossil CO_2 is readily swept away by the high rate of exchange. On the other hand, in places with low ventilation or exchange, the area of contamination is restricted in size.

ACKNOWLEDGMENTS

Thanks are due to Dr. U. Manze, Koeln, for ^{13}C assay, and to Prof. K. O. Münnich, Heidelberg, for valuable discussions. The author is indebted to Stiftung Volkswagenwerk for a grant enabling us to improve our ^{14}C dating equipment.

REFERENCES

Chatters, R. M., J. W. Crosby III, and L. G. Engstrand
    1969   Fumarole gaseous emanations: Their influence on carbon-14 dates. Tech. Ext. Serv. Washington State Univ. Circular 32:9.

DeVries, Hl.
    1958   Variation in concentration of radiocarbon with time and location on earth. Proc. Koninkl. Akad. Wetens. Amst. B 61:94-103.

Haxel, O.
    1958   Natürliche und künstliche Radioaktivität in der Atmosphäre. Deutsche Physikertagung Heidelberg. Mosbach (Physik Verlag). 22:134-152.

Houtermans, J., H. E. Suess, and W. Munk
    1967   Effect of industrial fuel combustion on the carbon-14 level of atmospheric CO_2. *In* Radioactive dating and methods of low-level counting. Wien. IAEA:57-68.

Libby, L. M., and W. F. Libby
    1972   Vulcanism and radiocarbon dates. Proc. 8th Int. Conf. on Radiocarbon Dating. New Zealand. 86-89.

MAK
    1971   Maximale Arbeitsplatzkonzentration gesundheitsschädlicher Arbeitsstoffe. Bundesarbeitsblatt Fachteil Arbeitsschutz. 9:250-255.

Münnich, K. O., and J. C. Vogel
    1958   Durch Atomexplosionen erzeugter Radiokohlenstoff in der Atmosphäre. Naturwiss. 45:327-329.

Münnich, K. O.
    1963   Der Kreislauf des Radiokohlenstoffs in der Natur. Naturwiss. 50:211-218.

Suess, H. E.
    1955   Radiocarbon concentration in modern wood. Science. 122:415-417.

Vogel, J. C.
    1972   Radiocarbon in the surface waters of the Atlantic Ocean. Proc. 8th Int. Conf. on Radiocarbon Dating. New Zealand. 267-279.

Zimen, K. E.
    1972   The future burden of the atmosphere and carbon-14 in the ethanol of wines. Proc. 8th Int. Conf. on Radiocarbon Dating. New Zealand. 99-105.

PART V
Shells and Water Plants

1

The Use of Marine Shells in Dating Land/Sea Level Changes

Joakim Donner and Högne Jungner

Marine shells have been dated in the University of Helsinki radiocarbon laboratory since 1973. Shell dates were first used in dating the deglaciation and land uplift in the area south of Disko Bugt in western Greenland. Here the marine sediments at various altitudes up to about 50 m contain shells of different species. Samples were collected by Donner in 1972 and the results were published by Donner and Jungner (1975). Different shell species from many sites were dated separately. The results are summarized in figure 1, including twenty-eight new dates and four earlier dates from the same area (K-1814 to K-1817). The wide scatter of these dates would seem initially to support the negative criticism directed toward the use of shell dates. Varying species from the same site have ages which differ considerably from each other; for instance, *Hiatella arctica* (Hel-345) and *Balanus* (Hel-436), dated at 8550 ± 190 B.P. and 6100 ± 160 B.P. respectively, or *Hiatella arctica* (Hel-437) and *Balanus* (Hel-364), dated at 8630 ± 200 B.P. and 6220 ± 160 B.P. These age differences can be explained. The beach deposits contain a mixture of shell material originating from shell species living in deep water as well as in shallow water. The shallow shell species were incorporated into the beach material last, before the deposit emerged from the sea as a result of the land uplift.

Figure 2 shows the distribution of *Hiatella* in an altitude-time diagram. The extensive spread of the dates is a result of the great depth amplitude of this species. Many of the dated shells were deposited below the intertidal zone and cannot be used for dating the uplift. *Mytilus edulis*, on the other hand, is a species living mainly in the shallow water of the intertidal zone, so it is suitable for dating the displacement of the shoreline (fig. 3). The date of this species also gives the date for the sea level position at the site from which the *Mytilus* shells were collected.

In order to test the validity of the results from Greenland, another area was chosen for similar studies. In 1974 shell samples were collected by Donner and Eronen in the Varangerfjord area in northern Norway, and in 1975 additional samples were taken from the outer coast of the Varanger peninsula. The datings are listed in tables 1 and 2. The

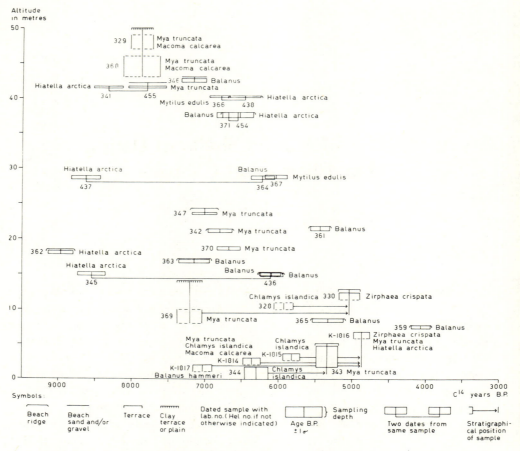

FIG. 1. Radiocarbon ages of dated shell samples in the area south of Disko Bugt plotted against altitude of samples (Donner and Jungner 1975, fig. 4).

results confirmed the conclusions reached on the basis of the samples from Greenland. It is seen in figure 4, which gives the results from the Varangerfjord area, that *Mytilus edulis* is used to determine the land/sea level changes (in two samples *Modiolus modiolus* was dated together with *Mytilus*). Therefore only *Mytilus edulis* and *Modiolus modiolus*) were dated in the samples from the Varanger peninsula, and the results show a regular relative regression of the sea level (fig. 5).

Sediments from two lakes studied by Eronen south of the Varangerfjord area were also dated. This is the same area from which shells were dated. Table 3 gives the ages obtained from these sediments. The threshold of Lake Mordvatnet is at 34 m and the age of the lowermost sediment sample (Hel-655) should approximately date the time when the sea level receded below this level and the basin became an independent lake. The age of the formation of Lake Vaervatnet with the threshold at 23 m can be dated at about 5600 B.P. (Hel-700).

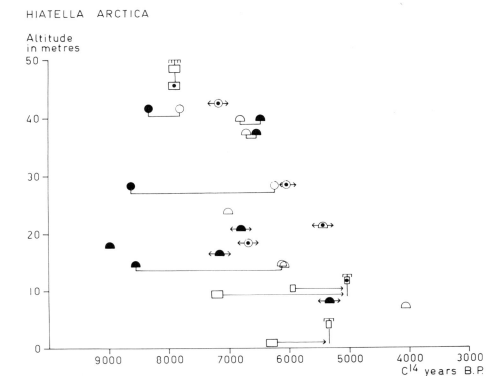

FIG. 2. Distribution of *Hiatella arctica* in dated shell samples. For symbols see fig. 3. (Donner and Jungner 1975, fig. 7).

A straight line through the five samples with *Mytilus edulis* and the two lowermost samples from the lakes gives the approximate regression in the Varangerfjord area. The straight line runs close to the dated samples. Of special interest is the good agreement between the date of the shell sample at 23.8 m and the date of the lake deposit from Vaervatnet. In addition it can be noted that the highest level at which pumice has been found in this area is 28 m. This pumice level is dated at approximately 7000 B.P. in Norway, a result agreeing well with the dates given in the diagram.

It can be concluded that accurate datings for land/sea level changes can be obtained by carefully choosing suitable shell species, such as *Mytilus edulis*. It can also be concluded that the shell dates agree well with dates from organic lake deposits.

REFERENCE

Donner, J., and H. Jungner
    1975   Radiocarbon dating of shells from marine Holocene deposits in the Disko Bugt area, West Greenland. Boreas. 4: 25-45.

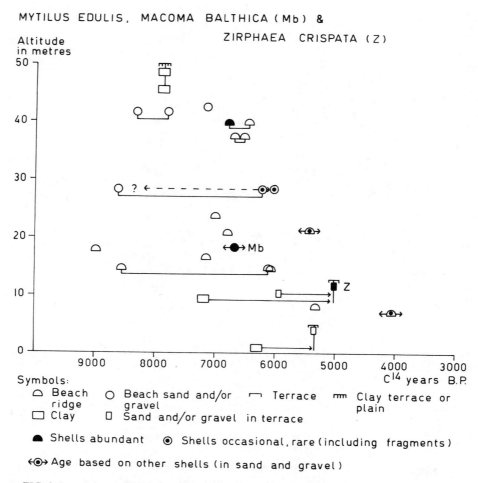

FIG. 3. Distribution of *Mytilus edulis*, *Macoma balthica*, and *Zirphaea crispata* in dated shell samples (Donner and Jungner 1975, fig. 3).

TABLE 1
Shell Samples from Varangerfjord

Locality	Altitude	Lab. n:o	Shells dated	Age B.P.
Nyelven	23.8 m	Hel-627	Mytilus edulis	5530 ± 150
		Hel-628	Mya truncata	8120 ± 170
Kariel	18.7 m	Hel-620	Mytilus edulis & Modiolus modiolus	4400 ± 130
Makviken	15.8 m	Hel-624	Mytilus edulis	4120 ± 140
		Hel-625	Mya truncata	6430 ± 150
Kariel	15.4 m	Hel-621	Mytilus edulis & Modiolus modiolus	3930 ± 130
		Hel-622	Arctica islandica	4300 ± 160
Krampenes	15.3 m	Hel-623	Mytilus edulis	3680 ± 140
Kariel	11.9 m	Hel-618	Mya truncata	4190 ± 130
		Hel-619	Arctica islandica	3820 ± 130
Makviken	11.2 m	Hel-626	Arctica islandica	3190 ± 120
Vestre Jakobselv	6.8 m	Hel-617	Arctica islandica	3830 ± 110

TABLE 2
Shell Samples from Outer Coast of Varanger Peninsula

Locality	Altitude	Lab. n:o	Shells dated	Age B.P.
Havningberg	14.1 m	Hel-747	Mytilus edulis	4300 ± 130
Sandfjord	12.4 m	Hel-749	Mytilus edulis	4290 ± 120
Kjölnes	11.4 m	Hel-751	Mytilus edulis	3820 ± 130
Store Molvik	9.0 m	Hel-750	Modiolus modiolus	2850 ± 120
Veines	8.5 m	Hel-753	Mytilus edulis	2850 ± 120
Sandfjordsbugten	4.1 m	Hel-752	Mytilus edulis	1340 ± 100
Havningberg	4.0 m	Hel-748	Mytilus edulis	520 ± 90

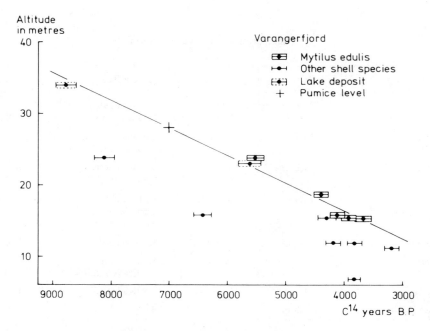

FIG. 4. Radiocarbon ages of dated samples in the Varangerfjord area, northern Norway.

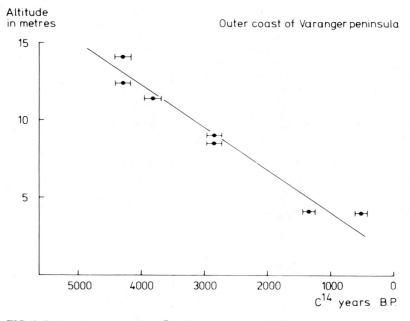

FIG. 5. Radiocarbon ages of dated shell samples from the outer coast of the Varanger peninsula, northern Norway.

TABLE 3

Lake Sediment Samples

Locality	Depth from lake surface (meters)	Lab. n:o	Age B.P.
Mordvatnet	6.80 – 6.85	Hel-655	7510 ± 220
Mordvatnet	7.16 – 7.21	Hel-656	8780 ± 180
Vaervatnet	4.22 – 4.25	Hel-699	5360 ± 110
Vaervatnet	4.25 – 4.28	Hel-700	5620 ± 190

2

The Suitability of Marine Shells for Radiocarbon Dating of Australian Prehistory

R. Gillespie and H. A. Polach

ABSTRACT

The problem of determining an appropriate initial ^{14}C activity for marine shells is discussed in terms of carbonate mineralogy, live-collected shells and charcoal/shell pair samples. New results for shells collected alive from Australian coastal waters before 1950 and in 1973 are presented, together with paired samples of charcoal and shell from coastal middens. From these results it is concluded that: (1) the current best estimate for the "apparent age" of Australian marine shells is 450 ± 35 years; (2) shells generally give a good indication of the age of midden deposits, and good agreement between charcoal and shell pairs can be expected; and (3) there were significant differences in the ^{14}C activity of 1973 shells from different regions, making contemporary samples unsuitable as modern ^{14}C dating reference standards.

INTRODUCTION

The use of marine shell carbonate for radiocarbon dating has met with criticisms (Olson 1963). This report will discuss some previous results obtained from marine shell samples and also will present new results from the Australian region, for which little work in this field has been reported. Much of the impetus to study shell dating has come from discussions with archaeologists working in Australia, who frequently find large shell middens near the coast. Most such sites have been dated using only the charcoal component, but shell is usually the major component of middens and is sometimes the only datable material present.

Three sets of samples are discussed here: (1) shells collected alive before 1950, (2) shells collected alive in 1973, and (3) charcoal/shell pairs from excavated middens. In some cases several different shell species from the same location have been measured, and the efficiency of the routine acid-etching pretreatment procedure has been studied in one set of excavated shells. The different methods used for age calculation are discussed and one is recommended.

Previous Work

MARINE SHELLS COLLECTED ALIVE BEFORE 1950

A search of the radiocarbon literature revealed sixty-eight measurements of the ^{14}C activity of marine shells collected between 1840 and 1950 (table 1). These were museum samples, assumed to have been collected alive. Recalculation of all measurements to $\nabla^{14}C$ values (normalized to $\delta^{13}C = 25^o/oo$ relative to PDB, age corrected and corrected for the Suess effect in surface ocean water by the method given in Mangerud and Gulliksen 1975) gives a mean "apparent age" or reservoir effect of 535 years (based on the mean value of $\nabla^{14}C = -64.5^o/oo$ relative to 0.95 NBS oxalic acid.)

It is clear from the data presented in table 1 that there are large and significant differences in ^{14}C concentration of surface ocean water samples as based on pre-1950 shell ^{14}C activity measurements (from table 1 maximum ∇ value = -42% and minimum = -114.6%). Since the differences were not adjusted by isotopic fractionation normalization, two other factors must be considered significant: namely, geographic location and local oceanic conditions. In general the age differences between surface ocean and atmosphere are greater at high latitudes than in the tropics (Broecker and Olson 1961, Mangerud and Gulliksen 1975). They are also greater in areas of restricted circulation and in regions of deep water upwelling (Berger et al. 1966; Taylor and Berger 1967; Rafter et al. 1972). These results indicate that it is mandatory to use local standards for marine shell ages in those regions where upwelling or restricted circulation occur and at high latitudes. Precise knowledge of local oceanographic data is also important, as demonstrated by Rafter et al. (1972).

POST-1950 MARINE SHELL MEASUREMENTS

Many measurements have been reported for the ^{14}C activity of ocean water bicarbonate to determine whether all marine shells have activities typical of their growth environment (Broecker and Olson 1961; Broecker 1963; Rafter et al. 1972), although Mook and Vogel (1968) have shown that shells are in equilibrium with their growth region water for the stable carbon and oxygen isotopes. Data from Rafter and O'Brien (1970, 1972) and Rafter (1968, personal communication) seem to indicate that marine shells follow the general trend of surface ocean water bicarbonate activity over the period 1954-73.

Erlenkeuser et al. (1975) report measurements on individual growth rings of marine shells, finding significant differences in ^{14}C activity which reveal shell carbonate as a sensitive indicator of the ^{14}C activity of the growth region. Local variations in shell carbonate activity may be due to variations in water temperature, salinity, and food supply as well as species-dependent metabolic differences (Andrews 1972; Weber 1974; Erlenkeuser and Willkomm 1973) but more data are required to determine the magnitude and significance of these effects.

TABLE 1

Known Age Shells Collected Alive Before 1950 and Measured at Other Laboratories

Geographic location	Collection year	∇C^{14} ‰ Age and industrial effect corrected	Reference
Jamaica	1929 ± 1	− 48.2 ± 5	Broecker and Olson (1961)
Jamaica	1884	− 43.4 ± 5	Broecker and Olson (1961)
Bahamas	1950	− 51.0 ± 5	Broecker and Olson (1961)
Bahamas	1885 ± 5	− 54.8 ± 7	Broecker and Olson (1961)
Iceland	1946	− 64.3 ± 6	Broecker and Olson (1961)
Iceland	1900	− 69.0 ± 6	Broecker and Olson (1961)
Iceland	1840	− 71.9 ± 6	Broecker and Olson (1961)
Hawaii	1840 ± 1	− 61.4 ± 6	Broecker and Olson (1961)
Moorea	1883 ± 3	− 48.2 ± 5	Broecker and Olson (1961)
Mexico	1939	− 77.6 ± 6	Berger et al. (1966)
Mexico	1935	−114.6 ± 6	Berger et al. (1966)
Mexico	1930	− 78.5 ± 6	Berger et al. (1966)
Gulf of California	1911	−112.7 ± 6	Berger et al. (1966)
Baja California	1932	− 90.9 ± 5	Berger et al. (1966)
Baja California	1938	− 79.5 ± 6	Berger et al. (1966)
Baja California	1939	− 70.0 ± 6	Berger et al. (1966)
Mexico	1938	− 77.6 ± 6	Berger et al. (1966)
Mexico	1938	− 71.9 ± 6	Berger et al. (1966)
California	1921	− 63.3 ± 6	Berger et al. (1966)
California	1878	− 60.4 ± 6	Berger et al. (1966)
Ecuador	1927	− 49.0 ± 5	Taylor and Berger (1967)
Ecuador	1927	− 61.4 ± 5	Taylor and Berger (1967)
Costa Rica	1935	− 81.4 ± 4	Taylor and Berger (1967)
Galapagos Is.	1934	− 63.3 ± 6	Taylor and Berger (1967)
Galapagos Is.	1934	− 86.1 ± 9	Taylor and Berger (1967)
Galapagos Is.	1934	− 54.8 ± 5	Taylor and Berger (1967)
Galapagos Is.	1932	− 47.1 ± 5	Taylor and Berger (1967)
Panama	1935	− 47.1 ± 6	Taylor and Berger (1967)
Panama	1935	− 59.5 ± 6	Taylor and Berger (1967)
Chile	1925	− 71.9 ± 4	Taylor and Berger (1967)
Chile	1935 ± 5	− 89.9 ± 9	Taylor and Berger (1967)
Peru	1935 ± 5	−128.9 ± 5	Taylor and Berger (1967)
Peru	1935 ± 5	− 81.4 ± 6	Taylor and Berger (1967)
Sweden	1935 ± 3	− 63.3 ± 9	Olsson et al. (1974)
Sweden	1930 ± 10	− 43.4 ± 7	Hakansson (1969)
Sweden	1930 ± 10	− 48.2 ± 6	Hakansson (1969)
Sweden	1901 ± 3	− 42.4 ± 6	Hakansson (1969)
Norway	1912	− 48.2 ± 9	Mangerud (1972)
Norway	1920	− 57.6 ± 9	Mangerud (1972)
Norway	1909	− 59.5 ± 10	Mangerud (1972)

TABLE 1 (*continued*)

Geographic location	Collection year	∇C^{14} ⁰/₀₀ Age and industrial effect corrected	Reference
Norway	1923	− 49.0 ± 9	Mangerud (1972)
Norway	1923	− 53.8 ± 9	Mangerud (1972)
Norway	1908	− 59.5 ± 9	Mangerud (1972)
Norway	1898	− 65.2 ± 10	Mangerud (1972)
Norway	1918	− 45.3 ± 11	Mangerud (1972)
Norway	1906	− 53.8 ± 9	Mangerud (1972)
Norway	1922	− 63.3 ± 9	Mangerud (1972)
Norway	1905	− 41.5 ± 9	Mangerud (1972)
Greenland	1900	− 70.9 ± 5	Hakansson (1973)
Greenland	1900	− 68.1 ± 5	Hakansson (1973)
Greenland	1899	− 70.0 ± 5	Hakansson (1973)
Greenland	1899	− 64.3 ± 5	Hakansson (1973)
New Zealand	1923	− 50.0 ± 5	Rafter (personal communication)
New Zealand	1925	− 45.3 ± 6	Rafter (personal communication)
New Zealand	1949	− 45.3 ± 5	Rafter (personal communication)
Norway	1906	− 56.7 ± 6	Mangerud and Gulliksen (1975)
Norway	1906	− 50.0 ± 6	Mangerud and Gulliksen (1975)
Norway	1857	− 55.7 ± 6	Mangerud and Gulliksen (1975)
Norway	1876	− 59.5 ± 9	Mangerud and Gulliksen (1975)
Norway	1857	− 55.7 ± 6	Mangerud and Gulliksen (1975)
Spitzbergen	1900	− 57.6 ± 6	Mangerud and Gulliksen (1975)
Spitzbergen	1926	− 55.7 ± 6	Mangerud and Gulliksen (1975)
Spitzbergen	1925	− 53.8 ± 6	Mangerud and Gulliksen (1975)
Spitzbergen	1878	− 66.2 ± 9	Mangerud and Gulliksen (1975)
Spitzbergen	1878	− 67.1 ± 9	Mangerud and Gulliksen (1975)
Canada	1899	− 87.1 ± 9	Mangerud and Gulliksen (1975)
Canada	1900	−101.3 ± 9	Mangerud and Gulliksen (1975)
Canada	1898	− 84.2 ± 9	Mangerud and Gulliksen (1975)

Values calculated by the authors from data given in the references.

CHARCOAL-SHELL PAIRS

For the determination of past relationships between atmosphere and ocean activities, the measurement of stratigraphically equivalent charcoal and shell pairs is important. A survey of *Radiocarbon* revealed forty-four such pairs (table 2) with a mean age difference of 507 years. This is in agreement with the reservoir effect of 535 years noted above, but it is possibly only fortuitous considering the wide range of localities concerned in the two groups of data. It does, however, indicate that the order of magnitude of the reservoir effect has not changed over the past 10,000 years.

TABLE 2
Charcoal/Shell Pairs Reported in *Radiocarbon*

Date list reference	Organic carbon date*	Shell carbonate date	Age difference + std. error
USGS I	2500 ± 200	2540 ± 200	−40 ± 283
USGS II	600 ± 200	580 ± 200	+20 ± 283
Lamont V	12,150 ± 250	12,350 ± 250	−200 ± 354
La Jolla I	960 ± 150	1060 ± 150	−100 ± 213
GSC I	12,200 ± 160	12,360 ± 140	−160 ± 213
UCLA II	1500 ± 80	1480 ± 80	+20 ± 113
UCLA II	1500 ± 80	1630 ± 80	−130 ± 113
La Jolla IV	1660 ± 200	1800 ± 300	−140 ± 361
La Jolla IV	2200 ± 200	2600 ± 200	−400 ± 283
La Jolla IV	5840 ± 200	6140 ± 250	−300 ± 320
Riken IV	3960 ± 175	3960 ± 130	0 ± 218
Riken IV	3370 ± 135	3260 ± 120	+110 ± 181
Groningen IX	10,030 ± 55	9780 ± 60	+250 ± 81
Groningen IX	9580 ± 85	9450 ± 55	+130 ± 101
U. Texas VII	2070 ± 110	2280 ± 90	−210 ± 142
U. Texas VII	1890 ± 100	1990 ± 100	−100 ± 141
U. Texas VII	1560 ± 100	1840 ± 100	−280 ± 141
U. Texas VII	2540 ± 110	2150 ± 60	+390 ± 125
U. Texas VII	1950 ± 80	2180 ± 90	−230 ± 120
U. Texas VII	1880 ± 90	2040 ± 90	−160 ± 127
U. Texas VII	1780 ± 100	1890 ± 90	−110 ± 135
U. Texas VII	1400 ± 110	1810 ± 90	−410 ± 142
U. Texas VII	1900 ± 90	2240 ± 90	−340 ± 127
U. Texas VII	2020 ± 80	2220 ± 80	−200 ± 113
U. Texas VII	2010 ± 90	1950 ± 70	+60 ± 114
U. Texas VII	1740 ± 100	2010 ± 90	−270 ± 135
Arizona VIII	2410 ± 110	2200 ± 500	+210 ± 512
Arizona VIII	3370 ± 150	3370 ± 110	0 ± 186
Birmingham VII	5450 ± 140	5315 ± 95	+135 ± 169
Riken VII	470 ± 105	335 ± 90	+135 ± 138
Uppsala XI	2560 ± 90	2920 ± 80	−360 ± 120
Pisa I	1863 ± 135	2160 ± 145	−297 ± 198
Chicago I	1951 ± 200	2285 ± 210	−334 ± 290
Michigan V	15,500 ± 700	13,900 ± 700	+1600 ± 990
Michigan V	9240 ± 500	9450 ± 400	−210 ± 640
Trondheim II	12,200 ± 350	12,700 ± 350	−500 ± 495
U. Georgia XIII	585 ± 60	665 ± 65	−80 ± 88
U. Texas X	720 ± 90	1100 ± 70	−380 ± 114
U. Texas X	1120 ± 110	1300 ± 70	−180 ± 130
U. Texas X	1120 ± 70	1740 ± 70	−620 ± 99

TABLE 2 (*continued*)

Date list reference	Organic carbon date*	Shell carbonate date	Age difference ± std. error
U. Texas X	1060 ± 90	1310 ± 60	−250 ± 108
U. Texas X	2170 ± 180	1500 ± 70	+670 ± 193
U. Texas X	180 ± 60	860 ± 50	−680 ± 78
U. Texas X	1650 ± 90	1870 ± 70	−220 ± 114

*Ages B.P. (5568 half-life) normalized to $\delta^{13}C = -25\%$

The specific question of the past relationship between atmosphere and surface ocean water activities has been discussed by Haynes et al. (1971) and Mangerud and Gulliksen (1975). It has been assumed that fluctuations in atmospheric activity would be attenuated by the buffering effect of the large oceanic reservoirs, particularly for changes within the residence time of atmospheric ^{14}C. It should be noted, however, that because of the local variations found in pre-1950 shells, any conclusions about past atmosphere/ocean results should be based on samples taken from closely similar environments. Most charcoal/shell pairs used in the following Australian study are from archaeological sites, where the deposition of both types of samples is assumed to be simultaneous.

Australian Work

SAMPLE PRETREATMENT

Charcoal samples were first cleaned by manual removal of rootlets and other foreign material, then boiled for 15 minutes in 1 molar phosphoric acid, cooled, and washed with distilled water. The residual charcoal was soaked overnight in a solution 0.1 molar sodium hydroxide and tetrasodium pyrophosphate (Goh and Molloy 1972), washed, boiled again in phosphoric acid, and washed to neutrality with distilled water.

Excavated shell samples were etched with an amount of dilute hydrochloric acid calculated to remove 10%-15% by weight of surface carbonate, then washed in distilled water. Shells from specimens which had been collected alive were soaked in 10% sodium hypochlorite for 48 hours, then washed in distilled water.

AGE CALCULATION

There are three basic methods used by radiocarbon dating laboratories for the calculation of radiocarbon ages for marine shell carbonate samples. It is important to recognize that different ages result, depending on which of these procedures is used. One procedure, used by Buddemeier and Hufen (1973), Kigoshi et al. (1973), Erlenkeuser and Willkomm (1973), and Vogel and Waterbolk (1972) makes no correction for isotopic fractionation or for the apparent age of marine shells. Vogel and Waterbolk (1972) aim to justify their method by stating, "no correction is applied to marine carbonates (sea shells) since the isotopic fractionation just compensates for the apparent age of surface ocean water." The second procedure, used by Lowdon et al. (1970) and Nydal et al. (1970), corrects all marine shells to a common $\delta C^{13} = 0°/oo$ relative to PDB. Since most marine shells have $\delta^{13}C = 0 \pm 2°/oo$ (Craig 1954, Mangerud 1972) this gives ages almost identical with those calculated by the first method, since these two procedures contain a correction factor of about 410 years for the apparent age of surface ocean waters. We show that this is correct only in certain oceanic regions. A third procedure, used by Mangerud and Gulliksen (1975) and Rafter et al. (1972), normalizes shell samples to a common $\delta^{13}C = -25°/oo$, then subtracts the apparent surface ocean water age appropriate to the particular region. Mangerud and Gulliksen (1975) use marine shell specimens, collected alive before 1950 and corrected for decay and for the Suess effect in surface ocean waters, to derive an apparent age appropriate for Scandinavian shell dating. Rafter et al. (1972) use a similar procedure in employing their extensive measurement of live-collected shell specimens and surface ocean water bicarbonate for New Zealand shell dating. It is clear from these respective studies that the apparent age or "reservoir effect" for marine shells varies with location, particularly at higher latitudes. It becomes imperative for valid archaeological dating to evaluate and to correct radiocarbon ages for geographical and local reservoir effects (apparent ages of surface ocean waters).

Since most terrestrial sample ages are related to 1890 wood standard by normalizing to $\delta^{13}C = -25°/oo$, and using 0.95 oxalic as the reference value, it would be a sensible approach to use this also for marine shells (and indeed all samples expressed as age B.P.). Even though $-25°/oo$, may not be the ideal choice (Olsson and Osadebe 1974), its use has become so well established as to make any change confusing and difficult. We recommend that all radiocarbon ages be calculated from $^{13}C/^{12}C$ fractionation corrected $D^{14}C$ values to the base of $-25°/oo$ w.r.t. PDB, and that the third calculation method be adopted in our respective laboratories. All our shell dates are reported as ages from which the local reservoir effect (age of surface ocean) has to be subtracted.

PRE-1950, LIVE-COLLECTED AUSTRALIAN SHELLS

Measurements of ^{14}C activity of marine mollusk shells collected alive at known dates between 1875 and 1950 are listed in table 4. These shells are from museum collections. Species names and mineralogy are given in table 3. Sampling locations are shown in figure 1. Results are given as $\Delta °/oo$, the millennial difference between 0.95 oxalic acid and sample ^{14}C activity age corrected for ^{14}C decay from year of live collection and 1950 corrected for $^{13}C/^{12}C$ isotopic fractionation, (i.e., normalized to base of $\delta^{13}C = -25°/oo$ relative to PDB; in absence of $\delta^{13}C$ measurements, the assumption is made that shell $\delta^{13}C = 0 \pm 2°/oo$).

Scrutiny of table 4 will show that the extreme Δ values are $-70°/oo$ and $-49°/oo$. By application of the "z" statistics (Polach, 1972) these can be shown to be in agreement ($z = 1.48$). Consequently all Δ values can be combined, giving an error-weighted mean $\Delta = -57.7 \pm 4°/oo$. After industrial effect correction (according to the method of Mangerud and Gulliksen 1975), this is equivalent to an apparent age or reservoir effect of 450 ± 32 years. 450 ± 35 years is the current best estimate for apparent age of Australian coastal surface waters relative to the pre-1950 atmosphere, and this value must be subtracted

TABLE 3

Species Names and Mineralogy of Mollusks Used in Australian Study

Code	Species	% Calcite (live) ±2%	Code	Species	% Calcite (live) ±2%
A	Anadara trapezia	0	N	Haliotis ruber	25
B	Austrocochlea constricta	4	O	Nerita atramentosa	40
C	Austrocochlea rudis	0	P	Patellanax peroni	70
D	Bankivia fasciata	0	Q	Pinctada margaritifera	16
E	Bembicium melanostomum	0	R	Pinna bicolor	0
F	Brachidontes rostratus	-	S	Pitar sp.	0
G	Cabestana spengleri	0	T	Plebidonax deltoides	0
H	Cellana solida	65	U	Proxichione laqueata	0
I	Conuber incei	0	V	Pyrazus ebinezus	0
J	Katelysia rhytiphora	0	W	Saccostrea cucculata	100
K	Littorina unifasciata	0	X	Thais orbita	75
L	Mactra obesa	95	Y	Turbo (Ninella) torquata	0
M	Mytilus edulis planulatus	2-20	Z	Turbo (Subninella) undulata	0

TABLE 4

Marine Mollusks Collected Alive Before 1950

Laboratory number	Location	Collection year	Species (table 3)	$\Delta°/oo$ (fractionation and age corrected to 1950)
SUA-354/1	Torres Strait (ca. 10°S, 143°E)	1875 ± 3	L	-58 ± 8
SUA-354/2	Torres Strait (ca. 10°C, 143°E)	1875 ± 3	R	-56 ± 10
SUA-357	Torres Strait (ca 10°S, 143°E)	1909	Q	-49 ± 10
SUA-355	Garden Is. W.A. (32°15'5, 115°40'E)	1930	U	-55 ± 10
SUA-393	Adelaide, S.A. (ca. 35°S, 139°E)	1937 ± 2	I	-70 ± 10
SUA-356	Narooma, N.S.W. (36°13'S, 150°07'E)	1950	J	-58 ± 10

Mean $\Delta = -57.7 \pm 4°/oo$; Industrial effect corrected mean, $\nabla^{14}C = 55 \pm 4\%$

from the ages of shells as reported by the Sydney and Australian National University Radiocarbon Dating laboratories.

1973: LIVE-COLLECTED AUSTRALIAN SHELLS

Measurements of the ^{14}C activity of marine shells collected alive in 1973 are listed in table 5. Species names are given in table 3. Results are given in terms of % Modern, the measured percentile difference in ^{14}C activity relative to 0.95 NBS oxalic acid, after the $^{13}C/^{12}C$ ratio of the samples has been normalized to the base of $\delta^{13}C = -25°/oo$ relative PDB (assuming all shells to have $\delta^{13}C = 0 \pm 2\%$). Sampling locations are shown in figure 1. These measurements appear to fall into two groups based on locality: those from

TABLE 5

Marine Mollusks Collected Alive in 1973

	East Coast of Australia		
Lab no.	Location	Species (see table 3)	^{14}C Activity % modern $\delta^{13}C$ corrected to base of $-25°/oo$
SUA-209/1	Moruya, N.S.W. 35° 54′S, 150° 07′E	T	107.1 ± 0.9
SUA-209/2	Moruya, N.S.W. 35° 54′S, 150° 07′E	D	104.1 ± 0.9
ANU-1091	Moruya, N.S.W. 35° 54′S, 150° 07′E	D	108.0 ± 0.6
SUA-23/1	Bass Pt., N.S.W. 34° 36′S, 150° 46′E	M	103.7 ± 0.6
SUA-23/2	Bass Pt., N.S.W. 34° 36′S, 150° 46′E	P	107.7 ± 0.6
SUA-23/3	Bass Pt., N.S.W. 34° 36′S, 150° 46′E	Y	107.4 ± 0.6
SUA-23/4	Bass Pt., N.S.W. 34° 36′S, 150° 46′E	G	105.2 ± 0.9
SUA-23/5	Bass Pt., N.S.W. 34° 36′S, 150° 46′E	W	108.7 ± 0.6
SUA-23/6	Bass Pt., N.S.W. 34° 36′S, 150° 46′E	O	109.6 ± 1.1
SUA-220	Broadwater, N.S.W. 29° 00′S, 153° 29′E	I	109.9 ± 0.9
SUA-218/1	Macleay Is. Q'ld	M	105.9 ± 0.8
SUA-218/2	Macleay Is. Q'ld	V	104.6 ± 0.8

Mean = 106.8 ± 2.1% Modern

TABLE 5 (*continued*)

	Southern Coast, Western Coast, and Northern Coast of Australia		
Lab no.	Location	Species (see table 3)	^{14}C Activity % modern δ^{13}C corrected to base of $-25^o/oo$
SUA-273/1	Swan Bay, Vic. 38° 17'S, 144° 40'E	J	118.6 ± 1.0
SUA-273/2	Swan Bay, Vic. 38° 17'S, 144° 40'E	Z	118.6 ± 0.9
SUA-300/1	Noarlunga, S.A. 35° 10'S, 138° 28'E	H	121.9 ± 1.0
SUA-300/2	Noarlunga, S.A. 35° 10'S, 138° 28'E	N	122.2 ± 1.0
SUA-273/3	Fremantle, W.A. 32° 03'S, 115° 47'E	E	124.0 ± 1.3
SUA-27/1	Cottlesloe, W.A. 32° 01S, 115° 45'E	C	116.9 ± 0.9
SUA-27/2	Cottlesloe, W.A. 32° 01S, 115° 45'E	K	113.4 ± 1.2
SUA-311	Pt. Stuart, N.T. 12° 13'S, 131° 53'E	W	117.6 ± 1.0
	Mean = 119.2 ± 3.5% Modern		
	East Coast of Tasmania		
SUA-294/1	Bruny Is., Tas., 43° 21'S, 147° 20'E	F	110.9 ± 0.9
SUA-294/2	Bruny Is., Tas., 43° 21'S, 147° 20'E	H	114.6 ± 0.9
SUA-294/3	Bruny Is., Tas., 43° 21'S, 147° 20'E	B	112.9 ± 1.2
	Mean = 112.8 ± 1.9% Modern		
	Estuarine Oyster Sample		
ANU-1173 C	Charon Pt., Q'ld 22° 23'S, 149° 49'E	Saccostrea succulata	116.0 ± 0.7

the eastern coast with mean values of %M = +106.8 ± 2.3 D^{14}C = 68 ± 22.3%) and those from the southern coasts with mean values of %M = +119.2 ± 3.5 (D^{14}C = +192 ± 34.5%). Both groups of shells can be considered as samples from normal populations, and the difference between the mean values is significant at the 0.1% level "z" and "t" test.

Since the activity of marine shell carbonate is generally representative of the activity of the surface ocean water bicarbonate in the growth region (Broecker and Olson 1961), it is necessary to consider why the waters on the eastern coast of Australia should be

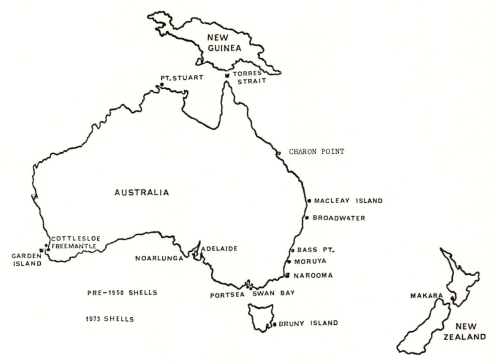

FIG. 1. Sampling locations for pre-1950 and 1973 shells.

significantly different from those on the southern coasts. Figure 2 (after Knox 1963, Maxwell 1968) shows that the major surface ocean currents affecting these regions are (1) the East Australian Current, which brings warmer subtropical water from the Coral Sea, and (2) the West Wind Drift, which brings cooler water from midsouthern latitudes of the Indian Ocean. The surface ocean water bicarbonate activity measurements of Fairhall et al. (1972), Linick and Suess (1972), Rafter and O'Brien (1972), and Vogel (1972) indicate that the increase in activity due to the nuclear explosions is greatest at midsouthern latitudes of the Indian and Pacific Oceans, with significantly lower activity in the tropical Pacific. The observed differences may be attributed to the different source waters for the two regions. It is of interest to note that two marine shell specimens collected alive in 1973 at Makra, New Zealand, had a mean value of $110.7 \pm 0.7\%$ Modern (Rafter, personal communication), which is midway between the two groups of Australian shells reported here. The discussions of Rafter and O'Brien (1970, 1972) indicate that Makra can be considered to be influenced by both the East Australian Current and the West Wind Drift.

Mention should be made here of the work of Erlenkeuser et al. (1975) who found, for marine shells collected alive in 1974, that the activity measured in individual growth rings showed significant differences. When whole shells are used, as in this work, the activity measured will be an integrated average over the growth period, in the manner applied to multiple tree rings. For shells which may be twenty or more years old when

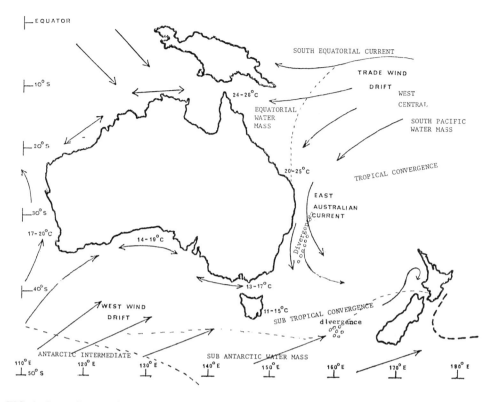

FIG. 2. Generalized surface ocean currents and mean water temperatures (Knox 1963, Maxwell 1968).

collected live, the measured activity will not be representative of the year of collection unless the last year's growth ring is isolated. While this is particularly important for post-1960 samples, it could also introduce errors in pre-1950 shells used to derive local modern reference standards. The useful fact remains that dating of whole marine shells provides a valuable indicator of current and past surface ocean water bicarbonate activity.

CHARCOAL/SHELL PAIRS

Measurements of charcoal and stratigraphically equivalent marine shells from Aboriginal middens on the Australian coast are listed in table 6. Radiocarbon ages for both charcoal and shell samples are calculated from $D^{14}C$ values, normalized to a common $\delta^{13}C = -25°/oo$ relative to PDB (in absence of $\delta^{13}C$ measurements the assumption was made that charcoal has a value $\delta^{13}C = -25 \pm 2°/oo$, the shell a value of $\delta^{13}C = 0 \pm 2°/oo$). In some cases the ^{14}C activity of more than one species of shell was measured (species names are as given in table 3). The mean age difference for twenty-nine pairs if +283 years (shells older than charcoal).

The group of samples from the Bass Point 50-60 cm level illustrate that acid etching pretreatment is both necessary and effective in removing younger surface contamination

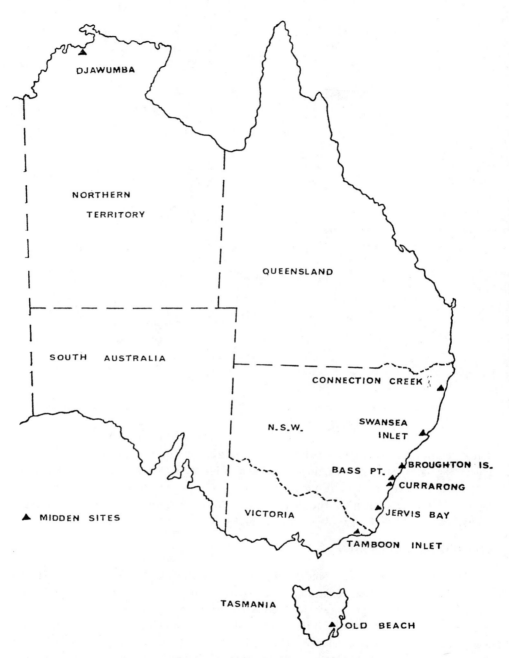

FIG. 3. Australian archaeological midden locations.

TABLE 6
Charcoal-Shell Pairs

Laboratory number	Site	Depth (cm)	Species	Age B.P. 5568 half-life $\delta^{13}C$ to $-25^o/oo$
SUA-47	Bass Point	10-15	Charcoal	320 ± 75
SUA-45/S1	Bass Point	10-15	M	725 ± 65
SUA-45/S2	Bass Point	10-15	P	680 ± 65
SUA-45/S3	Bass Point	10-15	Y	695 ± 65
SUA-45/S4	Bass Point	10-15	G	665 ± 65
SUA-146	Bass Point	30-35	Charcoal	985 ± 70
SUA-145/S1	Bass Point	30-35	M	1100 ± 65
SUA-145/S2	Bass Point	30-35	P	1250 ± 65
SUA-145/S3A	Bass Point	30-35	Y	1290 ± 65
SUA-145/S3B	Bass Point	30-35	Y	1290 ± 65
SUA-145/S3C	Bass Point	30-35	Y	1285 ± 65
SUA-145/S4	Bass Point	30-35	N	1270 ± 65
SUA-145/S5	Bass Point	30-35	G	1130 ± 65
SUA-145/S6	Bass Point	30-35	X	1190 ± 65
SUA-25	Bass Point	50-60	Charcoal	2650 ± 70
SUA-24/S1A	Bass Point	50-55	Y	2480 ± 70
SUA-24/S1B	Bass Point	50-55	Y	2870 ± 80
SUA-24/S2A	Bass Point	50-55	G	1990 ± 80
SUA-24/S2B	Bass Point	50-55	G	2180 ± 80
SUA-260C	Jervis Bay	65	Charcoal	390 ± 70
SUA-260/S1	Jervis Bay	65	M	800 ± 60
SUA-260/S2	Jervis Bay	65	P	720 ± 60
SUA-261C	Jervis Bay	104	Charcoal	1790 ± 90
SUA-261/S1	Jervis Bay	104	Y	970 ± 73
SUA-261/S2	Jervis Bay	104	M	970 ± 75
SUA-262C	Jervis Bay	148	Charcoal	910 ± 60
SUA-262/S1	Jervis Bay	148	Y	1285 ± 70
SUA-262/S2	Jervis Bay	148	O	1275 ± 70
SUA-262/S3	Jervis Bay	148	P	1125 ± 80
SUA-262/S4	Jervis Bay	148	M	1265 ± 70
SUA-238C	Swansea Inlet	0-4	Charcoal	1965 ± 90
SUA-238/S1	Swansea Inlet	0-4	W	2690 ± 90
SUA-238/S2	Swansea Inlet	0-4	A	2480 ± 90
SUA-264C	Djawumba	65-88	Charcoal	6350 ± 250
SUA-264S	Djawumba	65-78		
SUA-306	Old Beach	35	Charcoal	5800 ± 130
SUA-307	Old Beach	35	F	6010 ± 90
SUA-377C	Tamboon Inlet	40-50	Charcoal	360 ± 85
SUA-377S	Tamboon Inlet	40-50	F	660 ± 95
SUA-378C	Tamboon Inlet	10-20	Charcoal	220 ± 85
SUA-378S	Tamboon Inlet	10-20	F	420 ± 100

TABLE 6 (*continued*)

Laboratory number	Site	Depth (cm)	Species	Age B.P. 5568 half-life $\delta^{13}C$ to $-25^o/oo$
SUA-395C	Connection Ck. I	50-60	Charcoal	3720 ± 100
SUA-395S	Connection Ck. I	50-60	W	3750 ± 100
SUA-396C	Connection Ck. I	100-110	Charcoal	3400 ± 100
SUA-396S	Connection Ck. I	100-110	A	3790 ± 100
NSW-76	Currarong	37.5	Charcoal	1520 ± 100
SUA-241/S1	Currarong	30-40	W	2040 ± 70
SUA-241/S2	Currarong	30-40	V	1600 ± 70
SUA-402C	Broughton Is.	50-60	Charcoal	445 ± 170
SUA-402/S1	Broughton Is.	50-60	O	420 ± 85
SUA-402/S2	Broughton Is.	50-60	P	600 ± 85

from shells. This is not true for partly recrystallized shells, as shown by samples SUA-24/S2A and B, in which both pretreated and untreated shells are significantly different from the control charcoal sample. In the Jervis Bay series, the charcoal SUA-261C is clearly anomalous and may have been caused by the use of driftwood for burning (Lampert, personal communication).

The other charcoal/shell pairs show age differences ranging from -25 to +725 years, indicating that these middens are not ideal sites for the determination of past relationships between terrestrial and marine sample activities. Some of the problems of midden archaeology have been discussed by Ambrose (1967) and Hughes and Sullivan (1974), who point out the random accumulation and possible redeposition of both shell and charcoal components.

CONCLUSIONS

The data presented here suggest that despite the diversity and complexity of oceanic water mass movement around the Australian Continental Shelf, it is possible to derive a mean ocean surface water age based on pre-1950 shell collections. The coastal surface water depletion in ^{14}C relative to atmosphere and terrestrial wood can be expressed as a mean apparent age of 450 ± 35 years. If shell ^{14}C concentrations are corrected for isotopic fractionation to equate the activity of wood (i.e., $\delta^{13}C$ to the base of $-25^o/oo$ w.r.t. PDB), then the whole value of 450 must be subtracted from the reported age and the resultant error term combined according to: A-B ± $(a^2 + b^2)^{1/2}$, (where A ± a is age ± error of shell; B ± b is the age correction ± error). If the shell ^{14}C concentration is normalized to $\delta^{13}C = 0^o/oo$ w.r.t. PDB, or is not corrected at all, then only the value in excess of 410 years (i.e., 40 years ± 35 years) must be subtracted from the reported result.

To further ensure validity of results based on shell material, only those species which deposit their carbonate as aragonite should be used for dating because an X ray diffraction analysis will readily detect the presence of calcite recrystallization which might be indicative of contamination.

The collection and dating of contemporary (post-1950) shells, while a useful indicator of water mass circulation, cannot be used to validate geochronological studies because the diversity and complexity of distribution of a atom-bomb-produced ^{14}C in the surface ocean is too great to permit derivation of an appropriate age correction factor for a particular region of the coast. Large discrepancies from the observed and tabulated mean values, such as those observed in estuarine or Tasmanian coastal waters, will dictate caution and provide a useful indicator of present, but not necessarily of past, disequilibria. Large discrepancies in marine shell and charcoal pairs (i.e., of the order of 1000 years or greater) are likely to be due to misassociation of sample and event, for nothing in our study suggests oceanic disequilibria in coastal Australian waters of this order of magnitude.

ACKNOWLEDGMENTS

This work was carried out at the Sydney University Radiocarbon Laboratory and is based on the thesis submitted for attaining a Doctorate of Philosophy by Richard Gillespie, who would like to thank Dr. R. B. Temple (S.U.) for his encouragement, helpful advice, and critical evaluation. We would also like to thank Dr. J. P. White (S.U.) and Dr. J. Chappell (A.N.U.) for useful discussions, and the various Australian museums for donating shell samples. Laura O'Brien gave technical assistance in the Sydney Laboratory. R. J. Lampert (A.N.U.) and P. J. Hughes (A.N.U.) gave fully of their archaeological experience, and Maureen Powell (A.N.U.) typed the text.

REFERENCES

Ambrose, W. R.
    1967 Archaeology and shell middens. Archaeol. and Phys. Anthropol. *In* Oceania. 2:169-187.

Andrews, J. T.
    1972 Recent and fossil growth rates of marine bivalves. Paleogeogr. Paleochmatol. Paleoecol. 11:157-176.

Berger, R., R. E. Taylor, and W. F. Libby
    1966 Radiocarbon content of marine shells from the Californian and Mexican West coasts. Science. 153:864-866.

Broecker, W. S.
    1963 Carbon-14/carbon-12 ratios in surface ocean water. National Research Council Report No. 1075.

Broecker, W. S., and E. A. Olson
    1961 Lamont Radiocarbon Measurements VIII. Radiocarbon. 3:176-204.

Buddemeier, R. W., and T. H. Hufen
  1973   Hawaii Institute of Geophysics Radiocarbon Dates I. Radiocarbon. 15:70-74.

Craig, H.
  1954   Carbon-13 in plants and the relationship between carbon-13 and carbon-14 in Nature. J. Geology. 62:115-148.

Erlenkeuser, H., H. Metzner, and H. Willkomm
  1975   University of Kiel radiocarbon measurements VIII. Radiocarbon. 17:276-300.

Erlenkeuser, H., and H. Willkomm
  1973   University of Kiel radiocarbon measurements VII. Radiocarbon. 15:113-126.

Fairhall, A. W., A. W. Young, and P. A. Bradford
  1972   Radiocarbon in the sea. Proc. 8th Int. Cont. on Radiocarbon Dating. New Zealand. C2-16.

Goh, K. M., and B. P. J. Molloy
  1972   Reliability of radiocarbon dates from buried charcoals. Proc. 8th Int. Conf. on Radiocarbon Dating. New Zealand. G29-45.

Hakansson, S.
  1969   University of Lund Radiocarbon Dates II. Radiocarbon. 11:430.
  1973   University of Lund Radiocarbon Dates VI. Radiocarbon. 15:493.

Haynes, C. V., D. C. Grey, and A. Long
  1971   University of Arizona radiocarbon dates VIII. Radiocarbon. 13:1-18.

Hughes, P. J., and M. E. Sullivan
  1974   The redeposition of midden material by storm waves. J. Royal Soc. N.S.W. 107:6-10.

Keith, M. L., and G. M. Anderson
  1963   Radiocarbon dating: fictitious results with mollusc shells. Science. 141:634-636.

Kigoshi, K., N. Suzuki, and H. Fukatsu
  1973   Gakashuin natural radiocarbon measurements VIII. Radiocarbon. 15:42-67.

Knox, G. A.
  1963   The biogeography and intertidal ecology of the Australasian coasts. Oceanogr. Mar. Biol. Rev. 1:341-404.

Linick, T. W., and H. E. Suess
  1972   Bomb-produced radiocarbon in the surface waters of the Pacific Ocean. Proc. 8th Int. Conf. on Radiocarbon Dating. New Zealand. C87-93.

Lowdon, J. A., R. Wilmeth, and W. Blake jr.
  1970   Geological Survey of Canada radiocarbon dates X. Radiocarbon. 12:472-485.

Mangerud, J.
  1972   Radiocarbon dating of marine shells including a discussion of apparent age of recent shells from Norway. Boreas. 1:143-172.

Mangerud, J., and S. Gulliksen
  1975   Apparent radiocarbon ages of recent marine shells from Norway, Spitzbergen, and Arctic Canada. Quat. Research. 5:263-273.

Maxwell, W. G. H.
    1968   Atlas of The Great Barrier Reef. Elsevier.

Mook, W. G., and J. C. Vogel
    1968   Isotopic equilibrium between shells and their environment. Science. 159:874-875.

Nydal, R., K. Lovseth, and O. Syrstad
    1970   Trondheim natural radiocarbon measurements V. Radiocarbon. 12:205-237.

Olson, E. A.
    1963   The problem of sample contamination in radiocarbon dating. Ph.D. thesis. Columbia Univ.

Olsson, I. U., and F. A. N. Osadebe
    1974   Carbon isotope variations and fractionation corrections in C-14 dating. Boreas. 3:139-146.

Polach, Henry A.
    1972   Cross Checking of NBS Oxalic Acid and Secondary Laboratory Radiocarbon Dating Standards. *In* Proc. 8th Int. Conf. on Radiocarbon Dating. New Zealand. 2:688-717.

Rafter, T. A.
    1968   Carbon-14 variations in nature (3) measurements in the South Pacific and Antarctic Oceans. N.Z. J. Science. 11:551-589.

Rafter, T. A., H. S. Jansen, L. Lockerbie, and M. M. Trotter
    1972   New Zealand radiocarbon reference standards. Proc. 8th Int. Conf. on Radiocarbon Dating. New Zealand. H29-79.

Rafter, T. A., and B. J. O'Brien
    1970   Exchange rates between the atmosphere and the ocean. Proc. 12th Nobel Symp. Uppsala. 355-378.
    1972   Carbon-14 measurements in the atmosphere and in the South Pacific Ocean. Proc. 8th Int. Conf. on Radiocarbon Dating. New Zealand. C17-42.

Taylor, R. E., and R. Berger
    1967   Radiocarbon content of marine shells from the Pacific coasts of Central and South America. Science. 158:1180-1182.

Vogel, J. C.
    1972   Radiocarbon in the surface waters of the Atlantic Ocean. Proc. 8th Int. Conf. on Radiocarbon Dating. New Zealand. C43-55.

Vogel, J. C., and H. T. Waterbolk
    1972   Groningen radiocarbon dates X. Radiocarbon. 14:6-110.

Weber, J.
    1974   Carbon-13/carbon-12 ratios as natural isotopic tracers elucidating calcification processes in corals. Proc. 2d Int. Coral Reef Symp. Brisbane. 289-298.

3

Fraction Studies on Marine Shell and Bone Samples for Radiocarbon Analyses

R. E. Taylor and Peter J. Slota, Jr.

INTRODUCTION

The accuracy and reliability of radiocarbon values depend on the satisfaction of a set of basic assumptions relating both to the production and distribution of radiocarbon in nature and to the physical and chemical integrity of carbon isotope ratios in sample materials used in analyses. The assumption relating to the physiochemical integrity of sample materials states that there has been no alteration in a sample's carbon isotope ratio except by decay since it ceased to be a part of one of the carbon reservoirs — as, for example, at the death of an organism. Initial experiences suggested that this assumption apparently was being violated to different degrees in various sample materials. This led to a ranking of the reliability of samples in early ^{14}C literature. At the top of the list in the first edition of *Radiocarbon Dating* were charcoal and charred organic materials followed by well-preserved wood, grasses, cloth, and peat. At the bottom was well-preserved shell. Missing from the list was bone (Libby 1952:44).

Bone constitutes an important component of many archaeological deposits, especially as it relates to mortuary contexts and as it reflects subsistence patterns. Marine shell is the principal component of midden deposits in coastal sites. Laboratories concerned with archaeologically derived samples have been especially interested in scrutinizing more closely the exact nature of the problems associated with the use of such samples. The nature of some of the data generated as part of these studies will be briefly reviewed and the first results of recent work conducted at the University of California, Riverside, radiocarbon laboratory will be discussed.

Marine Shell Samples

Marine shell was quickly dubbed a relatively unreliable sample material. The earliest questions concerned (1) the possibility of isotopic exchange between the shell matrix and CO_2 dissolved in ground waters, and (2) the initial carbon isotope ratios for carbonates in relationship to terrestrial organics. The first question involved both exchange and possible recrystallization of the shell matrix after deposition; the second led to a concern with upwelling and other biological and geochemical effects of oceanic environments (Kulp et al. 1951, 1952; Suess 1954; Craig 1957).

The earliest published suite of marine shell carbonate/terrestrial organic pairs measured at the Chicago laboratory provided inconclusive data on the issue of reliability. Of five pairs measured, three were of statistically identical age, but two showed a significant discrepancy. The most serious was on a wood and shell pair from the South American site of Huaca Prieta (Libby 1955:132-133). The Lamont laboratory also reported in the early 1950's that certain shell carbonate values yielded radiocarbon ages from 1000-2000 years in excess of the associated terrestrial organics (Kulp et al. 1951, 1952; cf. Olson and Broecker 1961:148).

Several avenues of inquiry were pursued in an effort to clarify the problems involving marine shell. One approach continued to measure terrestrial organic/marine shell sample pairs. A review of the first seventeen volumes of *Radiocarbon* (1969-1975) yielded twenty-eight such pairs. The shell species were identified and, on the basis of species assignment, were unambiguously determined to derive from a normal marine or an open oceanic environment (cf. Gillespie and Polach, preceding paper.). In eighteen cases, the reported B.P. radiocarbon values of the pairs overlay at the one sigma level. Except for samples from the Peruvian area, all the remaining marine shell carbonate values exceeded their associated terrestrial control sample ages by only a small percentage. Other laboratories evaluated the problem of surface carbonate exchange in shell samples by measuring the ^{14}C activity in successive layers of the shell matrix. In cases where the carbonate age exceeded 10,000 years (i.e., Pleistocene age shell), some laboratories reported variations of up to 1000 years between the inner and outer fractions. In over 90% of the Pleistocene age samples, the difference between the inner and outer fraction did not exceed 5%. The outer shell fractions tended to yield slightly younger values, but this deviation, especially in Holocene age samples, rarely exceeded a small percentage (cf. Taylor 1970:35-36; Hakansson 1972; Shotton et al. 1975).

As part of a long-term archaeological dating program for the coastal areas of southern California and western Mexico, several projects were initiated to investigate the integrity of shell carbonate radiocarbon values for this oceanic region. The first project involved a study of contemporary radiocarbon concentrations in marine shells along the western continental margins of North and South America. The purpose of this study was to obtain an estimate of the magnitude and variability in upwelling effects (Berger et al. 1966; Taylor and Berger 1967; cf. Hubbs, et al. 1965; Gillespie and Polach, this volume).

An examination of the data in table 1 reflects the significant variation in ^{14}C activity in living mollusks ranging from a maximum of -8.5% along the Peruvian coast to a $+2.18\%$ value for the adjacent Ecuadorian coast. Four samples from the environment of the Galapagos Islands range from a little more than -4.0% to a little less than 0.5%. Such fluctuations in a small area emphasize the fact that broad generalizations about the magnitude of upwelling effects for any major oceanic region must be carefully scrutinized. Certainly any "worldwide average" figure could be very misleading if used as a corrective for marine shell ^{14}C values in specific regions. Problems that might be encountered

TABLE 1

Radiocarbon Content of Pre-bomb Marine Shells from the West Coasts of North and South America

UCLA Lab no.	Location and collection date	Shell species	Uncorrected $\delta^{14}C$ (%)	$^{13}C/^{12}C$ (per mille)	Corrected $\Delta^{14}C$ (%)
149	Monterey, Calif. (1878)	Mytilus californianus (Conrad)	−2.0 ± 0.65	−	−
1033	Seal Beach, Calif. (1921)	Lunatia lewisii (Gould)	−1.84 ± 0.56	−0.5	−1.74
963	Cedros Is., Baja, Calif. (1939)	Pecten (Lyropecten) subnodosus (Sowerby)	−2.53 ± 0.60	−0.2	−2.49
939	Magdalena Bay, Baja Calif. (1938)	Arca tuberculosa (sby)	−3.37 ± 0.62	−1.6	−3.06
916	Cape San Lucas, Baja Calif. (1932)	Strombus granulatus (Swainson)	−4.65 ± 0.52	−0.6	−4.54
917	Carmen Is., Gulf of Calif. (1911)	Strombus granulatus (Swainson)	−6.97 ± 0.60	+0.5	−7.06
914	Kino Bay, Sonora (1935)	Tivela bryonensis (Gray)	−7.05 ± 0.59	−0.4	−6.98
913	Mazatlan, Sinaloa (1939)	Anadara grandis (B&S)	−3.41 ± 0.56	−1.7	−3.08
915	Manzanillo, Colima (1930)	Ostrea fischeri (Dall)	−3.0 ± 0.58	+1.1	−3.21
940	Banderas Bay, Jalisco (1938)	S. granulatus (Swainson)	−2.55 ± 0.58	−0.8	−2.39
936	Isabel Island, Nayarit (1938)	Muricanthus regius (Wood)	−3.46 ± 0.58	−0.4	−3.38
938	Guatulco Bay, Oaxaca (1938)	Turritella leveostoma vgl.	−2.40 ± 0.58	+0.8	−2.56
1249A	Guayaquil, Ecuador (1927)	Cerithidea valida (Adams)	+1.01 ± 0.45	−5.72	+2.18
1249B	Guayaquil, Ecuador (1927)	Thais biserialis (Blainville)	−1.15 ± 0.53	+1.84	−1.51
1254	Port Parker, Costa Rica (1935)	Strombus granulatus (Swainson)	−3.11 ± 0.43	+1.74	−3.45
1255A	Santiago Is., Galapagos Is. (1934)	Kelletia kelleti (Forbes)	−1.48 ± 0.62	+0.35	−1.55
1255B	Santiago Is., Galapagos Is. (1934)	Astraea (Uvanilla) undosa (Wood)	−3.68 ± 0.94	+1.88	−4.04
1255C	Espanola Is., Galapagos Is. (1934)	Fasciolaria (Pleuroploca) princeps (Sowerby)	−0.42 ± 0.51	+1.33	−0.68
1255D	Santa Cruz Is., Galapagos Is. (1932)	Nerita (Ritena) scabricosta (Lamarck)	+0.22 ± 0.48	+2.88	−0.35

TABLE 1 (continued)

UCLA Lab no.	Location and collection date	Shell species	Uncorrected $\delta^{14}C$ (%)	$^{13}C/^{12}C$ (per mille)	Corrected $\Delta^{14}C$ (%)
1256A	Secas Is., Panama (1935)	Vasum caestus (Broderip)	+0.42 ± 0.61	+1.48	+0.12
1256B	Secas Is., Panama (1935)	Strombus galeatus (Swainson)	−0.90 ± 0.58	+1.30	−1.16
1277	Antofagasta, Chile (1925)	Concholepas concholepas (Bruguiere)	−2.61 ± 0.40	+0.09	−2.63
1278	Valparaiso, Chile (1930-1940)	Tequla aler (Lesson)	−4.09 ± 0.87	+1.32	−4.34
1279	Peru (1930-1940)	Oliva peruviana (Lamarck)	−8.29 ± 0.48	+1.15	−8.50
1282	Northern Peru (1930-1940)	Strombus peruvianus (Swainson)	−3.56 ± 0.57	−0.22	−3.52

(See Berger et al. 1966; Taylor and Berger 1967). Uncorrected $\delta^{14}C$ values expressed with respect to 0.95 NBS oxalic acid standard. Corrected $\Delta^{14}C$ values derived from $\Delta^{14}C = \{(1 + \delta^{14}C)/(1 + 2\delta^{13}C) - 1\} \times 100$.

under special oceanic conditions are highlighted in samples obtained from the Gulf of California (Sea of Cortez) where deviations as high as −7% are measured. Fortunately the values from the North American west coast do not exceed about −3.00% and vary by only a few percentage points.

A related issue raised as part of this study involved the nature of the constancy of the oceanic circulation patterns over the eastern Pacific. The question was to what degree values such as those cited in table 1 could be used to estimate oceanic upwelling effects along the continental margins of the Western Hemisphere, which dates further back into the early Holocene and terminal Pleistocene. Table 2 lists a series of paired terrestrial organics/marine shell samples from this coastal region obtained in stratigraphically associated contexts. These data suggest that oceanic circulation patterns, at least for the central coastal margin of North America, seem to have been constant as far back as early Holocene times.

The most recent approach to the use of marine shell as a sample material has been an examination of the ^{14}C concentrations in the carbonate as opposed to the total organic (principally conchiolin) fraction of single sample lots. Table 3 presents carbonate and organic fraction ^{14}C values for eight sample sets. The UCLA series represents values obtained by the senior author in 1969 and, except for two values, has not previously been reported. The carbonate fractions were obtained by acidification with 2N HCl. The organic/conchiolin fraction was obtained by combustion of the filtrate deposited on glass filter paper following total destruction of the carbonate matrix of the shell sample in 2N HCl. Because of the expected low organic yield, special care was exercised during the hydrolysis of the shell to prevent contact of the solution with possible contemporary

TABLE 2

Radiocarbon Determinations on Terrestrial Organics/Marine Shell Paired Samples from Coastal North America

Lab no.	Provenience	Material	Date
GSC-24	Vancouver Is., B.C.	Wood	12,200 ± 160
GSC-38		Shell	12,360 ± 140
L-391D	Vancouver Is., B.C.	Wood	12,150 ± 250
L-391E		Shell	12,350 ± 250
UCR-509	Newport Bay, Calif.	Charcoal	8,445 ± 280
UCR-510		Shell	8,045 ± 270
LJ-923	Baja Calif.	Charcoal	5,480 ± 200
LJ-924		Shell	6,140 ± 250
W-27	Baja Calif.	Charcoal	2,500 ± 200
W-26		Shell	2,540 ± 200
LJ-922	Baja Calif.	Charcoal	2,200 ± 200
LJ-925		Shell	2,600 ± 200
LJ-645	Baja Calif.	Charcoal	1,660 ± 200
LJ-611		Shell	1,800 ± 300
UCLA-187	Morett Site, Colima, Mexico	Charcoal	1,500 ± 80
UCLA-1035		Shell	1,480 ± 80
UCLA-1034		Shell	1,630 ± 80
LJ-85	Baja Calif.	Charcoal	960 ± 150
LJ-84		Shell	1,060 ± 150
W-155	La Jolla, Calif.	Charcoal	600 ± 200
W-154		Shell	580 ± 200

atmospheric ^{14}C contamination. From the UCR samples, before the isolation of the organic/conchiolin fraction was begun, approximately 50% by weight of the surface carbonate was removed separately followed by collection of the inner 50%-70% fraction. In three of the UCR samples and in one UCLA sample, separate measurements were conducted of the outer and inner carbonate fraction. In all cases the deviation between the two fractions did not exceed a small percentage, suggesting that little surface carbonate exchange occurred.

In six of the eight paired values, the differences in age between the organic/conchiolin fraction and the carbonate fraction tended to increase significantly with the increasing radiocarbon age of the samples. For the two samples with ages less than 1000 years, there were no statistical differences in the ages of the fractions. As the ages increase to a few thousand years, the differences become significant. With samples in the age range of 7000-8000 B.P., the differences were from 1000 to 2000 years. The comparisons were somewhat complicated by the fact that the UCLA series lacked stable isotope data and that the Santa Rosa Island values were not consistent with the other six sample sets. In this case the comparability of the carbonate and organics/conchiolin

TABLE 3

Radiocarbon Determinations on Carbonate and Organic/Conchiolin Fractions of Marine Shell Samples from the Southern California and West Mexican Coastal Archaeological Deposits

Sample number	Provenience	Sample type	Organic yield (%)	$\delta^{13}C$ ($^o/oo$)[1]	Uncorrected ^{14}C age (yrs. B.P.)[2]	Corrected ^{14}C age (yrs. B.P.)[3]
	A. UCLA Series						
UCLA-663B[4]	Santa Rosa Island,	Carbonate	–	–	7230 ± 120[6]	–
UCLA-663C	California	Conchiolin	~0.4	–	7460 ± 400[7]	–
UCLA-659C[4]	Santa Rosa Island,	Carbonate	–	–	6550 ± 150	–
UCLA-659B	California	Conchiolin	–	–	6850 ± 210[8]	–
UCLA-1681A[5]	Puerto Marquez,	Carbonate	–	–	3200 ± 80	–
UCLA-1681B	Mexico	Conchiolin	–	–	2400 ± 625	–
UCLA-1680A[5]	Puerto Marquez,	Carbonate	–	–	2490 ± 80	–
UCLA-1680B	Mexico	Conchiolin	–	–	1580 ± 390	–
	B. UCR Series						
UCR-446A	San Diego,	Carbonate[9]	–	+1.06	–	8085 ± 315
UCR-446B	California	Conchiolin	0.04	–23.03	–	5960 ± 270
UCR-431A	San Diego,	Carbonate[9]	–	+1.09	–	7160 ± 280
UCR-431B	California	Conchiolin	0.04	–22.63	–	6210 ± 240
UCR-441A	Point Mugu,	Carbonate[9]	–	+0.14	–	500 ± 130
UCR-441B	California	Conchiolin	0.05	–16.72	–	540 ± 130
UCR-440A	Point Mugu,	Carbonate[9]	–	+0.20	–	400 ± 140
UCR-440B	California	Conchiolin	0.06	–18.61	–	470 ± 120

[1] Expressed with respect to PDB standard. Error on values ± 0.2 $^o/oo$.
[2] Expressed with respect to 0.95 NBS oxalic acid standard. t½ = 5568 years.
[3] Corrected value normalized to -25.00 $^o/oo$ $\delta^{13}C$.
[4] Previously published value: Berger et al. 1964; Berger et al. 1965.
[5] Previously unpublished values.
[6] Inner fraction: outer fraction (UCLA-663A) 7120 ± 120 yrs. B.P.
[7] Published value 7210 ± 400 with 250 year correction factor applied.
[8] Published value 6500 ± 210 with 250 year correction factor applied.
[9] Inner fraction after removal of 40-50% of shell weight.

values seemed to be associated with a much higher organic yield. In the case of the UCR series values and the Puerto Marquez samples, external evidence (e.g., the archaeological contexts from which the samples were derived) supported the validity of the carbonate as against the organic/conchiolin values. Our present view is that the anomalous conchiolin values are associated with the assimilation of microorganisms into the conchiolin fraction of varying ^{14}C age. While the effect is not pronounced in material of relatively recent age, the contamination becomes marked apparently as a function of the time that the sample has been subject to attack by soil microorganisms. This process apparently does not affect the ^{14}C activity in the carbonate fraction.

Bone Samples

While shell received the lowest reliability rating in the earliest ranking of sample types, bone was not even listed. It was suggested, however, that it was "barely conceivable" that certain bone ^{14}C measurements under some circumstances might be reliable. It was assumed that bone would constitute a very poor sample type because of its low organic carbon content and generally porous structure. It should be noted that burned bone was highly recommended, but the actual sample was derived from carbonized skin and tissue, not from the bone structure itself.

Early ^{14}C values using the inorganic or carbonate fraction of bone were, in most cases, clearly erroneous (Tamers and Pearson 1965). For many sites the availability of usually reliable sample types such as wood or charcoal obviated the need to employ problematic samples such as bone. In some contexts, however, especially where stratigraphic discontinuities created a question of the association of skeletal material with, for example, charcoal, there was need to determine whether the organic or collagen fraction of bone structures had the isotopic integrity required to yield accurate ^{14}C values.

The first report of satisfactory agreement between bone collagen and associated charcoal ^{14}C values for a suite of samples was obtained by Berger et al. (1964). Contamination by noncollagenous organics remained a possibility (Haynes 1967), and several investigators suggested methods to ensure that a pure collagen fraction would be employed (Krueger 1965; Sellstedt et al. 1966; Berger et al. 1971; Longin 1971). Preparation of gelatin was suggested as a means to avoid potential contamination of the collagenous samples (Longin 1971; Protsch 1973). Since the collagen fraction in many bones, especially of Pleistocene age, was insufficient for routine radiocarbon analysis, Haynes (1968) attempted to investigate the reliability of using the bone apatite fraction in routine bone ^{14}C analysis. Hassan (1975) summarized the possible sources of anomalous results in ^{14}C dates, using both collagen and bone apatite. She concluded that physical contamination was the principal source of error in collagen work, while erroneous apatite dates could result from the exchange of carbon in apatite structures during recrystallization or surface exchange reactions. Stable isotope studies also suggested possible variability in the original ^{14}C activity in bone due to differences in taxonomic order, age, diet, or geochemical environment of the animal.

We have begun to study the reliability of bone ^{14}C values by conducting fraction studies on single bone sample lots to determine which of the organic fractions may be most susceptible to contamination. We have completed work on two duplicate sample suites from a single burial and we have measured the carbonate, collagen, and amino acid fraction in each case. Table 4 lists the results obtained to date. The carbonate fraction was obtained by acidification of a portion of each sample by 2N HCl in a closed system. The collagen fraction was obtained utilizing the method outlined in Berger et al. (1964). The amino acid fraction was prepared by refluxing the samples for 48 hours in 2N HCl and passing the solution through a Dowex ion exchange resin (2-X8). The eluted solution

TABLE 4

Radiocarbon Determinations on the Carbonate, Collagen, and Amino Acid Fractions of Human Bone Samples from a Pre-historic Burial in Northern California
(Site SJ0-112, Burial 36)

Sample number	Sample type	Organic carbon yield (%)	$\delta^{13}C^1$ (°/oo)	Corrected ^{14}C age^2
UCR-449A	Carbonate	–	−8.42	930 ± 140
UCR-449B	Collagen	6.95	−19.89	2765 ± 155
UCR-449C	Amino acids	1.26	−21.41	2930 ± 150
UCR-450A	Carbonate	–	−9.43	830 ± 100
UCR-450B	Collagen	6.49	−20.24	2835 ± 140
UCR-450C	Amino acids	1.12	−21.29	2960 ± 140

[1] Expressed with respect to PDB standard. Error on values ± 0.2 °/oo.
[2] Expressed with respect to 0.95 NBS oxalic acid standard (t½ = 5568 years) normalized to −25.00 °/oo $\delta^{13}C$.

was evaporated to dryness on a rotary evaporator and the solids were directly burned. The ^{14}C measurements were carried out as described in Taylor (1975).

Our data continue to support the early view that the carbonate fraction in bone yields highly unreliable results. In this case, the carbonate ^{14}C "ages" are nearly 2000 years younger than the collagen and amino acid fractions. By contrast the values obtained on all organic fractions are statistically identical and seem to reflect the actual age of the burial based on archaeological criteria including radiocarbon values on charcoal from other sites on the same time horizon.

CONCLUSION

The only generalization that can be supported concerning the reliability of either bone or marine shell ^{14}C values is that one cannot generalize. It is clear that both sample types can provide reliable values which can be compared directly with standard, terrestrial, organic sample types. Published data also exist which indicate that gross anomalies can and do occur. The question is which procedures to use in obtaining reliable values and which criteria to employ in evaluating the results obtained.

In the case of marine shell, it seems clear that the carbonate fraction from well-preserved marine shells from open ocean environments, where upwelling effects have been studied and where stable carbon isotope values are available, can yield radiocarbon values which are reliable. The highly negative view of marine shell carbonates generated in the early years of radiocarbon research was overly pessimistic because undue weight was placed on a small number of samples from a single region. Unfortunately the integrity of

^{14}C activity in the organic component of shell (conchiolin) can be seriously compromised and cannot be used as a check on the carbonate values.

Before marine shell can be used on a routine basis for any region, it is necessary to document the nature of the upwelling effects in sufficient detail so that any anomalous patterns can be identified and appropriate corrections can be established. The study of the upwelling effects in the eastern Pacific, for example, highlights the need to employ different precautions and procedures depending on the coastal region from which samples are derived. Along the southern California coast, an essentially constant −1.0% was noted. To the south, along most of the Mexican continental coastline, the effect increased to about −3.0%. In such cases it would be appropriate to add or subtract the relevant correction figure. In areas of relatively major upwelling (>5%) it would be necessary to add or subtract the correction value and, in addition, to increase proportionally the cited error value to compensate for the added uncertainty introduced by the marine environment. Such a step would be mandatory in areas where significant variation over short distances has been documented. It might also be appropriate to obtain duplicate sample analyses from the features, units, or sites to be dated. A suite of values which could be compared would quickly and accurately identify problems introduced by the use of samples from anomalous contexts.

Bone presents a more difficult problem. Since bone carbonates typically yield the most grossly variable values, the routine use of some portion or all of the bone organic fraction is indicated. Whether one is concerned with total organics, collagen, gelatin, or the component amino acids, the availability of sufficient sample to permit the ^{14}C assay to be made at an appropriate level of precision is a frequent problem. As previously noted, this problem becomes acute as the age of the sample increases. This situation points out the need to match research problems to appropriate instrumentation and counting procedures. ^{14}C detection systems can be operated so as to minimize the amount of sample required while maintaining the requisite statistical accuracy. The use of such systems permits the kind of fraction studies on bone samples which we have initiated. It is our view that bone samples can be routinely utilized in radiocarbon studies if proper pretreatment procedures are applied. We would suggest that the best insurance against unrecognized contamination would be the preparation of both the total organic or collagen fraction as well as the isolation of the component amino acids. If both organic fractions yield concordant ^{14}C values, the resultant value can be compared with determinations obtained from standard terrestrial organics.

REFERENCES

Berger, R., G. J. Fergusson, and W. F. Libby
    1965   UCLA Radiocarbon Dates IV. Radiocarbon. 7:336.
Berger, R., A. G. Horney, and W. F. Libby
    1964   Radiocarbon dating of bone and shell from their organic components. Science. 144:999-1001.
Berger, R., R. Protsch, R. Reynolds, C. Rozaire, and J. R. Sackett

1971 New radiocarbon dates based on bone collagen in California Paleoindians. Cont. of the Univ. of California Archaeological Research Facility. 12:43-49.

Berger, R., R. E. Taylor, and W. F. Libby
1966 Radiocarbon content of marine shells from the California and Mexican west coast. Science. 153:864-866.

Craig, H.
1957 The natural distribution of radiocarbon and the exchange time of carbon dioxide between atmosphere and sea. Tellus. 9:1-17.

Hakansson, S.
1972 University of Lund Radiocarbon Dates V. Radiocarbon. 14:380-400.

Hassan, A. A.
1975 Geochemical and mineralogical studies on bone material and their implications for radiocarbon dating. Dissertation. Southern Methodist University.

Haynes, C. V.
1967 Carbon-14 dating and early man in the New World. In Pleistocene extinctions: The search for a cause. P. Martin and W. Wright, jr., eds. New Haven, Yale University Press, 267-286.
1968 Radiocarbon analysis of inorganic carbon of fossil bones and enamel. Science. 161:687-688.

Hubbs, C. L., G. S. Bien, and H. E. Suess
1965 La Jolla natural radiocarbon measurements IV. Radiocarbon. 7:66-117.

Krueger, H. W.
1965 The preservation and dating of collagen in ancient bones. In Proc. 6th Int. Conf. Radiocarbon and Tritium Dating. (CONF-650652) 332-337.

Kulp, J. L., Feely, H. W., and L. E. Tryon
1951 Lamont natural radiocarbon measurements I. Science. 114:565-568.

Kulp, J. L., L. E. Tryon, W. R. Wickelman, and W. A. Snell
1952 Lamont natural radiocarbon measurements II. Science. 116:409-414.

Libby, W. F.
1952 Radiocarbon Dating. Chicago, University of Chicago Press.
1955 Radiocarbon Dating. 2d ed. Chicago, University of Chicago Press.

Longin, R.
1971 New method of collagen extraction for radiocarbon dating. Nature. 230:241-242.

Olson, E. A., and W. S. Broecker
1958 Sample contamination and reliability of radiocarbon dates. Trans. of the New York Acad. of Sciences. 20:593-604.
1961 Lamont Natural Radiocarbon Measurements VII. Radiocarbon. 3:141.

Protsch, R. R. R.
1973 The dating of upper Pleistocene subsaharan fossil hominids and their place in human evolution: With morphological and archaeological implications. Dissertation. University of California, Los Angeles.

Sellstedt, H., L. Engstrand, and N. G. Gejvall
1966 New application of radiocarbon dating to collagen residue in bones. Nature. 212:572-574.

Shotton, F. W., R. E. G. Williams, and A. S. Johnson
1975 Birmingham University radiocarbon dates IX. Radiocarbon. 17:255-275.

Suess, H. E.
    1954   U. S. Geological Survey Radiocarbon Dates I. Science. 120:467-473.
Tamers, M. A., and F. J. Pearson, jr.
    1965   Validity of radiocarbon dates on bone. Nature. 208:1053-1055.
Taylor, R. E.
    1970   Chronological problems in West Mexican Archaeology: An application of a dating systems approach in archaeological research. Dissertation. University of California, Los Angeles.
    1975   UCR radiocarbon dates II. Radiocarbon. 17:396-406.
Taylor, R. E., and R. Berger
    1967   Radiocarbon content of marine shells from the Pacific coasts of Central and South America. Science. 158:1180-1182.

4

Radiocarbon Activity in Submerged Plants from Various South Swedish Lakes

Sören Håkansson

ABSTRACT

The radiocarbon activity in submerged plants from lakes with more-or-less calcareous and noncalcareous surroundings has been determined and compared with the corresponding atmospheric activity. The depletion in ^{14}C compared to the atmosphere is of the same magnitude in the submerged plants from all but one of the lakes studied. The exception is a small, deep lake with noncalcareous surroundings, where the plants show a much larger depletion in ^{14}C than in the other lakes. This low activity may depend in part on a slow CO_2 exchange between the atmosphere and the lake water due to a large depth-to-surface ratio combined with good wind protection. Another possible explanation is that the lake may be fed mainly by old groundwater. Surface sediments from the deepest part of this lake have also been studied and samples from 1 to 7 cm below the sediment surface show a ^{14}C age of about 1400 years B.P.

The lack of significant differences in ^{14}C activity between the submerged plants from the different kinds of lakes studied indicates that the major part of the observed depletion in ^{14}C probably is caused by the relative slowness of the CO_2 exchange between the atmosphere and the lake water. This conclusion is based on the assumption that submerged plants in shallow, noncalcareous lakes show no significant apparent ^{14}C age under normal, pre-bomb conditions. The results also indicate that the apparent ^{14}C age for submerged plants from lakes with calcareous tills in the surroundings may be quite small or even absent (under pre-bomb conditions) if circumstances favor the exchange of CO_2 between the atmosphere and the lake water.

INTRODUCTION

The purpose of this study has been to gain information about the apparent age of organic lake sediments. It is well known that sediments from lakes in areas with calcareous bedrock near the surface or with highly calcareous till cover may be influenced by the so-called "hard water effect," yielding radiocarbon dates that are too old. In regions of known vegetation it is possible to determine the approximate extent of a hard water error by dating sediments from pollen-analytically determined characteristic levels, such as zone boundaries, comparing these with corresponding dates from terrestrial deposits (e.g., peat bogs). In lakes with undisturbed, continuous sediment deposition it is possible to date a sequence of samples from the youngest part of the sidement, extrapolating from the pre-bomb dates to the sediment surface in order to obtain the apparent age of the uppermost sediments. The water levels of many southern Swedish lakes were lowered within the last 100 years to create arable land. This lowering exposed older sediments to erosion by wave action. The uppermost sediments in such lakes are consequently a mixture of old and young material. In such cases it is not possible to extrapolate to the sediment surface in order to determine the apparent age. Total or partial deforestation of the area around a lake may have a similar effect on the exposed areas, with the increased wind exposure lowering the sedimentation level.

Other sources which have been used to determine the extent of hard water effects are submerged lake vegetation, algae, and carbon dioxide and bicarbonate in the water. These sources were used successfully in the beginning of the first radiocarbon decade (Deevey et al. 1954). Because of excessive ^{14}C in the atmosphere from nuclear weapon tests, it is no longer possible to use the activity values from these sources for a direct calculation of the apparent age, but measurement on such material may still give valuable information (Olsson et al. 1969: 541, Donner et al. 1971, Erlenkeuser and Willkomm 1971: 329; 1973: 114-115).

In order to make comparisons between the radiocarbon activity in submerged plants from lakes with hard and soft water, samples were collected from lakes with more or less calcareous versus noncalcareous surroundings. For the measurement of the corresponding atmospheric ^{14}C activity, suitable terrestrial plants were collected from some of the sites. For one particular lake (Lake Odensjön) the ^{14}C age of the top sediment from the deepest part was also determined.

Some facts relevant to the comparisons are presented in the following site descriptions.

Site Descriptions

LAKE AMMERN (OSTERGOTLAND)

There is calcareous till around the lake. The pH ca. 8 - 8.5 in the summer. Lake area 8.8 sq km. Maximum depth 18 m. Altitude 86.1 m. The lake water level was lowered in 1873. For radiocarbon dates on sediments, see *Radiocarbon* 1975, 17:180-181.

LAKE STRIERN (OSTERGOTLAND)

There is calcareous till around the lake. The pH ca. 8 - 9 in the summer. Lake area 1.5 sq km (1.7 x 1.1 km). Maximum depth 1.6 m. Altitude 87.3 m. The lake water level was lowered 2.5 m ca. 100 years ago. For radiocarbon dates on sediments, see *Radiocarbon* 1970, 12:541-543; 17:181-182.

LAKE VAN (OSTERGOTLAND)

The calcareous till around the lake has lower Ca content than the tills around Lake Ammern and Lake Striern. Lake area ca. 0.1 sq km (350 x 300 m). Maximum depth more than 4 m but probably less than 5 m. Altitude 92.4 m. For radiocarbon dates on sediments, see *Radiocarbon* 1974, 16:315-317. Information about the three lakes above was provided by H. Göransson (Dept. of Quaternary Geology, Univ. of Lund), who collected the samples from these lakes and the plant samples from Lake Odensjön.

LAKE HINNASJON (SMALAND)

There is no calcareous till or bedrock in the catchment area of the lake. The pH ca. 6.5. Lake area 0.3 sq km (1.6 x 0.3 km). Maximum depth 2.7 m. Altitude 170 m. For radiocarbon dates on sediments, see *Radiocarbon* 1975, v. 17:177-178. Information about this lake is from a study by Jensén (1974). The samples were collected by T. Persson (Dept. of Quaternary Geology, Univ. of Lund) and H. Göransson.

LAKE ODENSJON (NORTHWESTERN SCANIA)

The lake basin is eroded in crystalline bedrock (gneiss). The pH of surface water 7.6 in the middle of May 1976. The pH of water from the uppermost sediment from the deepest part of the lake 7.0. Lake area 0.015 sq km (165 x 130 m). Maximum depth ca. 20 m. Altitude 60 m. The lake has steep shores, more than 30 m high, along ca. 80% of the circumference, and is completely surrounded by forest. It is not exposed to the wind. The lake is fed almost entirely by groundwater. The surface sediment samples were collected from this lake with a 1 m Mackereth corer (Mackereth 1969).

HAKULLS MOSSE (NORTHWESTERN SCANIA)

This former lake is now a water-filled peat-cutting in a small bog (200 x 80 m). The till around the bog is sandy-silty and probably noncalcareous. The pH ca. 5.4 in summer (Müller, 1976). "Lake" area ca. 0.01 sq km. Maximum depth 1.3 m. Altitude 125 m. The biostratigraphy was studied by Berglund (1971) and samples for radiocarbon dating were collected in 1975.

LAKE BORRINGESJON, NORTHERN PART (SOUTHERN SCANIA)

The till around the lake and in the catchment area is mainly calcareous. The pH 9.1 near the outlet early in May 1976. Lake area 0.3 sq km (0.8 x 0.6 km). Altitude 49 m. The sample was taken just below the outlet in the rivulet connecting the two parts of the lake.

THE LAKE AT AMOSSEN (SOUTHERN SCANIA)

The till around the lake is calcareous. The pH 9.2 early in May 1976. Lake area ca. 0.03 sq km (300 x 100 m). Maximum depth ca. 3.5 m. Altitude 50 m.

Radiocarbon Activity

The results are given as a difference (Δ) from our radiocarbon standard (95% of the activity of NBS oxalic acid standard, age corrected to 1958):

$$\Delta = \delta^{14}C - (2\delta^{13}C + 50)\left(1 + \frac{\delta^{14}C}{1000}\right)$$

where $\delta^{14}C$ is the observed deviation from the radiocarbon standard in per mil and $\delta^{13}C$ the deviation from the PDB standard in per mil. For sediment samples the ^{14}C age B.P. (before 1950) is also calculated. The conventional half-life for ^{14}C of 5568 years is used.

RESULTS AND CONCLUSIONS

Results of the measurements are given in table 1 for plant samples and in table 2 and figure 2 for sediment samples (only Lake Odensjön). All Δ values for plant samples are presented in diagram form in figure 1, which shows that only Lake Odensjön deviates significantly from the rest. The low ^{14}C activity in this lake may depend in part on slow CO_2 exchange between the atmosphere and the lake water due to a large depth to surface ratio in combination with good wind protection. Another possible explanation is that the lake may be fed mainly by old groundwater. From the values for the surface sediments (table 2 and figure 2) it is obvious that we are here dealing with a considerable apparent ^{14}C age. The sediments from 1 to 7 cm below the sediment surface, relatively unaffected by the nuclear weapon test effect, have an average ^{14}C age of ca. 1400 years. The real age of these sediments is probably less than 200 years.

The difference $\Delta_2 - \Delta_1$ presented in table 3 and figure 3 is the difference in ^{14}C activity between terrestrial plants and submerged collected at about the same time. This difference represents approximately the ^{14}C depletion in submerged plants compared to the atmosphere for the year of collection. For those sites where no terrestrial plants are collected, the average Δ_2 value for the other sites is used for the calculation of this difference. For 1974 the average Δ_2 is 426.3 ± 6.5 (the somewhat deviating value for Lu-1027 omitted), and for 1975 it is 390.7 ± 6.5. In figure 3, where the deviating values for Lake Odensjön are omitted, it is clear that the depletion in ^{14}C compared to the atmosphere is of the same magnitude in the submerged plants from all the lakes studied except Lake Odensjön. There is no significant difference between lakes with calcareous and lakes with noncalcareous surroundings, but it must be remembered that none of the studied calcareous lakes is of the extreme hard water type found in areas with outcropping, calcareous bedrock. Each is situated in an area with only more or less calcareous till cover. Lack of significant difference between the studied calcareous and noncalcareous lakes indicates that the major part of the observed depletion in ^{14}C probably is caused by the relative slowness of the CO_2 exchange between the atmosphere and the lake water. This conclusion is based on the assumption that submerged plants in shallow, noncalcareous lakes show no significant apparent ^{14}C age under normal, pre-bomb conditions (Deevey et al. 1954: 287). The results also indicate that the apparent ^{14}C age for

TABLE 1

Radiocarbon Activity in Recent Plants from Various Swedish Lakes

Lab. no.	Locality lake	Lake area sq km	Approx. max. depth m	Type of lake	Approx. pH	Plant species	Date of coll.	$\delta^{13}C$ vs. PDB $^o/oo$	$\Delta_1 \pm \sigma$ Submerged plants $^o/oo$	$\Delta_2 \pm \sigma$ Terrestrial plants $^o/oo$	$\Delta_2 - \Delta_1{}^* \pm \sigma$ $^o/oo$
Lu-1184	Ämmern	8.8	18	calcareous	8.3	*Elodea canadensis*	Sept. 75	-17.2	264.9 ± 6.5		125.8 ± 9.2
Lu-1186	Ämmern					*Potamogeton perfoliatus*	Sept. 75	-17.4	263.9 ± 6.2		126.8 ± 9.0
Lu-1011	Striern	1.5	1.6	calcareous	8.5	*Myriophyllum spicatum*	Sept. 74	-20.0	297.8 ± 6.7		128.5 ± 9.4
Lu-1012	Striern					*Carex* sp.	Sept. 74	-26.3		426.3 ± 6.6	
Lu-1185	Striern					*Myriophyllum spicatum*	Sept. 75	-19.8	284.7 ± 6.3		106.0 ± 9.1
Lu-1015	Vån	0.1	>4	calcareous		*Elodea canadensis*	Sept. 74	-17.6	287.9 ± 6.2		137.9 ± 9.1
Lu-1016	Vån					*Carex elata*	Sept. 74	-27.4		425.8 ± 6.6	
Lu-1187	Vån					*Elodea canadensis*	Sept. 75	-18.2	260.3 ± 7.1		130.4 ± 9.6
Lu-1181	Hinnasjön	0.3	2.7	non-calcar.	6.5	*Myriophyllum alterniflorum*	Sept. 75	-24.3	270.4 ± 6.3		118.4 ± 9.1
Lu-1182	Hinnasjön					*Molinia caerulea*	Sept. 75	-27.9		388.8 ± 6.5	
Lu-1009	Odensjön	0.015	20	non-calcar.	7.5	*Myriophyllum alterniflorum*	Sept. 74	-18.7	153.2 ± 6.0		273.6 ± 8.9
Lu-1010	Odensjön					*Carex* sp.	Sept. 74	-27.4		426.8 ± 6.5	

TABLE 1 *(continued)*

Lab. no.	Locality lake	Lake area sq km	Approx. max. depth m	Type of lake	Approx. pH	Plant species	Date of coll.	$\delta^{13}C$ vs. PDB $^o/oo$	$\Delta_1 \pm \sigma$ Submerged plants $^o/oo$	$\Delta_2 \pm \sigma$ Terrestrial plants $^o/oo$	$\Delta_2 - \Delta_1{}^* \pm \sigma$ $^o/oo$
Lu-1183	Odensjön	0.015	20	non-calcar.	7.5	*Myriophyllum alterniflorum*	Sept. 75	−20.4	123.4 ± 5.8		267.3 ± 8.7
Lu-1162	Håkulls mosse	0.01	1.3	non-calcar.	5.4	submersed brown-mosses	Sept. 75	−25.9	247.7 ± 6.1		143.0 ± 8.9
Lu-1019	N. Börringesjön	0.3	–	calcareous	9.1	*Sium* sp.	Oct. 74	−29.5	283.4 ± 6.4		142.9 ± 9.1
Lu-1026	Åmossen	0.03	3.5	calcareous	9.2	*Ceratophyllum* sp.	Oct. 74	−24.5	287.4 ± 6.4		128.5 ± 9.1
Lu-1027	Åmossen					*Carex* sp.	Oct. 74	−27.4		415.9 ± 6.5	
Lu-1178	Åmossen					*Ceratophyllum* sp.	Oct. 75	−22.4	255.9 ± 6.2		137.2 ± 9.2
Lu-1179	Åmossen					*Carex* sp.	Oct. 75	−29.4		393.1 ± 6.8	
Lu-1180	Måryd					*Juncus* sp.	Oct. 75	−28.0		390.2 ± 6.5	

*For those sites where no measurements on terrestrial plants were made, the average Δ_2 value for the other sites was used for the calculation of $\Delta_2 - \Delta_1$.
For 1974 the average Δ_2 is 426.3 ± 6.5.
For 1974 the average Δ_2 is 390.7 ± 6.5.

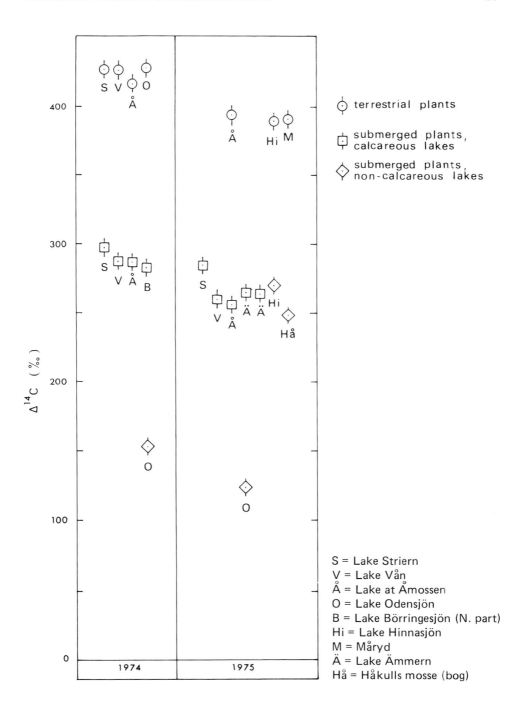

FIG. 1. ^{14}C activity above the standard level in recent plant samples.

TABLE 2
Radiocarbon Age of Surface Sediments from Lake Odensjön, Collected in Sept. 1974

Lab no.	Depth below sediment surface cm	$\delta^{13}C$ vs. PDB °/oo	$\Delta \pm \sigma$ °/oo	^{14}C age B.P. $\pm \sigma$ $T_{1/2}$ = 5568 yr.
Lu-1241	0-1	−29.4	135.8 ± 12.8	1170 ± 120
Lu-1242	1-3	−29.0	150.7 ± 10.8	1310 ± 105
Lu-1243	3-5	−29.4	164.6 ± 9.3	1440 ± 90
Lu-1244	5-7	−29.0	159.9 ± 9.9	1400 ± 95

The ^{14}C ages are corrected for deviations from $\delta^{13}C$ = −25.0°/oo in the PDB scale.

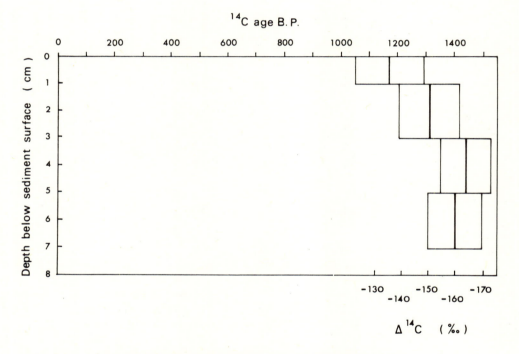

FIG. 2. Radiocarbon age of surface sediments from Lake Odensjön.

submerged plants from lakes surrounded by calcareous tills may be quite small or perhaps even absent (under pre-bomb conditions) if circumstances favor CO_2 exchange between the atmosphere and the lake water. If the tills are strongly calcareous, as they may be for some period after the deglaciation of an area, the submerged vegetation and the organic sediments in the lakes probably will be markedly influenced. The oldest organic sediments

TABLE 3

$\Delta_2 - \Delta_1$ Arranged According to Year of Collection and Type of Lake

1974					1975			
Calcareous surroundings		Non-calcar. surroundings			Calcareous surroundings		Non-calcareous surroundings	
Lake	$\Delta_2 - \Delta_1 \pm \sigma$ ⁰/₀₀	Lake	$\Delta_2 - \Delta_1 \pm \sigma$ ⁰/₀₀		Lake	$\Delta_2 - \Delta_1 \pm \sigma$ ⁰/₀₀	Lake	$\Delta_2 - \Delta_1 \pm \sigma$ ⁰/₀₀
Striern	128.5 ± 9.4	Odensjön	273.6 ± 8.9		Striern	106.0 ± 9.1	Odensjön	267.3 ± 8.7
Vån	137.9 ± 9.1				Vån	130.4 ± 9.6	Hinnasjön	118.4 ± 9.1
Åmossen	128.5 ± 9.1				Åmossen	137.2 ± 9.2	Håkulls mosse	143.0 ± 8.9
N. Börringesjön	142.9 ± 9.1				Ammern	125.8 ± 9.2		
					Ammern	126.8 ± 9.0	Average excluding Lake Odensjön	130.7 ± 17.4
Average	134.5 ± 7.2				Average	125.2 ± 11.7		

$\Delta_2 - \Delta_1$ is the difference in ^{14}C activity between terrestrial plants and submerged plants.

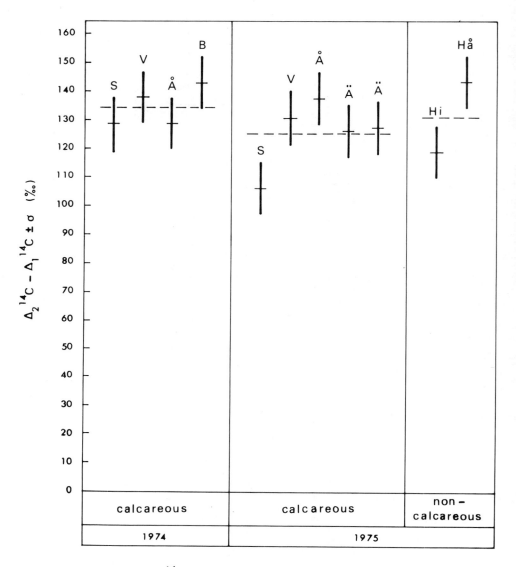

FIG. 3. The difference in ^{14}C activity ($\Delta_2 - \Delta_1 \pm \sigma$) between terrestrial plants and submerged plants from some Swedish lakes. The broken lines show the average values. The same site designations as in figure. 1.

in Lake Striern seem to be thus influenced (Håkansson 1970: 541-543), but it is difficult to determine the extent of the hard water error since the chronology for deglaciation of this area probably needs revision. In view of this, care must be exercised in generalizing from these results.

REFERENCES

Berglund, B. E.
  1971 Late-glacial stratigraphy and chronology in South Sweden in the light of biostratigraphic studies on Mt. Kullen, Scania. Geol. Fören. Stockholm Förh. 93:11-45.

Deevey, E. S., jr., M. S. Gross, G. E. Hutchinson, and H. L. Kraybill
  1954 The natural ^{14}C contents of materials from hard-water lakes. Proc. National Acad. Sci. 40:285-288.

Donner, J. J., H. Jungner, and Y. Vasari
  1971 The hard-water effect on radiocarbon measurements of samples from Säynäjälampi, northeast Finland. Comm. Physico-Mathematicae. Helsinki. 41:307-310.

Erlenkeuser, H. and H. Willkomm
  1971 University of Kiel Radiocarbon Measurements VI. Radiocarbon. 13:325-339.
  1973 University of Kiel Radiocarbon Measurements VII. Radiocarbon. 15:113-126.

Håkansson, S.
  1970 University of Lund Radiocarbon Dates III. Radiocarbon. 12:534-552.

Jensén, S.
  1974 Hinnasjön. Bottentopografi: makrovegetation och växtbiomassa. Medd. från Avd. för Ekologisk Botanik. Lunds Univ. 2 (8).

Mackereth, F. J. H.
  1969 A short core-sampler for sub-aqueous deposits. Limnol. Oceanogr. 14:145-151.

Müller, C.
  1976 Håkulls mosse och Mölle mosse: två "mossjöar" på Kullaberg. Aquannalen (Lund). 11 (1):37-39.

Olsson, I. U., S. El-Gammal, and Y. Göksu
  1969 Uppsala Natural Radiocarbon Measurements IX. Radiocarbon. 11:515-544.

PART VI
Oceanography

1

^{14}C Activity of Arctic Marine Mammals

Henrik Tauber

ABSTRACT

The ^{14}C activity of a series of recent bones of arctic marine mammals has been measured. When normalized to $\delta^{13}C = 0°/oo$, the apparent ages are 30-190 years. This is similar to, to, or slightly higher than, apparent ages of marine shells from the same regions when normalized in the same way. When not corrected for isotopic fractionation, ^{14}C dates of bones of arctic marine mammals are approximately 300 years too old compared to contemporaneous terrestrial plant material. Bones of human beings who subsisted on a mixed terrestrial and marine diet also show considerable apparent ages.

INTRODUCTION

The troposphere of each hemisphere is well mixed within a few months. This ensures a uniform content of naturally produced ^{14}C in contemporaneous terrestrial plants and animals, if correction is made for isotopic fractionation during the processes of carbon fixation and metabolism.

The mixing of the oceans, on the other hand, is a relatively slow process, permitting considerable differences in the ^{14}C content of various water masses. Nonetheless, except in certain regions characterized by upwelling of deep water, investigations have established that the bicarbonate of the mixed surface layers of the Atlantic, the Indian, and the Pacific north of about 40°S has a fairly stable ^{14}C content (Fonselius and Ostlund 1959, Broecker et al. 1960, Broecker and Olson 1961, Bien et al. 1963, Bien and Suess 1967). At the present rate of ocean circulation this ^{14}C content, if uncorrected for isotopic fractionation relative to terrestrial material, is close to the ^{14}C content of terrestrial plants. The increase in ^{14}C activity due to isotopic fractionation during the formation of bicarbonate, and the decrease due to mixing with deep water, almost cancel. This fact has long been utilized in the dating of marine shells which are in close isotopic equilibrium with bicarbonate of he surrounding waters.

Marine plants and marine animals at various stages of the biological food chain are not in isotopic equilibrium with the ocean bicarbonate, but show deviations in $\delta^{13}C$ of approximately $-10°/oo$ to $-20°/oo$ relative to the PDB-standard (Craig 1953, 1954). The ^{14}C dating of such materials needs some form of normalization in order to be strictly comparable to dates of terrestrial materials.

At arctic sites where wood is scarce, bones of marine mammals are often submitted to dating. However, only in a few instances have the ^{13}C and the ^{14}C content of bones of marine mammals been measured against samples of known age. Marine mammals may, moreover, migrate over large stretches of the oceans where variations in ^{14}C activity may have occurred. For these reasons a series of samples, including bones of seal, walrus, blue whale, and polar bear of known age and collected prior to the time of nuclear weapon tests was secured from the Zoological Museum, University of Copenhagen.

Methods

The samples are described in table 1. Also included are bone samples of two humans who lived on the southeastern coast of Greenland in the eighteenth century (Mathiassen 1933, 1936). These Eskimos had no contact with Europeans, could grow no plant crops, had no higher terrestrial animals to hunt, and so subsisted almost entirely on sea foods (i.e., marine animals, and possibly a few birds, most of which fed on fishes). Their isotopic composition, therefore, should be similar to that of marine mammals in the same region.

Collagen from the bones was isolated as gelatin by the extraction method of Longin (1971). After combustion, $\delta^{13}C$ was determined relative to the PDB standard by mass spectrometer analysis with a reproducibility better than $0.2°/oo$ and an overall uncertainty of about $1°/oo$. The age was corrected (T½ = 5730 yr). $\delta^{14}C$ was measured in a proportional gas counter relative to 0.95 times the activity of the NBS oxalic acid stan-

TABLE 1

Description of Samples

Sample	Provenience	Description
K-346	W-Greenland	Bones of Harp Seal (Pagophilus groenl.), died 1886. CN 13.
K-347	N-Greenland	Bones of Walrus (Odobenus rosmarus), died 1915 at Thule (76°33'N;68°46'W). CN 401.
K-348	E-Greenland	Bones of Polar Bear (Ursus maritimus), died 1932 at Kap Rink (68°21'N;28°44'W), Blosseville Coast. CN 2591.
K-349	Denmark	Baleen of Blue Whale (Sibbaldus muscul.), died 1931 at Aaro Sund (55°15'N;9°43'E), Lille Belt. CN 19.
K-350	E-Greenland	Bones of Eskimo (Homo sapiens), excavated 1932 at Grave 1, Kangartik (65°39'N;36°53'W), Angmagssalik, by Mathiassen (1933). Estimated age 1750 ± 50 A.D.
K-351	E-Greenland	Bones of Eskimo (Homo sapiens), excavated 1932 at Grave 3, Ruinnæsset (63° 16'N; 41° 7'W), Skjoldungen, by Mathiassen (1936). Estimated age 1750 ± 50 A.D.

dard in 1950. The ^{14}C deviations were then normalized to δ^{13}C = -25 ‰ by the equation (Broecker and Olson 1961)

$$\Delta = \delta^{14}C - 2(\delta^{13}C + 25)\left(1 + \frac{\delta^{14}C}{1000}\right) ‰ \qquad (1)$$

and Δ‰ values were finally corrected for the industrial lowering of the ^{14}C activity in sea water, which was assumed to proceed exponentially from 0 ‰ in 1890 to -10 ‰ in 1950 (Broecker and Olson 1961, Mangerud and Gulliksen 1975). These results are given as ∇ ‰ in table 2. It is seen that δ^{13}C values fall within the normal range for marine animals with a mean depletion relative to sea water of about -15‰. This also applies to the two samples of human bones.

From the ∇ ‰ values, apparent ages (T½ = 5570 yr.) at time of death have been calculated (table 3). For K-350 and K-351, a correction for secular variations in the atmospheric ^{14}C activity as given by the smooth curve by Damon et al. (1973), was applied before this calculation. In table 3 are included two bone samples of polar bear measured in Lund, Sweden, Lu-715 from Kapp Wijk (78°30' N, 15°0' E) at Spitzbergen, and Lu-779 from Kap Stephensen (68°25' N, 28°31' W) on the eastern coast of Greenland (Håkansson 1974). The last sample has here been corrected thirty years for the influence of industrial effect.

It is seen that, if normalized to δ^{13}C = -25‰, the apparent ages of the samples become 430-590 years. This is in accordance with the values found for marine shells from the same regions when normalized in the same way (Mangerud 1972, Hjort 1973, Mangerud and Gulliksen 1975).

In the great majority of ^{14}C laboratories ^{14}C measurements of marine shells are directly or indirectly normalized to δ^{13}C = 0 ‰, that is, to the mean value fround in ocean bicarbonate. In this case apparent ages of marine shells from the North Atlantic become 0-70 years, except for regions of northern Norway, the east coast of Greenland,

TABLE 2

Measurements on Recent Samples of Arctic Marine Mammals, and Humans

Sample	Year of death	δ^{13}C ‰ PDB	δ^{14}C ‰ age corr.	Δ ‰	∇ ‰ corr. ind. eff.
K-346	1886	-16.1	-41 ± 6	-58	-58 ± 6
K-347	1915	-10.5	-45 ± 6	-73	-71 ± 6
K-348	1932	-14.5	-36 ± 6	-56	-52 ± 6
K-349	1931	-17.2	-44 ± 6	-59	-55 ± 6
K-350	1750 ± 50	-13.0	-39 ± 8	-62	-62 ± 8
K-351	1750 ± 50	-12.6	-37 ± 6	-61	-61 ± 6

Measurements are relative to PDB-standard of ^{13}C and 0.95 the ^{14}C content of the NBS oxalic acid standard in 1950. ∇ ‰ are deviations corrected for industrial effect.

TABLE 3

Apparent Age of Samples of Arctic Mammals at Time of Death

Sample	Year of death	$\delta^{13}C$ ‰	Apparent age ($T_{1/2}$ = 5570)		
			Normalized to		Un-normalized
			$\delta^{13}C = -25$	$\delta^{13}C = 0$	for $\delta^{13}C$
K-346	1886	−16.1	480 ± 50	80 ± 50	340 ± 50
K-347	1915	−10.5	590 ± 50	190 ± 50	350 ± 50
K-348	1932	−14.5	430 ± 50	30 ± 50	260 ± 50
K-349	1931	−17.2	450 ± 50	60 ± 50	330 ± 50
K-350	1750 ± 50	−13.0	570 ± 65	170 ± 65	320 ± 65
K-351	1750 ± 50	−12.6	560 ± 50	160 ± 50	300 ± 50
Lu-715	1900 ± 50	−14.6	480 ± 70	80 ± 70	310 ± 70
Lu-779	1932	−16.2	465 ± 45	65 ± 45	320 ± 45

Measured ^{14}C activities are corrected for age of specimens relative to 1950, for industrial effect, and for secular variations in atmospheric ^{14}C activity (K-350 and K-351).

and northwest Greenland, where the outflow of water from the Arctic Ocean may increase apparent ages to 100-300 years (Mangerud 1972, Hjort 1973, Krog and Tauber 1974, Tauber and Funder 1975; Mangerud and Gulliksen 1975).

When normalized to $\delta^{13}C = 0$‰, the apparent ages at death of bones of marine mammals become 30-190 years with the highest ages at northern and eastern Greenland. A part of the apparent ages may reflect the age of the animal (or human) itself at death and an accumulated age of the marine plants and animals in the various steps of the food chain. This would hardly exceed 10-30 years for the animals. The samples of polar bear (K-348, Lu-715, Lu-779), though originating from eastern Greenland and Spitzbergen, do not show increased apparent ages as is usual in these waters. This may reflect the eating habits of polar bears. They subsist predominantly on seals, but they may occasionally eat berries or partake of a musk-ox.

At present, most ^{14}C dates calculated for bones of arctic marine mammals, as published in *Radiocarbon* and elsewhere, have not been corrected for isotopic fractionation. As suggested by the figures in table 3, these dates are approximately 300 years too old when compared to the age of contemporaneous terrestrial plant material.

So far no international agreement has been reached concerning a standardized mode of normalization of ^{14}C measurements of marine shells, plants, and animals. It is suggested that ^{14}C dates of all samples that have ultimately derived their ^{14}C from oceanic bicarbonate be normalized to $\delta^{13}C = 0$‰ PDB. This would preserve continuity with the majority of existing dates of marine shells, and it would minimize the apparent ages of samples of marine plants and animals.

Bones from humans who have subsisted mainly on sea food may give considerable apparent ages, up to approximately 500 years, if the measurements have been normalized to $\delta^{13}C = -25^{0}/oo$, which is the common procedure for this type of material. Samples from humans who had subsisted on a mixed terrestrial and marine diet may therefore show intermediate apparent ages. If Calvin photosynthesis has formed the basis for the terrestrial nutrients, the magnitude of these apparent ages may be estimated from the $\delta^{13}C$ values. If Hatch-Slack or CAM photosynthesis (Troughton 1973) has prevailed in the primary terrestrial assimilation, $\delta^{13}C$ values will be insensitive to the mixing ratio.

ACKNOWLEDGMENTS

Thanks are due to Ulrik Møhl and J. Balslev Jørgensen, University of Copenhagen, for supplying the samples, and to C. Vibe, University of Copenhagen, for information on the life habits of arctic marine animals.

REFERENCES

Bien, G. S., N. W. Rakestraw, and H. E. Suess
 1963 Radiocarbon dating of the deep water of the Pacific and Indian Oceans. *In* Radioactive Dating. Vienna, I.A.E.A., 159-173.

Bien, G. and H. Suess
 1967 Transfer and exchange of ^{14}C between the atmosphere and the surface water of the Pacific Ocean. *In* Radioactive dating and methods of low-level counting. Vienna, I.A.E.A., 105-115.

Broecker, W. S., R. Gerald, M. Ewing, and B. C. Heezen
 1960 Natural radiocarbon in the Atlantic Ocean. J. Geophys. Res. 66:2903-2931.

Broecker, W. S., and E. A. Olson
 1961 Lamont radiocarbon measurements VIII. Radiocarbon, 3:176-204.

Craig, H.
 1953 The geochemistry of the stable carbon isotopes. Geochim. et Cosmochim. 3:55-92.

Craig, H.
 1954 Carbon-13 in plants and the relationship between carbon-13 and carbon-14 variations in nature. J. Geol. 62:115-149.

Damon, P. E., A. Long, and E. I. Wallick
 1973 Dendrochronological calibration of the carbon-14 time scale. Proc. 8th Int. Conf. Radiocarbon Dating, New Zealand, A 38-A 43.

Fonselius, S., and G. Ostlund
 1959 Natural radiocarbon measurements on surface water from the North Atlantic and the Arctic Sea. Tellus, 1:77-82.

Hjort, C.
 1973 A sea correction for East Greenland. Geol. För. Stockholm Förh. 95:132-134.

Håkansson, S.
 1974 University of Lund Radiocarbon Dates VII. Radiocarbon, 16:307-330.

Krog, H. and H. Tauber
   1974   C-14 chronology of late and postglacial marine deposits in North Jutland. Geol. Survey Denmark, Yearbook 1973:93-105.

Longin, R.
   1971   New method of collagen extraction for radiocarbon dating. Nature, 230:241-142.

Mangerud, J.
   1972   Radiocarbon dating of marine shells, including a discussion of apparent age of recent shells from Norway. Boreas. 1:143-172.

Mangerud, J. and S. Gulliksen
   1975   Apparent radiocarbon ages of recent marine shells from Norway, Spitzbergen, and Arctic Canada. Quat. Research, 5:263-273.

Mathiassen, Th.
   1933   Prehistory of the Angmagssalik Eskimos. Medd. om Grønland. 92(4):155.
   1936   The former Eskimo settlements on Frederik VI's Coast. Medd. om Grønland, 109 (2):55.

Tauber, H. and S. Funder
   1975   C-14 content of recent mollusks from Scoresby Sund, Central East Greenland. Geol. Survey of Greenland, Report No. 75, Report of Activities 1974:95-99.

Troughton, J. H.
   1973   Carbon isotope fractionation by plants. Proc. 8th Int. Conf. Radiocarbon Dating, New Zealand, E 39-E 57.

2

Environmental Effects on Radiocarbon in Coastal Marine Sediments

Helmut Erlenkeuser

ABSTRACT

Coastal marine sediments from different environments were investigated for natural radiocarbon isotope concentration: from the Fladenground area (northern North Sea), from the tidal flats off the western coast of Schleswig-Holstein, northern Germany, from Kiel Bight (western Baltic Sea), and from the Bornholm Basin (southern Baltic). Sediment sampling, pretreatment, and wet oxidation technique with sedimentary organic carbon are outlined.

Organic matter as well as carbonates of coastal marine sediments are found to be composed of differently aged fractions. Sediment surface ages range between 800 years in the Baltic and 5000 years in the Fladenground area. In spite of these redistribution effects, however, the effective ^{14}C-age of the sediments is highly reproducible for a given environment. Minor variations are caused by varying supply of recent organic matter. In recent time dead carbon from fossil fuels and man-made ^{14}C from the nuclear weapon tests have entered the sedimentary environment ("ash-effect" and "bomb-effect").

For the Bornholm sediments the age composition of the organic matter could be analyzed in greater detail. The recent carbon fraction varies by a factor of 2.6 during the last 70 years, from 0.5% to 1.3% of sediment dry weight, and may be as low as 10% of total C-org content. The "eroded" fraction amounts to 3.6% of dry weight, the effective age being about 1500 years, and fossil fuel carbon is about 0.2% to 0.3% on dry weight basis.

INTRODUCTION

Ocean sediments have recorded much information about the history of environmental effects by the accumulation and composition of deposits. The geological and geochemical data reflect climatological and hydrographical aspects, tell about interactions between the sea and sea bottom, and, for the more recent sediment layers, document the impact of man on the sea. Furthermore, the sediment is a sink and a source for nutrients and carbon, withdrawn from and later released to the water layer above. Accumulation rates must thus be taken into account in ecological models. In recent years there has begun a steadily increasing engagement of widely different disciplines into the investigation of coastal marine sediments and estuaries, because it is vital to gain detailed knowledge about the response of the sea to the growing discharge of pollutants.

Some of the aspects for the postglacial period, can be studied from the sediments of the North Sea and the Baltic. Dating by natural radiocarbon is an appropriate means of establishing the sediment chronology. Good time resolution can be achieved because accumulation is generally fast, at rates of about 0.1 to 3 mm/yr for fine-grained, argillaceous deposits. High organic production coincident with a rapid fixation of the organic debris in the sediments produces appreciable organic carbon concentrations.

We have learned that organic matter of these sediments does not originate entirely from the autochthonous production of the year of deposition but is composed of different age-fractions which are resuspended elsewhere and redeposited along with the autogenic fraction. The radiocarbon age depends on the mixing ratio of the different fractions which might easily vary since mixing processes in sedimentary environments are slow and local factors may become dominating. We have studied sediment cores from the North Sea and the Baltic in greater detail to gain knowledge about the reliability and the geologic relevance of radiocarbon in coastal sediments.

Methods

The sediments were collected with box corers ranging between 50 cm and a few meters in length. Large coring barrel cross-sections, about 200 to 500 cm^2, were used in order to reduce the risk of strata distortion during coring. For better time resolution, thin, closely spaced sediment slices were cut from the cores, with each sample providing sufficient material for radiocarbon analysis. Surface sediments were collected with short box corers, which are more easily handled on shipboard than are long coring barrels. Each core was X rayed and samples were taken for different chemical and sedimentological studies prior to sectioning the core for ^{14}C dating. The sediment slices were about 5 to 20 mm thick, properly spaced according to the radiographic stratification pattern.

The rims of the sediment slices were discarded in order to avoid sample contamination by displaced materials adhering to the core liner walls. Samples were then stored in a freezer to minimize contamination from atmospheric CO_2 fixed by microbial activity (Geyh et al. 1974), a potential problem for sediments with low carbon content.

In order to remove carbonates — if these were not to be dated separately — the samples were pretreated with 10% hydrochloric acid in a water bath at 100°C for one hour, dialyzed for about a week up to slightly below neutrality, centrifuged, and dried at 70°C to constant weight (Suess and Erlenkeuser 1975). Initial wet weights were typically about 50 to 500 g. Depending on the organic carbon content, 50 to 300 g of dry sample was introduced into a 2 l reaction glass flask with round bottom, along with about 100 ml of distilled water and the appropriate quantity of potassium dichromate oxidant. The latter was added in amounts providing a 5 to 10-fold excess of oxygen over the organic carbon to be oxidized. A dropping funnel with 200 to 250 ml of degassed, concentrated sulfuric acid was attached to one of the inlets of the reaction vessel.

The sample was stirred moderately throughout the following procedure using a commercial, leak proof stirrer. High purity grade nitrogen gas was used as a carrier for removing air and CO_2 from the reaction flask. During oxidation, CO_2 was flushed into a dry ice trap for water removal and was carried to a liquid nitrogen cooled CO_2 trap. In order to prevent air from entering through any undetected small leaks, the pressure in the reaction line was kept slightly above atmosphere (at about 200 to 400 mm water column) by means of a self-controlling pressure-reduction valve situated at the inlet of the CO_2 trap. The CO_2 was frozen under vacuum while the carrier gas was pumped away.

For purification the CO_2 was vacuum-distilled into liquid nitrogen cooled active charcoal and back-distilled under vacuum at temperatures slowly increasing to room temperature. The CO_2 yield of about 97% can be further increased by heating the charcoal slowly up to 100°C within 10 minutes.

The wet oxidation technique reduces the concentration of electronegative impurities in the CO_2 gas, since most of the electronegative oxides which affect the counting characteristics of CO_2 remain dissolved in the liquid phase of the reaction mixture. We have used this method for about four years, especially with sediment samples. No difference in counter operation voltage has been observed between CO_2 from "dirty" samples and counting gases from carbonates and oxalic acid or tank CO_2. The carbon-to-CO_2 conversion has been found to be quantitative for widely differing types of samples, such as coal, graphite, wood, sediments, or algae.

We did not attempt to isolate chemical fractions from the sedimentary organic carbon. Rapid diagenesis of freshly deposited organic debris results in a complex cross-linking of the different constituents and functional groups of organic compounds. The chemical similarity of organic matter of different age suggests that age-specific carbon compounds cannot be extracted from sedimentary organic matter.

Results

Sediment cores were analyzed for radiocarbon from different locations of the North Sea and the Baltic (fig. 1): (A) from the Fladenground area in the northern North Sea, (B) from the tidal flats off the coast of Schleswig-Holstein, northern Germany, (C) from Kiel Bight, western Baltic Sea, and (D) from the Bornholm Basin in the Baltic proper. The distribution of radiocarbon with depth reflects three major effects: (1) the redistribution of old organic carbon (redistribution-effect); (2) the supply of dead carbon of fossil fuel residues (ash-effect); and (3) the supply of bomb-produced radiocarbon (bomb-effect).

THE REDISTRIBUTION EFFECT

Kiel Bight Sediments

Shallow water seas are characterized by vigorous wave action and by bottom currents which cause older deposits to be resuspended at one locality and redeposited at another

FIG. 1. Sediment cores from North Sea and Baltic. See text for key to locations A–D.

site. The organic carbon of coastal sediments, therefore, contains variously aged fractions yielding a ^{14}C age correspondingly higher than the actual depositional age.

This feature is demonstrated in figure 2 for two sediment cores from Kiel Bight. The cores were taken at more than 20m of water depth where bottom currents are usually weak and favor the accumulation of fine-grained, argillaceous deposits. The sediments are black muds, rich in hydrogen sulfide and organic carbon (3% to 5% on dry-weight basis). An irregular stratification in the lithofacies indicates varying strength of bottom currents and wave action in Keil Bight.

Sediment accumulation proceeds linearly with time toward the surface, at a rate of 1.4 mm/yr for core KI-620 from a local basin at 28m water depth, and at 0.45 mm/yr for the upper part of core KI-1105 from Vejsmäs submarine channel at 27 m water depth where the bottom current regime is more pronounced.

Neglecting for the moment the ^{14}C-age variations immediately below the depositional interface, the sediment surface age as obtained by extrapolation is about 800 years

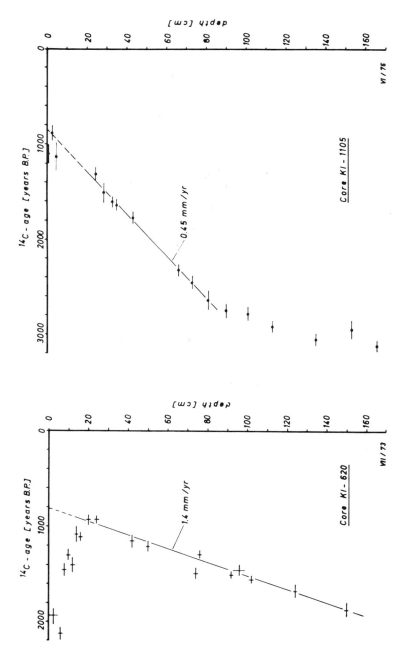

FIG. 2. Two sediment cores from Kiel Bight.

for both cores. These deposits are not as old as they appear to be from the ^{14}C dates. The low radiocarbon contents cannot be accounted for by fresh water effects. Detailed studies of water activities in the western Baltic and of shell segments of the slow-growing bivalve, *Cryprina islandica*, (Erlenkeuser et al. 1975) show that Keil Bight water activity in the preindustrial era was about 98% STD* for bottom water layer and 99% STD in the surface layer. The low radiocarbon content of the sediments demonstrates the existence of old organic carbon in modern deposits.

The scatter of ^{14}C dates around the mean sedimentation line is only slightly greater than calculated from counting statistics. It illustrates a high degree of reproducibility of radiocarbon concentrations in sediments.

Upon closer sample spacing some minor but significant ^{14}C variations became evident (fig. 3). These fluctuations occur both in the ^{14}C profile and in the organic carbon distribution (fig. 4). They result from a varying supply of recent organic carbon to the sediments, reflecting a varying rate of organic production in the water column or decomposition of recent matter at the bottom. Climatic and hydrographic variations may account for these phenomena (Niemisto and Voipio 1974). The variation of ^{14}C activity is related to the organic carbon content according to

$$\Delta A = \frac{A_{rec} - \hat{AC}}{\hat{c}} \Delta c$$

wherein

A_{rec} = recent activity in the water column

$\hat{AC}$ = measured activity, age corrected

$\hat{c}$ = total organic carbon content measured

provided Δc is the variation of the recent carbon fraction. For the present core ΔA amounts to 2% STD when the recent carbon fraction increases by 1% of sediment dry weight. Recent carbon variations appear to account for most small scale ^{14}C variations, although this rule cannot be applied unless the grain size distribution is known in order to allow for dilution effects by the minerogenic fraction.

Nordstrand Sediments

In order to better understand the depositional behavior of radiocarbon in the coastal marine environment, cores from the tidal flats of the North Sea, south of Nordstrand Isle near the coast of Schleswig-Holstein, were analyzed for carbon isotopes. Waves and tidal currents have caused the deposits to be thoroughly mixed and frequently redistributed. Marine sediments have been accumulating since Atlantic waters first entered this area.

*%STD: percentage of sample activity, normalized to $\delta^{13}C_{PDB}$ = −25 °/oo, as referred to 0.95 oxalic acid ^{14}C standard activity.

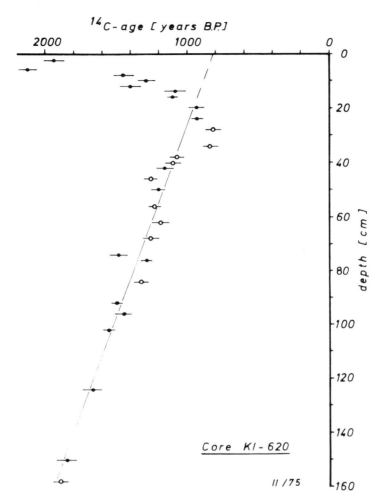

FIG. 3. Scatter of ^{14}C dates around mean sedimentation line.

The cores were drawn from two dredged pits situated one after the other, excavated in 1952 (core KI-973) and in 1962 (core KI-695), respectively. They silted up very quickly from the suspension load of tidal currents which flood the pits periodically. Tidal currents of varying strength produced a pronounced grain-sized stratification of winter versus summer layers of Nordstrand sediments. The annual accumulation rates can readily be determined from X ray photographs (Unsöld 1974).

Strong interaction between the sea and its deposits is reflected by high radiocarbon ages which are about 2500 to 3000 years for these modern deposits (figure 5). Residual variations of ^{14}C-age with depth may be understood on the basis of stable carbon isotope ratios and local sedimentation conditions. δ^{13}C values are between -19 °/oo and -23 °/oo

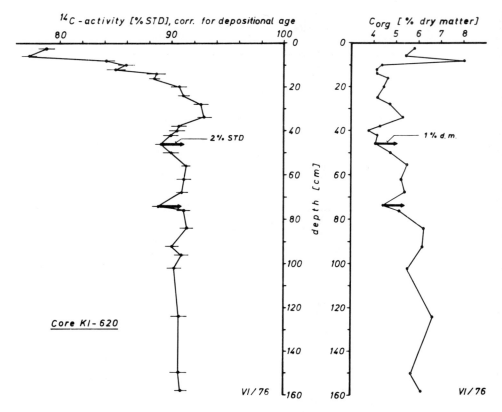

FIG. 4. ^{14}C profile (left) and C_{org} distribution (right).

on the PDB scale and are negatively correlated to ^{14}C activities, which tend to be lower where coarse grain deposition is favored. These findings indicate that the suspension load of the tidal currents contains coarsely-grained terrestrial plant litter of late Atlantic age, probably peat. It settles down in the pit at rates dependent on the local pattern of tidal currents, which have slowly changed as the sediment filling has risen to approach the mean high tide level. When corrected for terrigeneous additions and arranged for true depositional age instead of for depth, ^{14}C-age profiles result as shown in figure 6. The activities of samples of equal depositional age are about the same for both cores except for disturbances which occur in the early stage of the second pit dredged in 1962. The radiocarbon concentrations are strikingly uniform although, according to grain size distribution, widely differing weather conditions and tidal levels have occurred during the depositional history of these sediments.

A systematic activity difference of about 2.5% STD remains between the deeper sediment layers and the upper deposits from 1966 to 1971. This is probably due to the increasing supply of bomb-produced radiocarbon in the marine environment, particularly since 1963. Today water activities are about 122% STD. It is estimated, from the ^{14}C

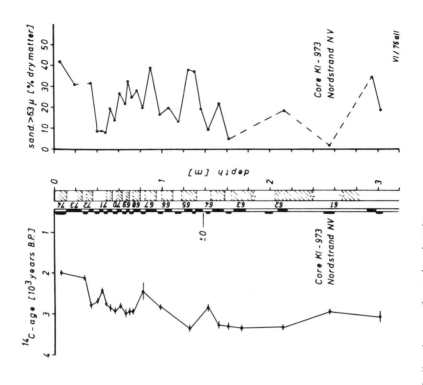
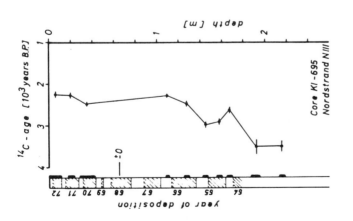

FIG. 5. Radiocarbon ages for modern deposits.

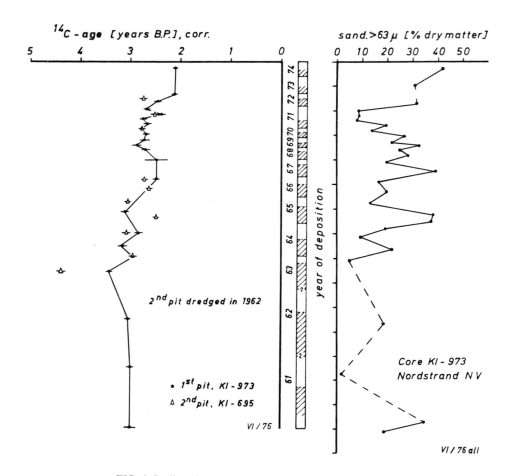

FIG. 6. Radiocarbon ages arranged for true depositional age.

increase in the sediments, that the annual contribution of recent organic matter comprises only about 10% of the total organic carbon of the sediments. In the uppermost layers of core KI-973 the recent carbon content is higher due to rootlets of pioneer plants at the sediment surface, which is now slightly above mean high tide level.

Fladenground Sediments

As further examples of the depositional behavior of ^{14}C in near-shore sediments, data from cores from two locations of the Fladenground area in the northern North Sea are presented. High sand contents indicate relatively strong prevailing bottom currents and low sedimentation rates (about 0.1 m/yr). The holocene sediment cover is extremely thin in core KI-958 (figure 7). Both carbonate and organic matter have been dated. The sediment surface age is about 1000 to 2000 years for the organic fraction and about 4000

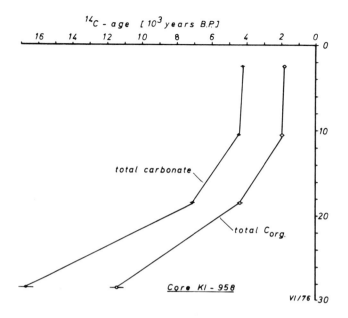

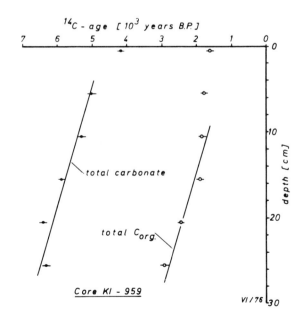

FIG. 7. Cores from Fladenground area, northern North Sea.

to 5000 years for the carbonate fraction although available measurements are insufficient for detailed analysis.

Radiocarbon dates obtained from the various sedimentary environments so far investigated give evidence for a well-developed geologic reproducibility of natural radiocarbon concentrations in the marine lithosphere. It appears that radiocarbon activities of the redistributed fraction are produced by numerous small-scale interactions between the sea and sea bottom. A statistically well defined mixing effect remains constant over long periods of depositional history. The statistical behavior of sediment redistribution seems better documented by carbon isotope composition than by grain size distribution, which is more readily affected and systematically varied by local hydrodynamic conditions.

The Ash-Effect

In the uppermost layers of Baltic Sea sediments, ^{14}C activities are greatly affected by the impact of man on the marine environment. Dead carbon from fossil fuels, along with a characteristic assemblage of certain heavy metals, dilutes the natural ^{14}C activities at rates gradually increasing toward the sediment surface (Erlenkeuser et al. 1974, Suess and Erlenkeuser 1975). The corresponding increase of the ^{14}C ages near the depositional interface is well documented in Kiel Bight sediments (figure 2) and occurs also, with some time delay, in the Bornholm Basin. In Kiel Bight the amount of dead carbon required to account for the observed increase of ^{14}C ages would be about 14% of the total organic carbon. This effect is too great to be ascribed to the well known dilution of atmospheric ^{14}C activity by release of gaseous CO_2 from fossil fuel burning (Suess- or Industrial-effect). We have suggested that particulate carbon and carbon residues of ash particles can be supplied to the coastal marine environments by washout from the atmosphere, by runoff from land, or by river discharge. The ash-effect may therefore be restricted to certain local environments, although transport processes by bottom currents may effect its dispersal over larger areas of the sea floor (Suess and Erlenkeuser 1975).

The Bomb-Effect

During the last decade, bomb-produced radiocarbon has entered the sea, providing an ideal tracer for identifying modern deposits. The complexity of radiocarbon activity distribution with depth in recent sediments is seen from a core (figure 8) taken from Bornholm Basin at 71 m water depth (Suess and Erlenkeuser 1975). The ^{14}C age distribution with depth demonstrates the combined influence of the redistribution-effect which makes the sediments about 1200 ^{14}C years old, of the ash-effect which starts developing at 14 cm of depth, and of the bomb-effect which is recognized in the uppermost 3 cm layer.

In order to analyze the ^{14}C age distribution observed, the total organic carbon is resolved into three fractions: (1) a recent carbon fraction, (2) an eroded carbon fraction, and (3) a fossil carbon fraction from fuel residues. The activity balance for a given sediment layer at the time of deposition is

$$c_{er}A_{er} + c_{rec}A_{rec} + c_{foss}A_{foss} = \hat{c}\hat{A}e^{T/\tau}. \qquad (2)$$

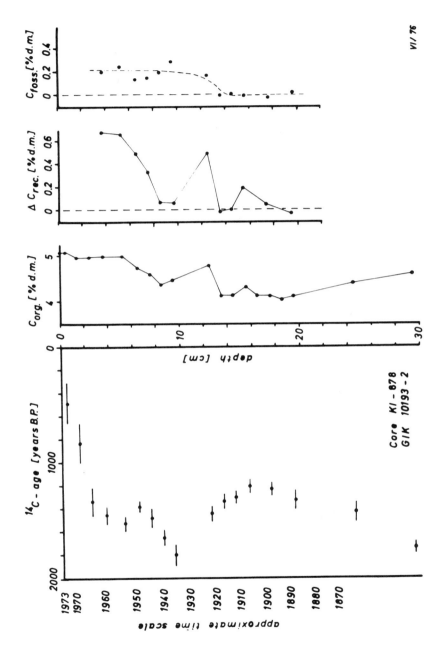

FIG. 8. Core from Bornholm Basin.

with

$$\hat{c} = c_{er} + c_{rec} + c_{foss}$$

c_{rec} = concentration, on dry weight basis, of recent organic carbon.

c_{er} = concentration of carbon, eroded elsewhere and redeposited.

c_{foss} = concentration of dead carbon from fossil fuels.

$\left. \begin{array}{l} A_{rec} \\ A_{er} \\ A_{foss} \end{array} \right\}$ = ^{14}C activities associated with the different constituents of the organic fraction, at the time of deposition. $A_{foss} = 0\%$ STD.

$\hat{A}$ = sediment activity as measured today.

T = depositional age of the sediment layer.

Geologic and sedimentological arguments may be used to derive additional assumptions which permit evaluation of equation (2). On the basis of isotopic reproducibility of organic matter redistribution processes, and in view of the fairly uniform facies of the Bornholm sediments, we may assume that the ^{14}C activity (A_{er}) and the concentration (c_{er}) of the eroded fraction are constant throughout the core and that total carbon variations are due to varying rates of recent and fossil carbon supply. Using carbon concentrations and ^{14}C activities at depths below the ash-effect for reference, the fossil carbon content (c_{foss}) and the excess (Δc_{rec}) over recent carbon at reference depth can be easily calculated. The fossil carbon content is about 4% or 5% of the total organic carbon (fig. 8) and remains constant above 10 cm of sediment depth. These results have been discussed in detail by Suess and Erlenkeuser (1975).

Recently the marine hydrosphere and the uppermost sediment layers have been affected by man-made radiocarbon from nuclear weapon tests in the atmosphere. Sedimentary response to the altered water activity may be used to estimate the molar fraction of recent carbon and, conversely, the activity of the eroded fraction. These data are of great interest from both geologic and biological points of view.

The analysis of sedimentary radiocarbon variations with time becomes difficult because the time scales of redistribution and the transport processes of particulate matter up to final fixation in the sediment are not known. The mobile, suspended particulate matter may form an intermediate reservoir, so the residence time of the organic carbon therein against final deposition has to be taken into account. As a first approximation, the time scales introduced by this hypothetical reservoir are assumed to be small as compared to the time scale of ^{14}C variation in the water during the bomb era. The activity of the sediment is assumed to be in a quasi-stationary state as to variations of the water activity and to follow the ^{14}C level in the hydrosphere without additional delay. Studies on modern activities of the Baltic Sea water and benthic samples (Erlenkeuser et al. 1975) suggest a residence time of about ten years for the CO_2 molecule in the surface water layer against exchange with the atmosphere. The box model calculations indicate a five-year water activity nearly constant at 142 %STD. Within this period of time the uppermost 10 to 15 mm of sediment were accumulated.

An additional assumption must be made with respect to the fossil carbon content of the upper bomb-era sediments. As suggested by the fossil carbon distribution (fig. 8), c_{foss} should remain essentially constant in the upper sediment layer at about 0.22% of dry weight. The bomb-effect in the sediments may then be analyzed in two ways. With the first method, represented in figure 9, the sediment at 4 cm of depth just below the bomb-effect is chosen for reference, obtaining the following results:

c_{rec} = 27% of $(c_{er} + c_{rec})$ = 1.3 % of dry weight

c_{er} = 73% of $(c_{er} + c_{rec})$ = 3.5 % of dry weight

c_{foss} (by extrapolation, see fig. 8) = 0.22% of dry weight

A_{er} = 82.8% STD   = 1510 years

These results indicate that the recent carbon is only a small fraction of the total organic matter in the sediments. The effective ^{14}C age of the eroded fraction is about 1500 years and seems to reflect a mixed activity of young deposits from the late holocene.

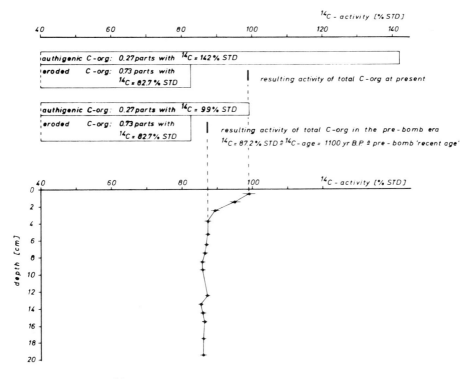

FIG. 9. Model of the ^{14}C-activities in different fractions of the organic matter in the sediments from the Bornholm Basin (Core KI-878 = GPI 10193-2)

With the second method, the bomb-era sediments are compared with the sediment layers at about 17 cm of depth which formed seventy years ago. The results are essentially the same as above, strongly supporting the model with regard to age composition of the sedimentary organic carbon.

c_{rec} (at 0-1 cm of depth) = 1.25% of dry weight
c_{rec} (at 17 cm of depth) = 0.5 % of dry weight
c_{er} = 3.6 % of dry weight
A_{er} = 83.9% STD = 1410 years.

It is interesting to note that the recent carbon fraction has varied by a factor of 2.6 during the last seventy years, from 0.5% of dry weight at about A.D. 1900 to 1.3% of dry weight in the modern deposits. A varying supply of recent organic carbon to Baltic Sea sediments has been proposed by Niemistö and Voipio (1974).

ACKNOWLEDGMENTS

This work was supported by the Deutsche Forschungsgemeinschaft within the program of the Sonderforschungsbereich 95. I wish to thank Dr. H. Willkomm for his stimulating interest and for our many discussions on problems of sediment dating. I am particularly indebted to colleagues from the Geologisch-Paläontologisches Institut und Museum of Kiel University, Erwin Suess, Gerhard Unsöld, and Friedrich Werner, for effective technical and scientific cooperation and stimulating discussion. The investigations on Flandenground sediments are part of the program of the Sonderforschungsbereich 94 of Hamburg University. I wish to acknowledge the expert assistance of staff members Monika Gumz, Heinz Finn, Horst Liebrenz, and Heidi Metzner, who processed the samples for carbon isotope analyses.
Contribution No. 132 of the Sonderforschungsbereich 95 "Interaction Sea/Sea Bottom," Kiel University.

REFERENCES

Erlenkeuser, H., E. Suess, and H. Willkomm
    1974   Industrialization affects heavy metal and carbon isotope concentrations in recent Baltic Sea sediments. Geochim. et Cosmochim. 38:823-842.

Erlenkeuser, H., H. Metzner, and H. Willkomm
    1975   University of Kiel radiocarbon measurements VIII. Radiocarbon. 17:276-300.

Geyh, M. A., W. E. Krummbein, and H. R. Kudrass
    1974   Unreliable ^{14}C dating of long-stored deep-sea sediments due to bacterial activity. Marine Geology. 17:M45-M50.

Niemistö, L., and A. Voipio
    1974   Studies on the recent sediments in the Gotland Deep. Merentutkimuslait. Julk./Havsforskningsinst. 238:13-23.

Suess, E., and H. Erlenkeuser
  1975 History of metal pollution and carbon input in Baltic Sea sediments. Meyniana. 27:63-75.

Unsöld, G.
  1974 Jahreslagen und Aufwachsraten in Schlicksedimenten eines künstlichen, gezeitenoffenen Sedimentationsbeckens (Wattgebiet südlich Nordstrand/Nordfriesland). (Annual layering and rates of accretion of muddy sediments in a man-made sedimentary basin of the tidal flats south of Nordstrand/Schleswig-Holstein). Meyniana. 26:103-111.

3

¹⁴C Routine Dating of Marine Sediments

Mebus A. Geyh

INTRODUCTION

Since publication of the first comprehensive paper on radiocarbon dating of marine sediments by Emery and Bray (1962), despite an increase in economic and scientific interest in marine research and in chronostratigraphic studies, dating methods have not improved. Only the problems of contamination of the carbonate fraction of marine sediments (Olsson and Eriksson 1965) and the influence of bacteria on the ^{14}C content of the organic carbon fraction (C-org) during storage of the cores have been studied (Geyh et al. 1974). A critical assessment seems overdue of the 279 available Hannover ^{14}C data of marine sediments obtained with routine runs. The results are compiled in table 1. These include designation of the cores and the water depths, coordinates of the sampling sites, dated fractions, core sections, and ^{14}C data. The $\delta^{13}C$ values, the C-org contents, and the sedimentation rates would complete the compilation if such data were available.

Sampling Sites and Description of the Analyses

The dated samples come from fifty-nine deep-sea cores taken with piston samplers of the gravity type from research ships F. S. Meteor, F. S. Valdivia, and Glomar Challenger during 1965-1974. Sampling areas (fig. 1) were the Atlantic Ocean between the Canary Islands and the coast of Portugal, the Mediterranean Sea, the Red Sea, the Persian Gulf, and the Arabian Sea. ^{14}C analyses were carried out on two fractions:

1. Carbonate fraction. The marine sediments were screened on 0.125 mm sieves using boiled distilled water. In some cases the organic carbon was removed by H_2O_2. The remaining carbonate fraction contained 50%-90% of foraminifers or coprolites.

2. (Acid-insoluble) organic carbon fraction. The carbonates of 100 to 1000 g of sediment were decomposed by HCl starting with a concentration of 10%. The acid-insoluble residue was dated. It represented ca. 30% of the total organic carbon of marine sediments.

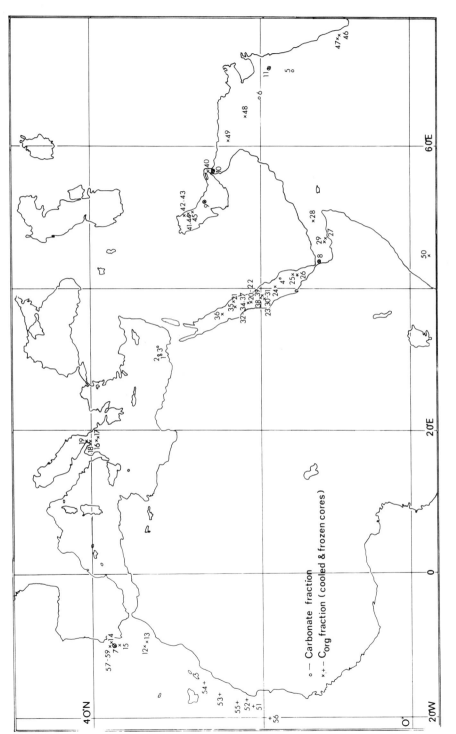

FIG. 1. Sampling sites of the cores. The numbers correspond to the core numbers of table 1.

TABLE 1

^{14}C-Dates and Estimated Sedimentation Rates of the Investigated Cores

No.	Name	Location	Coordinates	Hv	Frac.	Depth (cm)	$\delta^{13}C$ °/oo	Corg mg/g	RC date B.P.	Sed. rate mm/yrs	Top age B.P.
					A) ^{14}C-dates of the carbonate fraction						
1	N12 119 m	Mediter. Sea	31°02'N 30°39'E	1677	Fora.	5 8	+ 2.4		1995 ± 100	} 0.36 ± 0.11	+ 1820 ± 140
		Mediter. Sea		1678	Fora.	22 30	+ 2.2		2535 ± 135		
2	N14 328 m	Mediter. Sea	32°02'N 30°39'E	1674	Fora.	6 16	− 0.2		recent	0.36 ± 0.02	− 290 ± 110
		Mediter. Sea		1675	Fora.	92 103	+ 2.0		2645 ± 355		
		Mediter. Sea		1676	Fora.	180 190	+ 0.8		4730 ± 260		
3	N18 198 m	Mediter. Sea	32°02'N 31°48'E	1671	Fora.	1 4	+ 1.4		3545 ± 160	} 0.59 ± 0.13	+ 3500 ± 160
		Mediter. Sea		1672	Fora.	110 119			5545 ± 700		
		Mediter. Sea		1673	Fora.	191 195			6675 ± 870		
4	K71 1621 m	Red Sea	15°59'N 41°33'E	1665	Fora.	3 8	+ 1.4		1090 ± 185	0.25 ± 0.01	+ 900 ± 200
		Red Sea		1666	Fora.	58 63	+ 1.1		3040 ± 160		
		Red Sea		1667	Fora.	123 127	+ 1.3		5775 ± 185		
5	202 2700 m	Arabian Sea	13°33'N 70°50'E	2005	Fora.	60 70	+ 3.9		15120 ± 470	disturbed	
		Arabian Sea		2006	Fora.	169 183			13285 ± 1100		
		Arabian Sea		2007	Fora.	194 205			21460 ± 2060		
		Arabian Sea		2008	Fora.	294 306			14585 ± 405		
6	223 2650 m	Arabian Sea	20°02'N 66°51'E	1843	Fora.	30 40	+ 2.6		7125 ± 145	} 0.16 ± 0.02	+ 4950 ± 340
		Arabian Sea		1844	Fora.	87 95	+ 4.4		10610 ± 400		
		Arabian Sea		1845	Fora.	180 190	+ 3.2		20150 ± 780	0.09 ± 0.01	
		Arabian Sea		1846	Fora.	272 282	+ 3.1		16100 ± 1100		
		Arabian Sea		2002	Fora.	272 282			12490 ± 1075	contaminated	
		Arabian Sea		2004	Fora.	282 293			11890 ± 410		
		Arabian Sea		1847	Fora.	352 362	+ 0.9		25000 ± 1800		

B) Corresponding ^{14}C dates of the carbonate and Corg fractions of the same samples or cores

#	Station	Location	Coordinates	Sample	Fraction				Date	Ratio	+Date
7	8057B 2811 m	Atlantic Oc.	37°41'N 10°05'W	4139	Fora.	0	10		2465 ± 420	} 0.13 ± 0.03	+ 2080 ± 445
		Atlantic Oc.		4140	Fora.	140	150		13185 ± 2465	}	
		Atlantic Oc.		4141	Fora.	200	210		11390 ± 225	contaminated	
		Atlantic Oc.		4142	Fora.	280	290		8660 ± 1000		
	2900 m	Atlantic Oc.		3181	Corg.	0	3	1.4	2395 ± 420	} 0.15 ± 0.01	+ 1650 ± 340
		Atlantic Oc.		3183	Corg.	32	36	1.2	3515 ± 600	}	
		Atlantic Oc.		3184	Corg.	56	58	2.3	4890 ± 370	}	
		Atlantic Oc.		3185	Corg.	73	77	1.6	5450 ± 525	}	
		Atlantic Oc.		3186	Corg.	103	107	1.4	7950 ± 920		
		Atlantic Oc.		3187	Corg.	124	127	2.6	10650 ± 635		
		Atlantic Oc.		3188	Corg.	140	142	2.3	11160 ± 745		
		Atlantic Oc.		3190	Corg.	183	187	2.3	8955 ± 690		
		Atlantic Oc.		3858	Corg.	191	195	1.1	8610 ± 610	contaminated	
		Atlantic Oc.		3855	Corg.	219	223	1.9	12325 ± 495		
		Atlantic Oc.		3191	Corg.	284	287		10610 ± 725		
		Atlantic Oc.		3857	Corg.	299	303	3.8	13140 ± 650		
		Atlantic Oc.		3197	Corg.	344	348	1.7	8140 ± 460		
8	M42 960 m	Red Sea	12°03'N 43°46'E	1668	Fora.	1	10	+ 0.1	4885 ± 160	} 1.67 ± 0.32	+ 4870 ± 150
		Red Sea		1669	Fora.	92	98	+ 1.4	5470 ± 150	}	
		Red Sea		1670	Fora.	210	220	+ 1.3	6135 ± 175		
		Red Sea		3174	Corg.	1	5		950 ± 130	} 2.12 ± 0.44	+ 900 ± 110
		Red Sea		3175	Corg.	76	78		1220 ± 120	}	
		Red Sea		3176	Corg.	117	123		1445 ± 115		
		Red Sea		3177	Corg.	191	193		1840 ± 140		
9	1177 65 m	Persian Gulf	27°31'N 52°05'E	3143	Fora.	110	114		9975 ± 80	} influence of holes in the casing	
		Persian Gulf		3144	Fora.	110	114		10130 ± 95	}	
		Persian Gulf		3145	Fora.	110	114		10015 ± 65	}	
		Persian Gulf		3146	Fora.	110	114		10010 ± 110		
		Persian Gulf		3140	Fora.	254	269		10065 ± 65		
		Persian Gulf		3141	Fora.	325	340		10385 ± 70		

TABLE 1 (continued)

No.	Name	Location	Coordinates	Hv	Frac.	Depth (cm)		$\delta^{13}C$ ‰	C_{org} mg/g	RC date B.P.	Sed. rate mm/yrs	Top age B.P.
9		Persian Gulf		3142	Corg.	104	119		0.3	8295 ± 675	⎫	
		Persian Gulf		3139	Corg.	158	165			5975 ± 465	⎬ disturbed	
		Persian Gulf		3140	Corg.	254	269			2605 ± 1065	⎭	
10	1201	Persian Gulf	26°40'N 56°39'E	3147	Fora.	140	150			11805 ± 80	⎫	
	65 m	Persian Gulf		3151	Fora.	245	260			10450 ± 745	⎬ disturbed	
		Persian Gulf		3149	Fora.	310	325			8260 ± 375	⎭	
		Persian Gulf		3152	Corg.	40	44			2390 ± 330	⎫	
		Persian Gulf		3148	Corg.	190	195			6070 ± 400	⎬ 0.48 ± 0.01	+ 1685 ± 385
		Persian Gulf		3151	Corg.	245	260		0.5	8340 ± 1065	⎬	
		Persian Gulf		3149	Corg.	310	325		0.5	6810 ± 770	⎭	
11	210	Arabian Sea	17°41'N 70°57'E	1848	Copr.	30	40	+ 2.6		6560 ± 465	⎫	
	495 m	Arabian Sea		1849	Copr.	82	95			7460 ± 320	⎬	
		Arabian Sea		1850	Copr.	132	140			7260 ± 195	⎬ 0.78 ± 0.12	+ 5830 ± 315
		Arabian Sea		2009	Copr.	132	140			7450 ± 190	⎬	
		Arabian Sea		2010	Copr.	222	230			9440 ± 240	⎭	
		Arabian Sea		1851	Copr.	232	240	+ 4.2		7445 ± 425		
		Arabian Sea		2011	Copr.	270	278			22880 ± 670	⎫	
		Arabian Sea		1852	Copr.	332	337	+ 1.7		21895 ± 1425	⎬ 0.25 ± 0.05	
		Arabian Sea		2012	Copr.	386	395	+ 2.6		27420 ± 740	⎬	
		Arabian Sea		1853	Copr.	395	400		10.2	27750 ± 1485	⎭	
		Arabian Sea		2641	Corg.	2	3			4110 ± 125	⎫ 0.83 ± 0.08	+ 4230 ± 105
		Arabian Sea		2642	Corg.	50	60			5100 ± 125	⎭	
		Arabian Sea		2643	Corg.	172	180			6260 ± 145	⎫ 0.16 ± 0.01	
		Arabian Sea		3853	Corg.	402	407			19990 ± 635	⎭	

C) ^{14}C dates of the Corg fraction of long-term cooled, stored cores

No.	Name	Location	Coordinates	Hv	Frac.	Depth (cm)		$\delta^{13}C$ ‰	C_{org} mg/g	RC date B.P.	Sed. rate mm/yrs	Top age B.P.
12	8016D	Atlantic Oc.	33°50'N 9°43'W	3199	Corg.	5	8		1.1	128% mod ±7	⎫ 0.11 ± 0.02	− 540 ± 130
	4354 m	Atlantic Oc.		3201	Corg.	115	117		0.9	9140 ± 1200	⎭	
		Atlantic Oc.		3202	Corg.	226	229		0.5	>15000	⎫	
		Atlantic Oc.		3204	Corg.	286	289		0.7	3730 ± 765	⎬ disturbed	
		Atlantic Oc.		3251	Corg.	334	337		1.2	9155 ± 890	⎭	

#	Sample	Location	Lat	Lon								
13	8017B 3016 m	Atlantic Oc.	33°38'N	9°26'W	3196	Corg.	37	40	2.3	7050 ± 1000	}0.15 ± 0.05	+ 4620 ± 1690
		Atlantic Oc.			3856	Corg.	95	99	1.8	10745 ± 510		
		Atlantic Oc.			3179	Corg.	120	123	1.8	13630 ± 730	}0.07 ± 0.01	
		Atlantic Oc.			3180	Corg.	145	148	1.7	18685 ± 1625		
		Atlantic Oc.			3198	Corg.	177	180		6555 ± 1410		
14	8058B 1819 m	Atlantic Oc.	37°45'N	9°44'W	3182	Corg.	0	20	2.3	1820 ± 490	}0.46 ± 0.04	+ 1605 ± 505
		Atlantic Oc.			3859	Corg.	336	340		8875 ± 345		
		Atlantic Oc.			3192	Corg.	368	370		12960 ± 1250	}0.07 ± 0.02	
15	8066B 2710 m	Atlantic Oc.	37°00'N	9°48'W	3205	Corg.	0	10		2305 ± 300		+1950 ± 290
		Atlantic Oc.			3206	Corg.	74	76		4645 ± 490		
		Atlantic Oc.			3207	Corg.	144	146		8865 ± 530	}0.22 ± 0.02	
		Atlantic Oc.			3178	Corg.	214	216		9870 ± 630		
		Atlantic Oc.			3208	Corg.	268	270		14080 ± 1175	}0.12 ± 0.04	
16	OT28 930 m	Mediter. Sea	40°02'N	18°34'E	4000	Corg.	1	18	4.0	8305 ± 115	}0.13 ± 0.01	+ 7630 ± 130
		Mediter. Sea			4001	Corg.	65	75		13340 ± 180		
		Mediter. Sea			4002	Corg.	105	115		15815 ± 195	}0.05 ± 0.01	
		Mediter. Sea			4003	Corg.	130	152		21115 ± 160		
17	OT26 1050 m	Mediter. Sea	39°26'N	19°04'E	3999	Corg.	1	5		5035 ± 155	}0.55 ± 0.04	+ 4980 ± 160
		Mediter. Sea			3536	Corg.	173	188		8210 ± 185		
18	OT25 860 m	Mediter. Sea	39°24'N	18°33'E	\\\\	Corg.	0	10		4530 ± 140	}0.06 ± 0.004	+ 3750 ± 190
		Mediter. Sea			3997	Corg.	10	30		6885 ± 100		
		Mediter. Sea			3998	Corg.	92	112		11210 ± 165	}0.18 ± 0.01	
		Mediter. Sea			\\\\	Corg.	174	178		9640 ± 140		
19	OT5 826 m	Mediter. Sea	40°48'N	18°30'E	3994	Corg.	1	9		10170 ± 1120	slump masses	
		Mediter. Sea			3995	Corg.	27	35		10270 ± 1125		
		Mediter. Sea			3996	Corg.	102	110		10290 ± 1640		

TABLE 1 *(continued)*

No.	Name	Location	Coordinates	Hv	Frac.	Depth (cm)	$\delta^{13}C$ ‰	Corg mg/g	RC date B.P.	Sed. rate mm/yrs	Top age B.P.
20	6K 2200 m	Red Sea Red Sea Red Sea	21°17'N 38°03'E	5056 5057 5058	Corg. Corg. Corg.	96 196 296	−19.7	1.7 1.9 0.6	3645 ± 130 12690 ± 230 15185 ± 495	0.11 ± 0.01 0.40 ± 0.08	− 5040 ± 340
21	17K 2037 m	Red Sea Red Sea Red Sea	23°17'N 38°05'E	5751 5752 6590	Corg. Corg. Corg.	360 400 480 495 585 600	−20.1	0.1 0.3 0.3	1425 ± 545 3990 ± 385 3855 ± 640	0.87 ± 0.03	− 2920 ± 1900
22	79K 2066 m	Red Sea Red Sea Red Sea	21°18'N 38°05'E	5753 5754 5755	Corg. Corg. Corg.	276 309 767 833 1160 1180	−18.8 −17.6 −20.0	0.7 0.3 0.3	2510 ± 100 2950 ± 220 4020 ± 455	11.40 ± 6.27 3.50 ± 1.66	+ 2250 ± 200
23	108P 2049 m	Red Sea Red Sea Red Sea	19°34'N 38°39'E	4715 4716 4717	Corg. Corg. Corg.	21 76 253 283 410 438	−21.5 −21.4 −20.2	1.0 1.2 2.7	6840 ± 135 7585 ± 220 13265 ± 205	2.90 ± 0.99 0.27 ± 0.01	+ 6675 ± 170
24	137P 666 m	Red Sea Red Sea Red Sea Red Sea	18°10'N 40°17'E	4864 4718 4719 4720	Corg. Corg. Corg. Corg.	1 32 105 125 250 265 448 475	− 6.3 −20.3 −20.4 −20.3	1.0 1.6 2.4 1.9	17085 ± 1450 28215 ± 1100 23470 ± 740 29590 ± 1085	disturbed or contaminated	
25	167P 1124 m	Red Sea Red Sea Red Sea Red Sea	15°39'N 41°55'E	4865 4721 4722 4723	Corg. Corg. Corg. Corg	1 32 110 125 250 264 433 447	− 3.3 −14.0 −20.8 −18.1	3.3 2.7 5.5 3.4	21310 ± 510 26650 ± 1220 35695 ± 1585 32805 ± 1295	0.16 ± 0.02	+20000 ± 1360
26	172P 1128 m	Red Sea Red Sea Red Sea Red Sea	15°18'N 41°58'E	5047 5048 5049 5750	Corg. Corg. Corg. Corg.	142 239 441 711 728	−19.9 −13.7 −16.3	5.8 9.9 1.9 3.1	7940 ± 100 10150 ± 160 14670 ± 260 27285 ± 3145	0.44 ± 0.02 0.22 ± 0.06	+ 4740 ± 190
27	188P 1261 m	Red Sea Red Sea Red Sea Red Sea	17°24'N 46°06'E	5050 5051 5052 6591	Corg. Corg. Corg. Corg.	93 178 369 635 660	− 5.2 −24.2 −14.5 −22.4	3.5 1.5 3.1 1.0	24450 ± 480 14125 ± 560 13435 ± 260 11300 ± 830	disturbed	

No	Core / depth	Region	Coordinates	Lab No	Material	Depth (cm)		δ¹³C	‰	Age (yr BP)	Ratio	Δ
28	268 P 3075 m	Red Sea Red Sea Red Sea	13°00'N 49°29'E	5053 5054 5055	Corg. Corg. Corg.	85 185 385		−19.4 −20.8 −20.4	3.7 5.0 4.5	9035 ± 155 18395 ± 230 23260 ± 1195	{0.10 ± 0.01 0.41 ± 0.10	+ 1080 ± 350
29	278 P 1939 m	Red Sea Red Sea	11°29'N 47°15'E	4869 4870	Corg. Corg.	200 400	230 430	− 5.4 − 1.7	9.9 5.5	29650 ± 480 24875 ± 1300	contaminated	
30	332 K 2848 m	Red Sea Red Sea Red Sea	19°37'N 38°44'E	5770 5771 5772	Corg. Corg. Corg.	149 545 777	160 551 785	−21.6 −22.3 −28.4	1.8 1.1 0.8	1740 ± 115 2925 ± 120 10300 ± 400	{3.32 ± 0.47 0.31 ± 0.02	+ 1275 ± 170
31	334 K 2777 m	Red Sea Red Sea Red Sea Red Sea	19°39'N 38°45'E	5766 5767 5768 5769	Corg. Corg. Corg. Corg.	25 163 384 566	33 173 390 577	−22.6 + 0.1 − 8.3 −22.7	3.0 6.2 4.9 3.5	2080 ± 85 11200 ± 100 12200 ± 150 17300 ± 170	{0.15 ± 0.002 0.66 ± 0.02	+ 180 ± 100
32	447 K 2028 m	Red Sea Red Sea	21°25'N 38°04'E	5760 5761	Corg. Corg.	277 412	292 420	−22.0 −21.0	0.4 2.9	2640 ± 545 10200 ± 120	0.17 ± 0.01	−13700 ± 1700
33	451 K 1663 m	Red Sea Red Sea Red Sea Red Sea Red Sea	21°21'N 37°57'E	6592 6593 5059 5060 6594	Corg. Corg. Corg. Corg. Corg.	40 138 177 237 350	50 148 360	−26.8 −21.6 −17.2 −18.2 −19.1	2.8 2.2 4.3 2.0 0.8	7065 ± 125 7600 ± 220 10705 ± 105 21300 ± 640 16000 ± 740	1.83 ± 0.87 0.10 ± 0.01 0.05 ± 0.003	+ 6820 ± 210
34	454 K 2019 m	Red Sea Red Sea	21°21'N 38°06'E	6585 6587	Corg. Corg.	192 530	200 545	−22.9 −22.6	0.3 0.5	1615 ± 400 13450 ± 1010	0.28 ± 0.03	− 5200 ± 860
35	486 K 2440 m	Red Sea Red Sea Red Sea Red Sea	23°12'N 37°15'E	5762 5763 5764 5765	Corg. Corg. Corg. Corg.	0 108 238 390	10 114 245 400	−22.3 −22.0 −22.9 −18.3	1.2 0.8 2.5 0.2	2655 ± 120 13400 ± 640 5630 ± 90 10000 ± 550	{0.78 ± 0.04 0.35 ± 0.05	− 2590 ± 120
36	508 K 1550 m	Red Sea Red Sea Red Sea	24°43'N 36°17'E	6595 6596 6597	Corg. Corg. Corg.	27 150 370	40 160 380	−19.9 −22.0 −18.2	0.5 0.3 1.1	11050 ± 635 10080 ± 510 13100 ± 450	disturbed	

TABLE 1 (continued)

No.	Name	Location	Coordinates	Hv	Frac.	Depth (cm)		$\delta^{13}C$ °/oo	Corg mg/g	RC date B.P.	Sed. rate mm/yrs	Top age B.P.
37	575 K 2002 m	Red Sea Red Sea Red Sea Red Sea	21°24'N 36°04'E	5756 5757 5758 5759	Corg. Corg. Corg. Corg.	124 228 381 578	129 244 387 587	-10.8 -20.6 -20.0 -20.9	0.6 1.3 9.9 0.4	5640 ± 390 8355 ± 130 12160 ± 130 14600 ± 1100	} 0.40 ± 0.02	+ 2515 ± 330
38	609 K 2800 m	Red Sea Red Sea Red Sea Red Sea	20°04'N 38°30'E	6456 6457 6458 6459	Corg. Corg. Corg. Corg.	42 244 603 948	55 259 640 969	-22.7 -21.8 -25.1 - 6.4	2.4 2.0 3.5	12315 ± 405 12335 ± 380 4290 ± 250 11330 ± 400	disturbed	
39	69-37 2427 m	Red Sea Red Sea	20°07'N 38°29'E	6453 6455	Corg Corg	59 820	82 850	-22.2 -16.0	1.0 0.9	4115 ± 435 6300 ± 470	} 3.05 ± 0.64	+ 3570 ± 250
40	1056 D 65 m	Persian Gulf Persian Gulf Persian Gulf Persian Gulf Persian Gulf Persian Gulf Persian Gulf Persian Gulf Persian Gulf Persian Gulf Persian Gulf Persian Gulf	26°13'N 56°57'E	3792 3793 3794 3795 3796 3797 3798 3799 3153 3153 3800 3801 3154	Corg. Corg. Corg. Corg. Corg. Corg. Corg. Corg. Corg. Carb. Corg. Corg. Corg.	40 70 120 140 190 230 250 290 320 320 340 350 360	50 80 130 150 200 240 260 300 330 330 350 360 370		1.3 1.8 0.8 1.4 0.9 1.3 1.4 1.6 3.2 2.5 1.2	1845 ± 125 2765 ± 115 2740 ± 125 3610 ± 85 3840 ± 145 3615 ± 115 3615 ± 150 4035 ± 140 2915 ± 1125 4075 ± 155 4760 ± 165 4970 ± 190 8865 ± 425	} 1.36 ± 0.08 } 0.02 ± 0.003	+ 2070 ± 80
41	1060 29 m	Persian Gulf Persian Gulf Persian Gulf Persian Gulf	28°32'N 50°54'E	3171 3168 3169 3170	Corg. Corg. Corg. Corg.	0 30 231 380	7 35 241 390			975 ± 260 1560 ± 175 5695 ± 170 5840 ± 200	} 0.49 ± 0.03	+ 900 ± 160

42	1143 B 15 m	Persian Gulf Persian Gulf Persian Gulf Persian Gulf Persian Gulf	29°00'N 50°40'E	3414 4133 3415 4136 3416	Corg. Corg. Corg. Corg. Corg.	57 81 105 162 164	64 88 112 169 171	6.5 1.7 4.9 2.3	10870 ± 170 2170 ± 115 8660 ± 140 4155 ± 715 6245 ± 175	disturbed
43	1144 B 8 m	Persian Gulf Persian Gulf Persian Gulf	29°01'N 50°41'E	3417 3418 3419	Corg. Corg. Corg.	22 100 164	29 107 171	0.9 4.3 1.5	4015 ± 430 10940 ± 240 7910 ± 290	}0.11 ± 0.01 + 1750 ± 580
44	1160 29 m	Persian Gulf Persian Gulf Persian Gulf	28°32'N 50°53'E	4210 4211 4212	Corg. Corg. Corg.	87 127 333	96 135 338	3.0 1.9 1.8	2335 ± 120 3340 ± 195 5080 ± 185	{0.39 ± 0.09 1.17 ± 0.18 + 100 ± 600
45	1165 E 46 m	Persian Gulf Persian Gulf Persian Gulf Persian Gulf Persian Gulf Persian Gulf Persian Gulf	28°07'N 50°59'E	3420 3421 4233 3422 3423 3424 3425	Corg. Corg. Corg. Corg. Corg. Corg. Corg.	1 187 208 281 302 315 322	10 198 218 292 307 322 329	2.7 2.5 1.9 2.1 2.4 2.7	1970 ± 140 2250 ± 140 2195 ± 215 2625 ± 155 3060 ± 115 3740 ± 90 4255 ± 245	1.81 ± 0.16 + 1610 ± 130
46	187 212 m	Arabian Sea Arabian Sea Arabian Sea Arabian Sea	9°32'N 75°13'E	2638 3852 2639 2640	Corg. Corg. Corg. Corg.	2 85 196 282	20 90 210 294	3.6	2980 ± 110 1445 ± 320 17435 ± 425 17050 ± 430	disturbed + 2150 ± 120
47	188 988 m	Arabian Sea Arabian Sea Arabian Sea Arabian Sea	9°38'N 75°31'E	2644 2645 2646 2647	Corg. Corg. Corg. Corg.	2 68 167 365	12 80 169 375		1665 ± 65 4085 ± 130 9145 ± 110 12400 ± 270	{0.21 ± 0.003 0.62 ± 0.06 + 1250 ± 65

TABLE 1 *(continued)*

No.	Name	Location	Coordinates	Hv	Frac.	Depth (cm)		$\delta^{13}C$ ‰	Corg mg/g	RC date B.P.	Sed. rate mm/yrs	Top age B.P.
48	1048 2835 m	Arabian Sea Arabian Sea Arabian Sea Arabian Sea Arabian Sea	21°57'N 64°11'E	3158 3155 3159 3757 3156	Corg. Corg. Corg. Corg. Corg.	0 63 115 176 217	5 66 120 180 227			1910 ± 225 4995 ± 360 6860 ± 130 13215 ± 1105 17800 ± 530	0.23 ± 0.01 0.09 ± 0.004	+ 1870 ± 220
49	1052 3300 m	Arabian Sea Arabian Sea Arabian Sea Arabian Sea Arabian Sea Arabian Sea Arabian Sea Arabian Sea Arabian Sea	23°53'N 60°41'E	3167 3166 3165 3164 3163 3161 3157 3160 3758		0 90 160 186 260 300 320 350 357	2 95 170 191 265 310 325 360 363			3290 ± 310 3470 ± 220 2880 ± 120 2585 ± 155 3075 ± 210 3775 ± 445 9890 ± 520 4465 ± 125 26000 ± 1125	1.91 ± 0.24	+ 2220 ± 160
50	241-1 4084 m	Arabian Sea Arabian Sea Arabian Sea Arabian Sea Arabian Sea	2°22'S 44°41'E	5253 5254 5255 5256 5257		2 560 850 1150 1730	17 575 867 1165 1754		0.5 1.1 2.1 5.0 1.6	5015 ± 790 10130 ± 1140 13450 ± 805 22890 ± 580 21655 ± 1255	1.03 ± 0.08	+ 4930 ± 700

D) ^{14}C dates of the C org fraction of frozen cores

No.	Name	Location	Coordinates	Hv	Frac.	Depth (cm)		$\delta^{13}C$ ‰	Corg mg/g	RC date B.P.	Sed. rate mm/yrs	Top age B.P.
51	12309 2805 m	Atlantic Oc. Atlantic Oc. Atlantic Oc. Atlantic Oc.	26°50'N 15°07'W	6778 6779 6780 6781		1 30 70 330	5 40 80 340	-22.8 -21.1 -20.7	0.4 0.7 2.8 6.0	2295 ± 480 5700 ± 430 10480 ± 160 26040 ± 650	0.08 ± 0.004 0.16 ± 0.01	+ 1840 ± 430

52	12326 1035 m	Atlantic Oc. Atlantic Oc. Atlantic Oc. Atlantic Oc.	23°06'N 17°25'W	6769 6770 6771 6772	0 100 220 280	10 110 230 285	-20.5 -21.8 -20.1 -20.9	5.7 1.6 3.2 3.9	1255 ± 80 12320 ± 290 19650 ± 550 20800 ± 650	{0.09 ± 0.002 0.19 ± 0.01	+ 1200 ± 80
53	12328 2784 m	Atlantic Oc. Atlantic Oc. Atlantic Oc. Atlantic Oc.	21°09'N 18°34'W	6765 6766 6767 6768	0 150 300 520	10 160 310 530	-19.9 -20.6 -22.4 -20.4	1.5 3.5 4.5 1.1	3195 ± 250 13140 ± 250 25100 ± 770 >21000	}0.14 ± 0.004	+ 2710 ± 250
54	12329 3317 m	Atlantic Oc. Atlantic Oc. Atlantic Oc.	19°22'N 19°56'W	6782 6783 6784	1 120 250	10 130 260	-21.1 -19.9 -21.9	1.2 0.9 0.8	2045 ± 230 20050 ± 2325 >22800	{0.06 ± 0.01	+ 1215 ± 265
55	12379 2058 m	Atlantic Oc. Atlantic Oc. Atlantic Oc. Atlantic Oc.	23°08'N 17°45'W	6761 6762 6763 6764	100 220 450 545	110 230 460 555	-22.0 -23.5 -20.1 -19.9	2.6 3.0 9.9 3.8	8385 ± 210 10070 ± 1020 28050 ± 450 24050 ± 910	{0.71 ± 0.44 0.23 ± 0.02	+ 6910 ± 980
56	12392 2591 m	Atlantic Oc. Atlantic Oc. Atlantic Oc. Atlantic Oc.	25°10'N 16°51'W	6773 6775 6776 6777	0 105 195 306	7 110 200 312	-19.6 -21.0 -19.5 -20.9	0.5 4.9 2.7 5.7	710 ± 230 15145 ± 400 19000 ± 1020 28250 ± 1060	{0.07 ± 0.002 0.22 ± 0.07 0.12 ± 0.02	+ 225 ± 240
57	12393 2816 m	Atlantic Oc. Atlantic Oc. Atlantic Oc. Atlantic Oc.	37°40'N 10°06'W	4947 4948 4949 4950	76 226 296 426	86 236 306 436		1.1 0.8 0.6 0.4	8660 ± 230 14750 ± 395 14395 ± 355 15920 ± 580	0.12 ± 0.01	+ 2160 ± 925
58	123930 2800 m	Atlantic Oc. Atlantic Oc.	37°40'N 10°07'W	4771 4772 4773	116 156 196	126 166 200		4.4 5.1 3.5	11870 ± 300 14210 ± 340 15735 ± 365		
59	123931 2800 m	Atlantic Oc. Atlantic Oc.	37°51'N 10°05'W	5326 5327	356 416	366 426		3.0 8.0	18430 ± 365 21440 ± 490	{0.42 ± 0.03	

Most cores were frozen at 3°C for several years until the biostratigraphic investigation was completed. At present the cores are kept frozen for a few months before being subjected to ^{14}C analyses.

Radiocarbon measurements have been carried out with a 250 ml proportional counter at 1 atm C_2H_2 corresponding to a sample containing at least 100 mg of C. The background and net standard counting rates (oxalic acid × .95) are 0.7 and 4.9 cpm, respectively. ^{14}C analyses have generally been supplemented by bio- and lithostratigraphic studies, δ^{13}C determinations, analyses of the C-org content, and by U/Th datings.

Differential sedimentation rates have been calculated from ^{14}C data of adjacent core sections, and means have been determined from corresponding values. The top age has been extrapolated from the mean sedimentation rate of the uppermost core section. Compaction has not been considered.

Top ages of 0-2500 years appear to be associated with undisturbed sediments. Explanations for top ages exceeding 2500 years are (1) allochthonous material admixed within the dated core sections, (2) top material lost during sampling, or (3) sedimentation interrupted in the past. Negative top ages are explained by (1) an increasing sedimentation rate within the uppermost core section, or (2) the deposition of relocated material above the uppermost dated core sample.

^{14}C Data of the Carbonate Fraction of Marine Sediments

^{14}C dating of the carbonate fraction was begun despite contamination problems due to admixed fossil terrigenous material or isotopic exchange with atmospheric CO_2 (Olsson and Eriksson 1965). In 1965, our available marine samples were too small for the extraction of sufficient C-org. We dated twenty-one samples of six cores. The results have been in good agreement with the biostratigraphic findings.

The sedimentation rates of the Holocene cores (N12, N14, and N18) from the Mediterranean Sea (Einsele and Werner 1968) reflect decreasing material transport of the Nile River from .59 to .36 mm/a with increasing distance from the estuary. The top age of 3500 years of the core (N18) taken in the immediate vicinity of the estuary accounts for a noticeable deposition of fossil terrigenous material at this place.

The cores (202) at 3.5 m and 223 at 3.8 m from the Arabian Sea were biostratigraphically and sedimentologically investigated. A slight disturbance was found (Stackelberg 1972, Zobel 1973), so the core (202) shows no ^{14}C age/depth relation. The biostratigraphically determined Pleistocene/Holocene boundary of core (223) at 80 cm is in agreement with the ^{14}C data, but the exaggerated top age of 4900 years and the scattering of ^{14}C data below 80 cm confirm the presence of dislocated material. Rejuvenation by isotopic exchange might have occurred during the pretreatment of the marine samples. This contamination did not happen during storage, although holes were present in the core casing (table 1, core 9, Hv 3143-3141).

Although the ^{14}C data of the carbonate fraction do not contradict marine geologists' expectations, they contain more sedimentologic than chronologic information.

^{14}C Data of Corresponding Carbonate and Organic Carbon Fractions

Methodical investigation into the reliability of ^{14}C data of the C-org fraction has included comparison of corresponding ^{14}C data of the carbonate and the C-org fractions of cores (1048), (1177), (1201), and (8057). They differ by 2275 ± 365 years. Emery and Bray (1962) found 1700 years. The ^{14}C data of core (228) for the C-org fraction, ooids, and Foraminifera of 11,385 ± 320 B.P. (Hv 3854), 11,075 ± 145 (Hv 1859) and 9945 ± 140 B.P. (Hv 1858), respectively, are an exception because this sample was formed in flat water (Stackelberg 1972). In addition, five ^{14}C age/depth diagrams for both fractions were considered (fig. 2).

The 3.5 m core (8057) is reported (Kudrass and Thiede 1970, Thiede 1971) to have a Holocene sedimentation rate between .14 and .18 mm/a. This range has been verified by ^{14}C data of both fractions. Those which belong to Pleistocene sediments scatter widely and appear too small. The top ages of the two fractions agree with each other.

The 2.3 m core (M 42) has been biostratigraphically dated at about 1000 years. This is confirmed by the C-org ^{14}C data. The carbonate data are about 4000 years older but the sedimentation rates of both fractions agree at 2.12 ± .44 and 1.67 ± .32 mm/a respectively. Fossil carbonates rather than fossil C-org may have been admixed to the sediment.

The 3.4 m core (1177) has been biostratigraphically dated at 12,000 to 15,000 years. The ^{14}C data of the C-org fraction show an inverse age/depth relation. Those of the carbonate fraction are uniform. These findings would account for disturbed sedimentation although the biostratigraphic classification has not considered this possibility.

The 3.25 m core (1201) shows an inverse ^{14}C age/depth relation for the carbonate fraction and a regular one for the C-org fraction. This core correlates biostratigraphically with core (1056 D) (Diester 1972) and proves to be less than 8000 years.

Biostratigraphic dating of the 4.7 m core (210) has shown relocated material, verified by top ages of up to 5000 years, consisting of shelf sediments between 250 and 412 cm (Stackelberg 1972, Zobel 1973). Despite this disturbed sedimentation, the Holocene and Pleistocene sedimentation rates derived from ^{14}C data of both fractions are comparable. The corresponding top ages differ by 1700 ± 330 years.

The observation of agreement and disagreement between the radiocarbon data of both fractions obtained from disturbed sediments raises the question of possible mold contamination of the organic carbon fraction. Also at issue is the question of similar ages for the acid-soluble and acid-insoluble parts.

Mold need not change the ^{14}C content, as is proven in one case:

Core	Section	Hv	Age
1052	3.6–3.7 m	Hv 3161	3775 ± 445 B.P. without mold
		Hv 3162	3685 ± 360 B.P. with mold

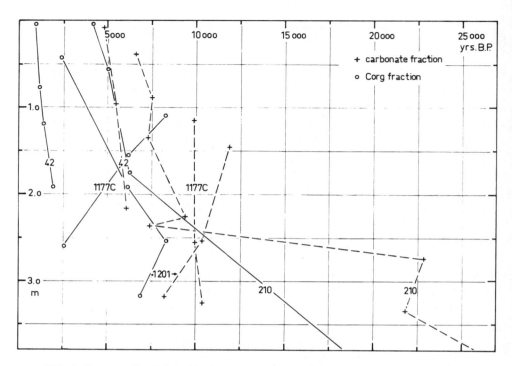

FIG. 2. Corresponding RC age/depth relations of the carbonate and the Corg fractions of cores M 42, 1177K, 1201, and 210.

On the other hand, the ^{14}C data of the corresponding acid-soluble and acid-insoluble organic carbon fractions differs unsystematically. The soluble organic carbon fraction was extracted by wet combustion.

Hv	Core	Section	Acid-soluble	Acid-insoluble
4136/4135	1145	1.62–1.69 m	4155 ± 715 B.P.	1595 ± 190 B.P.
4138/4137	10076	2.90–3.02 m	101.3 ± 2.5 % Mod.	1155 ± 105 B.P.

The sample of core (1145) has a biostratigraphic date of 2000 years. The ^{14}C age of the acid-soluble fraction appears too large while that of the second sample taken in 3 m depth should be too small. It is surprising that despite this confusing finding ^{14}C data of the C-org fraction obtained in different laboratories may agree (Diester 1972).

Core	Section	Hv	Age
1056	3.6–3.7 m	Hv 3154	8865 ± 485 B.P.
		Gr ????	8790 ± 100 B.P.

Beyond the explanations by Emery and Bray (1962), the ^{14}C age difference between the C-org and carbonate fractions of marine sediments can be understood through the results of Taylor (1976). Because he found a difference of up to 2500 years between the ^{14}C ages of the carbonate and the conchiolin of shells, the C-org fraction of marine sediments can be assumed to consist in part of conchiolin of Foraminifera.

Williams et al. (1969, 1975), investigating dissolved organic (DOC) and inorganic (DIOC) carbon in sea water, have found the DOC ^{14}C data to be larger than the corresponding DIOC ^{14}C data. The latter shows an increase at 2000 m water depth from 700 years to 3500 years. The same authors conclude that DOC is recycled in sea water. It would follow that DOC would not be sedimented in noticeable quantities. Otherwise the ^{14}C data of the carbonate fraction would be smaller than those of the C-org fraction, and the top ages would increase with water depth. Neither of these results were obtainable.

The ^{14}C data of corresponding C-org and carbonate fractions differ by 1700 to 2300 years, while the corresponding sedimentation rates of regularly sedimented cores are comparable. Disturbed sediments may show different ^{14}C age/depth relations for both fractions. Mold or holes in the core casing seem not to invalidate the ^{14}C data of either fraction, but the mode of pretreatment of long-stored sediments may be of importance for the ^{14}C C-org data. The ^{14}C data of the C-org fraction were considered to be reliable within the described methodical limits.

^{14}C Data of the C-org Fraction of Long-Stored Sediments

One-hundred-fifty-seven ^{14}C analyses were run of the C-org fraction of thirty-nine cores stored at 3°C for several years. The 3.4 m core (8016 B) with a ^{14}C content of 128% modern at the top had nuclear debris within the uppermost section (Thiede 1971). The scattering of the ^{14}C data indicated disturbed sedimentation. The biostratigraphical study did not consider this possibility. It classified the core sediments as Holocene based on findings of four ^{14}C data below 10,000 years. The sedimentation rate of the 1.8 m core (8017 B), expected to range from 0.06 to 0.15 mm/a (Thiede 1971), showed a rate of 0.15 mm/a. The top age of 4620 years and the anomalous ^{14}C age/depth relation of the deepest samples indicated contamination by allochthonous material. There was no paleontological assessment of disturbed sedimentation.

Four cores drilled in the Mediterranean Sea and intensively studied by biostratigraphers (Fabricius et al. 1970, Hesse and Rad 1972) had a high C-org content. The 1.1 m core (OT 5) consisted of slump masses, so the ^{14}C data were uniform. Sedimentation of the 1.8 m core (OT 25) evidenced past interruption as shown by a top age of 3750 years. This statement holds true for cores (OT 26) and OT 28) from the same sampling site. Various exaggerated top ages reflected an admixture of fossil material. This conclusion does not negate the biostratigraphic concept.

A corresponding investigation of Red Sea samples is in progress, so only a few particulars will be discussed. U/Th datings were made from the cores (6 K), (17 K), (344 K), (447 K), and (486 K). The ^{14}C and the U/Th data agree within the methodical limits (Geyh 1974, Geyh and Hoehndorf 1976). The Pleistocene sedimentation rate could not be shown to be larger than that of the Holocene, however. This may be due to contamination of the older sediments by bacterial activity.

High C-org contents of individual samples were associated with relatively large ^{14}C data as in (167 P), (188 P), (278 P), and (334 K). Possible explanations are the admixture of old organic carbon, or a uniform, absolute contamination of the core sediments, which is finally less effective for samples of high C-org contents.

An exception, core (188 P), was investigated biostratigraphically and isotopically by δ^{18}O (Schoell and Risch 1974). The core was assigned to the Pleistocene and the sediments were assumed to be undisturbed. The ^{14}C data scatter widely from 11,300 to 24,500 B.P. These appear to be reliable, as contamination problems should not exist in view of the relatively large C-org contents and the broad range of δ^{13}C values. Because there are sufficient indications of disturbed sedimentation, some scepticism is indicated as to the correctness of the biostratigraphic and isotopic interpretation.

Several samples of the Red Sea cores have δ^{13}C values exceeding $-18^{o}/oo$. It is probable that they have been hydrothermically influenced. The usual δ^{13}C values of the C-org fraction range from -18% to $-22^{o}/oo$, as in the conchiolin of shells.

The bottom and the top sections of the 3.7 m core (1056 D) from the Persian Gulf were dated twice. The ^{14}C data obtained immediately after coring (Diester 1972) and after three years of storage are comparable for the base sample, while the ^{14}C date of the top section after storage, is exaggerated by 1700 years. Diester assumed contamination and omitted all ^{14}C data obtained after the long storage, although the top age of 2070 years appears reasonable. A partial decomposition of the acid-insoluble organic carbon fraction during storage could explain some ^{14}C age shift.

Biostratigraphic investigation of five cores from the Persian Gulf was restricted to their chronological classification. Accordingly any information on disturbed sedimentation is missing except for that of the ^{14}C data. There is no admixed allochthonous material, since the top ages of the cores (1060), (1143 B), (1144 B), (1160), and (1165 E) are within the regular range.

The sediment of the 2.9 m core (187) from the Arabian Sea appears to be undisturbed except for the uppermost sample, which shows an exaggerated ^{14}C date (Stackelberg 1972, Zobel 1973). The assumed Holocene/Pleistocene boundary at 150 cm is acceptable by the ^{14}C data. The deepest sample may have been contaminated.

The 4.5 m core (188) contains undisturbed sediments as shown by the ^{14}C data and the biostratigraphic record (Zobel 1973). Both results are in agreement.

The 2.27 m core (1048) and 3.63 m core (1052) are undisturbed. Limited scattering of the ^{14}C data is due to the presence of some turbidity material. The Pleistocene sedimentation rate may be wrong in consequence of contamination by bacterial activity.

The 17.4 m core (241-1) shows in its sediment a faunal assemblage of mixed cool and warm water planktonic foraminifers. The sediment probably underwent intensive mechanical mixing during drilling (Zobel 1974). Despite this, results show a reasonable ^{14}C age/depth relation (fig. 3). The apparent discrepancy is explained by a decreasing downward-mixing of the slightly consolidated young top sediment.

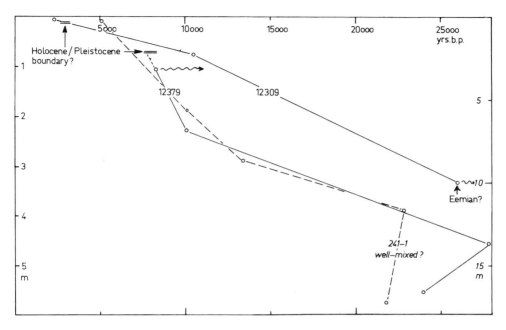

FIG. 3. RC age/depth relations of cores 241-1, 12379, and 12309, which seem to contradict the biostratigraphic concepts.

The ^{14}C data of the C-org fraction would have agreed with the biostratigraphic findings had careful and critical consideration been given both methods. The ^{14}C data of the C-org fraction of Pleistocene sediments often appear to be too young.

^{14}C C-org Data of Frozen Sediments

As reported earlier (Geyh et al. 1974), long-term stored sediments kept at 3°C may become contaminated by bacteria, as with the cores (8057 B) and 12393) (fig. 4). The C-org data exceeding 10,000 years scatter unsystematically while those of the short-term stored, frozen core (12393-1) show an acceptable ^{14}C age/depth relation (Thiede 1971). Out of twenty-three C-org ^{14}C data determined from six short-term stored, frozen cores taken in the Atlantic Ocean (12309), (12326), (12328), (12329), (12379), and (12392), the ^{14}C age/depth relations of cores (12326) 2.9 m, (12328) 5.3 m, and (12392) 3.1 m support the biostratigraphic concept (Diester 1975, personal communication). The sedimentation rates, the C-org contents, and the δ^{13}C values are comparable, but the ^{14}C data of the other three cores contradict the biostratigraphic classification based on the study of foraminifers and diatoms, the contents of radiolarians, carbonates, terrigene material, and the number of quartz grains.

The Holocene/Pleistocene boundary of the core (12309) 3.4 m should be at 12 cm and the Eemian should begin at 3.3 m (fig. 3). The ^{14}C data and the C-org content

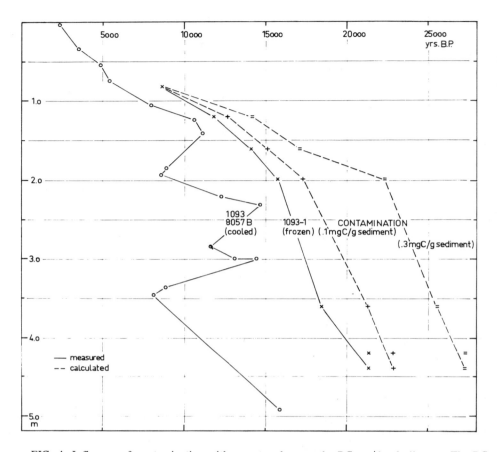

FIG. 4. Influence of contamination with recent carbon on the RC age/depth diagram. The RC data of the long-term stored cores 8057 B and 12393 seem too small. The short-term stored frozen core 12393-1 from the same sampling site shows a more reasonable RC age/depth relation. It becomes further smoothed if contamination with .1 mg recent carbon/g of sediment is assumed (dotted line).

increase from 2200 to 24,000 years and .4 to 6 mg of C/g of sediment, respectively. As the top age of 1845 years appears reliable, the biostratigraphic concept was questioned. A similar example gave the ^{14}C data of the 2.6 m core (12329). The 5.5 m core (12379) might have been disturbed, as reflected by the top age of 6900 years. The Holocene/Pleistocene boundary was expected at 70 cm, but was found at 2.2 m.

Reasonable top ages, corresponding sedimentation rates, and a small range of the $\delta^{13}C$ values support the reliability of the C-org ^{14}C data of short-term stored, frozen sediments. Conversely, the multiplicity of the sedimentologic analyses demands a more critical consideration of the ^{14}C data.

Provided that the biostratigraphic classification for the core (12309) is correct, and that the age difference in comparison with the ^{14}C data is due to contamination, about 0.4 ± 0.2 mg of recent organic carbon/g of sediment must have been present (table 2). A value of .3 ± .12 mg of recent C/g of sediment was found for core (12393) when the age

TABLE 2

Contamination with Recent Carbon Estimated From the Age Difference Between ^{14}C Data and Paleontologically Determined Ages for Core (12309)

Depth m	Biostr. age yrs. B.P.	^{14}C age yrs. B.P.	C-org %	Contamination	
				%	mg recent C/g
.02	2,300	2,300	.4	–	–
.35	17,000	5,700	.7	42	.3
.75	29,000	10,500	2.8	25	.7
3.35	70,000	26,000	6.0	4	.2

difference was considered between the long-term stored, cooled and the short-term stored, frozen sediments. As the latter might still have contained some recent carbon of unknown origin, the ^{14}C ages were calculated for a contamination with .1 and .3 mg of recent C/g of sediment (fig. 4). The best, smoothed ^{14}C age/depth relation was obtained for .1 mg of recent C/g of sediment.

According to these considerations, the age difference between the frozen core (12393-1) and the core (12393) can also be attributed to the difference in the C-org content if uniform contamination in depth is assumed.

The ^{14}C data of short-term stored, frozen Pleistocene sediments are still problematic. Freezing seems to preserve the acid-insoluble organic carbon somewhat better than cooling and diminishes contamination with recent carbon, which may range from .1 to .3 mg of recent C/g of sediment.

Statistics of the Accepted ^{14}C Data

The statistics of the accepted ^{14}C data distinguish between Holocene and Pleistocene samples and between opinions of the geochronologists and paleontologists, respectively. The reliability criterion of the former is a reasonable ^{14}C age/depth relation for at least two adjacent core sections beginning at the top. The judgment of the paleontologists was derived from their comments on the ^{14}C data (table 3).

Table 3

Statistics of Accepted ^{14}C Data Obtained With Different Techniques (in %)

	Number	Geochronologists		Palaeontologists	
		Holocene	Pleistocene	Holocene	Pleistocene
Carbonate fraction	58	92	42	79	32
C-org fraction at 3°C	192	82	57	66	48
C-org fraction at −3°C	29	95	95	55	70

The C-org ^{14}C data of the short-term stored, frozen Pleistocene sediments were more often accepted than the data of long-term stored sediments kept at 3°C, or the data of the carbonate fraction. In the case of Holocene sediments the opinion of the paleontologists seems to be contrary.

CONCLUSIONS

1. ^{14}C data/depth relations of marine sediments contain reliable information concerning conditions of the sedimentation but not concerning time marks or sedimentation rates. Consequently, isolated ^{14}C data are not suitable for chronological conclusions.

2. The C-org data of short-term stored, frozen sediments are more reliable than those of long-term stored sediments kept at 3°C, and both are more reliable than the ^{14}C data of the carbonate fraction, which are about 1700 to 2300 years larger than the corresponding C-org ^{14}C data. These statements refer in particular to Pleistocene samples.

3. C-org ^{14}C data of long-term stored sediments appear to be less reliable than those obtained immediately after coring. Freezing may protect the acid-insoluble organic carbon but may not prevent contamination. ^{14}C data of the C-org fraction depend upon the pretreatment technique.

4. ^{14}C data fitting a straight regression line must not be reliable.

5. The usual range of the δ^{13}C values of the C-org fraction is -18‰ to -22‰. Values exceeding this range may indicate hydrothermically influenced sediments.

ACKNOWLEDGMENTS

My colleagues H. R. Kudrass and A. Hoehndorf have supported this work with many suggestions. Dr. B. Zobel has given valuable comments on the judgment of the biostratigraphic results.

REFERENCES

Diester, L.
　1972　Zur spätpleistozänen und holozänen Sedimentation in zentralen und östlichen Persischen Golf. *Meteor. Forschungs Ergebnisse. Reihe C* (8):37-83.

Einsele, G., and F. Werner
　1968　Zusammensetzung, Gefüge und mechanische Eigenschaften rezenter Sedimente vom Nildelta, Rotem Meer und Golf von Aden. *Meteor. Forschungs Ergebnisse. Reihe C* (1):21-42.

Emery, K. O., and E. E. Bray
　1962　Radiocarbon dating of California basin sediments. Bull. Am. Ass. Pet. Geol. 46:1839-1856.

Fabricius, F., U. von Rad, R. Hesse, and W. Ott
　1970　Die Oberflächensedimente der Straße von Ortranto (Mittelmeer). Geol. Jb. 60:164-192.

Geyh, M. A.
1974 ^{14}C-Altersbestimmungen an der Corg Fraktion von Sedimenten der Valdivia Expeditionen. Valdivia. Wissenschaftliche Ergebnisse. I:14-20.

Geyh, M. A., and A. Höhndorf
1976 Complementary ^{14}C and Th/U analyses as a necessary contribution to stratigraphic investigations of marine sediments. Examples from the Red Sea. Geol. Jb. (In press.)

Geyh, M. A., W. E. Krumbein, and H. R. Kudrass
1974 Unreliable ^{14}C dating of long-stored deep-sea sediments due to bacterial activity. Marine Geol. 17:45-50.

Hesse, R., and U. Von Rad.
1972 Undisturbed Large Diameter Cores from the Strait of Otranto. The Mediterranean Sea. 645-653.

Kudrass, H. R., and J. Thiede
1970 Stratigraphische Untersuchungen an Sedimentkernen des ibero-marokkanischen Kontinentalrandes. Geol. Rundschau. 60:294-355.

Olsson, I. U., and K. G. Eriksson
1965 Remarks on ^{14}C dating of shell material in sea sediments. Progr. Oceanogr. 3:253-266.

Schoell, M., and H. Risch
1974 Isotopen-geochemische und mikro-paläontologische Untersuchungen an Sedimenten des Kernes 188P. Valdivia. Wissenschaftliche Ergebnisse. I:39-53.

Stackelberg von, U.
1972 Faziesverteilung in Sedimenten des indisch-pakistanischen Kontinentalrandes (Arabisches Meer). *Meteor.* Forschungs Ergebnisse. Reihe C (9):1-73.

Taylor, R. E.
1976 Radiocarbon dating of shells: Comparison of carbonate and conchiolin valves. Proc. 9th Int. Radiocarbon Conf. Univ. of Calif. Los Angeles and San Diego.

Thiede, J.
1971 Planktonische Foraminiferen in Sedimenten vom ibero-marokkanischen Kontinentalrand. *Meteor.* Forschungs Ergebnisse. Reihe C (7):15-102.

Williams, P. M., H. Oeschger, and P. Kinney
1969 Natural Radiocarbon Activity of the Dissolved Organic Carbon in the Northeast Pacific Ocean. Nature. 224 (5216):256-258.

Williams, P. M., and T. W. Linick
1975 Cycling of Organic Carbon in the Ocean: Use of naturally occurring radiocarbon as a long and short term tracer. *In* Isotope Ratios as Pollutant Source and Behaviour Indicators. Vienna, IAEA, 153-167.

Zobel, B.
1973 Biostratigraphische Untersuchungen an Sedimenten des indisch-pakistanischen Kontinentalrandes (Arabisches Meer). *Meteor.* Forschungs Ergebnisse. Reihe C (12):9-73.

1974 25th Quarternary and Neogene Foraminifera Biostratigraphy. Initial Reports on Deep Sea Drilling Project. XXV:573-578.

PART VII
Tree Rings and Other Known Age Samples

1

^{14}C in Modern American Trees

William F. Cain

ABSTRACT

^{14}C measurements by gas proportional counting of acetylene were made on six modern rural and urban trees from northeastern and northwestern United States. Urban trees are depressed below rural trees, presumably because of additional fossil carbon dioxide in the local atmosphere due to local air pollution. Oak trees systematically incorporate later carbon when sapwood is transformed to heartwood. This later carbon is not removed by acetone extraction followed by alkaline and acid washing. It is removed by bleaching to pure cellulose.

Proportions of stable carbon isotope enrichment can be calculated from the difference in Δ^{14}C between urban and rural trees, and from this the trend in local air pollution can be observed. Results for New York City are not in disagreement with carbon dioxide levels inferred from reports of carbon monoxide in the early 1950's.

INTRODUCTION

In 1954 Suess found that at the time of its growth the cellulosic carbon of wood from the nineteenth century A.D. had a higher ^{14}C content than wood from recent times. He attributed the difference to the dilution of the natural ^{14}C by the inactive fossil carbon dioxide added to atmosphere by man since the Industrial Revolution. This present work was undertaken in the hope that a similar effect could be measured on a local scale. The urban areas should show an even larger fossil fuel effect. By measuring ^{14}C in urban trees and comparing with that of rural trees, the trends in local air pollution could be obtained and then be available for correlation with health statistics for diseases associated with air pollution.

Tree-Samples and Methods of Treatment

Tree-samples were collected from three northeastern American cities: Washington, D.C., Boston, and New York City. Rural tree-samples were collected from about 100 km north of New York City and from western Oregon. Rings were counted, growth years assigned, and individual year-by-year samples were separated. During pretreatment these single year samples were splintered into toothpick fragments from 5 to 15 mm long, extracted with acetone for two hours, boiled with 2% aqueous NaOH, thrice rinsed with distilled H_2O, and then rinsed with 0.1 N aqueous HCl until the solution remained acidic. The wood was dried at 105°C overnight and then stored over $CaCl_2$ desiccant.

After this standard pretreatment a few samples were bleached to white, cellulosic fiber by alternate treatment with 3% aqueous NaOCl and 3% aqueous Na_2SO_3, according to the method of Norman and Jenkins (1967). These samples were then rinsed with distilled H_2O, dried at 105°C, and stored over $CaCl_2$. Wood samples were combusted and converted to C_2H_2 for gas proportional counting in Oeschger-type counters according to methods which have been in use at La Jolla and which have been described previously (Suess 1965).

The Results

The first tree is a red oak from Bear Mountain State Park (41°18'N, 74°0'W). This park is part of a large preserve of park land including Harriman State Park and West Point Military Reservation about 100 km north of metropolitan New York City. The ^{14}C measurements relative to the activity of a "standard" decade of rural fir (1871-1880) are plotted in figure 1. The ordinate is $\Delta^{14}C$ (°/oo). The activities of both the "standard" and the sample, after background subtraction, have been age-corrected to 1950 using the decay constant corresponding to the 5730 year half-life, and isotope-fractionation has been corrected to a $\delta^{13}C$ of -25.0 °/oo.

Since this is a rural tree, the negative values up to 1954 are attributed only to the worldwide fossil fuel effect. The large, positive values after 1954 correspond to the weapons testing ^{14}C of the nuclear era.

Figure 2 illustrates the relatively complex results for a turkey oak from Central Park, New York City (40°47'N, 73°58'W). The nuclear era shows the expected high levels corresponding to the atmospheric levels of the period. The values of the early part of the century are depressed below that of the (rural) Bear Mountain tree through 1915. Most of the points from 1920 to 1940 were taken from one radial segment of the tree, but the circled points were taken from a different sector. In this second sector the heartwood/sapwood boundary is about five years later than in the first. The elevated $\Delta^{14}C$ values in both sectors seem to correlate with a contamination associated with the heartwood. It seems that the process of conversion of sapwood to heartwood involves the deposition of acetone-insoluble materials from a later period. This is presumably material from the nuclear era, since there would seem to be no other time period within the lifetime of this tree (ca. 1850 to 1970) which could account for such high levels of ^{14}C.

To test the hypothesis, late heartwood values were measured from another tree, a white oak from western rural Oregon (Sheridan, Oregon; 45°7'N, 123°7'W). In the Central Park oak, the results had peaked in the first sector about 1926, which was six rings

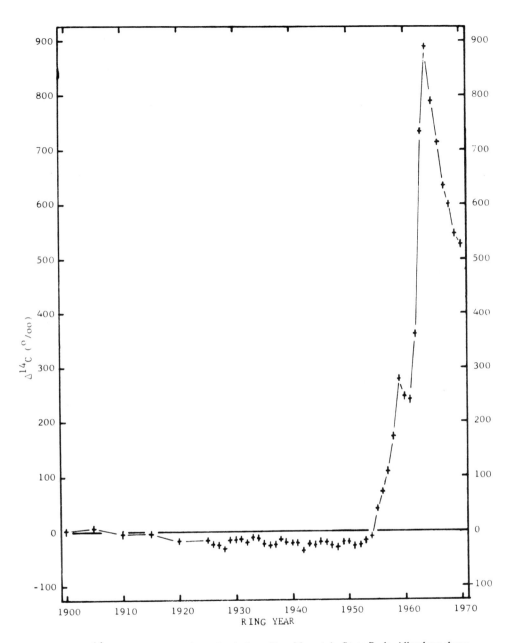

FIG. 1. $\Delta^{14}C$ versus ring year for red oak from Bear Mountain State Park. All values show one sigma error bars.

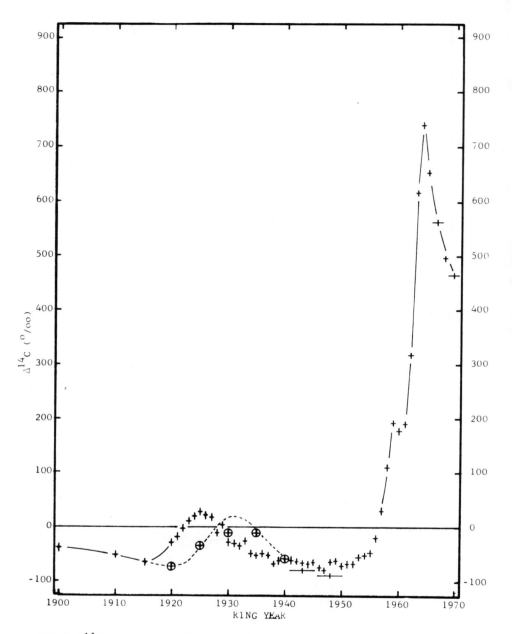

FIG. 2. $\Delta^{14}C$ versus ring year for turkey oak from Central Park, New York City. Circled points connected by dashed line are from radial sector where heartwood-sapwood boundary at time of cutting was approximately 1938. Other points are from sector where transition was about 1932. Two five-year samples (1941-45, 1946-50) are from bleached cellulose.

prior to the end of the heartwood (ca. 1932). In the second sector, the peak value would probably have appeared if measured about 1932 or 1933 (see dashed line in fig. 2). Such a position is consistent with the position of the heartwood/sapwood boundary in this sector.

Since the heartwood boundary in the Oregon oak occurs about 1945, and the tree was one year older when cut (1970), elevated values in the 1930's were expected, with the peak about 1938. Unfortunately the 1938 sample was lost during processing, but figure 3 shows an elevation similar to that in the Central Park tree, probably peaking in 1938. The values themselves are higher than the corresponding late heartwood values in the Central Park tree, due to the fact that this is a rural tree unaffected by additional local fossil carbon.

There is no apparent heartwood effect in the Bear Mountain tree, probably because it is entirely sapwood. About ninety years old, it is very small at about 25 cm in diameter, with growth rings up to 2 mm thick at most.

In the Central Park and Oregon oaks the transition from heartwood to sapwood is not abrupt but is distributed over several rings. A simple model of process was tried by correlating the heartwood values with running averages of the corresponding nuclear values. The correspondence was determined by an interval equal to the interval between the maximum in the contaminated years (1926 for Central Park and 1938 for Oregon) and the maximum nuclear year (1964). For Central Park the best correlation (+0.91) was obtained with five-year averages, and for Oregon the best correlation (+0.98) with three-year averages. (Nuclear era rings for Oregon were not measured, so the Bear Mountain values were used for correlation with the contaminated years.)

In figures 4 and 5 these correlations are plotted. A linear regression for the Central Park oak gives a slope of +0.134 and an intercept of -54.1 ‰. For the oak from Oregon the results are +0.084 and -17.7 ‰. This means that the measured value (Δ) for any ring in this anomalous period for the Central Park tree represents 0.866 of the actual value (Δ_o) and 0.134 of the corresponding later five-year average (Δ^*). That is, for the Central Park tree,

$$\Delta_o = \frac{\Delta - 0.134\ \Delta^*}{0.866}$$

and for the Oregon tree,

$$\Delta_o = \frac{\Delta - 0.084\ \Delta^*}{0.916}$$

The Δ_o values so calculated are plotted along with the measured values for the Central Park tree in figure 6 and for the Oregon tree in figure 7.

An urban tree from the National Arboretum in Washington, D.C., (38°55'N, 76°58'W) was analyzed for the period 1925 to 1970. As evident in figure 8, this tree seemed to show the least local fossil carbon increase of any of the urban trees studied. It also had the heartwood anomaly in the early 1950's, consistent with a transition from heartwood to sapwood about 1958. Much material was available for each ring, so an

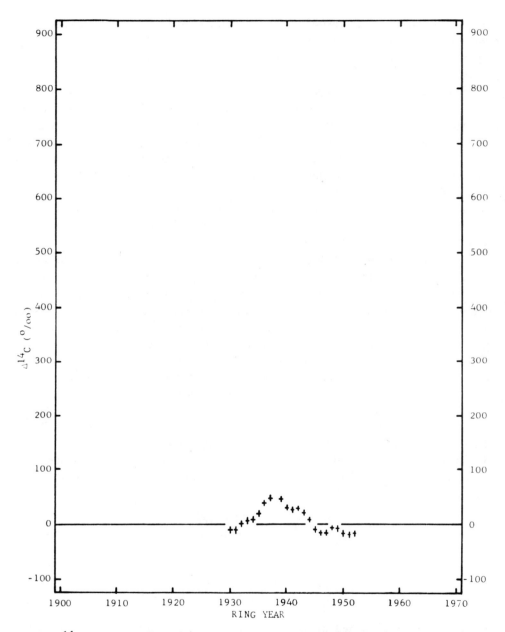

FIG. 3. $\Delta^{14}C$ versus ring year for a white oak from Sheridan Novitiate, Oregon. Heartwood-sapwood transition is approximately 1945.

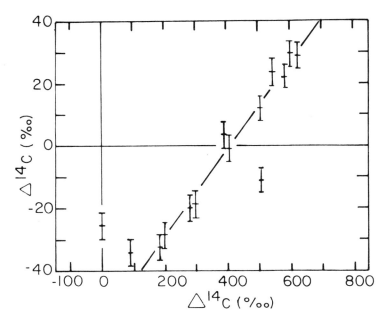

FIG. 4. Central Park Oak: $\Delta^{14}C$ of contaminated years (2920-33) versus corresponding five-year average of $\Delta^{14}C$ thirty-eight to forty-two years later (1958-70). Slope, 0.134; intercept, −54.1 °/oo; coefficient of correlation, +0.91.

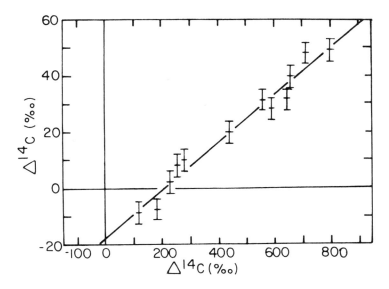

FIG. 5. Sheridan Novitiate Oak: $\Delta^{14}C$ of contaminated years (1930-42) versus corresponding three-year average of $\Delta^{14}C$ twenty-six to twenty-eight years later (1956-70). Slope, 0.084; intercept, −17.7 °/oo; coefficient of correlation, +0.98.

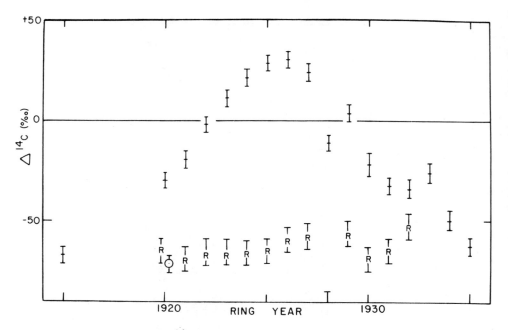

FIG. 6. Central Park Oak: $\Delta^{14}C$ versus ring year. "R" points have had bomb contamination subtracted as described in text. Circled point is an original uncontaminated value for 1920 for comparison with corrected R value at 1920.

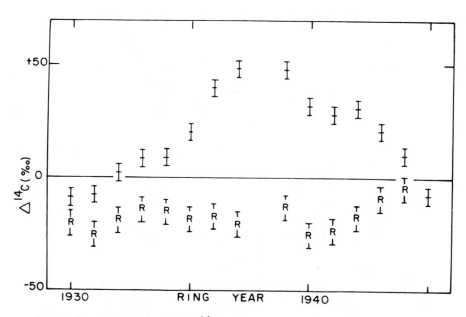

FIG. 7. Sheridan Novitiate Oak: $\Delta^{14}C$ versus ring year. "R" points have had bomb contamination subtracted as described in text.

additional sample for 1951 was bleached. The unbleached result was +3.0 ‰ while the bleached cellulose gave a result of −38.2 ‰. These points are included in figure 8.

There was insufficient material to run bleached samples for the Central Park tree through the whole anomalous period. There was, however, sufficient wood for a preliminary examination of the effect in the sapwood years. Two five-ring samples (1941-1945, and 1946-1950) gave results 11.6‰ and 14.0‰ lower, respectively than the average of the five-ring unbleached samples for the same periods. The bleached samples are included in figure 2.

Figure 9 plots the results for an elm from the Public Garden in Boston, Massachusetts (42°20'N, 71°4'W). This tree from a moderately polluted area does not seem to have any heartwood anomaly, which is in agreement with the absence of a distinctive heartwood/sapwood boundary. There is some change in coloration in the late 1930's and early 1940's, which may account for the slight elevation in $\Delta^{14}C$ values of the 1930's, or the elevation may be a real effect of the great economic depression.

Quantitative Estimate of Added Fossil Carbon

The enrichment in stable carbon isotopes (presumably from air pollution) (P) can be calculated,

$$P = \frac{\Delta^{14}C_r - \Delta^{14}C_u}{1000 + \Delta^{14}C_u} \times 100\%$$

where the subscripts "r" and "u" refer to rural and urban. For Central Park these results have been plotted in figure 10. The continuous line represents a rolling eight-year average of the values. Figure 11 gives a similar treatment of the data from Washington. Because of the heartwood contamination, the relative degree of air pollution is negative from 1948 to 1955. Presumably without the contamination the continuous rolling average would have followed the dashed line. Figure 12 shows the results from Boston. There were not enough data points on this tree for continuous averaging.

The Heartwood Effect: Radial Migration of Carbon?

Work reported by Wilson (1961), Willkomm and Erlenkeuser (1968), Fairhall and Young (1970), and Berger (1970), seemed to demonstrate that no significant amount of later carbon could appear in the solvent-extracted residue of earlier rings. Wilson found that, in a *Pinus radiata* felled in New Zealand in 1959, the activity of the cellulose and the "heartwood extractives removed by acetone and ethyl acetate" corresponded to the activity of the atmosphere when the cellulose was the outer ring of the tree. Willkomm and Erlenkeuser obtained results on an elm from Kiel between 1948 and 1964 which agreed well with atmospheric levels for the corresponding years. Fairhall and Young found that the acetone-soluble component of a heartwood ring of western red cedar was more than 60% above standard (unlike Wilson's results) but that the "[i]nsoluble residue from the acetone extraction show[ed] no excess ^{14}C." Berger injected ^{14}C-labeled sugar into a living tree but found no incorporation in the inner rings.

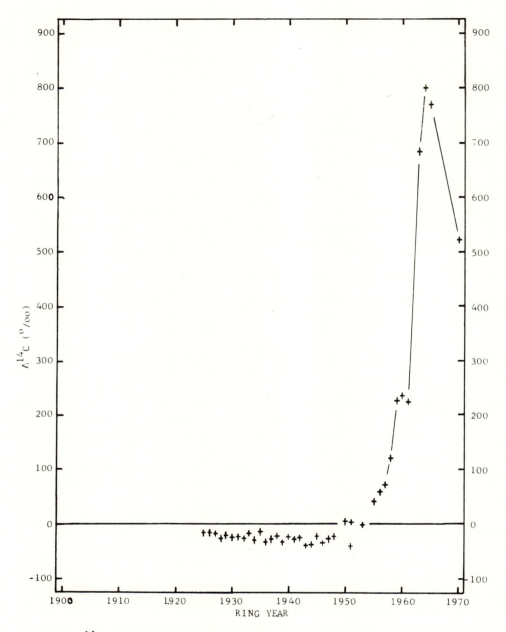

FIG. 8. $\Delta^{14}C$ versus ring year for white oak from U.S. National Arboretum, Washington, D. C. Heartwood-sapwood boundary about 1958. Lower point at 1951 is for bleached cellulose.

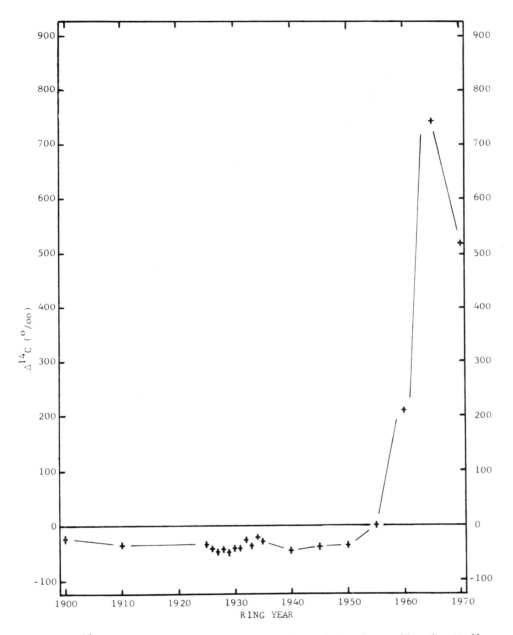

FIG. 9. $\Delta^{14}C$ versus ring year for American elm from Public Garden, Boston, Massachusetts. No truly distinctive heartwood-sapwood boundary.

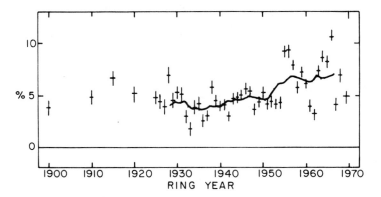

FIG. 10. New York City: Percentage of enrichment of stable isotopes (corresponds to local elevation of fossil original carbon dioxide) in Central Park Oak. Solid line represents eight-year rolling average.

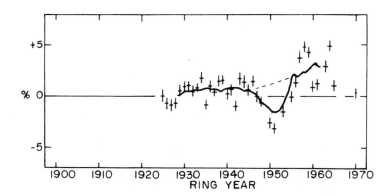

FIG. 11. Washington, D.C.: Percentage of enrichment of stable isotopes in National Arboretum Oak. Solid line is eight-year rolling average. Negative values around 1950 due to uncorrected heartwood effect.

Wilson's is the only work that seems to contradict directly the oak observations in this study. He was looking for a specific effect in the heartwood, but during the years (1957-1959) in which his rings became heartwood, the southern hemisphere was 5% to 15% above standard. The effect could well have escaped detection. Moreover, there seems to be a species dependency in the effect. It is pronounced in the oaks, but questionable in the Boston elm. One other tree, a spruce from Oregon, shows definite contamination, but not in a recognizable pattern (figure 13).

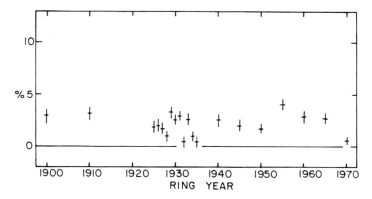

FIG. 12. Boston: Percentage of enrichment of stable isotopes in Public Garden Elm. No averaging.

In the oak from Oregon the effect was about 8% of later carbon. In the Central Park tree it was about 13%. For tree ring radiocarbon dating calibration work, this would not seem to be a significant source of cross-contamination in trees whose growth had ceased prior to the nuclear era. Within the normal lifetime of sapwood before the nuclear era, ^{14}C levels do not change by more than 2% to 3%. Even a 20% contamination by 3% higher ^{14}C would barely exceed statistical error in routine ^{14}C measurements. But studies of tree rings within the last two hundred years should be done only with bleached cellulose. This heartwood contamination and its implications, along with possible in situ production of ^{14}C at high altitudes induced by cosmic rays, is discussed in a report by Cain and Suess (1976).

Local Fossil Fuel Effect and Air Pollution Parameters

Air pollution records for the first sixty years of this century are scant, especially for urban carbon dioxide. Therefore it is difficult to calibrate the calculated stable isotope enrichments shown in figures 10-12, but some crude inferences are possible.

In 1952 Cholak reported that industrialized urban areas showed carbon *monoxide* concentrations up to 55 ppm, with averages between 5 and 10 ppm. The continuous air monitoring program (CAMP), conducted by the U.S. Public Health Service, reported that the annual average for six cities (Chicago, Cincinnati, Los Angeles, Philadelphia, San Francisco, and Washington) for the years 1962 to 1964 was 7.3 ppm (cited by Starkman 1971). Carbon dioxide can be related to carbon monoxide since, according to Starkman, the average emissions of an internal combustion gasoline engine before the days of emission controls averaged 5% carbon monoxide and 10% carbon dioxide by volume (city driving). Almost all carbon monoxide in the urban atmosphere is from automobiles. If we infer conservatively from Cholak a mean carbon monoxide concentration during the growing season in New York City of 5 ppm, then 10 ppm of carbon dioxide might be

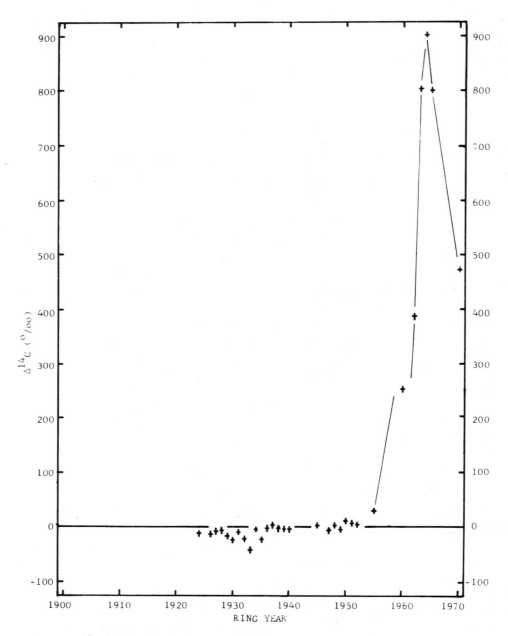

FIG. 13. $\Delta^{14}C$ versus ring year for spruce from Nestucca Bay, Oregon. Heartwood contamination is irregular.

expected from automotive emissions. Starkman estimates that 61% of the fossil fuel carbon in the urban atmosphere is from the automobile, so we can extrapolate to 16 ppm total excess carbon dioxide. If the "pure" air abundance of carbon dioxide were about 300 ppm in 1950, then 16 ppm excess fossil carbon dioxide would represent a stable isotope enrichment of about 5‰. This agrees with the levels shown in figure 10 for 1950.

ACKNOWLEDGMENTS

For supplying the tree specimens I thank Mr. C. M. O'Shea, Mr. P. Burke, and Mr. McDermott of the New York City Department of Horticulture; Mr. H. L. McGuire of the Boston Parks and Recreation Department; Dr. A. A. Piringer and Mr. Gene Eisenbeiss of the Washington National Arboretum; and Rev. G. Moreland, S. J., and Mr. M. Elia, S. J. of the Sheridan, Oregon Novitiate. The late Dr. George Bien, Mrs. C. Hutto, Mr. D. Hasha, Mr. R. L. Simpson, and Dr. T. Linick assisted in technical procedures and analyses. Prof. C. W. Ferguson of the University of Arizona advised and assisted in the dendrochronology. Mrs. K. Podvin, Mrs. E. Linck, and Mr. A. Briket of the laboratory of Prof. H. Craig provided many of the mass-spectrometric measurements. Prof. K. Marti and Dr. G. Lugmair advised and assisted in additional mass-spectrometric measurements. I am most grateful to Prof. Hans Suess who directed my doctoral dissertation submitted to the Department of Chemistry, University of California, San Diego, July 1975, from which this paper is excerpted.

I also thank the National Science Foundation for financial support through grant GA 21161.

REFERENCES

Berger, R.
    1970   Ancient Egyptian radiocarbon chronology. Phil. Trans. Roy. Soc. Lond. A. 269:23.
Cain, W. F., and H. E. Suess
    1976   Carbon-14 in tree rings. J. Geophys. Res. 81:3688.
Cholak, J.
    1952   The nature of atmospheric pollution in a number of industrial communities. Proc. 2d Nat. Air Pollution Symp. Los Angeles, Stanford Research Inst. 6.
Fairhall, A. W., and J. A. Young
    1970   Radiocarbon in the environment. Radionuclides in the Environment. ed. Gould. Washington, American Chemical Society:401.
Norman, A. G., and S. H. Jenkins
    1967   Biochem. J. 27:818. *In* B. L. Browning. Methods of Wood Chemistry. II. New York, Interscience, 406.
Starkman, E. S.
    1971   Vehicular emissions and control. Combustion-generated Air Pollution. New York, Plenum, 205.

Suess, H. E.
  1954 Natural radiocarbon and the rate of exchange of carbon dioxide between the atmosphere and the sea. Proc. Conf. on Nuclear Processes in Geolog. Settings. Williams Bay. NAS-NSF 52.
  1965 Secular variations of the cosmic-ray-produced carbon-14 in the atmosphere and their interpretations. J. Geophys. Res. 70:5937.

Willkomm, H., and H. Erlenkeuser
  1968 University of Kiel radiocarbon measurements III. Radiocarbon. 10:328.

Wilson, A.T.
  1961 Carbon-14 from nuclear explosions as a short-term dating system: use to determine the origin of heartwood. Nature. 191:714.

2

^{14}C Content of Nineteenth Century Tree Rings

M. J. Stenhouse

ABSTRACT

Individual tree rings spanning the years 1841-1890 were separated from a section of an oak tree grown in Washington D.C. After organic extraction the wood was bleached and the cellulose extract of each ring was analyzed for ^{14}C content in two independent counting systems. Results obtained by such analyses are discussed in terms of counter performance and possible short-term atmospheric ^{14}C fluctuations.

INTRODUCTION

Since the initial ^{14}C measurements made by de Vries (1958), which demonstrated a 2%-3% oscillation in atmospheric ^{14}CO$_2$ activity during the past 400 years, several intensive studies using dendrochronologically dated wood have resulted in a knowledge of the ^{14}C content of the atmosphere over the last eight millennia (Suess 1970a, Damon et al. 1972, Michael and Ralph 1972). Thus secular atmospheric ^{14}C variations — over periods of centuries as well as millenia — are now well documented. Although such variations violate the assumption of a uniform equilibrium concentration of atmospheric ^{14}C with time, first proposed by Libby (1955), calibration of the ^{14}C time scale with precisely dated wood has permitted the development of a more reliable radiocarbon dating technique. These long-term variations are not random but can be explained satisfactorily in terms of modulation of the ^{14}C production rate by systematic changes in the earth's magnetic field (Bucha 1970), solar activity, and climate (Suess 1968, Damon 1970, Suess 1970b).

Fluctuations over a lesser time scale, namely, annual variations of atmospheric ^{14}C activity, have been observed (Baxter and Walton 1971, Baxter and Farmer 1973), and have been correlated inversely with the eleven-year solar cycle. The latter correlation remains a stimulus for discussion, particularly with respect to the amplitude of the cyclic trend. Evidence may be presented to support (Lavrukhina et al. 1973, Dergachev 1975) and to contradict (Damon et al. 1973) these original findings. At issue is the lack of a suitable theory to justify a ^{14}C activity-sunspot relationship of such magnitude and on such a sensitive time scale (i.e., year to year). It is agreed that modulation of the cosmic ray flux alone, and hence of the ^{14}C production rate by the eleven-year sunspot cycle, is not sufficient to produce the 3% variations in tropospheric $^{14}CO_2$ activity measured by Baxter et al. In a mathematical treatment of this topic, Houtermans (1966) demonstrated that such solar-induced variations of the ^{14}C production rate are attenuated, by a factor of about 100, by the filtering effect of the atmospheric CO_2 reservoir. Therefore the final fluctuation of ^{14}C activity experienced in the troposphere is expected to be smaller than the statistical error of counting, typically 5°/oo. Other influential parameters, such as modification of the exchange rate of CO_2 with the oceans, may be significant in consideration of short-term oscillations; thus Cain and Suess (1976) attribute the eleven-year solar cycle effect to possible local meteorological conditions, such as the existence of an inversion layer which would be responsible for the reduced circulation of ambient CO_2. Nevertheless the major conclusion of the work of Baxter et al. (1971), which is often neglected in the solar cycle dependence debate, is the fact that annual fluctuations in atmospheric ^{14}C content were observed in excess of the statistical counting error, so that age analysis on correspondingly short-lived material is subject to an additional uncertainty.

Obviously short-term ^{14}C fluctuations require further investigation. If such variations exist, then the archaeological implications remain identical to those outlined previously (Baxter and Farmer 1973), while a geophysical/astrophysical explanation of the phenomenon is essential for a better understanding of the processes which govern the distribution of ^{14}C in nature. For this reason two new counting systems have been established at the La Jolla Radiocarbon Laboratory in order to measure the ^{14}C content of single tree rings. This paper gives an account of the preliminary data obtained.

Experimental

TREE RING PRETREATMENT

A section of oak tree (*Quercus robur*) grown in the National Arboretun, Washington D.C. (38° 55'N, 77° oo'W), and felled in January 1971, was selected for this investigation. Further information on this tree and its selection is provided elsewhere by W. F. Cain (1975), who collected the specimen. Sanding of the section facilitated identification of individual tree rings, typically 2-3 mm. thick, and rings were counted back to 1840. Ten-year sections from 1841-1890 was chopped off and each section was further subdivided into separate rings. Each annual sample was sliced into small fragments no larger than matchstick size, weighed, and sealed in labeled polyethylene bags. Whole wood was extracted overnight, using petroleum ether (40°-60° fraction) in a Soxhlet apparatus, and was subsequently dried in an oven at 110°C. Care was taken to ensure a drying time of at least 48 hours in view of the findings on solvent retention by Jansen (1972). The resultant wood was weighed again, and bleached using a $NaClO_2$/HCl solution (80 g. $NaClO_2$ dissolved in 110 ml. 1N HCl, in solution up to 3 liters) heated to 70°C. If necessary (if the cellulose was discolored) the process was repeated using a fresh bleaching solution. The cellulose residual yield was washed thoroughly with distilled H_2O, dried at 110°C, and weighed prior to combustion. Yields of cellulose were normally in the range 40%-50% by weight of whole wood.

PREPARATION AND COUNTING OF ACETYLENE

After pretreatment cellulose samples were combusted to CO_2 in a stream of oxygen within a silica tube and converted to acetylene via Li_2C_2; the purity of the final gas was verified by means of its triple point. Details of this now routine procedure of C_2H_2 preparation are given elsewhere (Linick 1977). Two sets of counting equipment were employed for ^{14}C activity measurement: (1) a quartz-shell counter, designated Q, with Fe cathode; of volume 2.5 liters and with separate guard ring (800 c.p.m.), and (2) a metal counter, designated M, constructed from stainless steel; of volume 2.0 liters and with separate guard ring (830 c.p.m.). Output pulses from each counter were passed through a voltage-sensitive preamplifier (x1) to electronic facilities suitable for two-channel monitoring of ^{14}C activity. The ^{14}C plateau of each counter was >500 volts with slope <1% per 100 volts change in working voltage. Filling pressures for the quartz and metal counters were 800 mm. and 900 mm. Hg pressure, respectively. Under these conditions, the net NBS oxalic acid standard ^{14}C activities were 25.08 ± 0.10 c.p.m. (Q), and 24.96 ± 0.11 c.p.m. (M); these values were obtained by linear least squares regression analysis performed on each set of individual counts. Background activity was measured using C_2H_2 prepared from anthracite or inactive calcium carbide. Over the analysis period, the background count rante of Q decreased from 8.20 c.p.m. at a rate of 0.20 c.p.m. per month (from statistical analysis, $r^2 = 0.95$; $s_{y.x} = \pm 0.06$), while a small but significant increase in metal counter background (8.50 c.p.m.) of 0.07 c.p.m. per month was detected ($r^2 = 0.80$; $s_{y.x} = \pm 0.09$). Despite this systematic background deviation, the net ^{14}C activities of the NBS oxalic acid, counted regularly over the analysis period, were constant in both counters within counting error, indicating unchanged efficiency of ^{14}C detection. Continuous monitoring of the coincidence as well as sample count rates in both channels was possible, so that small differences in gas gain from sample to sample could be corrected for counting efficiency variation. Each sample was counted for a two-day period in both counters so that a total of at least 70,000 counts was accumulated in each counting system. In this way two independent series of ^{14}C data were compiled for the same set of tree ring samples. The order in which tree rings were analyzed for ^{14}C content was nonconsecutive; in a ten-year batch, the odd-year rings were completed first. This approach was adopted to eliminate the effects of fluctuations which might occur in either counter performance. In addition the 1871-1880 section of Douglas fir used as a secondary standard at the La Jolla Radiocarbon Laboratory was routinely processed and counted as a further check on counter stability.

Net sample ^{14}C activities were corrected for isotopic fractionation via measurement of the $^{13}C/^{12}C$ ratio of CO_2 produced by back oxidation of a small portion of sample gas. After a further correction for radioactive decay since the year of growth, using the

5730-year half-life, activities were compared with 0.95 x activity of NBS oxalic acid. The latter count rate was corrected for decay since 1958, and normalized to $\delta^{13}C = -19.0 ‰$ with respect to PDB Belemnite.

Results

Table 1 contains the ^{14}C results of the tree rings analyzed in the Washington oak series. The data for each counting system are quoted separately and are also combined to produce the average values shown in the final column of table 1. When the sample C_2H_2 has been counted in one counter only, the single Δ value is added in parenthesis to the column of combined results. Figure 1 allows a visual comparison of the individual sets of data.

TABLE 1

^{14}C Content of Nineteenth Century Tree Rings

Washington Oak Series

Tree ring	Metal counter	Δ (age corrected) $\pm 1\sigma$ ‰ Quartz counter	Combined
1841	−0.7 ± 7.9	+7.2 ± 7.8	+3.3 ± 5.6
1842	+4.3 ± 7.9	−9.4 ± 7.7	−2.6 ± 5.5
1843	+3.2 ± 6.0	−3.8 ± 6.0	−0.3 ± 4.2
1844	—	−5.9 ± 7.9	(−5.9 ± 7.9)
1845	−8.7 ± 8.4	−15.9 ± 7.6	−12.2 ± 5.7
1846	+6.2 ± 8.1	−6.2 ± 7.7	0.0 ± 5.6
1847	−0.9 ± 8.1	−5.8 ± 7.7	−3.4 ± 5.6
1848	−6.7 ± 7.5	−2.2 ± 7.7	−4.5 ± 5.4
1849	+11.7 ± 8.1	−5.2 ± 8.0	+3.3 ± 5.7
1850	+0.8 ± 7.8	−1.7 ± 6.1	−0.5 ± 5.0
1851	+8.2 ± 7.6	—	(+8.2 ± 7.6)
1852	+1.1 ± 5.2	−4.4 ± 5.8	−1.7 ± 3.9
1853	−4.9 ± 7.9	−16.1 ± 7.7	−10.5 ± 5.5
1854	−5.4 ± 7.5	−22.6 ± 7.6	−14.0 ± 5.3
1855	+0.8 ± 8.1	−15.9 ± 8.0	−7.6 ± 5.7
1856	−4.1 ± 7.6	−11.0 ± 7.6	−7.6 ± 5.4
1857	—	−17.1 ± 10.1	(−17.1 ± 10.1)
1858	−10.7 ± 5.8	−9.7 ± 8.2	−10.2 ± 5.0
1859	−1.7 ± 8.3	−19.1 ± 8.4	−10.4 ± 5.9
1860	−3.9 ± 7.5	−6.2 ± 7.6	−5.1 ± 5.3
1861	+4.2 ± 8.3	−3.7 ± 5.1	+0.3 ± 4.9
1862	−10.3 ± 7.2	−19.8 ± 7.7	−15.1 ± 5.3
1863	+3.0 ± 7.8	−11.1 ± 5.7	−4.1 ± 4.8
1864	−11.0 ± 8.7	−13.9 ± 7.7	−12.5 ± 5.8
1865	−4.5 ± 8.0	—	(−4.5 ± 8.0)
1866	−2.0 ± 7.3	−10.7 ± 7.7	−6.4 ± 5.3
1867	+8.9 ± 8.1	−12.0 ± 8.3	−1.6 ± 5.8
1869	+4.7 ± 8.0	—	(+4.7 ± 8.0)

TABLE 1 (Continued)

Washington Oak Series

Tree ring	Metal counter	Δ (age corrected) ± 1σ °/oo		Combined
		Quartz counter		
1871	−0.5 ± 7.5	+0.1 ± 7.8		−0.2 ± 5.4
1872	−2.1 ± 7.1	−12.2 ± 7.7		−7.2 ± 5.2
1874	−24.8 ± 7.9	−18.0 ± 7.7		−12.5 ± 5.6
1876	−7.0 ± 8.1	−18.0 ± 7.7		−12.5 ± 5.6
1877	−12.4 ± 7.6	−11.5 ± 7.6		−12.0 ± 5.4
1878	−4.9 ± 7.9	−23.4 ± 5.9		−14.2 ± 4.9
1879	−13.1 ± 7.6	−21.5 ± 7.7		−17.3 ± 5.4
1880	+1.0 ± 6.2	−		(+1.0 ± 6.2)
1881	−3.7 ± 7.6	−4.8 ± 7.7		−4.3 ± 5.4
1882	−13.2 ± 7.2	−14.9 ± 6.8		−14.1 ± 5.0
1883	−26.8 ± 8.1	−22.1 ± 7.7		−24.5 ± 5.6
1884	−9.0 ± 5.6	−10.4 ± 7.7		−9.7 ± 4.8
1886	−12.6 ± 5.9	−		(−12.6 ± 5.9)
1887	−5.5 ± 8.1	−15.8 ± 7.2		−10.7 ± 5.4
1888	−2.1 ± 5.1	−14.3 ± 5.2		−8.2 ± 3.6
1889	−14.8 ± 5.8	−14.3 ± 7.5		−14.6 ± 4.7
1890	−3.7 ± 7.3	−2.2 ± 8.4		−3.0 ± 5.6

^{14}C activities were calculated as Δ, age corrected, using the Lamont VIII formulae (Broecker and Olson 1961). δ^{13}C values ranged from −23.25 °/oo to −26.39 °/oo. Years in which the amount of wood from the tree ring is insufficient for ^{14}C analysis are omitted from the table.

Discussion of Results

In any project designed to detect small differences in ^{14}C activity between samples, a stable counting system is essential. In this respect, the background count rates of both systems did not fulfill this condition so an additional element of uncertainty is incorporated in the final values. The 1σ counting error quoted (tyically ±8 °/oo) includes this uncertainty; combination of this error in each counting system yields a typical combined error of ±6 °/oo. In the absence of background deviation the corresponding ±1σ values would be ±6.5 °/oo and ±4.0 °/oo, respectively.

Linear least squares regression analysis was performed on each set of data, and the resultant regression lines are included in figure 1. Relevant statistical information associated with an analysis is compiled in table 2; in particular, the root mean square error, $s_{y \cdot x}$ (i.e. the standard deviation of estimates of y on x) is less than the individual counting error in each case. Thus the observed ^{14}C measurements can be explained by a straight line. While the slopes of the lines for the independent sets of data agree within error, the most significant feature indicated in figure 1 is the systematic difference in sample ^{14}C activities relative to NBS oxalic acid. This deviation amounts to ~8 °/oo from comparison

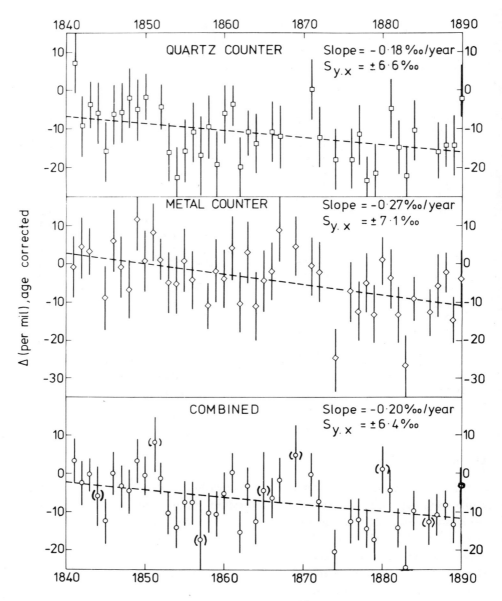

FIG. 1. Washington Oak Series — ^{14}C Data.

of the mean $\bar{x}$ (year) and $\bar{y}$ (Δ data) values in table 2, and can be attributed only to the uncertainty associated with the activity of the ^{14}C standard as measured in one or both counters. If one counting system alone should be responsible for this discrepancy, then lack of secondary counting facilities would result in a systematic error of ~60 years in any age determination performed during this period of study. Admittedly the latter

TABLE 2

Statistical Information on Least Squares Linear Regression Analysis

Output parameter	Q	M	Combined
x	1863.8	1865.2	1864.5
y	−11.14	−4.04	−7.14
r^2	−0.142	−0.242	−0.381
Slope/yr	−0.1816	−0.2675	−0.2010
Estimated variance of slope	0.0050	0.0054	0.0043
Intercept	327.3	494.9	367.7
Estimated variance of intercept	17436	18785	14861
RMS error	6.64	7.07	6.43

comment on accuracy is pessimistic, since statistical analysis of the NBS oxalic acid ^{14}C measurements suggests that the observed systematic error is shared by both counting systems. It is most essential to treat all ^{14}C activities, including those of background and standard, in a statistically acceptable manner so that the actual counting uncertainty associated with each sample may be assessed properly.

With respect to possible short-term ^{14}C fluctuations, the possibility of a cyclic trend is apparent for the set of combined measurements. In the combined series of the diagram the visual effect is accentuated by addition of tree ring data which were obtained in one counting system only. These particular samples are responsible for the pronounced maxima at 1851, 1869, and, to a lesser degree, 1880 (metal counter results). For these tree rings, substitution of the measured Δ values by their deviations from the appropriate regression line is a more realistic approach and greatly reduces the cyclic effect. Cyclic trends in the individual counter data are not so obvious, although the ^{14}C measurements are being subjected to more sophisticated mathematical computations designed to derive best fitting lines as objectively as possible (Stenhouse and Suess 1977, in preparation).

The overall sensitivity of ^{14}C analysis during this period is insufficient to detect ^{14}C fluctuations of 3%-4 °/oo amplitude. On the other hand, cyclic variations in annual atmospheric ^{14}C activity of 2%-3%, peak to trough, are not confirmed by this work. Additional measurements are needed on tree rings growing after 1890 in order to demonstrate the temporary atmospheric increase in ^{14}C content detected by other researchers (Lerman et al. 1970, Stuiver 1974, Lavrukhina and Alekseyev 1977).

CONCLUSIONS

The previous discussion deals with several factors of relevance to the accurate performance of a radiocarbon dating laboratory. Regular cross-checking of sample ^{14}C activities by an independent counting system is desirable if not essential. Intercalibration samples between different laboratories serve a similar purpose. Linear regression analysis by the method of least squares was performed on the ^{14}C data accumulated in this study; the results imply that the measured ^{14}C content of individual tree rings, plotted as a function of time, can be explained adequately by the regression lines. Further mathematical treatment is required to establish statistically whether or not these straight lines provide the best fit to the data.

ACKNOWLEDGMENTS

Excellent technical assistance, both in the preparation of acetylene and in mass spectrometric measurements, was provided throughout this work by Carol Hutto and Terri Broadwell, to whom the author is grateful. Professor H. E. Suess supplied valuable advice and encouragement throughout the project. Financial support through grants GA-25952 and DES75-05519 from the National Science Foundation is gratefully acknowledged.

REFERENCES

Baxter, M. S., and A. Walton
    1971   Fluctuations of atmospheric C-14 concentration during the past century. Proc. Roy. Soc. A321: 105-127.

Baxter, M. S., and J. G. Farmer
    1973   Radiocarbon: short-term variations. Earth and Planet. Sci. Lett. 20: 295-299.

Broecker, W. S., and E. A. Olson
    1961   Lamont VIII Radiocarbon Measurements. Radiocarbon. 3: 176-204.

Bucha, V.
    1970   Influence of the earth's magnetic field on radiocarbon dating. *In* Radiocarbon Variations and Absolute Chronology. Olsson, ed. Proc. 12th Nobel Symp. Uppsala. 501-511.

Cain, W. F.
    1975   Carbon-14 in tree rings of 20th century America. Dissertation. Univ. of Calif., San Diego.

Cain, W. F., and H. E. Suess
    1976   Carbon-24 in Tree Rings. J. Geophys. Res. 81 (21): 3688-3694

Damon, P. E.
    1970   Climatic versus magnetic perturbation of the atmospheric C-14 reservoir. *In* Radiocarbon Variations and Absolute Chronology. Olsson, ed. Proc. 12th Nobel Symp. Uppsala 571-594.

Damon, P. E., A. Long, and E. I. Wallick
    1972   Dendrochronologic calibration of the carbon-14 time scale. Proc. 8th Int. Conf. Radiocarbon Dating. New Zealand. A28-A43.
    1973   On the magnitude of the 11-year radiocarbon cycle. Earth and Planet. Sci. Lett. 20: 300-306.

Dergachev, V. A.
    1975   Variations in the solar activity and the radiocarbon abundance in the earth's atmosphere. Izvestiya Akad. Nauk SSSR, Seriya Fiz. 39(2): 325-333.

De Vries, H.
    1958   Variations in concentration of radiocarbon with time and location on earth. Proc. Koninkl. Ned. Akad. Wetenschap. Ser. B61: 94-102.

Houtermans, J. C.
    1966   On the quantitative relationship between geophysical parameters and the natural ^{14}C inventory. Z. Phys. 193: 1-12.

Jansen, H. S.
    1972   Transfer of carbon from solvents to samples. Proc. 8th Int. Radiocarbon Conf. New Zealand. B63-B68.

Lavrukhina, A. K., V. A. Alekseyev, and E. M. Galimov, et al
    1973   Radiocarbon in Sequoia growth rings. Doklady Akad. Nauk SSSR. 210: 238-240.

Lavrukhina, A. K., and V. A. Alekseyev
    1977   Laboratory of Cosmochemistry Radiocarbon Measurements I. Radiocarbon. 19(1):12-18.

Lerman, J. C., W. G. Mook, and J. C. Vogel
    1970   C-14 in tree rings from different localities. In Radiocarbon Variations and Absolute Chronology. Olsson, ed. Proc. 12th Nobel Symposium. Uppsala. 275-302.

Libby, W. F.
    1955   Radiocarbon Dating. 2d ed. Chicago, Univ. of Chicago Press, 175.

Linick, T. L.
    1977   La Jolla Radiocarbon Measurements VII. Radiocarbon. 19(1): 19-48.

Michael, H. N., and E. K. Ralph
    1972   Discussion of radiocarbon dates obtained from precisely dated sequoia and bristlecone pine samples. In Proc. 8th Int. Radiocarbon Conf. New Zealand. A12-A27.

Stuiver, M.
    1974   Natural radiocarbon in the 19th century. Proc. of the GSE Meeting. Geol. Soc. of Amer. Colorado.

Suess, H. E.
    1968   Climatic changes, solar activity, and the cosmic ray production rate of natural radiocarbon. Meteor. Monographs. 8(30): 146-150.
    1970a   Bristlecone-pine calibration of the radiocarbon time-scale; 5200 B.C. to the present. In Radiocarbon Variations and Absolute Chronology. Olsson, ed. Proc. 12th Nobel Symp. Uppsala. 303-311.
    1970b   The three causes of the secular C-14 fluctuations, their amplitudes and time constants. In Radiocarbon Variations and Absolute Chronology. Olsson, ed. Proc. 12th Nobel Symp. Uppsala. 595-605.

3

Subsurface Radar Probing for Detection of Buried Bristlecone Pine Wood

Henry N. Michael and Roger S. Vickers

INTRODUCTION

During the past three years, several experiments have been conducted in the use of subsurface radar probing for archaeological exploration. One of the experiments, reported here, has had the additional objective of locating remnants of the bristlecone pine, *Pinus aristata*, reclassified as *P. longaeva*, sp. nov., (Bailey 1970) buried beneath alluvial material in the drainage areas of the White Mountains of east-central California. This study has utilized a ground-penetrating radar system originally developed for use in locating underground utilities and shallow, geologically-related phenomena such as solution cavities, fractures, and tunnels.

This type of radar operates on simple principles, transmitting a short pulse of energy at a selected frequency into the ground and displaying any received echoes simultaneously on an oscilloscope screen and on a chart recorder. The major factors affecting its performance are soil conductivity (which is itself a function of soil moisture) and the dielectric contrast between the sought-after target and the material in which is is buried. An example of the radar losses due to soil conductivity in Chaco Canyon soils as a function of frequency is given in table 1. The losses are given in decibels per meter. A good figure-of-merit to use for a radar of this type is that it can detect return signals about 80 to 100 dB below the level of the transmitted signal. Putting in typical values for reflection coefficients and antenna coupling efficiencies, a typical system could tolerate only 40 to 60 dB of loss in the ground. From table 1, a soil with 6% moisture would cause a radar loss of about 19 dB/m and would therefore limit the useful two-way radar range to 2 to 2.5 m. The other major factor, the reflection coefficient of the target material, is dependent on the difference in the dielectric constant of the soil and target. The greater the difference, the more energy is reflected back to the radar receiver at the surface. Typical values for soil and target materials that might be encountered in archaeological surveys range from 1.6 to 7.0.

TABLE 1

Frequency (MHz)	Wet (as received, 6% moisture content by weight)			Dessicated (oven, 175°C)		
	C^*_p (pF)	R_p	α (dB/m)	C^*_p (pF)	R_p	α (dB/m)
5	8.2	1220	3.5	3.9	4700	1.9
10	5.9	1110	5.4	3.6	4250	2.3
15	5.1	1050	6.6	3.3	4000	2.7
20	4.6	980	7.8	3.2	3780	2.9
30	4.1	890	9.7	3.0	3400	3.5
50	3.6	755	13.0	2.9	2875	4.2
75	3.4	625	16.6	2.9	2500	4.9
100	3.6	520	18.9	3.0	2020	5.8
150	4.0	340	26.0	3.5	1250	8.1
200	5.0	205	34.4	4.0	700	12.6
250	7.3	104	46.5	5.2	370	18.3

*Chaco Canyon Pueblo Alto Soil (Depth 10 cm.). The two components of the impedance of the sample are C^*p, the capacitive component in picofarads, and Rp, the resistive component in ohms. The attenuation, α, is given in decibels per meter.

Early Uses

The radar system described in this paper was first used for an archaeological application in 1974, when a joint research team from Stanford Research Institute, the Museum Applied Science Center for Archaeology (MASCA) at the University of Pennsylvania, and the Chaco Center of the University of New Mexico combined to determine whether buried masonry structures could be detected with radar. As reported earlier (Vickers and Dolphin 1975a) the experiment was successful.

Typical output of the radar is demonstrated in figure 1, which shows the radar returns from a traverse across an outer wall at Hungo Pavi in Chaco Canyon. The horizontal axis corresponds to the horizontal traverse direction and is about eight meters long. The vertical axis represents the travel time of the electromagnetic pulse and is proportional to the depth into the ground from which the various echoes shown originated. In this particular example, anomalies such as the man-made structures show up as white shadows. The small area to the upper left was marked during the traverse and subsequently excavated. The uncovered target, shown in figure 2, was the upper surface of a well preserved wall about 16 cm below the surface. Despite the shallow burial, there was no discernible surface expression of the structure. Other areas were surveyed to determine the limitations of the system, including kivas that had been mapped and backfilled. The results were quite encouraging because, in soils with low radar losses, masonry targets could often be detected, although known adobe structures with modern fill material were

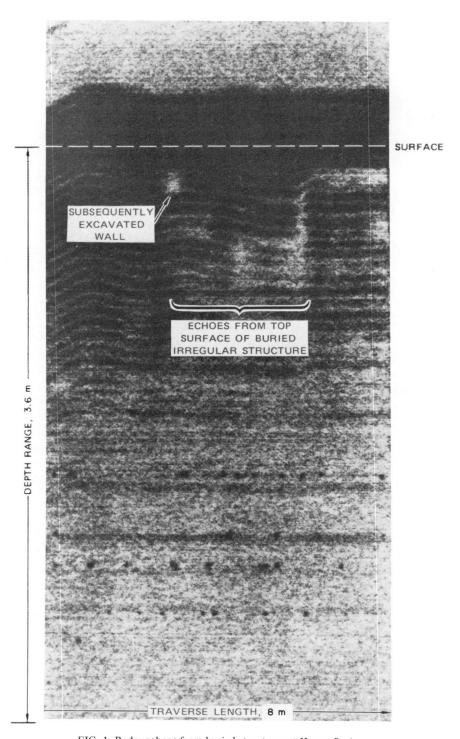

FIG. 1. Radar echoes from buried structures at Hungo Pavi.

FIG. 2. Excavated pueblo wall which had been located by soil-penetrating radar.

usually missed. In one case strong echoes were obtained over adobe walls, but it was not clear whether the echoes were due to the walls or due to the difference in soil compaction resulting from excavation procedures.

Use on Alluvial Fans

After this initial success, it was decided — in cooperation with MASCA — to test the system in detection of buried bristlecone pine wood in the western alluvial fans of the White Mountains. Because of the nature of depositions of materials on alluvial fans it was believed that wood might be found exceeding in age that presently known. Survey locations were selected (by H. N. Michael) on the basis of descriptions of the behavior of debris flows from the narrow canyon mouths during severe thundershowers over the canyon's drainage (Beaty 1963;1970). These tracts were about 2 m wide on the upper thirds of two fans — Silver Creek fan near Laws, California, and Milner Creek fan in Chalfant Valley, some 22 km to the north.

Numerous buried boulders in the alluvial material caused a dielectric constrast resulting in reflection characteristics similar to those expected from the logs. A number of traverses were made and locations where echoes appeared were marked for excavation. The radar losses in the fan alluvium are shown in figure 3. It can be seen that at 250 MHz, the attenuation limit of 60 dB would not be reached before 3 m depth, while at 100 MHz, 15 m penetration would be possible. Traverses were conducted at several frequen-

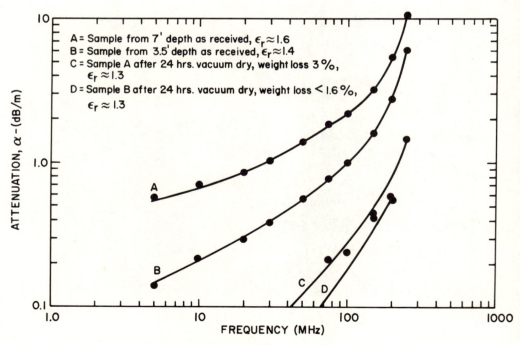

FIG. 3. RF losses in silt from Silver Creek alluvial fan site F.

cies between 70 and 300 MHz. From these it was determined that the best data were being obtained at about 100 MHz, which was standardized as the frequency for subsequent traverses made on the two fans. The data indicated that the radar signal was penetrating at least 10 m over most of the profiles. The positions of all anomalies with signatures close to those anticipated from buried wood were marked. Some of these were later excavated and each target was found to be a large boulder, similar to the one shown in figure 4. The several tracks surveyed did not yield wood. Possible explanations are that (1) human judgment of the locations was faulty; or that (2) the small size of the area surveyed — less than 1/2 hectare in an expanse of several dozen square kilometers — reduced all possibility of detecting wood to luck or coincidence; or that (3) no wood was preserved in the dry soils of the alluvial fans since, as it was deposited on the surface or buried shallowly during debris flows, it was destroyed by insects such as termites and carpenter ants.

Blind Digs

The surveys described above were completed in five days. Prior to the excavations the radar crew departed. Excavations were begun in the relatively moist soil at the apex of the Milner Creek fan, since the dry soils had yielded no wood. "Blind Dig No. 1," the first excavation, uncovered a buried logjam extending in depth from 1 m to 2 m. Samples of the various woods found, including that of bristlecone pine, are shown in figure 5. The radiocarbon date of the bristlecone pine was about 2300 B.P. Other blind digs yielded small pieces of wood, none of which proved to be of an age greater than 2000 years. These excavations demonstrated that wood is preserved in alluvial fans. In anticipation of future experiments and surveys in the area, specimens of woods (bristlecone pine and limber pine) were buried in the sterile excavations and their locations were noted. Part of the above-mentioned buried logjam was left unexcavated to facilitate future collection of data over these three locations with the results permitting discrimination between the radar signatures of boulders and those of buried wood.

Recent Developments

The work in Chaco Canyon (Vickers and Dolphin 1975b) and the White Mountains suggested improvement of the radar system and the associated recording and data reduction by removing the cables connecting the radar cart to the electronics and converting to an all-digital system. Without the umbilical cables, traversing can be made over greater distances and rougher terrain while avoiding problems of having to identify and remove cable reflections and reradiation of the impulse. A system has been developed and built in which the data can be transmitted via a telemetry link to a nearby vehicle equipped with a computer and display unit. A position-location system has been built into the radar cart providing an immediate capability at the computer for plotting the location of the radar traverses.

The new platform consists of a cordless, hand-towed cart (figure 6a) to which any type of sensing instrument for subsurface exploration can be readily attached. Such

FIG. 4. Boulder uncovered at the site of a radar echo.

FIG. 5. Samples of woods excavated from "Blind Dig no. 1".

FIG. 6. Second generation soil-penetrating radar system. (a) Cordless cart, (b) van with receiving equipment.

sensors can include ground-penetrating radar, presently available magnetometers, microgravity meters, or resistivity probes. Output from the sensing instrument is automatically transmitted by a radio link to a receiver located in a vehicle (figure 6b) parked at some convenient distance where the data are processed by a small computer, printed out within a few seconds, and digitally tape-recorded for permanent record. Simultaneously with the sensor operation, a position-location system records the cart position on the horizontal plan (to an accuracy of approximately 0.3 m) and plots this information at a suitable scale directly on a map or aerial photograph of the site being surveyed. Any detected anomalies can be superimposed on the map or aerial photograph of the site to guide the investigator in his subsequent exploration and excavation. Working in conjunction with him, the survey crew can provide markers and other on-the-spot information for his immediate use. Where terrain is too rugged to use the towed cart, the platform can be adapted for backpack operation by one or two men.

The automatic position-detection unit eliminates the need for time-consuming site-surveys and the laying out and staking of grids. Time and cost requirements for an archaeological survey are reduced and the site is left undisturbed. Any kink of grid including a random walk pattern can be chosen and, since results are available in real time, areas of the site showing unusual subsurface features can be immediately reexplored using a tighter grid spacing to obtain more details of subsurface features.

Work on construction and testing of the system was completed in early 1976. The system was field tested in the plaza area of Pueblo Alto, a hitherto unexcavated site that had been surveyed by the first generation system in 1975. A random walk survey pattern

FIG. 7. Map of results of radar survey at Pueblo Alto in 1975.

EXAMPLE OF POSITION-LOCATION SYSTEM OUTPUT

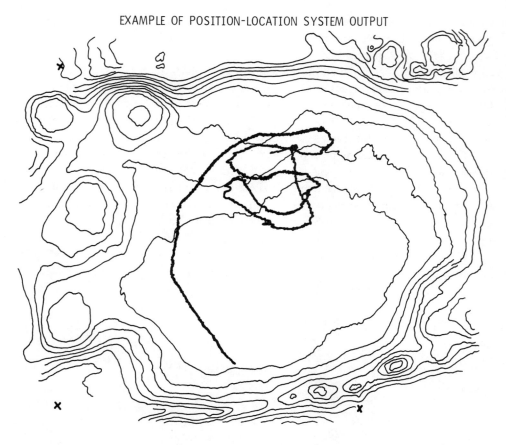

FIG. 8.

was used to avoid the bias previously experienced, in which apparent structures tended to align themselves with the survey grid. In several areas echoes were plotted that lay in a straight line, leaving little doubt that a linear feature was buried in those locations. The resultant plots are shown in figures 7 (1975 survey) and 8 (1976 survey).

ACKNOWLEDGMENTS

We acknowledge with gratitude the partial financial support of the National Science Foundation, Earth Sciences Division, through grant DES-74-22233 for the investigation of the alluvial fans and the radiocarbon dating of the samples in the Radiocarbon Laboratory, University of Pennsylvania, Dr. E. K. Ralph, Director.

Dr. L. T. Dolphin of the Stanford Research Institute directed the field operations with his usual enthusiasm. To him and to his assistants go our sincere thanks.

REFERENCES

Bailey, D. K.
 1970 Phytogeography and Taxonomy of *Pinus* Subsection *Balfourianae*. Annals Missouri Bot Garden. 57:210-249.

Beaty, C. B.
 1963 Origin of Alluvial Fans, White Mountains, California and Nevada. Annals Assoc. of Amer. Geog. 53 (4):516-535.
 1970 Age and Estimated Rate of Accumulation of an Alluvial Fan, White Mountains, California, U.S.A. Amer. J. Science. 268:50-77.

Vickers, R. S., and L. T. Dolphin
 1975a A Communication on an Archaeological Radar Experiment at Chaco Canyon, New Mexico. MASCA Newsletter. 11:1.
 1975b Subsurface Radar Sounding Experiments in Archaeology. Proc. URSI Annual Meeting, Colorado.

4

Radial Translocation of Carbon in Bristlecone Pine

A. Long, L. D. Arnold, P.E. Damon, C. W. Ferguson, J. C. Lerman, and A. T. Wilson

INTRODUCTION

Current knowledge about the variability of ^{14}C in past atmospheres, and the geophysical and astrophysical phenomena causing this variability, rests entirely on precise ^{14}C determinations on tree rings of known age (Damon et al. 1978). The need for greater analytical precision and accuracy has stimulated interest in a new ^{14}C analysis technology (Muller 1977, Bennet et al. 1977). Accuracy begins with the sample material itself and requires not only confidence in the age of tree ring(s), but assurance that all the carbon analyzed from the sample represents the year(s) of growth.

Assuming proper physical separation of the growth ring(s) from adjacent ones, two critical questions remain: Are the various chemical constituents of tree rings really of the same age and formed during the year in which the ring grew? If so, do the widely used sample pretreatment techniques remove noncontemporaneous carbon? We have attempted to answer these questions in the case of bristlecone pine (*Pinus longaeva*), the species most widely used for ^{14}C fluctuation determination (see also Berger, 1972).

Earlier work by Wilson (1961), Olsson et al. (1969, 1972), Fairhall and Young (1970), and Cain and Suess (1976) answered this question for *Pinus radiata*, a Swedish pine, an Argentine *Fitzroya cupressoides*, and a German oak, concluding that chemical separates from the same rings may have quite different $^{14}C/^{12}C$ values. The nearly universally-used wood pretreatment introduced by de Vries effectively removes humic acids and carbonates that a specimen may have accumulated while buried in soil, but the technique has not been tested for its ability to remove possible mobile constituents incorporated into the wood specimen while the tree was living.

Experimental Work

Nuclear weapon tests in the 1950's and 1960's provided the ^{14}C tracer spike necessary to obtain the required data. A tree which grew in Inyo National Park until 1975 had deposited its resinous heartwood up to the rings dating A.D. 1850 to 1895, varying with different radii. If mobile constituents were a reality in this tree, they would occur in the most recently deposited heartwood and would be recognizable by their bomb-era tag.

We sampled the following twenty-year samples: (1) A.D. 1855-1875 (heartwood), (2) A.D. 1930-1950 (pre-bomb sapwood), and (3) A.D. 1955-1975 (post-bomb sapwood); and we analyzed ^{14}C in the following: (1) untreated wood, (2) "de Vries treated" wood, (3) extracted wood, (4) solvent extracts, and (5) cellulose, all prepared using standard techniques (table 1). The results in terms of Δ (‰) in table I are plotted in figure 1. Δ is

TABLE 1

^{14}C Activity in Twenty Tree Ring Segments of Bristlecone Pine

Wood Fraction	Preparation Technique	^{14}C Activity in Δ (°/oo) Relative to Modern Standard		
		A.D. 1855-1875	A.D. 1930-1950	A.D. 1955-1975
Untreated		+27.03	+5.36	
de Vries Treated	Split wood into "match sticks", 2 to 3 mm diameter, 1 cm long. Treat with 4% HCl at 80 to 90° C overnight. Then 1% NaOH at 80 to 90° overnight. Finally, repeat HCl treatment and wash. Dry at 105°C.	+35.15		
Extracted	Grind to 20 mesh, extract in Soxhlet with Benzene-ethanol 2:1 for 20 hours, then in 95% ethanol for 5 hours, finally boil in H_2O and dry at 110°C.	+13.66		
Extractives	Evaporate benzene-ethanol mixture from above Soxhlet, add excess of water and evaporate. Repeat water step three times. Dry at 110°C.	+268.15		
Cellulose (Holocellulose)	Suspend extracted wood in solution of dilute acetic acid and NaClO, heat to 70°C, periodically adding fresh acetic acid and NaClO until pure white product forms. Filter, wash with H_2O, dry at 105°C.	+2.29	-19.88	+426.75

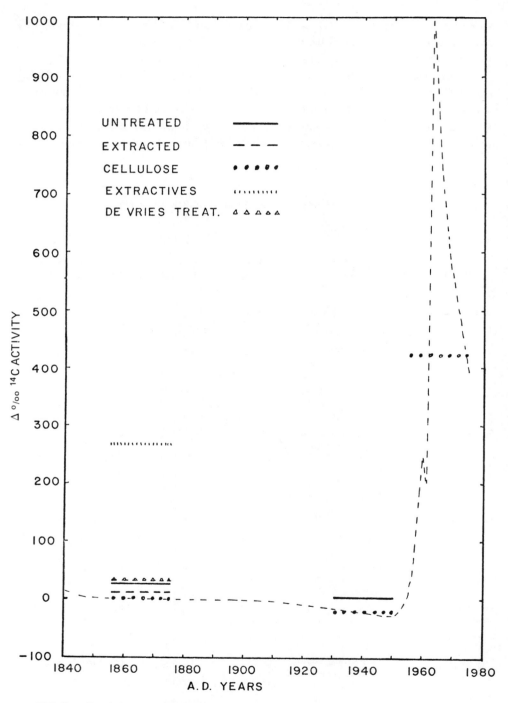

FIG. 1. Δ(°/oo) from table I plotted vs. age of tree ring specimens. Dashed curve is atmospheric Δ(°/oo) value derived from atmospheric sampling of Nydal et al. (1976) and Nydal and Lövseth (1970) and tree ring measurements of Lerman et al., (1970).

defined as the fractionation-corrected deviation of the sample ^{14}C activity in parts per mil from modern (or the A.D. 1890 ^{14}C activity, age corrected to A.D. 1950) as defined by Broecker and Olson (1961):

$$\Delta = \delta^{14}C - (2\delta^{13}C + 50)\left(1 + \frac{\delta^{14}C}{1000}\right)$$

where:

$$\delta^{14}C = \left[\frac{A_x}{A_o} - 1\right] \times 1000$$

and A_x and A_o are age corrected ^{14}C activities of sample and standard, respectively.

Discussion

Atmospheric $^{14}C/^{12}C$ values and wood Δ-values for the past few hundred years are well known. The last half of the nineteenth century is close to zero per mil by definition with respect to the carbon dating modern standard, and pre-bomb Δ-values are a few tens per mil below zero. During the bomb-era atmospheric and biospheric Δ-values rose sharply. They were briefly about 1000 °/oo in the northern hemisphere (see dashed curve in fig. 1).

The cellulose Δ-value on bomb-era tree rings is reasonable in that it averages the atmosphere over a twenty-year period. The other two cellulose values are also consistent with values obtained from trees and plants not living into the bomb-era, with the sapwood Δ depressed due to the industrial or Suess effect.

On the other hand, most strikingly, the material extracted from the heartwood contains ^{14}C activity well above the known atmospheric activity during the nineteenth century. Most of the extracted carbon could have formed only during the bomb-era; it must have radially traversed 100 rings while the late nineteenth century wood became heartwood. The Δ-value of untreated heartwood is consistent with the Δ-values of extractives and cellulose, considering that the extractives contained a measured 10.8% of the carbon in the untreated wood. Moreover, the de Vries treatment on the heartwood sample evidently did not remove the exotic carbon, yet as much as 50% of the sample material was destroyed during treatment. We will report elsewhere on lignin analyses in progress.

What does this signify for measurements of past atmospheric Δ-values through ^{14}C analyses of tree ring material? Few trees used for these studies grew into the bomb-era. Because the maximum variation in any 100-year A.D. period until 1900 is about 20 °/oo, and because extracted material composes only about 10% of the untreated wood, we expect a maximum effect of 2 °/oo on untreated bristlecone pine samples from this time period.

With the present state-of-the-art precision, ± 2 °/oo is not insignificant. With the possibility of a new generation of precision and accuracy in ^{14}C dating (Bennett et al. 1977), a new pretreatment, at least for bristlecone pine specimens, will be desirable. Holocellulose separation seems the most judicious choice at present.

CONCLUSIONS

1. A specimen of bristlecone pine heartwood contains carbon components which translocated through 100 rings of sapwood.

2. The standard de Vries HCl-NaOH-HCl pretreatment removes up to half of the wood, but not all of the translocated component.

3. For most tree ring calibration or atmospheric retrospective monitor samples, the probable error due to the phenomenon must have been small but not insignificant.

4. Present state-of-the-art ^{14}C detectors and possible future technology in ^{14}C measurement demand elimination of this possible cause of error.

5. We recommend solvent extraction to replace the de Vries treatment on all tree ring and plant material specimens collected for retrospective atmospheric monitoring. This is the compromise of choice between maximum removal of mobile carbon and minimum sample loss. If bomb-effect contamination is a possibility, or if sample size is not a limitation, we recommend the holocellulose separation.

ACKNOWLEDGMENTS

We are grateful to Ms. Linda Gonzalez for her excellent work in typing and proofing this manuscript. Our research was supported by National Science Foundation grant DES76-22629.

REFERENCES

Bennett, C. L., R. P. Beukens, M. R. Clover, H. E. Gove, R. B, Liebert,
A. E. Litherland, K. H. Purser, and W. E. Sondheim
    1977   Radiocarbon dating using electrostatic accelerators: negative ions provide the key. Science. 198:508-10.
Berger, R.
    1972   Tree-ring calibration of radiocarbon dates. Proc. 8th Internat. Conf. on Radiocarbon Dating, Wellington, New Zealand, 1:A97.
Broecker, W. S., and E. A. Olson
    1961   Lamont Radiocarbon Measurements VIII. Radiocarbon. 3:176-204.
Cain, W. F., and H. E. Suess
    1976   Carbon-14 in Tree Rings. J. Geophysical Res. 81:3688-3694.
Damon, P. E., A. Long, and J. C. Lerman
    1978   Temporal Fluctuations of Atmospheric ^{14}C: causal factors and implications. Ann. Rev. Earth Planet. Science. 6:457-494.
Fairhall, A. W., and J. A. Young
    1970   Radiocarbon in the Environment. Radionuclides in the Environment. *In* ed. Freiling. Adv. in Chem. 93. American Chemical Society, Washington, D.C.
Lerman, J. C., W. G. Mook, and J. C. Vogel
    1970 ^{14}C in tree rings from different localities. In Olsson, I.U., ed. Radiocarbon

Variations and Absolute Chronology. Proc. XII Nobel Symp. New York, Wiley Interscience, 652.

Müller, R. A.
  1977 Radioisotope dating with a cyclotron. Science. 196:489-494.

Nydal, R., K. Lövseth, and S. Gulliksen
  1976 A Survey of Radiocarbon Variation in nature since the test ban treaty. Proc. 9th Int. Radiocarbon Conf. Univ. of Calif., Los Angeles and San Diego.

Nydal, R., and K. Lövseth
  1970 Prospective decrease in Atmospheric Radiocarbon. J. Geophysical Res. 75 (12):2271-2278.

Olsson, I. U., M. Klasson, and A. Abd-el-Mageed
  1972 Uppsala Natural Radiocarbon Measurements XI. Radiocarbon. 14 (1): 247-271.

Olsson, I. U., S. El-Gammal, and Y. Göksu
  1969 Uppsala Natural Radiocarbon Measurements IX. Radiocarbon. 11 (2): 515-544.

Wilson, A. T.
  1961 Carbon-14 from Nuclear Explosions as a short-term dating system: Use to determine the origin of heartwood. Nature. 191 (4789):714.

5

The ^{14}C Level during the Fourth and Second Half of the Fifth Millennium B.C. and the ^{14}C Calibration Curve

Hans E. Suess

For the period from 4400 B.C. to 3000 B.C. a revised calibration is presented based on about 100 precision measurements of wood samples from "floating" tree ring chronologies of European oak and on 66 precision measurements of dendrochronologically dated wood samples of the California bristlecone pine. Also investigated is the "fine structure" in the time dependence of the ^{14}C level for the period from 4400 to 4200 B.C., when the magnetic dipole moment of the earth was minimal.

At the Uppsala Nobel Symposium on ^{14}C variations and the absolute radiocarbon time scale in 1969, I presented a preliminary ^{14}C calibration curve that permitted, within limits of error, conversion of conventional carbon dates into true historical age. The curve that I presented was based on some 300 individual measurements on tree ring dated wood obtained mostly from C. W. Ferguson of the University of Arizona tree ring laboratory. The curve, drawn by "cosmic schwung" showed many kinks and wiggles. Its purpose was to provide not only a convenient tool for deriving probable historical ages but also what was to be considered the most probable character if variations in the ^{14}C activity of atmospheric carbon dioxide during the past 8000 years. This character is of great interest for geophysics and geochemistry, as it illustrates secular variations of a well-defined geophysical parameter of the planet earth.

At the time of the Uppsala Conference several other laboratories had conducted measurements of samples of precisely known age. These measurements confirmed the general trend of Δ^{14}C with time, but they did not allow to recognize the presence of the irregularities, or "wiggles," demonstrated by the La Jolla measurements. This was in part due to the limited number of measurements of samples in part due to large experimental errors, and perhaps also in part due to possible differences in equipment calibration. Opinions at the Uppsala Nobel Symposium were divided. Some believed the wiggles of the type that I had described did not exist at all. Other accepted my preliminary curve in all its details and have since applied it to their own research problems.

At the time of publication of the Uppsala curve (Suess 1971), I specifically requested that no use be made of these early data for further evaluations, and in particular that they not be combined with data from other laboratories because of their incomplete nature. Despite this, several so-called "new" calibration curves have been published by other researchers based primarily on my 1969 data, but with some added results from other laboratories. In all cases, a much smoother curve has been derived than the one which was given by me at the Uppsala Symposium. Recently (1975) Malcolm Clark of Monash University, Australia, has published what he called "a rigorous statistical analysis" of all the published data on samples of known age, showing that the standard errors of all the data published, including those from La Jolla, are considerably larger than expected from counting statistics and assumed by the authors, and that, "the so-called wiggles in the calibration curve are statistically not justified." About 50% of the numerical values subjected to analysis have been derived from my pre-1969 measurements. The other 50% of the values consisted of results obtained since then by all the laboratories of the world combined. Figure 3 shows how the curve obtained by Clark compares with that given by me in Uppsala and with my empirical results obtained since then.

One fact which I believed was overlooked in Clark's analysis is that the experimentally obtained values do not generally follow a Gaussian distribution. Figure 1 illustrates this schematically by showing two extreme cases for the error distributions obtainable by carrying out many determinations from one and the same sample. Because of the high costs and the long time required, it is practically impossible to make the statistically required large number of measurements; but judging from the scatter of results for related measurements, results from an optimally working laboratory can be expected to follow a

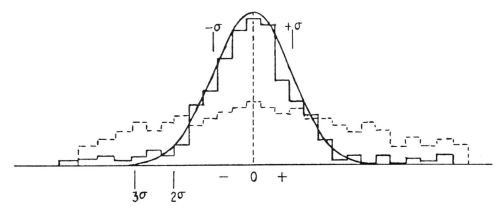

FIG. 1. Histograms of results that might be expected from two fictitious laboratories. The solid lines represent one operating well. The broken line represents one operating poorly. A Gaussian distribution and corresponding standard errors (1, 2, and 3 sigma) are indicated.

Gaussian distribution, except that some results will deviate to a much larger degree from an average value than would correspond to statistics. These large deviations may be due to human error in sample preparation, to sample contamination, or to malfunctioning of the equipment. The broken line in figure 1 indicates schematically the distribution of results to be expected from a poorly working laboratory. Here the spread may be three times that corresponding to counting statistics. Samples deviating too far from the expected values are often routinely omitted, so that the range of values obtained by the two types of laboratories does not seem to differ appreciably.

Since the Uppsala Conference more than 600 additional measurements have been conducted in La Jolla of samples of known age. Also I have objectively reevaluated the measurements on which the Uppsala curve was based, and I have omitted values that for experimental reasons appear doubtful. I could see that a slight memory effect of the counting system had affected results of measurements preceded by measurements of considerably higher or lower ^{14}C activity. This memory effect primarily affected background and standard values; many data had to be recalculated with revised values.

I will not present here another preliminary calibration curve subject to further revisions. Instead, I am presenting data for a selected time range that I consider to be, for all practical purposes, complete and final. Adjustments may be necessary by adding or subtracting on the B.P. scale some tens of years because, for a large part of the measurements, late nineteenth century wood and not the oxalic acid standard has been used as ^{14}C standard. The data cover the period from 4400 to 3000 B.C. All the results obtained are shown in figure 3; publication of numerical values shown here graphically, and of those for other periods of time, will be postponed until completion in La Jolla of a mathematical analysis of the data. Two types of samples have been used for the investigations: (1) dendrochronologically dated bristlecone pine wood, supplied by Prof. C. W. Ferguson of the University of Arizona, consisting always of ten-year segments, and (2) samples from oak trees from European riverbeds supplied by Dr. Bernd Becker of the University Stuttgart-Hohenheim, consisting frequently of single annual rings and never containing more than five annual rings. These latter samples belonged to so-called floating tree ring chronologies for which the absolute date had to be determined by matching the results with those obtained from Bristlecone pine measurements. The provenience of the material and some of the results were described in 1974 at the Mainz Conference and have been published by Suess and Becker (1977).

Figure 2 shows $\Delta^{14}C$ values obtained for samples from the Donau-8 series for the time from 4400 B.C. to 4200 B.C. The $\Delta^{14}C$ denotes the deviation of the ^{14}C content per mil from the calculated ^{14}C content relative to 0.944 times the NBS oxalic acid activity, taking the ^{14}C half-life to be 5730 years. A large number of samples for a relatively small time range has been measured in order to investigate variance and reproducibility of the results, and the general behavior of the $\Delta^{14}C$ values on a short time scale. This period of time was selected because the earth's magnetic field was close to a minimum, so that effects from solar activity would be more pronounced than at present, when the earth's magnetic field is higher by about a factor of three. When these measurements were begun, it was considered that an eleven-year cycle in the $\Delta^{14}C$ (Baxter and

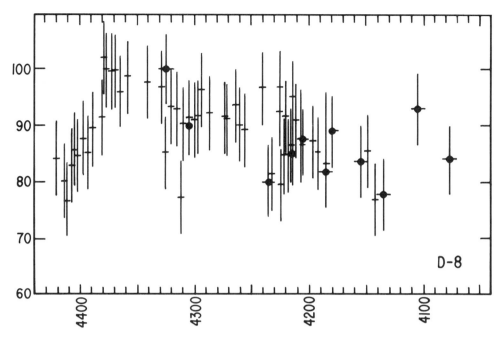

FIG. 2. Measured value of $\Delta^{14}C$ for the period from 4420 B.C. to 4100 B.C. in per mil, calculated with a ^{14}C half-life of 5720 years. Crosses refer to values obtained from a floating European oak chronology; circles refer to values for the Ferguson bristlecone pine samples.

Walton 1971) would be more pronounced at times of low magnetic dipole moment of the earth than at times of high geomagnetic moment. No indication for an eleven-year cycle was found, however. Figure 2 shows that in the late forty-fifth and the early forty-fourth centuries B.C. the ^{14}C level was rising rapidly by about 2%-3%. Thereafter, from about 4350 B.C. through the subsequent two or three centuries, it was decreasing slowly.

Single measurements for twenty year intervals or more cannot in principle allow discernment of events on a time scale of less than 100 years. At least eight measurements per century are necessary to make this possible, (Hartree 1952). This requirement is now practically met by our measurements for the time range from about 4400 B.C. to 3000 B.C. Figure 3 shows our experimental results. The filled-in circles indicate bristlecone pine measurements on precisely dendrochronologically dated wood obtained from Professor Ferguson. Some of the results are revised values of pre-1969 measurements. The squares are results on samples obtained from Dr. Becker from two floating tree ring chronologies, D-8 (squares) and D-7 (diamonds), as described by Suess and Becker (1977). Recently Dr. Becker has connected one of the floating chronologies with the tree ring sequence from Thayngenweiher and Burgäschisee in Switzerland, for which an absolute radiocarbon date by comparison with bristlecone pine wood has been published by Ferguson et al. (1966). The date for the arbitrary zero point of 3670 ±100 B.C. of this Swiss Lake-dweller sequence (Ferguson et al. 1966) and the zero point of 3500 B.C. of the chronology Donau-7, as reported in the proceedings of the Mainz symposium (Suess

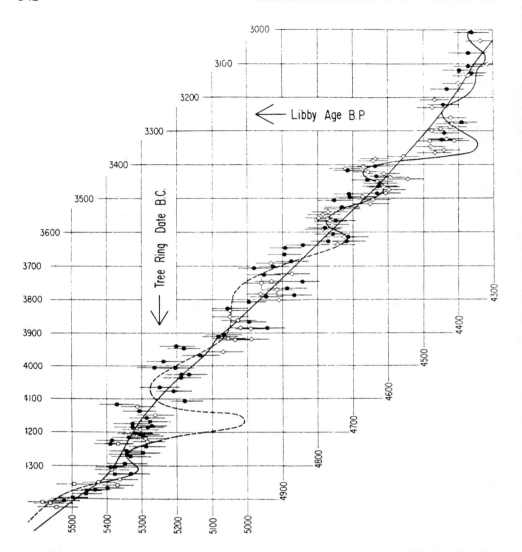

FIG. 3. Tree ring dates, namely, calendar dates B.C. versus conventional (Libby) radiocarbon dates ($T_{1/2}$ = 5568 yrs.) The irregular curve with wiggles is the one based on measurements carried out in La Jolla prior to 1969 (Suess 1970). Note that the broken part of the curve indicates particularly high uncertainty. The smooth curve is the one obtained by statistical analysis of "all" the available data (Clark 1975). Filled-in circles denote results on dendrochronologically dated bristlecone pine samples, squares and diamonds denote results from "floating" European oak chronologies as measured in La Jolla since 1969. (Solid squares are averages of results for two or more samples).

and Becker 1977), were found by Becker to be separated by 192 annual tree rings, 22 more than had been expected from the calibrated radiocarbon dates of the two series.

Figure 3 also shows the radiocarbon calibration curve as published in the Uppsala conference proceedings (Suess 1971). The reader will easily judge how well this curve described the true correlation between conventional radiocarbon dates and true calendar years. One characteristic feature is the steplike nature of the curve, which was clearly recognized in 1969. Details in the early curve may have to be revised, but its general character does not have to be changed. Wiggles of the type described in Uppsala do exist and, at least for the time range under consideration, my Uppsala curve better approximates an ideally correct curve than do any of the smooth curves derived by others.

The smooth line through figure 3 is the curve obtained by Clark (1975) by "rigorous statistical evaluation of all the results on tree-ring-dated samples" obtained by the world's radiocarbon laboratories, as discussed above. Clearly, it should not be possible to predict the outcome of future measurements on the basis of observations that are statistically not justified. Dr. Clark's calculations demonstrate conclusively that mixing the early La Jolla data with results obtained by other laboratories completely obscures details that can be recognized when considering the La Jolla data alone.

I would like to discuss the use of other so-called calibration curves. Such curves have been obtained by combining values from different laboratories and calculating "running means" of the experimental results. Curves obtained in this way may be within the limits of error of most experimental radiocarbon dates, but they give the erroneous impression of representing firmly established functions. They lead to an underestimation of the true uncertainties. They represent nothing more than noise passed through a low-pass filter so that smooth, random oscillations are obtained.

In figure 3 the reader will recognize several distinct oscillations that occurred during the one thousand years from 4400 to 3400 B.C. Determination of the geophysical causes of such features is, of course, the ultimate aim of this work.

Hopefully, in a year or two we shall have obtained an equally complete series of results for the third millennium B.C. In several more years we hope to have completed all the desirable measurements for the entire period for which calibrated dendrochronologically dated wood samples are presently available. By that time we can expect that tree ring scientists will have been able to prolong their series back to late Pleistocene times.

ACKNOWLEDGMENTS

As in all elaborate work on secular ^{14}C variations now in progress, its basis is provided by the Ferguson tree ring chronology, which dates back more than 8000 years. Dr. C. W. Ferguson supplied all the absolutely dated bristlecone pine samples for which results are reported here. Dr. Bernd Becker provided additional wood samples from his floating tree ring chronologies of European oak.

Technical operations of the La Jolla Radiocarbon Laboratory are carried out by Ms. Carol Hutto. They are supervised by Dr. Timothy Linick. Operation of the laboratory is financed by grant DES 74-22864 from the Earth Science Section of the National Science Foundation.

REFERENCES

Baxter, M. S., and A. Walton
    1971   Fluctuations of atmospheric carbon-14 concentrations during the past century. Proc. Roy. Soc. London. A. (321):105. (cf. P. E. Damon, A. Long, and E. I. Wallick (1973) Earth and Planet. Sci. Letters. 20:300-306.

Clark, M.
    1975   A calibration curve for radiocarbon dates. Antiquity. 252-72. (cf. H. Suess (1976) ibid. 61-63.)

Ferguson, C. W., B. Huber, and H. E. Suess
    1966   Determination of the age of Swiss lake dwellings as an example of dendrochronologically calibrated radiocarbon dating. Z. Naturfirschg. 21a:1173-1177.

Hartree, D. R.
    1952   Numerical Analysis. Oxford Univ. Press. 31:117-126.

Suess, H. E.
    1971   Bristlecone pine calibration of the radiocarbon time scale 5300 B.C. to the present. Proc. 12th Nobel Symp. on Radiocarbon Variations and Absolute Chronology. Uppsala. 595-605.

Suess, H. E., and B. Becker
    1977   Proc. Symp. Akad.d. Wiss. Mainz. 1974: Dendrochronologie und postglaziale Klimaschwankungen in Europa. Erdwiss. Forschung 13: 156-170.

6

Composite Computer Plots of ^{14}C Dates for Tree-Ring-Dated Bristlecone Pines and Sequoias

Elizabeth K. Ralph and Jeffrey Klein

Samples of archaeologically known ages were dated in this laboratory in 1952 and 1953 when it was found that contemporaneous modern woods had low counting rates (Ralph 1955). Subsequently, with the greater precision of gas proportional counting (CO_2), it was revealed that ^{14}C dates for samples representative of Egyptian dynasties were too young (Ralph and Stuckenrath 1960). Libby's original data (Arnold and Libby 1949) and calculations by Anderson and Libby (1951) had indicated that cosmic ray intensity had been constant to within about 10% for the past 20,000 years. With greater precision in counting techniques, however, we began to detect discrepancies of less than 10%.

With the financial support of the National Science Foundation and with the assistance of the University of Arizona Laboratory of Tree Ring Research, we obtained samples of tree-ring-dated *Sequoia gigantea* and later of *Pinus aristata* (or *longaeva*).

By 1972, at the time of the eighth International Conference in New Zealand, over 600 ^{14}C dates had been obtained by the Arizona, La Jolla, and Pennsylvania laboratories. As shown in figure 1 (reprinted from Ralph et al. 1973) there is good agreement among the three on the average, but the scatter of individual dates exceeds statistical expectations. Therefore, after experimenting with various techniques, we produced a third order polynomial fit for the average which illustrates the long-term deviation of ^{14}C dates. In addition we plotted nine-cell regression averages which retain the short-term oscillations. Both are illustrated in figure 2 (reprinted from Ralph et al. 1973).

By 1976 a total of 673 ^{14}C dates for dendro-dated samples had been obtained and released for publication. These are plotted in figures 3, 4, and 5, in which all outliers have been retained. In figures 3 and 4, the data are plotted without averaging. In figure 3 a fourth order least squares fit is illustrated, and in both figures 4 and 5, a sixth order polynomial least squares fit to the logarithm of dendro-ages is presented. The logarithm of the dendro-ages was employed to equalize the horizontal and vertical axes. It is reassuring to note that the sixth order polynomial crosses the ^{14}C deviation axis at

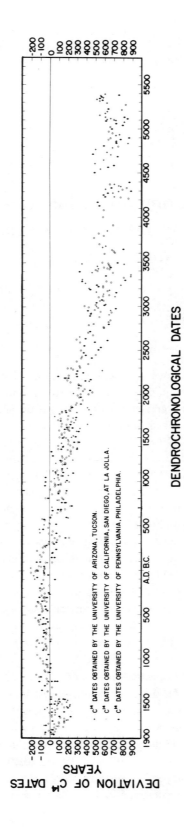

FIG. 1. Individual ^{14}C dates for dendro-dated samples, 5730 half-life.

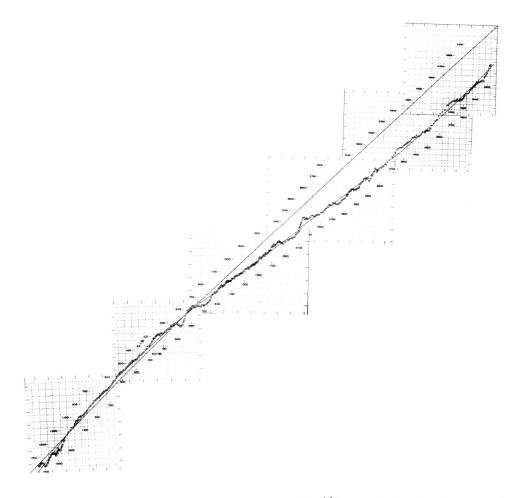

FIG. 2. Third order polynomial fit for the average of the ^{14}C versus dendro dates (dashed curve) and nine-cell regression averages (circles). Straight line represents 1:1 correspondence. ^{14}C scale is vertical; dendro scale, horizontal. 5730 half-life.

A.D. 1850, which is the age of our calibration sample, whereas the lesser order polynomials do not. Also, at the limit of dendro-dates, the sixth order polynomial curves upward less steeply.

An abbreviated explanation of the method employed is as follows: The function plotted is a sixth order polynomial least squares fit of the logarithm of the ^{14}C-date to the logarithm of the dendro-date, of the form:

$$y = a_0 + a_1 x + a_2 x^2 + a_3 x^3 + a_4 x^4 + a_5 x^5 + a_6 x^6 \qquad (1)$$

where y is the scaled logarithm, polynomial predicted "true" date, and x is the scaled

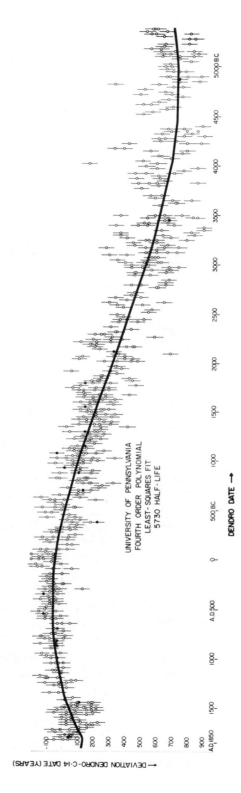

FIG. 3. Fourth order polynomial least squares fit and unaveraged dates. 5730 half-life.

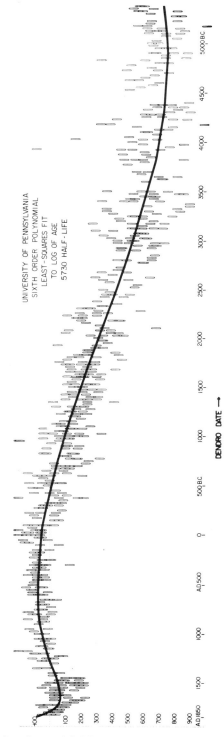

FIG. 4. Sixth order polynomial least squares fit to log of age and unaveraged dates. 5730 half-life.

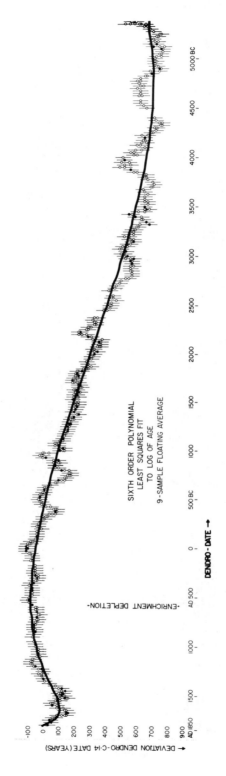

FIG. 5. Sixth order polynomial least squares fit to log of age and nine-cell regression averages. 5730 half-life.

logarithm, uncorrected ^{14}C date. To convert from years B.P. to "scaled logarithm", the following formula is used:

$$x = \frac{2\log_{10}(A) - (\log_{10}(105) + \log_{10}(7343))}{\log_{10}(7343) - \log_{10}(105)} \qquad (2)$$

where x = scaled logarithm of age; A = age in years B.P.

Consequently to convert from "scaled logarithm, polynomial predicted" age to age in years B.P.:

$$A_p = \exp\left[\frac{1.8447y + 5.8871}{2}\right] \qquad (3)$$

where y = scaled logarithm, polynomial predicted age.

A_p = polynomial predicted age in years B.P.

This basic method is described by Wolberg (1967).

Statistical measures, such as chi-square and the sum of the squares of residues of the data about the fit, indicate that there remain significant fluctuations, larger than expected by counting uncertainty, which are too short in period to be adequately fit by a low order polynomial. A fourier analysis of these oscillations was undertaken to determine their amplitude and frequency (prevalence), and to obtain some measure of the certainty of their existence (veracity). The nine-cell floating average, while limited in its usefulness as a quantitative indicator of these oscillations, is good qualitatively as a guide to regions where the sixth order polynomial is particularly deficient in its ability to follow the data. These areas were chosen first, as a test of this method of local fitting. (It should be noted that while it is somewhat awkward to use both polynomial and sinusoidal fits, the former converges more rapidly in fitting the longer-term variations, while the latter more effectively fits the short-term changes.) In figure 6 we have shown the region 1500 – 3000 B.C. in which the raw data, the nine-cell average, the sixth order polynomial, and the fourier analyzed variations about the polynomial fit are plotted. The sinusoidal fit was again done by least squares but, unlike the polynomial fit, outliers were rejected by use of a modified Chauvenet's criterion: points lying the greatest number of standard deviations from the fit were rejected one at a time, the fit was recalculated after each rejection, and all previous outliers were again tested for their distance from the fit. It was hoped by this means to minimize the effect of extremely deviant points. The region shown in figure 6 is a composite of three local fits, each of about 500 years. A check of the sensitivity of the local fit to the limits of the region under analysis was made by combining two neighboring regions and refitting to the entire region. Good agreement was obtained, indicating little sensitivity to the size of the region under consideration. It was required that the local fit match the polynomial at the ends of the region; this requirement was met by using a complete set of sine functions over the period of interest.

Work is in progress to complete the (Fourier) analysis of the oscillations about the long-term polynomial-approximated variations in the ^{14}C-inventory with the goal of

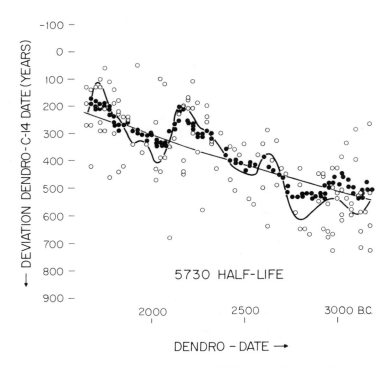

FIG. 6. Part of sixth order polynomial (thin curve), nine-cell averages (black circles) both taken from figure 5, raw data with outliers rejected (open circles), and sinusoidal local fits (heavy curves).

completing a table of corrections including a (detailed) analysis of the uncertainties of the fit. Examination of the nature of the variations with regard to minimum period and maximum rates of depletion and enrichment of ^{14}C in the atmosphere will lead to a better understanding of the basic mechanisms of changes.

In the meantime, we have presented the polynomial and nine-cell plots shown in figure 5 as our estimate of the true representation of the ^{14}C inventory over the past 7300 years.

ACKNOWLEDGMENTS

We acknowledge with gratitude financial support for the known-age dating program from the National Science Foundation, Earth Sciences Division, current grant DES 74-22233, and the cooperation of the Laboratory of Tree Ring Research, University of Arizona. We thank, especially, C. Wesley Ferguson for furnishing the tree-ring-dated bristlecone pines. Jeffrey Klein and Roger Warburton did the computer programming of the polynomial fits.

REFERENCES

Anderson, E. C., and W. F. Libby
    1951   Worldwide Distribution of Natural Radiocarbon. Physical Rev. 81:64.

Arnold, J. R., and W. F. Libby
    1949   Age Determinations by Radiocarbon Content: Checks with Samples of Known Age. Science. 110:678.

Ralph, E. K.
    1955   University of Pennsylvania Radiocarbon Dates I. Science. 121:149.

Ralph, E. K., and R. Stuckenrath, jr.
    1960   Carbon-14 Measurements of Known Age Samples. Nature. 188:185.

Ralph E.K., H. N. Michael, and M. C. Han
    1973   Radiocarbon Dates and Reality. MASCA Newsletter. 9 (1).

Wolberg, John R.
    1967   Prediction Analysis. Princeton, Van Nostrand.

7

Holocene Tree Ring Series from Southern Central Europe for Archaeologic Dating, Radiocarbon Calibration, and Stable Isotope Analysis

Bernd Becker

ABSTRACT

In order to construct a continuous ring series for the last 8700 years, tree rings of subfossil logs of oak trees in postglacial valley fillings of southern Central Europe are being analyzed. At its present stage the chronology has been completed from the present to 717 B.C. In addition, according to ^{14}C-calibrations by Suess, La Jolla, two large floating series have been dated from 1600 to 2900 B.C. (Bronze Age chronology) and from 2820 to 4060 B.C. (Neolithic chronology), respectively.

Several published calibrated dates of oak series from Neolithic Swiss lake-dwellings (Thayngen, Burgäschisee, and Niederwil) have been checked by cross – correlation with the Neolithic riverine oak chronology, Donau-7, for which an absolute calibrated date had been determined by the La Jolla laboratory.

Some aspects of future application of the postglacial oak tree ring sequence by stable isotope analyses (^{18}O/^{16}O, D/H) are discussed.

INTRODUCTION

Cooperation between German dendrochronologists and American radiocarbon laboratories has existed for more than a decade. The first remarkable result from this cooperation was the determination of absolute ages of Neolithic Swiss lake-dwellings using the late Professor Huber's floating tree ring chronology for Thayngen-Weier and Burgäschisee (Ferguson et al. 1966). A master chronology for German oak was established for medieval times back to 832 A.D. (Huber and Giertz 1969). For prehistoric times, only several short floating chronologies had existed from the Bronze Age and Neolithic times. An extension of the absolute oak tree ring chronology was achieved by using archaeological samples. In this way E. Hollstein succeeded in extending the Huber chronology to 299 A.D. He also constructed a floating tree ring series of 581 years from Roman and Preroman periods (Hollstein 1967, 1969, 1973).

After moving from Munich to Stuttgart-Hohenheim, our laboratory continued the work of Huber by analyses of subfossil oak trees buried in Holocene gravel fillings of rivers in southern Central Europe (Becker 1971). Huber had shown that it was possible to obtain remarkable correlations of Neolithic lake-dwellings (Huber et al. 1962, 1963, 1967) but that, because of the relatively short duration of the settlements, it was perhaps impossible to extend the chronology over long periods of time. This difficulty manifested itself again in attempts to correlate Bronze Age samples from Zug Sumpf with those from dwellings on the Zürichsee near Auvernier (Giertz, personal communication). Also, the successful cross-correlation by A. V. Munaut (Mook et al. 1972) did not result in significant extension of the Neolithic oak chronology. The deposition of oak logs, however, along our river valleys occurred continuously over more than 8500 years, affording the opportunity to correlate overlapping tree ring chronologies through several millennia. In this way the European oak tree ring series can in principle be constructed back to the time of the return migration of oaks into Central Europe after the last glaciation.

Numerous ^{14}C dates show that the oldest available logs from Holocene valley fillings have conventional radiocarbon ages of not more than 8700 years B.P. (Becker 1976, Delorme 1976, Schwabedissen et al. 1973). The oldest subfossile trees from fluvial deposits analyzed so far date from 9600 B.P. (Becker 1971). Together with birch tress (Betula verrucosa), Scotch pine trees (Pinus sylvestris) represent the forest type of the valley plains of Preboreal times in Europe. Recently, Furrer, Schweingruber, and Kaiser (Zurich) found pine and birch trees in a clay pit near Dättnau, Winterthur, (personal communication) which date from the Alleröd period (Suess, this volume). As was pointed out by Vogel et al. (1969), tree ring series of pine may be of special interest for radiocarbon calibration, especially in connection with floating pine chronologies from peat bogs in the Netherlands. The construction of a Holocene pine chronology is probably limited, however, by difficulties in obtaining suitable subfossil trees for the whole period.

In 1977 when our laboratory began tree ring analyses of subfossile oaks from fluvial gravel deposits, a group at the Belfast Paleoecology Laboratory began tree ring studies of subfossil oaks buried in peat bogs of Northern Ireland. At its present stage the Holocene oak chronology of Ireland consists of an absolute chronology for medieval times to 1000 A.D. (Baillie 1974), whereas five floating series extend to 4000 B.C., the longest of which has 2300 rings. There exist gaps of not more than 500 years within the past 6000 years; the largest gap exists between 750 and 1000 A.D. (Pilcher 1973, Pilcher and Baillie 1976).

Subfossil oak trees can be found in almost all the gravel pits in valleys where Holocene fluvial deposits are being dredged. We have sampled trees from more than 120 sites within the drainage basins of the rivers Rhein, Main, and Danube (figure 1). Great numbers of cross sections are available from this area for tree ring analyses. The logs are usually located several meters below groundwater level (in the upper Rhein Valley to a depth of 15 m), although occasional logs are found above the present groundwater table. The

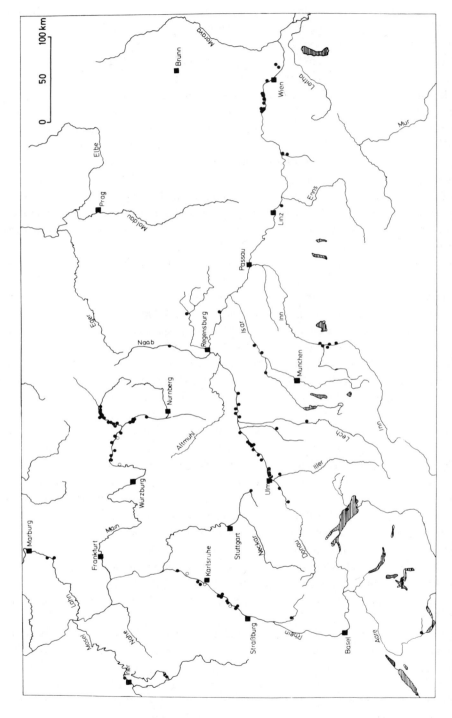

FIG. 1. Locations of gravel pits, from which subfossil oaks have been ^{14}C analyzed (black dots). The map shows the drainage system of the Rivers Main, Rhein, and Danube in southern Central Europe.

oak trees are very well perserved; the root systems and larger branches are often undamaged. Even after several millennia, heartwood is well preserved; in many cases only the sapwood is destroyed.

Since Huber began tree ring analyses in Central Europe, oak wood has been used almost exclusively. This is because the excellent durability of oak has made it the preferred timber for settlements since Neolithic times and the most common wood in archaeological excavations. Tree rings of oaks correlate over a remarkably large geographic area (Delorme 1973) and, contrary to the condition in coniferous wood, missing rings do not occur in oak tree rings. This is because at the beginning of each vegetation period the ring-porous oak develops rows of vessels with large diameters for water transportation. These vessels become inactive after the transition of sapwood to heartwood. Probably for this reason, each year ring-porous tree-species produce at least one growth layer of earlywood. During more than 25 years of tree ring studies of oak wood, no missing rings have been found in our laboratories.

Wood from oak and pine trees deposited under groundwater in fluvial gravels is practically of unlimited durability, and the environment of the deposit prevents significant contamination by foreign organic matter. The large ring width, frequently 5 mm or more, makes it possible to prepare single annual growth layers for radiocarbon measurements. To my knowledge, all earlier ^{14}C analyses are made on samples of about 10 rings.

Rings from near the center of a log were used for ^{14}C determinations only if they came from a height of more than 2 m above the roots, in order to avoid surface effects near the soil from CO_2 produced by decaying organic material, which may be lower in ^{14}C than contemporary material.

Absolute Master Chronology for Oak in Central Europe

Figure 2 shows the Holocene oak chronologies at their present stage. Two time scales are shown to compare dendrochronologically determined B.C. dates with radiocarbon dates B.P. of floating chronologies. The Ferguson bristlecone pine chronology, which now extends to about 6200 B.C. (Ferguson 1970, 1977), provides the basis for the calibration of all the floating chronologies as shown in the figure. "Wiggle matching" provides the possibility for precise dating of floating chronologies by comparing the radiocarbon data with those for the Ferguson bristlecone pine (Suess and Becker, 1977). As mentioned above, Huber's master chronology extended to A.D. 832. It was extended to A.D. 299 by Hollstein. The further tree ring series derived from Roman and Celtic settlements and from bridges in Germany and Switzerland. These tree ring series were established jointly with Huber and others (Hollstein, in press). Verification of the chronology to 717 B.C. was achieved by cross-correlating the Huber and Hollstein chronology with our series of subfossil oaks from the last two phases of widespread, fluvial activities from the Iron Age to Roman times and in the early Middle Ages (Becker 1976). The riverine oak chronologies 397 B.C. to A.D. 267, and A.D. 207-755 are the link between the historic and the prehistoric chronologies; it represents the oak tree ring patterns of Central

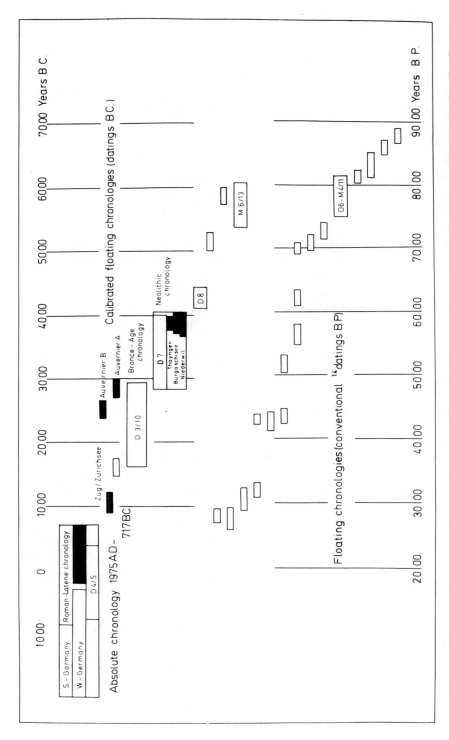

FIG. 2. Present stage of the postglacial oak tree ring chronology in southern Central Europe. Open squares denote subfossil oak series, closed squares chronologies of prehistoric settlements. The floating series are plotted versus the conventional ^{14}C-time scale B.P. (upper scale of the figure). Hatched squares give the position of the same series on a B.C. time scale (low part of the figure) as determined by the La Jolla Laboratory by comparison with the ^{14}C in bristlecone pine wood.

Europe for more than 2600 years from the present to the past. Hollstein has not yet published the complete data from 700 B.C. to A.D. 700; a final check of the B.C. chronology is still lacking. Cross-correlations of the published Hollstein and Huber chronologies demonstrate an important fact: oaks from fluvial deposits facilitate dendrochonological dating of archaeological samples grown elsewhere. In other words, irregularities in riverine oak tree ring patterns, caused by floods, groundwater oscillations, or other local factors within the valleys, are not sufficiently pronounced to obscure common patterns. A large number of tree ring cross-correlations from the period 700 B.C. to A.D. 700 demonstrates the validity of the master chronology for the whole area of the drainage basins of the rivers Rhein, Main, and Danube, and also for the area from the western Swiss lakes in the south to the German-Belgian border in the northwest as shown by archaeological samples.

Bronze Age Chronology

Professor M. A. Geyh, Hannover, has performed more than 100 radiocarbon datings on our subfossil oak samples, greatly facilitating cross-dating. The Bronze Age chronology consists of 1307 tree rings derived from 120 cross-dated trees of the first major period of log deposition along the rivers Rhein, Main, and Danube (Becker 1976). Twenty-two La Jolla ^{14}C-dates for this series indicate a chronology from about 2900 B.C. to 1600 B.C. The gap between the absolute chronology, which begins at 717 B.C., and the Bronze Age series which ends in 1605 B.C., is about 900 years. However, several floating chronologies fall within this time range, notably that of the prehistoric settlements Zug Sumpf – Zürichsee – Auvernier (V. Giertz, unpublished) dating from 1215 B.C. to 884 B.C. (Suess and Becker 1976). Recently we enlarged this series from 1447-832 B.C.

Neolithic Chronology

Considerable progress has recently been made in our floating Neolithic chronologies. Such chronologies from Swiss lake-dwellings have been ^{14}C-dated by several laboratories. Our most important accomplishment has been the cross-dating of our chronology Donau-7 of subfossil oaks from the rivers Main, Regnitz, and Danube with the three chronologies from the Neolothic settlements Thayngen-Weier (excavated by Guyan, analyzed by Huber) from Burgäschisee Süd and Südwest (Müller-Beck, Huber) and from Niederwil (Waterbolk, Munaut). Bristlecone pine calibrations have been recorded using "wiggle matching techniques" for the tree ring series from Thayngen-Burgashchisee (Ferguson et al. 1966), Niederwil (Mook et al. 1972), and for the oak series Donau-7 (Suess and Becker 1976).

This Neolithic chronology has been extended to 1235 tree rings dating from 4057 to 2823 B.C. by cross-correlating with another floating series (Main 5). A preliminary zero-point for the Main 5 series, determined by the La Jolla Laboratory (3275 ± 50 B.C.), differs from the now dendrochronologically fixed zero-point of the Donau-7 series (3356 B.C.) by 81 years.

The accuracy of these independently derived calibrated dates can now be compared by using the recently established tree-ring scale. Detailed information on the calibration of a complete set of Neolithic oak rings is given by Suess in this volume (Suess 1977). Here are listed only the dendrochronological positions of the series as obtained by the first number La Jolla ^{14}C measurements of the Donau-7 samples and their calibrations (Suess and Becker 1976). Figure 3 shows the Neolithic chronologies in their crossdated sequence. Furthermore, tree ring samples dated in La Jolla are indicated graphically in the figure.

The time span from the end of the Neolithic chronology to the beginning of the Bronze Age chronology is less than 300 years. After extension of the Neolithic chronology to 2823 B.C. both series overlap by more than 100 years. For this period there exist two chronologies for the Swiss lake-dwellings: Auvernier A and Auvernier B, dendrochronologically analyzed by V. Giertz, Munich, and calibrated by Suess (Suess and Strahm 1970). They date from 3000 to 2690 B.C. and from 2760 to 2550 B.C., respectively. It should soon be possible to connect the Neolithic and the Bronze Age series with a recently established 435-ring floating chronology from the Main valley (Main 10/12). 4800 years of the past 6000 years have been reconstructed with one absolute and two floating tree ring series. Several more existing floating chronologies should close the remaining two gaps, permitting dendrochronological dating, accurate to within one year, of the prehistoric sites Zug – Zürichsee, Thayngen-Burgäschissee-Niederwil, and Auvernier A and B.

Mesolithic Chronologies

Extension of the Neolithic chronology back to the fifth millennium B.C. depends on the feasibility of correlating a 700-ring floating chronology (Main 6/13) with the Donau-8 chronology, for which twenty-six radiocarbon datings have been completed (Suess and Becker 1976). Several earlier floating Holocene chronologies dating from the sixth and seventh millennia B.C. are also available for ^{14}C calibration measurement (figure 2). One of these has been extended since the conference by 797 tree rings and now covers a range of ^{14}C dates from 8280 to 7490 B.P. In addition, the Donan-8 chronology has been enlarged over 678 years, dating now 4791 to 4115 B.C.

Paleoecological and Paleoclimatological Studies on Holocene Oak Tree Rings

Changes of the water drainage system which caused valley fills and tree deposits during the Holocene can be recognized from dated phases of fluvial activity and inactivity, causing increased ore and decreased gravel redepositing, flood loam accumulation, and varying groundwater levels. These ecological factors controlling the growth of the subfossil oaks have been studied in cooperation with Professor W. Schirmer, Düsseldorf, who contributed valuable geological and paleopedological results. (Becker and Schirmer, 1978).

Some aspects of the paleoecological evidence based on the Holocene oak chronology are useful for paleoclimatological research by analyses of stable isotope ratios, such as $^{18}O/^{16}O$ and D/11. Results derived by stable isotope measurements on recent and his-

CROSSDATED MASTER-CHRONOLOGIES (OAK) OF NEOLITHIC SWISS LAKE DWELLINGS AND SUBFOSSIL RIVERINE OAKS OF THE DANUBE AND MAIN VALLEYS

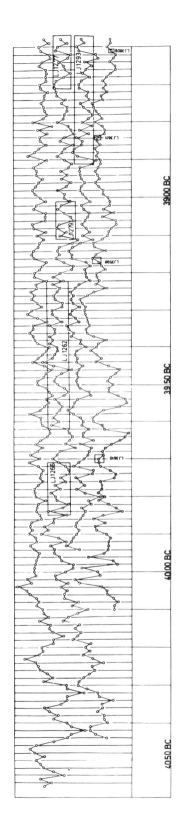
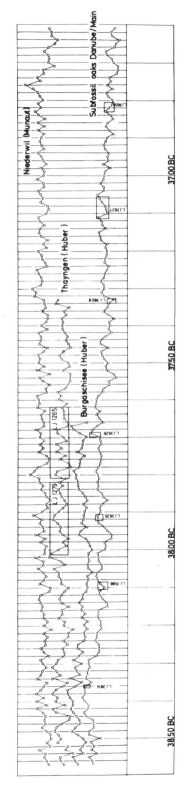

FIG. 3. The tree ring patterns of oak dwellings from the Neolithic sites Niederwil (Munaut), Thayngen, and Burgäschisee (Huber) and the curve for the subfossil Donau-7 series are plotted in their cross-dated position from 4057 B.C. to 3661 B.C. (±20 years). The tree ring samples used for earlier and recent calibration measurements by the La Jolla Laboratory are marked by rectangular blocks.

torical tree rings (Libby et al. 1976, Libby this volume) indicate possiblities for a reconstruction of past climate (temperature). Our paleoecological interpretations based on Holocene log deposits lead to the conclusion that ecological changes in the drainage systems predominate over effects from climatic variations on the subfossil tree ring patterns. This is illustrated by data derived from the Holocene oak tree ring patterns, such as mean ring width, correlation coefficients of cross-dated logs, and mean sensitivity from Holocene oak sites in the Rhein, Main, and Danube valleys (figure 4).

The ring width data provide the most accurate indications of changing ecological conditions within the valley flood plains. A significant trend exists in the growth rates of Holocene oak forests in all river systems studied. Wood layers from Boreal times are characterized by small growth rates, but the ring width increased during the early Atlantic. In contrast to this trend, the narrowest Holocene rings developed during the late Atlantic and the early Subboreal. This early part of the valley development during the Holocene appears to coincide with the following hydrological situations: (1) with the high groundwater levels of Boreal wet sites, (2) with increasingly favorable growth conditions in early Atlantic times caused by lowering of the groundwater tables due to down cutting of the riverbeds, and (3) with deterioration of the soils of the flood plains due to stagnant wetness, probably caused by a renewed raise of the groundwater level in late Atlantic times.

The rapid improvement in the ecological conditions, as evidenced by a significant incrase of the growth rates in Subboreal and Subatlantic oak sites, coincide with the major flooding and log-accumulation phases of the Neolithic-Bronze Age, the Iron Age-Roman period, and the early Medieval period. As a result of these river activities, extensive valley areas were redeposited and covered with fresh, fine-grained sediments. Since Roman times, alluvial loam sedimentation caused by forest clearances within the drainage basins has prevailed in the valley plains. The fast growth rates of riverine oak sites since that time demonstrate the improved soil fertility typical of our present valley plains.

Decreasing trends for cross-correlation and mean sensitivity (fig. 4) indicate changes of the Holocene valley ecosystem since the beginning of the Bronze Age. The widespread river activities changed, the sensitive and stressed growth conditions that had controlled the tree ring patterns of oaks during Boreal and Atlantic times became increasingly balanced by soil improvement through increased flood loam cover.

The tree ring patterns of our Holocene oak chronology do not correlate precisely with variations of Holocene climate. The improvement of the growth rates from late Atlantic to Subboreal times does not agree with paleobotanic evidence that the mixed oak forests reached their greatest extent in Central Europe, especially during the Atlantic period with its optimal climatic conditions.

Stable isotope analyses of subfossil oak wood will permit reconstruction of past climatic conditions, primarily conditions of temperature. Our resultant knowledge of the paleoecology of oak forests will then help us to understand the interactions of climatic hydrological variations and the influence of human activities upon the ecosystems of the Holocene valley flood plains.

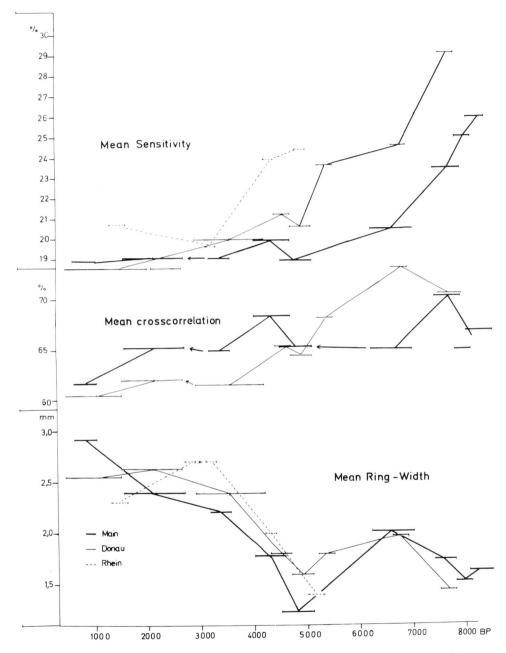

FIG. 4. Tree ring characteristics of absolutely dated or floating subfossil oak series of the past 8500 years, plotted against their dated growth periods: Mean sensitivity (%), mean crosscorrelation (%) after the "Gleichläufigkeitsmethode) and mean ring width (mm) are compared for the valleys of the rivers Rhine, Main, and Danube. In order to show existing trends the values are connected by arrows.

REFERENCES

Baillie, M. G. L.
 1974 Dendrochronolgy: An exercise in archeological dating. Irish Arch. Forum. 1:6-16.

Becker, B.
 1971 Möglichkeiten für den Aufbau einer absoluten Jahrringchronologie des Postglazials anhand subfossiler Eichen aus Donauschottern Ber. Deutsch. Bot. Ges. (85) 1-4:29-45.
 1977 Paläökologische Befunde zur Geschichte der postglazialen Flussauen im Südlichen Mitteleuropa. Proc. Symp. "Die Dendrochronologie des Postglazials". Akad.d.Wiss. Mainz. (In press.)

Becker, B., and W. Schirmer
 1978 Holocene valley development based on geological-pedological, dendrochronological and paleoecological investigations along the river Main. Boreas. 6: 303-321.

Delorme, A.
 1973 Über die geographische Reichweite von Jahrringchronologien unter besonderer Berücksichtigung mitteleuropäisher Eichenchronologien. Parehist. Zeitschr. (48) 2:133-143.
 1977 Möglichkeiten der Überbrückung regionaler Teilchronolgien zu einer überregionalen Postglazialchronologie der Eiche für Europa. Proc. of the Symposium "Die Dendrochronologie des Postglazials". Akad.d.Wiss. Mainz. (In press.)

Ferguson, C. W., B. Huber, and H. E. Suess
 1966 Determination of the Age os Swiss lake dwellings as an example of dendrochronologically-calibrated radiocarbon dating. Zeitschr.f.Naturforsch. (21a) 7: 1173-1177.

Ferguson, C. W.
 1970 Dendrochronology of Bristlecone Pine, Pinus aristata: Establishment of a 7484-year chronology in the White Mountains of Eastern Central California; Radiocarbon variations and absolute chronology. Proc. 12th Nobel Symp. Uppsala. Stockholm, Almquist and Wiksell, 237-259.

Hollstein, E.
 1967 Jahrringchronologien aus römischer and vorrömischer Zeit. Germania. (45) 1-2: 60-84.
 1969 Dendrochronologische Untersuchung der Einbäume in Historischen Museum der Pfalz. Festschr. In Hundert Jahre Hist.Mus.d.Pfalz. Speyer, pp. 191-204.
 1973 Jahrringkurven der Hallstattzeit.Trierer Zeitschr.f.Gesch.u.Kunst d.Trierer Landes. 36: 37-55.

Huber, B. and W. Merz
 1962 Jahrringchronologische Untersuchungen zur Baugeschichte der urnenfelderzeitlichen Siedlung Zug-"Sumpf".Germania, (40) 1: 44-56.
 1963 Jahrringchronologische Synchronisierung der jungsteinzeitlichen Siedlungen Thayngen-Weier und Burgäschisee Süd und -Südwest. Germania. (41) 1: 1-9.

Huber, B.
    1967   Seeberg,Burgäschisee Süd,Dendrochronologie. Acta Bernensia Bd.II,Seeberg-Burgäschisee Süd,Teil 4 Chronologie und Umwelt: 145-156.

Huber, B and V. Giertz-Siebenlist
    1969   Unsere tausendjährig Eichen-Jahrringchronologie durchschnittlich 57 (10-150) — fach belegt.Sitz.ber.d.Österr.Akad.d.Wiss., Mathem.Nat.wiss.K1.I, (178) 1-4: 37-42.

Libby, L. M., L. J. Pandolfi, and P. M. Payton, J. Marshall III, B. Becker and V. Giertz-Siebenlist
    1976   Isotopic tree thermometers.Nature. 261: 284-288.

Mook, W. G., A. V. Manaut, and H. T. Waterbolk
    1972   Determination of Age and Duration of Stratified Prehistoric Bog Settlements. Proc. 8th Intern. Conf. Radioc. Dating. New Zealand. F: 27-39.

Pilcher, J. R.
    1973   Tree-ring research in Ireland. Tree-Ring Bull. XXXIII: 1-5.

Pilcher, J. R. and M. G. L. Baillie
    1976   Belfast Radiocarbon Calibration. Antiquity. (In press.)

Schwabedissen, H., R. Schütrumpf, J. Freundlich and B. Schmidt
    1973   Pollenanalyse, Jahrringanalyse und C-14 Datierungen in ihrem Zusammenwirken für die urgeschichtliche Chronologie. Archäol. Norr. 1: 139-162.

Suess, H. E. and C. Strahn
    1970   The Neolithic Auvernier, Switzerland. Antiquity. 44: 91-99.

Suess, H. E. and B. Becker
    1977   Der Radiokarbongehalt von Jahrringproben aus postglazialen Eichenstämmen Südmitteleuropas. Proc. Symp. "Die Denrochronologie des Fostglazials". Akad.d.Wiss. Mainz. (In press.)

Vogel, J. C., W. A. Caspary, and A. V. Munaut
    1969   Carbon-14 trends in subfossil pine stubs. Science. 166: 1143-1145.

8

The Contribution of the Swiss Lake-Dwellings to the Calibration of Radiocarbon Dates

J. Beer, V. Giertz, M. Möll, H. Oeschger, T. Riesen, and C. Strahm

ABSTRACT

Studies of well preserved posts from the Neolithic lake-dwellings at Yverdon (Switzerland) have been combined to produce two continuous floating dendrochronologies. Thirty-eight samples from these chronologies have been radiocarbon dated. The results have been compared and fitted with spline-function-smoothed North American bristlecone pine data for calibration and determination of the absolute age. Differences between ages thus obtained and ages based on traditional chronology suggest a revision of some archaeological methods of dating.

INTRODUCTION

Organic material from lake-dwellings of the Neolithic and Bronze Age in Switzerland is extremely well preserved because it has been deposited in humid or wet layers. The village posts are ideally suited for dendrochronological study and ^{14}C analysis. With the excavations at Auvernier and other sites Suess and Strahm (1970) demonstrated that ^{14}C analysis of dendrochronologically determined samples and their comparison with the bristlecone pine chronology could facilitate determination of the absolute age of the settlements. During the recent excavation of the lake-dwellings at Yverdon, Switzerland (Strahm 1972/73), a new and more systematic project was planned in collaboration with Veronika Giertz, dendrochronologist of Munich, and the ^{14}C laboratory of the University of Bern. Some background information concerning the excavation will clarify the aims of the project.

Background Information

DESCRIPTION OF THE EXCAVATION

The site of Yverdon on the shores of lake Neuchâtel, Switzerland, reveals a typical lake-dwelling (fig. 1). Due to its location at the upper end of the lake, near the mouth of the river La Thièle, human traces and remains were covered early by sediments. This had had three effects: (1) the site has been well preserved under a covering layer; (2) due to sedimentation the shoreline has been pushed further into the lake, so today the site lies on solid ground; and (3) each short period of development has been stratigraphically separated by interrupting, water-borne deposits. This process has resulted in numerous, varied layers (fig. 2). Some layers consist of sand; others consist of organic material washed ashore by the lake. Other components, like clay lenses, stone heaps and construc-

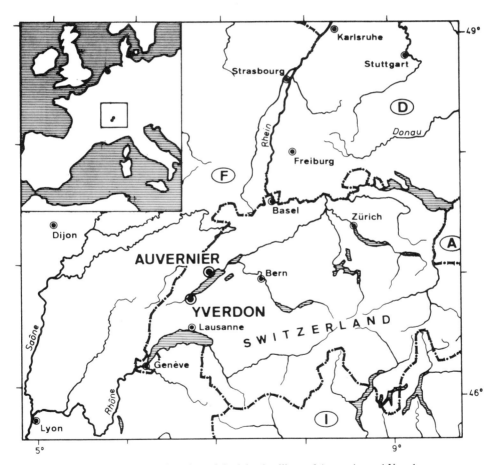

FIG. 1. Geographical situation of the lake-dwellings of Auvernier and Yverdon.

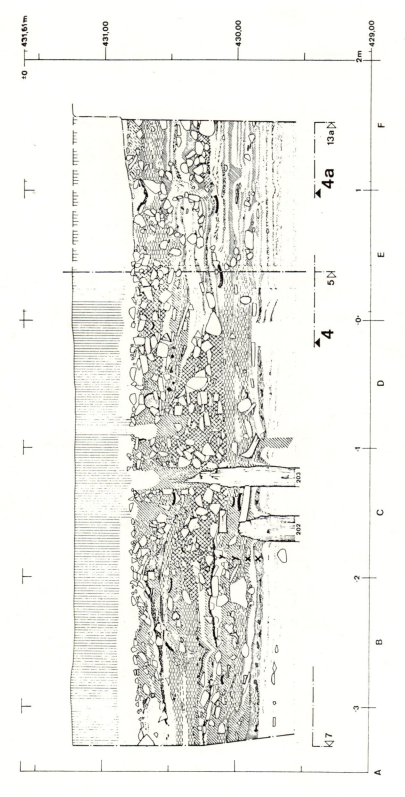

FIG. 2. The stratigraphy of Yverdon. X: layers pulled along by posts. Y: layers bordering on posts.

tional elements attest to human intervention. The layers are not extensive. Because they form in lenses and are sometimes very thin, it is difficult to distinguish one layer from another.

THE ARCHAEOLOGICAL MATERIAL

Several different artifacts lie in the fine strata. A very small selection is shown in figures 3 and 4. Based on these artifacts, the site of Yverdon is dated at the end of the Neolithic period in the so-called Auvernier group of the Saône-Rhône culture (Strahm 1975). Conventionally, this corresponds to the end of the third millennium, but it will be shown how this must be corrected by the calibration.

INTERRELATED CHRONOLOGY

The fine stratigraphic observations described provide a basis for checks of different chronological methods. Well separated layers of short duration, many stratified finds, stratigraphically dated posts for dendrochronological examination, and well defined samples for ^{14}C dating serve as the bases to establish a precise and detailed chronology. (In the following statements the term "chronology" is used strictly in the sense of time, not in the sense of evolution or cultural development). The results of these interrelated chronological elements aid in understanding short and subtle developments during the occupation of the site and in establishing a future time scale. To achieve this aim a variety of methods must be employed, applicable because of the special conditions of conservation in Yverdon. These methods include stratigraphy, cultural development, dendrochronology, and radiocarbon dating, each of which must support and complement the others. The resultant combination of different methods is termed "interrelated chronology" (fig. 5).

Methods of Analysis

Because the excavated materials at Yverdon could yield such important data for comprehensive chronology, all methods were used in order to gain as much information as possible. Yverdon thus serves to exemplify the following methods of interrelated chronology:

1. *Stratigraphy*: Stratigraphy shows the sequence of the sediment layers (fig. 2). It can be estimated that each layer was formed in a short time, some possibly in as little as one day.

2. *Cultural Development*: The cultural development is also a kind of chronology. By comparing the finds in their stratigraphic sequence, one can elaborate on the development of forms and objects. However, the time scale is in no way indicated.

    Stratigraphy and cultural development indicate the sequences and the changes of the finds. The time scale corresponding to these sequences is provided by dendrochronology and radiocarbon dating.

3. *Dendrochronology*: Dendrochronology can provide a floating time scale through

FIG. 3. Yverdon, Avenue des Sports. Examples of pottery.

FIG. 4. Yverdon, Avenue des Sports. A selection of tools.

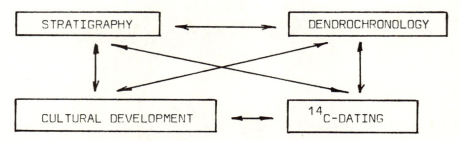

FIG. 5. The elements of the "interrelated chronology".

correlation of dendrochronologically classified posts and beams. If circumstances are favorable, it is possible to associate the vertically implanted posts with the layer in which they have been erected. This is because older layers are pushed downward by the implanted posts, while the more recent ones border on them (fig. 2).

4. *Radiocarbon Dating*: An initial point for a floating dendrochronological time scale must be determined by radiocarbon dating. However, the ^{14}C dates of well defined samples give only a "half absolute" time scale due to the difference between ^{14}C years and astronomical years. One aim of this report is conversion of the ^{14}C dates into absolute dates by comparison of ^{14}C results of floating chronologies with data of the bristlecone pine chronology.

In combination, these dating methods suggest the possibility of following the succession of the genesis of layers as well as of discerning the cultural development. This is important not only for the general understanding of the archaeological cultures, but also for the reciprocal control of each dating method, and it is productive of additional information on the ^{14}C variations.

The exhaustive dendrochronological determination of all the posts at Yverdon has resulted in a sequence of about 400 years, all documented by tree rings (fig. 6).

Results of the Dendrochronological Analysis

The stratigraphy of Yverdon indicates a continuous development without any significant interruption (fig. 2), which would suggest that the well preserved and stratigraphically dated posts would also give a continuous sequence with some overlapping. Dendrochronological examination indicates that the posts can be combined into two continuous floating chronologies, A and B (fig. 6), which do not overlap. Based on its stratigraphical position, chronology A is attributed to the lower layers, chronology B to the upper and younger layers. It is obvious that little time separates A and B.

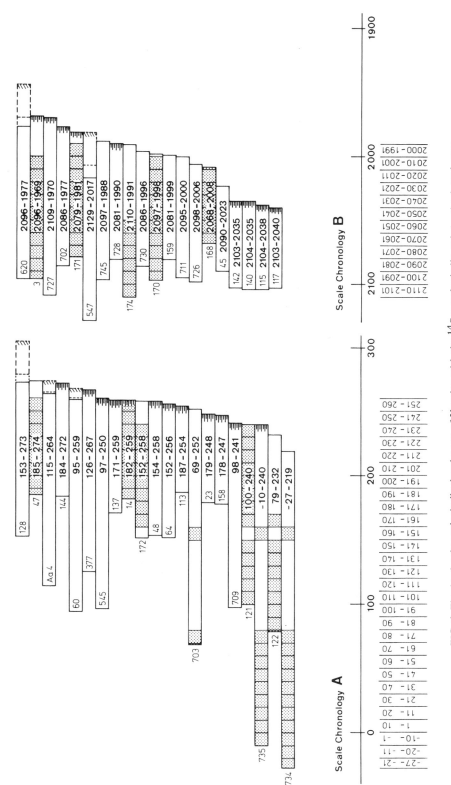

FIG. 6. The dendrochronologically dated posts of Yverdon with the ^{14}C-samples indicated by the dotted areas.

The Fitting of the Floating Chronologies

^{14}C MEASUREMENTS

Segments covering the same ten years were cut from different correlated posts to form a sample for ^{14}C dating. Thirty-eight samples were dated at the radiocarbon laboratory of the University of Bern. The samples from the floating chronologies A (290 years) and B (90 years) were measured partially in two similar counting systems with modern net count rates of ~15 cpm and backgrounds of ~1.27 cpm. The total counting periods for one sample ranged between 2500 and 10,000 minutes. All results were corrected for isotopic fractionation based on $\delta^{13}C$ measurements. The results are shown in table 1.

SMOOTHING OF THE BRISTLECONE PINE DATA

After several attempts to fit the data with curves published by different laboratories (Suess 1970, Ralph et al. 1973, Damon et al. 1974), it was decided to compare the Yverdon raw data directly with the bristlecone pine raw data. A list of raw data from the La Jolla, Pennsylvania, and Arizona laboratories was made available by E. K. Ralph. Only La Jolla data obtained prior to 1969 were included. Numerous, more precise La Jolla values obtained later do not appreciably change the results (H. E. Suess, personal communication). The bristlecone pine data were smoothed by a spline function (Reinsch 1967) and compared with the Yverdon data. In the spline-function-smoothing procedure, the optimal function $f(x)$ is composed of cubic parabolas which join at their common end points in such a way that $f(x)$, $f'(x)$, and $f''(x)$ are continuous. Hence the solution is a cubic spline. The smoothing function which must be constructed minimizes $x \int_0^{x_n} (f''(x))^2 dx$, and the sum of the quadratic deviations of the measured values $y_i(x_i)$ ($y_i = {}^{14}C$-age, x_i = bristlecone pine age) can be adapted to the sum of the quadratic statistical errors of the measurements. Figure 7 shows the bristlecone pine data determined by all three laboratories covering the range of dendro dates from 3800 to 5200 years B.P. together with the spline function. The ^{14}C ages were calculated with a half-life of 5730 years. The plateau from 4500 to 4800 dendro years corresponding to about 4200 ^{14}C years is characteristic. This plateau is probably in accordance with the younger part of chronology A. The older samples of chronology A seem to reflect the increase in ^{14}C age at approximatively 4,800 years B.P.

FITTING OF CHRONOLOGY A

To achieve an objective criterion for fitting, the square root of the average quadratic deviation $\sqrt{(\Sigma \sigma_i^2 / n)}$ ($\sigma_i^2 = \{y_i - f(x_i)\}^2$, n = number of values) of chronology A from the spline function was calculated for varying younger end points of chronology A. The results are shown in figure 8. The curve shows a well defined minimum at 4700 B.P. This represents the best fitting for the younger end point of chronology A. The agreement of the value of 67 years of $\sqrt{(\Sigma \sigma_i^2 / n)}$ at this minimum with the error of the measurements (40 to 70 years) is satisfactory, especially in view of error due to the uncertainty of the spline function fit. Because systematic error in the ^{14}C standard of the laboratory could

THE CONTRIBUTION OF THE SWISS LAKE-DWELLINGS

TABLE 1

Radiocarbon Dates of Floating Dendrochronologies Yverdon A and B

Tree ring sample no.	Laboratory dating no.	Radiocarbon age B.P.[1] (error 40-70 years)	
2020-2011	B-2682	4097	
2030-2021	B-2683	4218	
2040-2031	B-2684	4046	
2050-2041	B-2685	4172	
2060-2051	B-2686	4300	Chronology B
2070-2061	B-2687	4451	
2080-2071	B-2688	4176	
2090-2081	B-2689	4181	
2100-2091	B-2690	4184	
			Increasing age
251-260	B-2680	4298	
241-250	B-2679	4158	
231-240	B-2678	4289	
221-230	B-2677	4175	
211-220	B-2676	4232	
201-210	B-2675	4320	
191-200	B-2674	4141	
181-190	B-2673	4280	
171-180	B-2672	4232	
161-170	B-2671	4281	
151-160	B-2917	4167	
141-150	B-2669	4222	
131-140	B-2668	4289	
121-130	B-2667	4291	
111-120	B-2666	4269	Chronology A
101-110	B-2665	4339	
91-100	B-2664	4291	
81- 90	B-2663	4281	
71- 80	B-2916	4066[2]	
61- 70	B-2915	4447	
51- 60	B-2914	4457	
41- 50	B-2913	4437	
31- 40	B-2912	4476	
21- 30	B-2911	4517	
11- 20	B-2910	4537	
1- 10	B-2909	4463	
-9- 0	B-2908	4492	
-19- -10	B-2907	4484	
-27- -20	B-2006	4484	

[1] All age determinations are based on a ^{14}C half-life of 5730 years. The dates are corrected for isotopic fractionation. The result for each sample is calculated as mean value of two to four measurements.
[2] Sample B-2916 proved to be of poor counting gas-quality, and thus was rejected.

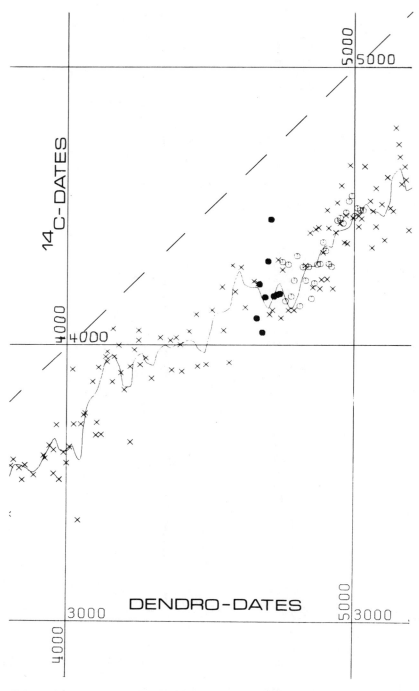

FIG. 7. Bristlecone pine raw data, spline function, and Yverdon data. Radiocarbon dates (based on a ^{14}C half-life of 5730 years) and dendrochronological dates in years before 1950. Younger end point of chronology A (circles) at 4750 B.P., younger end point of chronology B (filled-in circles) at 4660 B.P.

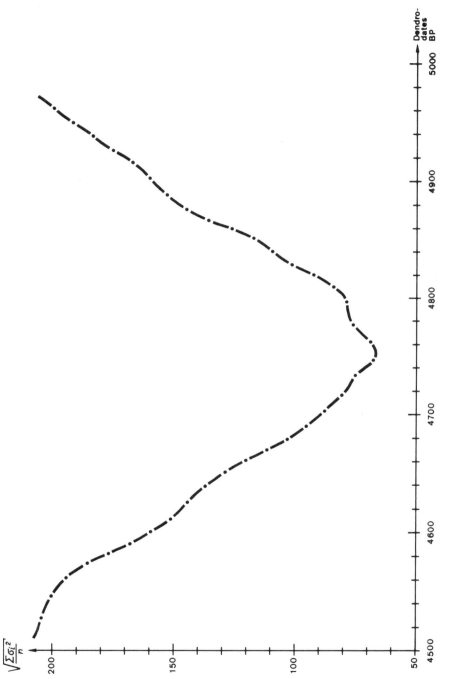

FIG. 8. Square root of average quadratic deviation of chronology A from spline-function-smoothed bristlecone pine data for varying younger end points of chronology A.

result in data too old or too young, the fitting procedures were repeated with all ^{14}C ages of chronology A decreased (and respectively increased) by Δy years, but no better fittings were obtained.

FITTING OF CHRONOLOGY B

The same process was applied to fit chronology B. The corresponding graph is shown in figure 9. In contrast to the results of A, there is no distinct minimum but rather a minimal range from 4520 to 4800 B.P. for the possible younger end points of chronology B. This is due to the samller number of data available for B, their greater scatter, and the fact that they lie in a region of the spline function with no characteristic trend, but additional archaeological arguments permit the reduction of the range of the possible younger end points of chronology B.

DISCUSSION

The ^{14}C data of chronology A were compared to a graph of the latest bristlecone pine data of the La Jolla laboratory (H. E. Suess, personal communication). The comparison agreed with the fit obtained by the mathematical procedure. The positive result confirms the feasibility of such fittings and at the same time indicates a common characteristic pattern of the ^{14}C calibration curve for both North America and Europe. Conversely, problems arising from the fit of chronology B point out the difficulties encountered when only a few data for the fitting procedure are available.

Importance of the Calibrated Dates from Yverdon

ARCHAEOLOGICAL ARGUMENTS FOR THE FITTING OF THE FLOATING CHRONOLOGIES

In considering the calibrated results, it has been shown that the younger end of the sample series of the older chronology (A) belongs unequivocally to about 4750 B.P. (2,800 B.C.). According to figure 9, the younger end of the younger chronology (B) falls into the time between 4800-4520 B.P. (2850-2570 B.C.). It must be remembered that chronology B itself embraces 90 years, and that both complete chronologies cannot overlap by more than 40 years since a longer overlap would have been recognized in the dendrochronological curves. Thus the range of the younger end of chronology B is limited to the period from 4660 B.P. (2710 B.C.) to 4520 B.P. (2570 B.C.) (fig. 10). The sample with the youngest possible absolute date (2570 B.C.) would be sample 2020-2011 of chronology B (see table 1). (The notation 2020-2011 is connected with the ^{14}C date of 2000 B.C., originally accepted as the average age of the settlement and arbitrarily coordinated with a specific point in chronology B). It should be emphasized that this youngest sample represents only the end of a whole series of well dated posts (figure 6) which produce information on several construction phases.

With reference to the occupation of Yverdon, the lowest layers which are combined with the posts of chronology A were formed at about 4750 B.P. (2800 B.C.), while the

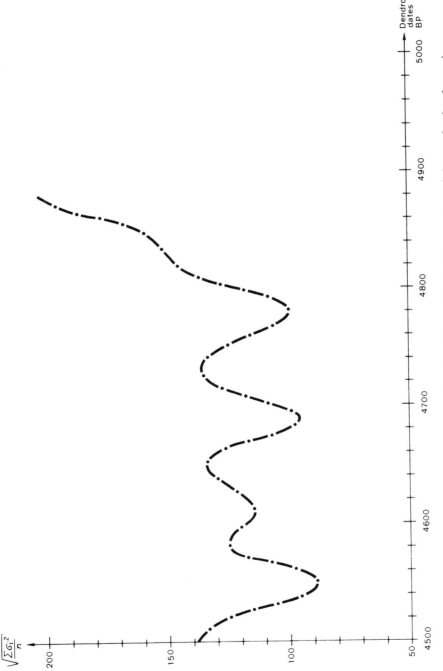

FIG. 9. Square root of average quadratic deviation of chronology B from spline-function-smoothed bristlecone pine data for varying younger end points of chronology B.

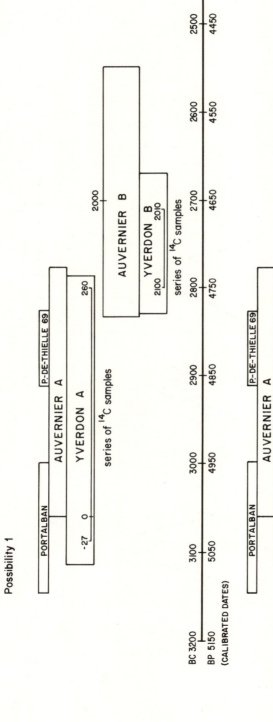

FIG. 10. The correlations of chronologies A and B of Yverdon and contemporary sites resulting from the fitting with the bristlecone pine data.

intermediary levels are contemporary with the younger end of chronology B at about 4520 B.P. (2570 B.C.). The curve of the tree ring samples is also confirmed by the earlier ^{14}C samples from Yverdon, collected in different archaeological layers. If these ^{14}C samples are put together according to their stratigraphical sequence, a very irregular, unsteady diagram (fig. 11) results, which is related to the graph of the tree ring samples. The initial incline and the abrupt decline correspond to the calibration curve for the dendro dates between 4800 and 4700 B.P. (2850 and 2750 B.C.) (Suess and Strahm 1970).

An interesting comparison can be made with an earlier calibration experiment in Auvernier. On the basis of his few data, H. Suess has established 2750 B.C. and 2550 B.C. as years corresponding to the younger ends of chronology A and chronology B. This differs only slightly from the fitting presented here. The curves determined by the two laboratories are similar in the case of chronology A. Chronology B, however, is not comparable because only four ^{14}C samples from Auvernier had been measured at that time.

HISTORIC SIGNIFICANCE

The absolute dates of prehistoric Europe, and especially of southeastern Europe, were based traditionally on the chronological records of ancient king lists in Mesopotamia (Milojčić 1949). These lists dated the archaeological context in which they were found. If an imported object from an adjoining culture lay in this context, this culture could also be dated. This process of dating was transferred to neighboring cultures, which were again correlated by other imports. By this so-called "chronological bridge", the dates of the third and second millenia B.C. were obtained for Yverdon. In this method of dating, retardation and other uncertainties were never allowed for. If the effect of retardation is considered, the middle European dates become even younger in comparison to the southeastern European dates; this means that the divergence existing between traditional and radiocarbon dates for Yverdon cannot result from retardation.

The uncertainty of the dates transferred from the Near East is revealed by the following consideration: If an object, well dated in the region I where it was produced, is found in a neighboring culture II, it cannot be determined whether it was imported at an initial or a later stage of this culture II. Even if the duration of culture II is estimated at only one hundred years, an uncertain period of two hundred years is obtained. For example, if the importation belongs to the final stage of culture II, culture II has begun one hundred years earlier, but if it belongs to the initial stage of II, the culture has continued after the importation for another one hundred years. If this method of dating (the chronological bridge) is continued, the uncertainty grows with each step. After four transfers an uncertainty of ± 400 years results (fig. 12). To get correct ages transferred, dates may have to be replaced by younger dates (in the case of retardation) or by older dates. The cumulative uncertainty produced by each transfer can be reduced by cross-datings, that is, by bringing back an object from a distant culture to one nearer to the Orient. Nevertheless, the chronological bridge has proved to be of putative precision.

582

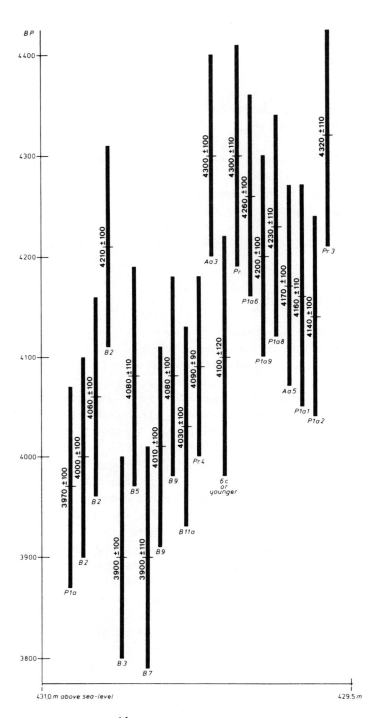

FIG. 11. The ^{14}C-dates of Yverdon from different layers.

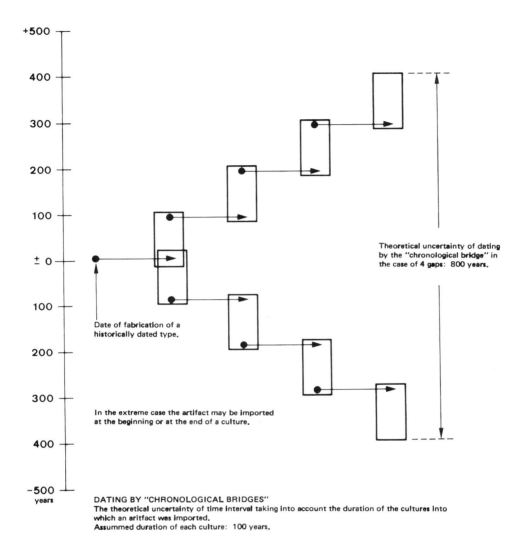

FIG. 12. Error produced by application of a "chronological bridge".

For Yverdon the transferred tranditional dates are inconsistent with the calibrated dates. According to the traditional chronology, a date of about 3950 B.P. (2000 B.C.) should be expected, but the calibrated date is approximately 4750-4520 B.P. (2800-2570 B.C.). This discrepancy may be explained by the uncertainty inherent in the traditional chronology, but there are additional arguments that call for the antedating of some cultures, of which Yverdon is one representative (Suess, Strahm 1970). Thus the calibration compels the archaeologist to reflect on his basis of dating.

SIGNIFICANCE FOR RESEARCH

The demonstrated results suggest the possibility of calibrating radiocarbon dates for a short period, and thus, of obtaining in future an absolute time scale, as demonstrated by H. Suess, E. Ralph and others. For archaeology, the results suggest revision of the traditional chronology, first for a short period, that is, the end of the third millenium B.C. Although the results have yet to be consolidated, they suggest that further research into analogous projects must continue. There is sufficient material in Switzerland to initiate comparable investigations (see Lambert and Orcel, this volume).

Since a number of samples from Swiss lake-dwellings have already been dated at the radiocarbon laboratory of the University of Bern, and since others will follow, further work on floating chronologies of the third and second millenniua B.C. will be conducted at the same institution. To avoid inter-laboratory differences, cross-checks with bristlecone pine samples of the corresponding period are anticipated.

REFERENCES

Damon, P. E., C. W. Ferguson, A. Long, and E. I. Wallick
    1974   Dendrochronologic Calibration of the Radiocarbon Time Scale. Amer. Antiquity. 39(2): 350-366.

Milojčić, V.
    1949   Die Chronologie der jüngeren Steinzeit Mittel- und Südosteuropas.

Ralph, E. K., H. N. Michael, and M. C. Han
    1973   Radiocarbon dates and reality. MASCA Newsletter. 9(1): 1-20.

Reinsch, C. H.
    1967   Smoothing by Spline Functions. Numerische Mathematik. 10: 177-183.

Strahm, Ch.
    1972/73   Les fouilles d'Yverdon. Jahrb. Ges. f. Ur- und Frühgeschichte. 57: 7-16.
    1975   Die Saône-Rhône-Kultur. Archäologisches Korrespondenzblatt. 5: 273-82.

Suess, H., and Ch. Strahm
    1970   The Neolithic of Auvernier, Switzerland. Antiquity. 44: 91-99.

Suess, H. E.
    1970   Bristlecone pine calibration of the radiocarbon time-scale 5200 B.C. to the present. Proc. 12th Nobel Symp. Uppsala. Stockholm, Almquist and Wiksell. New York, John Wiley and Sons, 303-311.

9

Dendrochronology of Neolithic Settlements in Western Switzerland: New Possibility for Prehistoric Calibration

G. Lambert and C. Orcel

The present study was initiated in October 1972, when the attention of Swiss authorities was drawn to the urgent problem of rescuing various archaeological sites directly endangered by the construction of the national highway 5 (RN5) (Schifferdecker et al. 1974). General management of the excavations in Neuchatel was entrusted to the archaeological museum. Excavations were made in the lake of Neuchatel, in the bay of Auvernier located at the foot of the eastern side of the Jura Mountains. These excavations revealed the existence of a large area now submerged in the lake that was once inhabited by prehistoric man. The settlements in this area succeed one another along the shoreline for a distance of 1000 m.

Among the settlements, which have been explored by divers and later excavated on "dry land" after the construction of dikes, the most recent are three attributed to the late Bronze Age. As these settlements do not directly concern the present subject, we will not go into further detail here.

For the Neolithic period a number of settlements have been excavated on a large scale with areas ranging from 600 to 1200 sq m. Among these the principal sites are Le Port, Le Boise-Lames (Boisaubert et al. 1974), Le Ruz-Chatruz, and La Saunerie. In addition to these sites numerous others have been located in the area. At least 27 main settlements have been found dating from 3200 B.C. to 1700 B.C. (5150 B.P. to 3650 B.P.) (Lundstrom et al. 1975). The longest stratigraphic sequence is to be found at the La Saunerie site (Vouga 1929, Jequier and Strahm 1965, Seuss and Strahm 1970, Valla 1972).

These lake sites have yielded thousands of posts, of which 5000 have corresponded to the Neolithic period. The preservation of these posts in a wet or humid environment is excellent, but deterioration follows rapidly upon excavation. A dendrochronological laboratory was established in Neuchatel to ensure maximum utilization of this fragile material. The program was initiated under the guidance of Mr. Schweingruber.

A sampling of the Neolithic material encountered to date in the bay of Auvernier shows that at present about 1000 posts have been measured. Of these, 150 have been synchronized. These correlations have been established by a visual method and have been placed in seven different sequences which at present cannot be definitely related to one another. These sequences, arranged chronologically from the earliest to the most recent, are:

First Sequence. Civilization of Cortaillod:

^{14}C date: Bern, 5150 B.P.
Duration: 230 years

> Three settlements:
>
>> Auvernier, Port; level V: 4 posts
>> Auvernier, Saunerie; lower levels: 22 posts
>> Twann, (Lake of Biel); lower levels: 11 posts

Second Sequence. Civilization of Cortaillod-évolué:

^{14}C date: Bern, 4400 B.P.
Duration: 164 years

> Three settlements:
>
>> Auvernier, Saunerie 2; lower levels: 28 posts
>> Montilier (canton de Fribourg): lower levels 5 posts
>> Twann, upper levels: 8 posts

Third Sequence. Horgen-Culture:

^{14}C date: none.
Duration: 236 years

> One settlement:
>
>> Auvernier, Graviers: 6 posts

Fourth Sequence. Lüscherz-Culture:

^{14}C date: 4050 B.P.
Duration: 266 years.

Four settlements:

> Auvernier, Saunerie; medium levels: 10 posts
> Auvernier, Brise-Lames: 20 posts
> Auvernier, Ruz-Chatruz: 12 posts
> Auvernier, Graviers: 10 posts

Fifth Sequence. Civilization of Saône-Rhône (Auverneir-Culture):

^{14}C date: ca. 3800 B.P.
Duration: 166 years.

One settlement:

> Auvernier, Saunerie 4; medium levels: 8 posts

Sixth Sequence. Cordée Civilization and early Bronze 1-2.

^{14}C date: none
Duration: 121 years.

One settlement:

> Auvernier, Saunerie 5; upper level: 4 posts

Seventh Sequence. Late Bronze.

^{14}C date: none.
Duration: 229 years.

Two settlements:

> Cudrefin (Canton de Vaud): 10 posts
> Morges, Les Roseaux (Canton de Vaud): 5 posts

If combined, these systems would represent a chronology of 1531 years ranging from 3650 B.P. to 5150 B.P. (conventional dates). According to present archaeological hypothesis, certain of these sequences probably should be combined.

Three computer programs have been developed to systematize our study. The first program (ROSE) is able to draw all the needed curves; the second program (ERICA) keeps a permanent file of all data; and the third program (IRIS) performs two kinds of statistical calculations: (1) the calculation of Bravais and Pearson coefficients on corrected curves (algorithms of Humbert, Centre de Calcul, Universitaire de Neuchatel) (De Martin 1974), and (2) the calculation of Eckstein coefficients (1969), based on La Place and Gauss demonstrations. Table 1 has been checked by this process.

TABLE 1

Dendrochronology of neolithic settlements in W. Switzerland.

Fourth Sequence: Lüscherz

A	B	N	A	B	N	A	B	N	A	B	N	A	B	N	A	B	N
1	325	1	51	165	2	101	155	9	151	90	33	201	75	30	251	75	21
2	325	1	52	155	2	102	145	10	152	95	33	202	75	29	252	80	20
3	230	1	53	135	2	103	145	10	153	105	33	203	75	28	253	75	20
4	320	1	54	155	2	104	190	12	154	125	33	204	80	28	254	65	19
5	250	1	55	150	2	105	180	12	155	100	33	205	80	27	255	60	19
6	310	1	56	195	2	106	140	12	156	100	33	206	85	27	256	80	19
7	165	1	57	170	2	107	140	13	157	105	33	207	85	27	257	85	19
8	175	1	58	160	2	108	130	13	158	110	33	208	85	27	258	90	17
9	275	1	59	130	2	109	140	16	159	110	32	209	95	26	259	70	16
10	175	1	60	145	2	110	120	17	160	140	32	210	105	26	260	65	15
11	225	1	61	125	2	111	125	17	161	115	31	211	80	26	261	80	14
12	210	1	62	115	2	112	100	17	162	100	31	212	105	26	262	80	14
13	160	1	63	110	3	113	125	17	163	105	31	213	95	26	263	95	14
14	150	1	64	115	3	114	160	18	164	80	31	214	110	26	264	85	14
15	225	1	65	145	3	115	175	20	165	85	31	215	90	26	265	80	13
16	265	1	66	125	3	116	165	20	166	80	31	216	100	26	266	85	13
17	220	1	67	150	3	117	155	21	167	95	31	217	70	26	267	100	12
18	215	1	68	130	3	118	100	22	168	95	30	218	80	26	268	90	11
19	190	1	69	120	3	119	135	22	169	105	29	219	90	26	269	90	11
20	165	1	70	160	3	120	145	23	170	95	29	220	85	26	270	75	11

A	B	N
301	45	3
302	55	3
303	70	3
304	65	2
305	65	1
306	80	1
307	60	1
308	65	1
309	65	1
310	65	1
311	55	1

A	B	N	A	B	N	A	B	N
21	165	1	71	160	3	121	185	23	171	105	29	221	95	26	271	75	11
22	125	1	72	135	3	122	135	23	172	95	28	222	85	26	272	70	11
23	160	1	73	105	3	123	145	24	173	85	28	223	105	26	273	75	11
24	125	1	74	135	3	124	175	24	174	75	28	224	80	26	274	75	10
25	195	1	75	130	3	125	155	24	175	105	28	225	85	26	275	70	10
26	175	1	76	125	3	126	185	24	176	95	28	226	70	26	276	70	9
27	180	1	77	110	3	127	185	24	177	95	28	227	80	26	277	80	9
28	150	1	78	135	3	128	140	25	178	85	28	228	75	26	278	80	8
29	150	1	79	115	3	129	150	25	179	90	28	229	75	26	279	65	8
30	185	1	80	100	3	130	120	25	180	80	28	230	85	26	280	50	8
31	185	1	81	120	3	131	140	25	181	75	28	231	85	26	281	65	8
32	210	1	82	120	3	132	135	26	182	70	29	232	80	26	282	65	8
33	200	1	83	120	3	133	120	26	183	85	29	233	100	26	283	65	8
34	230	1	84	110	3	134	135	26	184	80	29	234	85	26	284	55	8
35	190	1	85	120	3	135	135	26	185	85	29	235	100	26	285	65	8
36	160	1	86	145	4	136	120	27	186	80	30	236	90	26	286	55	7
37	160	1	87	120	4	137	110	28	187	80	30	237	95	26	287	60	7
38	150	1	88	135	4	138	110	28	188	90	30	238	85	26	288	70	7
39	120	1	89	115	4	139	100	28	189	75	31	239	90	25	289	70	7
40	120	1	90	125	4	140	120	28	190	95	29	240	95	25	290	70	7
41	135	1	91	160	5	141	120	28	191	95	29	241	95	25	291	65	7
42	125	1	92	150	5	142	130	28	192	95	31	242	85	24	292	60	6
43	175	1	93	145	6	143	110	29	193	85	31	243	80	24	293	60	5
44	150	1	94	145	7	144	105	30	194	70	31	244	75	24	294	50	5
45	125	1	95	150	7	145	135	30	195	80	31	245	60	24	295	40	5
46	115	1	96	125	7	146	130	31	196	85	30	246	75	23	296	40	5
47	100	1	97	145	8	147	135	31	197	90	30	247	60	23	297	40	5
48	170	2	98	155	8	148	135	32	198	110	31	248	80	23	298	45	5
49	165	2	99	155	8	149	135	32	199	90	31	249	70	22	299	40	5
50	170	2	100	160	9	150	110	32	200	90	31	250	90	21	300	40	5

A = years.
B = breadth of tree ring (average) 1/100 mm.
N = number of tree rings

REFERENCES

Boisaubert, J. L.
  1972- (In preparation.)
  75

Boisaubert, J. L., F. Schifferdecker, and P. Petrequin
  1974 Les villages néolithiques de Clairvaux (Jura, France) et d'Auverneir (Suisse). Problemes d'interpretation des plans. Paris, B.S.P.F., 355-382.

De Martin, P.
  1974 Analyse des cernes, dendrochronologie et dendroclimatologie. Paris, Masson, 78.

Eckstein, D.
  1969 Entwicklung und Anwendung der Dendrochronologie zur Alterbestimmung der Siedlung Haithabu. Thesis. Univ. of Hamburg.

Jequier, J. P., and C. Strahm
  1965 Les fouilles archeologiques d'Auvernier en 1964. Museé Neuchatelois. 2:78-88.

Lundstrom, K., G. Lambert, and F. Schifferdecker
  1975 Problèmes stratigraphiques et chronologiques du Neolithique d'Auvernier. Colloque du Groupe de Travail pour les Recherches pré-et proto-historiques en Suisse. 5.

Schifferdecker, F., P. Lenoble, and G. Lambert
  1974 Au bord du lac de Neuchatel, les stations littorales d'Auvernier. Archaeologia, Sept:58-65.

Suess, H. E., and C. Strahm
  1970 The Neolithic of Auvernier, Switzerland. Antiquity. XLIV:91-99.

Valla, F. R.
  1972 Les fouilles francaises à Auvernier (Neuchatel, Suisse) en 1948. Etude du materiel conservè au Musée de l'Homme, Paris, Archives Suisse d'Anthropologie Générale. XXXVI:1-79.

Vouga, P.
  1929 Classification du Néolithique Lacustre suisse. Indicateur d'Antiquities Suisses. 31:81-91, 161-180.

10

Archaeological Evidence for Short-term Natural ^{14}C Variations

R. Burleigh and A. Hewson

ABSTRACT

The subject of this report is the occurrence in the past of significant short-term ^{14}C variations as evidenced by the 300-year spread of radiocarbon ages obtained for an assemblage of deer antlers that had been rapidly sealed archaeologically. Other possible explanations, archaeological and geochemical implications, and plans for further work are discussed.

INTRODUCTION

The general trend and extent of natural ^{14}C deviation over the last seven to eight millennia is well established and there is considerable evidence also for short-term variations lasting up to a few hundred years. These facts are so well known that it is not necessary to discuss them here or even to cite the now familiar literature. At issue is whether there have also been substantial shorter-term fluctuations each having a periodicity of about ten to twenty years. There is some evidence for this effect from measurements of the ^{14}C activity of recent single tree ring series, in particular by Farmer and Baxter (1972). The present paper gives the results of measurements which were made originally for purposes of archaeological dating, but which support the idea that these shorter-term variations have also occurred in the more remote past.

Archaeological Background

In contrast to the main body of evidence now available for very short-term natural ^{14}C variations, the radiocarbon measurements presented here were obtained not from tree rings but from antlers of the red deer (*Cervus elaphus* Linn.). Since the antlers of deer are shed each year they represent potentially sensitive short-lived material which, in this particular case, happens to have been sealed archaeologically over a short period of time.

The antlers we have measured were obtained from excavations recently carried out by the British Museum at an extensive prehistoric flint mine site at Grime's Graves, Norfolk, England (52° 30' N, 0° 40' E). The location of the site is shown in figure 1. This figure also shows the location of the site in relation to the general distribution of the Upper Cretaceous Chalk formations in the region, in which bands of flint occur in situ.

In the later Neolithic period in Britain (ca. 2000 B.C. on the basis of radiocarbon dating) high quality flint used for the manufacture of tree-felling axes and other tools was mined on a large scale at Grime's Graves. The prehistoric miners sank roughly circular shafts some 13 m in depth and 3-5 m in diameter through the superficial deposits into the underlying chalk to reach a more or less continuous 20 cm layer of tabular flint, presum-

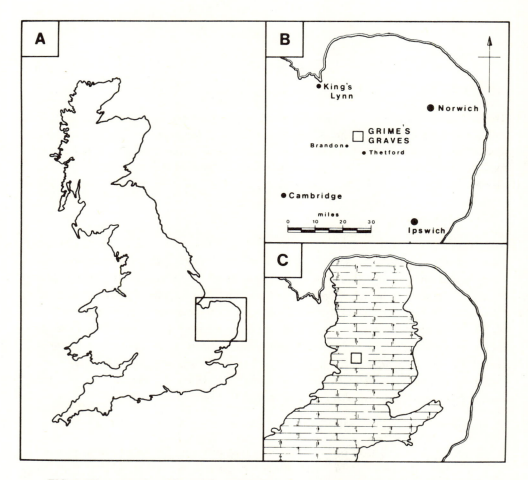

FIG. 1. Diagrammatic outline of the British Isles (A) and regional map of East Anglia (B), showing the location of Grime's Graves prehistoric flint mines. Map (C) shows the position of the site in relation to the cretaceous chalk formations of the region (hatched zone; the small square indicating the site itself is not to scale).

ably known to them from its occurrence as an outcrop at some point. At the base of the shafts, 4-7 horizontal galleries about 0.75-1.5 m high, 3-5 m wide, and 5-10 m in length were cut in order to win more flint. Figure 2 illustrates a typical gallery at Grime's Graves after clearing by modern excavation. There are other groups of similar prehistoric flint mines elsewhere in southern England and in various parts of continental Europe, notably in Belgium and the Netherlands (Clark 1952). The chronology of the mines has been reasonably well defined by radiocarbon dating (Burleigh 1976).

Deer antlers were used in large numbers by the Neolithic miners as levers or so-called "picks" (figure 3), and were apparently casually discarded when the points became too blunted for easy insertion into joints in the chalk. As a result these antlers became sealed in the chalk rubble backfilled into worked out galleries where they have since remained undisturbed in an environment entirely free from organic contamination.

The most effective way of working a mine once the shaft was completed would have been to backfill the chalk waste from a new gallery directly into a previously worked gallery (lifting only the waste from the first gallery to the surface). Stratigraphic evidence indicates that this method was used. Lack of appreciable weathering in the excavated shafts, coupled with the need for disposal of massive quantities of spoil from adjacent new shafts, strongly suggests that each pit was open for a relatively short time. Clark (1952) has estimated that a typical pit could have been fully worked out in six months. On this basis a reasonable estimate for the exploitation of the entire deep mining area at Grime's Graves, some 400 pits, would be less than 200 years.

Although any group of pits may have been open for a year or two, the overall picture is one of rapid sealing of at least partly "sorted", short-lived material. The possible significance of some of these factors manifested itself only as the radiocarbon measurements were assimilated, revealing an unexpectedly large age variation in material which had apparently been deposited and sealed over a short period of time.

Results

The first series of radiocarbon dates for the antlers from Pit 15 at Grime's Graves is listed in table 1. Following archaeological convention, the measurements are expressed as dates in the B.C. system but with the lower case notation "bc" used to indicate that these are uncalibrated radiocarbon ages based on the 5570-year Libby half-life for ^{14}C. These dates were obtained from collagen separated from antler by demineralization with dilute hydrochloric acid. Preservation of the antlers in a chalk environment precluded humic contamination. The collagen was converted to benzene for measurement of the ^{14}C activity by liquid scintillation counting. Reference to detailed laboratory procedures is given in a date list in *Radiocarbon* (Burleigh et at. 1976).

The measured ages span a period of about 300 radiocarbon years with a mean age of 1950 bc. The measured values tend to fall into two groups, one earlier than 1990 bc and the other later than 1940 bc. This appears to be related to the find positions of the antlers in the galleries (fig. 4). Taking into account the standard deviations of 79 years for the group as a whole and an average value of 47 years for the individual estimates, a time span of 120 years would be needed to explain the spread of the results. This is contrary

FIG. 2. Flint mine gallery at Grime's Graves (cleared by modern excavation; roof supports are modern; lighting is artificial). The seam of tabular flint is visible at the base of the gallery walls.

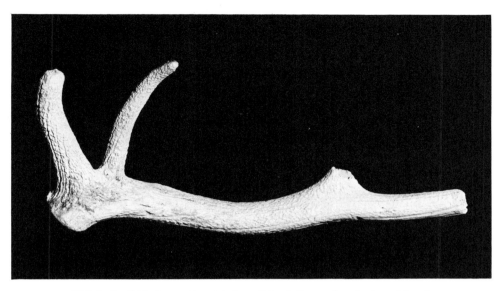

FIG. 3. Neolithic deer antler pick from Grime's Graves (length about 60 cm).

to all the available evidence which indicates that the pit system was worked for a very much shorter period of time.

There are two possible explanations for the spread of our results. Either the cause must lie somewhere in our laboratory procedures or it must be inherent in the sample material. The checks we maintain continuously on sample processing and measurement rules out the first of these possibilities and leads to the conclusion that the effect is real and must arise from the use of antler as dating material. These results and their interpretation as evidence for very short-term atmospheric ^{14}C variations are reported in *Nature* (Burleigh and Hewson 1976). A subsequent series of measurements is listed in table 2. These measurements form pairs with some of the samples listed in table 1 (see fig. 4 for the distribution of these pairs). The agreement between the two series is very close, perhaps more so than we deserve statistically. We interpret this second series of results both as supporting evidence for the model used to explain the first series and as further confirmation of the reproducibility of our laboratory procedures. Figure 5 is a composite plot of the two sets of results. All the measurements of both series were corrected for isotopic fractionation relative to PDB.

Discussion

We believe that our measurements provide corroborative evidence for very short-term ^{14}C variations during at least one period of the more remote past. We conclude this because of the special circumstances of physical association of the particular samples dated. Considerable evidence supports this hypothesis, but the remaining uncertainties raise interesting possibilities for further work.

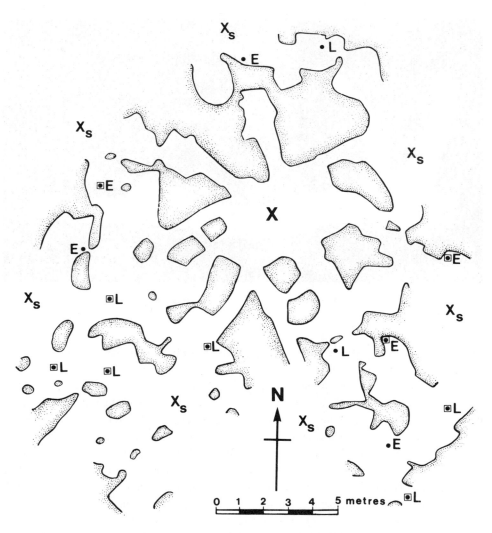

FIG. 4. Diagrammatic plan of the galleries of Pit 15 at Grime's Graves showing the original positions of the antler samples. "E" and "L" indicate whether the radiocarbon dates obtained fall early or late in the sequence (see text and table 1). Boxed dots represent paired samples. "X" marks the central shaft of Pit 15 and the x_s's indicate the approximate positions of the surrounding shafts.

Antler is short-lived, grows rapidly, is shed each year, and should reflect the atmospheric ^{14}C activity of the preceding spring and summer. Deer feed mainly on young, fresh growth in equilibrium with the prevailing atmospheric ^{14}C level. The results obtained can be explained by a variation in atmospheric ^{14}C activity of the order of 2% in less than 10 years, which is not incompatible with the findings of some previous work. It is an open question whether such variations may have occurred continuously. An alternative explanation that the antlers themselves were of widely different ages initially

TABLE 1
Radiocarbon dates for antlers from Pit 15

Age estimate (date ± σ)	Δ °/oo	Std. dev. of Δ	δ¹³C (°/oo)	Lab. No.
1786 ± 58	−12.7	4.6	−24.80	BM-980.
1870 ± 46	− 6.2	3.6	−24.99	BM-979.
1877 ± 45	− 5.6	3.5	−24.16	BM-973.
1899 ± 44	− 3.9	3.4	−23.03	BM-976.
1915 ± 44	− 2.7	3.4	−25.00	BM-978.
1918 ± 56	− 2.5	4.3	−23.32	BM-1001.
1932 ± 45	− 1.4	3.5	−21.16	BM-1002.
1937 ± 47	− 1.0	3.6	−24.13	BM-974.
1940 ± 42	− 0.8	3.2	−23.65	BM-996.
1990 ± 41	+ 3.1	3.1	−24.10	BM-975.
1999 ± 42	+ 3.7	3.2	−22.46	BM-1003.
2010 ± 56	+ 4.6	4.3	−24.87	BM-997.
2042 ± 45	+ 7.0	3.4	−23.00	BM-998.
2065 ± 61	+ 8.7	4.6	−24.53	BM-977.
2072 ± 57	+ 9.3	4.3	−23.25	BM-1000.

The dates are expressed in radiocarbon years bc and have been corrected for isotopic fractionation relative to the PDB Standard. The associated error terms are equivalent to one standard deviation and are derived from about 15 × 100 minute counts of each sample (totalling not less than 10^4 net counts). Δ Values have been calculated from the mean age for the series of 1950 bc. Figure 4 shows the general distribution of the dated antlers in the galleries of Pit 15.

TABLE 2
Comparison of dates for paired samples from Pit 15

Lab. No.	Original age estimate (see table 1)	Age estimate for duplicate sample	δ¹³C °/oo	Lab. No.
BM-980.	1786 ± 58	1790 ± 48	−23.80	BM-1056.
BM-979.	1870 ± 46	1918 ± 56	—	BM-1001.
BM-978.	1915 ± 44	1974 ± 47	−23.00	BM-1057.
BM-1002.	1932 ± 45	1926 ± 48	−22.90	BM-1058.
BM-974.	1937 ± 47	1954 ± 36	−22.22	BM-1054.
BM-996.	1940 ± 42	1884 ± 50	−23.30	BM-1053.
BM-975.	1990 ± 41	1937 ± 56	−23.20	BM-1051.
BM-1003.	1999 ± 42	2004 ± 43	−22.90	BM-1052.
BM-977.	2065 ± 61	2027 ± 47	−22.60	BM-1059.

Dates for pairs of antlers found together in Pit 15. The dates listed on the right provide a direct check on some of those from the first series (table 1) listed on the left of the table (BM-979 and BM-1001 were both in the first series but also formed a pair). As in table 1 all the dates are in radiocarbon years bc corrected for isotopic fractionation. The precision of measurement was comparable for both series.

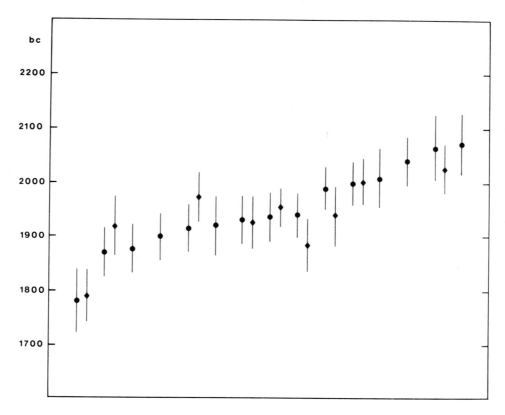

FIG. 5. Composite plot of all the dates listed in tables 1 and 2. The duplicate sample measurements are indicated by diamond symbols (dates are in radiocarbon years bc; error bars are equivalent to ±1 standard deviation).

is contrary to the stated archaeological evidence. The role of dietary factors as a possible cause of the apparent age differences is not supported by the $\delta^{13}C$ values listed in tables 1 and 2. All the antlers have been subjected to the same conditions of preservation in a humid but contamination-free environment at a temperature which has probably remained nearly constant at 7°C for more than 4000 years. Thus differential conditions of preservation or exposure to age contaminants cannot explain the spread of the results, and in situ ^{14}C production cannot explain the age differences. Producing a negligible effect in other materials at or near sea level (Harkness and Burleigh 1974), in situ ^{14}C production might be important in the case of antler which has a higher nitrogen content (Cain and Suess 1976). But Grime's Graves is less than 60 m above sea level, and for almost the whole of their existence the antlers have been shielded by at least 10 m of chalk. It has been suggested (Jope 1976) that antler horn may not reflect the true atmospheric ^{14}C level at the time due to its exceptionally rapid growth. If so, the age of the animal producing the antler might possibly be important. This factor could not be determined for the antlers we have dated. There appears to be scope here for a study of modern antler by means of bomb-carbon and as a matter of separate interest we plan to test antler for in situ ^{14}C production with reactor neutrons.

The assumption is made that only fresh antlers would be usable as tools and that their turnover by the miners would be rapid. The large numbers of discarded antlers found in the mines support to this point. It is not known whether the Grime's Graves area could have supported the number of deer necessary to supply the miners' demands or if antlers were traded over some distance, but it seems reasonable to suggest that the antlers found together in a particular pit system should represent the growth of one or a few, roughly successive seasons.

To test our hypotheses we plan a further series of age measurements based on a new group of antler samples from a separate pit at Grime's Graves. Apatite dates (Haynes 1968) are being obtained for comparative purposes from some antlers for which we already have collagen dates. No independent checks against charcoal are possible since sufficient has not been found in the mine galleries at Grime's Graves. It would be worth searching for similar ^{14}C variations in single tree rings of approximately the same age and in other suitable caches of archaeological material.

In conclusion, two main points can be made. First, from a practical point of view, whatever the true natural cause of the variations reported here, the inference is that radiocarbon ages for collagen from antler (but not bone) appear to be subject to substantial error. This may affect some British and European archaeological dates. Second, and more generally, ancient, short-lived material, especially when it forms a physical assemblage, may preserve a geochemical record uniquely valuable in determining the ultimate resolution that radiocarbon dating dating can achieve.

SUMMARY

Evidence is presented for the occurrence, about 4000 radiocarbon years B.P., of short-term radiocarbon variations similar to those previously reported in recent tree rings. This has been inferred from radiocarbon measurements of a series of red deer antlers utilized by Neolithic flint miners and sealed archaeologically over a short period of time. Whereas the antlers each represent a single year's growth and would all be expected to be the same age within a year or two, the measured ages have a spread of about 300 radiocarbon years. This is interpreted as evidence for an atmospheric ^{14}C variation of about 2% over 10 years. Other possible explanations are briefly discussed and some suggestions are made for further research. It is concluded that dates based on collagen from antler may be subject to considerable error. The potential importance of the record that may be preserved in all short-lived material is stressed.

ACKNOWLEDGMENTS

We thank Mr. G. de G. Sieveking of the Department of Prehistoric and Ramano-British Antiquities at the British Museum, who directed the underground explorations at Grime's Graves, for valuable advice and for supplying samples for measurement. Thanks are also due to the Trustees of the British Museum for the provision of funds enabling R. Burleigh to attend the Ninth International Radiocarbon Conference.

REFERENCE

Burleigh, R.
  1976   Radiocarbon dates for flint mines. 2d Int. Symp. on Flint. Maastrict. Nederlandse Geologische Vereniging, 89-91.

Burleigh, R., and A. Hewson
  1976   Evidence for short term atmospheric ^{14}C variations about 4000 B.P. Nature. 262:128-130.

Burleigh, R., A. Hewson, and N. Meeks
  1976   British Museum natural radiocarbon measurements VIII. Radiocarbon 18: 16-42.

Cain, W. F., and H. E. Suess
  1976   Carbon-14 in tree rings. J. Geophysical Research. 81:3688-3694.

Clark, J. G. D.
  1952   Prehistoric Europe: the economic basis. London, Methuen.

Farmer, J. G., and M. S. Baxter
  1972   Short-term trends in natural radiocarbon. Proc. 8th Int. Conf. on Radiocarbon Dating. New Zealand. 75-85.

Harkness, D. D., and R. Burleigh
  1974   Possible carbon-14 enrichment in high altitude wood. Archeometry. 16: 121-127.

Haynes, V.
  1968   Radiocarbon: analysis of inorganic carbon of fossil bone and enamel. Science. 161:687-688.

Jope, E. M.
  1976   Selection and preparation of protein material for radiocarbon dating. Proc. 9th Int. Radiocarbon Conf. Univ. of Calif. Los Angeles and San Diego.

11

Radiocarbon Variations Determined on Egyptian Samples from Dra Abu El-Naga

Ingrid U. Olsson and M. Farid A. F. El-Daoushy

ABSTRACT

Thirteen samples from Dra Abu El-Naga near Thebes in Egypt have been dated by the radiocarbon method. A comparison with dendrochronologically dated samples whose ^{14}C contents were measured in three other laboratories shows that the agreement between the ^{14}C content of these Egyptian samples and that of contemporaneous wood from different areas is good. A comparison with results from four other laboratories made on historically well-known samples, partly from the same graves, also shows good agreement.

INTRODUCTION

The present investigation was made as part of a joint ^{14}C project in which age determinations of well-known Egyptian samples were made in different laboratories. The aim of this paper is to give some technical information and the preliminary results, pending a joint analysis of all the dates. It also seeks to outline procedural methods found effective in obtaining reliable dates.

Samples

Samples were provided by H. N. Michael and E. Ralph from the Pennsylvania Radiocarbon Laboratory. They were collected between 13 and 15 April 1970 from the mudbrick pyramids at Dra Abu El-Naga South, Luxor, Egypt, by Lanny Bell. The reed samples are classified in three groups: (1) primary reeds, collected farthest from the surface and thus with minimal probability of external contamination, (2) secondary reeds, taken nearer the surface, and (3) tertiary reeds, taken nearest to the surface but not from the surface. These are the least preferred samples. Exposed reeds were discarded.

The samples belong to five tombs: (1) Nebwenenef, Tomb 157 (three reed samples), (2) Bekenkhons, Tomb 35 (one reed sample), (3) Roma-Roy, Tomb 283 (two *Tamarix* samples), (4) Tjanefer, Tomb 158 (two reed samples and one *Acacia* sample), and (5) Inhernakht, Tomb 282 (two reed samples).

In 1973, V. Täckholm analyzed part of the reed samples. Since she did not receive whole samples, there may have been a mixture of species, although the small samples she received were homogeneous except in one case. According to her analysis, all the reed samples are *Desmostachya bipinnata* (L.) Stapf., except the primary reeds from Tomb 157, Nebwenenef, which also contained some *Phoenix dactylifera* (L.). D. F. Cutler, of the Jodrell Laboratory in England, had analyzed the samples in 1971 with the same results.

Täckholm has also informed us that two grass species, *Desmostachya bipinnata* and *Imperata cylindrica* (L.) Beauv., both called "halfa", are used for making ropes and baskets.

The historical ages of these samples are closely related to the reign of Rameses II (Lanny Bell, personal communication). According to earlier statements, his reign began in 1304 B.C., but the present opinion is that it began in 1290 B.C. (Säve-Söderbergh, personal communication). For Nebwenenef, the historical age is 1290-1273 B.C., for Bekenkhons I, 1273-1223 B.C., for Roma-Roy, 1244-1196 B.C., for Tjanefer, 1267-1168 B.C., and for Inhernakht, 1290-1223, B.C.

Sample Treatment

The reed samples were broken into small pieces, the wood samples were cut into small pieces, and all samples were washed with distilled water to remove dust. The samples were boiled or simmered in 1% HCl overnight, centrifuged, treated repeatedly with distilled water, kept overnight at +80°C in a NaOH solution (1% NaOH), centrigued, again treated repeatedly with distilled water, and finally treated with HCl to remove any CO_2 absorbed before the samples were dried. The samples were then burned to produce carbon dioxide, which was purified in the modified way described by Olsson et al. (1972), to be used for the radioactivity measurement (Olsson 1958).

Radioactivity Measurement

The β-activity was measured in one, two, or three proportional counters made on principles described in the literature by de Vries, Stuiver, and Olsson. A new counter was used (PRF), described by El-Daoushy and Olsson (in press) in 1975 at the conference on low radioactivity measurements and applications at Tatranská Lomnica, Czechoslovakia.

This new counter had two electronic channels. The discriminators of these channels were set slightly differently, but, since they worked essentially in parallel, the mean value of the two channels was chosen and the statistical error was the same as for each of the channels or slightly lower. The choice of standards was such that one channel yielded slightly higher ages than the other when the counter was run at approximately 1.3 as well as 2.6 atm.

Since the counters were calibrated slightly differently, although the same oxalic-acid standard samples were used, the results were given separately for the three counters (PR1, PR4, and PRF) before they were averaged.

Standards, Normalization and Calculation of Uncertainty

All activity measurements were related to the international standard, 95% of the activity in 1950 of the oxalic acid from the NBS, with a $\delta^{13}C$ value of -19 °/oo on the PDB scale. Since our gas samples made from this oxalic acid all had $\delta^{13}C$ values very close to -19 °/oo (Olsson et al. 1972), no correction was needed for the present investigation.

All samples were measured at least twice for $\delta^{13}C$ values at the Karolinska Institute in Stockholm by Ryhage and his collaborators. In this case uncertainty was estimated at ca. 0.5 °/oo. The normalization to normal wood with $\delta^{13}C = -25$ °/oo was made according to principles accepted by ^{14}C laboratories (Olsson 1965, Olsson and Osadebe 1974).

Gas pressure was measured carefully and normalization for deviations from normal filling pressure, working voltage, etc., was made according to a modified computer program originated by Olsson (1965).

It should be mentioned that the given uncertainty includes not only statistical uncertainty from activity measurement, but also the uncertainties connected with all physical measurements performed in the laboratory, such as pressure measurements, $\delta^{13}C$ normalization, barometric-pressure correction of the background, and the standard value. The calculations for uncertainty have been detailed by Olsson (1965).

The ^{14}C ages were calculated with the half-life 5570 years. No error for the half-life was included in the uncertainty of the ^{14}C age. To facilitate comparisons with results calculated with the half-life 5730 years, final averages have been given in both half-lives (5570 and 5730 years).

To facilitate other comparisons, results have been given both in tabular form (table 1) and in diagrammatic form (fig. 1). The final mean values have been plotted in a diagram to allow comparison with tree ring measurements (fig. 2). In this preliminary report we have chosen to weight the results according to statistical uncertainty only, disregarding the fact that the samples have been classified as primary, secondary, and tertiary samples, (unless an unusual result had been obtained, as for tertiary reeds from Tomb 157).

Reliability of the Measurements

Possible error has several sources in determinations like the present ones. The pretreatment of each sample must be adequate. In one case, Bekenkhons I, Tomb 35, a test was made on the fraction soluble in the NaOH. Then the extraction *liquid* (1% NaOH) with extractives, was mixed with the first three portions of water from the washing. From this liquid, the humic acid and other compounds were precipitated by adding some conc. CHl. This fraction was dated at 1420 ± 80 B.P. (U-870). This value is significantly lower than that given by the fraction insoluble in NaOH. This demonstrates the importance of removing the fraction soluble in NaOH during pretreatment. In this investigation, the ratio of organic material in the fraction soluble in NaOH to that in the fraction insoluble in NaOH was about 1:2 to 1:4. Based on this ratio, a sample would appear about 315 years too young if 30% of the sample were a contaminant 1000 years younger than the sample.

TABLE 1

The results of the present investigation on samples from Dra Abu El Naga with the Pennsylvania dating numbers as sample numbers.

Sample no.	Description	Material	Species according to V. Täckholm	Stockholm Dating no. $\delta^{13}C$ ‰ PDB		Conventional radiocarbon age ($T_{1/2}$ = 5570 yr) (95% of oxalic acid AD 1950) $\delta^{13}C$ norm. to -25 ‰ PDB)			$T_{1/2}$ = 5730 $\delta^{13}C$ norm.
							MV	MV	
P-1730	Nebwenenef Tomb 157	Primary reeds	*Desmostachya bipinnata* (L.) Stapf + *Phoenix dactylifera*	-21.2	U-880 U-2460 U-5000	2815±140 2970±80 2865±110	2915±60	3015±40	3100±40
P-1731	Nebwenenef Tomb 157	Secondary reeds	*Desmostachya bipinnata* (L.) Stapf	-10.6	U-881 U-2461 U-5001	3060±125 3050±60 3240±110	3085±50		
P-1732	Nebwenenef Tomb 157	Tertiary reeds	*Desmostachya bipinnata* (L.) Stapf	-10.9	U-862 U-2462 U-5002	1390±135 1360±60 1480±110	1385±50		
P-1739	Bekenkhons I Tomb 35	Primary reeds	*Desmostachya bipinnata* (L.) Stapf	-10.7	U-869 U-2469 U-5004	2965±135 2975±70 2975±70	2975±50		3060±50
		SOL. fraction		-15.9	U-870	1420±80			

Lab #	Site	Material	Species	δ13C	Sample #	Dates	Combined	Combined	Combined
P-1735	Roma-Roy Tomb 283	Primary reeds	Desmostachya bipinnata (L.) Stapf	−12.4	U-2495 / U-5005	2915±80 / 2940±80	2925±55		
P-1736	Roma-Roy Tomb 283	Secondary reeds	Desmostachya bipinnata (L.) Stapf	−11.5	U-5016	(2595±105)			
P-1737	Roma-Roy Tomb 283	Tamarix		−30.8	U-2497 / U-5017	2925±80 / 3005±110	2955±65	2960±35	3045±35
P-1738	Roma-Roy Tomb 283	Tamarix		−25.8	U-5018	2640±135			
P-1738	Roma-Roy Tomb 283	Tamarix		−26.3	U-2498 / U-5019	3170±70 / 2735±140	3080±65		
P-1696	Tjanefer Tomb 158	Primary reeds	Desmostachya bipinnata (L.) Stapf	−11.1	U-866 / U-5026	2805±90 / (rejected)			
P-1698	Tjanefer Tomb 158	Secondary reeds	Desmostachya bipinnata (L.) Stapf	−11.3	U-868 / U-5028	2800±80 / 2790±130	2795±70	2815±35	2895±35

TABLE 1 *(continued)*

Sample no.	Description	Material	Species according to V. Täckholm	Stockholm Dating no. $°/oo$ $\delta^{13}C$ PDB		Conventional radiocarbon age ($T_{1/2} = 5570$ yr) (95% of oxalic acid AD 1950) $\delta^{13}C$ norm. to -25 $°/oo$ PDB		$T_{1/2} = 5730$ $\delta^{13}C$ norm.
						MV	MV	
P-1699	Tjanefer Tomb 158	*Acacia*		−25.2	U-2499 U-5029	2910±70 2755±55	2820±45	
P-1733	Inhernakht Tomb 282	Primary reeds	*Desmostachya bipinnata* (L.) Stapf	−11.0	U-863 U-5013	2830±95 2690±120	2780±75	2710±60
P-1734	Inhernakht Tomb 282	Secondary reeds	*Desmostachya bipinnata* (L.) Stapf	−11.9	U-864 U-5014	2855±115 2370±120	2625±85	2790±60

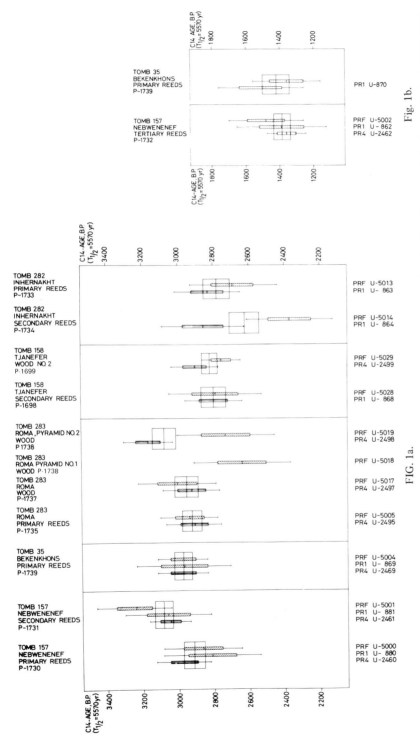

FIG. 1. The age determinations of the samples from Dra Abu El-Naga, expressed in conventional radiocarbon years (calculated with the half-life 5570 years, using 95% of the activity in 1950 of the oxalic-acid sample from the NBS as standard). The sample activities are normalized to −25 ‰ and the oxalic-acid standard to −19 ‰ in PDB. The lengths of the hatched rectangles indicate the range determined by ±1σ and the bars indicate the range determined by ±2σ. When two or more independent measurements are performed (for instance, in different counters), the mean value is calculated and the height of an open rectangle indicates ±σ of the mean value. In all cases, the fraction which remains insoluble in a 1% solution of NaOH after 1 night at +80°C is used. (b) One tertiary-reed sample, insoluble fraction, and the soluble fraction of another sample. The first result indicates a contaminated sample and the second result that the soluble fraction is much younger than the sample itself.

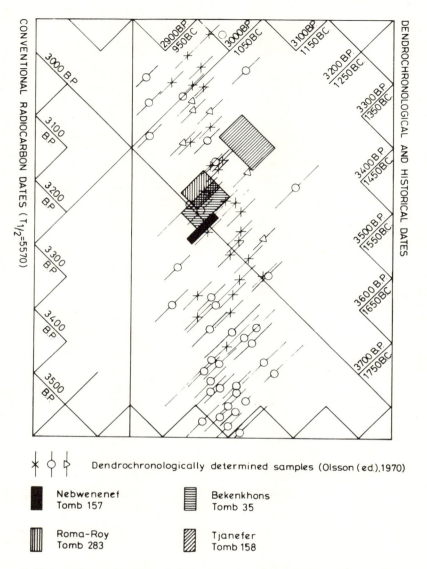

FIG. 2. The mean values of the present determinations in a diagram with ^{14}C determinations on dendrochronologically determined samples (from Plates I and IV in *Proceedings of the Twelfth Nobel Symposium*, Olsson (1970)).

Any malfunction of the counters or electronics may cause error. In the present investigation, one result (U-5026) was rejected from PRF, since the statistics clearly indicated an error.

For Inhernakht, Tomb 282, secondary reeds (U-864 and U-5014), results differed widely in the two counters, but, since the difference was <3σ, the two results were given. For Roma-Roy, Tomb 283, wood treatment 2, the difference was 3 σ. The lower result

(U-5019) 2735 ± 140 B.P. agreed with another date (U-5018) 2640 ± 135 B.P. Seven samples were dated in PR1 and PRF. Of these, three were dated at a lower age in PRF than in PR1; in one case the difference was almost 3 σ. Eight samples were dated in PR4 and PRF. Of these, three were dated at a lower age in PRF than in PR4. Four samples were dated in PR1 and PR4 (and PRF) with good agreement between the two counters PR1 and PR4. From these comparisons, it is difficult to exclude any result. One sample (U-5016) has not yielded entirely satisfactory statistics and the result should not be included in the final discussion. This sample (U-5016) has been remeasured, but the result is not yet available. As a check, the reeds from Inhernakht will also be remeasured, since the two measurements (U-5014) and (U-864) differ. A wood sample (U-5018) has been remeasured, but the result is not yet available. Since two samples (U-5016) and U-5026) were rejected because of bad statistics and because of measurement at 1.3 atmospheres using old electronics, it would seem better not to include other results from the same counter measured at the critical time when the results seemed less reliable. The Inhernakht sample has not been plotted in figures 2 and 3.

Since the fraction soluble in NaOH was dated at an appreciably lower age than the insoluble fraction, tests to check the pretreatment will be performed.

If present $\delta^{13}C$ values are compared with those from another laboratory dating the same samples, it must be remembered that the primary reeds from Tomb 157 consist of a mixture of two species having different photosynthetic pathways. If the propulsion between the two species should differ, the $\delta^{13}C$ values might also differ. It is natural that the fraction soluble in NaOH should show a more negative $\delta^{13}C$ value than the insoluble fraction, even though there are no contaminations, since the extract is usually deficient in ^{13}C. It should be remembered that different parts of a plant usually have slightly different $\delta^{13}C$ values (Olsson and Osadebe 1974).

Results

The results, as measured in the three counters, are indicated in fig. by rectangles indicating the range ±1σ and bars indicating ±2σ. The mean value for each sample is calculated. It will be seen that one sample described as tertiary reeds was probably contaminated. Since the soluble fraction of one sample yielded a much lower age than the insoluble fraction, special attention should be given to the pretreatment.

A comparison of the present results with dendrochronologically determined samples (fig. 2) indicates that there is a good agreement between the historically and dendrochronologically determined samples when used for studies of the ^{14}C variations. Some samples are reed samples and others are wood samples, but no sample is so well determined that short-term variations can be studied. A reed sample may, however, derive from one year and thus show deviations from an integrated curve obtained from wood samples if short-term variations exist, as suggested by some authors (Baxter and Farmer, 1973, Mook et al. 1973).

A comparison of the present results with some earlier Egyptian samples dated in four different laboratories indicates a good agreement between the present results and the earlier dates (Alessio and Bella 1965, Barker et al. 1971, Berger 1970, and Ralph 1959).

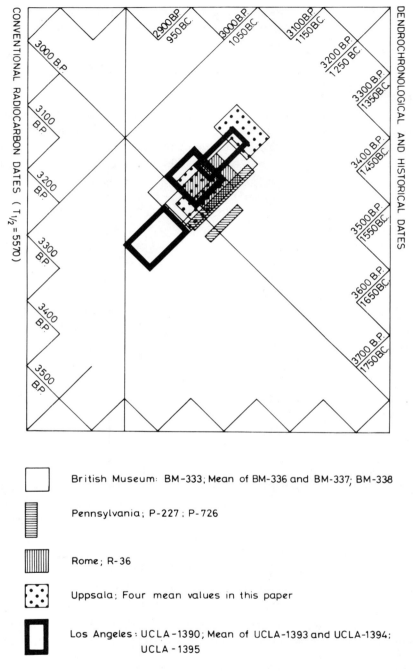

FIG. 3. The mean values of the present determinations in a diagram also containing other ^{14}C determinations on historically determined samples (measured in the laboratories at Los Angeles, Pennsylvania and Rome and at the British Museum).

The conclusion is that the Egyptian samples and the dendrochronological samples agree very well and that the ^{14}C activity of the atmosphere was significantly higher about 3100 to 3250 years ago than that of the present standard.

Note added in proof: It was later discovered that the sample Nebwenenef, Tomb 157, tertiary reeds was dated as U-869, U-2469 and U-5004 with the mean value 2975±50. The sample Bekenkhons I, Tomb 35, primary reeds was dated as U-862, U-2462 and U-5002 with the mean value 1385±50. Further explanations are given by Olsson and El-Daoushy (1978). This affects table 1 and the figures. Indeed, the two fractions of Bekenkhons I, Tomb 35, primary reeds agree well. Other dates in this series are also given by Olsson and El-Daoushy (1978).

ACKNOWLEDGMENTS

The authors gratefully acknowledge the financial support received from the Swedish Natural Science Research Council. One of us (M. F. A. F. El-Daoushy) was awarded a scholarship by the Swedish Institute. Special thanks are due to Professor V. Täckholm for identifying species, to Professor T. Säve-Söderbergh for valuable discussions, and to Dr. R. Ryhage and his co-workers for the ^{13}C/^{12}C determinations. Sincere thanks are due to Drs. E. Ralph and H. N. Michael, who provided the samples, to Professor K. Siegbahn, who made it possible for us to work at the Institute, and to Carina Sellin, Eva Haag-Lindström, and Maud Söderman, who helped us with the technical work in the laboratory.

REFERENCES

Alessio, M., and F. Bella
    1965   University of Rome carbon-14 dates III. Radiocarbon. 7:213-222.

Barker, H., R. Burleigh, and N. Meeks
    1971   British Museum natural radiocarbon measurements VII. Radiocarbon. 13:157-188.

Baxter, M. S., and J. G. Farmer
    1973   Radiocarbon: Short-term variations. Earth and Planetary Science Letters. 30:295-299.

Berger, Rainer
    1970   Ancient Egyptian radiocarbon chronology. Phil. Trans. Roy. Soc. Lond. A. 269:23-36.

El-Daoushy, M. F. A. F., and I. U. Olsson
    1976   An improved proportional counter for low-level counting with high efficiency. Proc. Int. Conf. on Low Radioactivity Measurements and Applications. Czechoslovakia. (In press.)

Mook, W. G., A. V. Munaut, and H. T. Waterbolk
    1973   Determination of age and duration of stratified prehistoric bog settlements. Proc. 8th Int. Conf. on Radiocarbon Dating. New Zealand. F27-F40.

Olsson, Ingrid
- 1958 A ^{14}C dating station using the CO_2 proportional counting method. Arkiv för Fysik. 13:37-60.
- 1965 Computer calculations of ^{14}C determinations. Proc. 6th Int. Conf. Radiocarbon and Tritium Dating. Pullman, Washington. 383-392.
- 1970 Radiocarbon Variations and Absolute Chronology. 12th Nobel Symp. Uppsala. Stockholm, Almqvist and Wiksell. New York, John Wiley and Sons.

Olsson, I. U., Martin Klasson, and Abdalla Abd-El-Mageed
- 1972 Uppsala natural radiocarbon measurements XI. Radiocarbon. 14:247-271.

Olsson, I. U., and F. A. N. Osadebe
- 1974 Carbon isotope variations and fractionation corrections in ^{14}C dating. Boreas. 3:139-146.

Ralph, E. K.
- 1959 University of Pennsylvania radiocarbon dates III. Radiocarbon. 1:45-58.

12

The Radiocarbon Contents of Various Reservoirs

Ingrid U. Olsson

ABSTRACT

Some potential systematic errors in the ^{14}C dating of samples near volcanoes, in limestone-rich areas, and on small islands in the oceans are discussed.

INTRODUCTION

One basic assumption in ^{14}C dating is that the activity of the sample is known when it is still in equilibrium with its surroundings. It is recognized that secular variations of the activity make the dates not very accurate on an absolute time scale, since the ages must be translated from radiocarbon years to sidereal years using a calibration curve. There is no unique way of drawing such a curve. A critical review of this problem is given by Clark (1975).

There may be complications due to local effects. Some determinations (Lerman 1974) suggest that there may be a somewhat lower activity in the atmosphere of the southern hemisphere than in that of the northern. Lower activity can be expected in the neighborhood of fumaroles because of release of old carbon dioxide.

Old, dissolved carbonate in groundwater and freshwater systems and in aged water causes a broad spectrum of "apparent ages" for the waters and for the submerged and floating plants taking their carbon from the dissolved CO_2 or bicarbonate (fig. 1).

In the oceans, apparent age is mainly determined by the exchange rates and the relationship between the reservoirs. The deep water is about 10^2 times larger than the surface water and both are regarded as being well mixed. At equilibrium, the proportion of carbon that circulates down into the deep water from the surface water is about 10^2 greater than that which rises up from the deep water to the surface water. Thus the radiocarbon in the deep water has more time to decay and the activity is less than that in the surface water. Due to this supply of low-activity carbon from below, the surface water has less activity than the atmosphere. The corresponding age is called the apparent age (fig. 1).

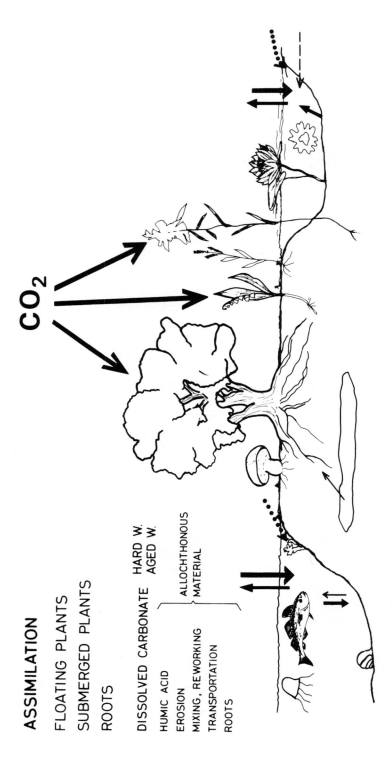

FIG. 1. A simplified drawing depicting some different types of materials used for ^{14}C dating. Arrows indicate exchange or transport of carbon either as carbon dioxide, bicarbonate, or as carbonaceous matter. An apparent age results when dissolved CO_2 or bicarbonate is used with a lower $^{14}C/^{12}C$ ratio than that of the atmosphere (e.g. shells, fishes, whales, submerged plants, and some floating plants). Sediments may be contaminated by erosion, reworking, etc. Humic acid transported to a lake will generally have a lower $^{14}C/^{12}C$ ratio than the atmosphere, but humic acid deposited in a sample may have a higher or lower $^{14}C/^{12}C$ ratio than the sample itself.

Roots penetrating down in a sediment constitute a material with a lower age than the sediment around the root. Some lakes are hard-water lakes and these usually have a high apparent age, but a soft-water lake can also have a low $^{14}C/^{12}C$ ratio of the dissolved CO_2 so that the water has an apparent age.

Datings of sediments such as gyttja often show a lower activity than is expected from the apparent age of the water, even if there is no graphite present. This is probably due to erosion and similar phenomena (fig. 1).

Volcanoes

It has earlier been shown (Chatters et al. 1969) that gases from fumaroles may affect neighboring plants, causing them to be dated at too low an activity. Recently, Šilar (1976) reported waters of great age in artesian aquifers. When the volcanic eruption started on Heimaey, Vestmannaeyjar, in Iceland, on 23 January, 1973, the author asked for some samples. Two air samples were submitted. One was collected between 2 and 4 February 1973 at Kirkjuvegur 59 in an open room. The tray containing NaOH was sheltered from precipitation and tephra. The second gas sample was collected between 20 and 23 February 1973 at Hradfrystistöd in a basement where gases from the volcano were concentrated. These two samples yielded excess activities of 410 ± 8 °/oo and -969.7 ± 3.5 °/oo, respectively. The latter value is equivalent to about 3% of the standard activity, indicating great contamination of the sample by inactive carbon dioxide from the volcano. Further details will be given later.

Two "grass" samples were submitted at the same time and two were submitted later. Only one of the latter two has been treated. One of the former grass samples was collected on 19 August 1973 from Breidabakka, yielding an excess activity of 533 ± 10 °/oo. The second sample was collected on 19 August 1973 from Hliskletti, yielding an excess activity of 425 ± 8 °/oo. The latter sample was collected at the end of 1974 on the tephra, yielding an excess activity of 405 ± 9 °/oo.

From earlier investigations and from current results, it is clear that dates determined on samples collected near volcanoes or hot springs must be treated with caution, since carbon dioxide containing less ^{14}C than the atmosphere may be assimilated by the plants. Further investigations are planned.

Assimilation by Roots

Since plants assimilate large amounts of water containing salts essential for the plant, the possibility must be considered of some carbon from this water being incorporated in the plants. For example, at Montezuma Well, Arizona, leaves from trees at the pond's edge, whose roots were perhaps under water, had lower activity than the grass, while the water had an extremely low ^{14}C content (Damon et al. 1964, Haynes et al. 1966). In 1963, the excess ^{14}C content of the leaves of *Populus frenonti* was 483 ± 5 °/oo; for the grass growing 1 to 2 ft above the water level it was 770 ± 25 °/oo; for the grass growing 75 ft above the water level it was 819 ± 28 °/oo; for the water it was -935.4 ± 4.2 °/oo.

In 1968 an experiment was started in Uppsala, growing tomato plants in buckets. To one bucket, ^{14}C was added in the form of carbonate from the half-life measurements. To another, ^{12}C was added in the form of limestone from Öland. In addition there were buckets to which no limestone was added. The first results were given by Olsson et al. (1972) and by Olsson (1972). No increase was apparent from the ^{14}C added, but the

NaOH-insoluble fraction of the roots of the tomato plant grown in the ^{12}C bucket seemed to have slightly lower activity than the others. The soluble fraction was analyzed, yielding significantly lower activity (fig. 2). Another tomato plant, for which neither ^{12}C nor ^{14}C was added, also yielded rather low activity for the insoluble fraction of the roots but not so low as the soluble fraction mentioned. In the next year, 1969, the insoluble fraction of the roots yielded a significantly higher activity for the plant grown in the bucket with ^{14}C than for that grown in the bucket with ^{12}C. Too few results are available for the following years, but further study seems warranted of this potential source of error.

A Possible Island Effect

The explanation given for the apparent age implies an island effect. Since some CO_2 with slightly lower ^{14}C activity than that of the normal atmosphere over the continents is given off from the surface layer of the ocean to the atmosphere, we should suspect that the ^{14}C activity of the air over the oceans would be slightly lower than that over the

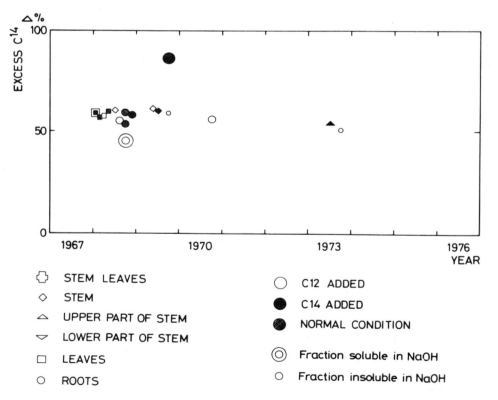

FIG. 2. The excess ^{14}C over the international standard for tomato plants grown in buckets in Uppsala. To one bucket, ^{14}C was added, to others, ^{12}C in the form of carbonates and to others, no extra carbonate. Different parts of the plants were used.

continents. This would probably be a very small effect, similar to that seen as a difference between the activities of the northern and the southern hemispheres. Some preliminary studies have been made, but no definite conclusions can be drawn since the values for the atmospheric samples have not yet been finally calculated. It would seem that the plant samples from the Faroes and the one from Greenland may have slightly lower ^{14}C activities than those from Abisko in the mountain region of Sweden. It was earlier noted that most samples from Svalbard had slightly lower activities than those from Abisko, but this could be explained as a latitudinal effect. If an island effect exists, the ^{14}C activity in the coastal areas of islands such as Svalbard, Iceland, and Greenland should be low. On the other hand, the proximity of large glaciers containing carbon dioxide, trapped and aged for a long time, may also affect the ^{14}C activity. The possibility of a lowering of the ^{14}C activity in the Norwegian fjords has been discussed by Mangerud (1972), who concludes that the glaciers in Norway and Greenland do not much influence the ^{14}C activity of the surface water.

The Influence of Food

Considering that a sample reflects the ^{14}C activity of the source of the carbon, not only will submerged plants show an apparent age in response to the surrounding water, but animals normally regarded as nonaquatic may show the apparent age of the ocean water. This has been demonstrated by two dates (Lu-715 and Lu-779) for polar bears given by Håkansson (1974). The bears were from Svalbard and East Greenland and had apparent ages of 480 ± 70 and 495 ± 45, respectively. A person who eats much fish will naturally have lower ^{14}C activity than a person who eats mostly terrestrial food.

ACKNOWLEDGMENTS

This investigation was financially supported by the Swedish Natural Science Research Foundation. Sincere thanks are due to Professor Kai Siegbahn for making available the facilities of the Institute of Physics in Uppsala. Special thanks are due to all my collaborators in the laboratory.

REFERENCES

Chatters, R. M., J. W. Crosby, III, and L. G. Engstrand
    1969   Fumarole gaseous emanations: Their influence on carbon-14 dates. Coll. of Engineering Circular 32. Washington State University.
Clark, R. M.
    1975   A calibration curve for radiocarbon dates. Antiquity. 59:251-266.
Damon, P. E., C. V. Haynes, and Austin Long
    1964   Arizona radiocarbon dates V. Radiocarbon. 6:91-107.
Håkansson, Sören
    1974   University of Lund radiocarbon dates VII. Radiocarbon. 16:307-330.
Haynes, C. V., jr., P. E. Damon, and D. C. Grey
    1966   Arizona radiocarbon dates VI. Radiocarbon. 8:1-21.

Lerman, J. C.
    1974   Les isotopes du carbone: Variations de leur abondance naturelle: application aux corrections des datations radiocarbone, à l'étude du métabolisme végétal et aux paléoclimats. Thesis. Université de Paris-Sud.

Mangerud, Jan
    1972   Radiocarbon dating of marine shells, including a discussion of apparent age of recent shells from Norway. Boreas. 1:143-172.

Olsson, I. U.
    1972   The pretreatment of samples and the interpretation of the results of ^{14}C determinations. *In* Climatic changes in Arctic areas during the last ten-thousand years. Symposium. Oulanka and Kevo, Finland. Acta Univ. Oul. A3, Geol. 1:9-37.

Olsson, I. U., Martin Klasson, and Abdalla Abd-El-Mageed
    1972   Uppsala natural radiocarbon measurements XI. Radiocarbon. 14: 247-271.

Šilar, Jan
    1976   Radiocarbon dating in groundwater research of the Bohemian Massif platform sediments. Proc. Int. Conf. on Low Radioactivity Measurement and Application. Czechoslovakia. (In press.)

13

Evaluation of Total Variability in Radiocarbon Counting

Stephen Meeks

Three possible major components account for the total variability in the counting statistics of a homogeneous source material in radiocarbon dating. The one component which is always present is the normal counting error due to the Poisson nature of the counts themselves. Further variability may be introduced from the laboratory preparation of the sample (Haas 1976). Finally the mean of the observable activity of the sample may fluctuate. This may be due to a number of causes, such as unstable counter characteristics perhaps due to voltage fluctuations, handling of the sample which may induce static electricity, exposure of the scintillator to light, or background fluctuations. These three components will be referred to as "the component due to counting", "the component due to the laboratory", and "the component due to normal error."

It seems reasonable to assume that the three components are independent or nearly so. Since the component due to counting can be evaluated only while counting a single laboratory sample preparation, the effect is said to be nested within the effect due to laboratory preparation. Two suitable statistical models for the situation described are given below. Following the two models is a further analysis of the normal counting error with comparisons of different estimators for the standard deviation of age estimates.

The basic model for which two cases will be developed can be described as follows

$$r_{ijk} = R + L_i + C_{i(j)} + E_{ijk} \quad \begin{array}{l} i = 1, \ldots, r \\ j = 1, \ldots, s_i \\ k = 1, \ldots, t_{ij} \end{array}$$

where

r_{ijk} = k^{th} count reading during the jth counting of the ith laboratory sample

R = mean observable count rate

L_i = random laboratory effect of ith sample preparation

$C_{i(j)}$ = random counting effect of jth counting of the ith laboratory sample

E_{ijk} = random normal error of kth reading during j^{th} counting of the ith laboratory sample

r = number of laboratory samples prepared

S_i = number of countings of ith laboratory sample

t_{ij} = number of readings during j^{th} counting of ith laboratory sample

n = total number of readings

$$= \sum_{i=1}^{r} \sum_{j=1}^{S_i} t_{ij} \left(\text{if } t_{ij} = 1 \text{ for all i,j then } n = \sum_{i}^{r} S_i \right)$$

It will be assumed that the random effects of the counting, laboratory, and normal error all follow normal (Gaussian) distributions with zero mean. The variances of these three will be denoted σ_ℓ^2, σ_c^2 and σ^2, respectively.

The analysis of this model will vary for $t_{ij} = 1$ for all i,j versus $t_{ij} > 1$ for some i,j. In the physical situation this translates into whether or not multiple readings during a single counting of one laboratory preparation can be considered replicate readings while the laboratory and counting effects are constant. As an example, if the characteristics of the counter can be assumed to be constant for ion can be considered replicate readings while the laboratory and counting effects are constant. As an example, if the characteristics of the counter can be assumed to be constant for periods of time during which several readings are obtained, then these readings might be considered replicates; otherwise they must be considered as unrelated readings. If replicate readings are not available, as is always the case for counters which give only a single total count, then $t_{ij} = 1$ for all i,j. The model with replicates supplies a test for whether the readings are indeed replicates.

Since the model without replicate readings assumes less about the physical situation than the model with replicate readings, it will be developed first. In either case the main question to be answered is whether there are significant laboratory or counting effects. In terms of the model, we wish to test if $\sigma_\ell^2 = \sigma_c^2 = 0$, or, in other words, if the laboratory or counting processes introduce extraneous variability. Under the hypothesis of no laboratory or counting effects, the model reduces to $r_i = R + E_i$. This simple model is characterized by the Poisson nature of the count readings so that $R = \sigma^2$, that is, the variance of the normal error is equal to the mean of the activity or, equivalently, the ratio $R/\sigma^2 = 1$. Estimates of R and σ^2 can be used to provide a test for equality of the two parameters. If sources of variability are present other than the normal error, the estimate of σ^2 will tend to be larger than the estimate of R, and the ratio R/σ^2 will tend to be smaller than 1. The purpose of the following analysis is to derive approximate criteria for answering the question of whether there are other sources of variability present, i.e., $\sigma_\ell^2 = \sigma_c^2 = 0$?

EVALUATION OF TOTAL VARIABILITY IN RADIOCARBON COUNTING

Case 1: $t_{ij} = 1$

$$\text{Hypothesis} \qquad r_i = R + E_i \qquad i = 1, \ldots, n$$

$$\text{vs alternative} \qquad r_{ij} = R + L_i + C_{i(j)} + E_{ij}$$

$$\bar{r} \equiv \frac{1}{n} \sum_{i=1}^{n} r_i$$

$$S^2 \equiv \frac{\sum_{i=1}^{n} (r_i - \bar{r})^2}{n-1}$$

$$X^2 \equiv (n-1) S^2 / \sigma^2$$

Under the hypothesis, $\bar{r}$ is distributed normally with mean R and variance σ^2/n. Also, X^2 is distributed as a central chi-squared with $n-1$ degrees of freedom. With these results

$$F \equiv \frac{n\bar{r}^2}{\sigma^2} \bigg/ \frac{X^2}{n-1} = n\bar{r}^2/S^2$$

has a noncentral F distribution with noncentrality parameter nR and with 1 and $n-1$ degrees of freedom. R is unknown so that the exact probability of a value of F given the value of R is not obtainable. In general n will be quite large so that $\bar{r}$ will be a very good estimate of R. To obtain approximate results R is replaced by $\bar{r}$ in further analysis.

The following result due to Severo and Zelen (1960) facilitates finding an approximate probability for F. Define X as below

$$X \equiv \frac{\left(\dfrac{F}{1+nR}\right)^{1/3} \left(1 - \dfrac{2}{9(n-1)}\right) - \left[1 - \dfrac{2(1+2nR)}{9(1+nR)^2}\right]}{\left[[2(1+2nR)/9(1+nR)^2] + \dfrac{2}{9(n-1)} \left(\dfrac{F}{1+nR}\right)^{2/3}\right]^{1/2}}$$

X is approximately normally distributed, so that $P(F < f) \doteq P(X < K_p)$ where K_p is a standard normal percentile.

If the hypothesis is rejected, estimates of σ_ℓ^2 and σ_c^2 are desirable. Using $\hat{}$'s to denote estimates,

$$\hat{\sigma}^2 = \bar{r}$$

$$\widehat{\sigma_c^2 + \sigma^2} = \sum_{i=1}^{r} \frac{\sum_{j=1}^{S_i}(r_{ij} - \bar{r}_i)^2}{n - r} \quad \text{where } \bar{r}_i = \frac{1}{S_i} \sum_{j=1}^{S_i} r_{ij}$$

$$\widehat{\sigma_c^2} = \widehat{\sigma_c^2 + \sigma^2} - \widehat{\sigma^2}$$

$$\widehat{\sigma_\varrho^2 + \sigma_c^2 + \sigma^2} = S^2$$

$$\widehat{\sigma_\varrho^2} = \widehat{\sigma_\varrho^2 + \sigma_c^2 + \sigma^2} - \widehat{\sigma_c^2 + \sigma^2}$$

While more complex analysis is possible, in a practical setting the estimates above should provide the investigator with sufficient information to search for specific areas of laboratory and counting techniques which are causing extra variability (Haas 1976).

Case II: $t_{ij} > 1$

The analysis proceeds as in case I except for estimation of the variance components and testing if the model is appropriate. The alternate model is now $r_{ijk} = R + L_i + C_{i(j)} + E_{ijk}$.

If replicate readings are available, there are two available independent estimates of σ^2

$$\hat{\sigma}_1^2 = \bar{r}$$

$$\hat{\sigma}_2^2 = \frac{\sum_{i=1}^{r}\sum_{j=1}^{S_i}\sum_{k=1}^{t_{ij}}(r_{ijk} - \bar{r}_{ij})^2}{n - s}, \quad \bar{r}_{ij} = \frac{1}{t_{ij}}\sum_{k=1}^{t_{ij}} r_{ijk}$$

$$S = \sum_{i=1}^{r} S_i$$

$\dfrac{(n-s)\hat{\sigma}_2^2}{\sigma^2/n}$ follows a central chi-squared with $n - s$ degrees of freedom.

$\bar{r}$ follows a normal distribution with mean R and variance σ^2.

$F \equiv n\bar{r}^2/\hat{\sigma}_2^2$ follows a non-central F distribution with non-centrality parameter nR with 1 and $n - s$ degrees of freedom.

$$X \equiv \frac{\left(\dfrac{F}{1+nR}\right)^{1/3}\left(1-\dfrac{2}{9(n-s)}\right) - \left[1-\dfrac{2(1+2nR)}{9(1+nR)^2}\right]}{\left[\dfrac{2(1+2nR)}{9(1+nR)^2} + \dfrac{2}{9(n-s)}\left(\dfrac{F}{1+nR}\right)^{2/3}\right]^{1/2}}$$

follows approximately a normal distribution.

Thus, $P(F < f) \doteq P(X < K_p)$ where K_p is a standard normal percentile A low probability suggests that readings within a single counting are not replicates and case 1 is appropriate.

Attempting to pool $\hat{\sigma}_1^2$ and $\hat{\sigma}_2^2$ into one estimator is not recommended. Rather $\hat{\sigma}^2 = \bar{r}$ is preferred since this is an unbiased estimator whether the correct model is case 1 or case 2.

$$\widehat{\sigma_\varrho^2 + \sigma_c^2 + \sigma^2} = S^2 \quad \text{as defined in case I}$$

$$\widehat{\sigma_c^2 + \sigma^2} = \frac{\displaystyle\sum_{}^{r}\sum_{}^{S_i}\sum_{}^{t_{ij}}(r_{ijk} - \bar{r}_i)^2}{n - r}, \quad \bar{r}_i = \frac{1}{t_i}\sum_{j=1}^{S_i}\sum_{k=1}^{t_{ij}} r_{ijk}$$

$$t_i = \sum_{j=1}^{S_i} t_{ij}$$

$$\hat{\sigma}_\varrho^2 = \widehat{\sigma_\varrho^2 + \sigma_c^2 + \sigma^2} - \widehat{\sigma_c^2 + \sigma^2}$$

$$\hat{\sigma}_c^2 = \widehat{\sigma_c^2 + \sigma^2} - \hat{\sigma}^2$$

Once the nature of the variability of the counting statistics is investigated, it is necessary to translate that variability into the variability of the age estimate.

Two common approaches to the derivation of confidence intervals for the age estimator involve transforming confidence intervals on the "modern to sample" ratio by the formula for the age estimator itself. Denoting the estimator of the modern to sample ratio by r, the age estimator by $\hat{A}$, and their variances by σ_r^2 and σ_A^2, the two approaches to confidence limits are:

1. $\hat{A} = \dfrac{1}{\lambda} \ln(\hat{r})$

$$\hat{\sigma}_A = \frac{1}{2}\left[\frac{1}{\lambda}\ln(\hat{r} + \hat{\sigma}_r) + \frac{1}{\lambda}\ln(\hat{r} - \hat{\sigma}_r)\right]$$

with the confidence limits given by

$$\hat{A} \pm K \cdot \hat{\sigma}_A$$

where K is the appropriate confidence level coefficient. This method will be referred to as the old symmetric method,

2. the confidence limits are given by

$$\frac{1}{\lambda} \ln (\hat{r} + k \cdot \hat{\sigma}_r)$$

and

$$\frac{1}{\lambda} \ln (\hat{r} - k \cdot \hat{\sigma}_r)$$

where k is again the appropriate confidence level coefficient. This method will be referred to as the asymmetric method.

Neither of these methods uses properties of the distribution of A and both can lead to a nonexistent confidence limit if the argument of the logarithm is a nonpositive number (i.e., $k \cdot \hat{\sigma}_r \geq \hat{r}$). To avoid this problem it is common to perform a test of hypothesis, either consciously or unconsciously, on whether the sample counted was dead. This is essentially the philosophy of giving only a minimum age if r is not greater than $2 \cdot \hat{\sigma}_r$.

As developed in Haas et al. (1976), the asymptotic variance-covariance structure for simultaneous estimation of the age (A), the counter efficiency (ρ), and the background mean (α) with mean sample activity (β) is

$$\begin{bmatrix} \text{Var}(\hat{A}) & \text{Cov}(\hat{A}, \hat{\rho}) & \text{Cov}(\hat{A}, \hat{\alpha}) \\ \text{Cov}(\hat{A}, \hat{\rho}) & \text{Var}(\hat{\rho}) & \text{Cov}(\hat{\rho}, \hat{\alpha}) \\ \text{Cov}(\hat{A}, \hat{\alpha}) & \text{Cov}(\hat{\rho}, \hat{\alpha}) & \text{Var}(\hat{\alpha}) \end{bmatrix}$$

$$V(A) = \frac{1}{n_1} \left[\frac{\rho\gamma + \alpha}{\lambda\rho\beta} \right]^2 + \frac{1}{n_2} \left[\frac{\rho\beta + \alpha}{\lambda\rho\beta} \right]^2 + \frac{1}{n_3} \left[\frac{\alpha(\gamma - \beta)}{\lambda\rho\gamma\beta} \right]^2$$

$$\text{Var}(\hat{\rho}) = \frac{1}{n_1} \left[\frac{\rho\gamma + \alpha}{\gamma} \right]^2 + \frac{1}{n_3} \left[\frac{\alpha}{\gamma} \right]^2$$

$$\text{Var}(\hat{\alpha}) = \frac{\alpha^2}{n_3}$$

$$\text{Cov}(\hat{A}, \hat{\rho}) = \frac{1}{n_1}\left[\frac{(\rho\gamma + \alpha)^2}{\lambda\rho\gamma^2}\right] - \frac{1}{n_3}\left[\frac{\alpha^2(\gamma - \beta)}{\lambda\rho\beta\gamma^2}\right]$$

$$\text{Cov}(\hat{A}, \hat{\alpha}) = \frac{1}{n_3}\left[\frac{\alpha^2(\gamma - \beta)}{\rho\lambda\gamma\beta}\right]$$

$$\text{Cov}(\hat{\rho}, \hat{\alpha}) = \frac{-1}{n_3}\left[\frac{\alpha^2}{\gamma}\right]$$

where λ = coefficient of 1n in the age estimator

n_1 = number of observed events while counting modern standard

n_2 = number of observed events while counting sample

n_3 = number of observed events while counting background

A new estimate of σ_A is the square root of the first element of the variance-covariance matrix with the parameters replaced by their estimates. This will yield a new method of determining symmetric confidence limits of the form $A \pm K \cdot \hat{\sigma}_A$. This method will be referred to as the new symmetric method. Note that this method gives finite confidence intervals for any positive estimate of the modern sample ratio. If there are sources of variability other than the normal Poisson error, Otlet (1976) generalizes the result above.

To compare the three methods, simulation was performed for a theoretical background mean count rate of 2.8 counts/min. and a counter efficiency of 0.72. Two confidence intervals were generated by each method; one at a 95% confidence level (k = 1.96) and one at a 70% confidence level (k = 1.04). For samples less than 30,000 B.P., there was little difference in the three methods, though the new symmetric intervals were consistently shorter but not with any practical significance.

Table 1 summarizes the percentage of intervals which covered the true age for 30,000 B.P. to 60,000 B.P.

At 70% confidence level, the new symmetric estimator was clearly superior to either alternative. At 95% confidence level only the asymmetric method maintained the appropriate coverage. The coverage of the new symmetric estimator gradually fell from 95% to about 87% or 88% at 60,000 B.P. with a 92% to 93% coverage at 40,000 B.P. The old symmetric method was not competitive; it often failed to provide any estimate of σ_A, and any existing coverage fell sharply. Figure 1 is a graph of the simulated coverage against age of the sample. Figure 2 is a histogram of simulated age estimates for a 57,000 B.P. sample, showing the asymmetric nature of the distribution of the age estimator which causes the decrease in coverage of the symmetric confidence intervals.

TABLE 1
Actual Coverage of Confidence Intervals from Simulation*.

Years B.P.	70% Confidence Level			95% Confidence Level		
	Asymmetric	New Symmetric	Old Symmetric	Asymmetric	New Symmetric	Old Symmetric
30,000	.6900	.6880	.6900	.9490	.9440	.9445
31	.6935	.6960	.6985	.9445	.9430	.9430
32	.6905	.6955	.6980	.9490	.9465	.9475
33	.6955	.7030	.7060	.9535	.9460	.9480
34	.7050	.7060	.7095	.9525	.9455	.9465
35	.7060	.7160	.7210	.9480	.9415	.9445
36	.7085	.7090	.7105	.9480	.9385	.9395
37	.7000	.6990	.7045	.9455	.9345	.9370
38	.7015	.7030	.7080	.9420	.9260	.9290
39	.6955	.7020	.7105	.9495	.9325	.9355
40	.7040	.7155	.7255	.9460	.9240	.9280
41	.7045	.7095	.7211	.9475	.9255	.9291
42	.6960	.7045	.7129	.9485	.9155	.9195
43	.7224	.7224	.7327	.9464	.9088	.9128
44	.7059	.7200	.7338	.9588	.9065	.9072
45	.7162	.7131	.7195	.9602	.8963	.8979
46	.7224	.7245	.7246	.9670	.8978	.8990
47	.7318	.7246	.7177	.9779	.9080	.9038
48	.7544	.7274	.7128	.9773	.9023	.9004
49	.7698	.7264	.7069	.9748	.8997	.8947

50	.7990	.7434	.9764	.9074	.8958
51	.8130	.7312	.9708	.8977	.8704
52	.8230	.7271	.9744	.8981	.8750
53	.8190	.7179	.9690	.8952	.8785
54	.8077	.7119	.9698	.8849	.8407
55	.8041	.7202	.9718	.8945	.8531
56	.8098	.7185	.9737	.9059	.8677
57	.8020	.7037	.9682	.8946	.8351
58	.7963	.7019	.9680	.8855	.8213
59	.7890	.6786	.9732	.8755	.7904
60	.7689	.6656	.9669	.8667	.7797

		.6983		
		.6709		
		.6521		
		.6535		
		.5892		
		.6095		
		.5925		
		.5816		
		.5361		
		.5204		
		.4991		

*Each entry is the result of generating 2000 confidence intervals for a background mean of 2.8 c/m and efficiency of 0.72.

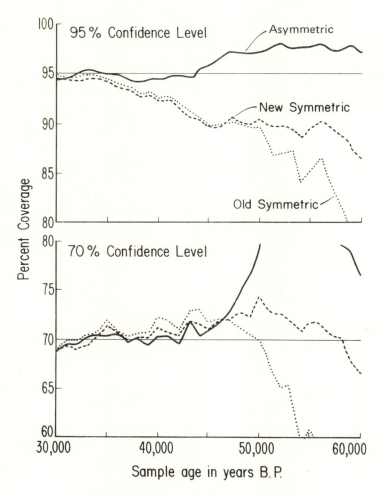

FIG. 1. Simulated coverage against age of samples

In conclusion, the new symmetric estimator is recommended. It always produces a finite confidence interval. If deviation of actual coverage from theoretical coverage is considered too severe, the method can be modified to produce asymmetric intervals to compensate without requiring extension of the interval to infinity, as is the case with the current asymmetric estimator. The old symmetric method is not competitive except on younger samples.

ACKNOWLEDGMENT

I wish to express my thanks to Miss Saadia Pankey for her abundant patience in preparing this manuscript.

EVALUATION OF TOTAL VARIABILITY IN RADIOCARBON COUNTING 629

```
FREQUENCY---------------------------------------------
  201    *      I                                                    *
  196    *      I                                                    *
  191    *      I I I                                                *
  186    *      I I I                                                *
  181    *      I I I I                                              *
  176    *      I I I I                                              *
  171    *      I I I I                                              *
  166    *      I I I I I                                            *
  161    *      I I I I I                                            *
  156    *      I I I I I                                            *
  151    *      I I I I I I                                          *
  146    *      I I I I I I                                          *
  141    *      I I I I I I                                          *
  136    *      I I I I I I I                                        *
  131    *      I I I I I I I                                        *
  126    *      I I I I I I I                                        *
  121    *      I I I I I I I                                        *
  116    *      I I I I I I I I                                      *
  111    *      I I I I I I I I                                      *
  106    *      I I I I I I I I                                      *
  101    *      I I I I I I I I                                      *
   96    *      I I I I I I I I                                      *
   91    *      I I I I I I I I                                      *
   86    *      I I I I I I I I I                                    *
   81    *      I I I I I I I I I                                    *
   76    *      I I I I I I I I I I                                  *
   71    *      I I I I I I I I I I                                  *
   66    *      I I I I I I I I I I                                  *
   61    *      I I I I I I I I I I                                  *
   56    *      I I I I I I I I I I I                                *
   51    *      I I I I I I I I I I I                                *
   46    *      I I I I I I I I I I I I                              *
   41    *    I I I I I I I I I I I I                                *
   36    *    I I I I I I I I I I I I I I                            *
   31    *    I I I I I I I I I I I I I I                            *
   26    *    I I I I I I I I I I I I I I                            *
   21    *    I I I I I I I I I I I I I I I I                        *
   16    *    I I I I I I I I I I I I I I I I                        *
   11    *    I I I I I I I I I I I I I I I I I I                    *
    6    *I I I I I I I I I I I I I I I I I I I  I   I               *
    1    *I I I I I I I I I I I I I I I I I I I I I I I I I   I I    *
---------------------------------------------------------------------
Class          5    10    15    20    25    30    30   35    40    45    50
True Age = 57000
```

FIG. 2. Simulated age estimates.

REFERENCES

Haas, Herbert
    1976   Specific Problems with Liquid Scintillation Counting of Small Benzene Volumes and Background Count Rate Estimation. 9th Int. Radiocarbon Conf. Univ. of Calif. Los Angeles and San Diego.

Haas, Herbert, et al.
    1976   Report on Preliminary Investigation into Adequacy of ^{14}C Counting Procedures. Report to Inst. for Study of Earth and Man. Southern Methodist Univ.

Otlet, R. L.
    1976   An Assessment of Laboratory Errors in Liquid Scintillation Methods of Carbon-14 Dating, Proc. 9th Int. Radiocarbon Conf.

Severo, N. C. and M. Zelen
    1960   Normal Approximation to the Chi-square and Non − Central F Probability Functions. Biometrika. 47:411.

PART VIII
Modeling Experiments

1

Prognosis for the Expected CO_2 Increase due to Fossil Fuel Combustion

H. Oeschger and U. Siegenthaler

ABSTRACT

Prognoses for the increase in atmospheric CO_2 due to fossil CO_2 input are calculated by means of a box diffusion model for the natural CO_2 exchange in nature. For a hypothetical CO_2 production stop in 1970, the CO_2 excess declines until the year 2000 to about 70% of its value in 1970. The diminution then gradually slows. For a constant CO_2 production rate, the atmospheric CO_2 content continually increases at a slightly diminishing rate. A future CO_2 production function is calculated that causes a logistic increase in the atmospheric CO_2 level to a final, constant excess of 50%. In this case the production rate increases to 160% until 2000 A.D., but in 2100 A.D. is only 70% of the 1970 rate.

INTRODUCTION

Atmospheric CO_2 increase due to fossil fuel combustion ranks among the most serious of environmental problems because of its enduring impact on the earth's radiation balance. To reliably forecast the atmospheric response to possible future CO_2 production functions, a thorough understanding of the dynamic characteristics of the CO_2 cycle is required. Important information on the natural CO_2 exchange is obtained from ^{14}C studies on samples from the atmosphere, hydrosphere, and biosphere:

1. The first ^{14}C depth profiles of samples from the Pacific (Bien et al. 1960) and Atlantic (Broecker et al. 1960) oceans, dating from 1957-1959, represent a steady situation with only minor influences of ^{14}C from nuclear weapons and CO_2 from fossil fuel.

2. Due to nuclear weapon tests, the average $^{14}C/C$ ratio in the ocean surface rose from 95% (pre-industrial atmospheric value = 100%) in 1957 to about 112% in 1970.

3. Until 1970, the integrated fossil CO_2 input amounted to about 19% of the pre-industrial mass of CO_2 in the atmosphere, with the result of a 10% increase in the atmospheric CO_2 concentration.

4. Until 1950, the atmospheric $^{14}C/C$ ratio had diminished by about 2% due to dilution by ^{14}C-free fossil CO_2.

Revelle and Suess (1957) and Craig (1957) investigated the CO_2 exchange between atmosphere, ocean, and biosphere by means of box models. Since then a number of papers on this topic have been published, most of them retaining the box model concept with a two-box ocean consisting of a mixed surface layer and of a deep-sea box below it. These models did not permit consistent description of the natural, steady-state ^{14}C distribution and the response to short-term disturbances of the system. The exchange parameters valid for short-term variations (uptake of bomb ^{14}C by the ocean) were usually quite different from those describing long-term phenomena (the preindustrial ^{14}C distribution). ^{14}C depth profiles from the ocean have shown that below the mixed layer, the $^{14}C/C$ ratio decreases gradually with depth. Therefore, the concept of a well-mixed deep-sea box drastically oversimplifies the actual situation. It is not surprising that such a model cannot satisfactorily simulate processes with different characteristic times unless artificial parameter adaptations are introduced for the individual processes. Oeschger et al. (1975) showed that these discrepancies do not occur if the deep sea is modeled as an eddy diffusive medium. Without parameter adaptation, all the above-mentioned phenomena related to the natural carbon cycle can consistently be described by the box diffusion model, briefly described below and detailed elsewhere (1975) by the authors.

The Model

Our box diffusion model consists of four reservoirs (fig. 1). The atmosphere and the mixed surface layer of the ocean are represented as well-mixed boxes. The continuous decrease of the $^{14}C/C$ ratio with depth is taken into account by considering the deep ocean as a medium with turbulent vertical mixing. The ocean is treated mathematically as a one-dimensional, eddy-diffusive medium or, in a numerical model version, as composed of a series of thin, well-mixed layers with exchange coefficients derived from the vertical eddy-diffusion coefficient, K (assumed constant with depth). K and the exchange parameter atmosphere (ocean, k_{am}) are determined from the natural ^{14}C distribution in atmosphere and ocean.

When the atmospheric CO_2 pressure changes, only the amount of dissolved CO_2 gas in the ocean varies proportionally. The corresponding variations of the HCO_3^- and the $CO_3^=$ contents (i.e., of the major part of total CO_2) depend on the sea water composition and are relatively smaller than the changes of p_{CO_2}. This effect can be taken into account by a buffer factor ξ. If the CO_2 partial pressure is increased by α%, the total CO_2 concentration of sea water in equilibrium with the atmosphere increases by α/ξ % only.

The biosphere is described in the model by a well-mixed box with 2.4 times the atmospheric carbon content and a mean residence time of 60 years. Some of the plants probably respond to an increased CO_2 supply with enhanced photosynthetic activity.

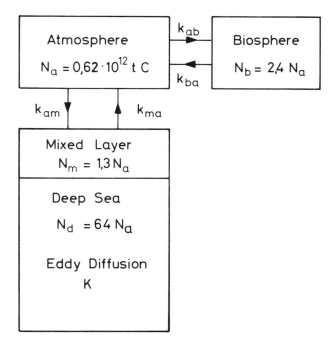

FIG. 1. The reservoirs of the box-diffusion model. The preindustrial amounts of carbon in each reservoir are indicated. Atmosphere, biosphere, and mixed layer of the ocean are modeled as well-mixed boxes; transport within the deep sea takes place by vertical eddy diffusion, K = eddy diffusion coefficient.

This effect is considered in the model by a biota growth factor ϵ. If the atmospheric CO_2 pressure is increased by α %, the flux from the atmosphere to the biosphere increases by $\epsilon\alpha$ %. According to Keeling (1973a) and our own model estimates, ϵ is probably between 0.0 and 0.4.

For calculation of atmospheric responses to different CO_2 input functions, we use the parameter values listed in table 1, with which the observed phenomena regarding the CO_2 cycle have been successfully modeled.

Prognoses of the Atmospheric CO_2 Increase for Different Input Functions

THE EQUILIBRIUM VALUE

If the integrated CO_2 input is α_o % of the preindustrial atmospheric CO_2 content, the new equilibrium, reached long after production has stopped, will be α_∞ % above the preindustrial value. The ratio α_∞/α_o can be calculated from the equation

$$\alpha_o N_a = \alpha_\infty N_a + \frac{\alpha_\infty}{\xi}(N_m + N_d) + \alpha_\infty \epsilon N_b$$

and

$$\frac{\alpha_\infty}{\alpha_o} = \frac{N_a}{N_a + (N_m + N_d)/\xi + \epsilon N_b}$$

with $\xi = 10$ and $\epsilon = 0.2$ we get

$$\frac{\alpha_\infty}{\alpha_o} = 0.125$$

One-eighth of the total CO_2 input remains in the atmosphere, so restoration of the original CO_2 concentration is unattainable. We must, however, recognize that such processes as the dissolution of carbonates from the ocean sediments or additional weathering are not considered in the model. These effects could become important for high CO_2 increases, particularly in the deep sea where they would manifest themselves only after much time because of slow vertical transport. We believe that such high CO_2 levels probably will not be reached, because man will have been forced to stop producing CO_2 in order to escape catastrophic climatic consequences.

DECREASE OF THE CO_2 LEVEL EXCESS AFTER A PRODUCTION STOP

The computed decrease of atmospheric CO_2 after a supposed CO_2 production stop in 1970 is plotted in figure 2. During the first few decades the decrease is relatively rapid and in the year 2000 the CO_2 excess is only about 70% of that in 1970. The diminution then becomes increasingly slow, tending toward a value of 0.22 times the 1970 excess; this ratio is roughly twice that calculated in the preceding paragraph, due to the fact that in 1970 almost half of the introduced CO_2 had entered the ocean and the biosphere.

TABLE 1

Model Parameters Used For Prognoses

Symbol	Meaning		Numical value
N_a	CO_2 in atmosphere		0.62×10^{12} t C
N_b	CO_2 in biosphere	pre-industrial amounts	$2.4\, N_a$
N_m	CO_2 in mixed layer		$1.3\, N_a$
N_d	CO_2 in deep sea		$64.2\, N_a$
k_{am}	Exchange coefficient atmosphere – mixed layer		1/7.7 y
k_{ab}	Exchange coefficient atmosphere – biosphere		1/25 y
K	Eddy diffusion coefficient (deep sea)		3987 m^2 s^{-1}
ξ	Buffer factor for excess CO_2		10
ϵ	Biosphere growth factor		0.2

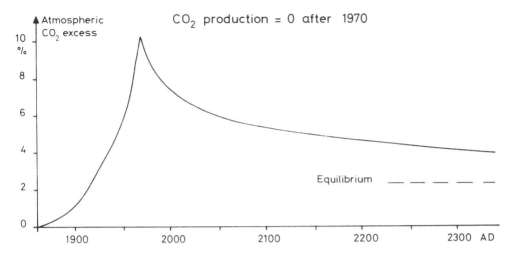

FIG. 2. Predicted atmospheric CO_2 excess after a fictive production stop in 1970. The pre-industrial level will never be reached again, but a new equilibrium value, 2.3% above it, will establish after a long time.

Consequently, if man should become aware of catastrophic climatic developments due to a high CO_2 level, this level would decline but slowly even after a total production stop. Even after many generations, the pre-industrial atmospheric CO_2 concentration would not be attained.

ATMOSPHERIC CO_2 INCREASE FOR A LOGISTIC CO_2 INPUT USING KNOWN FOSSIL FUEL RESOURCES

Zimen (1971) indicates that the resources of oil, coal, and gas are about twelve times the preindustrial CO_2 content. Assuming an integrated CO_2 production according to a logistic function, the production rate is given by

$$P(t) = \frac{d}{dt}\left(\frac{11.65 \, N_a}{1 + 61 \exp(-t/22)}\right)$$

t is the time in years; t = 0 in 1970.

The growth rate of P(t) immediately after 1970 is 4.6% per year (average rate 1960 to 1970: 5.0% per year).

The resulting CO_2 increase is shown in figure 3. Until about 2050 the atmospheric CO_2 excess would be about 55%-60% of the total of produced CO_2. At that time the concentration would be roughly three times more than today. The corresponding increase of the mean surface temperature can be estimated to about 3 to 4°C by extrapolating the results of Manabe and Wetherald (1975).

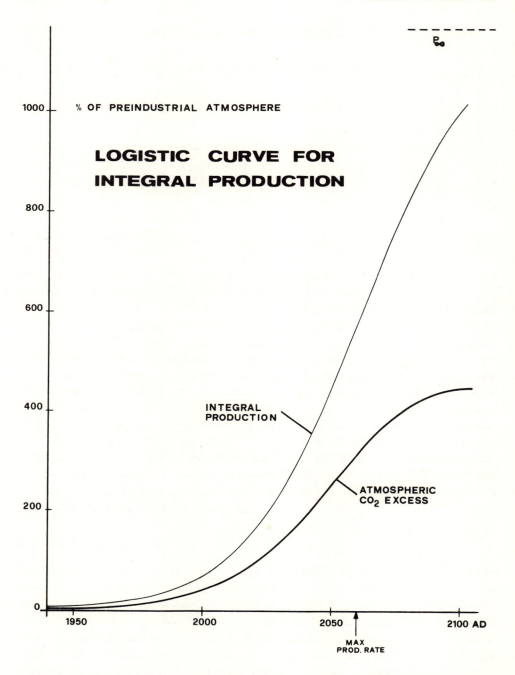

FIG. 3. Prediction for a logistic increase of the integral CO_2 production after 1970. $P\infty$ = the total amount of fossil fuel reserves which will have been burnt after infinite time.

ATMOSPHERIC CO_2 INCREASE WITH CONSTANT CO_2 INPUT

Due to the long residence time of the excess CO_2 in the atmosphere, and due to the fact that, even after an "infinite" time, one-eighth of the excess CO_2 remains in the atmosphere, no new equilibrium for the excess CO_2 is reached for a constant production rate. In this case, the atmospheric CO_2 content would steadily increase, though the slope would slightly decrease with time, as shown in figure 4. In 2050, 44% of the integral CO_2 production would still be in the atmosphere.

PRESCRIBED ATMOSPHERIC CO_2 EXCESS: LOGISTIC INCREASE ASYMPTOTICALLY TO 50%

A discussion of the climatic impact of fossil CO_2 may result in an upper tolerable limit for CO_2 excess. In the following we have arbitrarily assumed this limit to be 50%, which would cause a global warming of about 1°C according to the model computations of Manabe and Wetherald (1975). For our calculations we have assumed that this CO_2 excess (n_a) would be reached smoothly according to a logistic function after 1970:

$$n_a = \frac{0.5 \, N_a}{1 + 3.73 \exp(-t/23.5)}$$

If a CO_2 excess of 50% were tolerable, CO_2 production would be increased until the beginning of the next century by about 60%. Then energy production should be increasingly taken over by other energy sources. By the year 2100, the use of fossil fuel should be only 70% of that in 1970.

CONCLUSIONS

In discussing the potential danger of industrially produced CO_2 one must predict both the future concentrations and the implied climatic effects. The box diffusion model is able to reproduce the main features of the CO_2 cycle, so we have confidence in the prognoses presented. On the other hand, although sophisticated models exist for the climatic consequences, it has not yet been possible to include all possibly important feedbacks. Moreover, "experimental" verifications are difficult since the CO_2 effect is still masked by temperature fluctuations for varied reasons.

The models existing at present permit the following conclusions to be drawn: (1) a doubling of the preindustrial CO_2 level could cause a mean temperature rise of 2° to 3°C , probably involving drastic changes in the global weather pattern; (2) such a doubling could occur before the middle of the next century if consumption of fossil fuels should increase at the present rate; and (3) that the CO_2 increase is only partly reversible, and that any decrease, even after a total production stop, would proceed too slowly to escape catastrophic consequences once a critical CO_2 level had been exceeded.

In view of these alarming prospects, the CO_2 problem should be given much more attention in the public energy debate. At the same time, scientific efforts should continue for improving the relevant models and finding ways to test them. A paper is in preparation which discusses these problems in more detail, comparing different models.

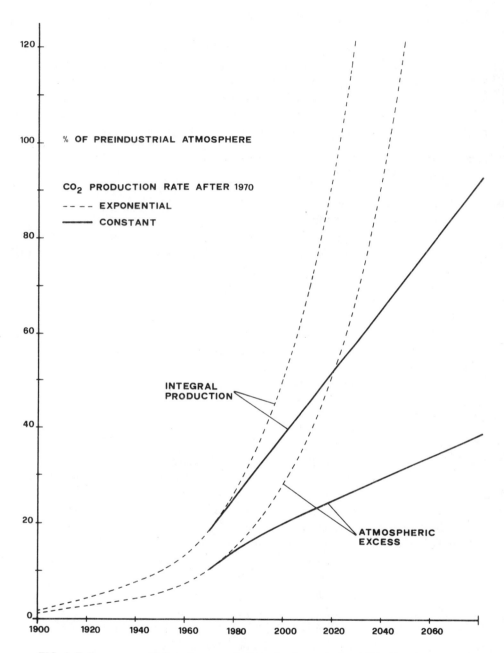

FIG. 4. Full curves: predictions for a constant production rate after 1970. Also shown are the curves (dashed) for an exponential increase after 1970 with an e-folding time of 35 years.

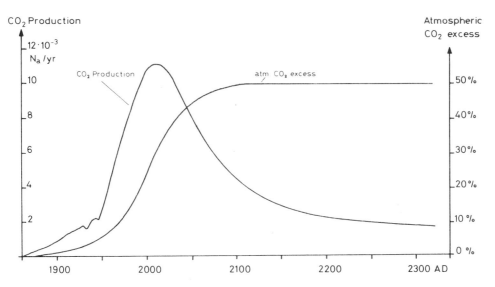

FIG. 5. Prescribed atmospheric CO_2 excess after 1970, logistically increasing to 50%. Production rate up to 1970 as compiled by Keeling (1973b), then computed so as to produce the prescribed CO_2 excess.

REFERENCES

Bien, G. S., N. W. Rakestraw, and H. E. Suess
    1960   Radiocarbon concentration in Pacific Ocean Water. Tellus. 12:436-443.

Broecker, W. S., R. Gerard, M. Ewing, M. , and B. C. Heezen
    1960   Natural radiocarbon in the Atlantic Ocean. J. Geophys. Res. 65:2903-2931.

Craig, H.
    1957   The natural distribution of radiocarbon and the exchange time of carbon dioxide between atmosphere and sea. Tellus. 9:1-17.

Keeling, C. D.
    1973a   The carbon dioxide cycle: reservoir models to depict the exchange of atmospheric carbon dioxide with the ocean and land plants. *In* Chemistry of the lower atmosphere. Rasool, ed. 251-329.
    1973b   Industrial production of carbon dioxide from fossil fuel and limestone. Tellus. 25:174-198.

Manabe, S., and R. T. Wetherald
    1975   The Effects of Doubling the CO_2 concentration on the climate of a General Circulation Model. J. Atmos. Sci. 32:3-15.

Oeschger, H., U. Siegenthaler, U. Schotterer, and A. Gugelmann
    1975   A box diffusion model to study the carbon dioxide exchange in nature. Tellus. 27:168-192.

Revelle, R., and H. E. Suess
    1957   Carbon dioxide exchange between atmosphere and ocean and the question of an increase of atmospheric CO_2 during past decades. Tellus. 9:18-27.

Zimen, K. E.
    1971   Nuclear Energy Reserves and Long-term Energy Requirements. Angewandte Chemie. Int. Ed. 10:1-11.

2

In Situ ^{14}C Production in Wood

C. J. Radnell, M. J. Aitken, and R. L. Otlet

ABSTRACT

It has been suggested that the cosmic ray thermal neutrons produced in the vicinity of bristlecone pine may explain the excess ^{14}C level indicated by the bristlecone pine calibration curve. Measurements have been made of the ^{14}C activity produced in pine and oak by reactor thermal neutrons at a level well above the natural ^{14}C level. Combined with estimates from other work of the cosmic ray thermal neutron flux for the altitude at which bristlecone pines grow, (3 km), the results indicate that, even for an 8000-year-old tree, the effect is not significant. The nitrogen content of the bristlecone pine sample, as derived from the observed increase in ^{14}C activity after neutron activation, is about 0.07%.

Investigation of possible contribution to ^{14}C activity from higher energy cosmic rays has also been made. The conclusion reached is that this effect is not significant.

The effect of various pretreatments on the reactor-produced ^{14}C production is not easily eliminated by pretreatment; substantial activity remains even in extracted cellulose.

INTRODUCTION

The bristlecone pine calibration curve (Suess 1970, Damon et al. 1972, Ralph et al. 1973) shows that the atmospheric ratio of ^{14}C to ^{12}C has varied over the past eight millennia. There appears to have been a higher concentration of ^{14}C (to 8%) prior to 2500 B.P., with minor oscillations in concentration (1% to 2%) throughout the last 8000 years, as shown in figure 1. Because the bristlecone pines have grown at high altitude (3 km) in the arid region of the White Mountains of California (geomagnetic latitude $\lambda = 44°N$), it has been questioned whether they give a valid record of worldwide atmospheric variations in the ^{14}C to ^{12}C ratio at ground level. Such discussion stems from the suggestion (RAMA 1969, unpublished) that a significant part of the ^{14}C activity in an old bristlecone pine could derive from continued in situ production of ^{14}C from nitrogen present in the wood by cosmic-ray-produced thermal neutrons (i.e., ^{14}N(n,p)^{14}C).

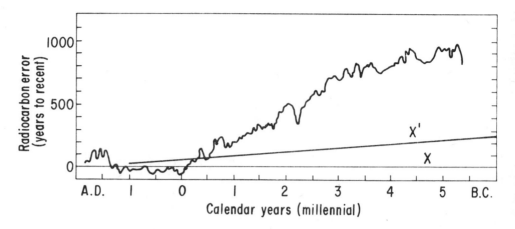

FIG. 1. Age error according to 9-sample averaging of tree ring measurements derived from data of Ralph et al. 1973). Use of the X line corresponds to the 5568 years half-life of ^{14}C and use of the X′ line to the revised half-life of 5730 years (from Aitken 1974).

A further possibility has been suggested of ^{14}C production by spallation reactions on the more abundant isotopic constituents of wood with the higher energy spectrum of cosmic rays (B. Rose, Archaeometry Conference 1973). Discussions with P. H. Fowler and J. R. Prescott (private communications 1973) suggest that significance would attach only to the spallation reaction, $^{16}O(n,^{3}He)^{14}C$, or to the less energetically favorable reactions $^{16}O(n,pd)^{14}C$ and $^{16}O(n,n2p)^{14}C$. Oxygen is the second most abundant element of wood, about 40% to 50% of the weight.

Neutron production by lightning (Libby and Lukens 1973, Fleischer et al. 1974, Fleischer 1975) and by solar flares are other possible sources of in situ ^{14}C production in bristlecone pine.

Since the majority of archaeological samples originate near sea level, any in situ effect produced by cosmic ray neutrons would be much less than that of the bristlecone pines, which grow at an altitude of about 3 km where the cosmic ray neutron flux is an order of magnitude greater than at sea level.

Harkness and Burleigh (1974) experimentally tested the idea of in situ ^{14}C in wood by irradiating bristlecone pine with a dose of neutrons estimated to be that received from cosmic ray neutrons, in the energy range thermal to 3 MeV by a 9000-year-old tree. Although they calculated, on the basis of a chemically determined value of 8.5% for the nitrogen content of their samples, that an excess activity of 0.12 dpm/gC should have been observed, no excess ^{14}C was detected. One possible explanation was that the ^{14}C produced by their irradiation could have been removed in the standard acid-alkali pretreatment they employed.

Following a theoretical examination of the production rates of in situ ^{14}C by thermal and high energy cosmic ray induced neutrons, we describe irradiation experiments using thermal neutron doses on wood which produce a measurable amount of in situ ^{14}C. The nitrogen content of bristlecone pine and oak is estimated from the results. An experiment to consider the ^{14}C production from high energy neutron spallation reactions with oxygen is reported, and a possible experiment is proposed to test for ^{14}C produced in bristlecone pine from neutrons generated by lightning.

The experimental work includes a controlled investigation into the extent to which in situ ^{14}C is removed by pretreatment. This is considered an important corollary to the in situ studies; the implications of the results are relevant to routine sample pretreatment processes.

In Situ ^{14}C Production Rate Theory

The ^{14}C activity (A_t) in tree rings after time (t) is governed by the simultaneous production and decay of ^{14}C within the wood. This can be described in its simplest form by the equation

$$A_t = A_o e^{-\lambda t} + P(1 - e^{-\lambda t})$$

where A_o is the original activity of the tree as fixed by the assimilation of carbon from the air, λ is the decay constant of ^{14}C, and P is the continual in situ production of ^{14}C in the tree rings by all contributing external sources. The latter can be written as

$$P = \Sigma N \sigma \varphi$$

where (N) is the number of target nuclei for each process involved, (σ) the activation cross section, and (φ) the neutron flux. Possible sources are discussed in more detail below.

THERMAL NEUTRONS

The most pertinent data for the thermal flux is considered to be that of Yamashita (1966), who measured the thermal and fast neutron fluxes at 3190 m altitude (about 700 g/cm^3) in the same geographic location as that of the bristlecone pine (the White Mountains of California). The thermal neutron flux measured 1 m above ground level was $(1.20 \pm 0.05) \times 10^{-2}$ n/cm^2/sec which, with his suggested correction for anistropy (about 15%), becomes about 1.4×10^{-2} n/cm^2/sec. This result is within the range of predicted values obtained by adjustment of data taken in other localities. It is considered the most reliable result because it is an experimentally determined value in the same geographic location and altitude being considered, but, as discussed in appendix A, the differences between all values quoted are less than an order of magnitude.

Two effects within the tree can be expected to enhance the incident thermal flux. These are (1) in situ production of neutrons by direct interaction of cosmic particles within the wood, and (2) the thermalization of higher energy neutrons in traversing the wood (B. Rose, Archaeometry Conference 1973). Yamashita's data implies a factor up to about 20 for the difference between the thermal neutron flux in the earth (near the surface) and in the air (near the ground). This is a higher value than implied from other data (see appendix A) and is therefore likely to represent an upper limit. Assuming this factor applicable to wood, the effective flux is then about 2.8×10^{-1} n/cm^2/sec, but in view of the 15 ± 5 value suggested in appendix A, the effective flux is more likely to be $(2.1 \pm 8) \times 10^{-1}$ n/cm²/sec.

Using a nitrogen content of 0.1% by weight, and a carbon content of 40%, the thermal neutron production rate is

$$P_t = N_N \sigma_t \varphi_t = (2.5 \pm 1.0) \times 10^{-3} \; ^{14}C \text{ atoms/min/gC}$$

where $\sigma_t = (1.81 \pm .05)$ b (Stehn et al. 1964).

This nitrogen content is chosen on the basis of its being of the same order as reported in this work and elsewhere (Libby and Lukens 1973; Damon et al. 1973); the value used by Harkness and Burleigh (1974) was two orders of magnitude greater.

HIGH ENERGY NEUTRONS

As stated earlier, the only significant source of ^{14}C production by anything order than thermal neutrons on ^{14}N would seem to be high energy neutrons on ^{16}O (oxygen being about 50% by weight of wood). These reactions are $^{16}O(n,^{3}He)^{14}C$, $^{16}O9n,pd)^{14}C$ and, $^{16}O(n,n2p)^{14}C$, the first being more energetically favorable: Q = -14.6 MeV, -20.1 Mev, and -22.3 MeV, respectively. The cross section for these reactions is taken as 10 mb and the integral flux for the neutrons most likely to produce the majority of ^{14}C from these reactions (i.e., in the region 40 to 160 MeV at 3 km altitude) as $(1.2 \pm 0.6) \times 10^{-2}$ n/cm²/sec. The basis of these figures is discussed in appendix B.

The ^{14}C production rate for these reactions in wood will be

$$P_h = N_o \sigma_h \varphi_h = 2.7 \times 10^{-4} \; ^{14}C \text{ atoms/min/gC}$$

with an uncertainty that is probably no worse than a factor of 2 (owing to the lack of experimental data for these cross sections).

Combining the thermal and high energy neutron ^{14}C production rates,

$$P = P_t + P_h$$

$$= (2.8 \pm 1.0) \times 10^{-3} \; ^{14}C \text{ atoms/min/gC}$$

Taking into account the decay over 8000 years, the in situ activity becomes

$$P(1 - e^{-\lambda t}) = (1.7 \pm 0.6) \times 10^{-3} \text{ dpm/gC}$$

or about 0.03% of the activity of 8000-year-old wood.

LIGHTNING-PRODUCED NEUTRONS

It has been suggested by Libby and Lukens (1973) that the neutrons generated in lightning bolts may have contributed to atmospheric ^{14}C production and also that they may have produced significant amounts of in situ ^{14}C in the bristlecone pines themselves, particularly on account of the vulnerability to lightning of high altitude trees on exposed ridges. They have suggested that this in situ effect might be significant and hence might help to explain certain short-term variations apparent in the bristlecone pine record.

Experimental

SIMULATED IN SITU PRODUCTION

Thermal Neutron Activation by $^{14}N(n,p)^{14}C$

Specimens of oak (*Quercus robur*) and bristlecone pine (*Pinus aristata*) were irradiated in the thermal column facility, J face of the light-water-moderated reactor, Herald, at AWRE, Aldermaston. This position provided an essentially thermal neutron flux for which the activation cross section of 1.81 ± 0.05 b could be expected to apply. Knowledge of this value, together with a measurement of the integrated thermal neutron dose from flux monitors irradiated with the sample, permitted direct measurement of the nitrogen content, important because of large differences in the reported estimates of nitrogen content. The nitrogen content was also checked chemically using the Kjeldahl (Gunning-Jodblauer) process (Strouts et al. 1962).

The assembly of the various flux monitors for the initial irradiation made to calibrate the position is shown in figure 2. Three types of monitors were used; (1) unshielded cobalt wires for the thermal flux, (2) cadmium-covered cobalt for an epithermal measurement, and (3) magnesium-nitrate-impregnated graphite rods from which to determine a supporting evaluation of the overall $^{14}N(n,p)^{14}C$ activation. Four cobalt wires were affixed to the vertical sides A, B, C, and D, and three more wires were fitted centrally within the impregnated graphite rods. The rods, 5 mm diameter, were inserted into

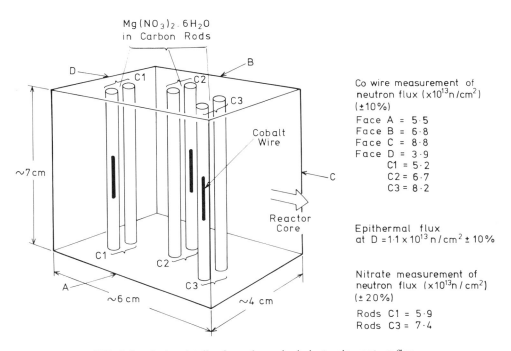

FIG. 2. Irradiation details of wood samples in isotropic neutron flux.

vertical holes cut into the body of the wood as shown. Impregnation was achieved by soaking the rods in a solution of magnesium nitrate for about one hour followed by oven drying. This was repeated until the final weighing indicated a total absorbed quantity of about 0.1 g, which was assumed to be $Mg(NO_3)_2 \cdot 6 H_2O$.

Following irradiation, the rods were crushed, burned to CO_2, and converted to benzene for ^{14}C counting using the standard techniques of this laboratory (Otlet and Slade 1974). Measurements of the ^{60}Co in the flux monitors and interpretation in terms of flux were carried out by P. Roscoe, AWRE, Aldermaston. Results of the flux calibration are given in figure 2. The apparent flux gradient along the length of the block (moving away from the reactor core) is possibly exaggerated by the local flux depression caused by placing the cadmium-covered monitor too close to the D-side uncovered cobalt monitor. The gradient is, however, only about a factor of 2. Since the average flux value determined from the mean of all the cobalt monitors is not significantly affected by the one suspect value, no ambiguity in interpretation should be caused. The results indicate a mean thermal neutron flux of about 3×10^{11} n/cm^2/sec[1].

Comparison of the activity induced in the carbon rods and the activity induced in the wood permits evaluation of the nitrogen content of the wood. Such evaluation is independent of any uncertainties about neutron energy spectrum in the reactor or its modification through attenuation in the wood itself.

In the first experiment, samples of oak and bristlecone pine, sized at about 6 cm $\times$ 7 cm $\times$ 4 cm, were irradiated for about 5 minutes and 50 minutes, respectively. After cooling, the samples were reduced to sawdust and each was mixed thoroughly for homogeneity. Nonirradiated, blank samples of each type of wood were similarly treated. Each was then divided into two parts giving two sets of identical samples. The first set were pretreated by boiling in 1M HCl (1 hour), washing with water, boiling in about 2% NaOH (1 hour), followed by washing and drying at 90°C overnight. The second set received no pretreatment.

All samples were then converted to benzene using the procedures mentioned (Otlet and Slade 1974) and counted in a Packard Model 3375 Tricarb liquid scintillation counter; the results are summarized in table 1. The implied nitrogen content of oak and pine from the activation measurement with samples not pretreated is in good agreement with the Kjeldahl determinations, allowing for a possible error of $\pm 100\%$ in the latter determinations; the pretreatment results are discussed later. This large error is due to the basic inefficiency of the Kjeldahl process in breaking down higher order nitrogen compounds. Bearing this in mind the results are in good agreement with the values published by Damon et al. (1973), 0.05%; Libby and Lukens (1973), 0.2%, but not with that of Harkness and Burleigh (1974), 8.5%.

The implied value is about half of the figure used so far in the above theoretical estimates of in situ production. Correcting the estimates to this new value gives the in situ contribution in 8000-year-old bristlecone pine as 10^{-3} dpm/gC, i.e., 0.02% of the natural ^{14}C activity.

TABLE 1

(a) Effects of pretreatment on the in-situ ^{13}C activity of pine and oak.

Pretreatment	Activity above natural ^{14}C level (dpm/gC) ± 5%		% nitrogen implied		$\delta^{13}C$ ‰ (relative to P.D.B.)	
	Pine ($7.2 \pm 0.6 \times 10^{14}$ n/cm² total neutron flux)	Oak ($7.7 \pm 0.6 \times 10^{13}$ n/cm² total neutron flux)	Pine (± 10%)	Oak (± 10%)	Pine (± 0.3‰)	Oak (± 0.3‰)
None	26.1	6.1	0.07	0.22	-20.0	-25.2
1M HCl + NaOH	18.0	5.2	0.06	0.19	-21.7	-26.4
1M HCl + NaOH + charring	21.1	6.6	0.05	0.24	-22.3	-26.5

(b) Chemical analysis of pine and oak.

	Carbon % (± 4)	Hydrogen % (± 1)	Oxygen % (± 5)	Nitrogen % (± 100%) (Kjeldahl)
Pine	36	3	45	.07
Oak	49	5	54	.09

Production by High Energy Reactions

The basis of the cross section used is the theoretical data of Reedy and Arnold (1973), discussed in appendix B. The purpose of the experiment now reported was to provide a check of the proton data used by Reedy and Arnold as a basis for their neutron cross section. Protons were used because of the difficulty in obtaining a neutron flux of appreciably intensity in the energy range 40 to 160 MeV. At the upper end of this range the interaction properties of neutrons and protons are similar, hence the cross section for p + ^{16}O → ^{14}C is comparable with that for n + ^{16}O → ^{14}C. At the lower end of the range the value for the proton cross section underestimates that for the neutron cross section.

The Harwell synchrocyclotron provided a convenient source of suitable high energy protons. Oxalic acid was chosen as target material in preference to wood to minimize the risk of conflicting activities from unknown trace impurities in the natural material. The details of the three irradiations and the results are given in table 2. Only minute quantities of material could be irradiated in the approximate 1 cm^2 circular area beam. The target thickness was kept at less than 0.5 g cm^{-2} to avoid excessive energy drop due to interactions in passing through the sample (0.5 g cm^{-2} corresponds to an energy drop of about 5 – 10 MeV).

After irradiation the three oxalic acid samples were left for two weeks to allow any short-lived by-products such as ^{11}C($t_{1/2}$ = 20.5 mins) and ^{13}N($t_{1/2}$ = 9.96 mins) to decay away (^{7}Be[$t_{1/2}$ = 53 days] is lost in the conversion of oxalic acid to C_6H_6 and the ^{11}C decays through about 1000 half-lives in two weeks). Following this initial cooling period, the irradiated material was carefully removed from its target capsule, weighted, and mixed into a larger quantity of nonirradiated oxalic acid carrier. It was then converted to benzene for counting using the normal procedures. The results are given in table 2; for 75 MeV protons the cross section is about 0.6 mb and for 160 MeV protons it is about 2.5 mb.

The values obtained are in reasonable agreement for present purposes with the experimental values of 2.6 ± 0.7 and 1.8 ± 0.5 mb, respectively, that were obtained by Tamers and Delibrias (1961); these values were used by Reedy and Arnold (1972).

TABLE 2

Proton irradiation of oxalic acid

Oxalic acid sample	Energy of proton beam (approx. MeV)	Total flux of protons (× 10^{14} p/cm^2)	Induced activity (dpm/gC)	Cross-section implied (mb)
TARGET 1	160	2.07 ± 0.01	0.4 ± 0.1	2.5 (± 20%)
TARGET 2	75	30.2 ± 0.3	1.8 ± 0.3	0.6 (± 40%)

Background samples have a count rate of 8.41 ± 0.36 dpm/gC.

EFFECT OF CHEMICAL PRETREATMENT

For this experiment, a piece of oak previously dated at 3500 B.C. was irradiated for one hour in the thermal column facility, J face of the light-water-moderated reactor, Herald, described earlier. After all short-lieved activities had been allowed to decay, the sample was reduced to sawdust, thoroughly homogenized, and divided into a number of discrete lots for separation pretreatments. As a control blank, a piece of nonirradiated wood, taken from an adjacent vertical section of the same oak log, was similarly reduced to sawdust and divided for separate pretreatments.

The pretreatments given are listed in table 1. With the exception of the 6M H_2SO_4 + NaOH, and the last cellulose extraction procedure (6M HNO_3 + NaOH), they represent the usual processes reported by various laboratories listed in *Radiocarbon*.

The acid-alkali processes are similar to one another except for the acid used, its concentration and, in one case, its order of application. The samples were boiled in the acid for an hour, washed repeatedly with distilled water until neutral, boiled in alkali for an hour, washed repeatedly with distilled water again until neutral, then placed overnight in a drying cabinet at about 90°C.

The "cellulose extration" process (HCl at $NaClO_2$) removes the ligning fraction of the wood. The wood shavings were heated at 70 to 80°C in acidified $NaClO_2$ solution (5 liters) until all chemical activity stopped, neutralizing the residue. This process was repeated until the reaction ceased completely and the remaining fraction was bleached white and pulpy. It was then dried, converted to benzene, and counted as before.

Details of the results, shown in table 3, are discussed elsewhere, but it is immediately evident both from table 3 and from table 1 that substantial reactor-induced activity

TABLE 3

Pretreatment effects on the in situ ^{14}C activity of irradiated wood
Total flux of thermal neutrons 7.7×10^{14} n/cm^2

Pretreatment	Specific Activity Excess (dpm/gC ± 0.1)	Fraction remaining (w/w % ± 2)	$\delta^{13}C$ (average) ± 0.3‰
None	31.6	100	−26.3
Charring	36.0	30	−27.9
Volatiles	27.1	70	−26.6
1M HCl + NaOH	15.9	52	−28.0
6M HCl + NaOH	15.1	52	−27.5
6M H_2SO_4 + NaOH	14.9	34	−27.2
NaOH + 6M HCl	19.1	55	−27.4
HCl + $NaClO_2$	10.9	30	−26.3
6M HNO_3 + NaOH	6.1	26	−25.4

remains in the sample after routine pretreatment. Even the cellulose extraction process leaves a significant amount of excess activity in the sample. A puzzling feature is the effect of charring; there is depletion of ^{14}C in the volatile component of the wood, driven off under charring, and enrichment of ^{14}C in the remaining charred component.

CONCLUSIONS

In situ ^{14}C production by thermal and high energy cosmic ray neutrons is insignificant compared to the ±½% counting error usually associated with a radiocarbon date for 8000-year-old bristlecone pine. An increase by a factor of 10 would be needed for the thermal neutron effect to become significant, and by a factor of at least 200 for an effect sufficient to explain the full 8% long-term excess indicated by the bristlecone pine calibration curve. The effect of higher energy neutrons is probably an order of magnitude lower than that of thermal neutrons. However, it should be noted that after 15,000 years the in situ ^{14}C level is about 0.1% of the natural ^{14}C level, and after 25,000 years the level is about ½%.

The thermal neutron conclusion is in general agreement with that of Harkness and Burleigh (1974). However, we differ diametrically in the explanation given for the absence of a significant effect. Those authors found a sufficiently high nitrogen content in bristlecone pine for a significant effect to be expected, and they concluded that its absence was due to removal of the reactor-produced ^{14}C during pretreatment. Our conclusion is that the nitrogen content is too low for a significant effect. As regards pretreatment, we find that substantial reactor-produced ^{14}C remains in the sample for all the pretreatments tried.

Significant in situ production of ^{14}C by lightning bolts remains a possibility. We hope to make measurements on a lightning-struck bristlecone pine in the near future, in parallel, we plan to insert a nitrogen-rich compound into a bristlecone pine and to measure it for ^{14}C activity after a few years of exposure.

ACKNOWLEDGMENTS

Many thanks to Dr. B. Rose of AERE, Harwell, for his original suggestion of possible spallation effects and for his helpful discussions; to Professor P. H. Fowler of Bristol and Professor J. R. Prescott of Adelaide for their helpful discussions and comments on the high energy neutron and proton cross sections of oxygen; to Dr. P. Roscoe of AWRE, Aldermaston, for help with thermal neutron irradiations and for the cobalt and cadmium measurements; to Drs. B. Syme and C. Whitehead of AERE, Harwell, for the proton irradiations; to Mr. E. H. Henderson of AERE, Harwell, for the Kjeldahl analyses; to Dr. Fletcher of the Research Laboratory for Archaeology and the History of Art at Oxford for the first sample of oak; to Dr. C. H. de Lassus of C.E.A., Saclay, for many helpful comments and cross section data; and to Richard Burleigh of the British Museum for the bristlecone pine (P-SW-INY-8a) and the calculations of RAMA. One of us (C. J. R.) gratefully acknowledges an E.M.R. grant from U.K.A.E.A.

APPENDIX A

In situ ^{14}C Production

For the thermal neutron in situ ^{14}C work, the most pertinent data sources are considered to be the works of Yamashita et al. (1966) and Gold (1968a and b).

The calculations of Harkness and Burleigh (1974) are based essentially on the observation by Gold (1968b) and the observation that in nitrogen gas at 2 atmospheres pressure in a proportional counter (200 m above sea level), the production rate of ^{14}C at ground level (geomagnetic latitude $\lambda = 53°N$) is $(4.75 \pm 0.19) \times 10^{-27}$ atoms/sec/nitrogen atom or $(1.22 \pm 0.05) \times 10^{-2}$ ^{14}C atoms/min/g nitrogen. If, for the purpose of calculation, wood is taken to be 0.1% N and 50% C (by weight), then the production rate (P) is $(2.45 \pm 0.10) \times 10^{-5}$ ^{14}C atoms/min/gC.

Altitude Variations

There is a strong variation in neutron flux with altitude. The neutron density rises exponentially with altitude to a maximum at an air pressure of about 100 g/cm^2 (about 16 km), after which it drops off sharply. Gold's data was taken at 200 m altitude, but bristlecone pines grow at about 3 km altitude. Harkness and Burleigh took the increase in ^{14}C production rate associated with this 3 km increase in altitude to be by a factor of 9.5; Yamashita measured the thermal neutron flux to be greater at 3190 m than at sea level by a factor of about 11 (and about 12 for 0.4 to 10^7 eV neutrons); and Lingenfelter's work (1963) implies a factor of about 8 for thermal neutrons. A factor of 10 ± 1 is used in the following calculations as a first estimate of the increase in neutron flux associated with this increase in altitude. As with most calculations in this work, the errors derived from the data used become rather insignificant as order of magnitude calculations tend to suffice overall.

Boundary Effects

Since wood is more dense than air it is to be expected that a larger thermalization of neutrons would occur in wood than in air. Gorschkov et al. (1964) calculate the neutron production by cosmic rays in granite as 1.8 times that in water, but the thermal neutrons resulting from this neutron production differ by about an order of magnitude. Yamashita et al. (1966) similarly find for measurements of fast neutron flux (0.4 to 10^5 eV) 1 m above the ground and in the air a difference of about 3 times and for thermal neutrons a difference of about 20 times. Yamashita also indicates that the thermal neutron flux in dry ground increases rapidly with depth and shows this effect by measuring the thermal fluxes on and in a concrete building, obtaining almost twice the thermal flux just inside the building. (The higher energy fluxes decrease with depth in the building.) The implication is that the thermal neutron flux may suffer similarly at the air/wood boundary, and that it does not necessarily decrease with depth in the wood but rather increases initially (and then decreases).

The absorption of thermal neutrons by hydrogen is small enough in wood not to alter the effective number of thermal neutrons encountered by the nitrogen. Since only thermal neutrons and a small number of epithermals were used in our experiments, little thermalization of neutrons could occur. The slight decrease in the thermal flux through the wood was due mainly to the decrease in flux while moving away from the reactor core. With cosmic rays, however, many more high energy neutrons exist, so thermalization would be expected to occur.

Geiger (1956) has shown experimentally that production of neutrons(q) may be approximated by

$$q \propto A^{1/3}$$

where A = atomic weight of material. If the earth is approximated as SiO_2 (as in Yamashita), or granite (as in Gorschkov et al. 1964), and the empirical formula for wood is taken as CH_2O, then the neutron production ratio would be about 0.8. About the same amount of neutrons would be produced in wood as would be produced in earth (SiO_2 or granite).

It is estimated that the air/wood boundary has the overall effect of increasing the effective thermal neutron flux by a factor of (15 ± 5).

Latitude Variations

Because of the large geomagnetic latitude variations in the cosmic neutron flux (10:1 at the top of the atmosphere, Miles 1964), Gold's data must be corrected before using it as relevant to the White Mountains of California ($\lambda = 37°N$). Gold's data was accumulated at $\lambda = 53°N$ and must be corrected for the different cosmic ray intensities associated with difference in latitudes (see Lingenfelter (1963) for an account of ^{14}C production in air associated with latitude and altitude, and also Holt et al. (1966)). Lingenfelter's data implies that Gold's values should be divided by 1.05. Holt et al. (1966) imply a value of about 1.2. The correction is taken as being about 1.2.

Solar Activity Variations

Gold took measurements in December 1966, about two years after a solar minimum. Because this corresponds closely to the mean of that solar cycle, this should be close to an average value of neutron intensity. Yamashita's experiments were made during a solar minimum in July-December 1964. Davison's (1967) data implies a factor of about 1.2 for the decrease in neutron flux about two years after a solar minimum. However, such corrections are swamped by differences between Yamashita's data, and that of Hess et al. (1957 and 1961) and Newkirk (1963), who find neutron production greater by about 3 times in air. The data of Yamashita, Gold, and Davison are taken as being the more accurate and hence are used in these calculations.

Combining all these effects with the in situ ^{14}C production rate data of $(2.45 \pm 0.10) \times 10^{-5}$ ^{14}C atoms/min/gC, the in situ ^{14}C activity produced over 8000 years in wood will be about $(1.5 \pm 1) \times 10^{-3}$ dpm/gC, or about 0.03% the natural activity of wood (c.f. the 0.03% implied by Yamashita's data).

APPENDIX B

High Energy Neutrons

CROSS SECTIONS

The cross section for the $^{16}O(p,3p)^{14}C$ is about 2 mb for proton energies above 200 MeV and about 3 mb at 80 to 90 MeV (Barbier 1969, Tamers and Delibrias 1961). For neutron cross sections, only theoretical values exist. From Audouze et al. (1967) the neutron cross section at 150 MeV can be calculated to be about 7 to 10 mb compared to Reedy and Arnold, whose cross section at 90 MeV is 13 mb, decreasing to 3 mb at 150 MeV. The data of Reedy and Arnold is based on experimental proton cross sections increased at low energies for the $^{16}O(n,^3He)^{14}C$ and $^{16}O(n,n2p)^{14}C$ reactions. A value of 10 ± 2 mb, in the 40 to 160 MeV energy range has been used in the main text as the neutron cross section. This value should be a maximum for the cosmic neutron flux covering this energy range.

FLUX CALCULATIONS

Extrapolating Davison's formulae from neutrons in the energy range 0.5 to 10 MeV to neutrons with energies between 40 and 160 MeV, the flux is calculated to be

$$\int_{40}^{160} 0.264 E^{-1.13} dE = 0.21 \text{ n/cm}^2/\text{sec}$$

the error being at least 20%. This value will tend to be higher than expected since most cosmic spectra imply a coefficient which becomes smaller (more negative) as energy increases.

From the spectra of Haynes (1964), the drop in cosmic flux from 11 km to 3 km is by a factor of about 15 ± 3. Korff et al. (1965) imply the drop in neutron flux from 11 km to 3 km to be by a factor of about 20 ± 5 at $\lambda = 42°N$; Lingenfelter implies about 15 at $\lambda = 40°N$; Holt et al. (1966) imply a factor of about 11 at $\lambda = 42°N$. These values are all extrapolations of differing energy range measurements. The decrease in neutron flux is taken as 18 ± 5.

For 40 to 160 MeV neutrons at 3 km an integral flux of $(1.2 \pm 0.6) \times 10^{-2}$ n/cm²/sec is obtained.

REFERENCES

Aitken, M. J.
    1974   Physics and Archaeology. 2d Ed. Clarendon Press.
Audouze, J., M. Epherre, and H. Reeves
    1967   Nucl. Phys. A97:144-163.
Barbier, M.
    1969   Induced Radioactivity. North Holland Publishing Co. 168-210.

Damon, P. E., A. Long, and E. I. Wallick
    1972   Proc. 8th Int. Conf. on Radiocarbon Dating. New Zealand. 44-59.
    1973   Earth and Planetary Sc. Letts. 20:311-314.
Davison, J. N.
    1967   PhD Thesis. Univ. of Bristol.
Fleischer, R. L., J. A. Plumer, and K. Crouck
    1974   J. Geophys. Res. 79 (33):4853-4858.
Flesicher, R. L.
    1975   Transactions. Am. Geophys. Union. 56 (6):368.
Geiger, K. W.
    1956   Can. J. Phys. 34:288-303.
Gold, R.
    1968a   Phys. Rev. 165 (5):1406-1411.
    1968b   Phys. Rev. 165 (5):1411-1414.
Gorschkov, G. V., V. A. Zyabkin, and O. S. Tsvetkov
    1964   Soviet J. At. En. English Trans. 17:492-496.
Harkness, D. D., and R. Burleigh
    1974   Archaeometry. 16:121-128.
Haynes, R. C.
    1964   J. Geophys. Res. 69:853.
Hess, W. N., H. W. Patterson, R. Wallace, and E. L. Chupp
    1959   Phys. Rev. 116 (2):445-457.
Hess, W. N., E. . Canfield, and R. E. Lingenfelter
    1961   J. Geophys. Res. 66:665-677.
Holt, S. S., R. B. Mendell, and S. A. Korff
    1966   J. Geophys. Res. 71 (21):5109-5116.
Korff, S. A., R. B. Mendell, and S. S. Holt
    1965   Proc. Int. Conf. Cosmic Rays.
Libby, L. M., and H. R. Lukens
    1973   J. Geophys. Res. 78 (26:5902-5903.
Lingenfelter, R. E.
    1963   Rev. Geophys. 1. (1):35-55.
Miles, R. F.
    1964   J. Geophys. Res. 69 (7):1277-1284.
Newkirk, L. L.
    1963   J. Geophys. Res. 68:1825-1833.
Otlet, R. L., and B. Slade
    1974   Radiocarbon. 16 (2):178-191.
Ralph, E. K., H. N. Michael, and M. C. Han
    1973   Radiocarbon dates and reality. MASCA Newsletter. 9:1-18.
RAMA
    1969   Unpublished communication with Richard Burleigh, British Museum Research Laboratory.
Reedy, R. C., and J. R. Arnold
    1972   J. Geophys. Res. 77 (44):537-555.

Stehn, J. R., M. D. Goldberg, B. A. Magurno, and R. Weiner Chasman
    1964   BNL-325, 2d ed. Supp. 2(1) Brookhaven National Laboratory, Neutron Cross Sections.

Strouts, C. R. N., H. N. Wilson, and R. T. Harr-Jones
    1962   Chemical Analysis. 3. Clarendon Press.

Suess, H. E.
    1970   Radiocarbon variations and absolute chronology. Olsson, ed. Stockholm, Almquist and Wiksell, 303-313.

Tamers, M. A., and G. Delibrias
    1961   Comples Rendu. 253:1202-1203.

Yamashita, M., L. D. Stephens, and H. W. Patterson
    1966   J. Geophys. Res. 71 (16):3817-3834.

PART IX
Radiocarbon and Climate

1

Isotopic Tree Thermometers: Anticorrelation with Radiocarbon

Leona Marshall Libby and Louis J. Pandolphi

ABSTRACT

We have obtained evidence that trees store the record of climate in their rings. In each ring, the ratios of the stable isotopes of hydrogen and oxygen vary in proportion to the air temperature when the ring was formed, because the isotopic composition of rain and atmospheric CO_2 varies with temperature. In this paper the stable isotope variations of hydrogen and oxygen in a Japanese cedar have been correlated with the secular variations of radiocarbon measured in bristlecone pines by Suess. We have found significant negative correlations for both isotope ratios over the last 1800 years. The inference is that the small scale (~1%) variations in ^{14}C concentration in tree rings are related to climate variations.

In our data we have found periodicities of 58, 68, 90, 96, 154, 174, 204, and 272 years. Because our samples were averaged over 5 years each, we were not able to detect the 21-year sun spot cycle in the present data. The Suess samples, averaged over about 25 years each, reveal a periodicity of 183 years in agreement with our periodicity of 174 years.

Principle

Trees subsist largely on atmospheric carbon dioxide and water. The isotopic compositions of atmospheric carbon dioxide, rain, and snow are known to vary in a manner strongly dependent on atmospheric temperature (Craig 1963, Environmental Isotope Data 1969-1975). A nonnegligible and temperature-dependent fractionation of isotopes in the formation of wood has been computed, assuming thermodynamic equilibrium, but it is small in comparison to that found in rain water (Libby 1972). Therefore a tree ring emerging from the outer, living layer of sapwood should represent atmospheric temperature at the particular time of emergence, except for possible isotope exchange of the heartwood with water. Isotope exchange between intact heartwood and water does not appear to occur, since it has been shown, using water-soluble, bio-organic dyes, that in many tree species water conduction remains limited to the outermost annual ring (Huber 1935). In addition it has been shown that radiocarbon sugars and radiocarbon dioxide, when injected into or ingested by a living tree, do not exchange with neighboring rings (Berger 1970), (Kigoshi and Endo 1961). Furthermore when *Sequoia gigantea* wood is soaked for 37 days in air saturated at 100% humidity with water which has been previously labeled to $\delta D_{SMOW} = 1170°/oo$, we have detected no deuterium exchange even though water saturation is optically evident (Libby et al. 1976).

Procedure

By milling a radial groove across a slice of tree, all the sawdust was collected from five or six rings at a time. The sample was thoroughly mixed, and about 5 mg were removed for each measurement. After treating $HgCl_2$ to produce CO_2 and HCl (Rittenburg and Pontecorvo 1956), the HCl was removed with quinoline and the O^{18}/O^{16} ratio was measured in a mass spectrometer. For measurement of the deuterium, the wood was burned to water and CO_2, the water was reacted with metallic uranium to produce hydrogen, and the D/H ratio was measured in a hydrogen mass spectrometer. The $\delta^{18}O$ measurements were based on the PDB standard and the δD measurements were based on the mean ocean vector (SMOW) standard, where

$$\delta O^{18} \equiv \{[(O^{18}/O^{16}) \text{ sample} - (O^{18}/O^{16}) \text{ std}] / (O^{18}/O^{16}) \text{ std}\} \times 10^3 \text{ °/oo}$$

$$\delta D = \{[(D/H) \text{ sample} - (D/H) \text{ std}] / (D/H) \text{ std}\} \times 10^3 \text{ °/oo}$$

Results

To test the method, the variations of stable isotope ratios in tree rings must be compared with temperature records from mercury thermometers. Only European trees are useful in the calibration and testing stages of this method of determining temperature histories, since only in Europe do mercury temperature records extend over two centuries (U.S. Department of Commerce 1966). We have compared O^{18}/O^{16} and D/H measurements for a German oak, *Quercus petraea*, the rings of which were counted by Huber and Siebenlist (1969), with temperature records from England (Manley 1959, 1961) back to the time of the thermometer's invention in the late seventeenth century. The temperature coefficients have been evaluated by least squares analysis (Libby and Pandolfi 1973, 1974). Also we have compared O^{18}/O^{16} for a Bavarian fir, *Abies alba*, the rings of which were counted by Becker et al. (1970) with temperature records made near where the tree grew (U.S.D.C. 1966) (see Libby et al. 1976). We have found that stable isotope ratios in these trees vary as does the local air temperature.

Discussion of General Applicability

Since the method would seem to have worked for European oak and fir, it would seem likely to apply wherever trees ingest rain water. Direct measurements have been made of the isotopic composition of atmospheric moisture (as rain and snow), and indirect measurements have been made of that of carbon dioxide, the O^{18} of which is probably in isotopic equilibrium with atmospheric moisture (Cohn and Urey 1938). These measurements are available in the publications of the International Atomic Energy Agency (IAEA 1969, 1970, 1971, 1973, 1975). They vary depending on air temperature, being larger during summer precipitation and smaller during winter precipitation. Isotopic composition of precipitation varies similarly with temperature in many places on the earth (Dansgaard 1964). It would seem that wherever trees depend on precipitation, they could be expected to store a record of temperatures in their rings.

Application to Old Europe

We had the temerity to measure the oxygen and hydrogen isotopic ratios and to infer the temperature in German oaks for that time period before thermometers were invented. These measurements indicated warm intervals at A.D. 1530, 1580, and 1650 and a very cold period at about A.D. 1700. These inferences agree with Bergthorsson's deductions of climate variations in Iceland (1962) (Libby et al. 1976).

Application to Japan

We measured a long-lived Japanese cedar, *Cryptomeria Japonica* or *Yaku-Sugi*, for both hydrogen and oxygen isotope ratios (fig. 1a). Carbon-14 in the rings of this tree had been measured by Kigoshi (1965). In Libby et al. (1976) its isotopic ratios of hydrogen and oxygen compared favorably with temperatures deduced from such evidence as blooming time of cherry trees and number of days per year of frozen lakes recorded in diaries. Such records have been evaluated for Japan (Yamamoto 1971) and for China (Chu 1973). We compared ratios in the modern part of the cedar (figs. 2a, 2b) with temperatures recorded at a nearby weather station, Miyazaki (Japan Meterological Agency 1975), and found good correlation of the respective variations. One should question whether the tree ring ages are correct in this tree (Kigoshi 1965), and whether the surrogate evidences given for comparison in the Yamamoto and Chu temperatures are either pertinent or correct. It may be possible to find other evidence for the past temperature changes indicated by the isotopic variations in the cedar. This would permit further indirect checks of the tree thermometer method, as for example, by records of the prices of wheat and other commodities.

We find by least squares analysis that $\delta D = 8.0\ \delta O^{18} + $ const. for the cedar, showing a slope of 8 which is consistent with that for worldwide rain water within experimental error (Dansgaard 1964). This would indicate that the tree is storing old atmospheric condensation (fig. 1).

We made a Fourier transform of the data in figure 1a to analyze for the power spectrum of periodicities. The results are shown in figures 3a, 3b, and in table 1. The same periods are found in both D/H and in O^{18}/O^{16} data within experimental error.

Because our samples integrate over a time period of approximately 5 years each, we

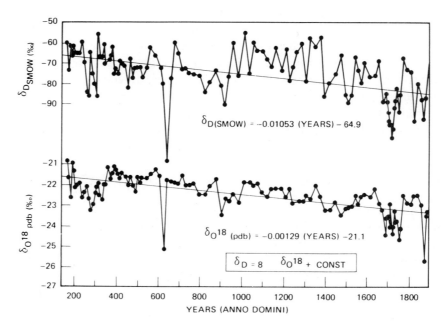

FIG. 1. The D/H ratio and the O^{18}/O^{16} ratio versus time. The dependence of δD on δO^{18} with a slope of 8.0, for the *Cryptomeria Japonica*.

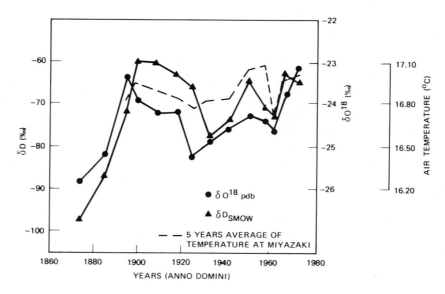

FIG. 2. Oxygen-18 and deuterium in modern *Cryptomeria Japonica* compared with average annual air temperatures measured at Miyazaki, Japan (Chu 1973). The Cryptomeria grew at 1350 meters in altitude, whereas Miyazaki is at sea level, about 150 miles northeast of Yaku Island. Averages over about eight years shown.

ISOTOPIC TREE THERMOMETERS 665

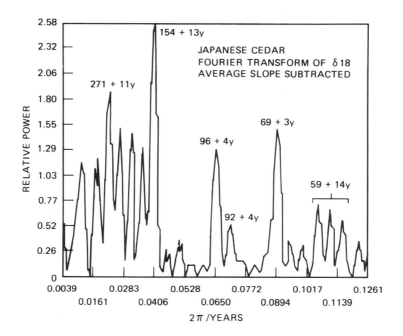

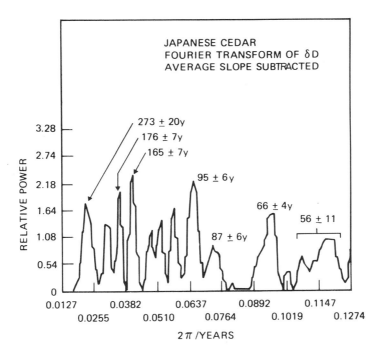

FIG. 3a.b. Power spectra obtained from Fourier transforms of the data in figure 1. The values of the periods so computed are not sensitive to various ways of truncating the data.

TABLE 1

Periods and their respective powers found in fourier transforms of
oxygen and hydrogen stable isotope data in Japanese cedar A.D. 140 - 1950

δ18		δD	
Period Years	Relative Power	Period Years	Relative Power
59 ± 14	0.67	56 ± 11	1.00
69 ± 3	1.50	66 ± 4	1.64
92 ± 2	0.52	87 ± 3	1.00
96 ± 2	1.29	95 ± 2	2.18
154 ± 6	2.58	153 ± 7	2.18
174 ± 7	1.29	174 ± 8	2.00
202 ± 7	1.45	205 ± 9	1.70
271 ± 11	1.80	273 ± 20	1.64

cannot expect to find evidence for periods of less than about 40 years in these data. This rules out the possibility of finding meaningful evidence for the 21-year sun spot cycle. There is some evidence for a 78-year solar cycle (Schove 1955). Three of the periods evidenced in the variations of the *Cryptomeria's* stable isotope ratios, namely 69 years, 92 years, and 96 years, when weighted with their respective powers, average to 83 years in agreement with Schove's 78-year period.

The sets of data span about 2000 years. We cannot expect to find evidence for periods of more than about 250 years, because the Fourier transform needs at least 8 or 10 cycles to yield meaningful results (Blackman and Tukey 1958).

Suess' bristlecone samples (Suess 1970), average over 25 years each, show a periodicity of 183 years in agreement with the period of 174 years in table 1 (fig. 4).

The absolute values of the deuterium ratios (ca. -70 SMOW) are about the same as in rain water. The absolute values of the oxygen ratios (ca. +20 SMOW) are halfway between that for rain (ca. -7 SMOW) and that for atmospheric CO_2 (ca. +40 SMOW). We conjecture, therefore, that carbon dioxide is in isotopic equilibrium with water both in the atmosphere and in the sap of the tree leaves.

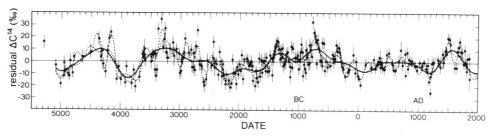

FIG. 4. Fluctuations in radiocarbon age about the chronological age measured by Suess (1970) in bristlecone pines.

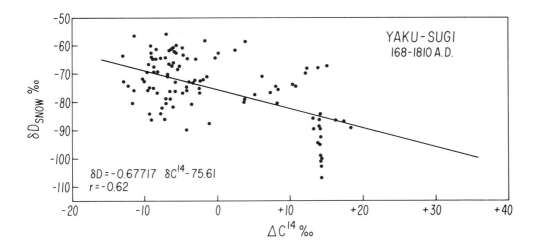

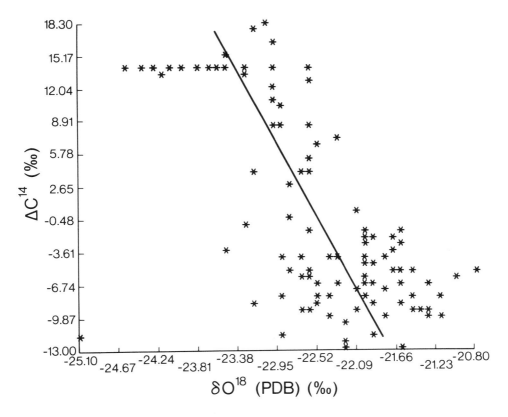

FIG. 5a.b. Correlation of fluctuations in radiocarbon age of bristlecone about the chronological age vs. a) δD and b) δO^{18} variations measured in Japanese cedar for 2000 years of tree rings.

We made least squares correlations of δD and δO^{18} measurements shown in figure 1 with the secular variations in ^{14}C concentration in the bristlecone pine sequence shown in figure 4, measured by Suess and Houtermans (Suess 1970). We found negative correlations for both sets of measurements, with correlation coefficients of 0.6 and 0.8, respectively (figs. 5a, b).

$$\delta D = -0.677 \delta^{14}C - 75.5 \text{ (SMOW)}; -0.62$$

$$\delta O^{18} = -0.0613 \delta^{14}C - 22.5 \text{ (PDB)}; -0.77$$

These correlations imply that the secular "wiggles" in radiocarbon age are related to variations of climate, a relationship adumbrated by Dansgaard et al. (1976). We plan to look for these anticorrelations in the isotope ratios of other trees. We plan to study trees felled by the advancing ice sheet at the Two Creeks site in Wisconsin 12,000 years B.P., expecting that the stable isotope ratios will indicate cooling as the glacier moved into the forest. It would seem likely, if the method is sound, that we could establish the history of climate back through many millennia. Of course further calibration with European trees will be necessary to substantiate the method.

ACKNOWLEDGMENTS

We are grateful to W. F. Libby for this guidance and advice regarding the chemistry. Our appreciation to V. Giertz-Sienbenlist, Bernd Becker, and K. Kigoshi, who supplied us with many of our wood samples and with dendrochronological dating, cannot be overemphasized. We deeply appreciate the laboratory assistance provided by John Marshall III, Sanford Ratner, Marion G. Weiler, and Patrick H. Payton, who designed and implemented our hydrogen chemistry and new vacuum line. Financial support was provided by the National Science Foundation.

REFERENCES

Becker, B., and V. G. Siebenlist
    1970   Flora. 159:310-346.
Berger, R.
    1970   Phil. Trans. Roy. Soc. London. Ser. A 269:23-26.
Bergthorsson, P.
    1962   Proc. Conf. on Climate 11th to 16th Centuries. Colorado. National Center for Atmospheric Research. Air Force Cambridge Research Laboratories.
Blackman, R. B., and J. W. Tukey
    1958   The Measurement of Power Spectra. New York, Dover Publications.
Chu, Ko-Chen
    1973   Scientica Sinica. 16 (2):226-250.
Cohn, M., and H. C. Urey
    1938   J.A.C.S. 60:679-687.

Craig, H.
    1963   Proc. Conf. on Nuclear Geology in Geothermal Areas. Spoleto. Consiglio Nazionalle delle Recherche. Italy.

Dansgaard, W.
    1964   Tellus. 15:436-468.

Dansgaard, W., S. J. Johnson, H. B. Clausen, and H. J. Langway
    1976   Nature. 227:482-483.

IAEA
    1969-75   Environmental Isotope Data Nos. 1-5. IAEA. Vienna.

Huber, B.
    1935   Berichte der Deutsche Botanischen Gesellschaft. 53:711-719.

Huber, B., and V. G. Siebenlist
    1969   Abteilung I. Östereichische Akademie der Wissenschaften. 178 (Band 1 bis 4): Heft 37-42.

Japan Meterological Agency
    1975   Air Temperatures for the World. Tokyo.

Kigoshi, K.
    1965   Proc. 6th Int. Conf. Radiocarbon and Tritium Dating. Pullman, Washington.

Kigoshi, K., and K. Endo
    1961   Chem. Soc. Japan Bull. 11:1738-1739.

Libby, L. M.
    1972   J. Geo. Res. 77:4310-4317.

Libby, L. M., and L. J. Pandolfi
    1973   Proc. Colloques Internationaux. Centre National de la Recherche Scientifique, Gif-sur-Yvette.
    1974   Proc. Nat. Acad. of Sciences. 71:2482-2486.

Libby, L. M., L. J. Pandolfi, P. M. Payton, J. Marshall III, B. Becker, and V. G. Siebenlist
    1976   Nature. 261:284-288.

Manley, G.
    1959   Archiv. Meterol. Geophys. Bioklimatol. Ser. V.A.K. 9:3-4.
    1961   Meterol. Mag. 90:303-310.

Rittenberg, D., and L. Pontecorvo
    1956   Int. J. Appl. Radiat. Isotop. 1:208-214.

Schove, D. J.
    1955   J. Geo. Res. 60:127-145.

Suess, H. E.
    1970   Radiocarbon Variations and Absolute Chronology. Proc. 12th Nobel Symp. ed. Olsson. Stockholm, Almquist and Wiksell, 595-605.

U. S. Department of Commerce
    1966   World Weather Records. 2. Europe. Environmental Science Services Administration, Environmental Data Services, Washington, D.C.

Yamamoto, T.
    1971   Geophys. Mag. 35 (2):187-206.

2

Causal Mechanisms in Climatic and Weather Changes as Revealed from Paleomagnetic Investigations of Samples Dated by the ^{14}C Method and from Geomagnetic Variations

Václav Bucha

The analyses of various types of geomagnetic and atmospheric manifestations have disclosed certain associations. The agreement in the occurrence of increased spectral densities as regards geomagnetic activity and the variations of atmospheric pressure over the geomagnetic pole proves the relationship between their periodicities. The results imply that changes in the intensity of corpuscular radiation, indicated by geomagnetic activity, affect the pressure patterns over the geomagnetic pole and polar region significantly, so that a pronounced modification of the general circulation may take place (Bucha 1976a, 1976b).

It is known that the last interglacial period began relatively suddenly approximately 10,500 years ago, and that it relieved the last glacial period in Europe and North America. The causes of this rapid and marked change which was accompanied by a considerable increase in temperature, particularly in Europe, have not yet been satisfactorily explained.

Paleomagnetic research in recent years has shown (Bucha 1976a) that the geomagnetic poles have displayed marked oscillations in the past; although the north geomagnetic pole was usually located in the vicinity of the geographic pole, it has been found that it deviated considerably from its equilibrium position during a certain period. Our latest paleomagnetic investigations of lake sediments dated by the ^{14}C method, of travertines and continental sediments (loesses and soils) covering the period of the last 40,000 years, have shown (fig. 1) that the north geomagnetic pole was located at lower geographic latitudes (about 60°N) from four to ten thousand years ago (fig. 2).

Radiocarbon dating of profile samples from Vracov lake was carried out at the ^{14}C laboratory, Berlin-Adlershof. The results are given in table 1 (Rybnickova and Rybnicek 1972). Marked deviations of magnetic directions were observed in the interval between

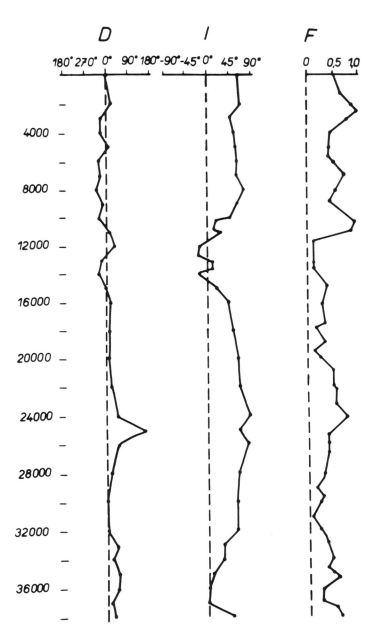

FIG. 1. Average variations of the magnetic declination, dip, and total intensity of the geomagnetic field over 38,000 years, investigated paleomagnetically for 9 parallel profiles of lake sediments and loesses.

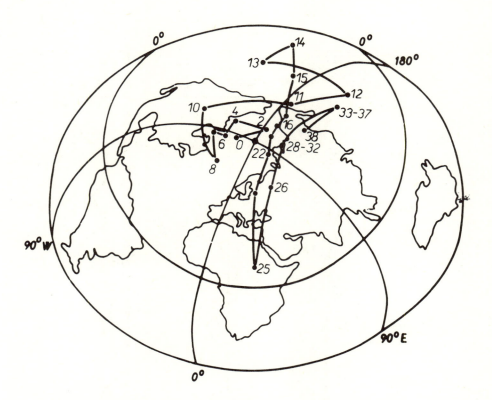

FIG. 2. Changes in the position of the virtual north geomagnetic pole over the last 38,000 years. Numbers indicate thousands of years.

twelve and fifteen thousand years ago, when the magnetic inclination in Central Europe decreased considerably into negative values of −20° and the virtual geomagnetic pole deviated into the region of the Pacific (fig. 2). Striking agreement was evidenced by comparing the sudden changes of climate and temperature in the interval twelve to ten thousand years ago (when the last period of glaciation terminated) with the marked changes in the positions of the geomagnetic pole from the Pacific to the North American continent (Bucha 1976a). However, it was first necessary to determine whether a connection exists between the positions of the geomagnetic pole and its role in forming the climate or weather, and to determine the function of the geomagnetic pole in the process of atmospheric circulation.

As regards the short-term changes, a unique dependence has existed between the C_i-indices (characterizing geomagnetic activity) and the last temperature variations (averages for the four winter months) in Prague during the last twenty-five years (fig. 3). After a sudden increase in geomagnetic activity, represented by the daily values of the K_p-indices, there has been a relatively sudden decrease of atmospheric pressure above the geomagnetic pole or its neighborhood at the 500 mb level, particularly during winter (fig. 4).

TABLE 1

Description of Layers of Profil Vracov, DU-1-B, (3 simplified)

Depth cm	Color	Lim. sup.	Structure	Lab No.	cm	Age
0-22	Reddish-brown	0	Stratum confusum			
22-28	Light reddish-brown	1	Moostorf, Carex-Torf			
28-40	Light reddish-brown	1	Carex-Torf, Moostorf			
40-50	Grey-brown	2	Phragmites-Carex-Torf, Moostorf			
50-54	Light grey	3	Phragmites-Carex-Torf			
54-56	Dark reddish-brown	2	Phragmites-Torf mit Ton			
56-61	Very dark brown	1	Moostorf, Phragmites-Torf			
61-69	Brown	1	Grobdetritusgyttja, Phragmites-Torf			
69-114	Dark brown	1	Moostorf			
114-122	Very dark brown	1	Grobdetritusgyttja, Moostorf			
122-147	Reddish brown	1	Grobdetritusgyttja	Bln 1009	140-145	3,430±100 B.P.
147-185	Grey brown	2	Carex-Phragmites-Torf, krümelig	Bln 1008	170-175	8,250±100 B.P.
185-200	Brown	1	Gyttja			
200-212	Green-Grey	3	Carex-Phragmites-Torf, krümelig	Bln 1007	205-212	6,495±100 B.P.
212-264			Gyttja	Bln 1006	225-232	10,765±140 B.P.
264-269	Light grey	3	Ton, Gyttja	Bln 1005	257-259	11,995±140 B.P.
269	Light grey	2	Ton, Sand	Bln 1004	264-269	11,993±250 B.P.

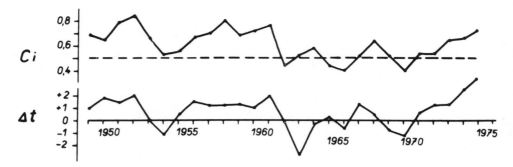

FIG. 3. Correlation of geomagnetic activity (C_i-indices) and the deviaitons of the temperature from the standard in Prague for the months of November-February (mean values) during the years 1949-1975.

Graphs representing gross agricultural production in some countries (FAO 1947-1974) (figs. 5-8) during years when an increased level of geomagnetic activity was recorded in the month of May (Bucha 1976b) (representing the main critical period for the growth of corn and other products in Central Europe [CSSR, FRG] and Canada) or in summer (April — July in the USSR), show a higher average gross agricultural production (e.g., in 1956-60, 1967-68, 1973-74). On the other hand, low geomagnetic activity was accompanied by a pronounced decrease in production (e.g., in 1954, 1962-65, 1970, 1972; figs. 5-8). The reasons for the decrease may vary for the individual regions of Europe and North America; for example, the decrease in agricultural output in SE North America in 1954 and 1972 was apparently due to excessive precipitation and floods, whereas in Central Europe (in Czechoslovakia, West Germany, and the USSR as well as in Canada) it may have been the result of drought.

It is not easy to explain fully the processes taking place above the geomagnetic pole where, as geomagnetic activity indicates, the kinetic energy of corpuscular radiation is apparently transformed into thermal energy which results in a decrease of atmospheric pressure. Several alternative explanations may be considered, with the most probable one being founded on the processes taking place in the magnetosphere.

As a result of the configuration of the sun's magnetic field, varying conditions are created in the course of solar cycles, allowing solar plasma to escape into interplanetary space. The plasma is observed to escape in the form of solar wind, sometimes at regular intervals with periods of 27, 13.5, 9, or 6.7 days, causing recurrent geomagnetic storms (Bucha 1976b).

The interplanetary magnetic field and the solar wind may increase the global instability of the magnetosphere. A magnetic storm is generated by a rapid sequence of magnetospheric substorms. If the interplanetary field is acting in the southernly direction for any length of time, a "neutral line of force" is formed in the plasma layer, along which the magnetic field is practically equal to zero. This is responsible for a highly unstable configuration of the plasma and field which changes abruptly to create a more stable

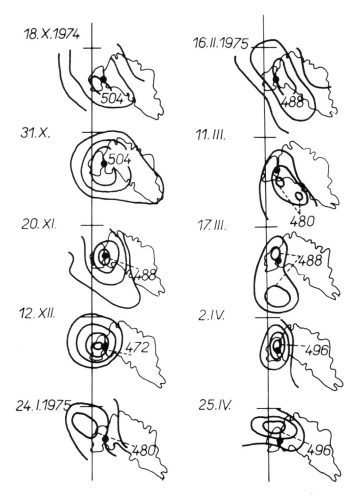

FIG. 4. Distribution of the cyclone over the geomagnetic pole (·) at the 500 mb level for 10 situations forming 3−6 days after a sudden marked enhancement of geomagnetic activity.

state. In this explosive process which, according to Roederer (1975), lasts for only 10-20 minutes, the accumulated energy of the magnetic field changes into the kinetic energy of the plasma. The plasma particles, which are ejected from the region of explosion, are accelerated in their motion towards the earth to a velocity of nearly 1000 km/s.

The particles diffuse toward the earth and penetrate along the magnetic lines of force into the upper atmosphere. Electric currents are generated by the gyration, oscillation, and drift motion of protons at energies on the order of $1 - 50$ keV, which share a significant portion of energy associated with the magnetospheric substorm (Akasofu 1968). The resulting electric field in the magnetosphere is communicated to the ionosphere and generates electric currents there. A concentrated electric current of intensity

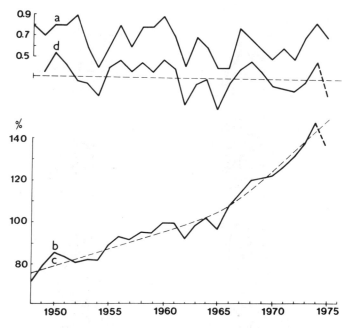

FIG. 5. Dependence of gross agricultural production in Czechoslovakia on geomagnetic activity. Curve (a) geomagnetic activity in May; curve (b) actual gross agricultural production; curve (c) average gross agricultural production; curve (d) deviations of the gross agricultural production from the average.

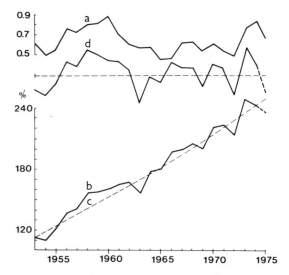

FIG. 6. Dependence of the gross agricultural production in the USSR on geomagnetic activity. Curve (a) geomagnetic activity as an average for the months of April to July; for significance of curves (b), (c), and (d), see figure 5.

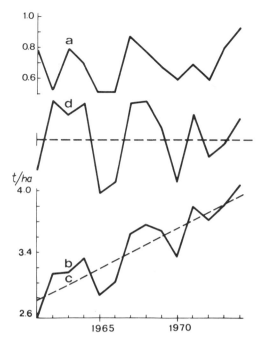

FIG. 7. Dependence of the average corn yields on geomagnetic activity in the FRG. Curve (a) geomagnetic activity in May; curve (b) actual corn yields in tons; curve (c) average corn yields; curve (d) deviations of the actual from the average corn yields.

up to 10^6 amperes, called the auroral electrojet, is induced along the auroral oval and causes a heating of the polar upper atmosphere and subsequent motion of atmospheric gas, particularly toward the center of the polar cap represented by the geomagnetic pole. According to our conclusions (Bucha 1976a, 1976b) this can contribute to the formation of the central cyclone mainly at the 500 mb level in the center of this region, that is, above the geomagnetic pole.

The function of the geomagnetic pole represents only a certain triggering mechanism in the whole process, in the course of which the warm masses of air coming from the subtropical regions, from the Gulf of Mexico along the east coast of the North American continent to the N, are channeled to the geomagnetic pole according to prevailing pressure conditions. We hope to point out several new associations which may serve to verify the existence of this proposed mechanism affecting the change of climate and weather.

It follows that changes in the position of the geomagnetic poles in the course of the past millennia could have been responsible for a shift of the whole region of low pressure (increase in cyclonal activity), the distribution of which for the present period and the winter is schematically shown in figure 9a.

With a view to the above conclusions, this means that the region of low pressure, in contrast to its distribution at present (fig. 9a), was displaced some 15-20° to the S five to

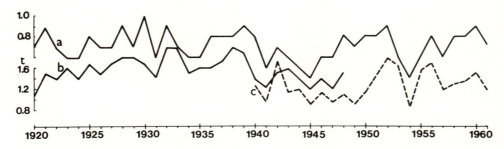

FIG. 8. Dependence of the average corn yields on geomagnetic activity in Czechoslovakia and Canada. Curve (a) geomagnetic activity in May; curve (b) yield in tons in Czechoslovakia; curve (c) the yield in Canada.

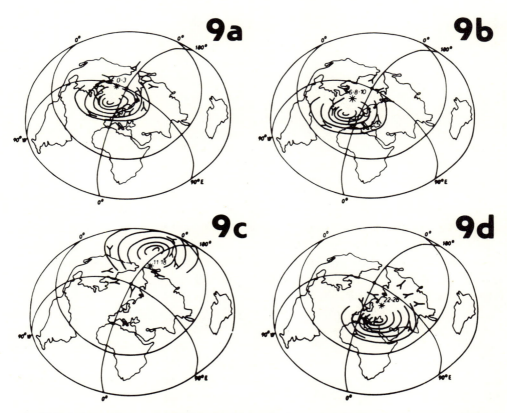

FIG. 9. Schematic expression of the low pressure region over the geomagnetic pole. * denotes the approximate mean position of the virtual geomagnetic pole, investigated paleomagnetically; (a) 0-3 for the last 3,000 years; (b) 6-8-10 for the interval between 5000 and 10,000 years ago; (c) 11-18 for the interval between 11,000 and 18,000 years ago; (d) 22-28 for the interval between 22,000 and 28,000 years ago). Large deviations from the mean position, marked *, did occur during these intervals. The arrows indicate prevailing wind directions.

ten thousand years ago, and in certain periods closer to Europe (fig. 9b) where there was a warmer climate than at present due to the steadier inflow of warm air from the SW. The influence of low pressure also affected the territories of North Africa and Asia Minor; resulting in a much higher proportion of precipitate days in these areas than applies now. After the low pressure area had moved to the N, North Africa became a desert.

In contrast, a relatively sudden change in the development of the climate took place in Europe and North America twelve to ten thousand years ago, for which period we found considerable change in the position of the north geomagnetic pole which moved from the Pacific (SW of Japan) to North America (fig. 2); the climate changed from the cold of the terminating glacial period of twelve thousand years ago to a very warm interglacial period. The data on the positions of the magnetic poles were determined by paleomagnetic investigation of sedimentary rocks, dated by the ^{14}C method. This change in climate did not affect northeastern Asia and Alaska; according to some investigations it was warmer there twelve thousand years ago than it is now.

As can be seen from the schematic illustration in figure 9c, the low pressure area was probably located in the Pacific at this time (eleven to eighteen thousand years ago) where the zonal flow was markedly intensified. On the other hand, Europe and North America, especially its eastern part, were under the influence of the cold northern flow. This could explain why these two continents were subject to glaciation for such a long period. The "little ice age", probably worldwide in character, seems to have been caused by the decreased intensity of solar wind (geomagnetic activity).

Arrhenius (1952) showed that changes in carbonate content due to different populations of foraminifera reflect the changes of world climate. In the Pacific the carbonate content increased in the glacial periods; in the North Atlantic it decreased. This conclusion can be interpreted much more naturally by saying that when glacial periods occurred in Europe and the North Atlantic, interglacial periods were taking place in the Pacific and vice versa. This interpretation of climatic changes does not require an explanation of the large changes in the intensity of solar energy during the Quaternary, which would be necessary if we were to explain the occurrence of significant warming and cooling simultaneously on the earth as a whole. According to our interpretation, the cooling occurs alternately at different places, either in the Atlantic and the adjacent territories, or in the Pacific and Asia. Long cold periods can be explained by predominating cold advection of the air masses into Europe and the eastern part of North America at times when the geomagnetic pole, and the predominating cyclone with it, were located in the Pacific. The interglacial periods, characterized by warming, on a long-range average as much as 15°C, occurred in Europe when W or SW flow predominated, and when the geomagnetic pole and the cyclones were located in the Atlantic and on adjacent territories, that is, in the European and eastern American parts of the Northern Hemisphere.

The main features of the development of atmospheric circulation in the region of the Atlantic display certain typical regularities. As synoptic maps show, tropical cyclones are mostly generated in the equatorial regions between Africa and Central America. They move to the W, and in the Caribbean and the Gulf of Mexico they turn towards the NE or N. They usually proceed along the Atlantic coast of North America (fig. 10) (Inst. Cubano

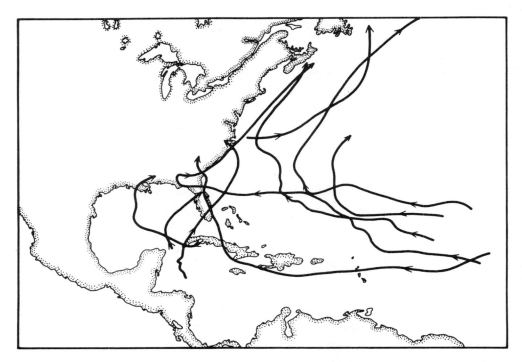

FIG. 10. Tracks of tropical cyclones and hurricanes in the Atlantic and the Caribbean in 1964 (8).

del Libro 1973) where they are capable of causing severe damage, including fatal casualties. At higher geographic latitudes the tropical cyclones usually change to extratropical cyclonal formations, emcompassing much larger areas; the flow velocity within them then decreases considerably.

The synoptic maps also indicate that the Caribbean represents the main source for the quasi-zonal Atlantic-European frontal zone, at first tending to the NE and then to the E, which reaches Europe only under certain conditions and with a different intensity. In considering some of the synoptic situations in the Atlantic in connection with differing values of geomagnetic activity, we find that marked differences occur in the nature of atmospheric circulation if we compare its development at the time of very low to zero geomagnetic activity with the distribution and trends of the pressure formations at the time of high geomagnetic activity.

In the Caribbean and the Gulf of Mexico, tropical cyclones are generated more frequently in summer as a result of the larger temperature instability of the air and the associated release of condensation heat. As shown in figures 11 and 12, *under low geomagnetic activity*, high pressure may be generated over the geomagnetic pole and Greenland, which, together with an extensive high pressure area over the Atlantic and possibly also over North America, will hinder the penetration of cyclonal formations from the Caribbean across the E of North America to the NE as well as to the E. As a result of this, the motion of the cyclonal formation from the Caribbean to the NE is

FIG. 11. Synoptic situations in the Northern Hemisphere between 19 and 25 November 1965. The very low geomagnetic activity corresponds to the pronounced high pressure over the geomagnetic pole (500 mb level top, sea level bottom).

FIG. 12. Synoptic situations between 18 and 24 June 1972, (low geomagnetic activity), representing meridional flow (500 mb level top, sea level below). (A) indicates position of hurricane "Agnes."

relatively slow, particularly in comparison to the motion of extra-tropical cyclones under zonal flow. For example, at a time of low geomagnetic activity over the Gulf of Mexico on 18 June 1972, the tropical cyclone (hurricane) "Agnes" was generated; this cyclone caused the largest damage recorded ever on the territory of North America. In the course of its very slow process to the NE (fig. 12) it devastated the eastern regions of North America; in its final phase (around 24 June 1972) it changed into an extratropical cyclone, which gradually decayed at higher geographic latitudes.

The low pressure region over the Gulf of Mexico and over North America was blocked by high pressure over the Atlantic and Greenland on its way to the NE. This resulted in extensive floods and damage to agricultural production due to excessive precipitation in the southeastern part of North America, apart from the cyclone mentioned above.

On the other hand in Central and Eastern Europe, into which the zonal flow had not penetrated as a result of the generation of blocking highs above the Atlantic and Greenland (fig. 12) high pressure areas were generated. These were associated with long-lasting droughts in the summers of 1970, 1972, and 1976, which were the cause of decreased yields. The same thing had happened in some previous years and seems also to be valid for Canada (figs. 5-8).

Under increased geomagnetic activity of longer duration, tropical cyclones from the Caribbean usually penetrate along the eastern coast of North America to the NNE and, as a result of the low pressure in the vicinity of the geomagnetic pole, they reach this region over northwestern Greenland without difficulty. Here the direction of flow changes and penetrates across Greenland, or along its southern coast to the E and over Iceland. If the increased level of geomagnetic activity is more permanent, a series of cyclones is generated and these push the Azores high to the S. The low pressure region is then distributed over most of the Atlantic, England, and Scandinavia. This opens a direct route to the E into Central and Southern Europe for the cyclones from the Caribbean and the east coast of North America. In Europe, this results in increased precipitation and temperatures in winter (e.g., 1973-74, fig. 13). These cyclones move at low geographic latitudes (between 30° and 45° N) toward Spain and sometimes, depending on their energy and the conditions over Europe, as far as Eastern Europe. Increased zonal atmospheric circulation seems to be valid for the whole northern hemisphere including the Pacific and the zonal flow, which increased precipitation across North America and Canada as well.

Model of Atmospheric Circulation and of its Changes in the Northern Hemisphere in Relation to Geomagnetic Activity

The earth and its atmosphere are heated by solar radiation, and they radiate approximately the same amount of heat into space. The equatorial regions accept more heat than they release, whereas in the polar regions the opposite holds true. For this reason, heat is gradually transferred from subtropical and tropical regions toward the poles, thereby creating the motion of the atmosphere called "general circulation." The principal atmospheric variables defining the state of the atmosphere are temperature, pressure, flow, humidity, cloud cover, and precipitation. If we consider worldwide circulation and if we

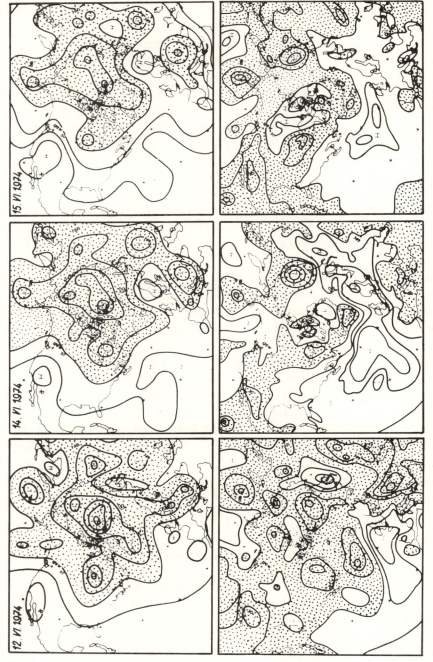

FIG. 13. Synoptic situations between 12 and 15 June 1974 (increased geomagnetic activity) (500 mb level top, sea level bottom), representing zonal flow.

include the effects of the oceans, continents, and mountain ranges, we should arrive at a fundamental pattern of the climatic conditions on the earth and the mechanism governing those processes in the atmosphere, implied, for example, by the dominating pattern of pressure formations in winter and summer. However, this pattern is frequently disrupted: the zonal flow changes to meridional flow, sudden penetration of Arctic air into Europe occurs, and the weather changes appreciably. A tentative model should help to explain the principal regularities in the changes of weather as well as the pronounced deviations in atmospheric circulation.

In the Atlantic the extensive Azores high predominates with its center over the Azores and temporarily over Europe and North America (Reil 1963) (fig. 14). In some cases the Azores high will unite with the high over Greenland to form a mighty ridge of high pressure over the whole Atlantic, which contributes to meridional flow. At other times it is pushed to the S by the low pressure centered in NE Canada and Greenland. Zonal flow from W to E is then intensified between these formations.

In summarizing our results, we will attempt to explain why the meridional flow changes to zonal (the former being responsible for the penetration of Arctic air into Europe following progressing cyclonal formations) and to explain the mechanism active in generating various types of atmospheric circulation in this region (Bucha 1976b).

Masses of air flow from the region of the Azores anticyclone and progress, together with an easterly flow of warm air, from Africa to the W and the Caribbean (fig. 14). This process also occurs along the paths of tropical cyclones and hurricanes which initially proceed due W (fig. 10) then change their course abruptly to the N and NE in the Caribbean.

FIG. 14. Worldwide distribution of sea level pressure and wind in July (9).

During transport of the heated air masses along the eastern coast of North America to the N, a heterogeneous pattern of pressure formations occurs in extratropical latitudes. For the sake of simplification we shall first consider the extremes of high and low geomagnetic activity and the corresponding pressure distributions.

When the level of geomagnetic activity is very low, a high pressure area (figs. 11 and 12) which merges temporarily with the Azores anticyclone occurs more frequently over the geomagnetic pole (fig. 15a). This creates a ridge of high pressure which prevents the flow's penetration from the coast of North America toward Europe. The masses of air proceed slowly along the eastern coast of North America toward the pole; if there is a pronounced high over the geomagnetic pole (under low geomagnetic activity), they may be deflected back into Europe and North America (penetration of Arctic air), or they

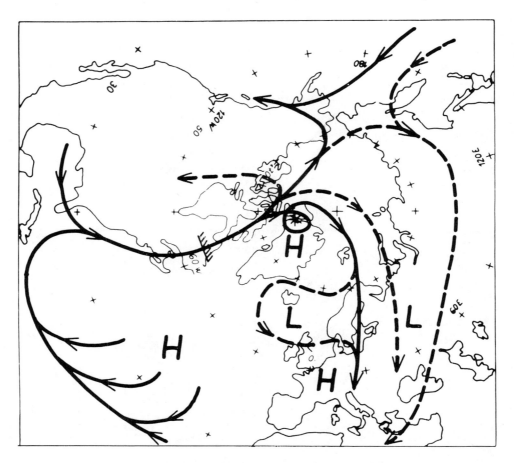

FIG. 15a. Schematic model representing the predominant air flow at the time of low geomagnetic activity in the Northern Hemisphere. The zonal flow penetrates from the region of the Azores anticyclone to the west into the Caribbean and from here to the north along the east coast of North America. Rapid progress is occasionally hindered by a high above eastern Canada (E). The air masses pass the North Pole and proceed to eastern Siberia or are deflected to the south (if there is a high above the geomagnetic pole) in the form of Arctic air penetrating into Europe and North America.

may penetrate slowly over NE Asia. This results in a drought or minimal precipitation in Europe; it creates unfavorable conditions for vegetation and is responsible for relatively cold winters. A schematic model of atmospheric circulation showing the predominant flow under conditions of low geomagnetic activity is presented in figure 15a.

During a prolonged increase in the level of geomagnetic activity, a low pressure area is generated over the geomagnetic pole, probably as a result of the processes previously mentioned; this low may persist over the region for a considerable length of time (fig. 13). It gradually increases and the atmospheric circulation intensifies, which results in the Azores high being pushed from N to S. The air flow near the geomagnetic pole around this central cyclone progresses (fig. 15b) and gradually extends further to the S into the Atlantic and over Europe until direct westerly flow is established between the eastern

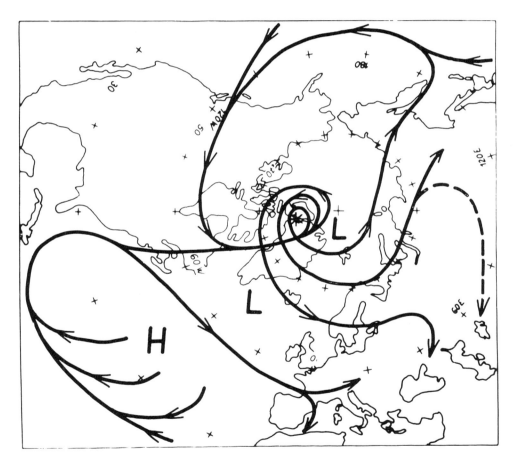

FIG. 15b. Schematic model of the air flow at the time of high geomagnetic activity in the Northern Hemisphere. As in figure 15a the circulation progresses from the Caribbean to the N and then into the region of the low over the geomagnetic pole. Here, the air stream is deflected to the SE. If enhanced geomagnetic activity persists, a low pressure area is maintained over the geomagnetic pole which gradually pushes the Azores high to the S until direct zonal flow is established from the east coast of North America into Europe. (*) represents the geomagnetic pole.

coast of North America and Europe, typical for zonal conditions. This results in mild winters, increased precipitation, rainy springs and summers, and favorable vegetative conditions. The schematic model of the intensified zonal atmospheric circulation at the time of increased geomagnetic activity is shown in figure 15b.

Many intermediate patterns of pressure formation also occur relative to geomagnetic activity and atmospheric circulation; moreover, the changes in geomagnetic activity are caused by extraterrestrial processes, in particular by the variation of the intensity of corpuscular radiation, whereas the changes in pressure are largely a reflection of general circulation of the earth's atmosphere connected directly with solar radiation. For this reason, a short increase in the level of geomagnetic activity may occur after a long period of low activity without markedly affecting the pressure pattern in the region of the geomagnetic pole (under conditions of a pronounced high) and vice versa.

Results of spectral analysis applied to a set of diurnal data for a four-year period (1962-65) (fig. 16), imply a relationship between the variations of geomagnetic activity (K_p-indices) and the variations of atmospheric pressure over the geomagnetic pole (at M_{P-O-PL} as well as at the 500 mb level, $M_{P-500-PL}$). As regards geomagnetic activity (M_{Kp}), figure 16 displays statistically significant spectral density periods of 27, 13.5, 9, and 6.7 days, with a maximum at 13.5 days (Bucha 1976b). A similar maximum is found on the spectral curve of atmospheric pressure vairations at sea level as well as at an altitude of 5 km. The subsidiary values of the increased spectral density in the vicinity of this maximum at 13.5 days suggest an occasional looser relationship between the variations of geomagnetic activity and pressure. The same is true of the periods of 27 days, which occur within the interval of 25.5 to 30.5 days with the 5 km pressure data. There is also an annual period which is particularly distinct for atmospehric pressure at higher levels.

The agreement in the occurrence of the increased spectral densities, particularly as regards the 13.5-day period, proves the relationship between the periodicity of the variations of geomagnetic activity and of atmospheric pressure, particularly at the time of solar minimum. The results in figure 16 imply that the changes in the intensity of corpuscular radiation indicated by geomagnetic activity significantly affect the pressure patterns over the geomagnetic pole and polar region, so that a pronounced modification of the general circulation may take place (figs. 15a, b). The correlation coefficient between the daily values of geomagnetic activity (K_p-indices) and the atmospheric pressure above the geomagnetic pole for several winter seasons has been found to be 0.6, which confirms the interrelationship. The idea of controlling the development and changes of weather is founded on these conclusions.

The function of the geomagnetic pole in the whole process of atmospheric circulation represents only a certain impulse in association with the generation of a cyclone over the geomagnetic pole (ohmic heating of air around the auroral oval at the time of increased geomagnetic activity), directing the spontaneous exchange of the air masses into a single point (the central cyclone) in which the incipient cyclonal activity continues to develop as a result of inflow of other air masses coming from the S (Bucha, in press).

Since this mechanism represents only a modulation of the existing processes, the energy required to generate it is considerably smaller than the energy of the existing

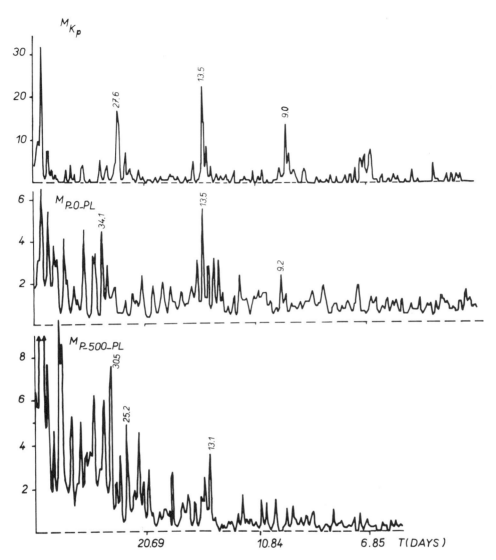

FIG. 16. Spectral function (maximum entropy) of geomagnetic activity (M_{Kp}) of atmospheric pressure over the geomagnetic pole at sea level (M_{P-O-PL}) and at the 500 mb level ($M_{P-500-PL}$) for the years 1962-1965. The most pronounced coinciding maximum (for a period of 13.5 days) and indicates the relationship between geomagnetic activity and atmospheric pressure over the geomagnetic pole.

atmospheric circulation as a whole (Bucha 1976a). The overall energy of the geomagnetic activity (and hence of solar wind particles) is 10^{-7} times smaller than the overall energy obtained by the earth from the sun, and over the polar cap it represents only a part concentrated into a single location.

These factors should be considered in proposing the control of weather. Provided it were possible to increase the temperature by 5-10°C and thereby to decrease the pressure over the geomagnetic pole it probably would be possible to create or strengthen the initial conditions for the intensification of cyclonal activity and zonal flow, which could contribute substantially to the improvement of climatic and environmental conditions on certain continents or their parts.

This outline of the possibilities of improving the climate and weather is only tentative. The procedures involved must be tested from other aspects and must be developed in detail with a view to the possible consequences of human intervention, considering means of creating conditions which would affect the weather only in the positive sense for peaceful purposes.

We have shown earlier (Bucha 1970) that the main deviations of the ^{14}C values from the standard have been caused by fluctuations of the geomagnetic moment during the past eight thousand years. The short term wiggles seem to have been related partly to changes in the intensity of the geomagnetic field (periods 60-2000 years) and to the variations of cosmic radiation (solar wind) indicated by geomagnetic activity.

REFERENCES

Akasofu, S.
    1968   Polar and Magnetospheric Substorms. Holland, D. Reidel.

Arrhenius, G.
    1952   Sediment Cores from the East Pacific. Swedish Deep-Sea Exped. (1947-48). 5:228.

Bucha, V.
    1970   Influence of the earth's Magnetic Field on Radiocarbon Dating. 12th Nobel Symp. Uppsala. Stockholm, Almquist and Wiksell.
    1976a   Variations of the Geomagnetic Field, the Climate and Weather. Studia geoph. et geod. 20:149.
    1976b   Changes in the Geomagnetic Field and Solar Wind: Causes of Changes of Climate and Atmospheric Circulation. Studia geoph. et geod.
    (in press)   Model of Changes in Atmospheric Circulation as Revealed from Geomagnetic Activity. Studia geoph. et geod.

FAO
    1947-74   Production Yearbooks.

Instituto Cubano del Libro
    1973   Trayectoria de huracanes y de perturbaciones ciclonica del oceano Atlantico, del mar Caribe, y del golfo de Mexico (1919-1969). Havana.

Reil, H.
    1963   Introduction to the Atmosphere. New York.

Roederer, J. G.
    1975   Planetary Plasmas and Fields. XVI. Gen. Assembly IUGG. Grenoble.

Rybníčková, E., and K. Rybníček
    1972   Erste Ergebnisse paläogeobotanischer Unterscuhungen des Moores bei Vracov. Südmähren. Folia Geobotanica et Phytotaxonomica. 7.

3

Sensitivity of Radiocarbon Fluctuations and Inventory to Geomagnetic and Reservoir Parameters

Robert S. Sternberg and Paul E. Damon

ABSTRACT

Long-term fluctuations in atmospheric $\Delta^{14}C$ were analyzed using the response of a two-box exchange model to a sinusoidal geomagnetic dipole moment. The model was used to calculate a theoretical value for the ^{14}C inventory and a ^{14}C fluctuation curve. Sensitivity of these calculations to values for seven independent geomagnetic and reservoir parameters was investigated. Calculated inventories were $\geqslant 122$ dpm/cm$_e^2$, somewhat greater than current estimates. There was no difficulty in producing model fluctuation curves with correct amplitudes, but there was a tendency for model curves to lag behind the data. Both the inventory and fluctuation curve analyses suggest that the maximum value of the geomagnetic dipole moment occurred at 2500 B.P. This is several hundred years earlier than is currently suggested by paleomagnetic data. To eliminate the lag problem, either the atmospheric residence time of a ^{14}C atom, or the ratio of total carbon in the sink reservoir to that in the ambient reservoir, must have a significantly higher value than those commonly used in the literature. Many plausible parameter sets are indicated by the analysis. The need for more paleomagnetic and radiocarbon data is indicated by the difficulty in predicting inventory and by the tendency for the model fluctuation curve and the data to diverge prior to 6000 B.P.

INTRODUCTION

It is now well established from measurements on tree rings, varves, and historical artifacts that there have been variations in atmospheric radiocarbon concentration over the last ten millennia (Olsson 1970). There are several a priori mechanisms that could cause the long-term (on the order of millennia) and short-term (decades) fluctuations (Grey 1969, Grey and Damon 1970, Lal and Venkatavaradan 1970, Lingenfelter and Ramaty 1970;

Suess 1970). Previous studies indicate that secular variation of the strength of the geomagnetic dipole moment is probably responsible for much of the long-term variation (Kigoshi and Hasegawa 1966, Bucha 1970, Damon 1970, Lal and Venkatavaradan 1970, Lingenfelter and Ramaty 1970, Suess 1970, Damon and Wallick 1972, Yang and Fairhall 1972). Damon (1970) found that if the maximum value of the dipole moment occurred near the beginning of the Christian era, as is generally suggested by paleomagnetic data (Bucha 1970, Kitazawa 1970), the theoretical atmospheric fluctuations curve predicted by a two-box model lagged behind the radiocarbon data by about 500 years. Damon suggested that four model parameters could be responsible for this discrepancy: the period of the assumed sinusoidal geomagnetic dipole moment, the total carbon content of the ambient reservoir, the exchange rate of radiocarbon from the atmosphere to the deep sea, and the activity of the deep sea at the reference year of A.D. 1890. Analysis at that time indicated that decreasing the value of any of these parameters might eliminate the lag problem. The purpose of this study is to investigate the sensitivity of a two-box exchange model to values for these and other parameters, and to see if the long-term radiocarbon trend curve can be modeled. We also investigated the ability of the model to evaluate the radiocarbon inventory.

Data

Figure 1 is a compilation of $\Delta^{14}C$ measurements on dendrochronologically dated tree ring samples covering the last 7500 years, excluding data since A.D. 1890. To determine the long-term trend of the data, least squares regressions were calculated for successively higher order degree polynomials. Data since A.D. 1890 were not included in these calculations since they are affected by dilution of ^{14}C by fossil fuel combustion, and more recent data are affected by bomb radiocarbon as well. After exclusion of these data, 670 points remain. Using the sequential F-test with a significance level of $\alpha = .05$, the fourth degree was the highest order polynomial term that was significant in the regression. Although some scatter of the data about the regression curve is evident in figure 1, the regression does account for 90% of the total variation of the data ($R^2 = .9$). The other 10% can be attributed to statistical scatter and to short-term radiocarbon fluctuations smoothed out by the regression.

For a multiple regression,

$$\frac{s(y)}{\sqrt{N}} < s(Y) < s(y) \qquad (1)$$

and

$$\bar{s}(Y) = \sqrt{\frac{k+1}{N}} s(y) \qquad (2)$$

(Daniel and Wood 1971) where $s(y)$ is the standard deviation of the data, $s(Y)$ is the standard error of estimate of the dependent variable of the regression, $\bar{s}(Y)$ is the average standard error of estimate, N is the number of data points, and k is the number of

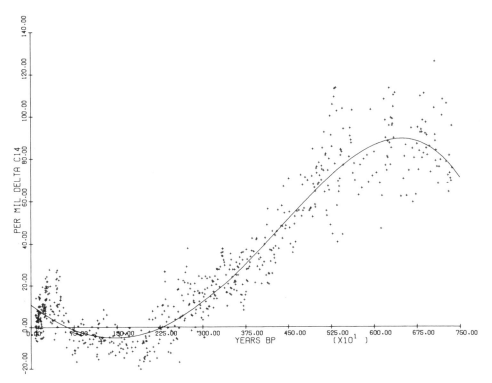

FIG. 1. $\Delta^{14}C$ (°/oo) vs. tree ring age in decades. Points represent data from six laboratories (Broecker and Olson 1959; Stuiver 1969; Damon *et al.* 1970; Lerman *et al.* 1970; Ralph and Michael 1970; data of Suess in Houtermans 1971; Damon *et al.* 1972). The curve is a fourth degree polynomial regression of $\Delta^{14}C$ on tree ring age.

independent variables (4 for a fourth degree polynomial). For the 670 data points,

$$.417°/oo < s(Y) < 10.8°/oo \text{ and } \bar{s}(Y) = .93°/oo.$$

The small values of the standard errors of estimate and the high value of R^2 indicate that the polynomial fit is a good indicator of the overall trend of the radiocarbon data. This polynomial fit curve will be used for comparison with model predictions.

The Model

The concept of box model studies involves dividing the earth into carbon reservoirs and modeling the input, exchange, and losses of the different reservoirs. Carbon is assumed to be randomly mixed within reservoirs. Depending upon the phenomenon investigated and the desired sophistication of a study, the number of reservoirs used will vary. Anywhere from one box (Grey and Damon 1970, Ralph 1972) to six boxes (Craig 1957, Plesset and Latter 1960, Bacastow and Keeling 1973, Ekdahl and Keeling 1973) have been used, although the majority of previous studies have involved two or three boxes (Arnold and Anderson 1957, Revelle and Suess 1957, Fergusson 1958, Wood and Libby 1964, Houtermans 1966, Kigoshi and Hasegawa 1966, Ramaty 1967, Damon 1970, Lingenfelter and Ramaty 1970, Yang and Fairhall 1972, Houtermans et al. 1973, Keeling 1973). For this investigation, a two-box model is adequate. For harmonic variation of radiocarbon production with periods on the order of 1000 years or more, a two-box model will give the same attenuation of atmospheric activity as a six-box model (Houtermans 1966, Ekdahl and Keeling 1973). Furthermore the nature of the sensitivity analyses to be discussed would be very cumbersome if a more complicated model were used.

Figure 2 is a schematic diagram of the two-box exchange model used. Radiocarbon is produced in the upper atmosphere with immediate injection into the ambient reservoir consisting of the atmosphere, biosphere, and mixed layer of the oceans. There is exchange between the ambient reservoir and the sink reservoir which consists of the deep oceans. Irreversible loss occurs through sedimentation as well as radioactive decay. All exchange rates (k_{as}, k_{sa}, k_{ds}) are assumed to be first order and linear. The inverse of an exchange rate k_{ij} is the mean residence time of an atom in reservoir i with respect to transfer to reservoir j. The exchange rates as well as N_a and N_s, the total carbon contents per square centimeter of the earth's surface of the ambient and sink reservoirs, respectively, are assumed constant with time. This is consistent with our neglecting of climatic effects in order to investigate geomagnetic effects, and with Damon's (1970) finding that total

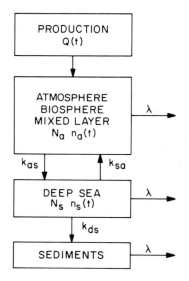

FIG. 2. Radiocarbon two-box exchange model. Radiocarbon produced at rate Q(t) is instantaneously injected into the ambient reservoir. Two-way exchange occurs between the ambient and sink reservoirs. N_i is the total carbon content of reservoir i; $n_i(t)$ is the ^{14}C content; k_{ij} is the exchange rate (inverse mean residence time) from reservoir i to j; λ is the ^{14}C decay constant.

climatic effects on radiocarbon fluctuations have been minimal for the last 8000 years. The number of ^{14}C atoms/cm$_e^2$ in the ambient and sink reservoirs, $n_a(t)$ and $n_s(t)$, vary in response to the time dependent production rate $Q(t)$, the forcing function for the model.

The two-box exchange model of figure 2 can be expressed mathematically by the following pair of differential equations:

$$\frac{d n_a(t)}{dt} = Q(t) - k_{as} n_a(t) + k_{sa} n_s(t) - \lambda n_a(t) \tag{3}$$

$$\frac{d n_s(t)}{dt} = k_{as} n_a(t) - k_{sa} n_s(t) - k_{ds} n_s(t) - \lambda n_s(t). \tag{4}$$

Since activity in a reservoir $A_i = \lambda n_i/N_i$, equation (3) can be multiplied by λ/N_a to obtain

$$\frac{d\left(\frac{\lambda n_a(t)}{N_a}\right)}{dt} = \frac{\lambda Q(t)}{N_a} - k_{as} \frac{\lambda n_a(t)}{N_a} + \frac{k_{sa}}{N_a} \lambda n_s(t) - \frac{\lambda^2 n_a(t)}{N_a} \tag{5}$$

$$\frac{d A_a(t)}{dt} = \frac{\lambda Q(t)}{N_a} - k_{as} A_a(t) + k_{sa} \frac{N_s}{N_a} \frac{\lambda n_s(t)}{N_s} - \lambda A_a(t) \tag{6}$$

$$= \frac{\lambda Q(t)}{N_a} - k_{as} A_a(t) + k_{sa} \frac{N_s}{N_a} A_s(t) - \lambda A_a(t). \tag{7}$$

Similarly,

$$\frac{d A_s(t)}{dt} = k_{as} \frac{N_a}{N_s} A_a(t) - k_{sa} A_s(t) - k_{ds} A_s(t) - \lambda A_s(t). \tag{8}$$

The function $Q(t)$ must be defined to solve equations (7) and (8). Elsasser et al. (1956) derived a relationship (hereafter referred to as Elsasser's relation) between the radiocarbon production rate $Q(t)$ and the strength of the geomagnetic dipole moment M, whereby $Q(t) \propto M(t)^{-\alpha}$, $\alpha = 0.52$. Using somewhat different approaches, Ramaty (1967) and Wada and Inoue (1966) derived similar relationships with slightly different values of α. The dipole moment $M(t)$ is assumed to vary sinusoidally with time. Compilations of paleomagnetic data from around the world (Cox 1968, Bucha 1970, Kitazawa 1970) are suggestive of a quasi-sinusoidal variation over the last 10,000 years (fig. 3). Thus,

$$M(t) = M_o + M_1 \sin \omega (T - t + \phi), \tag{9}$$

where M_o is the average value of the dipole moment, M_1 is the amplitude of the variation, T is the period of the oscillation, ϕ is a phase angle, and $\omega = 2\pi/T$ is the angular

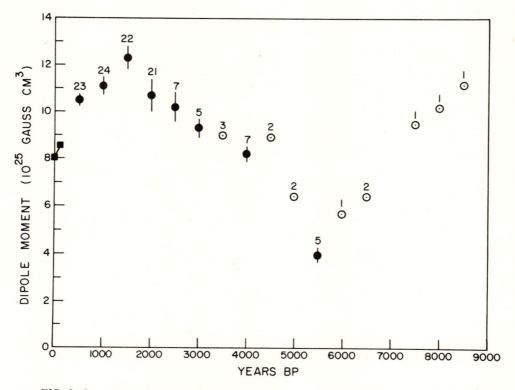

FIG. 3. Strength of the geomagnetic dipole moment vs. years B.P. (after Cox 1968). Variation during the last 130 years is known from observatory data. Other data is paleomagnetic (Smith 1967a, 1967b, 1967c). Averaging over 500 year intervals gives the points shown. The number of data averaged per interval is shown above the point. Vertical lines are standard errors of the mean for intervals with sufficient data for calculation.

frequency. The time t is in years before present so that $t' = T - t + \phi$ runs from past to present. For a sinusoidal variation of M(t), Elsasser's relation becomes

$$\frac{Q(t)}{Q_o} = \left(\frac{M(t)}{M_o}\right)^{-\alpha} \qquad (10)$$

where Q_o is the production rate corresponding to M_o. Using the expression of (9) for M(t) in equation 10),

$$Q(t) = Q_o \left(\frac{M_o + M_1 \sin \omega t'}{M_o}\right)^{-\alpha} \qquad (11)$$

$$= Q_o \left(1 + \frac{M_1}{M_o} \sin \omega t'\right)^{-\alpha} \qquad (12)$$

SENSITIVITY

Because all geomagnetic field models have $M_1 < M_o$, a binomial expansion can be used to give

$$Q(t) = Q_o \left[1 - \alpha \frac{M_1}{M_o} \sin \omega t' + \frac{\alpha(\alpha + 1)}{2} \left(\frac{M_1}{M_o} \sin \omega t' \right)^2 - \frac{\alpha(\alpha + 1)(\alpha + 2)}{6} \right.$$

$$\left. \left(\frac{M_1}{M_o} \sin \omega t' \right)^3 + - \ldots \right]. \tag{13}$$

This expression can now be used in equation (7), and one can obtain a term-by-term analytical solution of equations (7) and (8). Since M_1/M_o is generally about 0.5, the expansion for $Q(t)$ was truncated after the third power. Using the identities,

$$\sin^2 \theta = \frac{1 - \cos 2\theta}{2} \tag{14}$$

$$\sin^3 \theta = \frac{3}{4} \sin \theta - \frac{1}{4} \sin 3\theta \tag{15}$$

the expression for the forcing function of equation (7) becomes

$$\frac{\lambda Q(t)}{N_a} = C_1 + C_2 \sin \omega t' + C_3 \cos 2\omega t' + C_4 \sin 3\omega t' \tag{16}$$

where

$$C_1 = Q'_o \left[1 + \frac{\alpha(\alpha + 1)}{4} \left(\frac{M_1}{M_o} \right)^2 \right] \tag{17}$$

$$C_2 = Q'_o \left[-\alpha \frac{M_1}{M_o} - \frac{\alpha(\alpha + 1)(\alpha + 2)}{8} \left(\frac{M_1}{M_o} \right)^3 \right] \tag{18}$$

$$C_3 = Q'_o \left[\frac{-\alpha(\alpha + 1)}{4} \left(\frac{M_1}{M_o} \right)^2 \right] \tag{19}$$

$$C_4 = Q'_o \left[\frac{\alpha(\alpha + 1)(\alpha + 2)}{24} \left(\frac{M_1}{M_o} \right)^3 \right] \tag{20}$$

$$Q'_o = \frac{\lambda Q_o}{N_a}.$$

The particular solutions to equations (7) and (8) can now be obtained by the method of undetermined coefficents as

$$A_a(t) = \frac{-C_1 a_{22}}{D_o} + \frac{-a_{22} C_2 D_1 - \omega C_2 E_1}{D_1^2 + E_1^2} \sin \omega t' + \frac{-\omega C_2 D_1 + a_{22} C_2 E_1}{D_1^2 + E_1^2} \cos \omega t'$$

$$+ \frac{2\omega C_3 D_2 - a_{22} C_3 E_2}{D_2^2 + E_2^2} \sin 2\omega t' + \frac{-a_{22} C_3 D_2 - 2\omega C_3 E_2}{D_1^2 + E_2^2} \cos 2\omega t'$$

$$+ \frac{-a_{22} C_4 D_3 - 3\omega C_4 E_3}{D_3^2 + E_3^2} \sin 3\omega t' + \frac{-3\omega C_4 D_3 + a_{22} C_4 E_3}{D_3^2 + E_3^2} \cos 3\omega t' \qquad (21)$$

and

$$A_s(t) = \frac{C_1 a_{21}}{D_o} + \frac{C_2 a_{21} D_1}{D_1^2 + E_1^2} \sin \omega t' + \frac{-C_2 a_{21} E_1}{D_1^2 + E_1^2} \cos \omega t'$$

$$+ \frac{C_3 a_{21} E_2}{D_2^2 + E_2^2} \sin 2\omega t' + \frac{C_3 a_{21} D_2}{D_2^2 + E_2^2} \cos 2\omega t'$$

$$+ \frac{C_4 a_{21} D_3}{D_3^2 + E_3^2} \sin 3\omega t' - \frac{C_4 a_{21} E_3}{D_3^2 + E_3^2} \cos 3\omega t' \qquad (22)$$

where

$$a_{11} = -\lambda - k_{as} \qquad (23)$$

$$a_{12} = k_{sa} \frac{N_s}{N_a} \qquad (24)$$

$$a_{21} = k_{as} \frac{N_a}{N_s} \qquad (25)$$

$$a_{22} = -\lambda - k_{sa} - k_{ds} \qquad (26)$$

$$D_n = \lambda^2 + \lambda(k_{as} + k_{sa} + k_{ds}) + k_{as} k_{ds} - n^2 \omega^2 \qquad (27)$$

$$E_n = -n \omega (2\lambda + k_{as} + k_{sa} + k_{ds}). \qquad (28)$$

Since our concern is with atmospheric ^{14}C fluctuations, we will continue to work with equation (21) only. For comparison with the polynomial fit curve, we can easily convert equation (21) into an equation involving per mil $\Delta^{14}C$ through the relation

$$\Delta A_a(t) = \frac{A_a(t) - A_{ao}}{A_{ao}} \times 1000. \tag{29}$$

A_{ao}, the activity for a particular reference year, can be calculated from equation (21).

Note that $Q'_o = \lambda Q_o/N_a$ appears in every term of the solution (21) through the constants C_n. The $\Delta A_a(t)$ formulation thus enables us to use whatever units we wish for Q'_o or Q_o without affecting the resultant Δ value. Different units for Q'_o would introduce a multiplicative factor into every term of equation (21) and hence into the values for $A_a(t)$ and A_{ao}. This factor will then cancel out of the Δ formulation. To further simplify matters, we may arbitrarily set $N_a = 1$ and let $\nu = N_s/N_a = N_s$ be the relative amount of total carbon/cm$_e^2$ in the sink reservoir. N_s and N_a are always found in equation (21) as the ratio N_s/N_a (in a_{12} and a_{21}) except where N_a is associated with Q_o. Converting N_s and N_a to their true values would mean multiplying each by the same constant factor. This would not change the value of the ratio ν, and multiplying N_a by a constant in the Q'_o term would not change the value calculated for $\Delta A_a(t)$.

There are two complementary solutions to equations (21) and (22). Both involve exponential decay with time constants of about 50 and 8000 years. It is assumed that the dipole moment has varied sinusoidally for a time long enough so that these transients will have died out. After n time constants, the complementary solutions will be e^{-n} of their values when the forcing function was initiated. For example, after 32,000 years, the decay term with the shorter time constant will be negligible and the term with the longer time constant will be about 1.8% of its initial value. Cox (1968) postulated a steady-state dipole harmonic oscillator as part of a mechanism for geomagnetic polarity reversals, so it is reasonable to assume that the transients of the complementary solution can be neglected.

A theoretical calculation of the radiocarbon inventory can be made if the production function Q(t) is known. We are unaware of any previous attempts to do this. The inventory is simply the decay rate/cm$_e^2$ of all previously produced radiocarbon. The inventory can thus be expressed as

$$I = \frac{\int_{-\infty}^{0} Q(t)e^{-\lambda t}\, dt}{\int_{-\infty}^{0} e^{-\lambda t}\, dt} \tag{30}$$

$$= -\lambda \int_{-\infty}^{0} Q(t)e^{-\lambda t}\, dt \tag{31}$$

where t is years before present and Q(t) is the production rate/cm$_e^2$. Equation (31) shows that the inventory or decay rate is not in general equal to the average production rate. The production function is weighted by the exponential decay term, so that recently produced radiocarbon will contribute more to the inventory than that produced further in the past. Note that I depends only on λ and $Q(t)$. $Q(t)$ depends on $M(t)$, so the calculated value of I will be independent of the reservoir parameters.

Evaluation of the Model

To evaluate the equations for ambient and sink activities and radiocarbon inventory, values must be determined for the geomagnetic and reservoir parameters, none of which are precisely known. It therefore seems important to investigate the sensitivity of a model to these parameters by looking at the response of the model as the different parameters are independently varied over plausible ranges of values. Table 1 lists the model parameters that were varied, some minimum and maximum values suggested in the literature, and the minimum and maximum values used in the sensitivity analyses. All geomagnetic fields are constrained to have the known value of $M_{1950} = 8.0 \times 10^{25}$ gauss cm^3 at A.D. 1950. The maximum value of the sinusoidal dipole moment $M_{max} = M_o + M_1$. This maximum value for the field $M(t)$ of equation (9) occurs at time $T_{M_{max}}$ when

$$\frac{\pi}{2} = \omega\left(T - T_{M_{max}} + \phi\right) = \frac{2\pi}{T}\left(T - T_{M_{max}} + \phi\right),$$

or

$$T_{M_{max}} = \frac{3T}{4} + \phi + nT, \; n = \text{integer}. \tag{32}$$

M_{max}, $T_{M_{max}}$, T, and M_{1950} uniquely determine M_o, M_1, and ϕ, thus specifying the geomagnetic field. Values of M_{max}, $T_{M_{max}}$ and T were chosen such that the resultant field model would be in reasonable accord with the data of figure 3.

The wide range of reservoir exchange constants found in the literature is partly an inherent feature of the modeling process. Values for the exchange constants are generally found by using box models to simulate particular physical processes and by choosing constants that yield best agreement between the data and model calculations. The values determined for the exchange rates will then depend on the nature of the model (e.g., how many reservoirs) and what one is attempting to predict. This may be the value of a dependent variable at a particular point in time, or how this variable changes with time. As Oeschger et al. (1975) have emphasized, the nature of box models, at least those with only two or three reservoirs, is such that for phenomena of different characteristic frequencies, different exchange rates are required. It remains to be seen whether use of sophisticated box models or box diffusion models can be used to model phenomena ranging in characteristic times from years to millennia, without requiring changes of the exchange rates.

For steady-state exchange of ^{12}C between ambient and sink reservoirs

$$k_{as} N_a = k_{sa} N_s \tag{33}$$

SENSITIVITY

TABLE 1
Independent Parameters of the Sensitivity Analysis

Parameter	Units	Literature		Sensitivity analysis	
		Minimum	Maximum	Minimum	Maximum
M_{max}	10^{25} gauss cm^3			10.25	13.0
$T_{M_{max}}$	years B.P.			1500	2500
T	years			8000	10,000
a		0.47[1]	0.58[2]	0.45	0.60
ν		10[3]	33[4]	5	50
τ_a	years	⩽7[5]	75[6,7]	5	100
k_{ds}	years^{-1}	0.000004[8]	0.00001[9]	0.0	0.000025

Minimum and maximum values of M_{max}, $T_{M_{max}}$, and T used in the sensitivity analysis are based on figure 3. Values used for the other parameters are based on the given values from the literature: [1] Ramaty (1967); [2] Wada and Inoue (1966); [3] Arnold and Anderson (1957); [4] Wood and Libby (1964); [5] Fergusson (1958); [6] Houtermans et al. (1973); [7] Yang (1971); [8] Grey (1972); [9] Damon and Wallick (1972). (See text for definitions of the parameters).

and for steady-state exchange of ^{14}C

$$\frac{k_{as}}{k_{sa}} {}^{14}C = \alpha_{s/a} \frac{k_{as}}{k_{sa}} {}^{12}C \tag{34}$$

$$= \alpha_{s/a} \frac{N_s}{N_a} \quad \text{(Craig 1957)}$$

where $\alpha_{s/a}$ is the ratio of the fractionation factor of ^{14}C in the oceans relative to $-25\,^o/_{oo}$ wood divided by the fractionation factor of the atmosphere relative to wood. Analysis showed that activity calculations are barely affected for $1.0 < \alpha_{s/a} < 1.1$. Since $\alpha_{s/a}$ has a value of about 1.012 (Craig 1957), we can ignore fractionation effects and set $\alpha_{s/a} = 1.0$. So

$$\frac{k_{as}}{k_{sa}} {}^{14}C = \frac{N_s}{N_a} = \nu \tag{36}$$

and dropping the subscript

$$k_{sa} = \frac{k_{as}}{\nu}. \tag{37}$$

So, under the assumption of steady-state, k_{sa} is determined by k_{as} and ν. For a $\pm 50\,^o/_{oo}$

change in $\Delta^{14}C$ over 7500 years, the steady-state assumption should be reasonable for determining k_{sa}.

Analysis of the model was carried out in three phases as outlined below.

INVENTORY ANALYSIS

Inventory calculations were made according to equation (31) for the different geomagnetic fields specified by the parameters of table 1. Instead of Elsasser's relation between Q(t) and M(t), data of Lingenfelter and Ramaty (1970) were used which give the dependence of Q on both M and the state of the interplanetary magnetic field as characterized by the parameter η. The polar cosmic ray flux variation between solar minimum and maximum can be expressed as

$$\Phi(P,t) = \Phi(P,t_1)e^{\eta(t)/P\beta}$$

where Φ is the flux, P is the cosmic ray particle rigidity, β is the particle velocity, t is time, and t_1 is the time at which solar minimum occurs (Lingenfelter and Ramaty 1970, from Lockwood and Webber 1967). Figure 4 shows curves of Q vs M for three values of η. For the solar minimum of 1953-54 and the solar maximum of 1957-58, η had values of 0 and -2GV, respectively. Lingenfelter and Ramaty calculated the average production rate for 1937-1967 to be 2.2 atoms/cm$_e^2$/sec, corresponding to a constant value of η = -0.7GV over this period. The Q vs M curve for η = -0.7 was used for the inventory calculations, assuming the state of the interplanetary magnetic field for 1937-1967 to be representative for the period of time over which the inventory is calculated. Elsasser's relation is in good agreement with this curve and similar data of other investigators (Wada and Inoue 1966), except for values of M near the lower limit reached by some of our geomagnetic fields. Inventory was calculated using the rectangular rule for numerical integration with 100-year time intervals over 10 geomagnetic periods. The field was assumed to be uniformly sinusoidal over this time span, again consistent with Cox's model for a geomagnetic reversal mechanism.

SENSITIVITY ANALYSIS

The second phase involved an investigation of the sensitivity of the fluctuation curve generated by the model to both geomagnetic and reservoir parameters. Because of the wide ranges of values scanned for the seven independent parameters, it would have been impractical to look at the fluctuation curves predicted by each combination of parameter values. Instead time derivatives of the first harmonic terms of equation (21) were used to solve for the approximate maximum (Δ_{max}) and minimum (Δ_{min}) values of $\Delta A_a(t)$, and phase or lag time of the first harmonic response terms with respect to the sinusoidal geomagnetic field. The calculated lag time and the value of $T_{M\,max}$ determine the times at which Δ_{max} and Δ_{min} occur. This approach greatly simplifies the analysis by allowing quick inspection of three parameters — Δ_{max}, Δ_{min}, and lag time — to determine how well the curves predicted by a particular parameter set will match the regression curve.

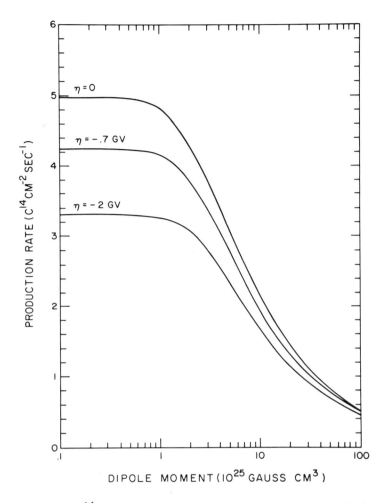

FIG. 4. ^{14}C production rate vs. strength of the geomagnetic dipole moment and the parameter η which characterizes the state of the interplanetary magnetic field (after Lingenfelter and Ramaty, 1970). For solar minimum (1953-54), $\eta = 0$; $\eta = -2$GV for solar maximum (1957-58); $\eta = -.7$GV to give the average ^{14}C production rate of 2.2 atoms/cm$_e^2$/sec from 1937-1967.

The first harmonic terms of equation (21) account for about 90% of the activity calculated for the first three harmonics, so that approach seems reasonable. For phases two and three, Q_o was determined for a particular M_o from the data of Lingenfelter and Ramaty for $\eta = -0.7$GV. In other words, Elsasser's relation, which was used in the solution of equation (21), was calibrated by using the data of figure 4.

For conversion of activities to Δ values, the time t_o at which A_{ao} was calculated was chosen as 625 B.P. (calendar years before present) rather than A.D. 1890 (60 B.P.). Although Δ^{14}C values for individual samples are calculated with respect to a reference activity of mid-nineteenth century wood, the regression curve through the Δ data passed through zero at 625 B.P., so using this year as t_o seems more representative of the entire data set.

PREDICTED TREND CURVES

Once phase 2 established the model sensitivity in calculations of Δ_{max}, Δ_{min}, and lag time, model fluctuation curves representing the entire solution of equation (21) were generated for selected parameter sets and compared with the polynomial fit curve.

Results

Table 2 lists parameters of interest in the inventory and sensitivity analyses and their values. Values for Δ_{max} at time $T_{\Delta_{max}}$ and Δ_{min} at time $T_{\Delta_{min}}$ were picked off the polynomial fit curve. From assays of the radiocarbon reservoirs, Suess (1965) and Grey (1972) both estimated the inventory to be 108 dpm/cm$_e^2$, and Grey suggested that the true value is very likely in the range of 105-117 dpm/cm$_e^2$. Figures 5 to 11 and tables 3 and 4 contain representative results from the inventory and sensitivity analyses. The results of these analyses were compared with the desired values of table 2.

The results of the inventory calculations are quite interesting if not altogether conclusive. Figure 12 shows the variation of Q(t) for a typical sinusoidal magnetic field, calculated using the data of Lingenfelter and Ramaty (1970). The nonlinearity of the relationship between Q and M, which is directly expressed in Elsasser's relation, makes Q(t) nonsinusoidal. Although more radiocarbon was produced in the half cycle with peak

TABLE 2

Data Values for ^{14}C Parameters Predicted in Sensitivity Analysis

Parameter	Value
Δ_{max}	90.0 °/oo
$T_{\Delta_{max}}$	6475 B.P.
Δ_{min}	−5.3° °/oo
$T_{\Delta_{min}}$	1450 B.P.
I	108 (105.6-117) dpm/cm$_e^2$

Δ_{max} and Δ_{min} are the extreme values for the curve in figure 1, and $T_{\Delta_{max}}$ and $T_{\Delta_{min}}$ are the times at which these points occur. The inventory value (I) was estimated by both Suess (1965) and Grey (1972). Grey also estimated the range shown in which the true value of I is likely to be.

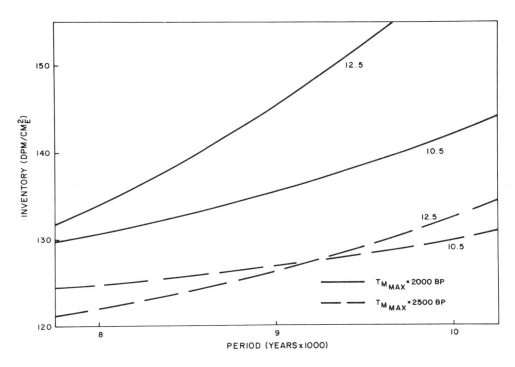

FIG. 5. Calculated values of inventory (dpm/cm$_e^2$) vs. geomagnetic period. Solid curves are for $T_{M_{max}}$ (time of maximum value of dipole field) = 2000 B.P.; dashed curves are for $T_{M_{max}}$ = 2500 B.P. Numbers along curves are values of the maximum dipole moment M_{max}.

Q_2, more of this radiocarbon has decayed away than that produced in the half cycle with peak Q_1. At the present time, $M \simeq M_o$ for most geomagnetic models and $Q \simeq Q_o$. Because of this and the higher ^{14}C production further in the past, values calculated for the inventory are usually quite close to Q_o. This is coincidental and will not be generally true for all times. As shown by figure 5, calculated values for the inventory are somewhat higher than the estimated value of 108 dpm/cm$_e^2$. The lowest values obtained of about 122 dpm/cm$_e^2$ are only 5 dpm/cm$_e^2$ higher than Grey's maximum estimate of 117 dpm/cm$_e^2$. Our model will yield an inventory value of 117 if $T_{M_{max}}$ is pushed back to 2900 B.P., but this does not seem reasonable. Required increases in M_{max} or decreases in T to obtain a value of 117 also seem incompatible with the paleomagnetic data. Thus we are left with either an incorrect evaluation of the inventory or a defect in our model. A weak point in the model could be the assumption of a steady sinusoidal geomagnetic dipole moment. It is unlikely that this assumption is strictly true. More data are needed before its validity can be more thoroughly investigated. Although our inventory analysis is not conclusive, it suggests a magnetic field with $T_{M_{max}}$ as far in the past as the data allow (2500 B.P.) and to a lesser extent, a low T of 8000-8500 years. This is significant since most magnetic models based on paleomagnetic data have $T_{M_{max}}$ occurring in the Christian era.

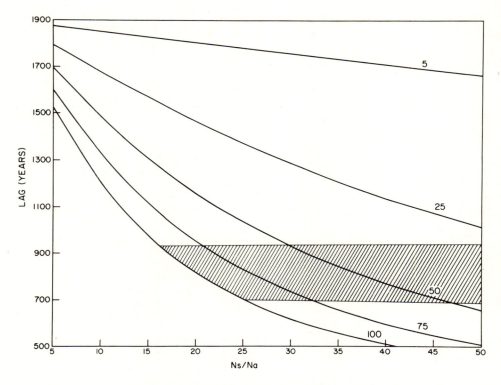

FIG. 6. Lag time of atmospheric $\Delta^{14}C$ vs. $\nu = N_s/N_a$, the ratio of total carbon in sink reservoir to total carbon in ambient reservoir. Curves are for indicated values of atmospheric residence time $\tau_a = k_{as}^{-1}$ in years. Hatched area represents the range of values that can be used when $T_{M_{max}} = 2500$ B.P. to give model ^{14}C fluctuation curves in reasonable agreement with the data.

Nongeomagnetic phenomena affecting radiocarbon production such as supernova explosions, antimatter meteorites, and solar activity will also affect the inventory. Supernova explosions and antimatter meteorites would increase Q and hence I, while variations in solar activity could cause either an increase or decrease in Q. We have assumed that solar activity and the interplanetary magnetic field over the last three solar cycles have been representative of the last 10 geomagnetic periods. This may not be true. Calculations have shown that if the average state of the interplanetary magnetic field can be described by the curve for $\eta = -2GV$ in figure 4, theoretical values for I are close to the estimated value of 108 dpm/gC when $T_{M_{max}} = 2500$ B.P. Similarly, although the last three solar cycles may accurately represent average conditions, the curve in figure 4 for $\eta = -0.7GV$ may not correspond to these conditions. Lingenfelter and Ramaty (1970) calculated the average production over the last three solar cycles to be $Q_{1950} = 2.2 \pm 0.4$ atoms/cm$_e^2$/sec or 132 ± 24 atoms/cm$_e^2$/minute. This value leads to the curve for $\eta = -0.7GV$. If the average production rate were actually 108 (= 132-24), calculated inventories would agree with the estimates.

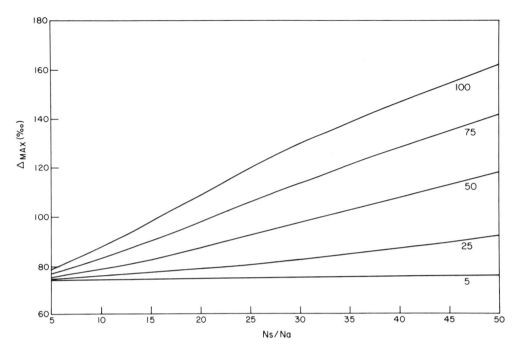

FIG. 7. Δ_{max} (maximum $^o/oo$ value of $\Delta^{14}C$) vs. $\nu = N_s/N_a$. Curves are for indicated values of τ_a.

Lingenfelter and Ramaty (1970) estimated that $0.75 \leq Q_{1950}/I \leq 1.61$, the spread largely reflecting the uncertainty in Q_{1950}. Using the parameter $R_o = Q_{1950}/I$ and paleomagnetic data rather than a geomagnetic field model to calculate theoretical ^{14}C fluctuation curves, they found that the radiocarbon data were fairly well bounded by curves for $R_o = 1.0$ and $R_o = 1.05$. These values for R_o imply that if the production rate calculations and assumptions are correct and $\overline{Q} = Q_{1950} = 132$ atoms/cm$_e^2$/minute, then $126 \leq I \leq 132$. Recalling our minimum inventory value of $I = 122$ in table 3 and figure 5, it appears that implicit in the analysis of Lingenfelter and Ramaty is an inventory problem similar to the one we have found. Conversely, if it is assumed that the estimated inventory of $I = 108$ is correct, the values for R_o imply that $108 \leq \overline{Q} \leq 113.4$, suggesting as we have above that either $\overline{Q}$ over the past few (<10) geomagnetic periods is substantially lower than Q_{1950} or that Q_{1950} is actually less than 132. Lingenfelter (personal communication) believes the recent production rate is probably no less than 126 atoms/cm$_e^2$/minute, but it still seems possible that the discrepancy between estimated and theoretical values for I may be more apparent than real, reflecting uncertainties in the data.

The sensitivity analysis shows that Δ_{min} is not nearly so sensitive to variation of parameter values as is Δ_{max}. This is because the model curves for $\Delta A_a(t)$ are forced to pass through zero at 625 B.P., which is fairly close to the desired Δ_{min} of -5.3 $^o/oo$ at 1450 B.P. Since the predicted curve pivots about the fixed point at 625 B.P., calculated

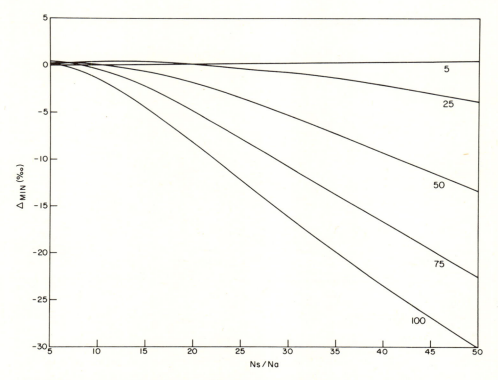

FIG. 8. Δ_{min} (minimum ^{o}/oo value of $\Delta^{14}C$) vs. $\nu = N_s/N_a$. Curves are for indicated value of τ_a.

activities will be most sensitive at times further removed from t_o. Hence the enhanced sensitivity of Δ_{max} at 6475 B.P. Another consideration in looking at the results of the sensitivity analysis is that although the sensitivity curves may change for different values of the fixed parameters, the slopes of the sensitivity curves remain nearly constant over the ranges of values investigated. This slope is what is actually being investigated in the sensitivity analysis.

It was mentioned in the introduction that Damon (1970) previously had problems with his model curves lagging the data by several hundred years. Our analysis of the lag time shows a functional dependence on T, k_{as}, and N_a consistent with Damon's (1970) earlier analysis. Tables 3 and 4 show that the lag time of the model curve behind the production function (or geomagnetic field) is not sensitive to the following parameters: M_{max}, $T_{M_{max}}$, α, k_{ds}. All these parameters, particularly M_{max} and $T_{M_{max}}$, significantly affect Δ_{max}. $T_{M_{max}}$ is fixed at 2500 B.P. as a result of the inventory analysis. Δ_{max} can still be changed considerably without affecting the lag time by changing α and k_{ds}, and to a lesser extent M_{max} and T. Δ_{min} is not particularly sensitive except for large changes in reservoir parameters. It is not generally difficult to obtain model fluctuation curves with the correct amplitudes, but initial models did show the predicted curves lagging the data. Pushing $T_{M_{max}}$ as far back as allowable, as also suggested by the inventory analysis, helps to reduce this lag. To eliminate the lag problem, it is necessary to use values for the

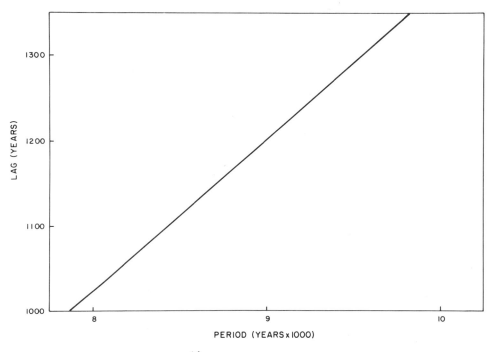

FIG. 9. Lag time of atmospheric $\Delta^{14}C$ vs. geomagnetic period. Lag time is independent of the value of the maximum dipole moment (M_{max}) and the time at which M_{max} occurs ($T_{M_{max}}$).

reservoir parameters somewhat higher than those usually suggested in the literature. The hatched area in figure 6 shows the general range of reservoir parameters ν and $\tau_a = k_{as}^{-1}$ which are needed to eliminate the lag problem for various models having $T_{M_{max}} = 2500$ B.P. Figure 13 compares the polynimial fit curve with one of our better model predictions. Agreement between the two is quite good.

The use of a relatively high value for τ_a is supported by steady-state calculations. In the nearly linear portion of the polynomial fit in figure 1 from about 2500-5500 B.P., the rate of change of $\Delta^{14}C$ is approximately 0.03 ‰/year, and this seems slow enough to justify steady-state considerations. From equation (7), steady-state implies that

$$k_{as} = \frac{\lambda(\overline{Q}/N_a - \overline{A}_a)}{\overline{A}_a - \overline{A}_s} \quad \text{or}$$

$$\tau_a = \frac{\overline{A}_a - \overline{A}_s}{\lambda(\overline{Q}/N_a - \overline{A}_a)} \tag{38}$$

where $\overline{A}_a$, $\overline{A}_s$, and $\overline{Q}$ are steady-state values for A_a, A_s, and Q. Equation (31) indicates that $\overline{Q} = I$. Figure 1 shows that $\Delta A_a(t)$ fluctuates about ± 50 ‰ about its mean, and use of the same parameter values in equation (22) as in equation (21) indicates a fluctuation

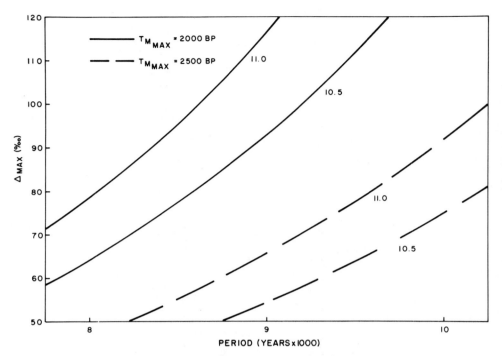

FIG. 10. Δ_{max} vs. geomagnetic period. Solid curves are for $T_{M_{max}}$ = 2000 B.P.; dashed curves are for $T_{M_{max}}$ = 2500 B.P. Numbers along curves give values of M_{max}.

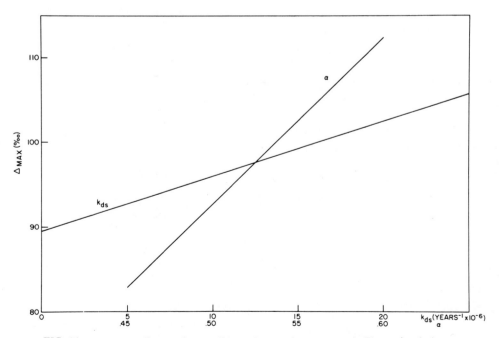

FIG. 11. Δ_{max} vs. sedimentation rate k_{ds} and vs. α, the exponent in Elsasser's relation.

TABLE 3

Values of C^{14} Dependent Variables for Minimum Values of the Model Independent Parameters*

Independent variable		Δ_{max} ⁰/₀₀	Δ_{min} ⁰/₀₀	Lag years	I dpm/cm²$_e$
T	8000	76.3	-5.3	950	122
M_{max}	10.25	43.2	-2.8	1033	126
$T_{M_{max}}$	2000	181.5	1.7	1033	140
a	0.45	83.0	-3.3	1033	124
τ_a	5	75.2	0.3	1779	124
ν	5	75.8	0.4	1700	124
k_{ds}	0.0	89.4	-3.3	1038	124

*As one independent parameter is varied, the others have the following fixed values: T = 8500, M_{max} = 12.75, $T_{M_{max}}$ = 2500, a = 0.5, τ_a = 50, ν = 25, k_{ds} = 0.000005. For the inventory analysis fixed values are: τ_a = 47.6, k_{sa} = 1/1500, ν = 31.5.

TABLE 4

Values of ^{14}C Dependent Variables for Maximum Values of the Model Independent Parameters*

Independent variable		Δ_{max} ⁰/₀₀	Δ_{min} ⁰/₀₀	Lag years	I dpm/cm²$_e$
T	10,000	164.6	0.8	1285	133
M_{max}	13.0	97.8	-3.5	1033	124
$T_{M_{max}}$	2500	92.7	-3.5	1033	124
a	.6	112.4	-3.9	1033	124
τ_a	100	119.5	-12.2	704	124
ν	50	118.5	-13.5	657	124
k_{ds}	0.000025	105.8	-4.4	1013	124

*As one independent parameter is varied, the others have the fixed values given for table 3.

in $\Delta A_s(t)$ of about ±35 ⁰/₀₀. Activities at A.D. 1890 are nearly equal to the minimum activities over the last 7500 years. Using A.D. 1890 values for A_a and A_s of 13.8 dpm/gC and 12.2 dpm/gC respectively (Damon 1970), good approximations to $\overline{A}_a$ and $\overline{A}_s$ are $\overline{A}_a$ = (1.05)(13.8) = 14.5 dpm/gC and $\overline{A}_s$ = (1.035)(12.2) = 12.6 dpm/gC. For $\overline{Q}$ = I = 108 dpm/cm²$_e$ and N_a = 0.351 gC/cm²$_e$ (Damon 1970), equation (38) gives τ_a = 53 years. Figure 6 indicates that if τ_a = 50, a value of ν = 30 will yield a theoretical fluctuation

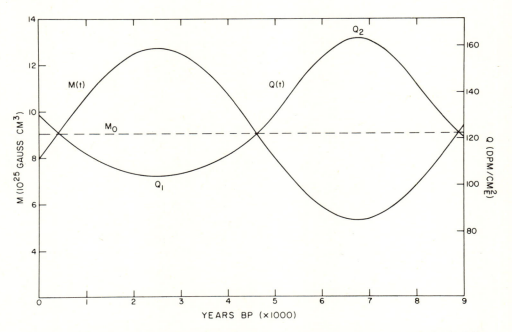

FIG. 12. Radiocarbon production rate Q(t) as a function of time for the geomagnetic dipole field $M(t) = 9.021 + 3.729 \sin 2\pi/8500 \, (t-375)$. $M_o = 9.021$. Note that Q(t) is asymmetrical about M_o.

curve in reasonable agreement with the data. This is consistent with other estimates of ν. Thus steady-state calculations support residence times derived through dynamic considerations. This apparently validates our assumption of steady-state in calculating values for k_{sa}. Equation (38) illustrates what must happen as more boxes are added to the model. As the carbon exchange system is more finely partitioned, total carbon content of the ambient reservoir (N_a) and activity differences between exchanging reservoir decrease, hence residence times must decrease as well. For example, let us separate the mixed layer from the ambient reservoir. Residence time in the ambient reservoir will then be with respect to exchange with the mixed layer. For $\overline{A}_a = 13.8$ dpm/gC as above, $\overline{A}_{m1} = .965 \, \overline{A}_a$, and $N_a = 0.126$ gC/cm$_e^2$ (Lal and Venkatavaradan 1970), equation (38) yields a residence time of $\tau = 5.0$ years. This strikingly demonstrates how residence times can be affected by reservoir partitioning.

CONCLUSIONS

Our analysis clearly indicates that our approach cannot be used to specify a unique set of parameters that will correctly predict both the inventory and fluctuation curve. It can be useful in delimiting parameter sets that give results in reasonable agreement with the data. Theoretical values calculated for the radiocarbon inventory are higher than inventory estimates. This suggests an error in the inventory estimates, our model, or our assumptions. The inventory analysis and the sensitivity analysis both suggest that if our model is

SENSITIVITY

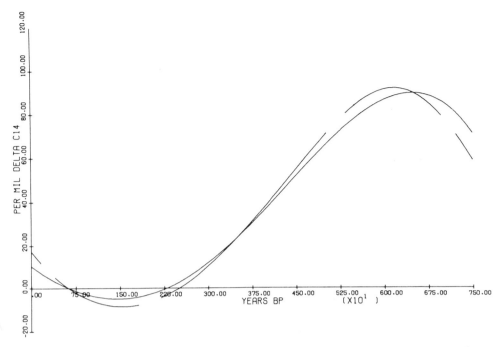

FIG. 13. Solid curve is the fourth degree polynomial regression of figure 1. Dashed curve is one of the better predictions of our model. Parameters are T = 8500, M_{max} = 12.5, $T_{M_{max}}$ = 2500, α = .45, τ_a = 75, ν = 25, k_{ds} = .000005.

reasonably valid, the dipole moment had its peak value several hundred years B.C., earlier than the current paleomagnetic data suggest. This is another example of the power of radiocarbon in integrating information from different disciplines.

There is a definite need for more data and for refinement of the modeling process. Extension of the radiocarbon chronology beyond 7500 B.P. would be quite helpful. Model fluctuation curves and the regression curve tend to diverge before 6000 B.P. It would be interesting to see if this trend continues into the past since it could be due to climatic effects of the last major glaciation. More paleomagnetic and archaeomagnetic data are also needed to determine if the geomagnetic field can be more accurately modeled and to see if the dipole moment has indeed varied sinusoidally. Spatial as well as temporal distribution of these data is important so that effects of secular variation due to activity of the nondipole field can be removed. Elucidation of the past behavior of the magnetic field and the radiocarbon production rate could be critical in clearing up the inventory problems of our analysis.

Use of box diffusion models and possibly more sophisticated box model studies to investigate long-term radiocarbon fluctuations could provide new insight, especially in determining the physical meaning of the exchange rates calculated by the various models for different phenomena. Models could also be used to study the combined effects of different causes of radiocarbon fluctuations.

ACKNOWLEDGMENTS

We would like to thank Paul Noren of the University of Arizona for helping with the mathematics, and Sandra Harralson for typing the manuscript. We are also grateful to Dr. Juan Carlos Lerman and Professors Austin Long and Robert Butler of the University of Arizona for reading and editing the manuscript. We accept full responsibility for any errors in the paper. This research was supported by National Science Foundation grant DES74-13362 to Paul E. Damon, and by the State of Arizona.

NOTE ADDED IN PROOF

The latest estimate of the ^{14}C inventory is 119 dpm/cm$_e^2$ (Damon et al. 1978) which agrees with our model predictions much better than previous estimates. A recent analysis of global paleointensity data from the past 8000 years by Duane Champion (thesis in preparation) of the U.S. Geological Survey, shows that the dipole moment had a maximum century-averaged value of 12.5 during the interval 2200-2300 B.P. Maximum entropy spectral analysis revealed the existence of a dominant periodicity having either a period of 7300 years (if all data are included) or 8700 years (if the eight oldest points, all from a single region, are excluded). The excellent agreement of Champion's figures with our geomagnetic parameters for figure 13 suggests our approach is valid, and illustrates the power of ^{14}C studies in delimiting geomagnetic field behavior.

REFERENCES

Arnold, J. R., and E. C. Anderson
    1957   The distribution of carbon-14 in nature. *Tellus.* 9:28-32.

Bacastow, R., and C. D. Keeling
    1973   Atmospheric CO_2 in the natural carbon cycle. II. Changes from A.D. 1700 to 2070 as deduced from a geochemical model. *In Carbon and the Biosphere. 24th Brookhaven Symp. in Biology.* New York. Nat. Tech. Inf. Serv. Springfield, VA. 86-135.

Broecker, W. S., and E. A. Olson
    1959   Lamont radiocarbon measurements VI. *Radiocarbon.* 1:111-132.

Bucha, V.
    1970   Influence of the Earth's magnetic field on radiocarbon dating. *In* Olsson ed. *Radiocarbon Variations and Absolute Chronology. 12th Nobel Symp.* Uppsala. Stockholm, Almqvist and Wiksell. New York, John Wiley and Sons. 501-511.

Cox, A.
    1968   Lengths of geomagnetic polarity reversal. *J. Geophys. Res.* 73:3247-3260.

Craig, H.
    1957   The natural distribution of radiocarbon and the exchange time of carbon dioxide between atmosphere and sea. *Tellus.* 9:1-17.

Damon, P. E.
    1970   Climatic versus magnetic perturbation of the atmospheric ^{14}C reservoir. *In* Olsson, ed. *Radiocarbon Variations and Absolute Chronology. 12th Nobel*

Symp., Uppsala. Stockholm, Almqvist and Wiksell. New York, John Wiley and Sons. 571-593.

Damon, P. E., J. C. Lerman, and A. Long
- 1978 Temporal fluctuations of atmospheric ^{14}C: Causal factors and implications: Ann. Rev. Earth Planet. Sci., 6: 457-494.

Damon, P. E., A. Long, and D. C. Grey
- 1970 Arizona radiocarbon dates for dendrochronologically dated samples. *In* Olsson, ed. *Radiocarbon Variations and Absolute Chronology. 12th Nobel Symp.* Uppsala. Stockholm, Almqvist and Wiksell. New York, John Wiley and Sons. 615-618.

Damon, P. E., A. Long, and E. I. Wallick
- 1972 Dendrochronologic calibration of the carbon-14 time scale. *Proc. 8th Int. Conf. on Radiocarbon Dating.* New Zealand. A28-A43.

Damon, P. E., and E. I. Wallick
- 1972 Changes in atmospheric radiocarbon concentration during the last eight millennia. *In Contributions to Recent Geochemistry and Analytical Chemistry.* Moscow, Nauka Publishing Office, 441-452 (in Russian; preprints in English).

Daniel, C., and F. S. Wood
- 1971 *Fitting Equation to Data.* New York, John Wiley and Sons, 55.

Ekdahl, C. A., and C. D. Keeling
- 1973 Atmospheric CO_2 in the natural carbon cycle. I. Quantitative deductions from records at Mauna Loa Observatory and at the South Pole. *In Carbon and the Biosphere. 24th Brookhaven Symp. in Biology.* Nat. Tech. Inf. Serv. Springfield, VA. 51-85.

Elsasser, W., E. P. Ney, and J. R. Winckler
- 1956 Cosmic-ray intensity and geomagnetism. *Nature. 178*:1226-1227.

Fergusson, G. J.
- 1958 Reduction of atmospheric radiocarbon concentration by fossil fuel carbon dioxide and the mean life of carbon dioxide in the atmosphere. *Proc. Roy. Soc. London. 243A*:561-574.

Grey, D. C.
- 1969 Geophysical mechanisms for ^{14}C variations, *J. Geophys. Res. 74*:6333-6340.
- 1972 *Fluctuations of Atmospheric Radiocarbon.* Dissertation. University of Arizona.

Grey, D. C. and P. E. Damon
- 1970 Sunspots and radiocarbon dating in the Middle Ages. *In* Berger, ed. *Scientific Methods in Medieval Archaeology.* Berkeley and Los Angeles, University of California Press. 167-182.

Houtermans, J.
- 1966 On the quantitative relationships between geophysical parameters and the natural ^{14}C inventory. *Z. für Physik. 193*:1-12.
- 1971 *Geophysical Interpretations of Bristlecone Pine Radiocarbon Measurements Using a Method of Fourier Analysis of Unequally Spaced Data.* Dissertation. University of Bern.

Houtermans, J. C., H. E. Suess, and H. Oeschger
    1973   Reservoir models and production rate variations of natural radiocarbon. *J. Geophys. Res. 78*:1897-1908.

Keeling, C. D.
    1973   The carbon dioxide cycle: reservoir models to depict the exchange of atmospheric carbon dioxide with the oceans and land plants. *In* Rasool, ed. *Chemistry of the Lower Atmosphere*. New York, Plenum Press, 251-329.

Kigoshi, K., and K. Hasegawa
    1966   Secular variation of atmospheric radiocarbon concentration and its dependence on geomagnetism. *J. Geophys. Res. 71*:1065-1071.

Kitazawa, K.
    1970   Intensity of the geomagnetic field in Japan for the past 10,000 years. *J. Geophys. Res. 75*:7403-7411.

Lal, D., and V. S. Venkatavaradan
    1970   Analysis of the causes of ^{14}C variations in the atmosphere. *In* Olsson, ed. *Radiocarbon Variations and Absolute Chronology. 12th Nobel Symp*. Uppsala. Stockholm, Almqvist and Wiksell. New York, John Wiley and Sons, 549-569.

Lerman, J. C., W. G. Mook, and J. C. Vogel
    1970 ^{14}C in tree rings from different localities. *In* Olsson, ed. *Radiocarbon Variations and Absolute Chronology. 12th Nobel Symp*. Uppsala. Stockholm, Almqvist and Wiksell. New York, John Wiley and Sons, 275-299.

Lingenfelter, R. E., and R. Ramaty
    1970   Astrophysical and geophysical variations in ^{14}C production. *In* Olsson, ed. *Radiocarbon Variations and Absolute Chronology. 12th Nobel Symp*. Uppsala. Stockholm, Almqvist and Wiksell. New York, John Wiley and Sons, 513-537.

Lockwood, J. A., and W. R. Webber
    1967   The 11-year solar modulation of cosmic rays as deduced from neutron monitor variations and direct measurements at low energies. *J. Geophys. Res. 72*:5977-5989.

Oeschger, H., U. Siegenthaler, U. Schotterer, and A. Gugelmann
    1975   A box diffusion model to study the carbon dioxide exchange in nature. *Tellus. 27*:168-191.

Olsson, I. U., ed.
    1970   *Radiocarbon Variations and Absolute Chronology. 12th Nobel Symp*. Uppsala. Stockholm, Almqvist and Wiksell. New York, John Wiley and Sons.

Plesset, M., and A. Latter
    1960   Transient effects in the distribution of carbon-14 in nature. *Proc. National Acad. Sci. 46*:232-241.

Ralph, E. K.
    1972   A cyclic solution for the relationship between magnetic and atmospheric C-14 changes. *In Proc. 8th Int. Conf. on Radiocarbon Dating*. New Zealand. A76-A84.

Ralph, E. K., and H. N. Michael
    1970   MASCA radiocarbon dates for sequoia and bristlecone pine samples. *In*

Olsson, ed. *Radiocarbon Variations and Absolute Chronology. 12th Nobel Symp.* Uppsala. Stockholm, Almqvist and Wiksell. New York, John Wiley and Sons. 619-623.

Ramaty, R.
  1967   The influence of geomagnetic shielding on ^{14}C production and content. *In* Hindmarsh, Lowes, Roberts, and Runcorn, eds. *Magnetism and the Cosmos.* NATO Advanced Study Inst. on Planetary & Stellar Magnetism, Univ. of Newcastle upon Tyne. Amer. Elsevier Publ. Co., 66-78.

Revelle, R., and H. E. Suess
  1957   Carbon dioxide exchange between atmosphere and ocean and the question of an increase of atmospheric CO_2 during the past decades. *Tellus. 9*:18-27.

Smith, P. J.
  1967a The intensity of the ancient geomagnetic field: a review and analysis. *Geophys. J. Roy. Astron. Soc. 12*:321-362.
  1967b Ancient geomagnetic field intensities. I. historic and archaeological data: sets H1-H9. *Geophys. J. Roy. Astron. Soc. 13*:417-419.
  1967c Ancient geomagnetic field intensities. II. Geologic data: sets G1-G21; historic and archaeological data: sets H10-H13. *Geophys. J. Roy. Astron. Soc. 13*:483-486.

Stuiver, M.
  1969   Yale natural radiocarbon measurements IX. *Radiocarbon. 11*:545-658.

Suess, H. E.
  1965   Secular variations of the cosmic-ray-produced carbon-14 in the atmosphere and their interpretations. *J. Geophys. Res. 70*:5937-5952.
  1970   The three causes of the secular ^{14}C fluctuations, their amplitudes and time constants. *In* Olsson, ed. *Radiocarbon Variations and Absolute Chronology, 12th Nobel Symp.* Uppsala. Stockholm, Almqvist and Wiksell. New York, John Wiley and Sons, 595-605.

Wada, M., and A. Inoue
  1966   Relation between the carbon-14 production rate and the geomagnetic moment. *J. Geomag. and Geolect. 18*:485-488.

Wood, L., and W. F. Libby
  1964   Geophysical implications of radiocarbon data discrepancies. *In* Craig, Miller, and Wasserburg, eds. *Isotopic and Cosmic Chemistry.* Amsterdam, North Holland Publishing Co., 205-210.

Yang, A., In-che
  1971   *Variations of Natural Radiocarbon During the Last 11 Millennia and Geophysical Mechanisms for Producing Them.* Dissertation. University of Washington.

Yang, A., In-che, and A. W. Fairhall
  1972   Variations of natural radiocarbon during the last 11 millennia and geophysical mechanisms for producing them. *In Proc. 8th Int. Conference on Radiocarbon Dating.* New Zealand. A44-A57.

PART X
Radiocarbon and Other Dating Methods

1

Dating of Lime Mortar by ^{14}C

Robert L. Folk and S. Valastro, Jr.

ABSTRACT

Development of a modified mortar dating technique by the ^{14}C method is described. This procedure may ve very valuable in the detailed dating of the construction of ancient buildings and may even resolve different events or phases of construction in the same structure.

Carbon-14 analysis of samples is commonly carried out on wood or charcoal derived from wood. The dates obtained from this type of sample often give ages which are too old because they may reflect the age of the tree and not the age of construction. Mortar dating, however, refers to the time of building construction because contemporary ^{14}C in the atmosphere is imbibed by the mortar during the setting or hardening process.

Earlier mortar dating attempts were complicated by dead carbon (carbon containing no contemporary ^{14}C isotope) contributed by limestone or marble contained in the aggregate. The new technique is dependent on a method which separates nearly all of the aggregate from the mortar and then counts only the first fraction of CO_2 evolved during the acidification process. Since the ^{14}C content of the mortar was totally absorbed as part of atmospheric CO_2, the radiocarbon assay of the sample will give an age equivalent to the setting or hardening process.

Five samples of mortar from Stobi were processed and assayed as a preliminary test of the technique. Stobi is an archaeological site in Macedonian Yugoslavia that was an important provincial town during Roman and early Byzantine times. A small sample (Tx-1431) taken from the narthex floor of the Central Basilica, dated archaeologically as late fourth or fifth century A.D., gave a radiocarbon date of A.D. 260 ± 180. Another sample (Tx-1944) from the same Basilica Floor 1 produced two dates with an average of A.D. 372 ± 60, which is in excellent agreement with the archaeological age estimate. Two samples (Tx-1941, Tx-1942) from the theater produced six dates averaging A.D. 233 ± 32; the archaeological age estimate places the theater in the late first to early second century A.D. A sample (Tx-1943) taken from the principal construction phase of the Episcopal Basilica, whose archaeological age estimate is about A.D. 400, gave an average age of A.D. 302 ± 60. In the fast city wall area, a layer of mortar with fourth century pottery above and below (Tx-1940) gave a date of 115 ± 60 B.C. This particular sample may have been contaminated due to annual flooding of the area from which it was taken. It is also possible that the mortar may have been redeposited. The geologic circumstances that can affect mortar evidently need to be investigated.

All of the dates above were subsequently corrected for ^{14}C half-life of 5730, dendrochronology, and were tentatively corrected for $\delta^{13}C$ fractionation.

The idea of dating mortar was conceived by G. Delibrias and J. Labeyrie (1965) who obtained excellent results with the technique in France. Dating was also attempted in England by M. S. Baxter and A. Walton (1970a) and in the United States by M. Stuiver and C. S. Smith (1965), but without success. We have now refined the technique to make it workable by (1) removing the aggregate and dating only the "live" mortar fraction, (2) using only the first fraction of CO_2 gas, which is evolved from the most reactive mortar, and (3) applying corrections with dendrochronology and $\delta^{13}C$.

Lime mortar and plaster have been used for over 3000 years, (Gourdin and Kingery 1975), and are made by the following process. Limestone ($CaCO_3$) is crushed and burned over a very hot wood fire at over 1000°C. The heat calcines the limestone, driving off the CO_2 and converting the limestone to CaO, quicklime, a white powder (Clark et al. 1940, White 1939). When the material is ready to be used in construction, a variable amount of "aggregate" is added. This consists of sand or gravel, often obtained from a local river, and is used to add bulk. Sometimes leftover, broken bits of brick or marble are thrown in. Water is added, the lime is mixed with the aggregate, and the material is then used to affix building stones or to plaster walls. As mortar hardens or "sets", it is reabsorbs CO_2 from the atmosphere (White 1939), converting back into cryptocrystalline calcite ($CaCO_3$) with a crystal size of 1 micron or less. On crystallization during the setting process, mortar absorbs the correct proportion of ^{14}C from the atmosphere and so can be used for dating the time of construction.

In using this technique, one must be sure that the mortar is indeed lime, $CaCO_3$. Some plaster utilized in buildings is really Plaster of Paris (gypsum), $CaCO_4 \cdot 2H_2O$, which is useless for radiocarbon dating since it contains no carbon. The two may be distinguished by the fact that weak acid will cause true lime mortar to effervesce violently, whereas gypsum mortar will not have a visible reaction. True cement, calcium aluminosilicate, also lacks carbon and cannot be used for dating.

Three complicating effects deserve mention. First, if the lime is improperly burned (too low a temperature, too coarse pieces of limestone), not all of the limestone will be converted to CaO, and the residual pieces will yield dead carbon and give too old a date (Stuiver and Smith 1965). Second, if the mortar is far inside a joint between two building stones, so that it is difficult for atmospheric CO_2 to be absorbed and harden the mortar, it may not set until some years after the time of construction. This is probably a very small effect, however, far less than the resolution error of ^{14}C dating. Most mortar has apparently set by the first 5 or 10 years (Baxter and Walton 1970b), except in some very thick castle walls where soft mortar still occurs after 100 years (White 1939).

A third and far more serious effect, which has nullified attempts to date mortar in England, is the aggregate problem (Stuiver and Smith 1965). If the aggregate used as filler in the mortar consists of quartz, chert, or feldspar sand, there is no problem; all the carbon will come from the live mortar and a correct date will be obtained. Luckily this was the condition in mortars successfully dated by Delibrias and Labeyrie (1965). If the aggregate contains sand or gravel-sized bits of reworked limestone (fig. 1) such as occur in some rivers draining terranes of carbonate rocks, or if the aggregate is made of leftover scraps of marble or limestone used in construction, there is a serious problem and unless special steps are taken, a large amount of dead carbon (old carbon depleted of all radioactive isotope ^{14}C) from the aggregate will be present along with the live carbon (containing ^{14}C) from the mortar itself, and a date obtained on this mixture will be uselessly old. Previous workers have not attempted to separate the mortar from the aggregate and have simply crushed the bulk material. This elementary step is the main modification we have made to render the technique productive.

Sample Preparation and Pretreatment

We have used a simple technique for eliminating the dead carbon contributed by carbonate rock chips in the aggregate (for more details see Valastro 1975). The mortar sample is rinsed to remove dust and then is dried at approximately 100°C. For full study, a piece of mortar about 2 Kg in weight should be collected.

The dried mortar lumps are gently broken with a rubber pestle to approximately 1 cm in size in order to separate the binding mortar powder, which is soft and friable, from the aggregate. Care should be taken not to crush sand grains or pebbles used as aggregate but to make as clean a separation as possible between aggregate and live white mortar powder by rubbing with a soft rubber pestle. Aggregate grains are usually much harder than the mortar itself, so that separation simply requires a little care and time. Next, the fine mortar powder is separated from the coarser aggregate by sieving through a series of 10, 30, 100, 200, and 230 mesh U.S. Standard sieves (respectively, 2.0, 0.59, 0.149, 0.074, and 0.0625 mm). Any size screen can be used if, on inspection with a binocular microscope, it is evident that aggregate sand grains are not going down into the "fines". The material that passes the screens should be inspected for presence of carbonate sand grains. This procedure is continued with freshly broken mortar lumps until approximately 300 gm or more of the powder that passes through the 230 mech screen (0.0625 mm) is accumulated.

A less laborious method of separating 300 gm of mortar powder involves placing approximately 400 gm of broken mortar fragments into a 2-liter Erlenmeyer flask. To the broken mortar is added enough water to afford a supernatant liquid. The liquid and the mortar fragments are continuously agitated until a thick suspension of the mortar powder is formed. The suspension is then passed through a 230 mesh screen and collected in a 3-liter beaker. The shaking is repeated in water until very little suspension is formed. The

FIG. 1. Mortar from Stobi as viewed in a petrographic thin-section (0.3 mm thick). The true, "live", $CaCO_3$ mortar is the featureless gray paste. Larger chunks are aggregate. The large elliptical fragment is a sand grain mode of marble, which provides a large amount of dead $CaCO_3$.

operation is repeated until it is estimated that 300 gm of the fine mortar powder has been accumulated. When collection is completed, the supernatant liquid is decanted from the beaker and the sample is dried in an oven at approximately 100°C. The cohesive cake is weighed, then stored and kept dry in an aluminum foil container until the sample is ready for the chemical preparation.

Chemical Preparation

It is desirable to have three full-size gas samples for ^{14}C determination, so a sufficient amount of mortar is required to generate three 6-liter volumes of CO_2 gas for eventual conversion into three 3-ml samples of benzene (C_6H_6) for counting. To determine the amount of mortar required, a 2-gm portion of each piece of mortar is chemically analyzed to determine the amount of C present. From this one can calculate the weight of mortar required to produce the three sufficient gas samples.

To determine the ^{14}C content, the proper amount of bulk mortar is weighed out and put into a 3-liter round-bottomed flask. To the contents is added enough water to cover the solids. Vacuum is applied by means of a mechanical pump until the chemical train is evacuated to a pressure of 0.01 Torr.

From a separatory funnel, altered to suit vacuum techniques and connected to the 3-liter round-bottomed flask by means of 24-40 vacuum joint, a 1:3 $HCl-H_2O$ acid solution is slowly added in a steady, dropwise flow while the contents of the flask are continuously agitated. Usually 155 ml of the acid solution are adequate to liberate enough CO_2 to yield 3 ml of benzene, which is then utilized in the liquid scintillation techniques.

Despite efforts at complete separation of mortar from aggregate, some pieces of dead limestone (from the aggregate or from unburned limestone pieces) may remain in the sample. If one were to collect all the CO_2 gas evolved from complete acidification of all the $CaCO_3$ present, one could get contamination from the dead limestone pieces. Fortunately the live mortar is very fine-grained, porous, and powdery, reacting quickly in acid. The dead limestone is harder, nonporous, and reacts more slowly. Thus by taking the first gas evolved by the acidification, one recovers CO_2 evolved from the most reactive live mortar. Most of the dead carbonate comes off in the later stages of effervescence. In our technique, the total amount of CO_2 evolved from each sample was divided arbitrarily into three portions. The first six liters of CO_2 to be evolved (designated as the "first fraction") is the essential one because it consists of the effervescence from the live mortar and gives the proper date. The second and third 6-liter quantities of gas are more contaminated with dead carbon. In our samples they have given spurious dates ranging from 200 to over 2000 years older than the first fraction.

Once the CO_2 preparation is completed, the normal procedure for liquid scintillation techniques as practised by the Radiocarbon Laboratory at the University of Texas at Austin is carried out (Pearson et al. 1966, Valastro et al. 1975).

The procedure is as follows: The CO_2 prepared is combined stoichimetrically with Li metal to form lithium carbide (Li_2C_2). The carbide form is hydrolyzed to form acetylene (C_2H_2). The acetylene is trimerized by means of silicon vanadium catalyst to form benzene; the final product is counted for residual radioactivity over a period of 24-48 hours and the age of the sample is then calculated. Dates are corrected for dendrochronology according to the calibration curve by Damon et al. 1972. The samples are also tentatively corrected for $\delta^{13}C$ fractionation (Damon et al. 1966) using the atmospheric value of CO_2, $-7\,^o/_{oo}$. All ^{14}C dates here are dated from reference year 1950. To obtain a correct date to coincide with archaeological dates, add 20 ± 5 years. In other words if a ^{14}C date is regarded here as A.D. 250, the true calendar date is A.D. 270.

Results

We have dated five samples from Stobi, a Hellenistic-Byzantine provincial capital in Yugoslavian Macedonia (Folk and Valastro 1976) (table 1, fig. 2). The first trial sample (Tx -1431) was collected in 1971 by the senior author from the narthex floor of the Synagogue Basilica complex. The estimated date based on archaeological evidence (imported pottery) is late fourth or fifth century A.D. (Wiseman, personal communication 1974). The sample was too small for the needs of the technique, so the date of A.D. 260 ± 180 involved a very large experimental error (all samples have been calculated on the basis of half-life 5730, NBS oxalic acid 95% acttivity, have been corrected for dendrochronology and tentatively corrected for $\delta^{13}C$). Nevertheless there was no statistical disagreement with the archaeological evidence, so we continued to work on mortar dating.

Two large samples were collected in 1973 from the theater by E. M. Davis, one (Tx-1941) from the foundation of the analemma in the east parodos, and the other (Tx-1942) from the south wall of the first radial corridor of the cavea, next to the west parodos. Archaeological evidence based on imported pottery points to an initial construction phase of the theater in the late first and early second centuries A.D., but it is not yet clear whether one building episode is represented, with modification of the plan as the work proceeded, or whether there was more than one separate stage of construction (Gebhard, personal communication 1974).

We split each of the two mortar samples from the theater into three parts, preparing and dating each part separately. The three dates from (Tx-1941) have an average of A.D. 233 ± 40, and those from (Tx-1942) have an average of A.D. 243 ± 50. These dates, being in excellent statistical agreement with each other, represent contemporaneity in radiocarbon dating terms. Since the archaeological evidence also indicates contemporaneity, all six dates from the two samples can be averaged, giving a mean age of A.D. 233 ± 32. (Long and Rippeteau 1974). This average implies that construction time of the theater may be later than the archaeological evidence suggests.

From the Episcopal Basilica, sample (Tx-1943) was obtained by E. M. Davis in 1973 from the socle foundation of the south wall of the principal construction stage. Architectural style, documentary evidence, and pottery dates this stage of construction as very close to A.D. 400 (Wiseman and Rodoševic, personnal communication 1974). As with the samples from the theater, (Tx-1943) was split into three parts which were prepared and counted independently of one another, resulting in an average of A.D. 302 ± 60. Charcoal

TABLE 1.
Age Determinations on Mortar Sample from Stobi

University of Texas Radiocarbon Laboratory cat no.	Stobi Project cat no.	Location	Assay no. sample mesh no. and mm size*	Age before present (Half-life 5568) age ref. 1950 uncorrected dates	Age before present (half-life 5730) age ref. 1950 corrected for Dendrochronologic value – Arizona Method	Date B.C.-A.D. corrected	Age before present (half-life 5730) age ref 1950 average date corrected for Dendrochronological value and corrected for $^{13}C/^{12}C$	Archaeological estimate of date
Tx-1931	Stobi-Folk	Central Basilica Narthex – Floor	(1) <200 0.074		1690 ± 180**	A.D. 260 ± 180		
Tx-1944	Mortar no. 1	Central Basilica Floor 1	(1) <230 0.0625 (2) Mortar pieces Avg	1628 ± 50 1560 ± 90 1594 ± 60	1615 ± 60 1542 ± 100 1578 ± 60	A.D. 335 ± 60 A.D. 408 ± 110 Avg A.D. 372 ± 60	$\delta^{13}C = -15.823$ 1510 ± 60 – 440 ± 60 A.D.	Late 4th or 5th century A.D.
Tx-1941	Mortar no. 4	Theater – E. Parodos Analemma Foundation	(1) <230 0.0625 (2) <230 0.0625 (3) <230 0.0625 Avg	1740 ± 60 1720 ± 60 1730 ± 60 1730 ± 34	1740 ± 70 1716 ± 70 1726 ± 70 1727 ± 40	A.D. 210 ± 70 A.D. 234 ± 70 A.D. 224 ± 70 Avg A.D. 223 ± 40	$\delta^{13}C = -9.320$ 1690 ± 40 – 260 ± 40 A.D.	– Early 2nd Century A.D.
Tx-1942	Mortar no. 9	Theater – S. Wall 1st Radial Corridor	(1) <230 0.0625 (2) <230 0.0625 (3) <230 0.0625 Avg	1738 ± 80 1690 ± 70 1715 ± 70 1713 ± 43	1730 ± 90 1680 ± 80 1710 ± 80 1707 ± 50	A.D. 220 ± 90 A.D. 270 ± 80 A.D. 240 ± 80 Avg A.D. 243 ± 50	$\delta^{13}C = -8.180$ 1700 ± 50 – 250 ± 50 A.D.	– Early 2nd Century A.D.
Tx-1943	Mortar no. 10	Episcopal Basilica S. Aisle Foundation of Principal Structure	(1) Mortar pieces (2) <230 0.0625 (3) <230 0.0625 (4) <230 0.0625 Avg	1680 ± 130 1627 ± 70 1670 ± 60 1640 ± 70 1654 ± 43	1670 ± 140 1614 ± 80 1660 ± 70 1629 ± 80 1643 ± 49	A.D. 278 ± 130 A.D. 336 ± 70 A.D. 290 ± 60 A.D. 321 ± 80 Avg. A.D. 306 ± 49	$\delta^{13}C = -19.207$ 1550 ± 50 – 400 ± 50 A.D.	A.D. 385-410
Tx-1940	Mortar no. 1	East City Wall Area – next to E. Wall of Casa Romana	(1) <230 0.0625 (2) <230 0.0625 Avg	2015 ± 80 2051 ± 60 2033 ± 50	2040 ± 90 2090 ± 70 2065 ± 60	B.C. 90 ± 90 B.C. 140 ± 70 Avg B.C. 115 ± 60	$\delta^{13}C = -16.220$ 1880 ± 60 – 70 ± 60 A.D.	4th century A.D.

*Each date is based on the first evolution of CO_2 gas; most samples were split into 2 or 3 subsamples for replicate analysis.

**Analysis on very small sample.

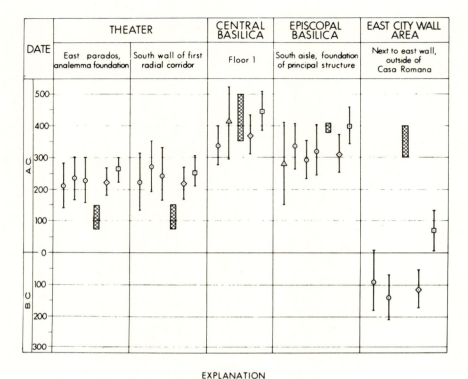

FIG. 2. Chart showing archaeological age estimate with ^{14}C mortar dates.

dates on this same structure give ^{14}C dates of A.D. 340 and A.D. 290, agreeing with the mortar dates and testifying to the validity of the mortar dating method (Davis et al. 1973), (fig. 3).

We have also dated a sample from the Casa Romana (Wiseman and Snively, personal communication 1974) with results that do not agree with archaeological evidence, but there are many possible sources of error here including a sample mixup (Folk and Valastro 1976). For more certainty, a mortar sample must be collected from the actual structural walls of the Casa Romana.

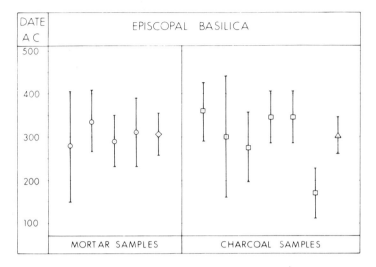

FIG. 3. Chart comparing mortar dates and charcoal dates from the Episcopal Basilica.

CONCLUSIONS

Radiocarbon dating of mortar is a relatively new technique, full development of which has been impeded by problems of contamination by dead carbon. If aggregates containing dead carbon are carefully removed from the samples, and if only the first evolution of CO_2 gas from the live mortar is used, promising results are obtained. From an ancient building it is easier to collect chunks of mortar, which give the correct date of actual construction, than to search for bits of charcoal, which give a date relevant to the time when the charcoal was part of a living tree.

In evaluating the radiocarbon dating of mortar, it is significant that the dates from split samples (in the case of the theater, from multisamples) were clustered. Mortar dates also agree with charcoal dates in one structure where a cross-check was obtained. Furthermore the mortar dates fall close to the archaeological dates, even though these latter are subject to continual fluctuation as new cultural discoveries are made. Many more samples from well dated structures are required so that patterns of date clusters can be used to calibrate the radiocarbon dates against the known dates. At such time one can expect the radiocarbon dating of mortar to provide reliable dates for mortar construction in buildings of unknown date.

If absolute dates through mortar dating should prove especially difficult to obtain, the technique may well provide relative ages for differentiating building episodes not distinguishable through other lines of evidence. Our experience with the two mortar samples from the theater suggests this likelihood, in which contemporaneity is attested to both by archaeological evidence and by radiocarbon dating. Mortar dating may also add confirmation to stylistic dating.

Furthermore dating of mortar and plaster offers much in the way of detailed dissection of the anatomy of construction of ancient buildings. Successive construction stages or alterations in a building are not always datable on stylistic or contextual grounds alone, and the application of mortar dating may well provide a valuable line of evidence for discerning such episodes.

As a hypothetical example of such a complex building, consider an ancient temple whose foundations were laid of rough rubble and mortar in 600 B.C. The temple superstructure was razed and converted to a Christian church about A.D. 100, when they made a mosaic floor and put up old style walls. After military destruction, new style walls were put up in A.D. 500, but the old floor was not modified. Perhaps repair work on the roof was done in A.D. 800, retaining the same stones and design as used in the A.D. 500 building. Plaster walls and frescos were added in A.D. 1000. Mortar dating would give several different dates for this one building, whereas an archaeological dating based on overall "style" might put the whole edifice at A.D. 500. Mortar dating can embellish and confirm the stylistic dating, providing many intriguing surprises (Delibrias and Labeyrie 1965).

ACKNOWLEDGMENTS

We are grateful to Dr. E. Mott Davis of the Anthropology Department, University of Texas at Austin, for collecting the samples and for the task of integrating the results of this work with the archaeological evidence at Stobi. We would like to thank Dr. James Wiseman of Boston University and director of the Stobi project for his assistance. Many thanks are due to Dr. Lynton S. Land of the University of Texas, Geology Department, for analyzing the samples for $\delta^{13}C$, and to Mrs. Alejandra G. Varela of the Radiocarbon Laboratory, University of Texas at Austin, for her assistance in the preparation of the samples for ^{14}C dating.

REFERENCES

Baxter, M. S., and A. Walton
    1970a Glasgow University Radiocarbon Measurement III. Radiocarbon. 12:496-502.
    1970b Radiocarbon dating of mortars. Nature. 225:937-938.

Clark, G. L., W. F. Bradley, and V. J. Azbe
    1940   Problem in lime burning; a new x-rays approach. Industrial and Engineering Chemistry. 32:972-976.

Damon, Paul, Austin Long, and E. I. Wallich
    1972   Dendrochronologic calibration of the Carbon-14 time scale. Proc. 8th Int. Conf. on Radiocarbon Dating. New Zealand. A28-43.

Damon, Paul, Austin Long, and Donald C. Gray
    1966   Revised ^{14}C dates for the reign of Pharoah Sesastris III. Geophys. Res. 71:1055-1063.

Davis, E. Mott, D. Srdoč, and S. Valastro.
    1973   Radiocarbon dates from Stobi, 1971 season Mano-Zissi and Wiseman, eds. Studies in the antiquities of Stobi. 1:23-36.

Delibrias, G., and J. Labeyrie
    1965   The dating of mortars by the Carbon-14 method. Proc. 6th Int. Conf. on Radiocarbon and Tritium Dating. Clearing House for Federal Scientific and Technical Information, U.S. Dept. of Commerce, 344-347.

Folk, Robert L.
    1973   Geologic Contributions to Archaeology and Dating Techniques. Yugoslavia. (abs.) Geol. Soc. America. 5:624.

Folk, Robert L., and S. Valastro, jr.
    1976   Radiocarbon dating of mortars at Stobi. Mano-Zissi, and Wiseman eds. Studies in the Antiquities of Stobi. II:29-41.
    (In Press).   Successful technique for dating of lime mortar by carbon-14. J. Field Arch.

Gebhard, Elizabeth
    1974   Personal communication to E. Mott Davis.

Gourdin, W. H., and W. D. Kingery
    1975   The beginning of pyrotechnology: Neolithic and Egyptian lime plaster. Field Arch. 2 (1/2):133-150.

Long, Austin, and Bruce Rippeteau
    1974   Testing contemporaneity and averaging radiocarbon dates. American Antiquity. 39(2):205-215.

Pearson, F. J., E. Mott Davis and M. A. Tamers
    1966   University of Texas radiocarbon dates IV. Radiocarbon. 8:453-466.

Stuiver, M., and C. S. Smith
    1965   Radiocarbon dating of a ancient mortar and plaster. Proc. 6th int. Conf. on Radiocarbon and Tritium Dating. Clearing House for Scientific and Technical Information, U.S. Dept. of Commerce, p. 338-343.

Valastro, S., jr., E. Mott Davis, and A. G. Varela
    1975   University of Texas at Austin dates X. Radiocarbon. 17:52-98.

Valastro, S., jr.
    1975   A new technique for the radiocarbon dating of mortar. M. A. Thesis. Univ. of Texas.

Valastro, S., jr., and Robert L. Folk
    1974   A new radiocarbon technique for dating mortar. (abs.) Geol. Soc. America. 6:993-994.

Wiseman, James
    1974   Personal communication to E. Mott Davis. Context storage lots 123 and 124 in Stobi records.

Wiseman, James, and Ž. Radoševic
    1974   Personal communication to E. Mott Davis. Lot 995 in Stobi records.

Wiseman, James, and C. S. Snively
    1974   Personal communication to E. Mott Davis. Lots 724, 725, and 726 in Stobi records.

White, Alfred H.
    1939   Engineering Materials. New York, McGraw Hill, 547.

2

Comparison of ^{14}C Ages and Ionium Ages of Corals

Kunihiko Kigoshi, Akira Tezuka and Hiroshi Mitsuda

ABSTRACT

The ionium age of coral can be compared directly to its radiocarbon age, but the reliability of the ionium age is not so high that it may be used to check the basic assumptions of radiocarbon dating.

An examination of coral samples which gave significantly discordant ionium and radiocarbon ages showed the existence of an appreciable amount of ionium adsorbed on the surface of the calcium carbonate in the coral during its preservation period. A method of cleaning secondary absorbed ionium described here provides an efficient means to make the ionium age reliable.

INTRODUCTION

Ionium datings on autogenic carbonates have contributed to the dating of geological events and sea level changes during the last 200,000 years, but the reliability of these dates is still in question. Some of the results of ionium datings have been directly compared with radiocarbon dates (Thurber et al. 1965, Osmond et al. 1970, Kaufman et al. 1965, 1971, Konishi et al. 1968) and some have been compared with protoactinium dates (Ku 1968, Konishi et al. 1968). The agreement of dates obtained by these different methods is not always good.

It is generally recognized that the ionium dates of coral samples are far more reliable than dates on shells, but even for the coral samples, disagreement between ionium and radiocarbon dates is not rare. This paper presents the results of an attempt to obtain reliable ionium dates on coral samples and to discuss the basic assumptions of both dating methods.

Analytic Method

The coral samples were sliced thin and washed with distilled water using an ultrasonic cleaner. The uranium, the thorium, and the ^{232}U spike were separated out from the solution of the coral in hydrochloric acid by means of anion exchange resin. The uranium was purified by solvent extraction using tributylphosphate and 8-oxyquinoline, then electroplated on a stainless steel disk. The amounts of uranium and thorium isotopes were determined by alpha spectroscopy using a surface-barrier type solid detector. The radiocarbon dates were measured on the carbon dioxide evolved by acid from the coral sample with normal dating techniques.

Disagreement of Ionium and Radiocarbon Ages

Many recent mollusks and corals have given concordant ionium and radiocarbon ages; but serious disagreements appear in older samples. For coral samples the disagreement has a general trend in which the ionium dates are older than the radiocarbon ages. The reason for this disagreement is believed to be either the loss of uranium in coral or the accretion of ^{230}Th to the coral during the preservation period. If we assume that the radiocarbon dates on the coral samples are correct, and that the discordant older ionium dates come from the adsorption of ^{230}Th onto the coral, the excess amount of ^{230}Th can be estimated for each coral sample.

The graph plotting the excess amount of ^{230}Th vs. radiocarbon age of the coral sample (fig. 1) indicates the elaboration of ^{230}Th during the preservation period. It was expected that the cleaning of surface contamination of ionium on coral samples would give concordant ionium and radiocarbon ages.

Decontamination Procedure and Results Obtained

To remove the adsorbed ^{230}Th from the surface of coral, it is necessary to find a condition in which the removed and dissolved ^{230}Th does not adsorb again onto the coral sample.

The adsorption of thorium on coral powder depends on the pH of the solution, as shown in figure 2. When the coral powder is put into an acidic solution, the pH change of the solution is so rapid that it is difficult to separate the solid phase from the washing solution with good, reproducible results. Decontamination with an acid solution might allow separation of the solid phase with reproducible results, but only when the pH of the solution rises above four, and the pH change slows. As would be expected from the adsorbed amount of ^{230}Th indicated in figure 2, a large volume of washing solution is necessary to prevent the readsorption of ^{230}Th onto the surface of the solid phase. It is better to add ^{232}Th as a holdback carrier in the cleaning solution to depress the degree of readsorption of ^{230}Th. In our preliminary experiments the decontamination was done without the ^{232}Th holdback carrier to avoid any disturbing effects it might impose on the analytic data of ^{232}Th in the coral.

For the decontamination by acid, the coral sample was powdered and sieved finer than 200-mesh, and powder which was too fine was removed by washing with distilled

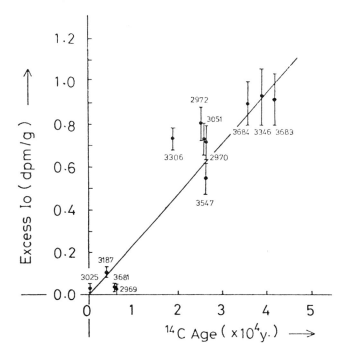

FIG. 1. Relation between excess ionium and ^{14}C age.

water. For one gram of coral powder, about one liter of 0.1 to 0.5 N hydrochloric acid was added. Within ten to twenty seconds the bubbling mixture of calcium carbonate and acidic solution was separated using a large Büchner funnel with rapid pumping. The pH of the separated filtrate was lower than three. The remaining amounts of the decontaminated coral powder were about half of the original amounts.

Table 1 shows a comparison of the analytic data on the decontaminated coral powder with the data obtained on the same sample cleaned only with distilled water using an ultrasonic cleaner. Before and after the decontamination process, the contents of ^{234}U and ^{238}U were not changed, but the content of ^{230}Th was lowered by this acid washing. This suggests that a part of the thorium in the coral was surface adsorbed. The calculated ionium ages of the decontaminated coral samples became younger and showed a good agreement with the corresponding radiocarbon ages with the exception of sample GaK-3684. The age of this last sample after decontamination remained higher than the radiocarbon age. Further decontamination processing may possibly bring the date closer to the radiocarbon date.

In contrast to the change of ^{230}Th before and after decontamination, the ^{232}Th content showed almost no change. This would suggest a very high $^{230}Th/^{232}Th$ ratio in the aqueous phase from which the coral adsorbed the thorium onto the skeletal surface in the preservation period.

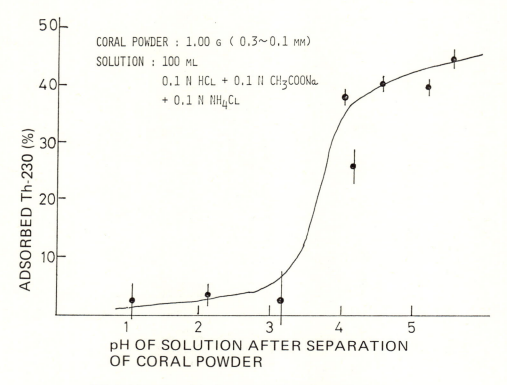

FIG. 2. Adsorption of thorium onto coral powder on pH of solution.

CONCLUSION

Ômura and Konishi (1970) have expressed the view that part of the uranium in coral may not be retained at the time of alteration from aragonite to calcite. Figure 3 shows the relation between the alteration of coral samples, expressed by the fraction of calcite in the coral, and the disagreement of ^{230}Th and ^{14}C ages. The disagreement of both datings is expressed by $\Delta_{Io-^{14}C}$ in the ordinate of figure 3. From figure 3, we could not find any correlation between the amount of calcite in the coral and the order of disagreement between the radiocarbon dates and ionium ages which were obtained without the acid cleaning process.

As indicated in table 1, the decontamination procedure made an appreciable change in aragonite/calcite ratios in the coral samples of GaK-3025 and GaK-3346. Both cases were similar to the other samples, however, in that the contents of uranium isotopes did not change after the decontamination procedure. This constancy of uranium content also supports the view that the discordant ionium ages in these samples were not the result of calcium carbonate alteration from aragonite to calcite in the preservation period. With respect to the cases reported here, it may be concluded that the disagreements between the radiocarbon dates and the ionium ages obtained without the acid cleaning process

TABLE 1

Comparison of Analytical Data Before and After the Acid Cleaning of Coral Powder

14C dating code no. GaK –	Mineral composition (wt %)		Uranium content d/m/g		Thorium content d/m/g		Io age (10^4 yr)	14C age (10^4 yr)
	aragonite	calcite	238U	234U	232Th	230Th		
2972	0	100	1.46 ± 0.05	1.53 ± 0.05	0.09 ± 0.001	1.11 ± 0.08	13.5 $^{+0.15}_{-0.14}$	2.54 $^{+0.15}_{-0.14}$
after cleaning	0	100	1.48 ± 0.04	1.56 ± 0.04				
3025	63	37	2.69 ± 0.09	2.99 ± 0.10	0.03 ± 0.02	0.044 ± 0.007	0.16 ± 0.11	0.02 ± 0.01
after cleaning	52	48	2.67 ± 0.09	2.96 ± 0.10	0.07 ± 0.02	0.07 ± 0.02	0.248 ± 0.06	
3306	100	0	0.66 ± 0.025	0.728 ± 0.03	0.04 ± 0.005	0.77 ± 0.056	(U < 10)	1.90 ± 0.045
after cleaning	100	0	0.69 ± 0.022	0.769 ± 0.025	0.04 ± 0.01	0.12 ± 0.04	1.89 ± 0.61	
3346	34	66	2.38 ± 0.06	2.68 ± 0.07	0.04 ± 0.008	1.70 ± 0.13	10.9 ± 1.6	3.93 $^{+0.31}_{-0.21}$
after cleaning	74	26	2.38 ± 0.09	2.66 ± 0.10	0.04 ± 0.006	0.762 ± 0.035	3.66 ± 0.25	
3547	100	0	2.05 ± 0.07	2.22 ± 0.08	0.07 ± 0.007	1.01 ± 0.07	6.53 ± 0.74	2.66 ± 0.14
after cleaning	100	0	2.12 ± 0.04	2.29 ± 0.05	0.07 ± 0.01	0.467 ± 0.06	2.45 ± 0.33	
3684	100	0	1.85 ± 0.07	2.00 ± 0.07	0.01 ± 0.003	1.44 ± 0.11	13.7 ± 2.1	3.61 ± 0.14
after cleaning	100	0	1.77 ± 0.06	1.86 ± 0.06	0.06 ± 0.01	0.545 ± 0.034	5.05 ± 0.56	

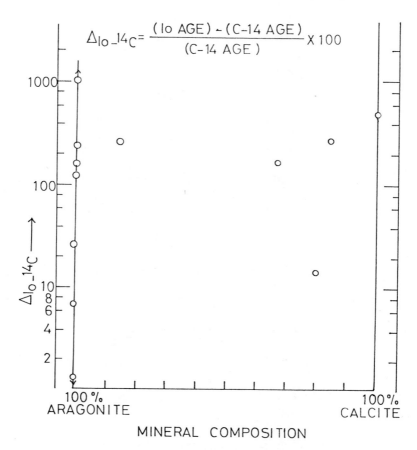

FIG. 3. Mineral composition and disagreement of ionium and radiocarbon ages.

TABLE 2

Sampling Site and Donor of Coral Sample

Code no. GaK –	Sampling site	Donor
2972	4° S 40° E Kenya, Africa	Nobuyuki Hori, Tokyo Metro. Univ.
3025	19°46′30″ N 72°5′30″ W Cape Heitien, Haiti	H.K. Brooks, Univ. of Florida
3306	4°28′ N 118°37′ E Semporna, Borneo	Shoji Fujii, Toyama Univ.
3346	31° S 159° E Lord Howe Isl, N.S.W.	C.V.G. Phipps, Univ. of Sydney
3547	13°07′ N 59°39′ W Bridgestone, Barbados	Arata Sugimura, Kobe Univ.
3684	21° S 166°57′W East coast of Cifou, New-Caledonia	J. Launay, ORSTOM Noumea

were the result of surface contamination of ionium on the calcium carbonate crystals in the coral samples during the preservation period.

If the discordant ages between the ^{230}Th and ^{14}C ages were caused only by surface contamination of coral samples, the proper decontamination process could open a way to the precise cross-checking of radiocarbon ionium datings which would give information complementary to tree ring checks of radiocarbon datings in the period older than 10,000 years.

ACKNOWLEDGMENTS

The authors wish to thank the donors of coral samples who permitted their use for this research.

REFERENCES

Kaufman, A., and W. S. Broecker
    1965   J. Geophys Res. 70:4039-4054.
Kaufman, A., W. S. Broecker, T. Ku, and D. L. Thurber
    1971   Geochim. et Cosmochim. 35:1155-1183.
Konishi, K., A. Ômura, and T. Kimura
    1968   Geol. Paleontology of Southeast Asia. 5:211-224.
Ku, T. L.
    1968   J. Geophys. Res. 73:2271-2276.
Ômura, A., and K. Konishi
    1970   J. Geol. Soc. Japan. 76:389-397.
Osmond, J. K., J. P. May, and W. F. Tanna
    1970   J. Geophs. Res. 75:469-479.
Thurber, D. L., W. S. Broecker, H. A. Potratz, and R. L. Blanchard
    1965   Science. 149:55-58.

3

The Dating of Fossil Bones using Amino Acid Racemization

Jeffrey L. Bada, Patricia M. Masters,
E. Hoopes, and D. Darling

ABSTRACT

Amino acid racemization dating is an important new technique which can be used to determine the ages of fossil bones which are either too old or too small to be dated by radiocarbon. The racemization method has an effective dating range beginning a few thousand years B.P. and extending back to several hundred thousand years, the actual range being dependent upon the temperature of the region where the bone was found. Only a few grams of bone are required in amino acid racemization dating.

In this paper we first describe the procedures used in racemization dating. We then present comparisons between racemization-derived ages and those estimated by some other, independent, dating technique, including radiocarbon. We also show that the racemization rates determined using well-dated fossil bones show good correlation with the estimated temperature exposure of the various samples. Finally, we list several criteria which can be used to judge the reliability of a racemization-deduced age.

INTRODUCTION

A new method of dating fossil materials based on the racemization reaction of amino acids has been developed within the last few years (see Bada and Schroeder 1975; Schroeder and Bada 1976 for reviews). This technique is based on the fact that the amino acids present in the proteins of living organisms consist only of the L-enantiomers. Over long periods of geologic time, however, these L-amino acids undergo slow racemization, producing the corresponding D-amino acids. Fossil materials have been found to contain both L- and D-amino acids, and the D/L amino acid ratio increases with the age of the sample. The extent of racemization of amino acids has been used to estimate the ages of various fossil materials including deep-sea sediments, shells, fossil bones, wood, and coprolites.

Amino acid racemization dating has been found to be particularly useful in estimating the ages of fossil bones (Bada and Helfman 1975). The method has an effective dating range beginning at a few thousand years B.P. and extending to several hundred thousand years B.P., the actual range being dependent upon the temperature of the region where the bone was found. Only a few grams of bone are required for amino acid racemization analysis.

Of the various amino acids, aspartic acid has been the most widely used in fossil bone dating. This amino acid has one of the fastest racemization rates of the stable amino acids (Bada 1972a, Bada, Kvenvolden, and Peterson 1973). At 20°C, in bone the half-life for aspartic acid racemization is approximately 15,000 years. Thus for most mid- or low-latitude sites, the racemization rate of aspartic acid is much slower than the decay rate of radiocarbon and therefore can be used to date bones which are too old to be dated by ^{14}C.

Since racemization is a chemical reaction, the rate of the reaction is temperature dependent. Thus in order to date a bone using racemization, it is necessary to evaluate the average temperature to which the bone has been exposed. This temperature can be evaluated using a procedure in which the in situ rate of amino acid racemization for a particular site is calculated by measuring the extent of racemization in a bone which has been previously dated, either by radiocarbon or by some other independent dating technique (Bada and Protsch 1973, Bada et al. 1974, Bada, Schroeder, and Carter 1974, Bada and Deems 1975, Bada and Helfman 1975). After this "calibration" has been carried out, other bones in this same general area can be dated, based on their extent of amino acid racemization. Ages determined using this calibration procedure have been shown to be in close agreement with ages deduced using ^{14}C or other chronological information (Bada et al. 1974, Bada and Deems 1975, Bada and Helfman 1975).

In this paper we first discuss the techniques used in amino acid racemization dating. We compare racemization-derived ages with those estimated by some other, independent dating technique, and we show that the aspartic acid racemization rates which we have determined from the extent of racemization in independently-dated bones correlate very well with the estimated temperature exposure of the various amples. Finally, we recommend a standard procedure for reproting amino acid racemization dates and list several criteria which can be used to judge the reliability of a racemization-deduced date.

Experimental Procedures Used In Racemization Analysis

An outline of the procedures used for determining the extent of racemization of several amino acids in a fossil bone sample is shown in figure 1. The general cleaning procedure used for racemization analysis is very similar to that used in the ^{14}C dating of bones (Olsson et al. 1974). Whereas in ^{14}C dating, collagen is extracted and used as the fraction, in amino acid racemization dating the total amino acid mixture from the fossil bone is isolated. Three separate fractions are obtained from a typical bone sample: an aspartic acid fraction, a glutamic acid fraction, and a fraction containing all of the remaining amino acids.

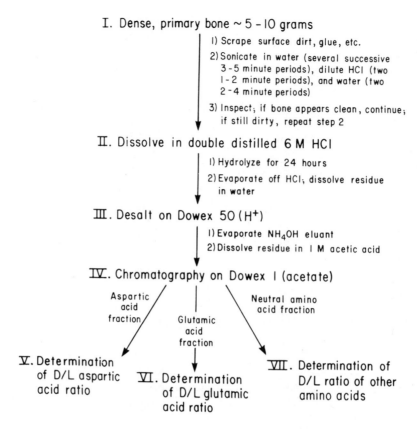

FIG. 1. An outline of the procedures used in amino acid racemization dating. More complete descriptions of the various steps can be found in the following references: step III, Bada, Luyendyk, and Maynard, 1970; step IV, Hirs, Moore, and Stein, 1954; step V, Bada and Protsch, 1973; and steps VI and VII, Hoopes, Peltzer, and Bada, 1978.

The extent of racemization of aspartic acid is determined using the diastereomeric dipeptide technique (Manning and Moore 1968, Bada and Protsch 1973). In this procedure, the isolated aspartic acid is reacted at 2°C and pH 10.2 with L-leucine-N-carboxyanhydride (L-leu-NCA), in order to synthesize two different dipeptides: L-leucyl-D-aspartic acid and L-leucyl-L-aspartic acid. These two dipeptides are then separated on an Automatic Amino Acid Analyzer (Bada and Protsch 1973). A sample chromatogram is shown in figure 2.

The extent of racemization of glutamic acid and the other amino acids is determined by gas chromatography after the synthesis of an appropriate derivative. There are several gas chromatographic techniques which could be used to determine the extent of racemization of amino acids. The one which we use (Hoopes, Peltzer, and Bada, 1978) involves first the synthesis of the methyl esters of the amino acids, which are then reacted

THE DATING OF FOSSIL BONES

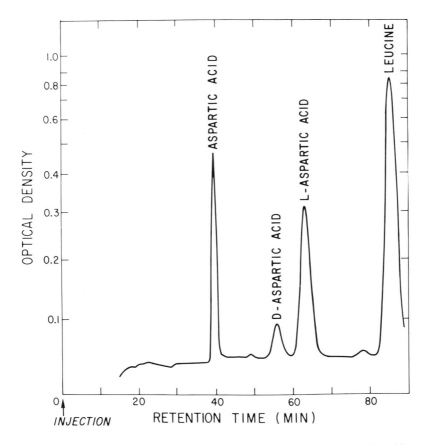

FIG. 2. Amino acid analyzer chromatogram showing diastereomeric dipeptides L-leucyl-D-aspartic acid (labeled D-aspartic acid) and L-leucyl-L-aspartic acid (labeled L-aspartic acid) for a human skeleton found at La Jolla, California in April 1976. The aspartic acid peak represents the material which did not react with the L-leu-NCA; the leucine is derived from the hydrolysis of L-leu-NCA.

with trifluoroacetyl-L-prolyl chloride (TPC). This particular procedure was chosen over the others because the TPC reagent can be synthesized from inexpensive starting materials and the gas chromatography analyses can be carried out on standard packed-glass columns. Thus this particular method does not require the use of any reagent which is difficult to obtain or the use of any specialized gas chromatographic columns. Sample chromatograms showing the analysis of the glutamic acid and neutral amino acid fractions using this technique are shown in Figures 3a and 3b, respectively.

It should be emphasized that great caution must be exercised in carrying out amino acid analysis in order to avoid contaminating a sample with more modern amino acids at some point in the sample workup (Hare 1965). Therefore all reagents are specially prepared. Water is double-distilled in an all-glass apparatus. Double-distilled 6M HCl is prepared by two successive azeotropic distillations. Ammonium hydroxide is prepared by

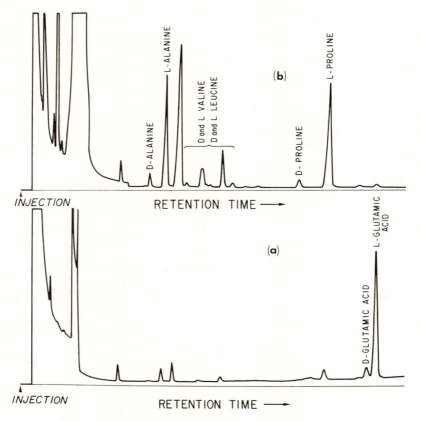

FIG. 3. Gas chromatographic trace showing the trifluoroactyl-L-prolyl amino acid methyl esters for a human skeleton found at La Jolla, California in April 1976. The glutamic acid derivatives isolated in step VI of figure 1 are shown in 3a; the neutral amino acid derivatives isolated from step VII are shown in 3b.

dissolving gaseous ammonia in double-distilled water on a vacuum line. All glassware is precleaned with potassium dichromate-sulfuric acid cleaning solution, and this is followed by extensive washing in double-distilled water.

Amino Acid Racemization-Deduced Ages

The amino acid racemization reaction can be written as

$$\text{L-amino acid} \underset{k_i}{\overset{k_i}{\rightleftharpoons}} \text{D-amino acid} \tag{1}$$

where k_i is the first-order rate constant for the interconversion of the L- and D-enantiomers for a particular amino acid. The kinetic equation for amino acid racemization is (see Bada and Schroeder 1972 for derivation)

$$\text{Ln}\left\{\frac{1+(D/L)}{1-(D/L)}\right\} - \text{Ln}\left\{\frac{1+(D/L)_{t=0}}{1-(D/L)_{t=0}}\right\} = 2 \cdot k_i \cdot t \quad (2)$$

where (D/L) is the measured enantiomeric ratio in the fossil bone, $(D/L)_{t=0}$ the enantiomeric ratio in a Modern bone sample carried through the same processing procedure, and t is the age of the sample. The $(D/L)_{t=0}$ term is not zero, since a small amount of racemization takes place during the acid hydrolysis step. For a 24-hour 6M HCl hydrolysis, the $(D/L)_{t=0}$ term for aspartic acid is 0.07 (Bada and Protsch 1973).

The value of k_i in the above equation for any particular bone will depend on the temperature history of the sample. In order to determine the value of k_i for a site, the extent of amino acid racemization is measured in a bone which has been dated by some other technique. After this calibration has been carried out, other samples from the area can be dated, based on their extent of amino acid racemization.

ASPARTIC ACID RACEMIZATION-DEDUCED AGES

Aspartic acid has the fastest racemization rate of any of the stable amino acids (Bada 1972a, Bada, Kvenvolden, and Peterson 1973). For most mid- and low-latitude sites, significant racemization of aspartic acid will have taken place in a bone in the time interval datable by radiocarbon. In areas such as Egypt, due to the very high temperatures, appreciable racemization of aspartic acid will be present in historically-dated samples. Because of the longer half-life of the aspartic acid racemization reaction compared to the decay rate of radiocarbon, aspartic acid racemization can also be used to date bones which are too old for radiocarbon dating. In table 1 are listed some comparisons between aspartic acid racemization-deduced ages and ages deduced by radiocarbon, from historical records, or from geological information.

The results listed in table 1 show that there is close agreement between the aspartic acid racemization ages and those deduced by other techniques, with the one exception of the Egyptian Sakkara sample. The disparate results for the Sakkara sample are difficult to explain. In the only other instance we have found so far where there was a disagreement between racemization and other independently-deduced ages, that is, Olduvai Gorge, Upper Ndutu Bed sample (Bada and Helfman 1975), it was clear that the racemization age was compatible with the age range suggested by other information and that the radiocarbone date of this sample was erroneous. The racemization ages of several other Egyptian bones (i.e., Tarkhan and Tura) agree very well with the historical ages. In fact, the analyses of three different Tarkhan skeletons have yielded essentially identical D/L aspartic acid values (average = 0.460 ± 0.010) even though the bones have differed greatly in their state of preservation and amino acid content (see table 1). Further work with Egyptian material should help elucidate the reason(s) for the anomalous Sakkara racemization age. It is important to note, however, that the Sakkara racemization age is too young, which suggests that if racemization ages are erroneous they will tend to be less than the true ages of the samples.

The results listed in table 1 for Nelson Bay Cave, Klasies River Mouth Cave (KRM), and Swartklip (SW) are particularly interesting. The bone samples from KRM and SW

TABLE 1

Comparisons between Aspartic Acid Racemization and Either Radiocarbon, Historical, or Geologically-Inferred Ages

Site	D/L Aspartic acid	Radiocarbon, historical or geologically-inferred age (yrs)	Aspartic acid age (yrs)
Muleta Cave, Mallorca, Spain (from Bada et al. 1974)	0.273 0.293 0.455	16,850 ±200 UCLA 1704D 18,890 ±200 UCLA 1704E 28,600 ±600 UCLA 1704A	$k_{asp}=1.25 \times 10^{-5} yr^{-1}$ 18,600 33,700
Murray Springs, Arizona (from Bada et al., 1974) Double Adobe, Arizona (from Bada and Helfman, 1975)	0.33 0.52 0.50	5,640 ±60 A-905A, B 11,230 ±340 A-805 10,420 ±100 A-1152	$k_{asp}=4.84 \times 10^{-5} yr^{-1}$ 10,500 9,900
EGYPTIAN SAMPLES			
Tarkhan, near Cairo (this report) F 1516 F 1691 F 1556	0.450*^{a} 0.475*^{b} 0.465*^{c} 0.453	1st and 2nd dynasty 3200-2700 B.C.	$k_{asp}=9.31 \times 10^{-5} yr^{-1}$ 3360 B.C. 3220 B.C.
Sakkara, near Cairo (this report)	0.137†	4th dynasty 2500-2400 B.C.	760 A.D. (?)
Tura, near Cairo (this report)	0.416	Middle Kingdom 2100-1600 B.C.	2030 B.C.
SOUTH AFRICAN COASTAL SITES			
Nelson Bay Cave (from Bada and Deems 1975)	0.151 0.167	16,700 ±240 I-6516 18,100 ±500 UW-185 18,660 ±110 GrN-5889	$k_{asp}=4.92 \times 10^{-6} yr^{-1}$ 20,000
Klasies River Mouth Cave 1 (from Bada and Deems 1975) Level^d 13 16 18 19	 0.370 0.467 0.474 0.548	 ~80,000 to 120,000^e	 65,000 89,000 90,000 110,000
Swartklip I (this report)	0.55	Beginning of last glacial? (i.e., ~80,000)^f	~110,000
Heuningsneskrans Shelter, South Africa (from Bada and Helfman 1975, and this report) Depth in Deposit 0.52 m 0.72 m 0.81 m	 0.54 0.62 0.71	 20,500 ±300 LJ 3136 23,900 ±800 LJ 3138 —	 $k_{asp}=2.64 \times 10^{-5} yr^{-1}$ 24,800 31,000

TABLE 1. (continued)

Site	D/L Aspartic acid	Radiocarbon, historical or geologically-inferred age (yrs)	Aspartic acid age (yrs)
Mt. Carmel Caves, Israel (from Bada and Helfman 1976)			
Sefunim layer 8-9	0.472	~27,000 [g]	$k_{asp}=1.64 \times 10^{-5} yr^{-1}$
Tabūn layer C, Square 5, Bed 22, T69	0.658	40,900 ± 1000 GrN 2729	44,000
Tabūn layer E, Square 14, Bed 49	0.742	~80,000 (?) [h]	54,000
Stránská Skála, near Brno, Czechoslovakia (this report)			
Layer No. 7 Depth 5.0-5.1 m	0.512	690,000 [i]	$k_{asp}=7.7 \times 10^{-7} yr^{-1}$
Layer No. 5 Depth 5.5-5.6 m	0.574		
Külna Cave, near Brno, Czechoslovakia (this report)			
Layer No. 7a Depth 3.8-4.0 m	0.072†	45,600 ±2850 GrN 6060 −2200	50,000
Layer No. 11 Depth 7.0-7.2 m	0.092†	Last Interglacial (?) [j]	80,000

Except as indicated, the samples were hydrolyzed 24 hours.
†Sample was hydrolyzed 4 hours. The D/L aspartic acid ratio for a modern bone hydrolyzed 4 hours $\simeq 0.023$ (Bada et al., in preparation); this is thus the $(D/L)_{t=0}$ value used in equation (2) for these samples.
*Sample was hydrolyzed 6 hours. The D/L aspartic acid ratio for a modern bone hydrolyzed 6 hours = 0.026 (Bada et al., in preparation); this is thus the $(D/L)_{t=0}$ value used in equation (2) for these samples.
[a] Aspartic acid content $\simeq 0.005$-0.01 mg/g of bone.
[b] Aspartic acid content $\simeq 0.05$-0.1 mg/g of bone.
[c] Aspartic acid content $\simeq 0.1$-0.2 mg/g of bone.
[d] The larger the level number, the lower the sample stratigraphically.
[e] See Klein, 1974.
[f] See Klein, 1975.
[g] A. Ronen, University of Haifa, personal communication.
[h] See discussion in Bada and Helfman, 1976.
[i] Brunhes/Matuyama paleomagnetic reversal (690,000 years) occurs between layers 5 and 7 (K. Valoch, Moravske Museum, Brno, personal communication).
[j] K. Valoch, personal communication.

which were analyzed came from levels which were too old for radiocarbon dating (Klein 1974, 1975). Independently derived information suggests, however, that the levels at both sites were deposited during or at the end of the last interglacial period (Klein 1974, 1975; also summarized by Bada and Deems 1975). Therefore the samples from KRM and SW should have ages in the range of 80,000 to 120,000 years (Shackleton and Opdyke 1973). This age range is in close agreement with the ages deduced from aspartic acid racemization. This correlation suggests that a bone which has a radiocarbon age of approximately 17,000 to 20,000 years (in this case, the Nelson Bay Cave sample) is a suitable calibration sample to use for dating bones which are much older, even those which are beyond the 40,000-year dating limit of radiocarbon.

Other important dates are those for Külna Cave, deduced using the Stránská skála calibration. At Stránská skála, the Brunhes/Matuyama magnetic reversal occurs between layers V and VII (Valock, personal communication). This provides an important datum point for calibrating the aspartic acid racemization reaction over a tremendously long time period. Because of the cool temperatures at the Stránská skála locality (present mean annual air temperature equals approximately 8°C; Valock, personal communication), the rate of aspartic acid racemization is considerably slower than for the other sites listed in table 1. The value of $k_{asp} = 7.7 \times 10^{-7}$ yr^{-1} deduced for the extent of racemization of aspartic acid in bones from layers V and VII represents the average value over the last 700,000 years. This calibration value was used to estimate the ages of samples from Külna Cave, an Upper Pleistocene site located close to Stránská skála. The Külna Cave racemization ages are in excellent agreement with those deduced from both radiocarbon and geologic evidence. The agreement between the racemization and ^{14}C ages for layer 7a is important because the radiocarbon age was extremely close to the upper dating limit of ^{14}C. Radiocarbon ages in the vicinity of 40,000 years should always be viewed with some caution. The excellent correspondence between the aspartic acid and the radiocarbon-deduced ages in this case suggests that the radiocarbon date is not greatly in error.

RACEMIZATION AGES DEDUCED USING OTHER AMINO ACIDS

It should be possible to determine racemization ages using several different amino acids. The reason that aspartic acid has been used extensively so far is that this amino acid has the fastest racemization rate. Easily detectable racemization of aspartic acid takes place in a time interval datable by radiocarbon. Alanine and glutamic acids have racemization rates which are about half that of aspartic acid, while the amino acids valine and leucine undergo racemization considerably more slowly (Bada 1972a, Bada, Kvenvolden, and Peterson 1973). Recently we have determined glutamic acid racemization ages for some samples which also had been dated by aspartic acid racemization. These preliminary results are listed in table 2. In general, glutamic acid racemization ages are compatible with those deduced from aspartic acid. The glutamic acid racemization ages are not as accurate as those deduced from aspartic acid racemization, however, because the D/L glutamic acid ratios determined by gas chromatography are less precise than the D/L aspartic acid ratios determined on the amino acid analyzer. In the future we plan to carry out extensive studies on using the extent of racemization of several amino acids to deduce a series of racemization ages for a single fossil bone sample.

The extent of racemization of several amino acids in a fossil bone could also be used to determine the degree of secondary amino acid contamination (Bada, Kvenvolden, and Peterson 1973). Elevated temperature kinetics experiments have shown that the extent of

TABLE 2

Racemization Ages Deduced Using Aspartic and Glutamic Acids: Murray Springs, Arizona. (The aspartic acid results are from Bada et al. 1974b and Bada and Helfman 1975).

Radiocarbon age (yrs)	D/L Ratio		Racemization age (yrs)†	
	Aspartic acid	Glutamic acid	Aspartic acid	Glutamic acid
5,640 ± 60, A-905A, B*	0.33	0.15	$k_{asp} = 4.86 \times 10^{-5} yr^{-1}$	$k_{glu} = 1.83 \times 10^{-5} yr^{-1}$
11,230 ± 340, A-805	0.52	0.21	10,500	~9,000
Double Adobe** 10,420 ± 600, A-1152	0.50	0.21	9,900	~9,000

*Sample used to calculate K values.
**Site located close to Murray Springs with similar climate.
†Calulated from equation 2. The $(D/L)_{t=0}$ term for glutamic acid = 0.05 for a modern bone hydrolyzed 24 hours (Hoopes, McCurdy, and Bada, unpublished results).

amino acid racemization in a bone should have the pattern D/L aspartic acid > D/L alanine ≃ D/L glutamic acid > D/L leucine. The presence of this pattern in a fossil bone sample provides evidence that the sample was not contaminated.

It should be emphasized that even though a sample may be contaminated with secondary amino acids, amino acid racemization still provides a minimum age (Bada 1974). Contamination introduces amino acids that are more modern (that is, ones that consist mainly of the L-enantiomers) than the indigenous collagen-bound amino acids. Contamination thus lowers the actual D/L amino acid ratio which should be present in the bone.

Correlation of Racemization Rates With Temperature

When using the amino acid racemization reaction to date fossil bones, it is generally assumed that the major factor which affects the extent of racemization in a sample is the average temperature to which the bone has been exposed. The calibration procedure discussed earlier is a method of evaluating this temperature, but since racemization is a chemical reaction, it has been suggested that factors other than temperature, such as pH, humidity, or leaching (Hare 1974, von Endt et al. 1975) could have an effect on racemization rates. It has also been argued that variations in these other environmental factors throughout geological time might affect the extent of racemization in a fossil bone in a manner perhaps not compensated by using the calibration technique (von Endt et al. 1975).

To test which environmental factors are the most important in determining the racemization rate in a fossil bone, we have examined the D/L aspartic acid ratios in Holocene radiocarbon-dated bones from various regions of the world (Bada and Helfman 1975). Since the general climate (and, by implication, temperature) of the earth has been more or less constant throughout the Holocene, the k_{asp} values determined from these bone samples should be roughly proportional to the present-day mean annual air temperatures of the sites where the bones are found. Our results have shown that there is generally good agreement between temperatures calculated from the k_{asp} values and the actual temperatures of the sites. We suggest that if other factors bear a major influence on racemization rates, these correlations would be much poorer than they are.

We now present additional correlations between racemization-deduced temperatures and the estimated temperatures to which Holocene and Pleistocene fossil bone samples have been exposed. One problem in the earlier calculations was our lack of a good reference sample, that is, a sample with an unequivocal temperature history. Recently, however, an excellent reference sample has become available. We have detected increasing racemization of aspartic acid in tooth enamel and dentine with increasing age in human beings (Helfman and Bada 1975, 1976). The value of k_{asp} of 8.1×10^{-4} yr^{-1} determined from the human tooth enamel and dentine studies thus corresponds to the value of k_{asp} at 37°C, the mammalian body temperature.

Before we can use the k_{asp} value determined from the teeth of living human beings, we must first show that the rate of racemization is the same in teeth and in bone. To demonstrate this, we have determined the D/L amino acid ratios in associated fossil bones and teeth. These comparisons are shown in table 3. The results demonstrate that the rate of racemization in teeth is essentially identical to that in bone.

We have used the k_{asp} value determined from the teeth of living human beings to calculate temperatures for Holocene and Pleistocene bones, obtaining the calculated temperatures from the equation

$$\text{Ln}\left[\frac{k_{asp}\text{ (tooth)}}{k_{asp}\text{ (fossil)}}\right] = \frac{E_a}{R}\left(\frac{1}{T_{calc}} - \frac{1}{T_{tooth}}\right) \quad (3)$$

where E_a = 33.4 kcal/mole (Bada 1972b), k_{asp} (tooth) = 8.1×10^{-4} yr^{-1}, k_{asp} (fossil) = the rate constant determined from the dated fossil bone sample, T_{tooth} = 37°C = 310°K, and T_{calc} = the calculated average temperature (°K) to which the fossil bone has been exposed. The temperatures calculated from the above equation have been compared not with the mean annual air temperature (T_a) but with the estimated exponential air temperature (T_{exp}). Because racemization is a chemical reaction, it is not a linear function of temperature. Based on the hydrolysis of sucrose in weakly acidic solutions, Lee (1969) has shown that the exponential temperature is related to the mean annual air temperature by the following equations:

For areas with no seasonal snow cover

$$T_{exp} \simeq (T_a + 0.226\ R_T - 2.23)/0.973 \quad (4a)$$

For areas with seasonal snow cover

$$T_{exp} \simeq (T_a + 0.161\ R_T + 1.23)/1.065 \quad (4b)$$

where R_T = the yearly range in temperature of the locality. Using this equation and the

TABLE 3

Comparisons of the Extent of Amino Acid Racemization in Associated Teeth and Bones

Site Location		Tooth	Bone
Riverside, California Human skeleton found?	D/L aspartic acid	0.457	0.437
La Jolla, California Human skeleton SDM 19241*	D/L aspartic acid	0.187	0.154
Human skeleton found 1976	D/L aspartic acid	0.184	0.223
Florisbad, South Africa Hippopotamus jaw (Bada, Protsch, and Schroeder 1973)	alloisoleucine/isoleucine	0.42	0.46

*San Diego Museum of Man notation.

T_a and R_T values obtained from meteorological records, we have calculated T_{exp} values and compared these with the temperatures calculated from the extent of racemization in well-dated fossil bones. The results of these correlations for Holocene bones are shown in figure 4. (The data on which figures 4 and 5 are based will be published elsewhere.) The results show that there is excellent agreement between the calculated and exponential temperatures. We feel that this agreement is remarkable, especially considering the diverse types of physical environment from which the various samples were obtained.

We have also carried out a similar comparison between the temperatures calculated from Pleistocene samples and present-day exponential temperatures. These results are shown in figure 5. As should be the case, since cooler temperatures generally prevailed throughout the Pleistocene (Emiliani 1972), the temperatures calculated from Pleistocene fossil bones are considerably lower than the present-day exponential temperatures.

We believe that the temperature correlations shown in figures 4 and 5 provide convincing evidence that temperature is the major environmental factor which determines the rate of racemization for any particular locality. As we have stated elsewhere (Bada and Helfman 1975), there are probably minor effects on amino acid racemization rates which are due to other environmental factors, but these are small compared to those which arise from variations in temperature.

Recommended Standard Procedure for Reporting Racemization Dates

With the increasing use of amino acid racemization dating, it is important to establish a set of criteria for assessing the reliability of a racemization-deduced age. In table 4 we have listed information which should be given when a racemization age is reported. The first two conditions should be considered a minimum requirement and always reported; otherwise, the racemization date should be considered meaningless. The reliability

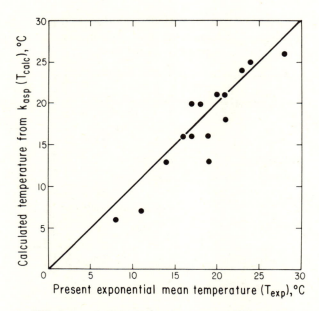

FIG. 4. A polot of T_{calc} vs. T_{exp} for Holocene bones. An exact agreement between the two temperatures is represented by the straight line.

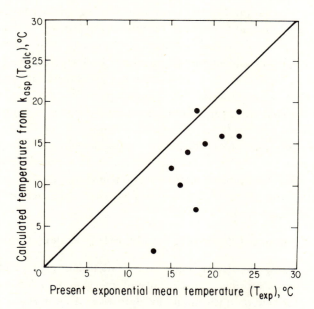

FIG. 5. A plot of T_{calc} vs T_{exp} for Pleistocene bones. An exact agreement between the two temperatures is represented by the straight line.

of the date increases with the number of conditions which are fulfilled. It should eventually be possible to satisfy an additional criterion and to report an average racemization age for a sample deduced using several different amino acids.

Of the various criteria listed in table 4, the second is by far the most important. As has been discussed elsewhere (Bada, in press), the D/L amino acid ratio for a particular bone can generally be unequivocally determined (condition 1). The conversion of this ratio into an age estimate can be done only if a suitable calibration value has been determined for the site or locality where the bone was found.

TABLE 4

Criteria for Assessing the Reliability of an Amino Acid Racemization Date

1.	The D/L aspartic acid ratio (or the D/L ratio for the amino acid being used) should be explicitly stated and the technique used to determine this ratio should be defined.
2.	The "calibration" constant (k_i) used to calculate an age should be reported, and method of deducing this value should be stated.
3.	The temperature calculated (T_{calc}) from the k_i "calibration" value should compare favorably with the estimated temperature history of the sample.
4.	The extent of amino acid racemization should follow the pattern D/L aspartic acid >D/L alanine ≃D/L glutamic acid >D/L leucine for an uncontaminated sample. If this pattern is not observed, the calculated age should be regarded as a minimum estimate.
5.	If sufficient material is available, the racemization analysis of a sample should be repeated several times to determine the reproducibility of the D/L ratio.

Future:

6.	The comparison of the racemization ages deduced from several amino acids should provide an estimate of the uncertainty of the racemization date.

CONCLUSION

The amino acid racemization dating technique provides an important new chronological indication which is especially useful for dating fossil bones which are undatable by radiocarbon or other radiometric techniques. The racemization method can be used to date bones which have ages ranging from a few thousand years to several hundred thousand years. Analyses can typically be carried out on only a few grams of bone.

Probably the most important applications of racemization dating will be in paleoanthropology and archaeology. The genus *Homo* is now thought to have existed for approximately two to three million years. Radiocarbon dating can thus be used only to date the last ~1%–2% of man's existence. Even if a human fossil is thought to be datable by radiocarbon, anthropologists are reluctant to do so because several hundred grams of bone have to be destroyed for ^{14}C dating. Because of these limitations, coupled with the fact that other radiometric techniques capable of dating periods too old for radiocarbon dating are not applicable to bone, most of the human fossils which have been found throughout the world have never been directly dated. In the future racemization dating should help to place at least the latter part of man's evolutionary history into a firmer chronological perspective.

ACKNOWLEDGMENTS

We thank the many contributors of samples for their generosity, especially R. Burleigh and G. Sieveking (Egyptian samples), R. Klein, J. Vogel, and Q. B. Hendy (South African coastal and Heuningsneskrans Shelter samples), W. W. Howells, T. Molleson, and A. Ronen (Mount Carmel samples) and K. Valoch (Czechoslovakian samples). This work was supported by grant EAR73-00320-A01 from the National Science Foundation.

REFERENCES

Bada, J. L.
- 1972a Kinetics of racemization of amino acids as a function of pH. J. Am. Chem. Soc. 95:1371-1373.
- 1972b The dating of fossil bones using the racemization of isoleucine. Earth Planet. Sci. Lett. 15:233-231.
- 1974 Reply to M. L. Bender: Reliability of amino acid racemization dating and paleotemperature analysis on bones. Nature. 252:379-381.
  Comment on the paper by R. Protsch. The absolute dating of upper Pleistocene sub-Saharan fossil hominids and their place in human evolution. J. Human Evolution. (In press.)

Bada, J. L., B. P. Leyendyk, and J. B. Maynard
- 1970 Marine sediments: Dating by the racemization of amino acids. Science. 170: 730-732.

Bada, J. L., and R. A. Schroeder
- 1972 Racemization of isoleucine in calcareous marine sediments: Kinetics and mechanism. Earth Planet. Sci. Letter. 15:1-11.
- 1975 Amino acid racemization reactions and their geochemical implications. Naturwissenschaften 62:71-79.

Bada, J. L., and R. Protsch
- 1973 Racemization reaction of aspartic acid and its use in dating fossil bones. Proc. Natl. Acad. Sci. U.S.A. 70:1331-1334.

Bada, J. L., R. Protsch, and R. A. Schroeder
- 1973 The racemization reaction of isoleucine used as a paleotemperature indicator. Nature. 241:394-395.

Bada, J. L., K. A. Kvenvolden, and E. Peterson
- 1973 Racemization of amino acids in bones. Nature. 245:308-310.

Bada, J. L., R. A. Schroeder, and G. Carter
- 1974a New evidence for the antiquity of man in North America deduced from aspartic acid racemization. Science. 184:791-793.

Bada, J. L., R. A. Schroeder, R. Protsch, and R. Berger
    1974   Concordance of collagen-based radiocarbon and aspartic acid racemization ages. Proc. Natl. Acad. Sci. U.S.A. 71:914-917.

Bada, J. L. and L. Deems
    1975   Accuracy of dates beyond the ^{14}C dating limit using the aspartic acid racemization reaction. Nature. 255:218-219.

Bada, J. L., and P. M. Helfman
    1975   Amino acid racemization dating of fossil bones. World Arch 7:160-173.
    1976   Applications of amino acid racemization dating in paleoanthropology and archaeology. *In* Colloque I: Datations absolues et analyses isotopiques en prehistorie; Methods et limites. Union Intern. Sci. Prehistorique Protohistorique. IX Congres. Paris, CNRS, 39-62.

Emiliani, C.
    1972   Quaternary hypothermals. Quaternary Res. 2:720-723.

Hare, P. E.
    1965   Amino acid artifacts in organic geochemistry. Carnegie Inst. Wash. Yearb. 64:232-235.
    1974   Amino acid dating of bone: the influence of water. Carnegie Inst. Wash. Yearb. 73:576-581.

Helfman, P. M., and J. L. Bada
    1975   Aspartic acid racemization in tooth enamel from living humans. Proc. Natl. Acad. Sci. U.S.A. 72:2891-2894.
    1976   Aspartic acid racemization in dentine as a measure of aging. Nature. 262:279-281.

Hirs, C. H. W., S. Moore, and W. H. Stein
    1954   The chromatography of amino acids on ion exchange resins: use of volatile acids for elution. J. Am. Chem. Soc. 76:6063-6065.

Hoopeo, E. A., E. T. Peltzer, and J. L. Bada
    1978   Determination of amino acid enantiomeric ratios by gas chromatography of the N-trifluoroacetyl-L-prolylpeptide methyl esters. J. Chromat. Sci. 16: 556-560.

Klein, R.
    1974   Environment and subsistence of prehistoric man in the Southern Cape Province, South Africa. World Arch 5:249-284.
    1975   Paleoanthropological implications of the non-archeological bone assemblage from Swartklip I, Southwestern Cape Province, South Africa. Quaternary Res. 5:275-288.

Lee, R.
    1969   Chemical temperature integration. J. Appl. Meteorology 8:423-430.

Manning, J. M., and S. Moore
    1968   Determination of D- and L-amino acids by ion exchange chromatography as L-D and L-L dipeptides. J. Biol. Chem. 243:5591-5597.

Olsson, I. U., M. F. A. F. El-Daousky, A. I. Abd-El-Mageed, and M. Klasson
    1974   A comparison of different methods for pretreatment of bones. I. Geologiska Foreningens i Stockholm Forhandlinger 96:171-181.

Schroeder, R. A., and J. L. Bada
    1976   A review of the geochemical applications of the amino acid racemization reaction. Earth Science Rev. 12:347-391.

Shackleton, J. J., and N. D. Opdyke
    1973   Oxygen isotope and paleomagnetic stratigraphy of Equatorial Pacific Core J28-238: Oxygen isotope temperatures and ice volumes on a 10^5 year and 10^6 year scale. Quaternary Res. 3:39-55.

von Endt, D. W., P. E. Hare, D. J. Ortner, and A. I. Stix
    1975   Amino acid isomerization rates and their use in dating archaeological bone. Soc. Amer. Archaeol. 40th Annual Meeting. Dallas. 66.

4

Amino Acid Racemization Dating of Fossil Shell from Southern California

Patricia M. Masters and Jeffrey L. Bada

ABSTRACT

The racemization kinetics of isoleucine were studied in both Total and Protein hydrolysates of *Chione* shell. Unlike the condition found in bone, racemization in *Chione* shell does not follow reversible first-order kinetics. Racemization rates also vary among different molluskan species.

Seven marine terrace deposits were dated by the extent of racemization of isoleucine in *Chione* shells. The "calibration" was provided by a uranium disequilibrium date on coral at the Nestor Terrace locality, SDSU 2577. Alleu/iso ratios for duplicate analyses were highly reproducible. Racemization analyses at five terraces in northern San Diego County have corroborated other 120,000-year terraces, as well as 100,000-year and ~220,000-year high sea stands.

A series of sixteen radiocarbon dates on shell from six archaeological sites were compared with isoleucine racemization dates on *Chione* from the same localities. Using shells from the site Ora 64 for the calibration sample, the calculated racemization dates for all of the sites except Del Mar differed by an average of 29% (vs. 6% for bone analyses) from the ^{14}C dates on shell carbonate. Reproducibility of alleu/iso ratios was poor for shells of the same provenience.

A new method of processing shell for a radiocarbon date is described. The amino acids, derived from shell proteins, are extracted by ion exchange chromatography and then are burned and counted. The ^{14}C date of 12,000 ± 1100 years produced by this method compares with a ^{14}C date on shell carbonate of 7380 ± 220 years for shell from the same levels. The 12,000-year date is compatible with the racemization age determined using the Ora 64 calibration. This discrepancy suggests that exchange processes are affecting the carbonate dates on shell from the Del Mar site.

The difference in average temperatures experienced by the interglacial Nestor Terrace shell and the Holocene Ora 64 shell was calculated to be 11°C. This $\overline{\Delta T}$ is in agreement with similar estimates for Florida.

INTRODUCTION

An important stage in the development of a new dating technique is evaluation of the reliability of dates derived from various fossil materials. The first applications of amino acid racemization as a fossil dating method were with calcareous deep sea sediments (Bada et al. 1970, Bada and Schroeder 1972) and with foraminifera (Wehmiller and Hare 1971). Subsequently the focus of racemization dating research shifted to fossil bone (Bada et al., this volume). Mitterer's work (1966, 1974, 1975) on *Mercenaria* suggested the feasibility of racemization dating of fossil mollusk shell. Due to our interest in a number of Indian shell midden sites along the southern California coast, we have attempted to apply the amino acid racemization dating technique to west coast mollusks.

Amino acid racemization dates on human bone from five sites along the coast of San Diego County suggest that human occupations occurred here some 30-50,000 years ago. Radiocarbon dates on collagen are not possible due to very low organic nitrogen in these skeletal remains. More than 100 ^{14}C dates, primarily on shell carbonate, exist for Indian sites in the county, but only one charcoal date (Davis 1976) approaches the time range suggested by the racemization ages on bone. Of particular interest is a series of nine shell carbonate ^{14}C dates on the middens which comprise SDM W-34 and W-34A, the locality where the skeletal remains of the Dek Mar Man were recovered in 1929. Three separate racemization analyses (Bada et al. 1974, Bada and Helfman 1975) on cranial and long bone fragments of the Del Mar Man fossil (SDM 16704) have yielded an age of 46,000 years B.P. This age cannot be reconciled with ^{14}C dates on the Del Mar shell, which range from 4590 ± 60 to 9260 ± 100 years, but we have many racemization dates on bone from other sites in the county which are consistent with this later occupation (Bada and Helfman 1975, Bada et al. 1974).

The discrepancy at the Del Mar site implies either that one group of dates is in error or that two different sets of occupations are being sampled. Perhaps it is to be expected that most ^{14}C dates in the county are young, since the later populations on this coast may have been more widespread and with a higher density than the earlier ones. The association of the fossil bones with stratigraphic positions within the sites will never be known because the bones were discovered eroding from the sea cliffs almost 50 years ago, and no excavations were undertaken at that time.

There has been a good deal of concern about the accuracy of radiocarbon dates on shell carbonate. Due to the tendency of the original aragonite to recrystallize into calcite, the more stable form of calcium carbonate, the carbon in shell carbonate is prone to exchange with nonindigenous carbon, thus contaminating the samples with carbonates dissolved in groundwater. Exchange has been shown to occur even with gaseous CO_2, which significantly decreases the apparent age of shells, particularly those near the limits of ^{14}C dating (Olsson 1965). Similarly, a rich organic environment such as a kitchen midden (containing food refuse and excrement) could seriously contaminate an underlying deposit containing older shell. Also, the isotopic environments in which the inorganic phase of the shell has been laid down can influence the ^{14}C ages. Marine shell may yield apparent ages ranging from 440-750 years older than the actual ages (Mangerud and

Gullicksen 1975, Gillespie and Polach, this volume). Similar problems exist with freshwater shell (Rubin et al. 1963).

Because of these considerations, we have taken two approaches to dating the organic component of shells. First, we have applied the amino acid racemization dating technique to Pacific Coast mollusk shells. We have discussed the racemization kinetics of isoleucine in *Chione* shell, and we have presented the results of racemization analyses on fossil *Chione* from seven interglacial terrace localities and six midden sites in southern California. Racemization dates on shell are evaluated in terms of reproducibility, kinetic complexity, temperature calculations, and comparisons with ages derived from independent dating techniques. Second, we have described a method for radiocarbon dating in which the amino acids extracted from *Chione* are processed in a manner similar to that used for collagen-based ^{14}C dates on bone.

Materials and Methods

Shells from the Del Mar site, SDM W-34 and W-34A, were recovered from two test pits during the summer of 1974 by P. M. Masters, C. Lee, and R. Tyson. The units were excavated in 10-centimeter levels from the surface to the underlying sterile sandstone. The unit excavated in the lower midden (W-34A) was 1.2 m in depth, and the unit in the upper midden (W-34) was 1.5 m in depth. The interglacial terrace shells were collected from two localities in San Diego (SDSU 2577 and 1854) by P. M. Masters, J. L. Bada, and S. Brown. Samples from the five marine shell deposites in northern San Diego County were supplied by P. Remeika and G. Kuhn.

Amino acid racemization analyses on shell were carried out in two ways: (1) sampling all the amino acids extracted from the shell (called Total) and (2) sampling only the protein-bound amino acids (referred to here as Protein). The first method is essentially identical to the procedure developed for analysis of fossil bone (Bada et al. 1974, Bada and Protsch 1973). In this case, all the amino acids within the shell, whether they occurred as free amino acids, in small peptides, or in proteins, were included in the final fraction to be analyzed. The second method separated the higher molecular weight proteins from the small peptides and free amino acids, thus yielding a more homogeneous fraction which was then broken down by $6\,M$ HCl hydrolysis into amino acids for analysis. Processing was begun with thoroughly cleaning a ~20-g shell by sonification in double-distilled water. The outer layers of the shell were dissolved away in $2\,M$ HCl, and then the shell was completely dissolved in $6\,M$ HCl. Because of the amount of calcium carbonate present in the shell, the pH of the solution was raised during this step. It was thus necessary to evaporate the samples to ~20 ml in a rotary evaporator. (Complete dryness was not possible, due to the hygroscopic property of the $CaCl_2$ formed.) The sample was next resuspended in 250 ml of $6\,M$ HCl and hydrolyzed for 24 hours. After bringing the sample nearly to dryness again in a rotary evaporator, it was neutralized by dilution with double-distilled water and desalted over a column of Dowex 50W-X8 cation exchange resin. The amino acids were eluted with NH_4OH, and the sample was then evaporated to dryness. The alloisoleucine/isoleucine (alleu/iso) ratio was determined on a Beckman-Spinco automatic amino acid analyzer.

For an analysis of the Protein fraction, the shell was cleaned as described above. It was then dissolved in 6 M HCl at 4°C in order not to disrupt the peptide bonds. This solution was then reduced to a small volume with the rotary evaporator at low temperature (<50°C), diluted 1:1 with double-distilled water, and dialyzed exhaustively against water. The dialysate was brought to dryness in the rotary evaporator and resuspended in 6 M HCl. Hydrolysis and subsequent processing of the sample followed the procedures described above.

The high temperature kinetics studies were performed on Recent *Chione* shells (*C. californiensis* and *C. undatella*) collected from Mission Bay in San Diego by JLB. Radiocarbon dates on two of these shells were 5360 ± 140 (LJ 3508) and 3980 ± 130 (LJ 3509). The shells were cleaned, broken into small fragments, and sealed under vacuum in tubes with double-distilled water. The sealed tubes were placed in a heating block and maintained in 135°C for various lengths of time. Subsequent analysis was the same as for the fossil shells.

As a control, an analysis was carried out on one valve of a freshly killed specimen of *Chione californiensis*.

The sample of amino acids submitted for a radiocarbon date was extracted from ~1.4 kg of *Chione* shells which had been excavated from the upper Del Mar midden (W-34) by R. Tyson. It was necessary to pool *Chione* from two decimeter levels (Dm 10 and Dm 11) in order to isolate a sufficient amount of amino acids. The shells were processed as described above for the Total fraction. After desalting, which required 14 liters of Dowex 50W-X8 resin, the yield of amino acids was 250 mg. A small aliquot was used for alleu/iso determination, and the remainder was burned and counted by M. Stenhouse at the La Jolla Radiocarbon Laboratory. The resin control was processed by T. Linick.

Kinetics of Isoleucine Racemization in Chione

The racemization reaction of isoleucine was used in the initial studies on foraminiferal deep-sea sediments and fossil bone. Racemization (or actually, epimerization) of L-isoleucine produces D-alloisoleucine, which is not found in the proteins of living organisms. The racemization reaction of isoleucine is written as

$$\text{L-isoleucine} \underset{k_{alleu}}{\overset{k_{iso}}{\rightleftharpoons}} \text{D-alloisoleucine} \tag{1}$$

where k_{iso} and k_{alleu} represent the first order rate constants for the interconversion of the D- and L-diastereomers. The integrated rate expression for the reversible racemization of isoleucine (see Bada and Schroeder 1972, for derivation) is

$$\ln\left[\frac{1 + (\text{alleu/iso})}{1 - K'(\text{alleu/iso})}\right] - \text{constant} = (1 + K') k_{iso} \cdot t \tag{2}$$

K' equals $1/K_{eq}$ where K_{eq} is the ratio of alloisoleucine to isoleucine at equilibrium. The integration constant, as determined from the analysis of the live-caught specimen of *C. californiensis*, is 0.033.

Before racemization dating can be attempted on any fossil material, it is necessary to determine to what extent the kinetics of the racemization reaction in this material is linear with time. The integrated first-order rate expression (eq. 2) can be used to calculate ages only within the linear range of the racemization reaction. Beyond the linear range, it is possible to deduce only minimum ages.

The results of heating *Chione* fragments for various lengths of time at 135°C are shown in figure 1. The racemization reaction of isoleucine in the Total shell hydrolysates approximates reversible first-order kinetics to an alleu/iso ratio of ~0.58. A value of K_{eq} = 1.05 has been established by heating a sample at 135°C for 2397 hours. This equilibrium constant is considerably lower than that determined for *Mercenaria* (Mitterer 1975) and foraminiferal tests (Wehmiller and Hare 1971), which suggests that the equilibrium alleu/iso ratio may be species-dependent.

Racemization of isoleucine in the Protein hydrolysates exhibits linear kinetics up to an alleu/iso ratio of ~0.46, and the K_{eq} for protein-bound isoleucine in *Chione* is 0.54. The low equilibrium ratio in the nondialyzable Protein fraction is probably one of the contributing causes of the abnormally low K_{eq} ratio in the Total amino acid fraction. The

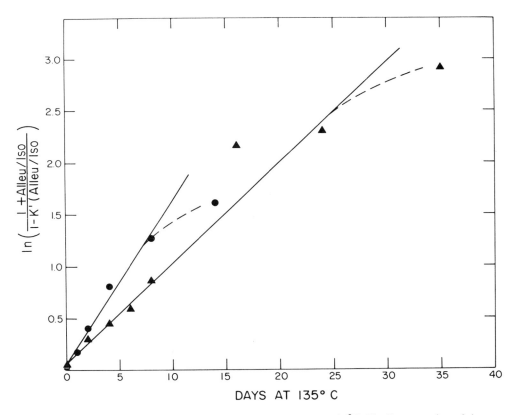

FIG. 1. The kinetics of racemization of isoleucine in bone at 135°C. The linear portion of the plot is a least-squares fit of the points. ●, analyses on total hydrolysate; ▲, analyses of Protein fraction.

fact that the alleu/iso K_{eq} ratio in the protein component is less than the value for free amino acids indicates that the configurational arrangement of amino acids in proteins may alter the equilibrium enantiomeric ratios as suggested by Petit (1974).

The nonlinear kinetics of the Protein fraction could be due to the presence of a number of different proteins in the conchiolin or organic matrix of *Chione* shell. Mollusk shell appears to be composed of a number of distinct proteins which probably vary in proportion and amino acid composition among different species (Gregoire et al. 1955, Bricteux-Gregoire et al. 1968). Since the amino acids in various proteins appear to racemize at different rates (Schroeder 1974), the nonlinear kinetics in *Chione* Protein may possibly indicate a complex, nonhomogeneous mixture of proteins and higher molecular weight peptides.

The more extensive linear range of isoleucine racemization in *Mercenaria* shell (Mitterer 1975) could be the fortuitous consequence of fewer types of proteins in this particular mollusk. At any rate, it is clear from figure 1 that racemization analyses on *Chione* can yield dates only when the alleu/iso ratios do not exceed ~0.6. Above this value, racemization ages are minimum estimates.

In contrast to the complicated kinetics in shells, reversible first-order kinetics are followed in bone. This probably is the result of a relatively homogeneous protein composition: 90% of the organic material in bone is collagen.

Species Differences in Racemization Rates

Five species of mollusk from two marine units, identified as last interglacial terraces by Ku and Kern (1974), were analyzed for alleu/iso ratios in both Total and Protein fractions. The results are presented in table 1. The similar alleu/iso ratios for the *Chione*

TABLE 1

Extent of Racemization of Isoleucine in Nestor Terrace Shells

Shell Type	SDSU Site No.	Alleu/Iso	
		Total	Protein
Chione	1854	0.389	0.113
Chione (1)	2577	0.365	0.088
Chione (2)	2577	0.361	–
Chione mean		0.372	0.101
Lucina (Epilucina)	2577	0.522	0.123
Mussel (Septifer)	2577	0.395	0.064
Abalone	2577	0.493	0.124
Olivella	1854	0.477	–
variance		0.0034	0.0006

specimens from the different localities are consistent with Ku and Kern's interpretation of the contemporaneity of the deposits.

The various mollusk species show considerable variation in the extent of racemization, although all the shells are of the same age. Differences in racemization rates among various taxonomic categories of mollusk shells have also been reported by Miller and Hare (1975). When the alleu/iso ratios in the Total and Protein fractions for the different species are compared, the variance of the Protein fractions is 1/6 that of the Total hydrolysates. This implies that the protein-bound isoleucine racemizes at similar rates in the different shells. The variance in the Total alleu/iso ratios may possibly be attributed to differing rates of diagenetic hydrolysis of constituent proteins resulting in different relative proportions of free amino acids, small peptides, and protein-bound amino acids (Schroeder and Bada 1976).

Racemization Dating of Shell from Interglacial Marine Terraces in San Diego County

Substituting the mean alleu/iso ratio for the two Total analyses of *Chione* from SDSU 2577 (i.e., 0.363) and the uranium disequilibrium date on coral of 120,000 ± 10,000 years (Ku and Kern, 1974) into equation (2) yields $k_{iso} = 3.0 \times 10^{-6}$ yr^{-1}. This k_{iso} value has been used to calculate ages for five additional interglacial terraces in northern San Diego County. All the terrace localities are mapped in figure 2. The alleu/iso ratios and deduced ages are listed in table 2. The reproducibility of the racemization results, as shown by duplicate analyses at four of the localities, is excellent.

Two terraces, UCLA 3458 and SDSU 318, are of particular interest because they have been designated as exposures of the late Pleistocene Nestor Terrace on geomorphic and paleontological grounds (Valentine 1960; Kern 1971). The alleu/iso ratios for these two localities yield dates consistent with the Nestor Terrace age of 120,000 ± 10,000 years determined by Ku and Kern (1974).

Radiometric evidence for glacial and interglacial events generally corroborate the racemization dates in table 2. Oxygen isotope data (Shackleton and Opdyke 1973) suggest a low ice volume (i.e., sea level maximum) at 100,000 years as well as at 120,000 years. Locality #14 may represent the 100,000-year event. Localities #1 and #130 have yielded racemization dates in the range of 180,000 to 220,000 years. These ages correspond to the boundary of oxygen isotope stages 6 and 7, which has an estimated date of 195,000 years (ibid). This boundary marks the end of the penultimate interglacial high sea stand.

The alleu/sio ratios for these latter two terrace exposures are close to the nonlinear portion of the kinetics plot (fig. 1). Therefore it is not possible to date older terraces in the San Diego area by the extent of isoleucine racemization in Total hydrolysates of *Chione*. Due to the slower kinetics of isoleucine racemization in the Protein fraction of *Chione*, dating of earlier interglacial episodes may yet be achieved with Protein analyses. Although *Chione* has been shown to have the slowest rate of isoleucine racemization of the five mollusk types studied at present (table 2), it is conceivable that some other species commonly found in these terraces may have a rate slower than *Chione*. This

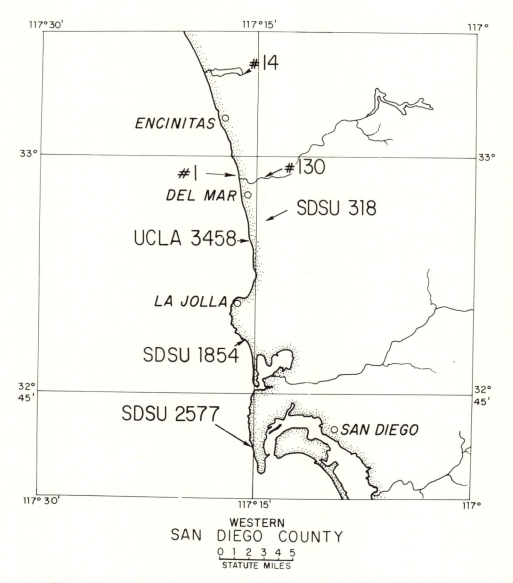

FIG. 2. Map of eight interglacial terrace fossil mollusk locations in San Diego County.

would provide another means to extend the racemization dating limit for marine mollusk in southern California beyond the third interglacial cycle.

Racemization Dating of Chione from Indian Middens

Chione shells from six archaeological sites located along the coast of southern California have been analyzed for Total alleu/iso ratios. The purpose of these analyses was (1) to establish a calibration constant for Holocene *Chione* along the southern California

TABLE 2

Isoleucine Racemization Dates on Total Hydrolyzates of Fossil *Chione* from Interglacial Marine Terraces in San Diego County

Terrace	Reference #	Total Alleu/Iso	Calculated* Age (x 10^3 yr)
14	a	0.304	98
130	a	0.595, 0.593†	220
1	a	0.570, 0.522	210, 180
SDSU 318	b	0.352, 0.359	120
UCLA 3458	c	0.338	110
SDSU 1854	d	0.389	130
SDSU 2577	d	0.365, 0.361	"calibration"

#References: (a) samples supplied by G. Kuhn and P. Remeika; (b) Kern 1971; (c) Valentine 1960; (d) Ku and Kern 1974.

*Ages deduced by using as a "calibration" the mean alleu/iso ratio (0.363) and a uranium disequilibrium date of 120,000 ± 10,000 years (Ku and Kern 1974) at SDSU 2577 to calculate $k_{iso} = 3.0 \times 10^{-6} yr^{-1}$.

†Duplicate analyses on separate shells.

coast and (2) then to compare the ^{14}C dates on shell carbonate with the *Chione* racemization ages at a number of sites. The sites were selected because they had either a large series of ^{14}C dates on shell (Ora 64, Great Western) or corroborative racemization dates on bone (Ora 370, Las Flores, Fairview, Ora 64). The results of the Total analyses are presented in tables 3A and 3B. Using the ^{14}C ages for each site or stratigraphic level and a value of 0.95 for K′, the rate constants (k_{iso}) were calculated for the various sites. The present-day average annual air temperatures (i.e., 16°C), are essentially the same for the various coastal localities. Therefore, the rate constants should be very similar for the six sites. The seven rate constants listed in table 3A vary between $1.4 \times 10^{-5} yr^{-1}$ and $3.2 \times 10^{-5} yr^{-1}$, more than a factor of two. This degree of variation occurs even within the same site between samples that are supposedly contemporaneous (see Fairview and Great Western). The deviation is even greater for four of the nine Del Mar k_{iso} values (table 3B).

Shell from Ora 64, near Newport Beach, California, was chosen as the calibration sample. At this site, a series of sixteen radiocarbon dates performed in three different laboratories on shell carbonate bracketed the occupation between 6000 and 7000 years. An aspartic acid racemization date of 6800 years on bone from the 60 to 70 cm level was calculated using a postglacial k_{asp} in bone of $1.5 \times 10^{-5} yr^{-1}$ (Bada et al. 1974). A *Chione* shell from 70 to 80 cm yielded a ratio of 0.149, which was used to calculate a k_{iso} of $2.0 \times 10^{-5} yr^{-1}$. The ages deduced from this k_{iso} value are listed in tables 3A and 3B and can be compared with the ^{14}C dates on shell carbonate. The average difference between the racemization dates and the dates in table 3A is 29%. The discrepancies between the radiocarbon and racemization ages are much larger for the Del Mar samples listed in table 3B; for these comparisons, the average difference is 88%.

The disagreement between the radiocarbon and racemization ages can readily be visualized in figure 3 where the alleu/iso ratios in the form of the Ln terms calculated

TABLE 3A

Isoleucine Racemization Results on *Chione* from Archeological Sites in Southern California

Site	Laboratory Number	Shell Carbonate ^{14}C Age (yrs B.P.)	Total Alleu/Iso	k_{iso}[a] $(\times 10^{-5} yr^{-1})$	Calculated Age (yrs B.P.) Ora 64 k_{iso}[b]
Ora 370	LJ 3515	1,370 ± 40	.050	2.4	1,670
Las Flores	LJ 3173	2,070 ± 50	.046	1.4	1,460
Fairview					
E9	UCLA 1496B	3,685 ± 100	.133	3.2	5,800
G2		3,685 ± 100(?)	.076	1.6	2,950
Ora 64	GAK 4137	6,790 ± 140	.149	2.0	–
Great Western					
J-15-6	LJ 3245	8,120 ± 90	.155	1.7	6,900
J-6-8	LJ 3160	8,270 ± 80	.230	2.6	10,800

TABLE 3B

Isoleucine Racemization Results on *Chione* from the Del Mar Site

Site	Laboratory Number	Shell Carbonate ^{14}C Age (yrs B.P.)	Total Alleu/Iso	k_{iso}[a] $(\times 10^{-5} yr^{-1})$	Calculated Age (yrs B.P.) Ora 64 k_{iso}[b]
W-34					
Dm 3	LJ 3175	4,590 ± 60	.291	6.1	14,000
Dm 7	LJ 3176	5,440 ± 70	.325 (.226)[e]	5.8 (3.9)	15,800 (10,600)
Dm 10-11	LJ 3507	7,380 ± 220[d]	.231	2.9	10,800
Dm 15	LJ 3177	9,260 ± 100	.295 (.378)	3.1 (4.1)	14,200 (18,800)
W-34A					
Dm 3	LJ 3221	8,040 ± 110	.170	1.9	7,700
Dm 6	LJ 3220	8,960 ± 120	.222	2.3	10,400
Dm 8	LJ 3262	9,080 ± 120	.164	1.6	7,400
Dm 10	LJ 3263	6,130 ± 100	.328 (.246)	5.2 (3.8)	16,000 (11,600)
Dm 11	LJ 3219	6,800 ± 100	.281	4.0	13,500
Dm 12	–	–	.417	–	21,000

[a] K_{iso} values calculated from equation 2 using $K' = 0.95$ and ^{14}C ages for each sample.

[b] Ages calculated from equation 2 using $k_{iso} = 2.0 \times 10^{-5} yr^{-1}$ from Ora 64.

[c] Calculated using $k_{iso} = 1.8 \times 10^{-5} yr^{-1}$ from Del Mar Dm 10-11, W-34.

[d] Radiocarbon date is 12,000 ± 1100 (LJ 3631) on amino acids extracted from *Chione* in Dm 10 and Dm-11. The k_{iso} calculated from this date is $1.8 \times 10^{-5} yr^{-1}$.

[e] Values in parentheses represent analyses on separate shells from the same decimeter level.

from equation 2 are plotted against the radiocarbon ages. The straight line is the least-squares fit to the points exclusive of the Del Mar samples. The correlation coefficient is 0.90. Some of the Del Mar results fall significantly above the line. This can be accounted for in either of two ways: (1) the radiocarbon dates on shell carbonate are anomalously young, or (2) the alleu/iso ratios are too high, probably as the result of past environmental heating.

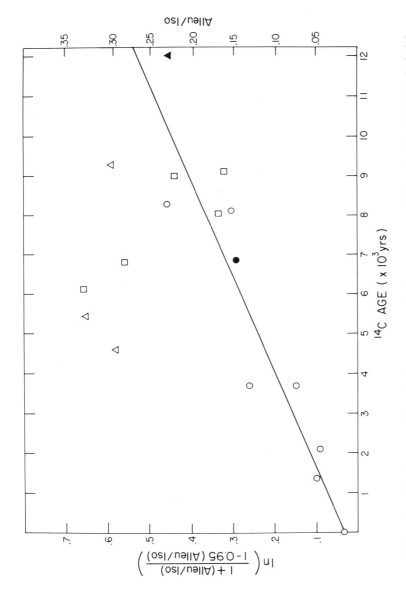

FIG. 3. Plot of the ln terms of the alleu/iso ratios against radiocarbon ages for *Chione* from the six archaeological sites listed in table 3. The straight line is the least-squares fit of all the points except those from Del Mar. O, t_o on live-caught *C. californiensis* and the five sites from table 3A; ●, Ora 64 "calibration"; △, Del Mar upper midden (W-34); ▲, Del Mar Dm 10-11 "calibration"; □, Del Mar lower midden (W-34A).

The greatest source of error in racemization dating would be temperature fluctuations not shared by the calibration sample (Bada and Helfman 1975). The basic assumption behind the calibration method is that the sample of known age has been exposed to the same average temperature as the sample which is being dated (Bada and Protsch 1973). When dealing with a food refuse material such as shell, this is not a particularly safe assumption, since the method of cooking the shellfish and the length of cooking time could have affected the degree of racemization. Consequently, an amino acid racemization date on food refuse shell from a human occupation site may differ from the actual age in an indeterminable manner. If it has been heated more than the calibration sample, the calculated age will be older than the actual age, and conversely, if it has experienced little or no cooking, the racemization date will be an underestimate. At this time, there is no adequate way to evaluate the degree of cooking or heating to which a shell may have been exposed.

Other factors may contribute to the observed differences between the carbonate ^{14}C dates and the deduced racemization dates. Mixing within the middens due to rodent or human disturbance could account for part of the discrepancies, as well as the fact that not one of the shells from sites listed in table 3A comes from the exact unit or level as the shell from which the ^{14}C date was derived. Only with the shells from the Del Mar middens were radiocarbon and racemization analyses performed on samples having the same provenience. Analyses of two different shells from the same level of the same unit are listed in parentheses in table 3B. The lack of reproducibility (and stratigraphic consistency) may be due either to reworking of the deposit or to differences in heating.

Radiocarbon Dating of Amino Acids Extracted from Chione Shell

To better evaluate the above explanations for the anomalously high alleu/iso ratios at Del Mar a second means of radiocarbon dating shell was attempted at this locality. Since bone collagen ^{14}C dates appear to be more reliable than ^{14}C dates on the nonapatite carbonate fraction from bone (Berger et al. 1964, Krueger 1965), it seemed reasonable to adapt a similar method to shell dating. The amino acids were isolated from shell from Dm 10 and 11 at Del Mar and submitted to the La Jolla Radiocarbon Laboratory. The radiocarbon date on the amino acids was 12,000 ± 1100 (LJ 3631), whereas the carbonates in shell from the same two levels at W-34 yielded an age of 7380 ± 220 (LJ 3507).

To test the possibility that contamination might account for the nonconcordant amino acid and carbonate radiocarbon ages, a small amount of the Dowex resin used to desalt the amino acids was also burned and counted. The resin was found to have an infinite age (LJ 3632), calling into question whether the amino acids isolated from the shell were contaminated by the resin. Evidence against resin contamination derives from the δ^{13}C analyses. The δ^{13}C for the amino acids isolated from the shells was –18.0 °/oo ± 2.0 °/oo. This value is consistent with the δ^{13}C value of –18 °/oo in the protein component from marine plankton (Degens 1969). The resin δ^{13}C was –27.5 °/oo. These δ^{13}C data suggest that resin contamination is not an important factor.

Additional evidence supporting the accuracy of the 12,000-year radiocarbon date on the amino acid fraction is provided by the fact that the k_{iso} value of 1.8×10^{-5} yr^{-1} calculated using this age is in good agreement with the Ora 64 calibration k_{iso} value. This correlation is illustrated in figure 3.

Paleotemperatures

The extent of amino acid racemization in a sample of known age can be used to calculate the average temperature exposure of the sample. Schroeder and Bada (1974) have demonstrated the use of aspartic acid racemization in bone to estimate glacial and postglacial temperature differences in the western Mediterranean and in equatorial Africa. In another study, Mitterer (1975) has used the racemization of isoleucine in *Mercenaria* shells to estimate paleotemperatures in Florida.

The Ora 64 k_{iso} value (i.e., 2.0×10^{-5} yr^{-1}) corresponds to the average rate of isoleucine racemization during the Holocene, while the Nestor Terrace k_{iso} value (i.e., 3.0×10^{-6} yr^{-1}) represents the average rate constant for the Upper Pleistocene. Using the Arrhenius equation

$$\ln \frac{k_2}{k_1} = \frac{E_a}{1.987 \text{ cal/mole}} \frac{\Delta \overline{T}}{T_1 T_2} \qquad (3)$$

and $E_a \simeq 28$ kcal mole^{-1} (Bada and Schroeder 1972, Mitterer 1975), the difference in average temperature ($\Delta \overline{T}$) to which the Ora 64 and Nestor Terrace shells have been exposed can be calculated. The estimated $\Delta \overline{T}$ is 11°C.* This value is in reasonable agreement with the $\Delta \overline{T}$ of 12-13°C calculated by Mitterer for southern and central Florida. Both $\Delta \overline{T}$ estimates suggest that the temperatures in coastal southern California and Florida were minimally 11-13°C cooler during the last glacial epoch because the interglacial shells reflect exposure to two warmer periods, one at the end of the last interglacial and the other during the Holocene. The inferred glacial temperatures seem much too cool for these regions. Due to the kinetic complexities of amino acid racemization in shells, the temperatures should be viewed with caution (Schroeder and Bada 1976). It is important to note, however, that the ratio of the Holocene to Upper Pleistocene k_{iso} values we have determined in shells from southern California is nearly the same as that reported by Mitterer for Florida. This should be the case, since temperature variations during the Pleistocene appear to have been similar to the two regions (Gates 1976).

Discussion

Early studies of amino acid diagenesis in fossil materials were carried out on mollusk shells (Abelson 1954, Hare and Mitterer 1966, Hare and Abelson 1967). At the time the present study was initiated, however, no absolute racemization dates had been published for mollusk shells, although considerable attention had been given to racemization dating

---

*The value of $\Delta \overline{T}$ is fairly insensitive to the actual values used to T_1 and T_2 (a change of 1°C in the value of T_1 or T_2 gives a change of 0.01°C in $\Delta \overline{T}$). Thus, approximate estimates for T_1 and T_2 will suffice for the calculation.

of other fossil materials (Bada et al. 1970, Wehmiller and Hare 1971, Bada and Schroeder 1972, Bada et al., this volume). In the course of this work, Mitterer (1975) reported the use of the racemization reaction of isoleucine in *Mercenaria* shells to estimate ages for several interglacial terrace deposits in Florida. He analyzed, but did not assign racemization ages to, radiocarbon-dated *Mercenaria* shells from seven archaeological sites in the southeastern United States and Mexico.

The purpose of this study was to examine the suitability of mollusk shells as a material for amino acid racemization dating and then to apply the method to shell from Indian midden and marine terrace deposits in southern California. We have found *Chione* shell to be a less reliable material than bone for several reasons. First, the more complex kinetics of racemization in *Chione* reduce the range over which absolute ages can be calculated. This limits the usefulness of the technique when dating Pleistocene events in temperature environments. Beyond the linear region of the reaction, only minimum ages can be calculated. Second, the differences in racemization rates among different mollusk species mean that racemization analyses can be compared from one locality to another only when the same species is available for dating or when the kinetics of racemization have been determined for each species. This is a more cumbersome process than racemization analyses of bone, where no species differences in racemization rates have been found. A third problem is apparent in the shell midden samples: reproducibility is poor in duplicate analyses of *Chione* specimens having the same provenience. Whether this is due to reworking or differential heating, the reliability of mollusk dating in cultural deposits is questionable. Reproducibility of alleu/iso ratios for the interglacial terrace shell is generally very good, however, and some confidence can probably be placed in racemization dates for geological deposits.

It has been stated that racemization dating of mollusk shells is far more reliable than racemization dating of bones (Hare 1974). If the ^{14}C dates on shell carbonate are assumed to be correct, the results reported here do not support this claim. More than half of the comparisons in table 3 are widely discrepant with the radiocarbon dates, while twenty-five published comparisons between ^{14}C and racemization dates on bone have shown only one nonconcordant result (and the amino acid date is probably the correct one [Bada and Helfman 1975]). The average difference between racemization dates and ^{14}C dates on bone is 6% (Bada and Helfman 1975), while the results presented here using *Chione* shell indicate an average difference of 29% (table 3A). The greatest discrepancy for bone comparisons is 18% (ibid). Among the five sites listed in table 3A, the largest difference is 57%. Divergences are even greater when considering the Del Mar results: ~200% in three of the nine comparisons. Both the Del Mar and Ora 64 calibrations are underestimating the ages of samples with alleu/iso ratios greater than 0.23 due to slowing of the k_{iso} under colder climatic conditions, so the discrepancies in the Del Mar and Great Western comparisons should be even greater.

A problem with the interpretation of the Del Mar results is that some of the alleu/iso ratios exceed the ratios from the Nestor Terrace sites. There are three possible explanations: (1) these particular shells may have been extensively heated, (2) the lower midden (W-34A) overlies a terrace deposit and some of these shells may have been mixed into the lowest level of the cultural deposit, and (3) contamination of the midden shell may have occurred by amino acids leached from an intermixed older shell deposit. The interglacial terrace locality #1 (fig. 2) is within 1700 feet of the Del Mar site, but this terrace is apparently much older than the lowest levels at Del Mar. Reexcavation of W-34A may help to resolve which of the above explanations is most reasonable.

Concerning the accuracy of amino acid racemization dates on shell relative to shell carbonate ^{14}C dates, it is not possible to decide on the basis of the results discussed here which dating method yields more reliable ages for mollusk shell. The agreement between the racemization date on W-34 Dm 10-11 (10,800 years) and the radiocarbon date on the extracted amino acids (12,000 ± 1100 years) is suggestive. If this observation were replicated in a series of comparisons, the reliability of ^{14}C dates on shell carbonate would be seriously in question, but the extensive work done on shell and charcoal pairs (see Gillespie and Polach, this volume) indicates that unusual conditions may be affecting the ^{14}C results at the Del Mar site. The nature of these conditions should be investigated.

Since the amino acid ^{14}C date correlates well with the racemization date at Del Mar, Dm 10-11, the other racemization ages at this site may be more accurate than the shell carbonate ^{14}C dates. This raises the possibility that W-34, W-34A, and perhaps other shell middens in San Diego County contain much older components than shell carbonate ^{14}C dates have indicated.

In summary, we can conclude that:

1. Racemization dates on mollusk shell from archaeological deposits probably will never approach the potential precision of radiocarbon dates.

2. Racemization analyses on mollusk shell may, however, provide some indication of contaminated samples, that is, exchanged ^{14}C in shell carbonates.

3. Racemization dates on bone are more reliable than on shell because of better reproducibility, the linear kinetics of racemization in bone, and lack of species differences in racemization rates.

4. The new method of radiocarbon-dating the amino acids extracted from shell appears to be promising and indicates a greater antiquity for certain midden deposits in San Diego County than had previously been considered.

5. Racemization dates on fossil *Chione* from interglacial terraces are highly reproducible and offer a reliable method for establishing late Pleistocene chronologies for the southern coast of California and elsewhere.

6. The slower kinetics of isoleucine racemization in the Protein fraction of *Chione* may permit the racemization dating limit for mollusks to be extended into the middle Pleistocene.

REFERENCES

Abelson, P. H.
  1954   Amino acids in fossils. Science. 119:576.

Bada, J. L., B. P. Luyendyk, and J. B. Maynard
  1970   Marine sediments: Dating by the racemization of amino acids. Science 170:730-732.

Bada, J. L., and P. M. Helfman
   1975 Amino acid racemization dating of fossil bones. World Arch. 7:160-173.
Bada, J. L., and R. Protsch
   1973 Racemization reaction of aspartic acid and its use in dating fossil bones. Proc. Natl. Acad. Sci. USA. 70:1331-1334.
Bada, J. L., and R. A. Schroeder
   1972 Racemization of isoleucine in calcareous marine sediments: Kinetics and mechanism. Earth Planet. Sci. Lett. 15:1-11.
Bada, J. L., R. A. Schroeder, and G. Carter
   1974 New evidence for the antiquity of man in North America deduced from aspartic acid racemization. Science. 184:791-793.
Bada, J. L., R. A. Schroeder, R. Protsch, and R. Berger
   1974 Concordance of collagen-based radiocarbon and aspartic acid racemization ages. Proc. Natl. Acad. Sci. USA. 71:914-917.
Berger, R., A. G. Horney A. G., and W. F. Libby
   1964 Radiocarbon dating of bone and shell from their organic components. Science. 144:999-1001.
Bricteux-Grégoire, S., M. Florkin, and Ch. Grégoire
   1968 Prism conchiolin of modern or fossil molluskan shells, an example of protein paleization. Comp. Biochem. Physiol. 24:567-572.
Davis, E. L.
   1976 Two dated La Jollan burials and their place in California prehistory: A review. Pac. Coast Arch. Soc. Quart. 12:1-42.
Degens, E. T.
   1969 Biogeochemistry of stable carbon isotopes. In Eglinton and Murphy, eds., Organic Geochemistry. Springer-Verlag, 304-329.
Gates, W. L.
   1976 Modeling the ice age climate. Science. 191:1138-1144.
Grégoire, Ch., Gb. Duchâteau, and M. Florkin
   1955 La trame protidique des nacres et des perles. Ann. Inst. Oceanogr. 31:1-36.
Hare, P. E.
   1974 Amino acid dating of bone: the influence of water. Carnegie Inst. Wash. Yearb. 73:576-581.
Hare, P. E., and P. H. Abelson
   1967 Racemization of amino acids in fossil shells. Carnegie Inst. Wash. Yearb. 66:526-528.
Hare, P. E., and R. M. Mitterer
   1966 Nonprotein amino acids in fossil shells. Carnegie Inst. Wash. Yearb. 63:362-364.
Kern, J. P.
   1971 Paleoenvironmental analyses of a late Pleistocene estuary in southern California. J. Paleontol. 45:810-823.
Krueger, H. W.
   1965 The preservation and dating of collagen in ancient bones. Proc. 6th Int. Conf. on Radiocarbon and Tritium Dating. Pullman, Washington. 332-337.

Ku, T. L., and J. P. Kern
  1974 Uranium-series age of the Upper Pleistocene Nestor Terrace, San Diego, California. Geol. Soc. America Bull. 85:1713-1716.

Mangerud, J., and S. Gulliksen
  1975 Apparent radiocarbon ages of Recent marine shells from Norway, Spitzbergen and Arctic Canada. Quaternary Res. 5:263-273.

Miller, G. H., and P. E. Hare
  1975 Use of amino acid reactions in some arctic marine fossils as stratigraphic and geochronological indicators. Carnegie Inst. Wash. Yearb. 74:612-617.

Mitterer, R. M.
  1966 Amino acid and protein geochemistry in mollusk shells. Dissertation. Florida State University.
  1974 Pleistocene stratigraphy in southern Florida based on amino acid diagenesis in fossil *Mercenaria*. Geology. 2:425-428.
  1975 Ages and diagenetic temperatures of Pleistocene deposits of Florida based on isoleucine epimerization in *Mercenaria*. Earth Planet. Sci. Lett. 28:275-282.

Olsson, I. U.
  1965 New experience on ^{14}C-dating of tests of foraminifera. Proc. 6th Int. Conf. on Radiocarbon and Tritium Dating. Pullman, Washington. 332-337.

Petit, M. G.
  1974 The racemization rate constant for protein-bound aspartic acid in woodrat middens. Quaternary Res. 4:340-345.

Rubin, M., R. C. Likins, and E. G. Berry
  1963 On the validity of radiocarbon dates from snail shells. J. Geol. 71:84-89.

Schroeder, R. A.
  1974 Kinetics, mechanism, and geochemical applications of amino acid racemization of various fossils. Dissertation. University of California, San Diego.

Schroeder, R. A., and J. L. Bada
  1973 Glacial-postglacial temperature difference deduced from aspartic acid racemization in fossil bones. Science 182:479-482.
  1976 A review of the geochemical applications of the amino acid racemization reaction. Earth Science Rev. 12:347-391.

Shackleton, N. J., and N. D. Opdyke
  1973 Oxygen isotope and paleomagnetic stratigraphy of Equatorial Pacific Core J28-238: Oxygen isotope temperatures and ice volumes on a 10^5 year and 10^6 year scale. Quaternary Res. 3:39-55.

Valentine, J. W.
  1960 Habitats and sources of Pleistocene mollusks at Torrey Pines Park, California. Ecology. 41:161-165.

Wehmiller, J., and P. E. Hare
  1971 Racemization of amino acids in marine sediments. Science. 1973:907-911.

PART XI
Appendix

A Calibration Table for Conventional Radiocarbon Dates

Hans E. Suess

Publication of this volume affords the opportunity to include up-to-date calibration data based on measurements carried out by the La Jolla Radiocarbon Laboratory before August 1978. The first La Jolla calibration curve presented in Uppsala in 1969 at the Twelfth Nobel Symposium (Suess 1970) was derived from about 300 measurements of wood dendrochronologically dated by C. W. Ferguson of the University of Arizona. The samples covered the time from A.D. 1300 to 5400 B.C. Since then the La Jolla laboratory has carried out some additional 600 ^{14}C measurements on dendrochronologically dated wood, including samples back to about 6000 B.C. It is now possible to derive from these results an improved curve for the functional relationship of conventional radiocarbon dates with true historical ages of terrestrial plant materials.

The purpose of the curve presented in Uppsala was not only to allow, within reasonable limits of error, the conversion of conventional radiocarbon dates to historical ages, but also to allow estimates of the age uncertainties which result from the irregularities, the so-called "wiggles," of the curve. The La Jolla radiocarbon dates on which the curve was based have been used by several authors to derive modified curves which, in general, are more regular and smoother than the La Jolla curve. (For references to original publications see Damon et al 1978.) The historical values obtained with such curves agree within limits of errors with those obtained from the La Jolla curve, but neglecting the irregularities results in underestimations of the intrinsic uncertainties in the conversion.

The radiocarbon dates, as reported by most radiocarbon laboratories, have for samples less then 8000 years old a statistical standard error of about 50 to 100 years. In figure 1 the 1969 La Jolla curve, as it was presented at the Uppsala Conference and published in its proceedings, is shown as a broken line. It was drawn free hand; time constants and amplitudes of the irregularities resulted from subjective guesses as it was intended to illustrate the general character rather than the fine structure and details of the curve. The solid line in figure 1 represents a spline function calculated and drawn by a computer, programmed by H. Kruse of the Max-Planck Institut für Chemie, Mainz. The degree of smoothing of the function was selected to minimize the loss of significant information, even though some of the experimental noise was undoubtedly retained.

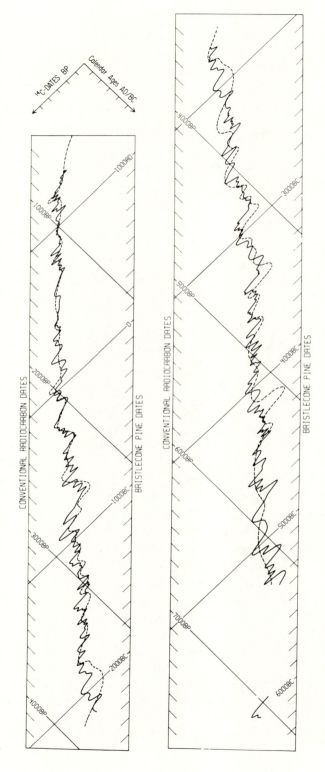

FIG. 1. Conventional radiocarbon dates versus calendar ages. Coordinates are tilted 45°. The solid line represents a computer drawn spline function based on La Jolla ^{14}C measurements carried out before August 1978. The broken line is the calibration curve of 1969 (Suess 1971), shown here for comparison.

Most of the data used to derive fig. 1 were results of measurements of Ferguson bristlecone pine samples. Those obtained prior to September 1977 have been published (Suess 1978). Values for two floating chronologies, D7 and D11, listed in this publication have been used also. Publication of additional data is in preparation.

As discussed at the conference on "Dendrochronology and Postglacial Climatic Variations" in 1974 (Suess and Becker 1977), the irregularities or wiggles in the ^{14}C content on which this colaboration is based of tree ring dated wood samples as a function of time are identical in wood grown in North America and in wood grown in Europe. This agrees with geophysical expectations, because the atmosphere of the earth mixes within a few years. Since 1974 several European laboratories have obtained results that agree with the La Jolla data when normalized to the same ^{14}C standard. Mook and De Jong in Groningen (personal communications) have measured European oak sequences from the 33rd and 34th centuries B.C. and have confirmed in a quantitative way the existence of a conspicuous wiggle during that period of time. By using larger amounts of wood, these authors were able to achieve a correspondingly higher precision than the La Jolla laboratory. The Belfast laboratory (Pearson et al. 1977), using relatively large samples of its own tree ring series of Irish oak, obtained very accurate values which show that during the late third millennium B.C. and the early second millennium B.C. the ^{14}C content of the atmosphere was more constant, as hence, the curve is more regular than the 1969 La Jolla curve. This is in perfect agreement with the more recent La Jolla measurements (fig. 1). Unpublished results of several other European laboratories, such as those in Heidelberg and Bern, also agree with the La Jolla measurements (Oeschger and Münnich, personal communications).

The early La Jolla results were obtained using late nineteenth century wood specimens as standards, before the decision to use 0.95 times the ^{14}C activity of the NBS oxalic acid as standard. It has been found that the radiocarbon content of the wood specimens corresponds to approximately 0.944 times that of the NBS oxalic acid. The La Jolla conventional age values calculated with this standard were therefore about 50 years younger than those obtained with the now accepted standard value. In order to facilitate comparison with the early results, both curves shown in figure 1 were derived with the early standard values. The data listed in the table, however, are calculated with the ^{14}C standard value of 0.95 × NBS oxalic acid, with $\delta^{13}C = -19$ °/oo; ($\delta^{13}C$ of plant material: -25 °/oo). Data for the last 600 years (ca. A.D. 1300 to the present) are not included because this time interval has been carefully investigated by M. Stuiver, Seattle, Washington (1978). Bristlecone pine data for the time interval from 5400 to 5900 B.C. are missing because Professor Ferguson finds it impossible to supply the La Jolla laboratory with the required quantities of twenty grams of wood per sample from this time range. Attempts are being made in La Jolla to bridge this period by measurements of a floating oak chronology from Europe.

The second column of table 1 lists calendar dates that, based on the La Jolla calibration measurements, are the most probable historical dates of samples with conventional radiocarbon ages as listed in column 1. For cases in which more than one or a whole range of dates appear equally probable, these dates are listed in column 2. These calendar dates

TABLE 1

The 1978 La Jolla Calibration Table for Radiocarbon Dates of Wood
Data in parentheses are improbable but not improbable,
÷ indicates range

Conventional* date	Most probable time of growth	Possible centuries of provenience
700 B.P.	A.D. 1230	13th A.D.
50	A.D. 1220	13th (12th) A.D.
800 B.P.	1220 ÷ A.D. 1070	(13th) 12th, 11th A.D.
50	1140 ÷ A.D. 1070	12th, 11th A.D.
900 B.P.	1130 ÷ A.D. 1070	12th, 11th A.D.
50	A.D. 1000	11th, 10th A.D.
1000 B.P.	A.D. 950	10th A.D.
50	A.D. 900	(10th) 9th A.D.
1100 B.P.	880 ÷ A.D. 800	9th A.D.
50	870 ÷ A.D. 700	9th, 8th A.D.
1200 B.P.	800 ÷ A.D. 700	8th A.D.
50	A.D. 680	7th A.D.
1300 B.P.	A.D. 650	7th A.D.
50	625 ÷ A.D. 550	(7th) 6th A.D.
1400 B.P.	A.D. 520	6th (5th) A.D.
50	A.D. 470	5th A.D.
1500 B.P.	A.D. 450	5th A.D.
50	425 ÷ A.D. 350	5th, 4th A.D.
1600 B.P.	A.D. 350	4th A.D.
50	330 ÷ A.D. 220	3rd, 4th A.D.
1700 B.P.	A.D. 220	3rd A.D.
50	220 ÷ A.D. 80	3rd, 2nd, 1st A.D.
1800 B.P.	210 ÷ A.D. 70	(3rd) 2nd, 1st A.D.
50	60 ÷ A.D. 60	1st A.D.
1900 B.P.	A.D. 30	1st A.D.
50	A.D. 25 ÷ 120 B.C.	1st A.D., 1st B.C.
2000 B.P.	0 ÷ 130 B.C.	1st, 2nd, B.C.
50	150 B.C.	2nd B.C.
2100 B.P.	180, 340 ÷ 370 B.C.	2nd (3rd) 4th B.C.
50	180 ÷ 375 B.C.	(2nd) 3rd, 4th B.C.

*Conventional (Libby) dates are based on: $T_{1/2}(^{14}C)$: 5568 years. Zero year: 1950. Contemporary standard is 0.95 NBS Oxidic Acid.

APPENDIX

TABLE 1 (Continued)

Conventional Date	Most probable time of growth	Possible centuries of provenience
2200 B.P.	390 B.C.	4th B.C.
50	400 ÷ 490 B.C.	5th B.C.
2300 B.P.	410 ÷ 580 B.C.	5th, 6th B.C.
50	510 ÷ 590 B.C.	6th B.C.
2400 B.P.	590 ÷ 760 B.C.	(6th) 7th, 8th B.C.
50	600 ÷ 770 B.C.	7th, 8th B.C.
2500 B.P.	790 B.C.	8th B.C.
50	810 B.C.	9th B.C.
2600 B.P.	820 ÷ 870 B.C.	9th B.C.
50	820 ÷ 880, 970 B.C.	9th (10th) B.C.
2700 B.P.	840 ÷ 980 B.C.	9th, 10th B.C.
50	900 ÷ 1070 B.C.	10th, 11th B.C.
2800 B.P.	900 ÷ 1100 B.C.	10th, 11th B.C.
50	1100 ÷ 1200 B.C.	12th B.C.
2900 B.P.	1110 ÷ 1310 B.C.	12th, 13th B.C.
50	1260 ÷ 1370 B.C.	13th, 14th B.C.
3000 B.P.	1310 ÷ 1390 B.C.	14th B.C.
50	1400 B.C.	14th, 15th B.C.
3100 B.P.	1400 ÷ 1480 B.C.	15th B.C.
50	1410 ÷ 1510, 1550 B.C.	15th (16th) B.C.
3200 B.P.	1520 ÷ 1580 B.C.	16th B.C.
50	1610 ÷ 1670 B.C.	17th B.C.
3300 B.P.	1620 ÷ 1670 B.C.	17th B.C.
50	1670 ÷ 1730 B.C.	17th, 18th B.C.
3400 B.P.	1740 B.C.	18th B.C.
50	1750 ÷ 1920 B.C.	18th, 19th, 20th B.C.
3500 B.P.	1830 ÷ 1940 B.C.	19th, 20th B.C.
50	1950 ÷ 2050 B.C.	20th, 21st B.C.
3600 B.P.	1960 ÷ 2100 B.C.	20th, 21st B.C.
50	2110 ÷ 2320 B.C.	22nd, 23rd, 24th B.C.
3700 B.P.	2120 ÷ 2340 B.C.	22nd, 23rd, 24th B.C.
50	2220 ÷ 2350 B.C.	23rd, 24th B.C.
3800 B.P.	2360 ÷ 2440 B.C.	24th, 25th B.C.
50	2370 ÷ 2450 B.C.	24th, 25th B.C.

TABLE 1 (Continued)

Conventional Date	Most probable time of growth	Possible centuries of provenience
3900 B.P.	2400 ÷ 2460 B.C.	25th B.C.
50	2420 ÷ 2540 B.C.	25th, 26th B.C.
4000 B.P.	2560 ÷ 2670 B.C.	26th, 27th B.C.
50	2570 ÷ 2840 B.C.	26th, 27th, 28th, 29th B.C.
4100 B.P.	(2600) 2800 ÷ 2880 B.C.	(27th) 29th B.C.
50	2820 ÷ 2890 B.C.	29th B.C.
4200 B.P.	2900 B.C.	29th, 30th B.C.
50	2910 ÷ 2970 B.C.	30th B.C.
4300 B.P.	2920 ÷ 3130 B.C.	30th, 31st, (32nd) B.C.
50	3010 ÷ 3240 B.C.	31st, 32nd, 33rd B.C.
4400 B.P.	3180 ÷ 3290 B.C.	32nd, 33rd B.C.
50	3340 B.C.	34th B.C.
4500 B.P.	3340 B.C.	34th B.C.
50	3350 (3420) B.C.	34th (35th) B.C.
4600 B.P.	3360 ÷ 3480 B.C.	34th, 35th B.C.
50	3490 B.C.	35th B.C.
4700 B.P.	3500 ÷ 3620 B.C.	36th, 37th B.C.
50	3560 ÷ 3630 B.C.	36th, 37th B.C.
4800 B.P.	3630 B.C.	37th B.C.
50	3640 ÷ 3780 B.C.	37th, 38th B.C.
4900 B.P.	3690 ÷ 3790 B.C.	(37th) 38th B.C.
50	3710 ÷ 3870 B.C.	38th, 39th B.C.
5000 B.P.	3820 ÷ 3880 B.C.	39th B.C.
50	3900 ÷ 3920 B.C.	40th B.C.
5100 B.P.	3930 B.C.	40th B.C.
50	3940 ÷ 4070 B.C.	40th, 41st B.C.
5200 B.P.	3980 ÷ 4080 B.C.	(40th), 41st B.C.
50	4090 ÷ 4240 B.C.	41st, 42nd, 43rd B.C.
5300 B.P.	4100 ÷ 4360 B.C.	42nd, 43rd, 44th B.C.
50	4290 ÷ 4370 B.C.	(43rd) 44th B.C.
5400 B.P.	4300 ÷ 4370 B.C.	44th B.C.
50	4380 B.C.	44th B.C.
5500 B.P.	4390 ÷ 4500 B.C.	(44th) 45th B.C.

TABLE 1 (Continued)

Conventional Date	Most probable time of growth	Possible centuries of provenience
50	4410 ÷ 4520 B.C.	45th, 46th B.C.
5600 B.P.	4460 ÷ 4550, 4700 B.C.	45th, 46th, 47th B.C.
50	4560 ÷ 4720 B.C.	46th, 47th, 48th B.C.
5700 B.P.	4570 ÷ 4730 B.C.	46th, 47th, 48th B.C.
50	4570 ÷ 4760 B.C.	46th, (47th) 48th B.C.
5800 B.P.	4770 B.C.	48th B.C.
50	4780 B.C.	48th B.C.
5900 B.P.	4790 ÷ 4940 B.C.	48th, 49th, 50th B.C.
50	4800, 4960, 5120 B.C.	49th, 50th, 51st (52nd) B.C.
6000 B.P.	4970 ÷ 5140 B.C.	50th, 51st, 52nd B.C.
50	4980 ÷ 5150 B.C.	50th, 51st, 52nd B.C.
6100 B.P.	4990 ÷ 5330 B.C.	(50th) 51st, 52nd, 53rd, 54th B.C.
50		
•		
•	Early and Mid 6th Millennium B.C.	
•		
7000 B.P.	6000 B.C.	59th, 60th, 61st, 62nd B.C.

do not necessarily imply that the samples must have come from the listed years. Because of the laboratory and field noise inherent in the conventional dates, the range for the time of origin of the samples is actually larger. Even if the dates given in column 1 were taken to be infinitely accurate, it would rarely be possible to establish more than the century of the provenience of the sample. All the centuries which might possibly represent the time of origin of the samples are listed in column 3. The statistical and experimental error of the conventional radiocarbon date will widen this time interval, which can easily be estimated by including in column 1 the dates that can be considered to fall within the range of errors of the measurements. For example, if a laboratory reports a conventional radiocarbon age of 5050 ± 50 years, the table shows that the sample probably dates from between 3820 and 3930 B.C. but it may well have grown at any time during the 39th or 40th century B.C. There is a limit as to how accurately a single sample of wood, charcoal, or some other plant material, can be dated by radiocarbon, no matter how accurately the radiocarbon determination is being carried out. The accuracy with which the age of other remains, such as samples containing marine organic and inorganic carbon, can be determined is much lower. More precise dates can only be obtained if series are available of samples whose age difference, or at least chronological order, is known. The best results can be obtained for so-called floating tree ring sequences covering several hundred years.

ACKNOWLEDGMENTS

I thank Mr. Hartwig Kruse and Dr. Timothy Linick for the programming of the computer-drawn calibration curve in figure 1. The curve is entirely based on the dendrochronological bristlecone pine sequence established by Professor Ferguson. The operations of the La Jolla Radiocarbon Laboratory are financed by the National Science Foundation, grant EAR76-22623. An award from the Humboldt Foundation provided the opportunity to obtain information quoted here as "personal communications."

REFERENCES

Damon, P. E., J. C. Lerman, and A. Long
    1978   Temporal Fluctuations of Atmospheric ^{14}C: Causal Factors and Implications. Ann. Rev. Earth Planet. Sci. 6:457-94.

Pearson, G. W., J. R. Pilcher, M. G. L. Baillie, and J. Hillam
    1977   Absolute radiocarbon dating using a low altitude European tree ring calibration. Nature. 270:25-28.

Stuiver, M.
    1978   Radiocarbon Timescale tested against magnetic and other dating methods. Nature. 273:271-274.

Suess, H. E.
    1971   Bristlecone pine calibration of the radiocarbon time scale 5200 B.C. to the present. 12th Nobel Uppsala. 303-313.
    1978   La Jolla Measurements of Radiocarbon in Tree-Ring-Dated Wood. Radiocarbon. 20:1-18.

Suess, H. E., and B. Becker
    1977   Dendrochronologie und Postglaziale Klimaschwankungen in Europa. Erdwissenschaftl. Forschg. 13:156.

Index

Abaj Takalik, 83
Africa: Acheulian, 8; Bantu, 25; Chronology, 10, 14; Iron Age, 9; later Stone Age, 8; middle Stone Age, 8
Allerød, 76
Altar de Sacrificios, 84
Altiplano, Cochabamba, 72
Animo acid dating, 740
ANU Sucrose, 115
Archaeology, underwater, 102
Arizona 1850 wood, 115
Arlington Springs Man, Santa Rosa Island, 4
Atmospheric hydrocarbons, 158
Australia, 404
Auvernier, 107, 566, 585
Ayacucho, Peru, 72

Background: reduction, 148, 176, 182, 208; reproducibility, 164
Barton Ramie, 85
Battleship curve, 90
Benzene preparation, 252
Bering Strait, 70
Bomb -^{14}C, 309, 313
Bone: dating, racemization, 740; modern, 4; samples, 428
Bristlecone pines, 520, 532, 538, 545, 643, 777
Bronze Age: chronology, Europe, 559; Swiss, 102
Bushman Rock Shelter, 71

Calibration table, ^{14}C dates, 777
California, southern, hominid remains, 73
Carbon, natural, urban particles, 172
^{13}C/^{12}C ratio of oxalic acid, 126
^{14}C: arterial samples, 346; bomb in humans, 324; calibration curve, 538, 545; cholesterol, 342; content, various reservoirs, 613; dates, Egypt, 601; dating, mortar, 732; dating of wood fractions, 140; gallstones, 346; Ionium Ages, 733; level, 4th and 5th millennium B.C., 538; production, wood, in situ, 643; racemization comparisons, 746; reservoir parameters, 633, 691; variations, 601, 691; variations, short-term, 591
Cave of Hearths, Makapansgat, 3
Cellular mutations, 355
Charcoal/shell pairs, 408, 426
Chevelon project, 90
China lake ^{14}CO$_2$ levels, 310
Climate: paleomagnetic investigations, 670; and uranium determinations, 100
Cluster analysis, 48, 92
CO$_2$: from car exhaust, 388; increase of, 633; stratospheric, 309
Collagen, 428
Comparison, charcoal, snails, seeds, 294
Corals, 733
Corn, 87
Corozal Project, 85
Cortaillod culture, 104
Counter: gas, small, 163; liquid, small, 163; to 60,000 years, 202
Counting efficiency, 197
Cuello, 84

Dättnau, Switzerland, 77
Dendrochronology, neolithic, western Switzerland, 566, 585
Diablo Canyon site, Calif., 5
Domestic stock, chronology, Africa, 21
Drift-gas deposits, 365
Dryas: Older, 81; Younger, 77

East Africa dates, 25
Egypt, upper, 18
Electronegative impurities, 185
Error: contaminated ^{14}C samples, 136; liquid Scintillation counting, 256, 266

Feldmeilen-site, 104
Flint mine, 591
Floating chronologies, 574
Fluorine method, 3
Fossil fuel effect, 507
F, U, N method, 4

Geomagnetic parameters, 691
Global history, 40
Greenland, 397
Grime's Graves, 592

Heartwood effect, 503
Hominids, fossil, America, 69
Horgen culture, 104
Howieson's Poort, 18
Humic acids, 277, 284

Indic religions, 38
Ionium/^{14}C ages, 733
Iron Age: East and South Africa, 27; dates, Bantu Africa, 26

Kalambo Falls, 13

Laguna Beach Woman, Calif., 5, 71
Lakes, Swedish, 433
Laser infrared, dating by, 148
Later Stone Age, South Africa, 18
Levallois technology, 16
Liquid Scintillation counting, 246, 256
Loessic soils, 281
Los Angeles Man, Woman, Calif., 5, 71
Low level counting improvements, 147, 202

Maize, domesticated, 87
Makalian Wct Phase, 8
Makapansgat, 3
Mammals, marine, arctic, 447
Man, anatomically modern, 69
Marine shells, 397, 404, 422
Marine terraces, shell racemization dating, 763
Mass-spectrometry, dating by, 148, 238
Master chronology, oak, 557
Maya lowlands chronology, 83
Measurement, small ^{14}C samples, 158
Mesolithic: Britain, 41, 66; chronology, Europe, 560; northern Europe, 41; Swiss, 108
Microliths, England, 41
Middle Paleolithnic: Europe, 13; Middle East, 13; North Africa, 13

Midland, Texas, 4
Modeling experiments, CO_2, 633
Mongoloids, America, 71

Nachikufan, 8
Natchez, Miss., 4
Neolithic, Swiss, 102
Neolithic chronology: Europe, 559; Swiss, 109
Neutrons: high energy, 646; Lightning 646; thermal, 645
Nile: Nubian, 18; Valley of, 18, 32, 35
North African Mousterian, 13

Ocean ^{14}C, 315
Old Kingdom, Egypt, 33
Olmec, 83
Oncogenesis, 358
Otovalo, Ecuador, 72
NBS Oxalic Acid, 115

Paleoclimitological studies with oak, 560
Paleoecological studies with oak, 560
Peat, 284
Pfyn culture, 104
Plants, aquatic, 433
Pottery, earliest American, 87
Pre-Aurignacian, Cyrenaica, 13
Pretreatment, wood or charcoal, 135

Radiation burden, human, 332
Radiocarbon dating, thermal diffusion, 216
Radon, error, 138
Residence times, ^{14}C, human tissue, 334, 351
Rise-time discrimination, 151

Sacaba, Cochabamba, 72
Sangoan, 13
Schnurkeramik culture, 104
Sea level: changes, 397; mean, 32, 34
Sediments: coastal marine, 453; marine 470; organic, 284
Seibal, 84
Seriation by radiocarbon dates, 89
Shell dating, racemization, 757
Shielding improvement, 176
60,000 year ^{14}C samples, 202
Snail shells, 294
Soil dating, 277, 284
Stone industries, England, early postglacial, 41
Stratospheric $^{14}CO_2$, 309
Subsurface radar, 520

Swasey phase, 85
Swiss Lake dwellers, 102, 566

Tepexpan, Mexico, 4
Treatment, standard of wood or charcoal, 139
Tree-rings: Holocene, Europe, 554; 19th century, 511
Tree thermometers, 661
Trees, modern American, 495
Translocation, radial in bristlecone pine, 532
Tropospheric $^{14}CO_2$, 309, 313
Two Creeks, 76

Uranium: in ocean cores, 97; and climate, 100

Variability, ^{14}C counting, 619
Volcanoes, 615

Weather, paleomagnetic investigations, 670
Western Eruopean chronology, 11, 14
Wet oxidation, oxalic acid, 125

Yverdon, Switzerland, 566

Zimbabwe ruins, 9